# Cytomegaloviruses
## Molecular Biology and Immunology

Edited by Matthias J. Reddehase

Assistant Editor: Niels Lemmermann

Caister Academic Press

Caister Academic Press
32 Hewitts Lane
Wymondham
Norfolk NR18 0JA
U.K.

www.caister.com

British Library Cataloguing-in-Publication Data

A catalogue record for this book is available from the
British Library
ISBN: 1-904455-02-6

Printed and bound in Great Britain

Cover image of cytomegalovirus designed by Andrew Townsend, courtesy of Daniel N. Streblow and
Jay A. Nelson (for greater detail, see Figure 5.1 in Chapter 5, p.92).

# Contents

# Contributors

**Shirley A. Aguirre**
Pfizer Global Research & Development
La Jolla Laboratories
San Diego, CA
USA

**Mark Bain**
Department of Medicine
University of Cambridge
Cambridge
UK

**Karl W. Boehme**
McArdle Laboratory for Cancer Research
University of Wisconsin-Madison
Madison, WI
USA

**William Britt**
Department of Pediatrics
University of Alabama
School of Medicine
Birmingham, AL
USA

WBritt@peds.uab.edu

**Cathrien A. Bruggeman**
Department of Medical Microbiology
Cardiovascular Research Institute Maastricht
University of Maastricht
Maastricht
The Netherlands

**Wolfram Brune**
Robert Koch Institute
Division of Viral Infections
Nordufer 20
Berlin
Germany

BruneW@rki.de

**Ivan Bubic**
University of Rijeka
Faculty of Medicine
Brace Branchetta 20
Rijeka
Croatia

**Ann E. Campbell**
Department of Microbiology and Molecular
Cell Biology
Eastern Virginia Medical School
Norfolk, VA
USA

Campbeae@evms.edu

**Andrew J. Carmichael**
Department of Medicine
University of Cambridge School of Clinical
Medicine
Cambridge
UK

**Teresa Compton**
McArdle Laboratory for Cancer Research
University of Wisconsin-Madison
Madison, WI
USA

tcompton@wiscmail.wisc.edu

**Christian Davrinche**
Institut National de la Santé et de la
Recherche Médicale
Centre Hopitalier Universitaire Purpan
Toulouse
France

davrinch@toulouse.inserm.fr

**Margarete Digel**
Institute of Medical Virology
University of Tuebingen
Elfriede-Aulhorn- Straße 6
Tuebingen
Germany

**Markus Eickmann**
Institut für Virologie
Philipps-Universität Marburg
Robert-Koch- Straße 17
Marburg
Germany

**Wade Gibson**
Department of Pharmacology and
Molecular Sciences
Johns Hopkins School of Medicine
Baltimore, MD
USA

wgibson@jhmi.edu

**Dorothee Gicklhorn**
Institut für Virologie
Philipps-Universität Marburg
Robert-Koch- Straße 17
Marburg
Germany

**Natascha K.A. Grzimek**
Institute for Virology
Johannes Gutenberg University Mainz
Obere Zahlbacher Straße 67
Mainz
Germany

**Gabriele Hahn**
Max von Pettenkofer Institute
Department of Virology
Ludwig-Maximilians-University
Munich
Germany

**Laura K. Hanson**
Department of Microbiology and Molecular
Cell Biology
Eastern Virginia Medical School
Norfolk, VA
USA

**Hartmut Hengel**
Institute for Virology
Heinrich-Heine-University Düsseldorf
Universitätstraße 1
Düsseldorf
Germany

hartmut.hengel@uni-duesseldorf.de

**Laura Hertel**
Department of Microbiology and Immunology
Health Sciences Addition
University of Western Ontario
London
Ontario
Canada

**Rafaela Holtappels**
Institute for Virology
Johannes Gutenberg University Mainz
Obere Zahlbacher Straße 67
Mainz
Germany

R.Holtappels-Geginat@uni-mainz.de

**Stipan Jonjic**
University of Rijeka
Faculty of Medicine
Brace Branchetta 20
Rijeka
Croatia

jstipan@medri.hr

**Suzanne J.F. Kaptein**
Department of Medical Microbiology
Cardiovascular Research Institute Maastricht
University of Maastricht
Maastricht
The Netherlands

Skap@lmib.azm.nl

**Ulrich H. Koszinowski**
Ludwig-Maximilians Universität München
Pettenkoferstr. 9a
München
Germany

koszinowski@mvp.uni-muenchen.de

**Astrid Krmpotic**
University of Rijeka
Faculty of Medicine
Brace Branchetta 20
Rijeka
Croatia

**Juan C. Lacayo**
Division of Infectious Diseases and
Immunology
University of Minnesota School of Medicine
Department of Pediatrics
Minneapolis, MN
USA

**Niels Lemmermann**
Institute for Virology
Johannes Gutenberg-University Mainz
Obere Zahlbacher Straße 67
Mainz
Germany

**Peter Lischka**
Institut für Klinische und Molekulare
Virologie
Friedrich-Alexander-Universität Erlangen-
Nürnberg
Schloßgarten 4
Erlangen
Germany

**Susan McDonagh**
Department of Cell and Tissue Biology
School of Dentistry
University of California
San Francisco, CA
USA

**Michael Mach**
Institut für Klinische und Molekulare
Virologie
Friedrich-Alexander-Universität Erlangen-
Nürnberg
Schloßgarten 4
Erlangen
Germany

Michael.Mach@viro.med.uni-erlangen.de

**Ekaterina Maidji**
Department of Cell and Tissue Biology
School of Dentistry
University of California
San Francisco, CA
USA

**Gerd G. Maul**
Gene Expression and Regulation Program
The Wistar Institute
Philadelphia, PA
USA

maul@wistar.org

**Jeffery L. Meier**
Department of Internal Medicine
University of Iowa
Iowa City, IA
USA

jeffery-meier@uiowa.edu

**Thomas Mertens**
Universitätsklinikum Ulm
Abteilung Virologie
Albert-Einstein-Allee 11
Ulm
Germany

thomas.mertens@medizin.uni-ulm.de

**Martin Messerle**
Institute for Virology
Hannover Medical School
Carl-Neuberg Straße 1
Hannover
Germany

messerle.martin@mh-hannover.de

**Detlef Michel**
Universitätsklinikum Ulm
Abteilung Virologie
Albert-Einstein-Allee 11
Ulm
Germany

**Edward S. Mocarski, Jr.**
Department of Microbiology & Immunology
D347 Fairchild Science Building
Stanford University School of Medicine
Stanford, CA
USA

mocarski@stanford.edu

**Michael W. Munks**
Molecular Microbiology & Immunology
Oregon Health Sciences University
3181 S.W. Sam Jackson Park Road
Portland, OR
USA

**Jay A. Nelson**
Vaccine and Gene Therapy Institute and
Department of Molecular Microbiology and
Immunology
Oregon Health Sciences University
Portland, OR
USA

nelsonj@ohsu.edu

**Satoshi Noda**
Laboratory for Infectious Diseases
Tokai University
School of Medicine
Isehara
Kanagawa
Japan

**Sandra Pepperl-Klindworth**
Institute for Virology
Johannes Gutenberg University Mainz
Obere Zahlbacher Straße 67
Mainz
Germany

**Lenore Pereira**
Department of Cell and Tissue Biology
School of Dentistry
University of California
San Francisco, CA
USA

pereira@itsa.ucsf.edu

**Bodo Plachter**
Institute for Virology
Johannes Gutenberg University Mainz
Obere Zahlbacher Straße 67
Mainz
Germany

plachter@uni-mainz.de

**Jürgen Podlech**
Institute for Virology
Johannes Gutenberg University Mainz
Obere Zahlbacher Straße 67
Mainz
Germany

**Klaus Radsak**
Institut für Virologie
Philipps-Universität Marburg
Robert-Koch- Straße 17
Marburg
Germany

radsak@staff.uni-marburg.de

**Matthias J. Reddehase**
Institute for Virology
Johannes Gutenberg University Mainz
Obere Zahlbacher Straße 67
Mainz
Germany

Matthias.Reddehase@uni-mainz.de

**Matthew Reeves**
Department of Medicine
University of Cambridge
Cambridge
UK

**Veronica Sanchez**
Department of Cellular and Molecular
Medicine
School of Medicine
University of California, San Diego
La Jolla, CA
USA

**Mark R. Schleiss**
Division of Infectious Diseases and
Immunology
University of Minnesota School of Medicine
Department of Pediatrics
Minneapolis, MN
USA

schleiss@umn.edu

**Christof K. Seckert**
Institute for Virology
Johannes Gutenberg University Mainz
Obere Zahlbacher Straße 67
Mainz
Germany

**Thomas Shenk**
Department of Molecular Biology
Princeton University
Princeton, NJ
USA

tshenk@molbio.princeton.edu

**Christian O. Simon**
Institute for Virology
Johannes Gutenberg University Mainz
Obere Zahlbacher Straße 67
Mainz
Germany

**John Sinclair**
Department of Medicine
University of Cambridge
Cambridge
UK

js@mole.bio.cam.ac.uk

**Christian Sinzger**
Institute of Medical Virology
University of Tuebingen
Elfriede-Aulhorn- Straße 6
Tuebingen
Germany

christian.sinzger@med.uni-tuebingen.de

**J. G. Patrick Sissons**
Department of Medicine
University of Cambridge School of Clinical
Medicine
Cambridge
UK

jgps10@medschl.cam.ac.uk

**Barry Slobedman**
Centre for Virus Research
Westmead Millennium Institute and The
University of Sydney
Westmead, NSW
Australia

**Richard D. Smith**
Biological Sciences Division
Pacific Northwest National Laboratory
Richland, WA
USA

**Deborah H. Spector**
Department of Cellular and Molecular
Medicine
School of Medicine and School of Pharmacy
and Pharmaceutical Sciences
University of California, San Diego
La Jolla, CA
USA

dspector@ucsd.edu


**Thomas Stamminger**
Institut für Klinische und Molekulare
Virologie
Friedrich-Alexander-Universität Erlangen-
Nürnberg
Schloßgarten 4
Erlangen
Germany

Thomas.Stamminger@viro.med.uni-
erlangen.de


**Frank R.M. Stassen**
Department of Medical Microbiology
Cardiovascular Research Institute Maastricht
University of Maastricht
Maastricht
The Netherlands


**Mark F. Stinski**
Department of Microbiology
University of Iowa
Iowa City, IA
USA

mark-stinski@uiowa.edu


**Daniel N. Streblow**
Vaccine and Gene Therapy Institute and
Department of Molecular Microbiology and
Immunology
Oregon Health Sciences University
Portland, OR
USA


**Takako Tabata**
Department of Cell and Tissue Biology
School of Dentistry
University of California
San Francisco, CA
USA


**Qiyi Tang**
Gene Expression and Regulation Program
The Wistar Institute
Philadelphia, PA
USA


**Susan M. Varnum**
Biological Sciences Division
Pacific Northwest National Laboratory
Richland, WA
USA


**Cornelis Vink**
Department of Medical Microbiology
Cardiovascular Research Institute Maastricht
University of Maastricht
Maastricht
The Netherlands


**Markus Wagner**
Bavarian Nordic GmbH
Virology Research
Fraunhofer Straße 13
Martinsried
Germany


**Kirsten Lofgren White**
Gilead Sciences Inc.
San Mateo, CA
USA


**Mark R. Wills**
Department of Medicine
University of Cambridge School of Clinical
Medicine
Cambridge
UK

**Jiake Xu**
School of Surgery and Pathology
University of Western Australia
Nedlands, WA
Australia

**Albert Zimmermann**
Robert Koch-Institut
Division of Viral Infections
Nordufer 20
Berlin
Germany

# Abbreviations

**A**

| | |
|---|---|
| aa | amino acid |
| AAF | alpha-activated factor |
| ACD | amino conserved domain |
| ACV | aciclovir |
| ADAM | a disintegrin and a metalloprotease |
| ADAR | RNA-specific adenosine desaminase/adenosine deaminases that act on RNA |
| ADCC | antibody-dependent cellular cytotoxicity |
| AIDS | acquired immune deficiency syndrome |
| AP | assembly protein |
| AP1 | ATF-2/c-Jun |
| APC | anaphase-promoting complex |
| APC | antigen-presenting cell |
| apoE | apolipoprotein E |
| Arp2/3 | actin-related protein 2/3 |
| ART | active retroviral therapy |
| ATCC | American Type Culture Collection |
| ATP | adenosine triphosphate (also ADP, AMP, CTP, UDP, etc.) |
| AV | anchoring villi |

**B**

| | |
|---|---|
| BAC | bacterial artificial chromosome |
| BM | bone marrow |
| BMC | bone marrow cell |
| BMT | bone marrow transplantation |
| bp | base pair(s) |

**C**

| | |
|---|---|
| CAK | Cdk-activating kinase |
| CBP | CREB-binding protein |
| CCD | carboxyl conserved domain |
| CCIC | circulating cytomegalic infected cells |
| CCMV | chimpanzee CMV |
| CCR | CC chemokine receptor |
| Cdk | cyclin-dependent kinase |
| CDV | cidofovir |
| ChIP | chromatin immuno-precipitation |
| CID | cytomegalic inclusion disease |
| CIITA | class II transactivator |
| CK | chemokine |

| | |
|---|---|
| CKII | casein kinase II |
| CLR | C-type lectin receptors |
| Clr | C-type lectin-related protein |
| CLT | CMV latency-associated transcript |
| CMI | cell-mediated immunity |
| CMP | common myeloid progenitor |
| CMV | cytomegalovirus |
| CNS | central nervous system |
| Con A | concanavalin A |
| COX2 | cyclooxygenase 2 |
| CPE | cytopathogenic effect |
| CRS | cis-repression signal |
| CT | cytoplasmic tails |
| CTB | cytotrophoblast |
| CTD | C-terminal domain |
| CTE | constitutive transport element |
| CTL | cytolytic (or cytotoxic) T lymphocyte |
| CTLL | cytolytic (CD8) T lymphocyte line(s) |

**D**

| | |
|---|---|
| Daxx | Death domain-associated protein 6 |
| DBs | dense bodies |
| DC | dendritic cell(s) |
| DC-SIGN | DC-specific ICAM3 grabbing non-integrin |
| DISC | death-inducing signalling complex |
| DN | dominant negative |
| DNA | deoxyribonucleic acid |

**E**

| | |
|---|---|
| E | early (phase of cytomegaloviral gene expression) |
| E/RE | early/recycling endosome |
| EBNA | Epstein–Barr virus nuclear antigen |
| EC | endothelial cell(s) |
| ECM | extracellular matrix |
| EGFP | enhanced green fluorescent protein |
| EGF-R | epidermal growth factor receptor |
| eIF-2$\alpha$ | eukaryotic translation initiation factor 2$\alpha$ |
| EJC | exon junction complex |
| ELISA | enzyme-linked immunosorbent assay |
| ELISPOT | enzyme-linked immunospot |
| EM | electron microscope (microscopy) |
| ER | endoplasmic reticulum |
| ERF | ETS2-repressor factor |
| ERGIC | endoplasmic reticulum Golgi intermediate compartment |

**F**

| | |
|---|---|
| FACS | fluorescence-activated cell sorter |
| FAK | focal adhesion kinase |
| FasL | Fas ligand |
| FcRn | neonatal Fc receptor |
| FRT | FLP recombinase target |
| FTICR | Fourier transform ion cyclotron resonance |
| FV | floating villi |

**G**

| | |
|---|---|
| GAF | gamma-activated factor |
| gB | glycoprotein B |
| GBP | GTP-binding proteins |

| | |
|---|---|
| gc | glycoprotein complexes |
| GCV | ganciclovir |
| Gfi-1 | growth factor independence-1 |
| GFP | green fluorescence protein |
| gH | glycoprotein gH |
| GM-CSF | granulocyte/macrophage colony-stimulating factor |
| GMP | granulocyte–macrophage progenitor cells |
| Golgi | Golgi apparatus |
| gp | glycoprotein |
| GPCMV | guinea pig cytomegalovirus |
| GPCRs | G protein-coupled receptor |
| GPI | glycosylphosphatidylinositol |
| gpt | guanosine phosphoribosyl-transferase |
| GSK-3 | glycogen synthase kinase 3 |
| GvH | graft-versus-host |
| GvHD | graft-versus-host-disease |

## H

| | |
|---|---|
| H-2 | histocompatibility complex of the mouse |
| HAART | highly active antiretroviral therapy |
| HAT | histone acetyltransferase |
| HCG | human chorionic gonadotropin |
| HCMV | human cytomegalovirus |
| HCV | hepatitis C virus |
| HDAC | histone deacetylases |
| HEK | human embryonic kidney |
| HF | human fibroblast |
| HHV | human herpesvirus |
| HIV | human immunodeficiency virus |
| HLA | human histocompatibility leukocyte antigen |
| HMT | histone methyltransferase |
| HP1 | heterochromatin protein-1 |
| HPL | human placental lactogen |
| HSC | hematopoietic stem cell |
| HSCT | hematopoietic stem cell transplantation |
| Hsp | heat shock protein |
| HSPG | heparan sulfate proteoglycan |
| HSV | herpes simplex virus |
| HuMIG | human inducer of IFN-$\gamma$ |
| HUVEC | umbilical vein endothelial cell |
| HvG | host-versus-graft |
| HvGD | host-versus-graft disease |

## I

| | |
|---|---|
| i.c. | intracardiac |
| i.n. | intranasal |
| i.p. | intraperitoneal(ly) |
| i.v. | intravenous(ly) |
| IDO | IFN-$\gamma$ induced indoleamine 2,3 dioxygenase |
| IE | immediate-early (phase of cytomegaloviral gene expression) |
| IFN | interferon (e.g. IFN-gamma) |
| IFNAR1 | IFN-$\alpha/\beta$ receptor 1 |
| IFNGR1 | IFN-$\gamma$ receptor 1 |
| IFN$\gamma$ or IFN-$\gamma$ | interferon gamma |
| Ig | immunoglobulin |
| IHC | immunohistochemistry |
| IKK$\epsilon$: | I$\kappa$B-kinase $\epsilon$ |
| IL | interleukin (e.g. IL-2) |

| | |
|---|---|
| IMT | intimal-medial thickening |
| INM | inner nuclear membrane |
| iNOS | inducible nitric oxide synthase |
| IRAK | IL-1R-associated kinase |
| IRF | interferon regulatory factor |
| ISG | interferon stimulated gene |
| ISGF3 | IFN-stimulated gene factor 3 |
| ISRE | IFN-stimulated response element |
| ITAM | immunoreceptor tyrosine-based activation motif |
| ITIM | immunoreceptor tyrosine-based inhibitory motif |

## J

| | |
|---|---|
| JNK | Janus kinase |

## K

| | |
|---|---|
| Kan | kanamycin |
| kb | kilobase(s) |
| kbp | kilobasepair(s) |
| kDa | kilodalton |
| KIR | killer cell Ig-like receptors |
| ko | knockout |
| KSHV | Kaposi´s sarcoma associated herpesvirus |

## L

| | |
|---|---|
| L | late (phase of cytomegaloviral gene expression) |
| LAP | lamina associated polypeptides |
| LBR | lamin B receptor |
| LC | Langerhans cell(s) |
| LC | liquid chromatography |
| LC-MS/MS | LC-tandem mass spectrometry |
| LDL | low density lipoprotein |
| LDLrec | LDL receptor |
| LIM | latently infected mouse |
| LMP | late myeloid progenitor |
| LMP | latent membrane protein |
| LPS | lipopolysaccharide |
| LRR | leucine rich repeat |
| LSS | latent start site |
| LTα | lymphotoxin alpha |

## M

| | |
|---|---|
| Mφ | macrophages |
| M | molar |
| MAb | monoclonal antibody |
| MAP | mitogen-activated protein |
| MAPK | mitogen-activated protein kinase |
| mC-BP | mCP-binding protein |
| MCK | MCMV chemokine |
| MCM | mini-chromosome maintenance |
| MCMV | mouse (murine) cytomegalovirus |
| MCP | major capsid protein |
| mCP | minor capsid protein |
| MDM | monocyte-derived macrophages |
| mFasL | membrane bound FasL |
| MHC | major histocompatibility complex |
| MIC | MHC class I chain-related molecule |
| MIE | major immediate-early |
| MIEP | major immediate early promoter |

| MIP | macrophage inflammatory protein |
|---|---|
| MMP | matrix metalloproteinase |
| MNC | mononuclear cell |
| MOI | multiplicity of infection |
| MPMV | Mason–Pfizer monkey virus |
| MRF | modulator recognition factor |
| mRNA | messenger RNA |
| mRNP | ribonucleoprotein complex |
| MS | mass spectrometry |
| MudPIT | multidimensional protein identification technology |
| MULT-1 | murine UL-16 binding protein-like transcript 1 |
| MVA | modified vaccinia virus Ankara |
| MyD88 | myeloid differentiation factor 88 |

**N**

| NA | nucleoside analogue |
|---|---|
| NCR | natural cytotoxicity receptor |
| ND | nuclear domain (e.g. ND10) |
| NE | nuclear envelope |
| NES | nuclear export signal |
| NF | nuclear factor (e.g. NF-κB) |
| NIEP | non-infectious enveloped particle |
| NK cell | natural killer cell |
| NKCR | NK-cell receptor(s) |
| Nkrp-1 | NK receptor protein-1 |
| NKT cells | natural killer T cells |
| NL | nuclear lamina |
| NLS | nuclear localization signal |
| NO | nitric oxide |
| NP | nuclear pore |
| NPC | nuclear pore complex |
| NPM | nuclear pore membrane |
| NT-Ab | virus-neutralizing antibody |
| NTP | nucleoside triphosphate |

**O**

| OAS | $2',5'$-oligoadenylate synthetases |
|---|---|
| Ocil | osteoclast inhibitory lectin |
| ONM | outer nuclear membrane |
| ORC | origin recognition complex |
| ORF | open reading frame(s) |
| oriP | latent origin of DNA replication |

**P**

| p.i. | postinfection |
|---|---|
| P/CAF | p300/CBP-associated factor |
| PACS-1 | phosphofurin acidic cluster sorting protein 1 |
| PAGE | polyacrylamide gel electrophoresis |
| PAMP | pathogen associated molecular pattern |
| pAP | assembly protein precursor |
| PB | peripheral blood |
| PBMC | peripheral blood mononuclear cell |
| PCR | polymerase chain reaction |
| pDC | plasmacytoid DC |
| PFU | plaque-forming unit(s) |
| PGE | ProGlyGlu sequence |
| PI(3)K | phosphatidylinositol 3,4,5 triphosphate kinase |
| PIC | pre-initiation complex |

| | |
|---|---|
| PKA | protein kinase A |
| PKC | protein kinase C |
| PKR | protein kinase R |
| PML | promyelocytic leukemia |
| PMN | polymorphonuclear leukocyte |
| PNC | perinuclear cisterna |
| poly(A)+ RNA | polyadenylated RNA |
| pp | phosphoprotein |
| PP | pocket proteins |
| pPR | maturational protease precursor |
| pre-RC | pre-replication complex |
| PRR | pathogen recognition receptor |

**Q**

| | |
|---|---|
| qcPCR | quantitative competitive PCR |

**R**

| | |
|---|---|
| R | receptor (e.g IL-2R) |
| r | recombinant (e.g. rIFN-gamma) |
| RAE-1 | retinoic acid early inducible genes 1 |
| RCMV | rat cytomegalovirus |
| RER | rough endoplasmic reticulum |
| RhCMV | rhesus macaque cytomegalovirus |
| RIP1 | receptor interacting protein 1 |
| RNA | ribonucleic acid |
| RNAi | RNA interference |
| RRE | Rev-response element |
| RT | reverse transcription |
| RT-PCR | reverse transcriptase polymerase chain reaction |
| rVV | recombinant vaccinia virus |

**S**

| | |
|---|---|
| s.c. | subcutaneous(ly) |
| SAGE | serial analysis of gene expression |
| SBP | silencing binding protein |
| SCID | severe combined immunodeficiency |
| SCP | smallest capsid protein |
| SCT | stem cell transplantation |
| SDS | sodium dodecyl sulfate |
| SG | salivary glands |
| SGV | salivary gland-derived MCMV |
| SH2 | Src homology 2 domain |
| SHP1 | Src homology phosphatase 1 |
| siRNA | short interfering RNA |
| SIV | simian immunodeficiency virus |
| SMC | smooth muscle cell(s) |
| STAT1 | signal transducer and activator of transcription1 |
| STB | syncytiotrophoblast |
| StoS | stop-to-stop |
| SUMO | small ubiquitin modifier |
| SV | secretory vesicles |
| Syk | spleen tyrosine kinase |

**T**

| | |
|---|---|
| TAA | transplant-associated arteriosclerosis |
| TAF | TBP-associated factor |
| TAFIID | transcription-associated factor IID |
| TAFs | TBP, TFIIB, and TBP-associated factors |

| | |
|---|---|
| TAP | nuclear export factor 1 |
| TAP | transporter associated with antigen processing |
| TBK1 | TANK-binding kinase 1 |
| TBP | TATA-binding protein |
| TC | tissue culture |
| TCM | central memory T cell |
| TCR | T cell receptor for antigen |
| TEM | effector memory T cell |
| tet | tetracycline |
| TGF | transforming growth factor |
| TGN | trans-Golgi network |
| Th | CD4-positive T helper |
| TIMP-3 | tissue inhibitor of metalloproteinases 3 |
| TIR | Toll/IL-1 receptor |
| TLR | toll-like receptor |
| TNF | tumor necrosis factor |
| TPR | tetrapeptide repeat |
| TRAF6 | TNF-receptor-associated factor 6 |
| TRAIL | TNF-related apoptosis-inducing ligand |
| TREX | transcription/export complex |
| TRIF | TIR–domain-containing adaptor protein inducing IFN–$\beta$ |
| ts | temperature sensitive |
| TSA | Trichostatin A |
| TVS | transplant vascular sclerosis |

## U

| | |
|---|---|
| UL | unique long domain |
| ULBP | UL binding protein |
| US | unique short domain |
| UV | ultraviolet (light) |

## V

| | |
|---|---|
| V region | variable region of Ig or of TCR |
| VAK | virus activated kinases |
| ValGCV | valganciclovir |
| VCAM-1 | vascular endothelial adhesion molecule-1 |
| VEGF | vascular endothelial growth factor |
| Viperin | virus inhibitory protein, endoplasmic reticulum-associated, and interferon-inducible |
| VIPR | viral protein interfering with antigen presentation |

## W

| | |
|---|---|
| WT | wild-type |

## X

| | |
|---|---|
| X-Gal | 5-bromo-4-chloro-3-indolyl-beta-D-galactopyranoside |

## Y

| | |
|---|---|
| YY1 | Yin Yang-1 |

## Z

| | |
|---|---|
| ZAP70 | 70 kd zeta-associated protein |

## Miscellaneous

| | |
|---|---|
| $\beta_2$m | $\beta_2$ microglobulin |
| 2':5'-OAS | 2'5'-oligoadenylate synthetase |
| 3D | three-dimensional |

# Preface

## From protozoan to proteomics

On September 23, 2003, I happened to catch an infected e-mail that had passed through our university's firewall, spam filters and virus alert. It contained a particularly insidious and malicious virus that occupied a great deal of my time – and actually still does! As you may have already guessed, I'm talking about cytomegalovirus. On that fateful day when Annette Griffin introduced herself as an Acquisitions Editor with Horizon Scientific Press to ask whether I would be interested in editing a book on CMV, I was of course hesitant and colleagues warned me of the enormity of the task. What finally convinced me to undertake this project, however, was the proposed concept of the book. The aim was a publication that 'details the cutting-edge research and future potential in this increasingly important field', a publisher's view likely to meet with general approval in the scientific CMV community. I also agreed completely with the publisher's idea that the book 'should bring together recent research on human and animal cytomegaloviruses with chapters on infection models, pathogenic mechanisms, interactions with the host, genomics and molecular biology, latency and reactivation, and strategies for control'. At a time when parallel sessions are planned by conference organizers according to discipline rather than on the basis of pending problems that need to be solved by integrating different views and approaches, there is a risk that we will fail to understand CMV disease in all its complexity because molecular biologists and immunologists no longer listen to each other. Following my initial training in immunology, as a graduate student in the Koszinowski lab I profited a great deal from my mentor's strategy of combining the ideas and techniques of immunologists and molecular biologists. By editing this book, I have profited again from reading more than a thousand manuscript pages covering a broad spectrum of CMV research, and I certainly had plenty of new ideas for my own lab's to-do-list! There are many good monographs, articles in review series, and journal special issues on selected topics. Yet, to the best of my knowledge, this book is without precedent in that it provides an overview of current opinion and cutting-edge research on literally all aspects of CMV infections, although with a focus on basic science that was intended by the publisher. Rather than being asked to write a lecture book chapter or a comprehensive review of literature, the authors were encouraged to give opinions and to put forward hypotheses.

What I liked most was Horizon Press's plan to 'bring together recent research on human and animal cytomegaloviruses'. One cannot deny that there is some distance between these two groups. On the one hand, basic research in animal models sometimes ignores the pending clinical questions; on the other hand, information from animal models is sometimes ignored and rediscovered years later, often without fair referencing. This seems to have become something of a tradition. When I was a young postdoctoral scientist visiting Dr. Monto Ho in the Cathedral of Learning at the University of Pittsburgh, he told me an anecdote that can also be found in his classic book *Cytomegalovirus Biology and Infection* (Ho, 1982). It actually dates back to a very honest report given by Dr. Weller on the history of the isolation of human CMV (Weller, 1970): Dr. Margaret Smith, well known in MCMV research for the isolation of the Smith strain, used the mouse salivary gland virus as a model to establish the methods for isolation of the agent that causes cytomegalic inclusion disease (CID) in humans. Her pioneering accomplishment in the mouse model was published in 1954 (Smith, 1954). It is less well known that she almost simultaneously succeeded in growing the corresponding agent from the salivary gland of an infant. She even discovered that the human agent does not produce cytopathic lesions in mouse tissue nor, indeed, the mouse one in human tissue. She was convinced – and rightly so, as became clear soon after – that each of the two agents produces cytopathic lesions only in homologous tissue, which reflects the species specificity of CMVs. Sadly, her submitted manuscript on the human agent was initially rejected by the journal's editors on the grounds that she might have propagated the mouse virus in human tissue. It therefore took another two years until human CMV became known to medical virologists as a result of back-to-back publications written by Rowe and colleagues and by Margaret Smith (Rowe et al., 1956; Smith, 1956), followed one year later by a publication from the Weller lab (Weller et al., 1957). Only recently have we begun to understand the molecular basis of the species specificity of CMVs, and, again, the mouse model is about to pave the way. It appears that the introduction of a single antiapoptotic gene into the genome of MCMV confers the ability to replicate in human cells on the virus (Wolfram Brune, pers. comm.; and Jurak and Brune, Abstract at the ASM Conference on Viral Immune Evasion, Acapulco, March 2005).

The search for the cradle of cytomegalovirus research takes us to Bonn, Germany, to a meeting of 14 members of the medical section of the Natural History Society of Prussian Rhineland and Westphalia that was chaired by Dr. Leo on June 27, 1881 (Andrä, 1881). During this meeting, the pathologist Dr. Ribbert gave a case report on a stillborn infant with Lues (syphilis)-like symptoms and interstitial nephritis associated with the presence of tremendously enlarged 'cytomegalic' cells that were also characterized by an enlarged nucleus. It was not until 1904 (Jesionek and Kiolemenoglou) that these 'curious cells' were observed again in the kidneys, lungs, and liver of a luetic fetus. Although the cytomegalic cells were then mistaken for protozoa, specifically for gregarines, the precision with which the cytopathic effect was described and illustrated deserves our highest respect (Figure 1). Unfortunately, when Ribbert saw the report by Jesionek and Kiolemenoglou on the protozoan-like cells, he reinterpreted his original observation in support of the protozoan hypothesis (Ribbert, 1904). This led Monto Ho (1982) to conclude that 'he [Ribbert] was unable to interpret his observation until he saw the report by Jesionek and Kiolemenoglou who had noted for the first time the presence of protozoan-like cells in lungs, kidneys and liver'. As a native German speaker, I have had the opportunity of reading the original minutes from the meeting of June 27, 1881, and for me there is no doubt at all that Ribbert was the first to correctly note the enlargement of cells and cytopathological alterations that are distinctively characteristic of CMV disease.

In 'homage' to the intranuclear inclusion body that is pathognomonic of CMV, we have compared the highly skilled sketch produced by Jesionek and Kiolemenoglou in 1904 (Figure 1) with 'modern' in situ hybridization photographic images taken a century later (Figure 2). Hybridization with virus-specific DNA probes identifies the intranuclear inclusion body as the site at which viral DNA is concentrated, and we know today that this is the site of viral DNA packaging into nucleocapsids (see Chapter 12 by Wade Gibson). After an interval of a hundred years, it is instructive to compare the margination of cellular chromatin that is condensed in 'polar bodies', which, as we can show today, do not usually contain viral genomes.

There is another short scientific anecdote relating to Figure 2. Here, the murine model was used to study in vivo coinfection with MCMV-WT and a mutant virus in which the gene of interest was deleted by replacement (Cicin-Sain et al., 2005; for a commentary, see Tremp, 2005). In the normal mouse liver, it is a relatively frequent event that lack of cytokinesis after mitosis leads to hepatocytes that possess two nuclei. The two viruses happened to coinfect such a binuclear hepatocyte; interestingly, MCMV-WT conquered one nucleus and the mutant virus the other. This finding implies that, in each of the two nuclei, a single viral genome molecule was successful in evading the host cell's epigenetic defense mechanisms (see Chapter 7 by Qiyi Tang and Gerd G. Maul).

This example brings us back to the original purpose of the book – to 'CMV today'. We leave behind the protozoan and turn instead to proteomics, genomics, and all the other topics of cutting-edge science.

Matthias J. Reddehase

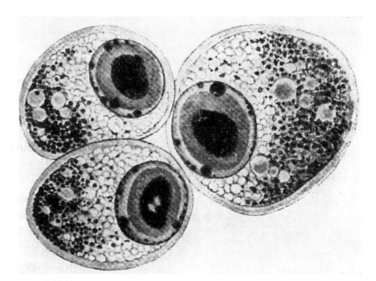

**Figure 1** Intranuclear inclusion body as seen in the year 1904. Protozoan-like "owl's eyes" cells from the left kidney of an alleged luetic fetus (Jesionek and Kiolemenoglou, 1904) portrayed from the Zeiss microscopic image after staining with hematoxylin and eosin. Reproduced by permission of the Münchner Medizinische Wochenschrift (MMW)-Fortschritte in der Medizin.

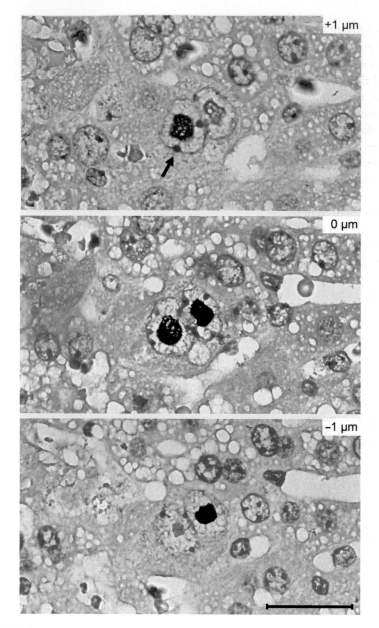

**Figure 2** 100th anniversary of the intranuclear inclusion body in the year 2004. Binuclear hepatocyte in the liver of an immunocompromised mouse co-infected with MCMV-WT (red) and an M36 gene (Menard et al., 2003) deletion mutant (black). Shown are serial 1-μm sections analyzed by two-color in situ hybridization. (Top) DNA probe specific for MCMV-WT. The arrow points to displaced cellular chromatin that is condensed in a so-called polar body. (Middle) DNA probes specific for MCMV-WT and the M36 gene deletion mutant. (Bottom) DNA probe specific for the mutant. Bar represents 25 μm. Courtesy of J. Podlech, Institute for Virology, Johannes Gutenberg-University, Mainz, Germany

## Acknowledgments

In the first place I have to thank Annette Griffin, Horizon Press's Acquisitions Editor, for her patience with me and all the other authors from the 'peculiar CMV community'. I have appreciated the cooperation of all the authors; they have invested their valuable time and were willing to share their scientific ideas with us. Special thanks go to Ulrich H. Koszinowski, whose lab I joined in 1980 as a student of immunology and whom I left in 1994 as a full professor and chair of virology. I wouldn't be editor of this book without him and the members of the 'early Koszinowski lab' in Tübingen: Günther Keil, Angelika Ebeling-Keil, our electron microscopist Frank Weiland, Stipan Jonjic, Margarita Del Val, Martin Messerle, Brigitte Bühler, Konrad Münch, Mathias Fibi, Wolfgang Mutter, and so many others who accompanied me on the way. Finally, this book would not have been possible without the superb technical help of my Assistant Editor Niels Lemmermann.

By defraying in part the costs for color figures, printing of this book is supported by the Deutsche Forschungsgemeinschaft, Sonderforschungsbereich (Collaborative Research Grant) 490 'Invasion and Persistence in Infections'.

## References

Andrä, C.J. (ed.). (1881). Verhandlungen des Naturhistorischen Vereines der preussischen Rheinlande und Westfalens. Achtunddreissigster Jahrgang. Vierte Folge: 8. Jahrgang. (Bonn: Max Cohen & Sohn), pp. 161–162.

Cicin-Sain, L., Podlech, J., Messerle, M., Reddehase, M.J., and Koszinowski, U.H. (2005). Frequent co-infection of cells explains functional in vivo complementation between cytomegalovirus variants in the multiply-infected host. J. Virol. 79, 9492–9502.

Ho, M. (1982). Cytomegalovirus. Biology and Infection. (New York: Plenum Medical Book Company).

Jesionek, A., and Kiolemenoglou, B. (1904). Ueber einen Befund von protozoënartigen Gebilden in den Organen eines hereditär-luetischen Fötus. Münchener Medizinische Wochenschrift 51, 1905–1907.

Menard, C., Wagner, M., Ruzsics, Z., Holak, K., Brune, W., Campbell, A.E., and Koszinowski, U.H. (2003). Role of murine cytomegalovirus US22 gene family members in replication in macrophages. J. Virol. 77, 5557–5570.

Ribbert, H. (1904). Über protozoenartige Zellen in der Niere eines syphilitischen Neugeborenen und in der Parotis von Kindern. Zbl. Allg. Pathol. 15, 945–948.

Rowe, W.P., Hartley, J.W., Waterman, S., Turner, H.C., and Huebner, R.J. (1956). Cytopathogenic agent resembling human salivary gland virus recovered from tissues cultured of human adenoids. Proc. Soc. Exp. Biol. Med. 92, 418–424.

Smith, M.G. (1954). Propagation of salivary gland virus of the mouse in tissue cultures. Proc. Soc. Exp. Biol. Med. 86, 435–440.

Smith, M.G. (1956). Propagation in tissue cultures of a cytopathogenic virus from human salivary gland virus (SGV) disease. Proc. Soc. Exp. Biol. Med. 92, 424–430.

Tremp, A. (2005). Fatal alliance. Nature Rev. Microbiol. 3, 669.

Weller, T.H. (1970). Cytomegaloviruses: the difficult years. J. Infect. Dis. 122, 532–539.

Weller, T.H., Macauley, J.C., Craig, J.M., and Wirth, P. (1957). Isolation of intranuclear inclusion producing agents from infants with illnesses resembling cytomegalic inclusion disease. Proc. Soc. Exp. Biol. Med. 94, 4–12.

# Human Cytomegalovirus Infections and Mechanisms of Disease

*William Britt*

## Abstract

The pathogenesis of infections with human cytomegalovirus (HCMV) have been modeled in small animals and primates utilizing the respective CMVs. In most cases, acute infection is associated with significant levels of virus replication and dissemination to multiple organs. In the immunocompetent animal, such infections are rapidly controlled by a number of effector functions of the innate and adaptive immune response. The pathogenesis of acute HCMV infections can be readily explained by the control of virus replication and the resolution of virus-induced cytopathology. There appears to be a linkage between levels of virus replication, organ dysfunction, and disease in patients as well as in experimental models with acute CMV infections. In contrast, chronic infections with CMV have as a major component of their pathogenesis a bidirectional relationship between viral gene expression and the host inflammatory response such that viral persistence is facilitated by the host inflammatory response and the host inflammatory response is fueled by the presence of the virus. In these cases, disease can be attributed to both viral and host functions. The viral gene products that appear to play a role in chronic inflammation have evolved with CMVs and are likely unimportant for replication in vitro. As such, defining the role of these viral functions in disease associated with chronic HCMV infections almost certainly will require relevant animal models.

## Introduction

Clinical syndromes associated with human cytomegalovirus (HCMV) infections have been described for nearly a century. Although the pathological consequences of fetal infection with HCMV were among the first reported in the medical literature, HCMV was recognized as an opportunistic pathogen in immunocompromised patients soon after widespread introduction of allograft transplantation (Ho, 1991; Rifkind, 1965; Rubin, 2002; Singh et al., 1988). In the recent past, HCMV was the most frequent opportunistic pathogen in HIV-infected patients and, although severe HCMV infections in patients with AIDS have become less commonplace in patients treated with active retroviral therapy (ART), the long-term consequences of HCMV infection in AIDS patients remain undefined. The disease syndromes associated with HCMV in immunocompromised patients can be related to virus replication. In contrast, disease syndromes that have been linked to chronic or persistent HCMV infection are less well described and in some cases, remain only interesting associations. Similarly, recent speculations into the role of HCMV in human malignancy are consistent with the modifications of the cell cycle, cellular stress and survival responses, and apoptotic responses that follow HCMV infection (Arnoult et al., 2004; Castillo and Kowalik, 2002; Castillo and Kowalik, 2004; Fortunato et al., 2000; Fortunato and Spector, 1998; Goldmacher et al., 1999; Kalejta et al., 2003; Michaelis et al., 2004; Mocarski, 2001; Skaletskaya et al., 2001; Wang et al., 2001; Yu and Alwine, 2002; Zhu et al., 1995). Yet, when examined closely, available literature has provided only limited definitive information about specific pathogenic mechanisms during acute HCMV syndromes. Much less is known about the pathogenesis of diseases associated with chronic or persistent HCMV infection.

Although the pathogenic mechanisms of HCMV infections remain poorly understood, arbitrarily separating disease syndromes into those associated with acute infection and those with chronic or persistent infection can simplify the discussion of the pathogenesis of HCMV infections. The virologic and immunologic characteristics of acute and chronic HCMV syndromes are listed in Table 1.1. Differences between acute and chronic disease syndromes in terms of virus replication and immunological responses are less distinct than suggested in Table 1.1; however, these are the more general characteristics and point to key aspects of the pathogenesis of HCMV infections. Virus replication has been correlated with the risk of virus dissemination, end-organ disease, and the risk of morbidity and mortality from HCMV infection in a number of different clinical settings in the immunocompromised host (Chemaly et al., 2004; Cope et al., 1997; Emery, 1999; Emery et al., 2000; Erice et al., 2003; Gor et al., 1998; Limaye et al., 2001; Nichols et al., 2001; Razonable et al., 2003; Spector et al., 1999; Spector et al., 1998). In these clinical populations, intervention with antiviral drugs has been shown to prevent dissemination (prophylaxis), limit virus dissemination and disease (pre-emptive therapy), and morbidity and mortality from HCMV infection (treatment) (Badley et al., 1997; Boeckh, 1999; Emery, 1999; Ljungman, 2001; Paya et al., 2002; Razonable et al., 2003; Schmidt et al., 1991; Singh, 2001; Spector et al., 1993; Winston et al., 2003; Wreghitt et al., 1999). Thus, it appears that the pathogenesis of these HCMV associated disease syndromes can be related to the failure of the host to limit virus replication and spread, perhaps secondary to the failure of components of the adaptive immune response, including both antiviral antibodies and cytotoxic T-lymphocytes (Alberola et al., 2001; Boppana et al., 1995; Chou et al., 1987; Komanduri et al., 2001; Li et al., 1994; Pass et al., 1983; Rasmussen et al., 1994; Reusser et al., 1997; Schoppel et al., 1997; Schoppel et al., 1998). However, with the exception of the most severely immunocompromised hosts, such as bone marrow allograft recipients and patients with AIDS that have failed ART, only a subset of patients who are at risk develop invasive HCMV disease. Furthermore, the course of infection in most patient populations can be highly variable, even in the most immunosuppressed patients. Thus, it appears that as yet undefined aspects of HCMV biology contribute significantly to the pathogenesis of disease in the group of patients with acute HCMV infection.

In contrast, the pathogenesis of disease manifestations that have been associated with chronic or persistent HCMV infection is not well understood. These disease syndromes have not been consistently related to levels of virus replication, and virus-associated end-organ disease is usually limited to a single organ or organ system. Although virus replication is required for induction of these clinical manifestations, studies have provided conflicting

**Table 1.1** Characteristics of acute and chronic disease associated with HCMV

| | Virus replication[1] | Multiple organ disease[2] | Mechanisms of immune control[3] | Immune-associated disease[4] |
|---|---|---|---|---|
| **Acute disease** | | | | |
| Mononucleosis syndrome | High | Yes | Innate/adaptive | No |
| Congenital infections; perinatal infections in premature infants | High | Yes | Unknown | Yes[5] |
| CMV syndromes in allograft recipients | High | Yes | Adaptive | No[5] |
| Disseminated Infections In AIDS patients | High | Yes | Adaptive | No[5] |
| **Chronic disease** | | | | |
| Congenital infection | Low | No | Unknown | Yes[6] |
| Vascular disease in transplant recipients | Low | No | Adaptive | Yes |
| Vascular disease in normal hosts | Low | No | Adaptive | Yes |
| Inflammatory disease and malingnacy[7] | Low | No | Unknown | Yes |

[1]Increased levels of virus replication determined by virus titration and/or viral burden assayed by quantitative PCR in peripheral compartments such as blood or within infected organs.
[2]Multiorgan disease include infection in liver/spleen, lung, gastrointestinal tract, CNS (congenital), eye, and in allograft recipients in the transplanted organ.
[3]Innate functions include interferons, NK cell responses. Adaptive responses include antiviral antibodies and CD4+/CD8+ T-lymphocyte responses.
[4]Immune-mediated disease defined by organ dysfunction associated with inflammatory cell infiltrates in virus infected tissue, but specific immunopathological mechanisms have not been identified.
[5]Studies in a small number of humans and in experimental models of CMV retinitis suggest that disease is associated with apoptosis of retinal epithelial cells, perhaps secondary to expression of Fas ligand (Buggage et al., 2000; Cinatl et al., 2000; Scholz et al., 2003; Zhang and Atherton, 2002)
[6]Congenitally infected infants exhibit pathologic findings attributable to virus replication; however, persistent replication in infected infants and late manifestations such as hearing loss suggest that other mechanisms, including immunopathologic damage, may account for late onset disease.
[7]HCMV genetic material and viral proteins have been detected in human malignancies together with inflammatory cellular infiltrates. HCMV has been found in biopsy specimens from patients with inflammatory diseases of the gastrointestinal tract.

information as to whether virus replication is required for continued disease activity (De La Melena et al., 2001; Lautenschlager et al., 1997; Lemstrom et al., 1997; Streblow et al., 2003; Zhou et al., 1999). The immune-mediated control during persistent HCMV infection is assumed to include both innate and adaptive components, yet neither function appears sufficient to eradicate virus persistence.

## Acute HCMV infections: clinical and pathogenic aspects

HCMV infections characterized by high levels of virus replication that are observed in immunocompromised hosts often include a set of clinical abnormalities similar to that observed in experimental animals infected with their CMV. Infections in normal, non-immunosuppressed hosts are infrequently associated with clinically apparent disease, but when symptomatic, these patients often develop clinical findings that are similar to those observed in immunocompromised patients. Several features of infection in normal hosts are exaggerated in the immunocompromised host including the level and duration of virus replication, laboratory evidence of organ involvement such as hepatitis, and clinical findings such as fever. A comparison of clinical and virologic parameters of HCMV infection in human populations with HCMV disease is provided in Table 1.2.

Interestingly, animal models of CMV infection including the mouse and guinea-pig model recapitulate many aspects of acute HCMV infection (Bia et al., 1983; Griffith and Aquino-de Jesus, 1991; Kern, 1999; Krmpotic et al., 2003; Reddehase et al., 1994; Reddehase et al., 2002; Trgovcich et al., 1998; Woolf, 1991). In addition, recent studies in the rhesus macaque have provided further evidence for the role of virus replication in the acute disease associated with HCMV infection (Chang et al., 2002; Kaur et al., 2003; Lockridge et al., 1999; Sequar et al., 2002; Tarantal et al., 1998). In each of these animal models, the level of virus replication correlates with disease activity, and organ damage and disease can be attributed to viral cytopathology. Most studies in animal models have utilized subcutaneous, intraperitoneal, or intravenous inoculation, thus bypassing mucosal sites of virus replication and local immune responses. Early infection of the liver and spleen is observed in these animals and the secondary viremia that follows, seeds sites of persistence such as the salivary glands and bone marrow (Griffith and Aquino-de Jesus, 1991; Kern, 1999; Lockridge et al., 1999; Reddehase et al., 1994; Reddehase et al., 2002).

Both innate and adaptive immune responses have been shown to be involved in the control of virus replication and disease resolution (Baroncelli et al., 1997; Chatterjee et al., 2001; Harrison and Caruso, 2000; Jonjic et al., 1994; Kaur et al., 1996; Koszinowski et al., 1991; Podlech et al., 2000; Polic et al., 1998; Reddehase et al., 1988; Reddehase et al., 1987; Reddehase et al., 1985; Sequar et al., 2002; Steffens et al., 1998). Although adaptive immune responses, particularly virus-specific, CD8[+] CTL responses, more efficiently eliminate virus infected cells than innate responses, the latter responses have been shown to provide a significant but incomplete level of protective immunity in MCMV-infected mice lacking adaptive immunity (Krmpotic et al., 2002; Polic et al., 1998). Similar clinical and laboratory findings have been reported in rhesus macaques inoculated intravenously with RhCMV, including the development of hematological abnormalities and evidence of hepatocellular disease (Lockridge et al., 1999). In addition, rhesus macaques infected with RhCMV by mucosal routes also exhibit spread to the liver and spleen presumably following a viremia that occurs after virus replication in local tissue, including regional lymphoid tissue. However, in these animals, laboratory abnormalities in liver function and hematologic parameters are less often observed than in those animals inoculated intravenously suggesting that local control of virus replication can suppress virus replication and limit the level of virus inoculum reaching the liver and spleen (Lockridge et al., 1999). Findings consistent with those in RhCMV-infected animals have been reported in mice given MCMV by oral routes (S. Jonjic, personal communication).

**Table 1.2** Clinical and laboratory findings in acute CMV infections

| Population | Symptoms | Laboratory findings | Virologic findings | Immune deficit |
|---|---|---|---|---|
| Normal host (mononucleosis; transfusion associated infections) | Fever, fatigue, adenopathy, hepatitis, splenomegaly | Leukopenia, thrombocytopenia, hepatocellular damage | Viremia, viruria; persistent virus excretion (months) | Unknown; innate and adaptive responses intact but delayed |
| Fetus; premature infant | Hepatosplenomegaly, microcephaly, jaundice | Thrombocytopenia, hepatocellular damage, CNS damage | Viremia, viruria; persistent virus excretion (months–years) | Unknown; delayed adaptive responses; innate responses present |
| Organ transplant recipients | Fever, fatigue, adenopathy, hepatitis, splenomegaly, pneumonitis | Leukopenia, thrombocytopenia, hepatocellular damage, acute graft dysfunction | Viremia, viruria, persistent virus excretion (months) | Depressed or delayed adaptive immune responses |
| AIDS | Fatigue, hepatitis, colitis, retinitis | Hepatocellular damage | Viremia, viruria | Loss of adaptive immune responses |

These data argue that local control of virus replication plays an important role in the level of virus replication and ultimately, the quantity of virus that reaches secondary targets of infection such as the liver and spleen. Reduction in the inoculum size that reaches organs such as the liver and spleen presumably also alters viral burden during persistence by decreasing the quantity of virus dissemination to sites of viral persistence (Reddehase et al., 1994). Although the control of virus replication at sites of infection has not been experimentally defined, it is likely that innate immune responses play a critical role in limiting virus replication during the early phases of virus infection (see also Chapter 15). Innate immune responses, including both cellular functions such as NK cells and soluble mediators such as interferons likely limit virus replication within local sites prior to the recruitment of immune effector functions of the adaptive immune system (Bukowski et al., 1984; Salazar-Mather et al., 1998). It must be assumed that these same innate effector functions also limit virus replication and spread within regional tissue as well as within organs such as the liver and spleen, thus providing some degree of protective immunity at each step of virus amplification and dissemination.

Although HCMV infection is nearly universal in most populations, only rarely do normal individuals exhibit the symptoms described in Table 1.2. Thus, the immune response to HCMV can be considered as being protective in normal individuals, even in the absence of any preexisting immunity. The term "protective immunity" must then be qualified by defining protective immunity as protection from symptomatic disease and clinically apparent organ dysfunction. In the small number of cases of acute HCMV infection that have been studied, prolonged virus replication, viremia, and viruria have all been reported in asymptomatic individuals (Drew et al., 2003; Horwitz et al., 1986; Jordan et al., 1973; Pannuti et al., 1985; Zanghellini et al., 1999). These results argue that protective immunity is sufficiently robust to limit end-organ disease but relatively ineffective in limiting virus replication and spread.

An obvious question that has not been well studied is why do these immune responses fail to protect some normal immunocompetent individuals from disease, such as those patients with community-acquired, mononucleosis syndromes? Several obvious explanations have been suggested including a difference in inoculum size such that individuals with symptomatic infections were infected with larger amounts of virus leading to more rapid spread to the liver and spleen. Although there is no direct evidence for this proposed mechanism of disease in HCMV infections, studies in experimental animals point to the relationship between inoculum and organ involvement (Bernstein, 1999; Bia et al., 1983; Kern, 1999; Krmpotic et al., 2003; Lockridge et al., 1999; Shanley et al., 1993; Trgovcich et al., 2000). Furthermore, the course of disease, including the interval to symptomatic disease following exposure to infected blood products, in patients with transfusion-acquired HCMV infection is consistent with exposure to a larger effective inoculum secondary to a blood-borne infection (Adler, 1983; Bowden, 1995; Lang and Hanshaw, 1969; Preiksaitis et al., 1988; Yeager, 1974). Another possibility that could account for the rare failures in controlling the development of disease in normal populations is the presence of specific deficits in immune responsiveness in a small subset of individuals within the population. The role of the MHC in control of protective responses to virus infection is a fundamental principle of immunology and individuals with deficits in NK function have been shown to be at risk for severe HCMV infections (Biron and Brossay, 2001; Biron et al., 1989). Recent studies have demonstrated the influence of MHC genes on the recognition of immuno-dominant antigens of HCMV, including the role of MHC polymorphism in the CD8[+] CTL recognition of a dominant viral antigen (Zaia et al., 2001). However, it is unlikely that MHC polymorphisms could account for restricted recognition of HCMV antigens considering the plethora of potential antigens encoded by this virus, a possibility that has recently been addressed in comprehensive studies utilizing overlapping peptides to map antigenic epitopes in the entire HCMV genome (L. Picker, personal communication).

More intriguing possibilities that could account for differences in the phenotype of infection among normal individuals from an outbred population such as humans, include infection with more virulent strains of HCMV or more efficient expression of viral immune evasion functions secondary to polymorphisms in the genotype of the human host. Studies utilizing MCMV deletion mutants have clearly demonstrated that replication-competent viruses can vary greatly in their replicative capacity when assayed in vivo as well as in their in vitro cellular tropism (Brune et al., 2001b; Hanson et al., 2001; Menard et al., 2003; Saederup et al., 2001). Thus, it is possible that strains of HCMV with expanded cellular tropism or strains with more virulent phenotypes circulate within the population.

Furthermore, it is well recognized that rodent CMVs including guinea-pig CMV acquire a more virulent phenotype when passaged through their natural host. Several different explanations have been offered for this observation, ranging from propagation of viruses with increased virulence whose titers are amplified following in vivo passage to a cytokine storm that may arise following infection with virus in salivary gland homogenate (Trgovcich et al., 2000).

Regardless of the explanation for this alteration in viral phenotype, the observation that in vivo passaged virus can exhibit increased virulence raises the possibility that variation in the virulence of infecting strains of HCMV secondary to their passage through specific hosts and host tissue could account for the differences in clinical outcomes of HCMV patients. Studies in MCMV have documented co-infections with multiple strains of virus can serve to complement viral functions, including potential virulence functions, during infection with non-clonal mixtures of viruses (U. Koszinowski, personal communication). It is also a near certainty that humans are also infected with mixtures of HCMVs and depending on the phenotypic composition of these viral mixtures, could exhibit markedly different responses to HCMV infection independent of the genetic make-up of the host.

Similarly, differences in the interplay between host responses and virus-encoded immune evasion gene products could result in phenotypic differences in outcomes in individuals infected with HCMV. Studies of the pathogenesis of MCMV infections have demonstrated that the host genotype can be a determinant in the activity of viral-encoded immune evasion genes (Wagner et al., 2002). As an example of how MHC polymorphism can influence the outcome of viral immune evasion functions is the relative insensitivity of the H-2K[b] allele to the MHC down-regulation by MCMV *m04*, *m06*, and *m152* genes (Wagner et al., 2002). Thus, it is possible that certain alleles would be less sensitive to immune evasion functions encoded by MCMV. Other investigators have demonstrated that the HCMV US2 gene product binds specific products of the HLA-A locus and not to some alleles in the HLA-B locus, thus exhibiting some HLA specificity in its proposed immune evasion functions (Gewurz et al., 2001). Therefore, effectiveness of immune evasion activity could be MHC allele-specific and account for differential levels of virus replication and possibly even symptomatic infection in some individuals.

## Acute infections in immunocompromised hosts

The pathogenesis of acute HCMV infections in immunocompromised hosts such as allograft recipients and patients with AIDS has been related to the level of virus replication in several studies (Cope et al., 1997; Emery, 1999; Gewurz et al., 2001; Spector et al., 1999). A similar relationship exists in small animal models and non-human primate models of HCMV infection in immunocompromised hosts (Brune et al., 2001a; Krmpotic et al., 2003; Lockridge et al., 1999). In fact, in HCMV infections in solid-organ allograft recipients, it has been argued that the viral burden is the most reliable predictor of invasive infection with HCMV (Cope et al., 1997; Emery, 1999). However, more careful examination of the reported data has suggested that invasive disease cannot routinely be predicted from an absolute level of viral burden, but that the rate of increase in the viral burden is perhaps best correlated with the development of invasive, end-organ infections (Emery, 1999; Emery et al., 2000). In the majority of these studies, viral burden has been assayed in the blood compartment and not in target organ tissue raising the possibility that blood viral burden is not reflective of organ viral burden. Results from studies in an experimental murine model of CMV infections have indicated that monitoring the viral burden in the blood compartment only indirectly correlates with viral replication in infected organs (Brune et al., 2001a). This experimental finding provides a possible explanation for the lack of a linear correlation between viral burden and disease and suggests that increasing viral burdens are of more prognostic value in immunocompromised hosts. It could be argued that an increasing viral burden is evidence of uncontrolled virus replication with a high risk of dissemination to other organs.

The loss of adaptive immune responses has been shown in both animal models and in immunocompromised humans to lead to uncontrolled virus replication and widespread viral dissemination. Studies in immunocompromised mice infected with MCMV have detailed the role of various adaptive immune effector functions in the control and resolution of MCMV replication and dissemination. Both CD8$^+$ and CD4$^+$ T-cells have been proposed to play a major role in resistance to disseminated infection in this model of human disease (Jonjic et al., 1988; Koszinowski et al., 1991; Krmpotic et al., 2003; Polic et al., 1996; Steffens et al., 1998; Volkmer et al., 1987). MCMV-specific CD8$^+$ CTLs have been shown in several different experimental systems to provide protective immunity to immunocompromised mice infected with MCMV (Jonjic et al., 1988; Reddehase et al., 1988; Steffens et al., 1998). Interestingly, MCMV-specific virus neutralizing antibodies were also demonstrated to play a key role in limiting virus dissemination in this mouse model of human disease (Jonjic et al., 1994; Rapp et al., 1993). Similar findings have been described for immunocompromised humans with HCMV infections (Schoppel et al., 1997; Snydman et al., 1987; Snydman et al., 1993; Winston et al., 1987; Winston et al., 1993). Early studies provided correlative associations between the loss of T-lymphocyte reactivity for HCMV and development of disease (Reusser et al., 1991). Immunotherapy with HCMV-specific CD8$^+$ T-cells provided passive protection in bone marrow transplant recipients (Walter et al., 1995). Furthermore, long-term protection was not achieved in these patients unless an endogenous HCMV-specific CD4$^+$ T-cell response was reconstituted (Boeckh et al., 2003; Walter et al., 1995). This result suggested that protective immunity to this persistent virus infection requires virus-specific CD4$^+$ T-cell responses, a finding that has subsequently been demonstrated in studies of other persistent virus infections (Dittmer et al., 1999; Sun and Bevan, 2003). In the setting of high viral loads and appropriate CD4+ and CD8+ T-cell responses, the role that viral-encoded evasion functions or viral genes encoding expanded cellular tropism may play in the control of virus replication and disease has not been well studied. Although the role of immune evasion functions in the resistance to immune-mediated viral clearance has not been defined, it could be argued that expression of these viral genes is responsible for the phenotypic variability in the severity of HCMV infections in the post-transplant period.

## Acute HCMV infection in the fetus and premature newborn infant: the unique situation of intrauterine infection

Infection of the developing fetus in utero (congenital or present at birth) and, less frequently, severe infection in the premature newborn following exposure to breast milk or transfused blood products are both characterized by high levels of virus replication and multiorgan involvement (Anderson et al., 1996; Boppana et al., 1997; Boppana et al., 1992; Hamprecht et al., 2001; Stagno et al., 1980; Vochem et al., 1998; Yeager, 1974; Yeager et al., 1981). These infants can present with severe infections characterized by organ-threatening and, less frequently, life-threatening disease. Prominent among the clinical findings are hepatitis, splenic enlargement, increased intravascular clearance of blood cells, and bone marrow depression as manifested by anemia and decreased platelet count in the peripheral blood, and in the case of congenitally infected infants, central nervous system (CNS) involvement.

The pathogenesis of the disease in infected infants is most readily explained by virus replication and cellular damage secondary to viral cytopathology (Becroft, 1981). However, one notable exception is the disease in the CNS. The pathology of CNS disease in infants with congenital HCMV infection has been reported for only a limited number of patients and a comprehensive description of pathological changes in the brain of a large number of infected infants is lacking. A number of abnormalities have been described, including findings consistent with migration deficits, calcifications and loss of cellularity, and areas of focal cerebritis and cerebellar hypoplasia (Malm et al., 2000; Perlman and Argyle, 1992). In most cases the abnormalities are symmetrical and are consistent with involvement of the neuroepithelium of the subventricular zone and/or disease secondary to vasculitis in the CNS (Perlman and Argyle, 1992).

Whether disease is secondary to virus-induced cytopathology in cells of the neuroepithelium and/or supporting cells including the glia or, the loss of vascularity is not known. Alternatively, disease could be secondary to inflammation within the CNS, a hypothesis consistent with the finding of inflammatory cells in the CNS of infants with congenital HCMV infection (Balcarek et al., 1993). In general, CNS disease remains static in the postnatal period with the exception of involvement of the auditory system. Auditory abnormalities develop in nearly 8% of infants with congenital HCMV infection and can develop after delivery at a time when peripheral diseases such as hepatitis have resolved (Dahle, 2000; Fowler et al., 1997; Williamson et al., 1992). In addition, auditory abnormalities can progress in the first several years of life suggesting ongoing disease in the auditory system (Dahle, 2000; Fowler et al., 1997; Williamson et al., 1992).

Owing to the absence of fundamental pathologic information, the pathogenesis of the auditory disease in congenital HCMV infection is even less well understood and has been suggested to be secondary to viral cytopathology. To date less than 12 temporal bones from HCMV infants have been studied, limiting the development of any testable hypothesis that could account for the course of the disease and the pathological findings. Animal models of the CNS disease associated with congenital CMV infection have been of limited value with the possible exception of the rhesus macaque (Tarantal et al., 1998). Murine models have required intracerebral inoculation, thus bypassing virus replication in the liver and spleen and perhaps more importantly, the host immune response (Kosugi et al., 1998; Kosugi et al., 2002; van Den Pol et al., 1999; van Den Pol et al., 2002). Similarly, CNS disease in the guinea-pig CMV model of congenital CMV has not been consistently demonstrated. However, it is of interest that intracerebral inoculation of murine CMV in mice can result in a cerebritis and virus replication in several different cell types in the choroid plexus and the subventricular epithelium, but the resulting disease is more focal and occurs primarily in areas of virus replication as compared to findings in infants with congenital HCMV infection (Kosugi et al., 1998; Kosugi et al., 2002; van Den Pol et al., 1999; Van Den Pol et al., 2000).

It has been assumed that the susceptibility of the fetus and newborn infant to HCMV infection is secondary to a developmental immaturity of the immune system. Yet there is little understanding of fetal immunity and it is a curious finding that some infants can present with severe infections, including hepatitis and bone marrow depression, but little or no CNS disease, whereas others can have primarily CNS damage. Furthermore, infants infected in the perinatal period can exhibit signs and symptoms of severe HCMV infection yet rarely if ever do these infants have CNS involvement. Thus, a unifying explanation for the variability of disease presentation in congenitally infected infants has not been formulated. Infants with congenital HCMV infection have both antiviral antibodies at birth and virus-specific CD8[+] and CD4[+] T-cell responses, albeit at levels reduced when compared to responses in older infants (Gibson et al., 2004; Tu et al., 2004). Antiviral antibodies are presumably passively acquired from the mother and reflect the response of the mother to HCMV infection. For many years it was claimed that congenitally infected infants had undetectable CD4[+] and, by inference, CD8[+] T-cell responses to HCMV and that this observation could be correlated with persistent virus excretion (Pass et al., 1984). More recent methodology has allowed detection of CD4[+] and CD8[+] T-cell responses in these infants, but at levels that are significantly reduced compared to responses in young children (Ogawa-Goto et al., 2003; Tu et al., 2004).

Importantly, there appears to be no difference in the CD4[+] and CD8[+] T-cell responses in congenitally infected infants with and without symptomatic disease (Gibson et al., 2004; Pass et al., 1984). In fact, infants infected at the time of delivery and who remain without symptoms of infection also have a similar depression in HCMV-specific CD4[+] responses, providing additional evidence that this in vitro measure of immunity may offer little insight into the pathogenesis of congenital HCMV infections (Gibson et al., 2004; Pass et al., 1984; Tu et al., 2004). Thus, although most investigators believe that the developmental immaturity in the immune response to HCMV contributes significantly to the pathogenesis of congenital HCMV infection, firm evidence is lacking and the mechanism(s) that are responsible for the increased susceptibility of the developing fetus to damaging infections with HCMV are incompletely defined.

Finally, the susceptibility of the developing organ system in the fetus is clearly an important determinant in the pathogenesis of congenital HCMV infection. Although this aspect of congenital infection remained unstudied because of the lack of an appropriate animal model, pathogenic mechanisms could include direct effects of HCMV infection on organogenesis in early stages of development secondary to virus-induced cytopathology as well as downstream effects related to the susceptibility of progenitor cells present in fetal tissues to HCMV infection and damage.

The unique relationship between the immunocompetent mother and the immunologically immature fetus results in a biological system in which immunity that limits virus replication and disease is provided by the mother. However, in some cases protective maternal immunity has little or no effect on the infection in the developing fetus. Furthermore, it is questionable if evolutionary pressures have driven the maternal immune response toward a response that specifically modifies the outcome of a fetal infection. Thus, any protection afforded to the fetus by the maternal immune response can be explained either indirectly by a modification of the maternal infection with resulting modification of the fetal infection or directly through passive acquisition of maternal immunity (antibody?) by the fetus. It has been argued that modification of maternal infection can also modify the course of fetal infection, yet there is little direct evidence to support this claim. Clearly, the virological course of maternal infection can be modified by virus-specific immunity, but it is unclear if this effect in pregnant women can be directly translated into similar modifications of the disease in the fetus. It is generally argued that antiviral antibodies cannot prevent HCMV infection of the fetus but can modify disease based on their capacity to reduce the effective viral burden. Pre-existing immunity in a pregnant woman could reduce infection of the placenta and perhaps even decrease the effective inoculum reaching the fetus. Somewhat different results have been obtained in studies of a rhesus macaque model system. The results of primate studies have suggested that disease can develop and progress in the face of maternal antibody responses, independent of the size of the viral inoculum (Chang et al., 2002; Tarantal et al., 1998).

A unique characteristic of perinatal HCMV infections is that infection of the infant or fetus follows exposure to a viral inoculum that is rarely if ever associated with overt disease in the mother. Thus, congenital infections are transmitted in the presence of an effective immune response that has provided protection to the pregnant women or, in the case of transfusion-associated infection, to the blood product donor. This observation has profound implications for the development of protective vaccines for prevention of damaging intrauterine and perinatal HCMV infections. This is perhaps best illustrated by early observations indicating that a significant percentage of infants with damaging congenital HCMV infection acquired their infection from a mother with preexisting protective immunity, a finding that suggested that definitions of protective immunity in the normal host may have little relevance to protective immunity for the developing fetus.

Additional evidence supporting this concept comes from a primate model of CMV disease. From these studies in the rhesus macaque model of congenital CMV infection it can be concluded that regardless of the maternal immune status, if virus reaches the fetus, disease ranging from focal microscopic damage of the CNS to severe CNS disease with loss of normal brain architecture can occur (Chang et al., 2002; Tarantal et al., 1998). Therefore, if maternal immunity exerts protective activity in the prevention of intrauterine infection, it most likely provides this by modification of virus transmission to the fetus, presumably at the maternal–placental interface (see also Chapter 2). From studies of the natural history of congenital HCMV infections, this protection is far from complete and may reduce transmission by at most 20- to 30-fold (Stagno et al., 1982b; Stagno et al., 1986). Although this level of protection from fetal infection appears at first glance to be significant, data from natural history studies have indicated that within a population of women with a high prevalence of HCMV infection, a similar number of damaging congenital infections will be derived from infants born to women with pre-existing "protective" immunity as will be identified from women initially infected with HCMV during pregnancy (Stagno et al., 1977; Stagno et al., 1982a). Conversely, in a population of women in which the prevalence of HCMV infection is low, most affected infants are derived from infections acquired during pregnancy as would be expected because most maternal infections are initial or primary infections in this group. This projection suggests that even if a vaccine induced similar levels of protective immunity as natural infection, it will limit disease secondary to intrauterine HCMV infection by less than 50% in the population of women with a high seroprevalence of HCMV infection and likely only be efficacious for a selected subpopulation of women in the total population. Infants can also acquire HCMV during the perinatal period secondary to ingestion of virus containing breast milk or transfusion with HCMV-infected blood products. These infections can be severe, particularly in premature infants, presumably secondary to the lack of passively acquired maternal immunity (Yeager et al., 1981).

As has been demonstrated in the murine model, antiviral antibodies are thought to function by reducing the magnitude of the infecting inoculum as well as by limiting virus dissemination (Jonjic et al., 1994; Reddehase et al., 1994). Similar to findings in experimental animals, passively acquired antiviral antibodies in humans do not prevent infection (Snydman et al., 1987; Yeager et al., 1981). A corollary of this interpretation is that passively acquired antiviral antibodies could protect fetuses from damaging intrauterine infections. In animal models of congenital CMV infection such as the guinea-pig model (see also Chapter 24) it appears that virus-neutralizing antibodies could limit both intrauterine disease and perinatal infection and disease (Bratcher et al., 1995; Chatterjee et al., 2001). However, this system differs from congenital HCMV infection in that antiviral antibodies are passively transferred in close temporal proximity to viral challenge and virus is given as a cell-free preparation by subcutaneous or intraperitoneal injections. Moreover, the same virus strain that was used to induce the antibody response in donor animals has been used to challenge the pregnant animals.

Regardless of these shortcomings, these studies have provided in vivo evidence that passively acquired antiviral antibodies can modify fetal infection. Although these data provide support for a hypothesis that is consistent with the protective activity of antiviral antibodies in intrauterine HCMV infection, there is little definitive support for this mechanism of immune protection of the developing human fetus. Natural history studies of congenital HCMV infections have documented severe infections in infants born to women with documented immunity prior to pregnancy, including anti-HCMV antibodies (Ahlfors et al., 1981; Ahlfors et al., 2001; Boppana et al., 2001; Yamamoto et al., 2001). Thus, the pathogenesis of intrauterine infection is difficult to unify into a single mechanism. Disease in the fetus undoubtedly represents a combination of immunological deficits together with the susceptibility of developing organ systems to virus infection and virus-induced cellular and organ damage.

## Persistent HCMV infections and chronic disease: general considerations

The role of HCMV infection in chronic disease, particularly chronic inflammatory disease, has been proposed for several decades (Britt and Alford, 1996). More recently, the role of HCMV in chronic vascular disease has become an area of intense interest (see also Chapter 24). The continual improvements in the management of acute graft rejection in allograft recipients has extended survival in the vast majority of these patients and, as a result, long-term complications of allograft transplantation such as chronic graft rejection have become a major determinant of the clinical outcome of these patients. The development of a vasculopathy characterized by a vascular sclerosis in the transplanted organ has become a significant clinical problem of cardiac transplantation and is a major contributor to graft loss in these patients (Grattan et al., 1989; Hosenpud, 1999; Libby and Zhao, 2003). In addition, vascular disease together with tubular sclerosis has been linked to graft dysfunction in HCMV-infected renal allograft recipients (Lopez et al., 1974; Peterson et al., 1980; Soderberg-Naucler and Emery, 2001; Yilmaz et al., 1996). In both cases, HCMV infection has been implicated as at least a co-factor in these diseases and possibly even as the etiology.

Vascular disease in the allograft recipient is thought to represent the extreme of spectrum of vascular diseases associated with persistent HCMV infection, both in terms of the pathologic damage and the rapidity of development and progression of disease. In contrast, vascular disease in normal individuals including atherosclerotic coronary artery disease and peripheral arterial disease has also been linked to persistent HCMV infection within the walls of medium sized and large arteries (Epstein et al., 1996; Hendrix et al., 1989; Hendrix et al., 1991; Pampou et al., 2000; Vossen et al., 1996). This group of diseases as well as others associated with HCMV have in common a persistent infection coupled with chronic inflammation (Libby, 2002; Libby, 2003; O'Connor et al., 2001; Streblow et al., 2001). This association has led to a concept that chronic diseases associated with HCMV infection arise secondary to a local inflammatory response such that the virus persists in the presence of host inflammatory cells and soluble inflammatory mediators. A key component of the proposed mechanism of disease is the contribution of HCMV to the induction of inflammatory mediators from hematopoietic cells, endothelial and epithelial cells, and in some cases from smooth muscle cells. It has been proposed that virus infected cells and inflammatory cells together engage in a bidirectional stimulation creating a circuit that favors virus persistence at the cost of vascular damage secondary to the local inflammatory reaction.

Studies in several systems have clearly demonstrated the unique relationship between CMVs and cells of the inflammatory response. Examples of these interactions include activation of cellular genes by viral infection and the expression of immediate-early viral gene products, activation of viral transcription by cellular transcription factors, induction of pro-inflammatory and inflammatory cytokine, chemokine, and adhesion molecule expression by virus-infected cells, and expression of viral-encoded chemokine receptors that function as GPCRs in response to soluble mediators released by inflammatory cells (Billstrom

Schroeder and Worthen, 2001; Craigen et al., 1997; Dengler et al., 2000; Grundy et al., 1998; Knight et al., 1999; Koskinen, 1993; Mocarski, 2001; Streblow et al., 1999; Yilmaz et al., 1996). In addition, it appears that CMVs can utilize mononuclear cells not only as cellular sites of persistence but also as vehicles for trafficking to sites of disease and dissemination in the infected host (Saederup and Mocarski, 2002). Finally, latent virus can be reactivated from mononuclear cells in responses to inflammatory cytokines (Hummel et al., 2001; Soderberg-Naucler et al., 1997a; Soderberg-Naucler et al., 1997b). Thus, there is a considerable body of evidence that link HCMV to the development and progression of diseases in which inflammation underlies the basic pathologic process.

## HCMV infections and vascular disease in transplant recipients and normal individuals

The role of HCMV in vascular disease, particularly in cardiac allograft recipients, has been appreciated for over 20 years (Gyorkey et al., 1984; Melnick et al., 1993). Epidemiological associations between HCMV infection and coronary atherosclerotic heart disease have been reported in the early 1980s and more recent studies have further strengthened this link (Adam et al., 1987; Epstein et al., 1999; Nieto et al., 1996; O'Connor et al., 2001). In addition, studies of atherosclerotic lesions from patients with coronary artery disease and from cardiac allografts with vascular disease have demonstrated HCMV nucleic acids within the vessel walls (Epstein et al., 1996; Gyorkey et al., 1984; Hendrix et al., 1991; Speir et al., 1994). Treatment of cardiac allograft recipients with ganciclovir, an inhibitor of HCMV replication (see also Chapter 27), decreased the frequency of vascular disease compared with untreated control patients (Valantine et al., 1999). Perhaps the most convincing evidence linking HCMV to vascular disease have been studies in animal models (Chapter 24). These models of cardiac allograft rejection utilize heterotopic transplantation of histocompatibility mismatched cardiac allografts in rodent models (Lemstrom et al., 1993; Lemstrom et al., 1995). Infection of transplanted animals with the respective CMV results in more rapid progression of graft rejection (De La Melena et al., 2001; Lemstrom et al., 1993; Lemstrom et al., 1995; Streblow et al., 2003).

Some groups investigating CMV-associated allograft rejection in animal models have suggested that infection of the transplanted organ is necessary for accelerated rejection whereas others have argued that infection of the host alone is sufficient to cause more rapid rejection of the allograft (De La Melena et al., 2001; Koskinen et al., 1999; Lemstrom et al., 1995; Lemstrom et al., 1997; Streblow et al., 2003; Tikkanen et al., 2001; Zhou et al., 1999). Rejection can be modified by treatment of the infected animals with antivirals that limit virus replication, but only if treatment is initiated early after infection (Lemstrom et al., 1997). Histological changes in the allograft are consistent with similar pathology seen in human allografts and reveal concentric narrowing of the lumen associated with smooth muscle cell infiltration and deposition of foam cells (De La Melena et al., 2001 Zhou, 1999; Lemstrom et al., 1995).

A number of mechanisms have been proposed to account for the pathogenesis of this vascular disease; however, the mechanism that is most consistent with available data is the migration of virus-infected smooth muscle cells into the intima of the vessel wall in response to the inflammatory endothelitis associated with graft rejection. It has been shown that smooth muscle cells (SMCs) can be infected with CMVs and that HCMV infection of human SMCs and RCMV infection of rat SMCs results in expression of a functional cellular GPCR, the US28 (R33) gene product, on the cell surface (Streblow et al., 1999; Streblow et al., 2003). Chemokines released by infiltrating monocytes can result in the formation of a gradient that directs migration of CMV-infected SMCs into the areas of inflammatory cell infiltrates (Streblow et al., 1999). Migration of these cells into the intima results in subintimal thickening and obliteration of the vessel lumen. Furthermore, infected SMCs together with infected endothelial cells can recruit additional activated mononuclear cells into the area of inflammation and further compromise the functional integrity of the graft. Thus, once initiated, the bidirectional interaction between inflammatory cells and CMV-infected smooth muscle and endothelial cells can propagate this pathogenic response by recruitment of additional mononuclear cells, stimulation of cytokine/chemokine expression in infiltrating cells, and induction of viral gene expression and virus replication in response to pro-inflammatory and inflammatory cytokines. A more complete discussion of the association between CMV infection and vascular disease is presented in Chapter 24.

The role of HCMV in vascular disease in humans such as coronary artery disease has been proposed for several decades. Epidemiological studies have provided both data in support of the hypothesis as well as findings that suggest there is no identifiable association between HCMV and vascular disease in normal individuals (Adler et al., 1998; Blum et al., 1998; Blum et al., 2003; Borgia et al., 2001; Epstein et al., 1996; Epstein et al., 1999; Grahame-Clarke et al., 2003; Melnick et al., 1993; Nieto et al., 1996; O'Connor et al., 2001). These conflicting data are difficult to reconcile, but the high prevalence of both coronary artery disease and HCMV infection in the population make such studies difficult to sufficiently power for meaningful statistical analysis. Furthermore, in most cases these studies attempted to link HCMV infection and vascular disease by an indirect measure of HCMV infection of the vessel wall such as serum antibody titers. Only a few studies have extended the analysis to include vascular tissue specimens from patients with coronary artery disease and in these studies, evidence has been presented supporting a link between HCMV infection of the vessel wall and disease. Although these data cannot be viewed as definitive evidence for a role of HCMV in vascular disease in normal individuals, the findings from studies of allograft vascular disease together with the consensus view that coronary artery disease is an inflammatory disease provide more than sufficient support to justify further investigations into the mechanism(s) by which HCMV could contribute to vascular disease in the non-immunocompromised host.

Years of observational research have shown that known host factors such as family history, genetic susceptibility, and lifestyle influence the development of coronary artery disease but these risk factors do not represent the totality of risk for the development of disease. If HCMV infection is a major contributor to disease, and based on the fact that it is a ubiquitous agent in most populations, then variations in disease penetrance could be secondary to as yet undefined interactions between the virus and the host or alternatively, secondary to variations of viral genotype resulting in different phenotypic behavior between different strains of HCMV.

The observation that strains of HCMV differ in their cellular tropism, particularly their capacity to grow in endothelial cells and monocyte-derived macrophages, has fueled interest in the role of HCMV in vascular disease and also raised the possibility that specific viral genes could be responsible for endothelial tropism and therefore vascular disease following HCMV infection. Thus, the endothelial tropism of some strains of HCMV could provide at least one explanation of why not all HCMV-infected individuals develop clinically significant coronary atherosclerotic artery disease. Attempts to identify viral genes or viral gene products that define the endothelial cell tropism of HCMV have as yet been unsuccessful; however, the M45 gene of MCMV has been shown to confer endothelial cell tropism by virtue of its capacity to limit an apoptotic response in endothelial cells infected with MCMV, and this provides a precedent for the role of a viral gene in determining cellular tropism of CMVs (Brune et al., 2001b). It is also possible that HCMV endothelial cell tropism will be determined by multiple viral genes whose differential expression may add another layer of specificity to the cellular tropism and phenotypic behavior of HCMV.

At another level, at least three envelope glycoproteins have been shown to exhibit considerable sequence variation, raising the possibility that different strains of HCMV could have different phenotypic behavior in vivo based entirely on sequence variation of viral proteins that interact with different cell types during the initial phase of virus infection (Pignatelli et al., 2003; Rasmussen et al., 2003; Spaderna et al., 2002). Other viral genes have been shown to exhibit similar sequence variation, for instance UL144, an ORF encoding a TNF receptor-like molecule (Lurain et al., 1999). Therefore, it may be difficult to define the relationship between replication, cellular tropism, and in vivo virulence utilizing only a limited number of prototypic viruses in cell culture systems. (Sinzger et al. 1999) Consistent with this potential limitation of current experimental approaches has been a report that described the recovery of multiple genotypes of virus from autopsy tissue derived from fatal cases of congenital HCMV infection (Arav-Boger et al., 2002). The identification of these viruses as being genetically unique was based on the polymorphism in the UL144 gene. More provocative were the findings that different viral genotypes could be recovered from specific tissues, suggesting that multiple viral genotypes were present in the inoculum and that individual genotypes determined the tissue tropism of HCMV (Arav-Boger et al., 2002). Although definitive evidence of strains of HCMV that exhibit more virulent in vivo behavior has not been reported, there are clearly strains with expanded cellular tropism and at least in some cases, this expanded tropism results in the efficient infection of the endothelium.

These findings would argue that vascular disease can develop in models of CMV vascular disease secondary to interactions with the endothelium and that some strains of HCMV could more efficiently infect endothelial cells and lead to disease. Other phenotypic differences between disease producing and non-producing strains could be the expression of immune evasion functions and/or the activation of cellular responses such as induction of adhesion molecule expression, chemokine expression, and maintenance of persistent infection. Any of these differences in the in vivo phenotypic behavior, either individually or in combination, could provide a selective advantage for growth in endothelial cells and contribute to a disease-inducing phenotype in a subset of viruses.

## HCMV persistence and disease: a potential role of HCMV in chronic inflammatory diseases

Although the pathogenesis of acute HCMV infections has been described simply in terms of virus replication and host control of virus replication and dissemination, the pathogenesis of HCMV infections associated with chronic diseases cannot be described within a simple model of virus replication and antiviral immunity. Several lines of evidence suggest that HCMV and other CMVs can target cells at sites of inflammation, and an extensive literature indicates that pro-inflammatory cytokines and other soluble mediators of inflammation can induce HCMV replication as well as reactivate virus from latency (Hummel et al., 2001; Smith et al., 1992; Soderberg-Naucler et al., 1997a; Soderberg-Naucler et al., 1997b). In addition, CMVs have been shown to replicate in mononuclear cells and monocyte-derived macrophages and to utilize these cells to disseminate within the host (Fish et al., 1995; Guetta et al., 1997; Rice et al., 1984; Riegler et al., 2000; Sinclair and Sissons, 1996; Soderberg-Naucler et al., 1998; Taylor-Wiedeman et al., 1991; Waldman et al., 1995). Taken together, these findings support the concept that in some cases of chronic inflammatory disease associated with CMV infection, the virus and the host inflammatory response have a co-dependent relationship that favors virus persistence even in the presence of a host inflammatory response.

In these cases, persistence of CMVs could accentuate and possibly prolong the host inflammatory response, thus contributing to the pathologic process. Aspects of this strategy of persistence are utilized by several successful intracellular microbial agents including HIV, *Mycobacterium tuberculosis*, and *Toxoplasma gondii*. In the last two cases, persistent infection can lead to an exaggerated host inflammatory response that can be significant cause of tissue damage and organ dysfunction. HCMV appears to exhibit similar biologic behavior by persisting in the presence of host immune effector functions, perhaps secondary to the expression of viral immune evasion functions. These viral functions presumably not only limit the capability of the host to control of virus persistence but also likely modulate the intensity of the inflammatory response. By limiting the nature and the intensity of the inflammatory response, HCMV infection could facilitate the development of chronic inflammation. In such cases, CMVs can be viewed as a co-factor of chronic disease rather than as an underlying etiologic agent of disease.

The pathogenesis of these diseases is likely not directly attributable to virus-induced cytopathology but more correctly described by immune-mediated mechanisms. An extension of arguments suggesting a role of CMVs in chronic inflammatory diseases could include a role for this virus in the malignant behavior of certain human cancers that express many of the hallmarks of chronic inflammation or wound healing such as infiltration with mononuclear cells and angiogenesis. Clearly, the involvement of HCMV in human tumors remains speculative at this time but recent findings of HCMV nucleic acids and virus-encoded proteins in a glioblastomas, a tumor characterized by its vascularity and by histological evidence of inflammatory infiltrates, has prompted a reexamination of the role of HCMV in human cancers (Cobbs et al., 2002). The mechanism(s) by which HCMV contributes to the malignant phenotype is unknown, but possibilities include the induction of angiogenic factor expression such as vascular endothelial growth factor (VEGF) from either tumor cells, supporting cells, or infiltrating inflammatory cells infected with HCMV. Alternatively, HCMV may contribute by its effects on the cell cycle or apoptotic pathways (see also Chapter 11). However, the role, if any, of HCMV in human malignancies remains to be defined, and current findings at this stage represent only intriguing and somewhat provocative observations.

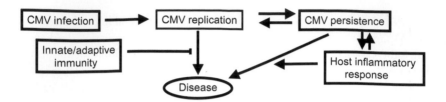

**Figure 1.1** Host immunity and CMV infection/disease.

## Conclusions

With few exceptions, until relatively recently the pathogenesis of HCMV infections has been discussed in terms of a conventional paradigm of viral replication and viral cytopathology. Resolution of acute virus infection and clinical disease is thought to require the concerted actions of both innate and adaptive immunity. In contrast to the role of host immunity in the resolution of acute CMV infection, disease in the immunocompromised host with chronic CMV infection appears to be secondary to both virus replication and the inflammatory response of the host. The results of studies of HCMV infection in allograft recipients have been used to model the association of HCMV with chronic inflammatory diseases in the normal population. A proposed mechanism of disease includes the bidirectional interactions between HCMV and cells of the inflammatory response leading to chronic inflammation and viral persistence within the host, and in some cases within a population. When viewed together, a simplified view of HCMV pathogenesis can be postulated (Figure 1.1). In this reductionistic model, the contribution of potentially key considerations such as in vivo cell tropism of individual CMVs and variation in host and viral genotypes are ignored. Under this premise, the primary function of host immunity is to limit disease associated with relatively high levels of virus replication but not to eradicate persistent virus infection. Viral persistence can be achieved and maintained by several mechanisms, including latency and persistence of a productive infection (chronic infection). In chronic infections, virus persistence can be aided by the host inflammatory response. In some cases such interactions between the host and the virus are bidirectional, facilitating infection and chronic inflammation in local sites and virus spread from these local sites to new sites or even to previously infected tissue. This unique relationship between CMVs and the host inflammatory response would argue for an expanded role of CMVs in human diseases, particularly those associated with chronic inflammation.

### Acknowledgments

The author would like to acknowledge thoughtful discussions with Drs. Michael Mach, Jay Nelson, and Michael Jarvis during the preparation of this chapter. The author's laboratory is supported by funding through National Institutes of Health, NIAID.

### References

Adam, E., Melnick, J.L., Probtsfield, J.L., Petrie, B.L., Burek, J., Bailey, K.R., McCollum, C.H., and DeBakey, M.E. (1987). High levels of cytomegalovirus antibody in patients requiring vascular surgery for atherosclerosis. Lancet 2, 291–293.
Adler, S.P. (1983). Transfusion-associated cytomegalovirus infections. Rev. Infect. Dis. 5, 977–993.
Adler, S.P., Hur, J.K., Wang, J.B., and Vetrovec, G.W. (1998). Prior infection with cytomegalovirus is not a major risk factor for angiographically demonstrated coronary artery atherosclerosis. J. Infect. Dis. 177, 209–212.

Ahlfors, K., Harris, S., Ivarsson, S., and Svanberg, L. (1981). Secondary maternal cytomegalovirus infection causing symptomatic congenital infection. N. Engl. J. Med 305, 284.

Ahlfors, K., Ivarsson, S.A., and Harris, S. (2001). Secondary maternal cytomegalovirus infection—A significant cause of congenital disease. Pediatrics 107, 1227–1228.

Alberola, J., Tamarit, A., Cardenoso, L., Estelles, F., Igual, R., and Navarro, D. (2001). Longitudinal analysis of human cytomegalovirus glycoprotein B (gB)-specific and neutralizing antibodies in AIDS patients either with or without cytomegalovirus end-organ disease. J. Med. Virol. 64, 35–41.

Anderson, K.S., Amos, C.S., Boppana, S., and Pass, R. (1996). Ocular abnormalities in congenital cytomegalovirus infection. J. Am. Optom. Ass. 67, 273–278.

Arav-Boger, R., Willoughby, R.E., Pass, R.F., Zong, J.C., Jang, W.J., Alcendor, D., and Hayward, G.S. (2002). Polymorphisms of the cytomegalovirus (CMV)-encoded tumor necrosis factor-alpha and beta-chemokine receptors in congenital CMV disease. J. Infect. Dis. 186, 1057–1064.

Arnoult, D., Bartle, L.M., Skaletskaya, A., Poncet, D., Zamzami, N., Park, P.U., Sharpe, J., Youle, R.J., and Goldmacher, V.S. (2004). Cytomegalovirus cell death suppressor vMIA blocks Bax- but not Bak-mediated apoptosis by binding and sequestering Bax at mitochondria. Proc. Nat. Acad. Sci. U. S. A. 101, 7988–7993.

Badley, A.D., Seaberg, E.C., Porayko, M.K., Wiesner, R.H., Keating, M.R., Wilhelm, M.P., Walker, R.C., Patel, R., Marshall, W.F., DeBernardi, M., et al. (1997). Prophylaxis of cytomegalovirus infection in liver transplantation: a randomized trial comparing a combination of ganciclovir and acyclovir to acyclovir. NIDDK Liver Transplantation Database. Transplantation 64, 66–73.

Balcarek, K.B., Oh, M.K., and Pass, R.F. (1993). Maternal viremia and congenital CMV infection. In Multidisciplinary Approach to Understanding Cytomegalovirus Disease, S. Michelson, and S.A. Plotkin, eds. (New York, Excerpta Medica), pp. 169–173.

Baroncelli, S., Barry, P.A., Capitanio, J.P., Lerche, N.W., Otsyula, M., and Mendoza, S.P. (1997). Cytomegalovirus and simian immunodeficiency virus coinfection: longitudinal study of antibody responses and disease progression. J. Acquir. Immune. Defic. Syndr. Hum. Retrovirol. 15, 5–15.

Becroft, D.M.O. (1981). Prenatal cytomegalovirus infection: epidemiology, pathology, and pathogenesis. In Perspective in pediatric pathology, H.S. Rosenberg, and J. Bernstein, eds. (New York, Masson Press), pp. 203–241.

Bernstein, D.I., Bourne, N. (1999). Animal models for cytomegalovirus infection: Guinea-pig CMV. In Handbook of Animal Models of Infection, M.S.O. Zak, ed. (London, Academic Press), pp. 935–941.

Bia, F.J., Griffith, B.P., Fong, C.K., and Hsiung, G.D. (1983). Cytomegaloviral infections in the guinea-pig: experimental models for human disease. Rev. Infect. Dis. 5, 177–195.

Billstrom Schroeder, M., and Worthen, G.S. (2001). Viral regulation of RANTES expression during human cytomegalovirus infection of endothelial cells. J. Virol. 75, 3383–3390.

Biron, C.A., and Brossay, L. (2001). NK cells and NKT-cells in innate defense against viral infections. Curr. Opin. Immunol. 13, 458–464.

Biron, C.A., Byron, K.S., and Sullivan, J.L. (1989). Severe herpesvirus infections in an adolescent without natural killer cells. N. Engl. J. Med. 320, 1731–1735.

Blum, A., Giladi, M., Weinberg, M., Kaplan, G., Pasternack, H., Laniado, S., and Miller, H. (1998). High anti-cytomegalovirus (CMV) IgG antibody titer is associated with coronary artery disease and may predict post-coronary balloon angioplasty restenosis. Am. J. Cardiol. 81, 866–868.

Blum, A., Peleg, A., and Weinberg, M. (2003). Anti-cytomegalovirus (CMV) IgG antibody titer in patients with risk factors to atherosclerosis. Clin. Exp. Med. 3, 157–160.

Boeckh, M. (1999). Current antiviral strategies for controlling cytomegalovirus in hematopoietic stem cell transplant recipients: prevention and therapy. Transpl. Infect. Dis. 1, 165–178.

Boeckh, M., Leisenring, W., Riddell, S.R., Bowden, R.A., Huang, M.L., Myerson, D., Stevens-Ayers, T., Flowers, M.E., Cunningham, T., and Corey, L. (2003). Late cytomegalovirus disease and mortality in recipients of allogeneic hematopoietic stem cell transplants: importance of viral load and T-cell immunity. Blood 101, 407–414.

Boppana, S.B., Fowler, K.B., Vaid, Y., Hedlund, G., Stagno, S., Britt, W.J., and Pass, R.F. (1997). Neuroradiographic findings in the newborn period and long-term outcome in children with symptomatic congenital cytomegalovirus infection. Pediatrics 99, 409–414.

Boppana, S.B., Pass, R.F., Britt, W.J., Stagno, S., and Alford, C.A. (1992). Symptomatic congenital cytomegalovirus infection: neonatal morbidity and mortality. Pediatr. Infect. Dis. J. 11, 93–99.

Boppana, S.B., Polis, M.A., Kramer, A.A., Britt, W.J., and Koenig, S. (1995). Virus specific antibody responses to human cytomegalovirus (HCMV) in human immunodeficiency virus type 1-infected individuals with HCMV retinitis. J. Infect. Dis. 171, 182–185.

Boppana, S.B., Rivera, L.B., Fowler, K.B., Mach, M., and Britt, W.J. (2001). Intrauterine transmission of cytomegalovirus to infants of women with preconceptional immunity. N. Engl. J. Med. *344*, 1366–1371.

Borgia, M.C., Mandolini, C., Barresi, C., Battisti, G., Carletti, F., and Capobianchi, M.R. (2001). Further evidence against the implication of active cytomegalovirus infection in vascular atherosclerotic diseases. Atherosclerosis *157*, 457–462.

Bowden, R.A. (1995). Transfusion-transmitted cytomegalovirus infection. Hematol. Oncol. Clin. North Am. *9*, 155–166.

Bratcher, D.F., Bourne, N., Bravo, F.J., Schleiss, M.R., Slaoui, M., Myers, M.G., and Bernstein, D.I. (1995). Effect of passive antibody on congenital cytomegalovirus infection in guinea-pigs. J. Infect. Dis. *172*, 944–950.

Britt, W.J., and Alford, C.A. (1996). Cytomegalovirus. In Fields Virology, Third Edition, B.N. Fields, D.M. Knipe, and P.M. Howley, eds. (New York, Raven Press), pp. 2493–2523.

Brune, W., Hasan, M., Krych, M., Bubic, I., Jonjic, S., and Koszinowski, U.H. (2001a). Secreted virus-encoded proteins reflect murine cytomegalovirus productivity in organs. J. Infect. Dis. *184*, 1320–1324.

Brune, W., Menard, C., Heesemann, J., and Koszinowski, U.H. (2001b). A ribonucleotide reductase homolog of cytomegalovirus and endothelial cell tropism. Science *291*, 303–305.

Bukowski, J.F., Woda, B.A., and Welsh, R.M. (1984). Pathogenesis of murine cytomegalovirus infection in natural killer cell-depleted mice. J. Virol. *52*, 119–128.

Castillo, J.P., and Kowalik, T.F. (2002). Human cytomegalovirus immediate-early proteins and cell growth control. Gene *290*, 19–34.

Castillo, J.P., and Kowalik, T.F. (2004). HCMV infection: modulating the cell cycle and cell death. Int. Rev Immunol. *23*, 113–139.

Chang, W.L., Tarantal, A.F., Zhou, S.S., Borowsky, A.D., and Barry, P.A. (2002). A recombinant rhesus cytomegalovirus expressing enhanced green fluorescent protein retains the wild-type phenotype and pathogenicity in fetal macaques. J. Virol. *76*, 9493–9504.

Chatterjee, A., Harrison, C.J., Britt, W.J., and Bewtra, C. (2001). Modification of maternal and congenital cytomegalovirus infection by anti-glycoprotein b antibody transfer in guinea-pigs. J. Infect. Dis. *183*, 1547–1553.

Chemaly, R.F., Yen-Lieberman, B., Castilla, E.A., Reilly, A., Arrigain, S., Farver, C., Avery, R.K., Gordon, S.M., and Procop, G.W. (2004). Correlation between viral loads of cytomegalovirus in blood and bronchoalveolar lavage specimens from lung transplant recipients determined by histology and immunohistochemistry. J. Clin. Microbiol. *42*, 2168–2172.

Chou, S., Kim, D.Y., Scott, K.M., and Sewell, D.L. (1987). Immunoglobulin M antibody to cytomegalovirus in primary and reactivation infections in renal transplant recipients. J. Clin. Microbiol. *25*, 52–55.

Cobbs, C.S., Harkins, L., Samanta, M., Gillespie, G.Y., Bharara, S., King, P.H., Nabors, L.B., Cobbs, C.G., and Britt, W.J. (2002). Human cytomegalovirus infection and expression in human malignant glioma. Cancer Res. *62*, 3347–3350.

Cope, A.V., Sabin, C., Burroughs, A., Rolles, K., Griffiths, P.D., and Emery, V.C. (1997). Interrelationships among quantity of human cytomegalovirus (HCMV) DNA in blood, donor-recipient serostatus, and administration of methylprednisolone as risk factors for HCMV disease following liver transplantation. J. Infect. Dis. *176*, 1484–1490.

Craigen, J.L., Yong, K.L., Jordan, N.J., MacCormac, L.P., Westwick, J., Akbar, A.N., and Grundy, J.E. (1997). Human cytomegalovirus infection up-regulates interleukin-8 gene expression and stimulates neutrophil transendothelial migration. Immunology *92*, 138–145.

Dahle, A.F., Fowler, K.B.; Wright, J.D.; Boppana, S.B.; Britt, W.J.; Pass, R.F. (2000). Longitudinal Investigation of Hearing Disorders in Children with Congenital Cytomegalovirus. J. Am. Acad. Audiol. *11*, 283–290.

De La Melena, V.T., Kreklywich, C.N., Streblow, D.N., Yin, Q., Cook, J.W., Soderberg-Naucler, C., Bruggeman, C.A., Nelson, J.A., and Orloff, S.L. (2001). Kinetics and development of CMV-accelerated transplant vascular sclerosis in rat cardiac allografts is linked to early increase in chemokine expression and presence of virus. Transplant. Proc. *33*, 1822–1823.

Dengler, T.J., Raftery, M.J., Werle, M., Zimmermann, R., and Schonrich, G. (2000). Cytomegalovirus infection of vascular cells induces expression of pro-inflammatory adhesion molecules by paracrine action of secreted interleukin-1beta. Transplantation *69*, 1160–1168.

Dittmer, U., Brooks, D.M., and Hasenkrug, K.J. (1999). Requirement for multiple lymphocyte subsets in protection by a live attenuated vaccine against retroviral infection. Nature Med. *5*, 189–193.

Drew, W.L., Tegtmeier, G., Alter, H.J., Laycock, M.E., Miner, R.C., and Busch, M.P. (2003). Frequency and duration of plasma CMV viremia in seroconverting blood donors and recipients. Transfusion *43*, 309–313.

Emery, V.C. (1999). Viral dynamics during active cytomegalovirus infection and pathology. Intervirology *42*, 405–411.

Emery, V.C., Sabin, C.A., Cope, A.V., Gor, D., Hassan-Walker, A.F., and Griffiths, P.D. (2000). Application of viral-load kinetics to identify patients who develop cytomegalovirus disease after transplantation. Lancet *355*, 2032–2036.

Epstein, S.E., Speir, E., Zhou, Y.F., Guetta, E., Leon, M., and Finkel, T. (1996). The role of infection in restenosis and atherosclerosis: focus on cytomegalovirus. Lancet *348*, 13–17.

Epstein, S.E., Zhou, Y.F., and Zhu, J. (1999). Infection and atherosclerosis: Emerging mechanistic paradigms. Circulation *100*, 20–28.

Erice, A., Tierney, C., Hirsch, M., Caliendo, A.M., Weinberg, A., Kendall, M.A., Polsky, B., and Team, A.C.T.G.P.S. (2003). Cytomegalovirus (CMV) and human immunodeficiency virus (HIV) burden, CMV end-organ disease, and survival in subjects with advanced HIV infection (AIDS Clinical Trials Group Protocol 360). Clin. Infect. Dis. *37*, 567–578.

Fish, K.N., Stenglein, S.G., Ibanez, C., and Nelson, J.A. (1995). Cytomegalovirus persistence in macrophages and endothelial cells. Scan. J. Infect. Dis. Supp. *99*, 34–40.

Fortunato, E.A., McElroy, A.K., Sanchez, I., and Spector, D.H. (2000). Exploitation of cellular signaling and regulatory pathways by human cytomegalovirus. Trends Microbiol. *8*, 111–119.

Fortunato, E.A., and Spector, D.H. (1998). p53 and RPA are sequestered in viral replication centers in the nuclei of cells infected with human cytomegalovirus. J.Virol. *72*, 2033–2039.

Fowler, K.B., McCollister, F.P., Dahle, A.J., Boppana, S., Britt, W.J., and Pass, R.F. (1997). Progressive and fluctuating sensorineural hearing loss in children with asymptomatic congenital cytomegalovirus infection. J. Pediatr. *130*, 624–630.

Gewurz, B.E., Wang, E.W., Tortorella, D., Schust, D.J., and Ploegh, H.L. (2001). Human cytomegalovirus US2 endoplasmic reticulum-lumenal domain dictates association with major histocompatibility complex class I in a locus-specific manner. J. Virol. *75*, 5197–5204.

Gibson, L., Piccinini, G., Lilleri, D., Revello, M.G., Wang, Z., Markel, S., Diamond, D.J., and Luzuriaga, K. (2004). Human cytomegalovirus proteins pp65 and immediate-early protein 1 are common targets for CD8[+] T-cell responses in children with congenital or postnatal human cytomegalovirus infection. J. Immunol. *172*, 2256–2264.

Goldmacher, V.S., Bartle, L.M., Skaletskaya, A., Dionne, C.A., Kedersha, N.L., Vater, C.A., Han, J.W., Lutz, R.J., Watanabe, S., Cahir McFarland, E.D., Kieff, E.D., Mocarski, E.S., Chittenden, T. (1999). A cytomegalovirus-encoded mitochondria-localized inhibitor of apoptosis structurally unrelated to Bcl-2. Proc. Natl. Acad. Sci. USA 96, 12536–12541.

Gor, D., Sabin, C., Prentice, H.G., Vyas, N., Man, S., Griffiths, P.D., and Emery, V.C. (1998). Longitudinal fluctuations in cytomegalovirus load in bone marrow transplant patients: relationship between peak virus load, donor/recipient serostatus, acute GVHD and CMV disease. Bone Marrow Transplant. *21*, 597–605.

Grahame-Clarke, C., Chan, N.N., Andrew, D., Ridgway, G.L., Betteridge, D.J., Emery, V., Colhoun, H.M., and Vallance, P. (2003). Human cytomegalovirus seropositivity is associated with impaired vascular function. Circulation *108*, 678–683.

Grattan, M.T., Moreno-Cabral, C.E., Starnes, V.A., Oyer, P.E., Stinson, E.B., and Shumway, N.E. (1989). Cytomegalovirus infection is associated with cardiac allograft rejection and atherosclerosis. JAMA *261*, 3561–3566.

Griffith, B.P., and Aquino-de Jesus, M.J. (1991). Guinea-pig model of congenital cytomegalovirus infection. Transplant. Proc. *23*, 29–31.

Grundy, J.E., Lawson, K.M., MacCormac, L.P., Fletcher, J.M., and Yong, K.L. (1998). Cytomegalovirus-infected endothelial cells recruit neutrophils by the secretion of C-X-C chemokines and transmit virus by direct neutrophil-endothelial cell contact and during neutrophil transendothelial migration. J. Infect. Dis. *177*, 1465–1474.

Guetta, E., Guetta, V., Shibutani, T., and Epstein, S.E. (1997). Monocytes harboring cytomegalovirus: interactions with endothelial cells, smooth muscle cells, and oxidized low-density lipoprotein. Possible mechanisms for activating virus delivered by monocytes to sites of vascular injury. Circ. Res. *81*, 8–16.

Gyorkey, F., Melnick, J.L., Guinn, G.A., Gyorkey, P., and DeBakey, M.E. (1984). Herpesviridae in the endothelial and smooth muscle cells of the proximal aorta in arteriosclerotic patients. Exp. Mol. Pathol 40, 328–339.

Hamprecht, K., Maschmann, J., Vochem, M., Dietz, K., Speer, C.P., and Jahn, G. (2001). Epidemiology of transmission of cytomegalovirus from mother to preterm infant by breastfeeding. Lancet 357, 513–518.

Hanson, L.K., Slater, J.S., Karabekian, Z., Ciocco-Schmitt, G., and Campbell, A.E. (2001). Products of US22 genes M140 and M141 confer efficient replication of murine cytomegalovirus in macrophages and spleen. J. Virol. 75, 6292–6302.

Harrison, C.J., and Caruso, N. (2000). Correlation of maternal and pup NK-like activity and TNF responses against cytomegalovirus to pregnancy outcome in inbred guinea-pigs. J. Med. Virol. 60, 230–236.

Hendrix, M.G., Daemen, M., and Bruggeman, C.A. (1991). Cytomegalovirus nucleic acid distribution within the human vascular tree. American Journal of Pathology 138, 563–567.

Hendrix, M.G., Dormans, P.H., Kitslaar, P., Bosman, F., and Bruggeman, C.A. (1989). The presence of cytomegalovirus nucleic acids in arterial walls of atherosclerotic and nonatherosclerotic patients. Am. J. Pathol. 134, 1151–1157.

Ho, M. (1991). Observations from transplantation contributing to the understanding of pathogenesis of CMV infection. Transplant. Proc. 23, 104–109.

Horwitz, C.A., Henle, W., Henle, G., Snover, D., Rudnick, H., Balfour, H.H., Mazur, M. H., Watson, R., Schwartz, B., and Muller, N. (1986). Clinical and laboratory evaluation of cytomegalovirus-induced mononucleosis in previously healthy individuals. Report of 82 cases. Medicine 65, 124–134.

Hosenpud, J.D. (1999). Coronary artery disease after heart transplantation and its relation to cytomegalovirus. Am. Heart J. 138, 469–472.

Hummel, M., Zhang, Z., Yan, S., DePlaen, I., Golia, P., Varghese, T., Thomas, G., and Abecassis, M.I. (2001). Allogeneic transplantation induces expression of cytomegalovirus immediate-early genes in vivo: a model for reactivation from latency. J. Virol. 75, 4814–4822.

Jonjic, S., del Val, M., Keil, G.M., Reddehase, M.J., and Koszinowski, U.H. (1988). A nonstructural viral protein expressed by a recombinant vaccinia virus protects against lethal cytomegalovirus infection. J. Virol. 62, 1653–1658.

Jonjic, S., Pavic, I., Polic, B., Crnkovic, I., Lucin, P., and Koszinowski, U.H. (1994). Antibodies are not essential for the resolution of primary cytomegalovirus infection but limit dissemination of recurrent virus. J. Exp. Med. 179, 1713–1717.

Jordan, M.C., Rousseau, W.E., Stewart, J.A., Noble, G.R., and Chin, T.D.Y. (1973). Spontaneous cytomegalovirus mononucleosis: clinical and laboratory observations in nine cases. Ann. Intern. Med. 79, 153–160.

Kalejta, R.F., Bechtel, J.T., and Shenk, T. (2003). Human cytomegalovirus pp71 stimulates cell cycle progression by inducing the proteasome-dependent degradation of the retinoblastoma family of tumor suppressors. Mol. Cell. Biol. 23, 1885–1895.

Kaur, A., Daniel, M.D., Hempel, D., Lee-Parritz, D., Hirsch, M.S., and Johnson, R.P. (1996). Cytotoxic T-lymphocyte responses to cytomegalovirus in normal and simian immunodeficiency virus-infected rhesus macaques. J. Virol. 70, 7725–7733.

Kaur, A., Kassis, N., Hale, C.L., Simon, M., Elliott, M., Gomez-Yafal, A., Lifson, J.D., Desrosiers, R.C., Wang, F., Barry, P., Mach, M., Johnson, R.P. (2003). Direct relationship between suppression of virus-specific immunity and emergence of cytomegalovirus disease in simian AIDS. J. Virol. 77, 5749–5758.

Kern, E.R. (1999). Animal models for cytomegalovirus infection: Murine CMV. In Handbook of Animal Models of Infection, O. Zak, Sande, M., ed. (London, Academic Press), pp. 927–934.

Knight, D.A., Waldman, W.J., and Sedmak, D.D. (1999). Cytomegalovirus-mediated modulation of adhesion molecule expression by human arterial and microvascular endothelial cells. Transplantation 68, 1814–1818.

Komanduri, K.V., Feinberg, J., Hutchins, R.K., Frame, R.D., Schmidt, D.K., Viswanathan, M.N., Lalezari, J.P., and McCune, J.M. (2001). Loss of cytomegalovirus-specific CD4+ T-cell responses in human immunodeficiency virus type 1-infected patients with high CD4+ T-cell counts and recurrent retinitis. J. Infect. Dis. 183, 1285–1289.

Koskinen, P.K. (1993). The association of the induction of vascular cell adhesion molecule-1 with cytomegalovirus antigenemia in human heart allografts. Transplantation 56, 1103–1108.

Koskinen, P.K., Kallio, E.A., Tikkanen, J.M., Sihvola, R.K., Hayry, P.J., and Lemstrom, K.B. (1999). Cytomegalovirus infection and cardiac allograft vasculopathy. Transplant. Infect. Dis. 1, 115–126.

Kosugi, I., Kawasaki, H., Arai, Y., and Tsutsui, Y. (2002). Innate immune responses to cytomegalovirus infection in the developing mouse brain and their evasion by virus-infected neurons. Am. J. Pathol. 161, 919–928.

Kosugi, I., Shinmura, Y., Li, R.Y., Aiba-Masago, S., Baba, S., Miura, K., and Tsutsui, Y. (1998). Murine cytomegalovirus induces apoptosis in non-infected cells of the developing mouse brain and blocks apoptosis in primary neuronal culture. Acta Neuropathol. 96, 239–247.

Koszinowski, U.H., Reddehase, M.J., and Jonjic, S. (1991). The role of CD4 and CD8 T-cells in viral infections. Curr. Opin. Immunol. 3, 471–475.

Krmpotic, A., Bubic, I., Polic, B., Lucin, P., and Jonjic, S. (2003). Pathogenesis of murine cytomegalovirus infection. Microbes Infect. 5, 1263–1277.

Krmpotic, A., Busch, D.H., Bubic, I., Gebhardt, F., Hengel, H., Hasan, M., Scalzo, A.A., Koszinowski, U.H., and Jonjic, S. (2002). MCMV glycoprotein gp40 confers virus resistance to CD8+ T-cells and NK cells in vivo.. Nature Immunol. 3, 529–535.

Lang, D.J., and I Ianshaw, J.B. (1969). Cytomegalovirus infection and the post-perfusion syndrome: recognition of primary infections in four patients. N. Engl. J. Med. 280, 1145–1149.

Lautenschlager, I., Soots, A., Krogerus, L., Kauppinen, H., Saarinen, O., Bruggeman, C., and Ahonen, J. (1997). CMV increases inflammation and accelerates chronic rejection in rat kidney allografts. Transplant. Proc. 29, 802–803.

Lemstrom, K., Koskinen, P., Krogerus, L., Daemen, M., Bruggeman, C., and Hayry, P. (1995). Cytomegalovirus antigen expression, endothelial cell proliferation, and intimal thickening in rat cardiac allografts after cytomegalovirus infection. Circulation 92, 2594–2604.

Lemstrom, K., Persoons, M., Bruggeman, C., Ustinov, J., Lautenschlager, I., and Hayry, P. (1993). Cytomegalovirus infection enhances allograft arteriosclerosis in the rat. Transplant. Proc. 25, 1406–1407.

Lemstrom, K., Sihvola, R., Bruggeman, C., Hayry, P., and Koskinen, P. (1997). Cytomegalovirus infection-enhanced cardiac allograft vasculopathy is abolished by DHPG prophylaxis in the rat. Circulation 95, 2614–2616.

Li, C.R., Greenberg, P.D., Gilbert, M.J., Goodrich, J.M., and Riddell, S.R. (1994). Recovery of HLA-restricted cytomegalovirus (CMV)-specific T-cell responses after allogeneic bone marrow transplant: correlation with CMV disease and effect of ganciclovir prophylaxis. Blood 83, 1971–1979.

Libby, P. (2002). Inflammation in atherosclerosis. Nature 420, 868–874.

Libby, P. (2003). Vascular biology of atherosclerosis: overview and state of the art. American J. Cardiol. 91, 3A–6A.

Libby, P., and Zhao, D.X. (2003). Allograft arteriosclerosis and immune-driven angiogenesis. Circulation 107, 1237–1239.

Limaye, A.P., Huang, M.L., Leisenring, W., Stensland, L., Corey, L., and Boeckh, M. (2001). Cytomegalovirus (CMV) DNA load in plasma for the diagnosis of CMV disease before engraftment in hematopoietic stem-cell transplant recipients. J. Infect. Dis. 183, 377–382.

Ljungman, P. (2001). Prophylaxis against herpesvirus infections in transplant recipients. Drugs 61, 187–196.

Lockridge, K.M., Sequar, G., Zhou, S.S., Yue, Y., Mandell, C.P., and Barry, P.A. (1999). Pathogenesis of experimental rhesus cytomegalovirus infection. J. Virol. 73, 9576–9583.

Lopez, C., Simmons, R.L., Mauer, S.M., Najarian, J.S., Good, R.A., and Gentry, S. (1974). Association of renal allograft rejection with virus infection. Am. J. Med. 56, 280–289.

Lurain, N.S., Kapell, K.S., Huang, D.D., Short, J.A., Paintsil, J., Winkfield, E., Benedict, C.A., Ware, C.F., and Bremer, J.W. (1999). Human cytomegalovirus UL144 open reading frame: sequence hypervariability in low-passage clinical isolates. J. Virol. 73, 10040–10050.

Malm, G., Grondahl, E.H., and Lewensohn-Fuchs, I. (2000). Congenital cytomegalovirus infection: a retrospective diagnosis in a child with pachygyria. Pediatr. Neurol. 22, 407–408.

Melnick, J.L., Adam, E., and Debakey, M.E. (1993). Cytomegalovirus and atherosclerosis. Eur. Heart. J. 14, 30–38.

Menard, C., Wagner, M., Ruzsics, Z., Holak, K., Brune, W., Campbell, A.E., and Koszinowski, U.H. (2003). Role of murine cytomegalovirus US22 gene family members in replication in macrophages. J. Virol. 77, 5557–5570.

Michaelis, M., Kotchetkov, R., Vogel, J.U., Doerr, H.W., and Cinatl, J., Jr. (2004). Cytomegalovirus infection blocks apoptosis in cancer cells. Cell. Mol. Life. Sci. 61, 1307–1316.

Mocarski, E.S., Tan Courcelle, C. (2001). Cytomegaloviruses and their replication. In Fields Virology, D.M.K.a.P.M. Howley, ed. (Philadelphia, Lippincott Williams and Wilkins), pp. 2629–2673.

Nichols, W.G., Corey, L., Gooley, T., Drew, W.L., Miner, R., Huang, M., Davis, C., and Boeckh, M. (2001). Rising pp65 antigenemia during preemptive anticytomegalovirus therapy after allogeneic hematopoietic stem cell transplantation: risk factors, correlation with DNA load, and outcomes. Blood 97, 867–874.

Nieto, F.J., Adam, E., Sorlie, P., Farzadegan, H., Melnick, J.L., Comstock, G.W., and Szklo, M. (1996). Cohort study of cytomegalovirus infection as a risk factor for carotid intimal-medial thickening, a measure of subclinical atherosclerosis. Circulation 94, 922–927.

O'Connor, S., Taylor, C., Campbell, L.A., Epstein, S., and Libby, P. (2001). Potential infectious etiologies of atherosclerosis: a multifactorial perspective. Emerg. Infect. Dis. 7, 780–788.

Ogawa-Goto, K., Tanaka, K., Gibson, W., Moriishi, E., Miura, Y., Kurata, T., Irie, S., and Sata, T. (2003). Microtubule network facilitates nuclear targeting of human cytomegalovirus capsid. J. Virol. 77, 8541–8547.

Pampou, S., Gnedoy, S.N., Bystrevskaya, V.B., Smirnov, V.N., Chazov, E.I., Melnick, J.L., and DeBakey, M.E. (2000). Cytomegalovirus genome and the immediate-early antigen in cells of different layers of human aorta. Virchows Arch. 436, 539–552.

Pannuti, C.S., Vilas Boas, L.S., Angelo, M.J., Amato Neto, V., Levi, G.C., de Mendonca, J.S., and de Godoy, C.V. (1985). Cytomegalovirus mononucleosis in children and adults: differences in clinical presentation. Scandinavian J. Infect. Dis. 17, 153–156.

Pass, R.F., Britt, W.J., Stagno, S., and Alford, C.A. (1984). Specific cell mediated immunity in congenital and perinatal CMV infection. In Herpesvirus (New York, Alan R. Liss, Inc.), pp. 197–209.

Pass, R.F., Griffiths, P.D., and August, A.M. (1983). Antibody response to cytomegalovirus after renal transplantation: comparison of patients with primary and recurrent infection. J. Infect. Dis. 147, 40–46.

Paya, C.V., Wilson, J.A., Espy, M.J., Sia, I.G., DeBernardi, M.J., Smith, T.F., Patel, R., Jenkins, G., Harmsen, W.S., Vanness, D.J., and Wiesner, R.H. (2002). Preemptive use of oral ganciclovir to prevent cytomegalovirus infection in liver transplant patients: a randomized, placebo-controlled trial. J. Infect. Dis. 185, 854–860.

Perlman, J.M., and Argyle, C. (1992). Lethal cytomegalovirus infection in preterm infants: clinical, radiological, and neuropathological findings. Ann. Neurol. 31, 64–68.

Peterson, P.K., Balfour, H.H., Jr, Marker, S.C., Fryd, D.S., Howard, R.J., and Simmons R.L. (1980). Cytomegalovirus disease in renal allograft recipients: a prospective study of the clinical features, risk factors and impact on renal transplantation. Medicine 59, 283–300.

Pignatelli, S., Dal Monte, P., Rossini, G., Chou, S., Gojobori, T., Hanada, K., Guo, J.J., Rawlinson, W., Britt, W., Mach, M., and Landini, M.P. (2003). Human cytomegalovirus glycoprotein N (gpUL73-gN) genomic variants: identification of a novel subgroup, geographical distribution and evidence of positive selective pressure. J. Gen. Virol. 84, 647–655.

Podlech, J., Holtappels, R., Pahl-Seibert, M.F., Steffens, H.P., and Reddehase, M.J. (2000). Murine model of interstitial cytomegalovirus pneumonia in syngeneic bone marrow transplantation: persistence of protective pulmonary CD8-T-cell infiltrates after clearance of acute infection. J. Virol. 74, 7496–7507.

Polic, B., Hengel, H., Krmpotic, A., Trgovcich, J., Pavic, I., Luccaronin, P., Jonjic, S., and Koszinowski, U.H. (1998). Hierarchical and redundanT-lymphocyte subset control precludes cytomegalovirus replication during latent infection. J. Exp. Med. 188, 1047–1054.

Polic, B., Jonjic, S., Pavic, I., Crnkovic, I., Zorica, I., Hengel, H., Lucin, P., and Koszinowski, U.H. (1996). Lack of MHC class I complex expression has no effect on spread and control of cytomegalovirus infection in vivo. J. Gen. Virol. 77, 217–225.

Preiksaitis, J.K., Brown, L., and McKenzie, M. (1988). Transfusion-acquired cytomegalovirus infection in neonates. Transfusion 28, 205–209.

Rapp, M., Messerle, M., Lucin, P., and Koszinowski, U.H. (1993). In vivo protection studies with MCMV glycoproteins gB and gH expressed by vaccinia virus. In Multidisciplinary Approach to Understanding Cytomegalovirus Disease, S. Michelson, and S.A. Plotkin, eds. (Amsterdam, Excerpta Medica), pp. 327–332.

Rasmussen, L., Geissler, A., and Winters, M. (2003). Inter- and intragenic variations complicate the molecular epidemiology of human cytomegalovirus. J. Infect. Dis. 187, 809–819.

Rasmussen, L., Morris, S., Wolitz, R., Dowling, A., Fessell, J., Holodniy, M., and Merigan, T.C. (1994). Deficiency in antibody response to human cytomegalovirus glycoprotein gH in human immunodeficiency virus-infected patients at risk for cytomegalovirus retinitis. J. Infect. Dis. 170, 673–677.

Razonable, R.R., van Cruijsen, H., Brown, R.A., Wilson, J.A., Harmsen, W.S., Wiesner, R.H., Smith, T.F., and Paya, C.V. (2003). Dynamics of cytomegalovirus replication during preemptive therapy with oral ganciclovir. J. Infect.Dis. 187, 1801–1808.

Reddehase, M.J., Balthesen, M., Rapp, M., Jonjic, S., Pavic, I., and Koszinowski, U.H. (1994). The conditions of primary infection define the load of latent viral genome in organs and the risk of recurrent cytomegalovirus disease. J. Exp. Med. 179, 185–193.

Reddehase, M.J., Jonjic, S., Weiland, F., Mutter, W., and Koszinowski, U.H. (1988). Adoptive immunotherapy of murine cytomegalovirus adrenalitis in the immunocompromised host: CD4-helper-independent antiviral function of CD8-positive memory T-lymphocytes derived from latently infected donors. J. Virol. 62, 1061–1065.

Reddehase, M.J., Mutter, W., Munch, K., Buhring, H.J., and Koszinowski, U.H. (1987). CD8-positive T-lymphocytes specific for murine cytomegalovirus immediate-early antigens mediate protective immunity. J. Virol. 61, 3102–3108.

Reddehase, M.J., Podlech, J., and Grzimek, N.K. (2002). Mouse models of cytomegalovirus latency: overview. J. Clin. Virol. 25, S23–36.

Reddehase, M.J., Weiland, F., Munch, K., Jonjic, S., Luske, A., and Koszinowski, U.H. (1985). Interstitial murine cytomegalovirus pneumonia after irradiation: characterization of cells that limit viral replication during established infection of the lungs. J. Virol. 55, 264–273.

Reusser, P., Attenhofer, R., Hebart, H., Helg, C., Chapuis, B., and Einsele, H. (1997). Cytomegalovirus-specific T-cell immunity in recipients of autologous peripheral blood stem cell or bone marrow transplants. Blood 89, 3873–3879.

Reusser, P., Riddell, S.R., Meyers, J.D., and Greenberg, P.D. (1991). Cytotoxic T-lymphocyte response to cytomegalovirus after human allogeneic bone marrow transplantation: pattern of recovery and correlation with cytomegalovirus infection and disease. Blood 78, 1373–1380.

Rice, G.P.A., Schrier, R.D., and Oldstone, M.B.A. (1984). Cytomegalovirus infects human lymphocytes and monocytes: virus expression is restricted to immediate-early gene products. Proc. Natl. Acad. Sci. USA 81, 6134.

Riegler, S., Hebart, H., Einsele, H., Brossart, P., Jahn, G., and Sinzger, C. (2000). Monocyte-derived dendritic cells are permissive to the complete replicative cycle of human cytomegalovirus. J. Gen. Virol. 81, 393–399.

Rifkind, D. (1965). Cytomegalovirus infection after renal transplantation. Arch. Intern. Med. 116, 554–558.

Rubin, R. (2002). Clinical approach to infection in the compromised host. In Infection in the Organ Transplant Recipient, R. Rubin, Young, LS, ed. (New York, Kluwer Academic Press), pp. 573–679.

Saederup, N., Aguirre, S.A., Sparer, T.E., Bouley, D.M., and Mocarski, E.S. (2001). Murine cytomegalovirus CC chemokine homolog MCK-2 (m131–129) is a determinant of dissemination that increases inflammation at initial sites of infection. J. Virol. 75, 9966–9976.

Saederup, N., and Mocarski, E.S., Jr. (2002). Fatal attraction: cytomegalovirus-encoded chemokine homologs. Curr. Top. Microbiol. Immunol. 269, 235–256.

Salazar-Mather, T.P., Orange, J.S., and Biron, C.A. (1998). Early murine cytomegalovirus (MCMV) infection induces liver natural killer (NK) cell inflammation and protection through macrophage inflammatory protein 1alpha (MIP-1alpha)-dependent pathways. J. Exp. Med. 187, 1–14.

Schmidt, G.M., Horak, D.A., Niland, J.C., Duncan, S.R., Forman, S.J., and Zaia, J.A. (1991). A randomized, controlled trial of prophylactic ganciclovir for cytomegalovirus pulmonary infection in recipients of allogeneic bone marrow transplants; The City of Hope-Stanford-Syntex CMV Study Group. N. Engl. J. Med. 324, 1005–1011.

Scholz, M., Doerr, H.W., and Cinatl, J. (2003). Human cytomegalovirus retinitis: pathogenicity, immune evasion and persistence. Trends Microbiol. 11, 171–178.

Schoppel, K., Kropff, B., Schmidt, C., Vornhagen, R., and Mach, M. (1997). The humoral immune response against human cytomegalovirus is characterized by a delayed synthesis of glycoprotein-specific antibodies. J. Infect. Dis. 175, 533–544.

Schoppel, K., Schmidt, C., Einsele, H., Hebart, H., and Mach, M. (1998). Kinetics of the antibody response against human cytomegalovirus-specific proteins in allogeneic bone marrow transplant recipients. J. Infect. Dis. *178*, 1233–1243.

Sequar, G., Britt, W.J., Lakeman, F.D., Lockridge, K.M., Tarara, R.P., Canfield, D.R., Zhou, S.S., Gardner, M.B., and Barry, P.A. (2002). Experimental coinfection of rhesus macaques with rhesus cytomegalovirus and simian immunodeficiency virus: pathogenesis. J. Virol. *76*, 7661–7671.

Shanley, J.D., Biczak, L., and Formon, S.J. (1993). Acute murine cytomegalovirus infection induces lethal hepatitis. J. Infect. Dis. *167*, 264–269.

Sinclair, J., and Sissons, P. (1996). Latent and persistent infections of monocytes and macrophages. Intervirology *39*, 293–301.

Singh, N. (2001). Preemptive therapy versus universal prophylaxis with ganciclovir for cytomegalovirus in solid organ transplant recipients. Clin. Infect. Dis. *32*, 742–751.

Singh, N., Dummer, J.S., Kusne, S., Breinig, M.K., Armstrong, J.A., Makowka, L., Starzl, T.E., and Ho, M. (1988). Infections with cytomegalovirus and other herpesviruses in 121 liver transplant recipients: transmission by donated organ and the effect of OKT3 antibodies. J. Infect. Dis. *158*, 124–131.

Sinzger, C., Schmidt, K., Knapp, J., Kahl, M., Beck, R., Waldman, J., Hebart, H., Einsele, H., and Jahn, G. (1999). Modification of human cytomegalovirus tropism through propagation in vitro is associated with changes in the viral genome. J. Gen. Virol. *80*, 2867–2877.

Skaletskaya, A., Bartle, L.M., Chittenden, T., McCormick, A.L., Mocarski, E.S., and Goldmacher, V.S. (2001). A cytomegalovirus-encoded inhibitor of apoptosis that suppresses caspase-8 activation. Proc. Natl. Acad. Sci. USA *98*, 7829–7834.

Smith, P.D., Saini, S.S., Raffeld, M., Manischewitz, J.F., and Wahl, S.M. (1992). Cytomegalovirus induction of tumor necrosis factor-alpha by human monocytes and mucosal macrophages. J. Clin. Inves. *90*, 1642–1648.

Snydman, D.R., Werner, B.G., Dougherty, N.N., Griffith, J., Rubin, R.H., Dienstag, J.L., Rohrer, R. H., Freeman, R., Jenkins, R., Lewis, D., et al. (1993). Cytomegalovirus immune globulin prophylaxis in liver transplantation. A randomized, double-blind placebo-controlled trial. Ann. Intern. Med. *119*, 984–991.

Snydman, D. R., Werner, B. G., Heinze-Lacey, B., Berardi, V. P., Tilney, N. L., Kirkman, R. L., Milford, E. L., Cho, S. I., Bush, H. L., Levey, A. S., et al. (1987). Use of cytomegalovirus immune globulin to prevent cytomegalovirus disease in renal transplant recipients. N. Engl. J. Med. *317*, 1049–1054.

Soderberg-Naucler, C., and Emery, V.C. (2001). Viral infections and their impact on chronic renal allograft dysfunction. Transplantation *71*, SS24–30.

Soderberg-Naucler, C., Fish, K.N., and Nelson, J.A. (1997a). Interferon-gamma and tumor necrosis factor-alpha specifically induce formation of cytomegalovirus-permissive monocyte-derived macrophages that are refractory to the antiviral activity of these cytokines. Journal of Clin. Invest. *100*, 3154–3163.

Soderberg-Naucler, C., Fish, K.N., and Nelson, J.A. (1997b). Reactivation of latent human cytomegalovirus by allogeneic stimulation of blood cells from healthy donors. Cell *91*, 119–126.

Soderberg-Naucler, C., Fish, K.N., and Nelson, J.A. (1998). Growth of human cytomegalovirus in primary macrophages. Methods (Duluth) *16*, 126–138.

Spaderna, S., Blessing, H., Bogner, E., Britt, W., and Mach, M. (2002). Identification of glycoprotein gp-TRL10 as a structural component of human cytomegalovirus. J. Virol. *76*, 1450–1460.

Spector, S.A., Hsia, K., Crager, M., Pilcher, M., Cabral, S., and Stempien, M.J. (1999). Cytomegalovirus (CMV) DNA load is an independent predictor of CMV disease and survival in advanced AIDS. J. Virol. *73*, 7027–7030.

Spector, S.A., Weingeist, T., Pollard, R.B., Dieterich, D.T., Samo, T., Benson, C.A., Busch, D.F., Freeman, W.R., Montague, P., Kaplan, H.J., and et al. (1993). A randomized, controlled study of intravenous ganciclovir therapy for cytomegalovirus peripheral retinitis in patients with AIDS. AIDS Clinical Trials Group and Cytomegalovirus Cooperative Study Group. J. Infect. Dis. *168*, 557–563.

Spector, S.A., Wong, R., Hsia, K., Pilcher, M., and Stempien, M.J. (1998). Plasma cytomegalovirus (CMV) DNA load predicts CMV disease and survival in AIDS patients. J. Clin. Invest. *101*, 497–502.

Speir, E., Modali, R., Huang, E.S., Leon, M.B., Shawl, F., Finkel, T., and Epstein, S.E. (1994). Potential role of human cytomegalovirus and p53 interaction in coronary restenosis. Science *265*, 391–394.

Stagno, S., Dworsky, M.E., Torres, J., Mesa, T., and Hirsh, T. (1982a). Prevalence and importance of congenital cytomegalovirus infection in three different populations. J. Pediatr. *101*, 897–900.

Stagno, S., Pass, R.F., Cloud, G., Britt, W.J., Henderson, R.E., Walton, P.D., Veren, D.A., Page, F., and Alford, C.A. (1986). Primary cytomegalovirus infection in pregnancy. Incidence, transmission to fetus, and clinical outcome. JAMA 256, 1904–1908.

Stagno, S., Pass, R.F., Dworsky, M.E., Henderson, R.E., Moore, E.G., Walton, P.D., and Alford, C.A. (1982b). Congenital cytomegalovirus infection: The relative importance of primary and recurrent maternal infection. N. Engl. J. Med. 306, 945–949.

Stagno, S., Reynolds, D.W., Huang, E.S., Thames, S.D., Smith, R.J., and Alford, C.A. (1977). Congenital cytomegalovirus infection: occurrence in an immune population. N. Engl J. Med. 296, 1254–1258.

Stagno, S., Reynolds, D.W., Pass, R.F., and Alford, C.A. (1980). Breast milk and the risk of cytomegalovirus infection. N. Engl. J. Med. 302, 1073–1076.

Steffens, H.P., Kurz, S., Holtappels, R., and Reddehase, M.J. (1998). Preemptive CD8 T-cell immunotherapy of acute cytomegalovirus infection prevents lethal disease, limits the burden of latent viral genomes, and reduces the risk of virus recurrence. J. Virol. 72, 1797–1804.

Streblow, D.N., Kreklywich, C., Yin, Q., De La Melena, V.T., Corless, C.L., Smith, P.A., Brakebill, C., Cook, J.W., Vink, C., Bruggeman, C.A., Nelson, J.A., and Orloff, S.L. (2003). Cytomegalovirus-mediated up-regulation of chemokine expression correlates with the acceleration of chronic rejection in rat heart transplants. J. Virol. 77, 2182–2194.

Streblow, D.N., Orloff, S.L., and Nelson, J.A. (2001). Do pathogens accelerate atherosclerosis? J. Nutr. 131, 2798–2804.

Streblow, D.N., Soderberg-Naucler, C., Vieira, J., Smith, P., Wakabayashi, E., Ruchti, F., Mattison, K., Altschuler, Y., and Nelson, J.A. (1999). The human cytomegalovirus chemokine receptor US28 mediates vascular smooth muscle cell migration. Cell 99, 511–520.

Sun, J.C., and Bevan, M.J. (2003). Defective CD8 T-cell memory following acute infection without CD4 T-cell help.. Science 300, 339–342.

Tarantal, A.F., Salamat, M.S., Britt, W.J., Luciw, P.A., Hendrickx, A.G., and Barry, P.A. (1998). Neuropathogenesis induced by rhesus cytomegalovirus in fetal rhesus monkeys (Macaca mulatta). J. Infect. Dis. 177, 446–450.

Taylor-Wiedeman, J., Sissons, J.G., Borysiewicz, L.K., and Sinclair, J.H. (1991). Monocytes are a major site of persistence of human cytomegalovirus in peripheral blood mononuclear cells. J. Gen. Virol. 72, 2059–2064

Tikkanen, J., Kallio, E., Pulkkinen, V., Bruggeman, C., Koskinen, P., and Lemstrom, K. (2001). Cytomegalovirus infection-enhanced chronic rejection in the rat is prevented by antiviral prophylaxis. Transplant. Proc. 33, 1801.

Trgovcich, J., Pernjak-Pugel, E., Tomac, J., Koszinowski, U.H., and Jonjic, S. (1998). Pathogenesis of murine cytomegalovirus infection in neonatal mice. In CMV-Related Immunopathology, M. Scholz, H.F. Rabenau, H.W. Doerr, and J. Cinatl, eds. (Basel, Karger), pp. 42–53.

Trgovcich, J., Stimac, D., Polic, B., Krmpotic, A., Pernjak-Pugel, E., Tomac, J., Hasan, M., Wraber, B., and Jonjic, S. (2000). Immune responses and cytokine induction in the development of severe hepatitis during acute infections with murine cytomegalovirus. Arch. Virol. 145, 2601–2618.

Tu, W., Chen, S., Sharp, M., Dekker, C., Manganello, A.M., Tongson, E.C., Maecker, H.T., Holmes, T.H., Wang, Z., Kemble, G., Adler, S., Arvin, A., and Lewis, D.B. (2004). Persistent and selective deficiency of CD4+ T-cell immunity to cytomegalovirus in immunocompetent young children. J. Immunol. 172, 3260–3267.

Valantine, H.A., Gao, S.Z., Menon, S.G., Renlund, D.G., Hunt, S.A., Oyer, P., Stinson, E.B., Brown, B.W.,Jr., Merigan, T.C., and Schroeder, J.S. (1999). Impact of prophylactic immediate post-transplant ganciclovir on development of transplant atherosclerosis: a post hoc analysis of a randomized, placebo-controlled study. Circulation 100, 61–66.

van Den Pol, A.N., Mocarski, E., Saederup, N., Vieira, J., and Meier, T.J. (1999). Cytomegalovirus cell tropism, replication, and gene transfer in brain. J. Neurosci. 19, 10948–10965.

van den Pol, A.N., Reuter, J.D., and Santarelli, J.G. (2002). Enhanced cytomegalovirus infection of developing brain independent of the adaptive immune system. J. Virol. 76, 8842–8854.

Van den Pol, A.N., Vieira, J., Spencer, D.D., and Santarelli, J.G. (2000). Mouse cytomegalovirus in developing brain tissue: analysis of 11 species with GFP-expressing recombinant virus. J. Comp. Neurol. 427, 559–580.

Vochem, M., Hamprecht, K., Jahn, G., and Speer, C.P. (1998). Transmission of cytomegalovirus to preterm infants through breast milk. Pediatr. Infect. Dis. J 17, 53–58.

Volkmer, H., Bertholet, C., Jonjic, S., Wittek, R., and Koszinowski, U.H. (1987). Cytolytic T-lymphocyte recognition of the murine cytomegalovirus nonstructural immediate-early protein pp89 expressed by recombinant vaccinia virus. J. Exp. Med. 166, 668–677.

Vossen, R.C., van Dam-Mieras, M.C., and Bruggeman, C.A. (1996). Cytomegalovirus infection and vessel wall pathology. Intervirology 39, 213–221.

Wagner, M., Gutermann, A., Podlech, J., Reddehase, M.J., and Koszinowski, U.H. (2002). Major histo-compatibility complex class I allele-specific cooperative and competitive interactions between immune evasion proteins of cytomegalovirus. J. Exp. Med. 196, 805–816.

Waldman, W.J., Knight, D.A., Huang, E.H., and Sedmak, D.D. (1995). Bidirectional transmission of in-fectious cytomegalovirus between monocytes and vascular endothelial cells: an in vitro model. J. Infect. Dis. 171, 263–272.

Walter, E.A., Greenberg, P.D., Gilbert, M.J., Finch, R.J., Watanabe, K.S., Thomas, E.D., and Riddell, S.R. (1995). Reconstitution of cellular immunity against cytomegalovirus in recipients of allogeneic bone marrow by transfer of T-cell clones from the donor. N. Engl. J. Med. 333, 1038–1044.

Wang, J., Belcher, J.D., Marker, P.H., Wilcken, D.E., Vercellotti, G.M., and Wang, X.L. (2001). Cytomegalovirus inhibits p53 nuclear localization signal function. J. Mol. Med. 78, 642–647.

Williamson, W.D., Demmler, G.J., Percy, A.K., and Catlin, F.I. (1992). Progressive hearing loss in infants with asymptomatic congenital cytomegalovirus infection. Pediatrics 90, 862–866.

Winston, D.J., Ho, W.G., Bartoni, K., and Champlin, R.E. (1993). Intravenous immunoglobulin and CMV-seronegative blood products for prevention of CMV infection and disease in bone marrow transplant recipients. Bone Marrow Transplant. 12, 283–288.

Winston, D.J., Ho, W.G., Lin, C.H., Bartoni, K., Budinger, M.D., Gale, R.P., and Champlin, R.E. (1987). Intravenous immune globulin for prevention of cytomegalovirus infection and interstitial pneumonia after bone marrow transplantation. Ann. Intern. Med. 106, 12–18.

Winston, D.J., Yeager, A.M., Chandrasekar, P.H., Snydman, D.R., Petersen, F.B., Territo, M.C., and Valacyclovir Cytomegalovirus Study, G. (2003). Randomized comparison of oral valacyclovir and in-travenous ganciclovir for prevention of cytomegalovirus disease after allogeneic bone marrow trans-plantation. Clin. Infect. Dis. 36, 749–758.

Woolf, N.K. (1991). Guinea-pig model of congenital CMV-induced hearing loss: a review. Transplant. Proc. 23, 32–34.

Wreghitt, T.G., Abel, S.J., McNeil, K., Parameshwar, J., Stewart, S., Cary, N., Sharples, L., Large, S., and Wallwork, J. (1999). Intravenous ganciclovir prophylaxis for cytomegalovirus in heart, heart-lung, and lung transplant recipients. Transpl. Int. 12, 254–260.

Yamamoto, A.Y., Mussi-Pinhata, M.M., Cristina, P., Pinto, G., Moraes Figueiredo, L.T., and Jorge, S.M. (2001). Congenital cytomegalovirus infection in preterm and full-term newborn infants from a popu-lation with a high seroprevalence rate. Pediatr. Infect. Dis. J. 20, 188–192.

Yeager, A.S. (1974). Transfusion-acquired cytomegalovirus infection in newborn infants. Am J. Dis. Child. 128, 478–483.

Yeager, A.S., Grumet, F.C., Hafleigh, E.B., Arvin, A.M., Bradley, J.S., and Prober, C.G. (1981). Prevention of transfusion-acquired cytomegalovirus infections in newborn infants. J. Pediatr. 98, 281–287.

Yilmaz, S., Koskinen, P.K., Kallio, E., Bruggeman, C.A., Hayry, P.J., and Lemstrom, K.B. (1996). Cytomegalovirus infection-enhanced chronic kidney allograft rejection is linked with intercellular ad-hesion molecule-1 expression. Kidney Int. 50, 526–537.

Yu, Y., and Alwine, J.C. (2002). Human cytomegalovirus major immediate-early proteins and simian virus 40 large T antigen can inhibit apoptosis through activation of the phosphatidylinositide 3′-OH kinase pathway and the cellular kinase Akt. J. Virol. 76, 3731–3738.

Zaia, J.A., Gallez-Hawkins, G., Li, X., Yao, Z.Q., Lomeli, N., Molinder, K., La Rosa, C., and Diamond, D.J. (2001). Infrequent occurrence of natural mutations in the pp65(495–503) epitope sequence pre-sented by the HLA A*0201 allele among human cytomegalovirus isolates. J. Virol. 75, 2472–2474.

Zanghellini, F., Boppana, S.B., Emery, V.C., Griffiths, P.D., and Pass, R.F. (1999). Asymptomatic primary cytomegalovirus infection: virologic and immunologic features. J. Infect. Dis. 180, 702–707.

Zhou, Y.F., Shou, M., Guetta, E., Guzman, R., Unger, E.F., Yu, Z.X., Zhang, J., Finkel, T., and Epstein, S.E. (1999). Cytomegalovirus infection of rats increases the neointimal response to vascular injury without consistent evidence of direct infection of the vascular wall. Circulation 100, 1569–1575.

Zhu, H., Shen, Y., and Shenk, T. (1995). Human cytomegalovirus IE1 and IE2 proteins block apoptosis. J. Virol. 69, 7960–7970.

# Routes of Human CMV Transmission and Infection at the Uterine–Placental Interface

## 2

*Lenore Pereira, Ekaterina Maidji, Susan McDonagh, and Takako Tabata*

### Abstract

Congenital CMV infection affects 1–3% of babies in the United States annually, causing mortality and permanent disabilities. CMV infection of the placenta precedes virus transmission to the fetus and is linked to unusual cytotrophoblast interactions at the uterine–placental interface. Differentiating cytotrophoblasts invade the uterus, establish blood flow to the placenta, and share properties with endothelial and immune cells. Routes of virus transmission indicate that CMV infects the uterine wall and vasculature and spreads to invasive cytotrophoblasts. We summarize our understanding of the routes of infection and dysregulated functions in the developing human placenta.

## CMV infects specialized cell types in the human placenta

The human placenta is not an effective barrier to CMV, as indicated by histopathology reports that virus replicates in placenta with or without fetal transmission. Approximately 15% of women with primary infection abort spontaneously (Griffiths and Baboonian, 1984). In this case the placenta, not the fetus, shows evidence of infection, which suggests that placental involvement is important in its own right and precedes virus transmission to the fetus (Benirschke et al., 1974; Hayes and Gibas, 1971). Immunohistochemical analysis of placentas from pregnancies with congenital CMV infection and evidence of villitis found viral proteins in cells within chorionic villi including trophoblasts, stromal fibroblasts, macrophages, infiltrating lymphocytes and fetal blood vessels (Benirschke and Kaufmann, 2000; Mostoufi-Zadeh et al., 1984; Muhlemann et al., 1992; Nakamura et al., 1994; Pereira et al., 2003; Sinzger et al., 1993). CMV can cause premature delivery and, in 25% of affected infants, intrauterine growth restriction (Istas et al., 1995), outcomes that are often associated with placental pathologies. In vitro, CMV replicates in cytotrophoblasts isolated from early- and late-gestation placentas (Fisher et al., 2000; Halwachs-Baumann et al., 1998; Hemmings et al., 1998). The routes of virus transmission and the types of immune responses elicited and functions impaired are likely linked to the unusual nature of cytotrophoblast interactions with maternal cells at the uterine–placental interface (Figure 2.1).

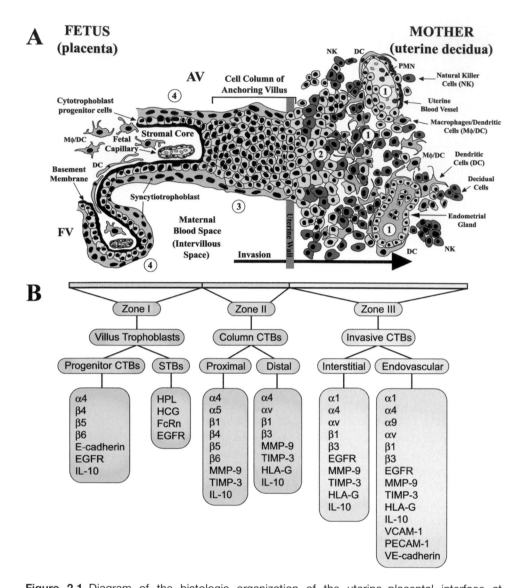

**Figure 2.1** Diagram of the histologic organization of the uterine–placental interface at midgestation and expression of stage-specific differentiation molecules. (A) The basic structural unit of the placenta is the chorionic villus, composed of a stromal core with blood vessels, surrounded by a basement membrane and overlain by cytotrophoblast progenitor cells. As part of their differentiation program, cytotrophoblasts detach from the basement membrane and adopt one of two lineage fates. They either fuse to form the syncytiotrophoblasts that cover floating villi or join a column of extravillous cytotrophoblasts at the tips of anchoring villi. The syncytial covering of floating villi mediates nutrient, gas, and waste exchange and passive transfer of IgG from maternal blood to the fetus (Zone 1). The anchoring villi, through the attachment of cytotrophoblast columns, establish physical connections between the fetus and the mother (Zone II). Invasive cytotrophoblasts penetrate the uterine wall up to the first third of the myometrium (Zone III). A portion of the extravillous cytotrophoblasts breach uterine spiral arterioles and remodel these vessels by destroying their muscular walls and replacing their endothelial linings. The diagram was modified from (Zhou et al., 1997). (B) Zones indicate the stage-specific development of trophoblasts and differentiation molecules expressed. Integrins:

# Development of the hemochorial human placenta

The embryo's acquisition of a supply of maternal blood is a critical hurdle in pregnancy maintenance. The mechanics of this process are accomplished by cytotrophoblasts, which are specialized epithelial cells. The histology and differentiation processes at the maternal–fetal interface, diagrammed in Figure 2.1, have been studied intensively by the Fisher laboratory (Damsky and Fisher, 1998; Damsky et al., 1994; McMaster et al., 1995; Zhou et al., 1997). Placentation is a stepwise process whereby cytotrophoblast progenitors leave the basement membrane to initiate blood flow to the placenta. This entails trophoblast differentiation along two independent pathways, depending on their location. In floating villi, they fuse to form a multinucleate syncytial covering attached at one end to the tree-like fetal portion of the placenta (Figure 2.1, zone I). The rest of the villus floats in a stream of maternal blood, which optimizes exchange of substances between the mother and fetus across the placenta. In the pathway that gives rise to anchoring villi (zone II), cytotrophoblasts aggregate into columns of nonpolarized mononuclear cells that attach to and penetrate the uterine wall. The ends of the columns terminate within the superficial endometrium and give rise to invasive cytotrophoblasts. A subset of these cells, either individually or in clusters, commingle with resident decidual, myometrial and immune cells. During endovascular invasion, masses of cytotrophoblasts open the termini of uterine arteries and migrate into the vessels, thereby diverting blood flow to the placenta (zone III). Cytotrophoblasts replace the endothelial lining and partially disrupt the muscular wall, whereas in veins, they are confined to the portions of the vessels near the inner surface of the uterus. Together, the two components of cytotrophoblast invasion anchor the placenta to the uterus and permit a steady increase in the supply of maternal blood that is delivered to the developing fetus.

# Cytotrophoblast differentiation along the invasive pathway

Cytotrophoblasts undergo a novel differentiation program, switching from an epithelial to an endothelial phenotype that resembles vasculogenesis, controlled through the coordinated actions of numerous interrelated factors (Figure 2.1). The cells express novel adhesion molecules, integrins, Ig superfamily members and proteinases that enable attachment and invasiveness, and immune-modulating factors for maternal tolerance of the hemiallogeneic fetus (Cross et al., 1994; Norwitz et al., 2001). Interstitial invasion requires down-regulation of integrins characteristic of epithelial cells ($\alpha 6 \beta 4$) and novel expression of $\alpha 1 \beta 1$, $\alpha 5 \beta 1$, $\alpha v \beta 3$ (Damsky et al., 1994). Integrins $\alpha 4$ and $\alpha 9$, characteristic of leukocytes, promote cell–cell and cell–matrix adhesion during invasion (Palmer et al., 1993; Tabata, submitted; Zhou et al., 1997). Epidermal growth factor receptor (EGFR), expressed by cytotrophoblasts, interacts with maternal ligands that up-regulate the cells' invasiveness in early gestation (Bass et al., 1994). Endovascular cytotrophoblasts express vasculogenic fac-

---

$\alpha 1$, $\alpha 4$, $\alpha 5$, $\alpha 9$, $\alpha v$, $\beta 1$, $\beta 3$, $\beta 4$, $\beta 5$, $\beta 6$ (Damsky et al., 1992; Tabata, manuscript submitted; Zhou et al., 1997). Hormones: human placental lactogen (HPL); human chorionic gonadotropin (HCG) (Kovalevskaya et al., 2002). Immune molecules: MHC class I HLA-G (McMaster et al., 1995); IL-10 (Roth and Fisher, 1999); neonatal Fc receptor (FcRn) (Simister et al., 1996). Proteinases and inhibitors: matrix metalloproteinase 9 (MMP-9) (Librach et al., 1991); tissue inhibitor of metalloproteinases 3 (TIMP-3) (Bass et al., 1997). Epidermal growth factor receptor (EGFR) (Bass et al., 1994). Sites proposed as natural routes of CMV infection at the uterine–placental interface are numbered 1 to 4.

tors and receptors, including VE (endothelial) cadherin and vascular endothelial adhesion molecule-1 (VCAM-1) and platelet endothelial cell growth factor (Damsky and Fisher, 1998; Zhou et al., 1997). To degrade the extracellular matrix (ECM) of the uterine stroma, cytotrophoblasts up-regulate matrix metalloproteinase 9 (MMP-9) (Librach et al., 1991) and the tissue inhibitor of metalloproteinases 3 (TIMP-3), which regulates invasiveness (Bass et al., 1997). Immune molecules – nonclassical MHC class Ib molecule HLA-G (Kovats et al., 1990; McMaster et al., 1995) and IL-10 (Roth and Fisher, 1999; Roth et al., 1996) – enable maternal tolerance. Cytotrophoblasts share properties of immune cells and express chemokine receptors, suggesting that chemokine networks at the maternal–fetal interface and paracrine actions of chemokine ligands in the villus stroma and the decidua could contribute to aspects of placental development and cytotrophoblast invasion (Drake et al., 2004; Red-Horse et al., 2004).

## Routes of placental infection and virus transmission to the fetus

The cellular organization of the placenta and recent studies of early-gestation tissues suggest routes by which CMV infection spreads from the uterus to the placenta and then the embryo/fetus (Fisher et al., 2000; Pereira et al., 2003). One route of transmission could occur during invasion into the uterine wall (Figure 2.1, sites 1 and 2). Interstitial cytotrophoblasts could encounter infected endometrial glands, decidual cells and fibroblasts, and decidual granular leukocytes. Endovascular cytotrophoblasts remodeling the uterine vasculature replace maternal endothelial and vascular smooth muscle cells that could harbor persistent infection. Infected cytotrophoblasts (site 2) could spread virus in a retrograde manner through the cell columns to the anchoring chorionic villi (site 3). Another route of transmission is hematogenous across the syncytiotrophoblast layer that covers floating chorionic villi (site 4). These placental cells, in direct contact with maternal blood, express the neonatal Fc receptor (FcRn), a molecule that facilitates maternal IgG transfer and passive immunization of the fetus (Simister et al., 1996). In syncytiotrophoblasts that accumulate IgG–virion complexes pinocytosed from maternal blood, FcRn could transport complexes to cytotrophoblast progenitors that become infected in the presence of low-avidity antibodies (Fisher et al., 2000; Maidji et al., submitted; Pereira et al., 2003). Focal cytotrophoblast infection might spread to stromal fibroblasts and fetal blood vessels. Infected endothelial cells and possibly neutrophils carry virus via the placental circulation to the fetus. Villus core macrophages phagocytose transcytosed IgG–virion complexes and virions from infected foci. Accordingly, neutralizing antibodies and innate immune cells that suppress infection at the uterine–placental interface protect the fetus from infection.

## CMV proteins expressed in cytotrophoblast progenitor cells in chorionic villi infected in utero and in vitro

Clues about potential routes of virus transmission emerged from analysis of naturally infected placentas and villus explants infected in vitro (Figures 2.2A and 2.3). Tissue sections of chorionic villi were double-stained with anti-cytokeratin to identify trophoblast cells and with a monoclonal antibody to CMV IE proteins 1 and 2 to identify infected cells. Naturally infected placentas showed clusters of cytotrophoblast progenitor cells underlying the syncytium that were productively infected and expressed CMV IE proteins (Figure 2.3A and B). Often, these clusters were the only cells infected. These patterns of infection were

**A. Villus Explant**
**plated on Matrigel**

**B. CTBs plated on Matrigel**

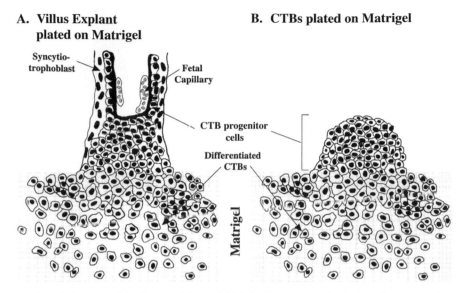

**Figure 2.2** Culture models for studying CMV infection using chorionic villus explants and purified cytotrophoblasts. (A) Diagram of an anchoring villus explant attached to a Matrigel substrate by cytotrophoblasts (CTBs) from the cell column. (B) Diagram of purified differentiating cytotrophoblasts cultured on Matrigel. The cytotrophoblast progenitors cells aggregate, invade the matrix, and express stage-specific molecules upon differentiation. For infection, CMV is added to media on the villus explants and cells.

virtually indistinguishable from those found after infection in vitro. Using a model tissue culture system, chorionic villi are plated on filters coated with Matrigel (an extracellular matrix), infected with CMV and then cultured (Figure 2.2A). These experiments revealed an unexpected infection pattern (Fisher et al., 2000). Notably, the syncytiotrophoblast that cover the villus surface were not infected and failed to stain for CMV IE proteins, whereas nuclear staining of small, isolated clusters of underlying cytotrophoblast progenitor cells was observed (Figure 2.3C and D). In some explants, CMV IE proteins were also detected in cytotrophoblasts in the cell columns of anchoring villi (Figure 2.3E and F). These studies suggested that in vitro infection models the initial steps in placental infection, and that virus is transmitted from trophoblasts to other cell types in the villus core in naturally infected tissues.

## CMV and pathogenic microorganisms detected in the decidua and adjacent placenta

Intrauterine CMV infections do not occur in isolation and often include other viral and bacterial pathogens (Collier et al., 1990; Coonrod et al., 1998; Romero et al., 2002). Analysis of biopsy specimens from the decidua and adjacent placentas of 265 uncomplicated pregnancies found CMV DNA in 69% of specimens, and CMV with bacteria in 38% (McDonagh et al., 2004; Pereira et al., 2003). When found in isolation, CMV was detected in 27% of placental samples. Other pathogens included herpes simplex virus type 1 (HSV-1) in 3%, HSV-2 in 9%, and more than one bacterial species in 15% of specimens. Only sixteen percent of placental samples were negative for these pathogens, suggesting

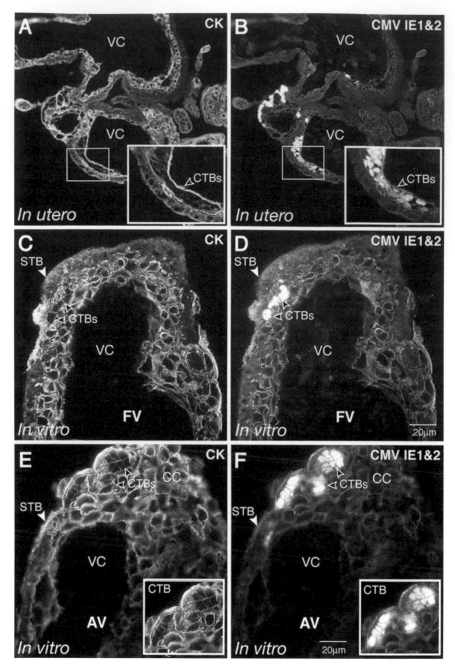

**Figure 2.3** CMV infects underlying cytotrophoblasts in naturally infected chorionic villi and villus explants infected in vitro. Naturally infected chorionic villi stained with antibodies to (A) cytokeratin (CK) and (B) CMV IE1&2 proteins. (C and E) Cytokeratin-stained floating villi (FV) and anchoring villi (AV) showed syncytiotrophoblasts (STB) that cover the surface and the underlying cytotrophoblasts (CTBs). (D and F) CMV IE1&2 proteins expressed by underlying clusters of infected cytotrophoblasts. The inner stromal villus cores (VC) were negative for infected cells. CMV protein expression was also detected in CTBs in cell columns (CC) of anchoring villi (AV). Insets show infected CTBs at higher magnification.

that early-gestation placentas frequently contain DNA from viral and bacterial pathogens. Detailed analysis of paired first-trimester decidual and placental biopsy specimens from individual pregnancies showed that some pathogens were present in both. CMV DNA was detected in 89% of the decidual samples and 63% of the placentas. When CMV was found in isolation in the decidua (40%), virus was also sometimes present in the placenta (26%). In contrast, bacterial DNA was detected in the placenta (11%) and less frequently in the decidua (6%). Together these results suggest that CMV can be selectively transferred from the decidua to the adjacent placenta and that infection can occur with or without pathogenic bacteria.

## Patterns of CMV infection in the uterine wall mirrored in the adjacent placenta

Examination of CMV proteins in paired decidual and placental biopsy specimens from seropositive donors showed three staining patterns related to neutralizing titers (Pereira et al., 2003). In the first pattern, islands in both decidual and placental compartments stained strongly for CMV-infected cell proteins (Figure 2.4A). This pattern predominated in samples from donors with low CMV neutralizing titers and a few with intermediate titers and other pathogens. In the decidua, cytokeratin-positive glandular epithelial cells, endovascular cytotrophoblasts in remodeled uterine blood vessels, and interstitial cytotrophoblasts were sometimes positive. Resident decidual cells strongly stained for viral proteins, suggesting that these cells were permissive for viral replication. In the adjacent portions of the placenta, floating villi contained cytotrophoblast expressing CMV-infected cell proteins that localized to the nuclei and cytoplasm. Abundant vesicles amassed close to the plasma membrane of the villus surface and contained the virion envelope glycoprotein gB. In regions with infected syncytiotrophoblasts, fibroblasts and fetal capillaries in the villus core expressed infected-cell proteins. Invasive cytotrophoblasts in developing cell columns that anchor the placenta to the uterine wall also stained. In contrast, macrophages (Mϕ) and dendritic cells (DCs) within the villus stromal cores contained infected-cell proteins in cytoplasmic vesicles but not in the nuclei, suggesting phagocytosis.

In the second pattern, the number of cells that stained for CMV-infected-cell proteins was reduced in the decidua, and occasional focal infection was found in the placenta (Figure 2.4B). This pattern predominated in samples from donors with low to intermediate neutralizing titers, several of which contained other pathogens as well. In the decidua, CMV replication was detected in some glandular epithelial cells and decidual cells. In the interstitium, Mϕ/DCs were abundant throughout, especially near infected glands and blood vessels. These cells contained gB-positive cytoplasmic vesicles but were not infected. Sometimes the adjacent placentas contained small clusters of cytotrophoblast progenitor cells that expressed CMV-infected-cell proteins. Isolated gB-containing vesicles were present in the syncytiotrophoblast layer. In the villus core, Mϕ/DCs containing CMV gB-positive vesicles were often observed.

In the third pattern, few cells stained for CMV-infected-cell proteins in the decidua, and none were found in the placenta (Figure 2.4C). This pattern predominated in samples from donors with intermediate to high neutralizing titers, several of which contained other pathogens as well. In the decidua, neutrophils with viral proteins were found in uterine blood vessels near infected cells. In the adjacent portions of the placenta, syncytiotropho-

**A**

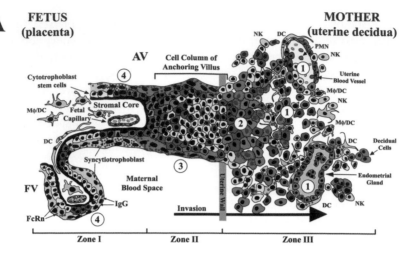

**B**

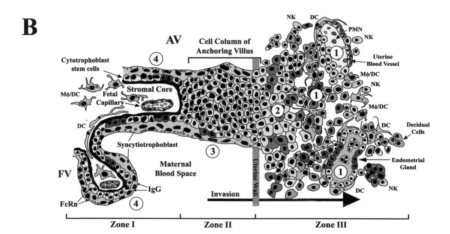

**C**

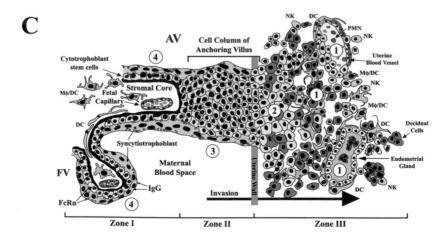

blasts contained numerous CMV gB-positive vesicles but were not infected. In villus core MΦ/DCs, gB accumulated in large cytoplasmic vesicles. When placentas were stained for IgG, syncytiotrophoblasts contained many positive vesicles and a subset contained gB. In villus core MΦ/DCs, some gB-staining vesicles colocalized with the more abundant IgG-positive vesicles. In some tissues, the presence of CMV virions was confirmed by electron microscopy.

To determine whether multiple CMV strains infect the uterine–placental interface, a region of the gB gene with characteristic nucleotide differences was sequenced. Sequence analysis of selected CMV-positive samples revealed that the gB genotypes were similar to variants in groups 1, 2, and 3 (Chou and Dennison, 1991). In these cases, one genotype was present in both the decidua and adjacent placenta, as indicated by genotype sequence. The one exception was tissue from a seropositive donor lacking detectable neutralizing antibodies: this specimen contained a mixture of gB genotypes, suggesting different strains could colonize the pregnant uterus early in the course of maternal infection.

In summary, CMV is commonly present at the maternal–fetal interface, one possible explanation for why pregnant women shed virus from the cervix (Collier et al., 1995; Shen et al., 1993; Stagno et al., 1975). Bacteria were found in some donors with intermediate to high CMV neutralizing titers whose uninfected placentas contained virion proteins, suggesting limited viral replication in the decidua (Pereira et al., 2003). Reactivation from decidual MΦ/DCs might occur as a consequence of inflammatory responses to pathogenic bacteria and could depend on the number of latently infected MΦ/DCs that infiltrate the uterus (Cook et al., 2002; Hahn et al., 1998; Soderberg-Naucler et al., 2001). Placentas from healthy pregnant donors contained isolated areas of infection that were a small part of the whole tissue. Since these tissues were from normal pregnancies, placental infection that leads to serious congenital infection likely involves the decidual and placental components that stained for infected-cell proteins, i.e., an exacerbation of the situation found in samples from women with the lowest CMV neutralizing titers and some with intermediate titers as well as bacterial pathogens.

## Hematogenous route of transmission: FcRn-dependent transcytosis of IgG–virion complexes to underlying cells

The unusual pattern of infection in cytotrophoblast progenitors in chorionic villi and villus explants suggested that virions could be transported across the syncytium and infect the

**Figure 2.4** (opposite) Schematic of patterns of CMV infection at the uterine–placental interface. (A) Acute CMV infection in the decidua associated with an infected placenta. CMV-infected cells (red) and gB-containing vesicles (red) in MΦ/DCs are seen at the uterine–placental interface. Infected cells were found in endometrial glands, uterine blood vessels and invasive cytotrophoblasts. Infection was transmitted to portions of the adjacent placenta, as indicated by expression of viral replication proteins by trophoblasts and fetal capillaries. Some MΦ/DCs contained cytoplasmic vesicles with CMV gB, suggesting phagocytosis without productive infection. (B) Reduced infection in the decidua mirrored in the placenta. Focal infection found in cytotrophoblast progenitor cells and virion phagocytosis by villus core MΦ/DCs. (C) Suppressed infection in the decidua and placenta with gB present in vesicles in syncytiotrophoblasts and in villus core MΦ/DCs without viral replication. FV, floating villi; AV, anchoring villi; FcRn, neonatal Fc receptor; IgG, immunoglobulin G; PMN, neutrophil; NK, natural killer. Sites proposed as routes of infection are numbered 1 to 4.

underlying cells (Figure 2.3). If this occurred via immune complexes of virions and IgG, then infection could depend on FcRn transcytosis. Immunostaining of early-gestation tissue confirmed the presence of FcRn in apical microvilli of syncytiotrophoblasts that interface with the maternal blood and also contain vesicles with IgG and gB (Leach et al., 1996; Leitner et al., 2002; Maidji et al., submitted; Pereira et al., 2003). As predicted, maternal IgG, purified from placental biopsy specimens, reacted with CMV-infected fibroblasts (by immunofluorescence) and with viral replication proteins and gB (by immunoblot). Accordingly, chorionic villi showed focal CMV infection when tissue-associated IgG was of low avidity (see Figure 2.3F) (Pereira et al., 2003). Figure 2.5 shows an illustration modeling the early steps in hematogenous spread of CMV infection in a chorionic villus based on analysis of FcRn-dependent transport of IgG–virion complexes in polarized epithelial T-84 cells, naturally expressing the receptor (Dickinson et al., 1999).

Experiments first assessed whether virion transport in T-84 cells is dependent on CMV-specific IgG and FcRn binding. When cells were infected with IgG–virion complexes, they were transcytosed from the apical to the basal membrane domain regardless of whether antibodies had high or low neutralizing titers (Figure 2.6A and C), but virions alone were not transcytosed. Quantification of CMV DNA showed that a small fraction of the input complexes from the apical compartment were transported to the basal compartment. When cells were pretreated with chicken Fc IgY and F(ab')$_2$ IgG fragments that fail to bind FcRn, virion complexes were transcytosed (Figure 2.6A). In contrast, when cells were pretreated with human Fc IgG fragment or the immune complexes were pretreated with Staphylococcus protein A, which binds Fc, virion complexes were not transcytosed, suggesting competition for FcRn could preclude binding and transport of virion complexes (Figure 2.6B). Infectivity of transcytosed complexes in fibroblasts showed that virion complexes with low-neutralizing IgG were infectious (Figure 2.6D), whereas those with high-neutralizing IgG were not.

This study demonstrated that polarized epithelial cells expressing FcRn transcytose IgG–virion complexes that could transmit infection to underlying cells in the presence of low-avidity IgG. The findings could explain the pattern of infection in cytotrophoblast progenitors in placentas from women with low neutralizing titer (Fisher et al., 2000), the presence of CMV DNA and virion proteins in uninfected placentas from immune donors (Pereira et al., 2003), and the association between primary maternal infection, low-avidity IgG, and a high incidence of congenital CMV (Boppana and Britt, 1995; Fowler et al., 2003; Revello et al., 2002; Stagno et al., 1982). Conceivably, FcRn-dependent IgG–virion transport offers a mechanism for virus transmission to organs composed of epithelial cells expressing this receptor – lung, kidney, intestine and breast – in the course of natural infection (Cianga et al., 2003; Haymann et al., 2000; Israel et al., 1997; Kobayashi et al., 2002; Spiekermann et al., 2002).

## CMV replicates in placental cytotrophoblasts in vitro

Several groups have reported that cytotrophoblasts isolated from early-gestation (Fisher et al., 2000; Hemmings et al., 1998) and term placentas (Halwachs-Baumann et al., 1998) are susceptible to CMV infection. A detailed examination of the viral life cycle was done using an in vitro model of cytotrophoblasts purified from chorionic villi plated as a monolayer on Matrigel (see Figure 2.2B). Under these conditions the cells form aggregates, analogous to

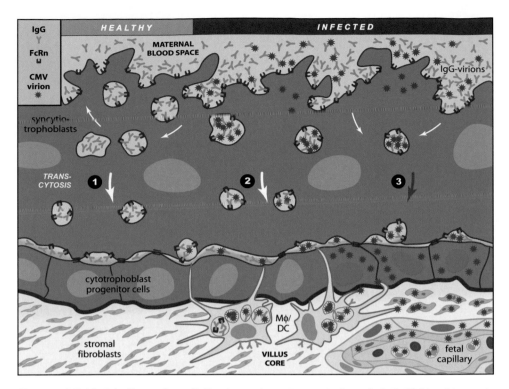

**Figure 2.5** Model illustrating FcRn-dependent transcytosis of IgG-CMV virions in syncytiotrophoblasts covering chorionic villi in early gestation. Pathway 1: Healthy placenta; IgG pinocytosed from maternal blood space binds to FcRn in endocytic vesicles and is recycled or transcytosed across the syncytium to the basal membrane and released. Pathway 2: Suppressed placental infection in an immune mother; pinocytosed IgG–virion complexes of high-avidity IgG bind to FcRn in endocytic vesicles and are stored or transported and released into the intercellular space where they contact cytotrophoblast progenitors. Virion complexes, internalized by immature Mφ/DCs in the villus core across cytotrophoblast cell junctions, are sequestered in vacuoles (Pereira et al., 2003). Pathway 3: Active infection in placenta; pinocytosed low-avidity IgG–virion complexes bind to FcRn in endocytic vesicles and are transcytosed, released, and infect underlying cytotrophoblasts. Virion nucleocapsids are sometimes detected close to the maternal blood space in syncytiotrophoblasts. Focal infection spreads to villus core stromal fibroblasts, fetal capillary cells and leukocytes. Symbols: IgG, FcRn and virions are shown at the left. Mφ/DCs are villus core macrophage/dendritic cells. Model based on (Maidji et al., submitted).

cell columns, and differentiate along the invasive pathway. In CMV-infected cells, nuclear staining for IE proteins was detected by 24 h and cytoplasmic staining for gB was detected by 72 h in 20–40% of the cells. There was an increase in the titers of intracellular and progeny virions during the culture period, confirming that differentiating/invasive cyto-trophoblasts are fully permissive for CMV replication. Infected cytotrophoblasts showed considerable changes, evidenced by dysregulated expression of stage-specific adhesion and immune molecules and functions, described next.

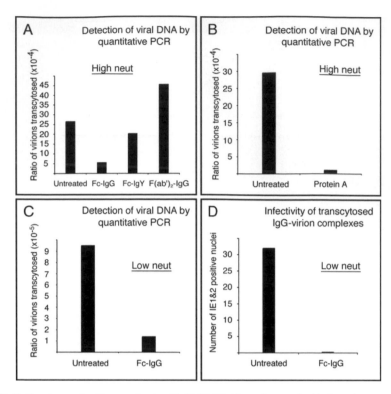

**Figure 2.6** FcRn-dependent transcytosis of IgG-CMV virions from apical to basal compartments across polarized T-84 cells (Maidji et al., submitted). Purified IgG from a highly immune donor: (A) cells treated with Fc-IgG, Fc-IgY and F(ab')₂-IgG; (B) immune complexes treated with protein A. Purified IgG from a donor with low CMV neutralizing activity: (C) cells treated with Fc-IgG; (D) infectivity of transcytosed IgG–virion complexes in fibroblasts. Viral DNA was detected by quantitative PCR.

## CMV-infected cytotrophoblasts down-regulate expression of class Ib molecule HLA-G

In healthy placentas, the nonclassical MHC class Ib molecule HLA-G is expressed in differentiating cytotrophoblasts, particularly cells in anchoring villi with an increasing gradient of expression in the distal columns that is maintained once the cells enter the uterine wall (McMaster et al., 1995). Immunolocalization experiments showed that CMV infection of differentiating cytotrophoblasts down-regulates HLA-G expression (Fisher et al., 2000) (Figure 2.7). At late times after infection when high levels of CMV gB were detected, staining for HLA-G was lost, whereas uninfected cells stained with anti-HLA-G.

Several CMV genes down-regulate expression of classical MHC class Ia molecules (reviewed in Ploegh, 1998). Studies using mutants in which all of the genes known to down-regulate cell-surface expression of MHC class Ia molecules are deleted (Jones and Muzithras, 1992) showed that HLA-G expression was not rescued, suggesting a novel mechanism for dysregulation (Fisher et al., 2000). Others reported that HLA-G is resistant to the effects of US11, which binds to class I heavy chains and mediates their dislocation to the cytosol and subsequent proteasomal degradation (Schust et al., 1998). Subsequent

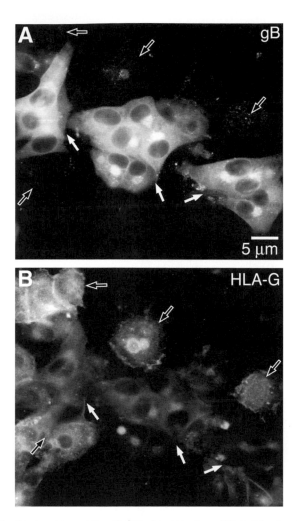

**Figure 2.7** CMV infection impairs cytotrophoblast expression of HLA-G in vitro. Cytotrophoblasts were purified and infected. At 72 h, the cells were immunostained for (A) gB and (B) HLA-G. Cells that did not express gB (black arrows) expressed HLA-G, and staining for gB (white arrows) was associated with a marked reduction in HLA-G expression.

analyses using chimeric molecules of MHC class Ia and Ib showed that the degradation efficiency depended on sequences in the heavy-chain cytosolic tail that HLA-G lacks (Barel et al., 2003). Analysis of HLA-G expression in cytotrophoblasts infected with the clinical strain VR1814 suggested that down-regulated expression occurs at the transcriptional level and does not involve viral glycoproteins that alter class Ia expression (S. McDonagh and L. Pereira, unpublished).

## CMV infection down-regulates α1β1 integrin expression
Congenital CMV infection is associated with abnormal placentation at a morphological level and intrauterine growth restriction (Benirschke et al., 1974), likely related to impaired remodeling of uterine arterioles. This prompted examination of the expression of the lam-

inin/collagen receptor integrin α1β1 in the context of CMV infection. Co-localization of CMV gB (Figure 2.8A and C) and integrin α1 expression (Figure 2.8B and D) showed that cells that did not stain for gB (Figure 2.8A) expressed integrin α1 in a plasma membrane-associated pattern (Figure 2.8B). Diffuse cytoplasmic staining for gB in infected cytotrophoblasts was also correlated with integrin α1 expression (see cell marked with an asterisk in Figure 2.8C and D), but accumulation of gB in vesicles (Figure 2.8C) at late times after infection was associated with the absence of staining for integrin α1 (Figure 2.8D). In contrast, analysis of another integrin that is up-regulated as the cells invade showed that α5, which inhibits invasion, was not affected. Recent studies showed corresponding transcriptional changes in VR1814-infected cytotrophoblasts (S. McDonagh and L. Pereira, unpublished).

## CMV impairs cytotrophoblast invasiveness in vitro

The impact of CMV infection on cytotrophoblast invasion was examined using an in vitro assay (see Figure 2.2B) (Fisher et al., 2000). This functional assay tests the ability of isolated cytotrophoblasts plated on the upper surfaces of Matrigel-coated filters to penetrate the surface, pass through pores in the underlying filter, and emerge on the lower surface of the membrane. The ability of cells infected with CMV to invade was dramatically impaired, as compared with control uninfected cells. Interestingly, the effect on invasion was greater than could be accounted for by the number of CMV-infected cells, suggesting that the pres-

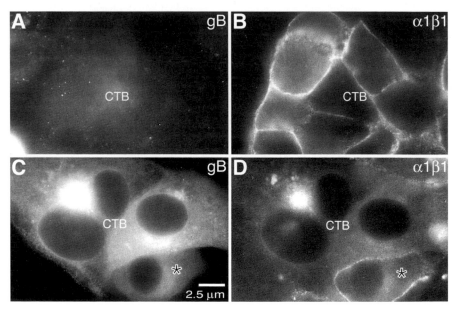

**Figure 2.8** CMV infection in vitro eventually down-regulates cytotrophoblast expression of integrin α1. Purified cytotrophoblasts were infected with CMV in vitro. At 72 h, the cells were fixed and stained for expression of gB and integrin α1. Cytotrophoblasts that did not express gB (A) displayed prominent staining for integrin α1 in a plasma membrane-associated pattern (B). Likewise, cells that stained in a diffuse cytoplasmic pattern for gB (C) also reacted with the anti-integrin antibody (D; cell marked with an asterisk). However, when gB was localized in a vesicular pattern, integrin staining was not detected (D).

ence of infected cells in the invading aggregates influences the behavior of the population as a whole and could be exacerbated by a secreted viral factor, addressed in the next section.

## CMV infection down-regulates MMP-9 protein levels via cmvIL-10, which impairs cell–cell and cell–matrix interactions

In early gestation, invasive cytotrophoblasts secrete relatively large amounts of MMP-9 that remodel the extracellular matrix; later, when invasion is complete, MMP-9 levels fall (Librach et al., 1991). The inactive proenzyme is activated by cleavage and removal of an inhibitory domain. Activated MMP-9 is absolutely required for invasion, whereas pro-MMP-9 is associated with noninvasive cells. Cytotrophoblasts infected with the clinical strain VR1814, but not with a laboratory strain, reduced MMP-9 production and activity (Figure 2.9A) (Yamamoto-Tabata et al., 2004). Cytokines and growth factors, including human IL-10, that are produced by cytotrophoblasts and impair the cells' invasion regulate MMP-9 expression and activity (Roth and Fisher, 1999; Roth et al., 1996). CMV exploits the IL-10 signaling pathway, expressing an IL-10 homolog (Kotenko et al., 2000). Although cmvIL-10 shares only 27% sequence identity with human IL-10 (hIL-10), the proteins have essentially identical affinity for the receptor, IL-10R1, and similarly reorganize the cell-surface receptor complex (Jones et al., 2002). MMP-9 activity was reduced in differentiating cytotrophoblasts treated with purified recombinant cmvIL-10 or hIL-10 in a dose-dependent fashion, equivalent to AD169 infection, but was reduced even more in VR1814-infected cells, suggesting that clinical strains have a detrimental effect on function (Yamamoto-Tabata et al., 2004) (Figure 2.9A and B). In addition, cytotrophoblasts treated with cmvIL-10 alone up-regulated the cells' production of the cytokine (Figure 2.9C). The effect of reduced proteinase activity was assessed by the frequency with which cytotrophoblasts passed through narrow pores in a Matrigel-coated filter, quantified using the in vitro functional assay (Figure 2.3B). Significantly fewer cells traversed the filter pores after treatment with cmvIL-10 alone (equivalent to hIL-10-treated cells), as compared with control untreated cells (Yamamoto-Tabata et al., 2004). These studies suggest that CMV also exploits an immune mechanism to dysregulate cytotrophoblast invasiveness that impairs extracellular matrix remodeling, thereby compounding changes resulting from integrin misexpression in infected cells.

## Outstanding questions about intrauterine CMV infection

We have begun to decipher the routes of CMV transmission during pregnancy, patterns of intrauterine infection, and dysregulated functions of the developing placenta. Considering the hypoimmune-responsive environment at the maternal–fetal interface, CMV may have developed the capacity for persistent infection or reactivation in cycling women. Many outstanding questions remain to be answered with regard to transmission in the uterine wall and intervillous blood space.

1   Do seropositive women have intrauterine CMV infection that is suppressed by innate and specific immune responses?
2   Can latently infected Mφ/DC progenitors that populate the pregnant uterus infect the decidua in women with recent seroconversion?

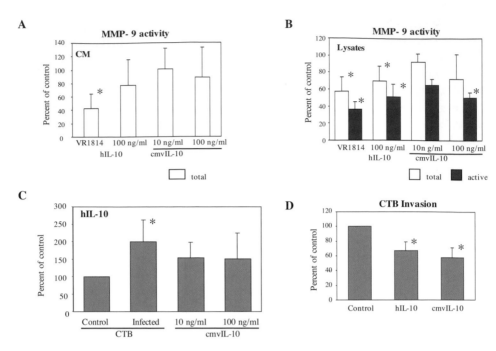

**Figure 2.9** Reduced MMP-9 activity in differentiating cytotrophoblasts infected with VR1814 or treated with cmvIL-10, and impaired invasiveness (Yamamoto-Tabata et al., 2004). Cytotrophoblasts (CTB) were left untreated, infected, treated with conditioned medium (CM), or treated with recombinant human or CMV interleukin-10 (hIL-10 and cmvIL-10, respectively). Matrigel-coated wells without cells served as a control. CM and cell lysates were then analyzed for MMP-9 activity by gelatin zymography (A and B). CMV infection, hIL-10 treatment and cmvIL-10 (100 ng/ml) treatment all decreased MMP-9 (92-kDa proenzyme; 86-kDa activated form) in CM (A) and cell lysates (B). Cytotrophoblasts were infected or treated with hIL-10 or cmvIL-10 for 4 days, and the CM (A) and lysates (B) were analyzed for MMP-9 activity, which was quantified as percent of control. Up-regulation of hIL-10 expression in differentiating cytotrophoblasts after VR1814 infection and cmvIL-10 treatment (C). CM from VR1814- and mock-infected control or cmvIL-10-treated cytotrophoblasts was collected on day 1 and the cells were cultured for 24 to 36 h. hIL-10 was then quantified using ELISA. Results are expressed as percent of control. CM from untreated mock-infected cells served as a control. Asterisks indicate a significant difference between control and infected or treated cells ($P < 0.05$). Bars, SD. (D) Impaired cell motility in functional assays of endothelial cell wound healing and cytotrophoblast invasion of Matrigel in vitro. Differentiating/invading cytotrophoblasts from first-trimester placentas, plated on Matrigel-coated filters, were treated with cmvIL-10 or hIL-10, fixed at 48 h and stained with cytokeratin-specific antibody. Cells that reached the filter underside were counted. Values (percent of control) are expressed as the mean ± SD of three experiments. Asterisks indicate a significant difference between control and IL-10-treated cells ($P < 0.05$).

3    Do pathogenic microorganisms trigger inflammation and local reactivation or persistent intrauterine infection?

4    Does reinfection occur at the uterine–placental interface in immune women?

5    Do women clear intrauterine infection following delivery? These difficult questions are now being addressed in order to understand the causes and reduce the incidence and damage due to congenital CMV infection.

## Acknowledgments

We are grateful to members of the Pereira and Fisher laboratories for inspiring discussions; Zoya Kharitonov, Hsin-Ti Chang and James Nachtwey for technical assistance, Doug Beckner and Keith Jones for preparing the illustrations, and Mary McKenney for editing the manuscript. This project was supported by US Public Health Service grants AI46657, AI53782 and EY13683 from the National Institutes of Health and grants from the March of Dimes Birth Defects Foundation and the University of California Academic Senate.

## References

Barel, M.T., Pizzato, N., van Leeuwen, D., Bouteiller, P.L., Wiertz, E.J., and Lenfant, F. (2003). Amino acid composition of alpha1/alpha2 domains and cytoplasmic tail of MHC class I molecules determine their susceptibility to human cytomegalovirus US11-mediated down-regulation. Eur. J. Immunol. 33, 1707–1716.

Bass, K.E., Li, H., Hawkes, S.P., Howard, E., Bullen, E., Vu, T.K., McMaster, M., Janatpour, M., and Fisher, S.J. (1997). Tissue inhibitor of metalloproteinase-3 expression is up-regulated during human cytotrophoblast invasion in vitro. Dev. Gen. 21, 61–67.

Bass, K.E., Morrish, D., Roth, I., Bhardwaj, D., Taylor, R., Zhou, Y., and Fisher, S.J. (1994). Human cytotrophoblast invasion is up-regulated by epidermal growth factor: evidence that paracrine factors modify this process. Dev. Biol. 164, 550–561.

Benirschke, K., and Kaufmann, P. (2000). Pathology of the Human Placenta, Fourth Edition eds (New York, Springer).

Benirschke, K., Mendoza, G.R., and Bazeley, P.L. (1974). Placental and fetal manifestations of cytomegalovirus infection. Virchows. Arch. B. Cell Pathol. 16, 121–139.

Boppana, S.B., and Britt, W.J. (1995). Antiviral antibody responses and intrauterine transmission after primary maternal cytomegalovirus infection. J. Infect. Dis. 171, 1115–1121.

Chou, S.W., and Dennison, K.M. (1991). Analysis of interstrain variation in cytomegalovirus glycoprotein B sequences encoding neutralization-related epitopes. J. Infect. Dis. 163, 1229–1234.

Cianga, P., Cianga, C., Cozma, L., Ward, E.S., and Carasevici, E. (2003). The MHC class I related Fc receptor, FcRn, is expressed in the epithelial cells of the human mammary gland. Hum. Immunol. 64, 1152–1159.

Collier, A.C., Handsfield, H.H., Ashley, R., Roberts, P.L., DeRouen, T., Meyers, J.D., and Corey, L. (1995). Cervical but not urinary excretion of cytomegalovirus is related to sexual activity and contraceptive practices in sexually active women. J. Infect. Dis. 171, 33–38.

Collier, A.C., Handsfield, H.H., Roberts, P.L., DeRouen, T., Meyers, J.D., Leach, L., Murphy, V.L., Verdon, M., and Corey, L. (1990). Cytomegalovirus Infection in Women Attending a Sexually Transmitted Disease Clinic. J. Infect. Dis. 162, 46–51.

Cook, C.H., Zhang, Y., McGuinness, B.J., Lahm, M.C., Sedmak, D.D., and Ferguson, R.M. (2002). Intra-abdominal bacterial infection reactivates latent pulmonary cytomegalovirus in immunocompetent mice. J. Infect. Dis. 185, 1395–1400.

Coonrod, D., Collier, A.C., Ashley, R., DeRouen, T., and Corey, L. (1998). Association between cytomegalovirus seroconversion and upper genital tract infection among women attending a sexually transmitted disease clinic: a prospective study. J. Infect. Dis. 177, 1188–1193.

Cross, J.C., Werb, Z., and Fisher, S.J. (1994). Implantation and the placenta: key pieces of the development puzzle. Science 266, 1508–1518.

Damsky, C.H., and Fisher, S.J. (1998). Trophoblast pseudo-vasculogenesis: faking it with endothelial adhesion receptors. Curr. Opin. Cell Biol. 10, 660–666.

Damsky, C.H., Fitzgerald, M.L., and Fisher, S.J. (1992). Distribution patterns of extracellular matrix components and adhesion receptors are intricately modulated during first trimester cytotrophoblast differentiation along the invasive pathway, in vivo. J. Clin. Invest. 89, 210–222.

Damsky, C.H., Librach, C., Lim, K.H., Fitzgerald, M.L., McMaster, M.T., Janatpour, M., Zhou, Y., Logan, S.K., and Fisher, S.J. (1994). Integrin switching regulates normal trophoblast invasion. Development 120, 3657–3666.

Dickinson, B.L., Badizadegan, K., Wu, Z., Ahouse, J.C., Zhu, X., Simister, N.E., Blumberg, R.S., and Lencer, W.I. (1999). Bidirectional FcRn-dependent IgG transport in a polarized human intestinal epithelial cell line. J. Clin. Invest. 104, 903–911.

Drake, P.M., Red-Horse, K., and Fisher, S.J. (2004). Reciprocal chemokine receptor and ligand expression in the human placenta: implications for cytotrophoblast differentiation. Dev. Dyn. *229*, 877–885.

Fisher, S., Genbacev, O., Maidji, E., and Pereira, L. (2000). Human cytomegalovirus infection of placental cytotrophoblasts in vitro and in utero: implications for transmission and pathogenesis. J. Virol. *74*, 6808–6820.

Fowler, K.B., Stagno, S., and Pass, R.F. (2003). Maternal immunity and prevention of congenital cytomegalovirus infection. JAMA. *289*, 1008–1011.

Griffiths, P.D., and Baboonian, C. (1984). A prospective study of primary cytomegalovirus infection during pregnancy: final report. Br. J. Obstet. Gynaecol. *91*, 307–315.

Hahn, G., Jores, R., and Mocarski, E.S. (1998). Cytomegalovirus remains latent in a common precursor of dendritic and myeloid cells. Proc. Natl. Acad. Sci. USA *95*, 3937–3942.

Halwachs-Baumann, G., Wilders-Truschnig, M., Desoye, G., Hahn, T., Kiesel, L., Klingel, K., Rieger, P., Jahn, G., and Sinzger, C. (1998). Human trophoblast-cells are permissive to the complete replicative cycle of human cytomegalovirus. J. Virol. *72*, 7598–7602.

Hayes, K., and Gibas, H. (1971). Placental cytomegalovirus infection without fetal involvement following primary infection in pregnancy. J. Pediatr. *79*, 401–405.

Haymann, J.P., Levraud, J.P., Bouet, S., Kappes, V., Hagege, J., Nguyen, G., Xu, Y., Rondeau, E., and Sraer, J.D. (2000). Characterization and localization of the neonatal Fc receptor in adult human kidney. J. Am. Soc. Nephrol. *11*, 632–639.

Hemmings, D.G., Kilani, R., Nykiforuk, C., Preiksaitis, J., and Guilbert, L.J. (1998). Permissive cytomegalovirus infection of primary villous term and first trimester trophoblasts. J. Virol. *72*, 4970–4979.

Israel, E.J., Taylor, S., Wu, Z., Mizoguchi, E., Blumberg, R.S., Bhan, A., and Simister, N.E. (1997). Expression of the neonatal Fc receptor, FcRn, on human intestinal epithelial cells. Immunology *92*, 69–74.

Istas, A.S., Demmler, G.J., Dobbins, J.G., and Stewart, J.A. (1995). Surveillance for congenital cytomegalovirus disease: a report from the National Congenital Cytomegalovirus Disease Registry. Clin. Infect. Dis. *20*, 665–670.

Jones, B.C., Logsdon, N.J., Josephson, K., Cook, J., Barry, P.A., and Walter, M.R. (2002). Crystal structure of human cytomegalovirus IL-10 bound to soluble human IL-10R1. Proc. Natl. Acad. Sci. USA *99*, 9404–9409.

Jones, T.R., and Muzithras, V.P. (1992). A cluster of dispensable genes within the human cytomegalovirus genome short component: IRS1, US1 through US5, and the US6 family. J. Virol. *66*, 2541–2546.

Kobayashi, N., Suzuki, Y., Tsuge, T., Okumura, K., Ra, C., and Tomino, Y. (2002). Fcrn-mediated transcytosis of immunoglobulin G in human renal proximal tubular epithelial cells. Am. J. Physiol. Renal. Physiol. *282*, F358–365.

Kotenko, S.V., Saccani, S., Izotova, L.S., Mirochnitchenko, O.V., and Pestka, S. (2000). Human cytomegalovirus harbors its own unique IL-10 homolog (cmvIL-10). Proc. Natl. Acad. Sci. USA *97*, 1695–1700.

Kovalevskaya, G., Genbacev, O., Fisher, S.J., Caceres, E., and O'Connor, J.F. (2002). Trophoblast origin of hCG isoforms: cytotrophoblasts are the primary source of choriocarcinoma-like hCG. Mol. Cell. Endocrinol. *194*, 147–155.

Kovats, S., Main, E.K., Librach, C., Stubblebine, M., Fisher, S.J., and DeMars, R. (1990). A class I antigen, HLA-G, expressed in human trophoblasts. Science *248*, 220–223.

Leach, J.L., Sedmak, D.D., Osborne, J.M., Rahill, B., Lairmore, M.D., and Anderson, C.L. (1996). Isolation from human placenta of the IgG transporter, FcRn, and localization to the syncytiotrophoblast: implications for maternal–fetal antibody transport. J. Immunol. *157*, 3317–3322.

Leitner, K., Ellinger, A., Zimmer, K.P., Ellinger, I., and Fuchs, R. (2002). Localization of beta 2-microglobulin in the term villous syncytiotrophoblast. Histochem. Cell. Biol. *117*, 187–193.

Librach, C.L., Werb, Z., Fitzgerald, M.L., Chiu, K., Corwin, N.M., Esteves, R.A., Grobelny, D., Galardy, R., Damsky, C.H., and Fisher, S.J. (1991). 92-kD type IV collagenase mediates invasion of human cytotrophoblasts. J. Cell Biol. *113*, 437–449.

Maidji, E., McDonagh, S., Genbacev, O., Duke, G., and Pereira, L. Neonatal Fc-receptor mediated transcytosis of IgG–virion complexes of maternal antibodies modulate cytomegalovirus infection in the human placenta. (manuscript submitted)

McDonagh, S., Maidji, E., Ma, W., Chang, H.T., Fisher, S., and Pereira, L. (2004). Viral and bacterial pathogens at the maternal–fetal interface. J. Infect. Dis. *190*, 826–834.

McMaster, M.T., Librach, C.L., Zhou, Y., Lim, K.H., Janatpour, M.J., DeMars, R., Kovats, S., Damsky, C., and Fisher, S.J. (1995). Human placental HLA-G expression is restricted to differentiated cytotrophoblasts. J. Immunol. *154*, 3771–3778.

Mostoufi-Zadeh, M., Driscoll, S.G., Biano, S.A., and Kundsin, R.B. (1984). Placental evidence of cytomegalovirus infection of the fetus and neonate. Arch. Pathol. Lab. Med. *108*, 403–406.

Muhlemann, K., Miller, R.K., Metlay, L., and Menegus, M.A. (1992). Cytomegalovirus infection of the human placenta: an immunocytochemical study. Hum. Pathol. *23*, 1234–1237.

Nakamura, Y., Sakuma, S., Ohta, Y., Kawano, K., and Iashimoto, T. (1994). Detection of the human cytomegalovirus gene in placental chronic villitis by polymerase chain reaction. Hum. Pathol. *25*, 815–818.

Norwitz, E.R., Schust, D.J., and Fisher, S.J. (2001). Implantation and the survival of early pregnancy. N. Engl. J. Med. *345*, 1400–1408.

Palmer, E.L., Ruegg, C., Ferrando, R., Pytela, R., and Sheppard, D. (1993). Sequence and tissue distribution of the integrin alpha 9 subunit, a novel partner of beta 1 that is widely distributed in epithelia and muscle. J. Cell Biol. *123*, 1289–1297.

Pereira, L., Maidji, E., McDonagh, S., Genbacev, O., and Fisher, S. (2003). Human cytomegalovirus transmission from the uterus to the placenta correlates with the presence of pathogenic bacteria and maternal immunity. J. Virol. *77*, 13301–13314.

Ploegh, H.L. (1998). Viral strategies of immune evasion. Science *280*, 248–253.

Red-Horse, K., Drake, P.M., and Fisher, S.J. (2004). Human pregnancy: the role of chemokine networks at the fetal-maternal interface. Expert. Rev. Mol. Med. *2004*, 1–14.

Revello, M.G., Zavattoni, M., Furione, M., Lilleri, D., Gorini, G., and Gerna, G. (2002). Diagnosis and outcome of preconceptional and periconceptional primary human cytomegalovirus infections. J. Infect. Dis. *186*, 553–557.

Romero, R., Espinoza, J., Chaiworapongsa, T., and Kalache, K. (2002). Infection and prematurity and the role of preventive strategies. Semin. Neonatol. *7*, 259–274.

Roth, I., Corry, D.B., Locksley, R.M., Abrams, J.S., Litton, M.J., and Fisher, S.J. (1996). Human placental cytotrophoblasts produce the immunosuppressive cytokine interleukin 10. J. Exp. Med. *184*, 539–548.

Roth, I., and Fisher, S.J. (1999). IL-10 is an autocrine inhibitor of human placental cytotrophoblast MMP-9 production and invasion. Dev. Biol. *205*, 194–204.

Schust, D.J., Tortorella, D., Seebach, J., Phan, C., and Ploegh, H.L. (1998). Trophoblast class I major histocompatibility complex (MHC) products are resistant to rapid degradation imposed by the human cytomegalovirus (HCMV) gene products US2 and US11. J. Exp. Med. *188*, 497–503.

Shen, C.Y., Chang, S.F., Yen, M.S., Ng, H.T., Huang, E.S., and Wu, C.W. (1993). Cytomegalovirus excretion in pregnant and nonpregnant women. J. Clin. Microbiol. *31*, 1635–1636.

Simister, N.E., Story, C.M., Chen, H.L., and Hunt, J.S. (1996). An IgG-transporting Fc receptor expressed in the syncytiotrophoblast of human placenta. Eur. J. Immunol. *26*, 1527–1531.

Sinzger, C., Müntefering, H., Löning, T., Stöss, H., Plachter, B., and Jahn, G. (1993). Cell types infected in human cytomegalovirus placentitis identified by immunohistochemical double staining. Virchows Archiv A. Pathol. Anat. Histopathol. *423*, 249–256.

Soderberg-Naucler, C., Streblow, D.N., Fish, K.N., Allan-Yorke, J., Smith, P.P., and Nelson, J.A. (2001). Reactivation of latent human cytomegalovirus in CD14(+) monocytes is differentiation dependent. J. Virol. *75*, 7543–7554.

Spiekermann, G.M., Finn, P.W., Ward, E.S., Dumont, J., Dickinson, B.L., Blumberg, R.S., and Lencer, W.I. (2002). Receptor-Mediated Immunoglobulin G Transport across Mucosal Barriers in Adult Life: Functional Expression of FcRn in the Mammalian Lung. J. Exp. Med. *196*, 303–310.

Stagno, S., Pass, R.F., Dworsky, M.E., Henderson, R.E., Moore, E.G., Walton, P.D., and Alford, C.A. (1982). Congenital cytomegalovirus infection: The relative importance of primary and recurrent maternal infection. N. Engl. J. Med. *306*, 945–949.

Stagno, S., Reynolds, D., Tsiantos, A., Fuccillo, D.A., Smith, R., Tiller, M., and Alford, C.A., Jr. (1975). Cervical cytomegalovirus excretion in pregnant and nonpregnant women: suppression in early gestation. J. Infect. Dis. *131*, 522–527.

Yamamoto-Tabata, T., McDonagh, S., Chang, H.T., Fisher, S., and Pereira, L. (2004). Human cytomegalovirus interleukin-10 down-regulates matrix metalloproteinase activity and impairs endothelial cell migration and placental cytotrophoblast invasiveness in vitro. J. Virol. *78*, 2831–2840.

Zhou, Y., Fisher, S.J., Janatpour, M., Genbacev, O., Dejana, E., Wheelock, M., and Damsky, C.H. (1997). Human cytotrophoblasts adopt a vascular phenotype as they differentiate. A strategy for successful endovascular invasion? J. Clin. Invest. 99, 2139–2151.

# Human Cytomegalovirus Genomics

3

Thomas Shenk

Abstract

This chapter provides an overview of the application of genomic technology, specifically DNA microarray technology, for the analysis of the HCMV-host cell interaction. The relatively small number of reports using microarrays that monitor global HCMV or cellular RNA accumulation are reviewed, and the implications of the experimental results are considered. These initial studies have only just begun to harness the potential of microarrays, and the chapter concludes with a discussion of future challenges and opportunities.

## Introduction

Cytomegaloviruses infect a wide range of animal species, and they are members of the *Cytomegalovirus* genus of the *Beta-herpesvirinae* subfamily of the *Herpesviridae* (reviewed in Roizman and Pellett, 2001). The *Beta-herpesvirinae* include a second genus, *Roseolovirus*, which comprises HHV-6A, HHV-6B and HHV-7. Members of the *Beta-herpesvirinae* are related in their gene content and organization, and they all exhibit a restricted host range and relatively long replication cycles.

Cytomegaloviruses have been identified that infect many different host species, including human and non-human primates and rodents. Viruses that infect different species share genes in common, but also contain genes that are unique to the virus infecting a specific host species. The host range of these viruses is restricted, e.g., a human virus will not infect a rodent and vice versa.

Human cytomegalovirus (HCMV) is a complex virus. Its 235,000 bp genome contains more than 200 ORFs likely to encode proteins, which are conserved in the five clinical isolates of the virus that have been sequenced (Dolan et al., 2004; Murphy et al., 2003). A genomic map of an HCMV clinical isolate, FIX (VR1814) (Grazia Revello et al., 2001; Hahn et al., 2002), is shown in Figure 3.1. Many of the ORFs have not yet been shown to encode proteins, but the FIX ORFs displayed in Figure 3.1 are also found in the Toledo, PH and TR strains of the virus (Murphy et al., 2003). After infection of cultured fibroblasts, the viral ORFs are expressed in a highly organized and regulated cascade of IE, E and L transcription.

HCMV replicates in and spreads through many cell types within an infected host. In other cell types it becomes quiescent and enters latency. It does not induce a global shut down of cellular mRNA accumulation. Rather, HCMV expresses its mRNAs as the host cell continues to synthesize and process its own mRNAs.

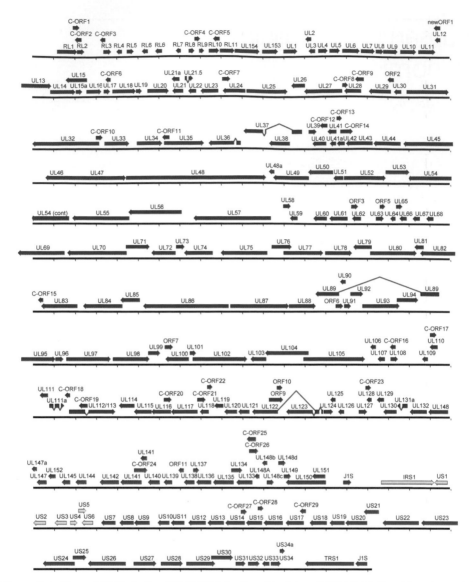

**Figure 3.1** Map of ORFs in the FIX (VR1814) clinical isolate of HCMV. ORFs ≥80 amino acids that are conserved in the Toledo, PH and TR strains of HCMV (see Murphy et al., 2003) are displayed. Arrow heads mark the C-terminus of coding regions. ORFs shown in yellow are those present in AD169 at this position, because this region was deleted during the cloning of the FIX genome into a bacterial artificial chromosome. The red ORFs are not present in the AD169 laboratory strain. Caret symbols mark spliced ORFs. Two adjacent, homologous ORFs are designated RL6, and UL141 contains a stop codon and is shown as UL141 and 141.1.

Given the relatively large size of the viral genome and the complexity of its interaction with an infected host, genomic technologies are proving to be especially useful analytical tools in HCMV studies. The ability to simultaneously monitor the accumulation of all virus and host cell mRNAs can provide a comprehensive view of the interaction of HCMV with different host cells under a wide range of physiological conditions.

The application of genomics to pathogen-infected cells and organisms is still in its infancy. Nevertheless, genomic studies have begun to provide intriguing new insights to the pathogen-host cell interaction (for a recent overview see Bryant et al., 2004). Here I will relate what we have learned from genomic studies of HCMV-infected cells. I will overview the relatively small set of HCMV genomic studies that have been reported, consider the implications of the results, and discuss several important issues to be addressed in the future.

## HCMV dramatically modulates the host cell transcriptome

HCMV does not institute a global block to cellular gene transcription or mRNA accumulation, but one would predict that infection has a profound effect on the transcriptional activity of cellular genes. The gB and gH virion glycoproteins induce the activity of cellular transcription factors as they contact cell-surface targets (Yurochko et al., 1997); virion proteins, such as UL82-coded pp71, direct the degradation of cellular transcriptional repressor proteins (Kalejta et al., 2003) and activate transcription (Liu and Stinski, 1992); viral proteins produced very rapidly after infection, such as UL123-coded IE1 and UL122-coded IE2, regulate the activity of a variety of promoters (Malone et al., 1990; Pizzorno et al., 1988; Stenberg et al., 1990), at least in part by antagonizing the activity of histone deacetylases (Nevels et al., 2004); and infection re-programs cell cycle progression (Bresnahan et al., 1996; Jault et al., 1995; Lu and Shenk, 1996; see also Chapter 11) with inevitable changes in cell gene expression.

Despite the strong prediction, relatively few changes in the accumulation of cellular mRNAs after HCMV infection were reported before the availability of genomic technologies (reviewed in Mocarski and Courcelle, 2001). One noteworthy experiment demonstrated the induction of a number of cellular mRNAs after infection of monocytes with the Towne laboratory strain of HCMV (Yurochko and Huang, 1999). The authors' hypothesis predicted that a set of cellular genes might be induced, and candidate mRNAs were tested by Northern blot assays. The induced mRNAs included IL-1β, A20, NF-κB and IκB, and their identities raised the possibility that HCMV infection leads to monocyte activation.

### HCMV impact on fibroblast mRNA accumulation

The first study designed to search on a global scale for virus-induced changes in cellular gene expression (Zhu et al., 1997) employed differential display analysis (Liang and Pardee, 1992) to examine fibroblasts infected with the AD169 laboratory strain of HCMV. Fifteen cellular RNAs were identified that accumulated to substantially higher levels at 8 h after infection of quiescent as compared to uninfected quiescent cells. UV-irradiated virions, which were unable to express viral genes, also induced the accumulation of this set of mRNAs, arguing that the accumulation was caused either by contact of the virion with the cell or by a virion protein after fusion of the virus envelope with the plasma membrane. All of the overexpressed cellular RNAs proved to be induced by type I interferon, and the possibility that interferon was present in the infecting virus preparations or that the cytokine was rapidly produced after infection of fibroblasts was ruled out in control experiments. As a result, this experiment provided an early indication that HCMV infection induces an innate antiviral response during the first hours after infection of fibroblasts.

Subsequent to the differential display analysis, an Affymetrix array that contained oligonucleotides corresponding to 6,600 human mRNAs was utilized to assay for changes in fibroblast mRNA levels after infection with AD169 (Zhu et al., 1998). The assay identified 258 cellular mRNAs that changed by a factor of $\geq$ 4, including 24 type 1 interferon-responsive genes. Perhaps the most interesting observation to come from this exercise was that RNAs encoding constituents of the pathway that converts arachidonic acid to prostaglandin $E_2$ were substantially modulated. Lipocortin 1, cytosolic phospholipase $A_2$ and cyclooxygenase 2 (COX2) mRNA levels were altered to favor the enhanced production of prostaglandins, which act as second messengers and are able to elicit numerous physiological responses in cells and tissues. Nonspecific COX2 inhibitors, such as aspirin and indomethacin, interfere with HCMV replication (Spier et al., 1998; Tanaka et al., 1988; Zhu et al., 2002). High doses of COX2-specific inhibitors reduce the production of progeny virus in fibroblasts by a factor of > 100 (Zhu et al., 2002). The block is relieved by addition of exogenous prostaglandin $E_2$ to the drug-treated cultures, demonstrating that the drugs do, indeed, antagonize HCMV replication by inhibiting COX2 activity. As yet the mechanism by which prostaglandin stimulates viral replication is not clear. It is noteworthy that RhCMV, which infects rhesus macaques, encodes a homolog of COX2 (Hansen et al., 2003) that is required for RhCMV replication in endothelial cells (Rue et al., 2004). Further, pseudorabies virus, an alpha-herpesvirus that infects pigs, recently has been shown to induce COX2 (Ray and Enquist, 2004) and, like HCMV, its replication is inhibited by COX2 inhibitors (Ray et al., 2004).

A second cataloging exercise employed an Affymetrix array containing 12,626 unique oligonucleotide probe sets to assay in a fairly detailed time course for the impact of AD169 infection on fibroblast RNA levels (Browne et al., 2001). A total of 1,425 mRNAs were modulated by a factor of $\geq$3 in at least two consecutive samples during the time course, and about the same number of RNAs were elevated in amount as were depressed. Most of these changes occurred by 24 h after infection, during the IE and E phases of the replication cycle. Thus, HCMV infection rapidly causes profound changes in cellular RNA levels, well before the onset of viral DNA replication; indeed, changes were detected at the first time point tested, 30 min after infection. Very recently, a detailed analysis of AD169 effects on fibroblasts during L stages of the replication cycle has appeared (Hertl and Mocarski, 2004). Notably, this study described the elevated expression of mitosis-related cellular RNAs, which were associated with the formation of abnormal mitotic spindles during the L phase of infection.

In addition to the time course noted above, Browne et al. (2001) compared infection with UV-irradiated virus to replication-competent virus at 6 h after infection. Although the two viruses down-regulated a similar set of RNAs, the UV-irradiated virus elevated the level of many more cellular RNAs than did the replication-competent virus. The cell RNAs elevated after infection with the UV-irradiated virus included RNAs corresponding to proinflammatory chemokines and proteins that are induced by type 1 interferon. Subsequently, it was shown that a mutant virus, which is unable to produce UL83-coded pp65, also allows the accumulation of numerous RNAs encoding antiviral products (Browne et al., 2003). The mutant phenotype might result from a massive induction of NF-κB, much greater than the induction that has been documented after infection with wild-type virus. NF-κB is known to participate in the activation of many antiviral genes. Microarray experiments with herpes simplex virus type 1 revealed a similar induction of antiviral RNA accumulation after infection with mutant viruses unable to express the ICP0 IE protein (Eidson et al., 2002; Mossman et al., 2001).

The HCMV pp65 protein also was shown to influence cellular RNA accumulation in microarray experiments using arrays comprising sequence-verified cDNAs (Abate et al., 2004). A total of 220 cDNAs were differentially expressed by wild-type as compared to pp65-deficient virus during the first 4 h after infection. This study concluded that pp65 prevented the activation of interferon regulatory factor 3 (IRF3) and did not effect NF-κB, based on a comparison between wild-type and mutant viruses. This conclusion is in conflict with Browne et al. (2003), who reported just the opposite: IRF3 DNA binding is activated to a similar extent by mutant and wild-type viruses, whereas NF-κB is activated to a much greater extent after infection with a pp65-deficient mutant than after infection with wild-type virus. At this point the conflict is unresolved.

An array experiment utilizing 8,942 cDNAs printed on glass slides demonstrated that the UL55-coded glycoprotein B (gB) and type I interferon elicited essentially the same increases in fibroblast RNA levels as did infection with AD169 (Simmen et al., 2001). At 24 h after infection, 52 RNAs were induced by virus infection and 51 of them were induced by either purified recombinant gB or type 1 interferon. This result argues that most of the changes induced by HCMV in fibroblasts result from the interaction of gB with cell-surface receptors, and a subsequent study has shown that the addition of recombinant gB activates interferon regulatory factor 3 (IRF3), as does the intact, wild-type virus (Boehme et al., 2004). Activation of IRF3 could explain the ability of gB to induce an interferon response. What is the cell-surface receptor that is contacted and activated by gB? The viral glycoprotein has been reported to bind to the epidermal growth factor receptor (Wang et al., 2003) and multiple integrins (Feire et al., 2004), and both of these receptors activate signal transduction pathways. Further, HCMV activates an inflammatory response through Toll-like receptor 2 (Compton et al., 2003), and it is conceivable that gB interacts with this receptor as well.

The observation that recombinant gB induces the accumulation of nearly the same set of cellular RNAs as does infection with replication competent virus (Simmen et al., 2001) appears to be inconsistent with the finding that UL83-coded pp65 blocks the induction of numerous cell RNAs (Browne and Shenk, 2003). Given the role of pp65, one would anticipate that gB alone would induce more cellular RNAs than AD169 virions, which would deliver pp65 to the infected cell. It is possible that the different results stem from technical differences between experiments. Alternatively, the activity of additional viral gene products, which would be expected to influence results after infection with virions, might be confounding the interpretation.

Virus-coded products other than gB can modulate the expression of cellular genes. The UL122-coded IE2 protein is a promiscuous transcriptional activating protein (Malone et al., 1990; Pizzorno et al., 1988; Stenberg et al., 1990) that functions at least in part by antagonizing the activity of histone deacetylases (Nevels et al., 2004), and the protein drives quiescent G0 fibroblasts into G1 (Castillo et al., 2000; Murphy et al., 2000; Sinclair et al., 2000). Affymetrix arrays were used to identify cellular genes that were modulated in response to infection with an adenovirus vector expressing the HCMV IE2 protein as compared to a control adenovirus vector expressing GFP (Song and Stinski, 2002). A total of 64 cellular genes were induced by a factor of 4 or greater in response to the IE2 protein, and half of the induced cellular RNAs encoded proteins with roles in cell cycle progression or DNA replication. This analysis beautifully correlates the ability of IE2 protein to activate cell cycle progression with the induction of a set of mRNAs that would be activated as a consequence. Further, it demonstrates that a viral protein expressed after the HCMV genome reaches the nucleus can activate a substantial number of cellular genes. Thus, although the majority of cellular RNAs modulated by HCMV appear to result from the interaction of gB with cell-surface proteins, other viral proteins also alter the levels of cellular mRNAs.

### HCMV impact on mRNA accumulation in myeloid cells

HCMV resides in a latent state in monocytes and their progenitors (Stanier et al., 1989; Taylor-Wiedeman et al., 1991), and cultured granulocyte–macrophage progenitors (CD14+/CD15+/CD33+) have been explored as a model host cell for HCMV latency (first described by Kondo et al., 1994). Viral DNA persisted after infection of these cells, and the viral genome did not exhibit its normal cascade of viral gene expression. Rather, a small set of transcripts derived from the vicinity of the IE1-coding UL123 ORF, which do not encode IE1 protein, were detected in these cells. Importantly, virus replication was reactivated after co-cultivation of the infected granulocyte–macrophage progenitors with fibroblasts. The failure to express the normal set of viral genes for a prolonged period of time, together with the retention of viral DNA and the ability to reactivate, argue that this is a valid model for HCMV latency.

   To probe the host cell response to latent HCMV, an array containing cDNAs corresponding to 7,642 sequence-verified human genes, as well as an assortment of additional expressed sequence tags, was used to assay RNAs prepared on day 14 after infection of human granulocyte–macrophage progenitor cells with the Toledo strain of HCMV (Slobedman et al., 2004). Toledo has been passaged to a limited extent in fibroblasts, and, in contrast to the AD169 and Towne laboratory strains, some isolates of Toledo have retained the ability to replicate in a variety of cell types other than fibroblasts. The analysis identified 30 genes whose RNA was elevated by a factor of ≥ 2.0 and six genes whose RNA was reduced by a factor of ≥ 2.7 in infected versus uninfected cultures. The modulated RNAs fall into multiple functional categories (transcription factors, host defense, cell growth, signaling), and lead to intriguing hypotheses. For example, RNA coding the POU2F2 transcription factor is elevated, and the factor has been reported to repress IE transcription of both herpes simplex virus and varicella zoster virus (Lillycrop et al., 1999; Patel et al., 1998). This leads to the suggestion that POU2F2 might serve to repress HCMV IE transcription during latency in granulocyte–monoocyte progenitors. If this proves to be the case, the next challenge will be to understand how POU2F2 becomes elevated in latently infected cells. Microrray experiments also have identified changes in RNA levels in murine sensory neurons harboring latent herpes simplex virus type 1 (Kramer et al., 2003).

## Assessment of the HCMV transcriptome in multiple host cells

As noted earlier in this review, HCMV is a complex virus with more than 200 ORFs (Figure 3.1), which are conserved in the five clinical isolates of the virus that have been sequenced to date (Dolan et al., 2004; Murphy et al., 2003). Given the complexity of the virus, it is not practical to assay individual RNAs by Northern blot or RT-PCR, if it is important to assess the activity of the entire genome. Consequently, HCMV-specific arrays have been generated that make it possible to simultaneously monitor the products of the entire HCMV transcriptome.

## HCMV RNA accumulation in fibroblasts

The first HCMV-specific array was described by Chambers et al. (1999). The array comprised 75-base oligonucleotides printed on glass slides. It was used to characterize the temporal expression of 151 HCMV ORFs in fibroblasts infected with the Towne laboratory strain of HCMV. This study demonstrated the reliability and sensitivity of arrays when used to monitor viral gene expression. It also "raised the bar" for future studies by demonstrating that it is now feasible to directly examine the full set of viral mRNAs rather than examine a small selected subset and extrapolate the result to the entire genome as had been done in the past.

Another microarray was generated using PCR-amplified DNA fragments corresponding to approximately 190 unique HCMV ORFs that were spotted onto a nitrocellulose membrane (Bresnahan and Shenk, 2000a). This array was used to monitor the expression of viral RNAs in fibroblasts infected with a virus lacking the UL82 ORF, which encodes the pp71 tegument protein (Bresnahan and Shenk, 2000b). The analysis demonstrated conclusively that pp71 enhances the accumulation of virus-coded RNAs at the very start of infection. The array has been used subsequently in the phenotypic analysis of additional viral variants (Adamo et al., 2004; Browne and Shenk, 2003; Heider et al., 2002), facilitating definitive assessments of the effects of various mutations on the HCMV transcriptome.

## HCMV RNA accumulation in CD34$^+$ stem cells

In addition to the granulocyte–monoocyte progenitor cell system discussed above, a second cell culture system for HCMV latency has been developed using more primitive hematopoietic stem cells. HCMV DNA is found in CD34$^+$ cells from human bone marrow as a result of natural infection (Khaiboullina et al., 2004 and references therein). Are viral RNAs expressed in these cells? The PCR-generated HCMV-specific cDNA array (Bresnahan and Shenk, 2000a) was printed on glass slides and used to monitor the accumulation of viral RNAs as a function of time after infection of CD34$^+$ bone marrow cells (Goodrum et al., 2002). The infected CD34$^+$ cultures appear to faithfully mimic HCMV latency: viral gene expression was transient, viral DNA was maintained, and virus replication was reactivated by co-culture of the CD34$^+$ cells with permissive fibroblasts. The microarray analysis demonstrated that the pattern of viral RNA accumulation in the CD34$^+$ population was strikingly different than in fibroblasts. Not only was RNA expressed transiently, but the pattern of RNAs expressed did not correspond to the normal IE, E, and L cascade. A subset of the complete genome was expressed and it included members of all three temporal classes of viral RNA at the earliest time tested. Subsequent analysis identified a subpopulation of CD34$^+$ cells (CD34$^+$/CD38$^-$) that exhibited the properties of latently infected cells, including the transient accumulation of a subset of the RNAs encoded by the HCMV transcriptome (Goodrum et al., 2004). The use of arrays has revealed that HCMV mRNA accumulation is markedly different in CD34$^+$ hematopoietic cell populations than in permissive fibroblasts.

## Looking to the future

The utility of microarrays in the investigation of HCMV biology is clear; they facilitate a global assessment of the interaction of the pathogen with its infected host cell. It is also evident that the use of microarrays as a tool for study of HCMV is in its infancy. Indeed, in this short review, I have discussed essentially all studies that have utilized microarrays in HCMV studies. The initial studies have raised many more questions than they have answered. What are the challenges going forward?

### Extract biologically Informative information from array data

Much of the work with microarrays to this point has cataloged RNAs whose expression is altered under a specified condition. Then the catalog is "cherry picked" using good biological sense, and potentially informative changes are explored in greater detail. This is how the requirement for COX2 in the HCMV (Zhu et al., 1998; Zhu et al., 2002) and pseudorabies (Ray and Enquist, 2004; Ray et al., 2004) replication cycles was identified. Numerous laboratories are working to extract substantially more information from array data. Indeed, the computational analysis of the large data sets has rapidly become markedly more sophisticated. The field is moving from the identification of lists of elevated and reduced RNAs to an appreciation of the cellular pathways that are induced or repressed and the prediction of how alterations to multiple pathways interact and impact on overall cellular physiology (for a recent review see Hackl et al., 2004). Work spanning many fields of biology is advancing this technology, and virology is no laggard. Technical advances in data collection (Glasby and Ghazal, 2003) and analysis (Challacombe et al., 2004) have been reported for HCMV microarray studies.

### Consider limitations to interpretations

Some of the limitations to the interpretation of microarray data are obvious. For example, an array only can quantify the relative steady-state levels of RNAs. It divulges nothing of the physiological basis for changes in RNA concentrations, i.e., changes in transcription rate versus changes in half-life. Fortunately, the mechanism underlying the change in level of an individual RNA can be readily investigated, when necessary.

Microarray data also do not distinguish direct from secondary effects. In the case of HCMV, many of the changes might well be secondary to other changes in cellular gene expression. This is especially likely for changes that occur later during the replication cycle, although the activation of a new late viral gene can certainly induce a new wave of primary consequences to the host transcriptome. Unfortunately, there is no global method that can be applied to identify the set of primary changes in RNA levels. Changes that occur very soon after application of a specific stimulus are, of course, candidates for direct effects, but many of the changes that occur rapidly after HCMV infection are, in fact, secondary to changes in the activity of several transcription factors. For example, contact of the virus with cell-surface receptors activates NF-κB (Yurochko et al., 1997), and NF-κB then participates in the rapid induction of a cellular antiviral response. Application of the stimulus in the presence of cycloheximide can give some indication of direct effects on RNA levels, but interpretation of this type of experiment is complicated by indirect effects of the drug, which can impact on RNA levels (Browne et al., 2001). An improved understanding of the signal transduction pathways that are induced when the virus contacts its host cell, together with an understanding of RNA changes in the context of pathways will be very helpful in distinguishing direct from indirect effects of HCMV infection.

Finally, as is again obvious, the level of a protein does not necessarily track the level of its cognate mRNA. A microarray determination that an RNA is elevated or unchanged after infection does not guarantee that the same is true for the protein that it codes. A dramatic example of this has come from work with herpes simplex virus. Analysis of cellular RNA levels using Affymetrix arrays demonstrated that herpes simplex virus infection enhanced the accumulation of about 2% of the RNAs assayed relative to the levels in mock-infected cells (Taddeo et al., 2002). Subsequently, one of the elevated RNAs, encoding the stress-inducible protein IEX-1, was studied in detail (Taddeo et al., 2003). Northern blot analysis confirmed the microarray result; IEX-1 RNA is elevated, reaching a peak at 3–7 h after infection. However, IEX-1 protein was not detected beyond 1 h after infection. The IEX-1 RNA was cleaved in a UL41-dependent manner, but nevertheless accumulated in the cytoplasm in the environment of an infected cell. This result is intriguing in terms of herpes simplex virus biology, but it also demonstrates a limitation to interpretation of microarray data. Even though the microarray reported that IEX-1 RNA was elevated, it was cleaved and its protein was not produced. Microarray data is more predictive of protein expression if polysome-associated RNA is separated from free cytoplasmic ribonucleoproteins prior to analysis (Johannes et al., 1999). Ultimately, it will be useful to combine high throughput proteomic analysis with DNA arrays to generate a complete picture of changes in gene expression after infection.

## Determine which changes are important and why

There are undoubtedly three categories of changes to RNA accumulation that occur within HCMV-infected cells. Some changes represent attempts on the part of the cell to respond to infection and combat the replication and/or spread of the virus. The induction of RNAs encoding interferon-responsive proteins and pro-inflammatory chemokines after HCMV infection (Browne et al., 2001; 2003) fall into this category. Other changes are critical to the virus. The expression of these cellular RNAs must be altered for successful replication and/or spread of the virus. The induction of COX2 mRNA is an example of this type of change. If the elevation of COX2 activity is blocked by a COX2 inhibitor, virus replication is inhibited (Zhu et al., 2002). The final category, and perhaps the largest category, might be termed "bystander" changes. These are changes that occur as a consequence of HCMV replication, but provide no advantage to either the pathogen or its host.

Although in many instances we can guess whether a change in gene expression is likely beneficial to the virus, host or to neither, it will be important to definitively determine the consequences of numerous changes. The effect of a reduction in level of an RNA can generally be probed by forced overexpression of the RNA. This often can be accomplished by using a construct containing a promoter that is insensitive to viral regulation and lacking mRNA destabilizing sequences, etc. It is more challenging to investigate the role that an overexpressed cellular RNA might play in the HCMV replication cycle. Since virus replication is restricted to human cells, the numerous available mouse cell variants are not helpful in HCMV studies. However, the RNAi field is advancing rapidly (for recent reviews see Novina and Sharp, 2004; Sherr and Eder, 2004), and the ability to use RNAi as a genetic tool to specifically inhibit the accumulation and expression of cellular mRNAs will undoubtedly contribute importantly to our understanding of the consequences of elevated RNAs identified in microarray experiments.

## Issues specific to HCMV

Most of the microarray studies discussed in this review employed the AD169 or Towne laboratory strains of HCMV. These strains were developed as vaccine candidates by extensive serial passage of clinical isolates in cultured fibroblasts, and they are known to differ substantially – both genetically and biologically – from clinical isolates of HCMV (discussed in Cha et al., 1996; Davison et al., 2004; Murphy et al., 2003). The laboratory strains have been adapted to fibroblasts where they replicate much more efficiently than clinical isolates of HCMV. However, the laboratory strains have lost the ability to infect and replicate in a variety of cell types that normally host HCMV as it replicates and spreads within an infected individual, such as epithelial cells, endothelial cells and macrophages.

Does an HCMV clinical isolate elicit the same cellular response as the artificially evolved laboratory strains of the virus? Does the virus elicit the same response in a variety of different host cell types or do fibroblasts, endothelial cells, epithelial cells, macrophages, etc., each react to HCMV infection with a unique transcriptome signature? In the future, it will be important to extend the microarray studies that have been performed with laboratory strains in fibroblasts to clinical isolates in a variety of cell types.

## Acknowledgments

I thank many colleagues for stimulating and helpful discussions that helped me to design this chapter.

## References

Abate, D.A., Watanabe, S., and Mocarski, E.S. (2004). Major human cytomegalovirus structural protein pp65 (ppUL83) prevents interferon response factor 3 activation in the interferon response. J. Virol. 78, 10995–11006.

Adamo, J.E., Schroer, J., and Shenk, T. (2004). Human cytomegalovirus TRS1 protein is required for efficient assembly of DNA-containing capsids. J. Virol. 78, 10221–10229.

Boehme, K.W., Singh, J., Perry, S.T., and Compton, T. (2004). Human cytomegalovirus elicits a coordinated cellular antiviral response via envelope glycoprotein B. J. Virol. 78, 1202–1211.

Bresnahan, W.A., Boldogh, I., Thompson, E.A., and Albrecht, T. (1996). Human cytomegalovirus inhibits cellular DNA synthesis and arrests productively infected cells in late G1. Virology 224, 150–160.

Bresnahan, W.A., and Shenk, T. (2000a). A subset of viral transcripts packaged within human cytomegalovirus particles. Science 288, 2373–2376.

Bresnahan, W.A., and Shenk, T.E. (2000b). UL82 virion protein activates expression of immediate-early viral genes in human cytomegalovirus-infected cells. Proc. Natl. Acad. Sci. USA 97, 14506–14511.

Browne, E.P., and Shenk, T. (2003). Human cytomegalovirus UL83-coded pp65 virion protein inhibits antiviral gene expression in infected cells. Proc. Natl. Acad. Sci. USA 100, 11439–11444.

Browne, E.P., Wing, B., Coleman, D., and Shenk, T. (2001). Altered cellular mRNA levels in human cytomegalovirus-infected fibroblasts: viral block to the accumulation of antiviral mRNAs. J. Virol. 75, 12319–12330.

Bryant, P.A., Venter, D., Robins-Browne, R., and Curtis, N. (2004). Chips with everything: DNA microarrays in infectious diseases. Lancet Infect. Dis. 4, 100–111.

Castillo, J.P., Yurochko, A.D., and Kowalik, T.F. (2000). Role of human cytomegalovirus immediate-early proteins in cell growth control. J. Virol. 74, 8028–8037.

Cha, T.A., Tom, E., Kemble, G.W., Duke, G.M., Mocarski, E.S., and Spaete, R.R. (1996). Human cytomegalovirus clinical isolates carry at least 19 genes not found in laboratory strains. J. Virol. 70, 78–83.

Challacombe, J.F., Rechtsteiner, A., Gottardo, R., Rocha, L.M., Browne, E.P., Shenk, T., Altherr, M.R., and Brettin, T.S. (2004). Evaluation of the host transcriptional response to human cytomegalovirus infection. Physiol. Genomics *18*, 51–62.

Chambers, J., Angulo, A., Amaratunga, D., Guo, H., Jiang, Y., Wan, J.S., Bittner, A., Frueh, K., Jackson, M.R., Peterson, P.A., Erlander, M.G.,and Ghazal, P. (1999). DNA microarrays of the complex human cytomegalovirus genome: profiling kinetic class with drug sensitivity of viral gene expression. J. Virol. *73*, 5757–5766.

Compton, T., Kurt-Jones, E.A., Boehme, K.W., Belko, J., Latz, E., Golenbock, D.T., and Finberg, R.W. (2003). Human cytomegalovirus activates inflammatory cytokine responses via CD14 and Toll-like receptor 2. J. Virol. *77*, 4588–4596.

Dolan, A., Cunningham, C., Hector, R.D., Hassan-Walker, A.F., Lee, L., Addison, C., Dargan, D.J., McGeoch, D.J., Gatherer, D., Emery, V.C., Griffiths, P.D., Sinzger, C., McSharry, B.P., Wilkinson, G.W., and Davison, A.J. (2004). Genetic content of wild-type human cytomegalovirus. J. Gen. Virol. *85*, 1301–1312.

Eidson, K.M., Hobbs, W.E., Manning, B.J., Carlson, P., and DeLuca, N.A. (2002). Expression of herpes simplex virus ICP0 inhibits the induction of interferon-stimulated genes by viral infection. J. Virol. *76*, 2180–2191.

Esclatine, A., Taddeo, B., Evans, L., and Roizman, B. (2004). The herpes simplex virus 1 UL41 gene-dependent destabilization of cellular RNAs is selective and may be sequence-specific. Proc. Natl. Acad. Sci. USA *101*, 3603–3608.

Feire, A.L., Koss, H., and Compton, T. (2004). Cellular integrins function as entry receptors for human cytomegalovirus via a highly conserved disintegrin-like domain. Proc. Natl. Acad. Sci. USA *101*, 15470–15475.

Glasbey, C.A., and Ghazal, P. (2003). Combinatorial image analysis of DNA microarray features. Bioinformatics *19*, 194–203.

Goodrum, F.D., Jordan, C.T., High, K., and Shenk, T. (2002). Human cytomegalovirus gene expression during infection of primary hematopoietic progenitor cells: a model for latency. Proc. Natl. Acad. Sci. USA *99*, 16255–16260.

Goodrum, F. D., Jordan, C.T., Terhune, S.S., High, K., and Shenk, T. (2004). Differential outcomes of human cytomegalovirus infection in primitive hematopoietic cell subpopulations. Blood *104*, 687–695.

Grazia Revello, M., Baldanti, F., Percivalle, E., Sarasini, A., De-Giuli, L., Genini, E., Lilleri, D., Labo, N. and Gerna, G. (2001) In vitro selection of human cytomegalovirus variants unable to transfer virus and virus products from infected cells to polymorphonuclear leukocytes and to grow in endothelial cells. J. Gen. Virol. *82*, 1429–1438.

Hackl, H., Cabo, F.S., Sturn, A., Wolkenhauer, O., and Trajanoski, Z. (2004). Analysis of DNA microarray data. Curr. Top. Med. Chem. *4*, 1357–1370.

Hahn, G., Khan, H., Baldanti, F., Koszinowski, U.H., Revello, M.G. and Gerna, G.. (2002) The human cytomegalovirus ribonucleotide reductase homolog UL45 is dispensable for growth in endothelial cells, as determined by a BAC-cloned clinical isolate of human cytomegalovirus with preserved wild-type characteristics. J, Virol. *76*, 9551–9555.

Hansen, S.G., Strelow, L.I., Franchi, D.C., Anders, D.G., and Wong, S.W. (2003). Complete sequence and genomic analysis of rhesus cytomegalovirus. J. Virol. *77*, 6620–6636.

Heider, J.A., Bresnahan, W.A., and Shenk, T.E. (2002). Construction of a rationally designed human cytomegalovirus variant encoding a temperature-sensitive immediate-early 2 protein. Proc. Natl. Acad. Sci. USA *99*, 3141–3146.

Hertel, L., and Mocarski, E.S. (2004). Global analysis of host cell gene expression late during cytomegalovirus infection reveals extensive dysregulation of cell cycle gene expression and induction of Pseudomitosis independent of US28 function. J. Virol. *78*, 11988–12011.

Jault, F.M., Jault, J.M., Ruchti, F., Fortunato, E.A., Clark, C., Corbeil, J., Richman, D.D., and Spector, D.H. (1995). Cytomegalovirus infection induces high levels of cyclins, phosphorylated Rb, and p53, leading to cell cycle arrest. J. Virol. *69*, 6697–6704.

Johannes, G., Carter, M.S., Eisen, M.B., Brown, P.O., and Sarnow, P. (1999). Identification of eukaryotic mRNAs that are translated at reduced cap binding complex eIF4F concentrations using a cDNA microarray. Proc. Natl. Acad. Sci. USA *96*, 13118–13123.

Kalejta, R.F., Bechtel, J.T., and Shenk, T. (2003). Human cytomegalovirus pp71 stimulates cell cycle progression by inducing the proteasome-dependent degradation of the retinoblastoma family of tumor suppressors. Mol. Cell. Biol. *23*, 1885–1895.

Khaiboullina, S.F., Maciejewski, J.P., Crapnell, K., Spallone, P.A., Dean Stock, A., Pari, G.S., Zanjani, E.D., and Jeor, S.S. (2004). Human cytomegalovirus persists in myeloid progenitors and is passed to the myeloid progeny in a latent form. Br. J. Haematol. *126*, 410–417.

Kondo, K., Kaneshima, H., and Mocarski, E.S. (1994). Human cytomegalovirus latent infection of granulocyte–macrophage progenitors. Proc. Natl. Acad. Sci. USA *91*, 11879–11883.

Kramer, M.F., Cook, W.J., Roth, F.P., Zhu, J., Holman, H., Knipe, D.M., and Coen, D.M. (2003). Latent herpes simplex virus infection of sensory neurons alters neuronal gene expression. J. Virol. *77*, 9533–9541.

Liang, P., and Pardee, A.B. (1992). Differential display of eukaryotic messenger RNA by means of the polymerase chain reaction. Science *257*, 967–971.

Lillycrop, K.A., Dent, C.L., Wheatley, S.C., Beech, M.N., Ninkina, N.N., Wood, J.N., and Latchman, D.S. (1991). The octamer-binding protein Oct-2 represses HSV immediate-early genes in cell lines derived from latently infectable sensory neurons. Neuron *7*, 381–390.

Liu, B., and Stinski, M.F. (1992). Human cytomegalovirus contains a tegument protein that enhances transcription from promoters with upstream ATF and AP-1 cis-acting elements. J. Virol. *66*, 4434–4444.

Lu, M., and Shenk, T. (1996). Human cytomegalovirus infection inhibits cell cycle progression at multiple points, including the transition from G1 to S. J. Virol. *70*, 8850–8857.

Malone, C.L., Vesole, D.H., and Stinski, M.F. (1990). Transactivation of a human cytomegalovirus early promoter by gene products from the immediate-early gene IE2 and augmentation by IE1: mutational analysis of the viral proteins. J. Virol. *64*, 1498–1506.

Mocarski, E.S., and Courcelle, C.T. (2001). Cytomegaloviruses and their replication, In Fields Virology, D.M. Knipe, and P.M. Howley, eds. (Philadelphia: Lippincott Williams & Wilkins), pp. 2629–2673.

Mossman, K.L., Macgregor, P.F., Rozmus, J.J., Goryachev, A.B., Edwards, A.M., and Smiley, J.R. (2001). Herpes simplex virus triggers and then disarms a host antiviral response. J. Virol. *75*, 750–758.

Murphy, E.A., Streblow, D.N., Nelson, J.A., and Stinski, M.F. (2000). The human cytomegalovirus IE86 protein can block cell cycle progression after inducing transition into the S phase of permissive cells. J. Virol. *74*, 7108–7118.

Murphy, E., Yu, D., Grimwood, J., Schmutz, J., Dickson, M., Jarvis, M.A., Hahn, G., Nelson, J.A., Myers, R.M. and Shenk, T.E. (2003). Coding potential of laboratory and clinical strains of human cytomegalovirus. Proc. Natl. Acad. Sci. USA *100*, 14976–14981.

Nevels, M., Paulus, C., and Shenk, T. (2004). Human cytomegalovirus immediate-early 1 protein facilitates viral replication by antagonizing histone deacetylation. Proc. Natl. Acad. Sci. USA *101*, 17234–17239.

Novina, C.D., and Sharp, P.A. (2004). The RNAi revolution. Nature *430*, 161–164.

Patel, Y., Gough, G., Coffin, R.S., Thomas, S., Cohen, J.I., and Latchman, D.S. (1998). Cell type specific repression of the varicella zoster virus immediate-early gene 62 promoter by the cellular Oct-2 transcription factor. Biochim. Biophys. Acta. *1397*, 268–274.

Pizzorno, M.C., O'Hare, P., Sha, L., LaFemina, R.L., and Hayward, G.S. (1988). Trans-activation and autoregulation of gene expression by the immediate-early region 2 gene products of human cytomegalovirus. J. Virol. *62*, 1167–1179.

Ray, N., Bisher, M.E., and Enquist, L.W. (2004). Cyclooxygenase-1 and -2 are required for production of infectious pseudorabies virus. J. Virol. *78*, 12964–12974.

Ray, N., and Enquist, L.W. (2004). Transcriptional response of a common permissive cell type to infection by two diverse alpha-herpesviruses. J. Virol. *78*, 3489–3501.

Roizman, B. and Pellett, P.E. (2001). The family herpesviridae: a brief intoduction, In Fields Virology, D.M. Knipe, and P.M. Howley, eds. (Philadelphia: Lippincott Williams & Wilkins), pp. 2381–2397.

Rue, C.A., Jarvis, M.A., Knoche, A.J., Meyers, H.L., DeFilippis, V.R., Hansen, S.G., Wagner, M., Fruh, K., Anders, D.G., Wong, S.W., Barry, P.A., and Nelson, J.A.. (2004). A cyclooxygenase-2 homologue encoded by rhesus cytomegalovirus is a determinant for endothelial cell tropism. J. Virol. *78*, 12529–12536.

Scherr, M., and Eder, M. (2004). RNAi in functional genomics. Curr. Opin. Mol. Ther. *6*, 129–135.

Simmen, K.A., Singh, J., Luukkonen, B.G., Lopper, M., Bittner, A., Miller, N.E., Jackson, M.R., Compton, T., and Fruh, K. (2001). Global modulation of cellular transcription by human cytomegalovirus is initiated by viral glycoprotein B. Proc. Natl. Acad. Sci. USA *98*, 7140–7145.

Sinclair, J., Baillie, J., Bryant, L., and Caswell, R. (2000). Human cytomegalovirus mediates cell cycle progression through G(1) into early S phase in terminally differentiated cells. J. Gen. Virol. *81*, 1553–1565.

Slobedman, B., Stern, J.L., Cunningham, A.L., Abendroth, A., Abate, D.A., and Mocarski, E.S. (2004). Impact of human cytomegalovirus latent infection on myeloid progenitor cell gene expression. J. Virol. 78, 4054–4062.

Song, Y.J., and Stinski, M.F. (2002). Effect of the human cytomegalovirus IE86 protein on expression of E2F-responsive genes: a DNA microarray analysis. Proc. Natl. Acad. Sci. USA 99, 2836–2841.

Speir, E., Yu, Z.X., Ferrans, V.J., Huang, E.S., and Epstein, S.E. (1998). Aspirin attenuates cytomegalovirus infectivity and gene expression mediated by cyclooxygenase-2 in coronary artery smooth muscle cells. Circ. Res. 83, 210–216.

Stanier, P., Taylor, D.L., Kitchen, A.D., Wales, N., Tryhorn, Y., and Tyms, A.S. (1989). Persistence of cytomegalovirus in mononuclear cells in peripheral blood from blood donors. BMJ 299, 897–898.

Stenberg, R.M., Fortney, J., Barlow, S.W., Magrane, B.P., Nelson, J.A., and Ghazal, P. (1990). Promoter-specific trans activation and repression by human cytomegalovirus immediate-early proteins involves common and unique protein domains. J. Virol. 64, 1556–1565.

Taddeo, B., Esclatine, A., and Roizman, B. (2002). The patterns of accumulation of cellular RNAs in cells infected with a wild-type and a mutant herpes simplex virus 1 lacking the virion host shutoff gene. Proc. Natl. Acad. Sci. USA 99, 17031–17036.

Taddeo, B., Esclatine, A., Zhang, W., and Roizman, B. (2003). The stress-inducible immediate-early responsive gene IEX-1 is activated in cells infected with herpes simplex virus 1, but several viral mechanisms, including 3′ degradation of its RNA, preclude expression of the gene. J. Virol. 77, 6178–6187.

Tanaka, J., Ogura, T., Iida, H., Sato, H., and Hatano, M. (1988). Inhibitors of prostaglandin synthesis inhibit growth of human cytomegalovirus and reactivation of latent virus in a productively and latently infected human cell line. Virology 163, 205–208.

Taylor-Wiedeman, J., Sissons, J. G., Borysiewicz, L.K., and Sinclair, J.H. (1991). Monocytes are a major site of persistence of human cytomegalovirus in peripheral blood mononuclear cells. J. Gen. Virol. 72, 2059–2064.

Wang, X., Huong, S.M., Chiu, M.L., Raab-Traub, N., and Huang, E.S. (2003). Epidermal growth factor receptor is a cellular receptor for human cytomegalovirus. Nature 424, 456–461.

Yurochko, A.D., and Huang, E.S. (1999). Human cytomegalovirus binding to human monocytes induces immunoregulatory gene expression. J. Immunol. 162, 4806–4816.

Yurochko, A.D., Hwang, E.S., Rasmussen, L., Keay, S., Pereira, L., and Huang, E.S. (1997). The human cytomegalovirus UL55 (gB) and UL75 (gH) glycoprotein ligands initiate the rapid activation of Sp1 and NF-kappaB during infection. J. Virol. 71, 5051–5059.

Zhu, H., Cong, J.P., Mamtora, G., Gingeras, T., and Shenk, T. (1998). Cellular gene expression altered by human cytomegalovirus: global monitoring with oligonucleotide arrays. Proc. Natl. Acad. Sci. USA 95, 14470–14475.

Zhu, H., Cong, J.P., and Shenk, T. (1997). Use of differential display analysis to assess the effect of human cytomegalovirus infection on the accumulation of cellular RNAs: induction of interferon-responsive RNAs. Proc. Natl. Acad. Sci. USA 94, 13985–13990.

Zhu, H., Cong, J.P., Yu, D., Bresnahan, W. A., and Shenk, T. E. (2002). Inhibition of cyclooxygenase 2 blocks human cytomegalovirus replication. Proc. Natl. Acad. Sci. USA 99, 3932–3937.

# Manipulating Cytomegalovirus Genomes by BAC Mutagenesis: Strategies and Applications

4

*Wolfram Brune, Markus Wagner, and Martin Messerle*

## Abstract

The generation of CMV mutants has been a difficult task especially because of the large genome size and the slow replication cycle of the CMVs. The recent cloning of CMV genomes as infectious bacterial artificial chromosomes (BAC) in *E. coli* opened new horizons for the construction of mutant CMVs by utilizing the methods of bacterial genetics. This chapter gives an overview of BAC cloning and the various mutagenesis techniques that allow targeted as well as random manipulations of CMV genomes. Selected examples give an impression of the power of the reverse and forward genetic procedures. The new techniques provide the basis for a comprehensive analysis of CMV gene functions as well for vaccine development.

## Introduction

Our knowledge about cytomegalovirus (CMV) gene functions is advancing rather slowly, although the DNA sequence of the HCMV laboratory strain AD169 has been known since 1990 (Chee et al., 1990) and those of rodent CMVs have also been available for several years (Rawlinson et al., 1996; Vink et al., 2000). Reasons for the overall scarcity of data are of course the large number of proteins encoded and the size of the genomes with associated difficulties in generating viral mutants. CMVs have the largest genomes among the herpesviruses, ranging in size from 221 (rhesus CMV) to 241 kbp (chimpanzee CMV) (Davison et al., 2003; Hansen et al., 2003). Coding potentials of up to 208 and 254 proteins were predicted for the HCMV laboratory strain AD169 and for clinical isolates of HCMV, respectively (Chee et al., 1990; Murphy et al., 2003b). Based on a more stringent evaluation, the genetic content of the AD169 strain was estimated at 145 genes and that of wild-type HCMV at between 164 and 167 genes (Dolan et al., 2004). The majority of the proteins encoded by HCMV has not yet been characterized. Even for those proteins about which we have some functional knowledge, e.g. the envelope glycoproteins, a clear understanding of the mode of action is usually missing. A number of ORFs has been subcloned in plasmid expression vectors and their functions studied independently of the viral infection and separately from other viral genes, for instance by transient transfection experiments. However, many viral proteins have to interact with other viral or cellular proteins in order to exert their effects and many genes display a distinct temporal expression pattern during the infection cycle. Thus, it is desirable to study the role of genes in the context of the viral infection.

In order to assign defined properties of the virus to certain genes, attempts to generate variants of CMVs with altered phenotypes were already being made in the seventies. Since the replication machinery of herpesviruses is highly accurate, spontaneous mutants emerge only rarely and chemical (treatment with nitrosoguanidine) or physical methods (irradiation with UV light) had to be used to introduce mutations into CMV genomes (Ihara et al., 1978; Yamanishi and Rapp, 1977). Restricted growth at an elevated temperature was a convenient read-out and selection principle. A series of temperature-sensitive HCMV and MCMV mutants were generated by these means and on the basis of their ability to complement each other in cell culture they could be divided into groups (Akel and Sweet, 1993). Rescue experiments and the assignment of mutations to specific genes had to await the cloning of subgenomic DNA fragments and the determination of the DNA sequences of CMV genomes. But even then the connection of a mutation with a certain phenotype turned out to be difficult, because the respective CMV genomes often contained several mutations (Kumura et al., 1990).

The introduction of insertion and deletion mutagenesis by Spaete and Mocarski (1987) paved the way to reverse genetics of CMVs. This was the method of choice for the disruption of CMV genes that are non-essential for replication in cell culture and also for generating recombinant CMVs expressing reporter genes in order to follow the course of an infection in animal hosts (reviewed by Mocarski and Kemble, 1996). Although this was a major step forward, the method was afflicted with a number of limitations that precluded rapid genetic analyses of CMVs. For instance, non-homologous recombination is usually favored in eukaryotic cells and homologous recombination events are relatively rare. Accordingly, the mutagenesis method was rather inefficient and the formation of illegitimate recombinants was often observed. Separation of mutant and wild-type viruses required lengthy and time-consuming plaque purification or limiting dilution procedures. Isolation of viral mutants displaying a growth deficit remained an even more difficult task. The application of selection methods, e.g. by employing the *gpt*, *neo* or *puro* gene, facilitated the isolation of recombinant viruses (Abbate et al., 2001; Greaves et al., 1995; Wolff et al., 1993) but did not overcome the intrinsic limitations of the approach.

The regeneration of CMVs from overlapping cosmids represented a new principle in that recombination between the overlapping fragments led to the reconstitution solely of mutant virus, obviating the need to select against non-recombinant wild-type virus (Ehsani et al., 2000; Kemble et al., 1996;). The method was used successfully to generate interstrain variants of HCMV (Kemble et al., 1996) as well as an HCMV ie1 mutant, which displayed an impaired growth phenotype and had to be grown on a complementing cell line (Mocarski et al., 1996).

The genetic analysis of small DNA viruses profited from the cloning of the complete viral genomes into plasmid vectors and the manipulation of the viral DNA sequences with the powerful molecular techniques in vitro or in *E. coli*. For many years the sheer size of the CMV genomes represented an insurmountable barrier, which prevented their cloning as a single entity. Only the development of new cloning vectors, e.g. the *E. coli* F-factor-derived bacterial artificial chromosome (BAC) that offered a cloning capacity of up to 300 kbp (Shizuya et al., 1992), made this vision become real.

## Cloning of CMV genomes as bacterial artificial chromosomes

At first sight cloning of CMV genomes may appear a trivial task, which can be achieved easily by ligation of the linear viral genome into a BAC cloning vector followed by transformation into an E. coli host. In practice however, such an approach has not been successful. This is probably due to the fact that large DNA molecules are prone to shearing and extensive handling leads to their destruction. For this reason the viral DNA and the BAC vector were joined the other way round, i.e. the BAC vector sequences were introduced into the viral genome by homologous recombination utilizing the cellular recombination machinery (Figure 4.1A). To this end, a recombination plasmid was constructed that contained the BAC vector flanked by viral DNA sequences homologous to the intended insertion site in the viral genome. The linearized recombination plasmid was transfected into fibroblast cells followed by infection with the respective CMV strain (Borst et al., 1999; Messerle et al., 1997). Recombinant viruses carrying the BAC vector integrated into their genomes were amplified in the presence of mycophenolic acid and xanthine using the guanosine phosphoribosyl transferase (gpt) gene for selection as described by Greaves et al. (1995). Fol-

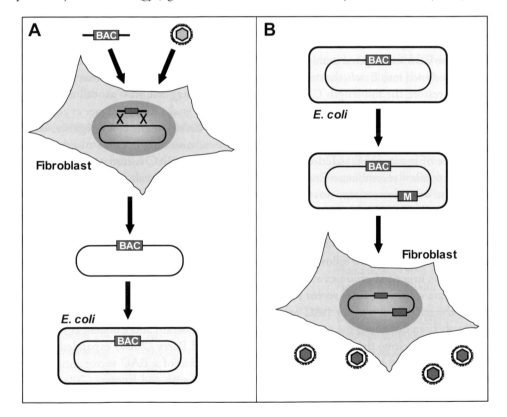

**Figure 4.1** (A) Cloning of a CMV genome as a BAC. The BAC replicon (blue) flanked by viral sequences is transfected into fibroblasts, which are subsequently infected with CMV. Viral genomes that have incorporated the BAC cassette by homologous recombination are isolated and transferred to E. coli. (B) The principle of generating mutant CMVs using BAC technology. The CMV genome maintained as a BAC in E. coli is used as a substrate to introduce a mutation (M, red). The mutant BAC is then transfected into fibroblasts in order to reconstitute mutant virus.

CMV BAC molecules. However, others had already established sophisticated methods for the manipulation of large DNA molecules or the *E. coli* chromosome and these methods could be adapted for the manipulation of CMV BACs. Fortunately, mouse and human geneticists became interested in the manipulation of BACs around the same time, so one could profit from their efforts. Meanwhile, there is a wealth of techniques that facilitate the introduction of any kind of mutation into the BAC-cloned CMV genomes with high efficiency and within a very short time frame.

### *RecA*-mediated allelic exchange

Manipulation of CMV genomes usually aims at the generation of different alleles of a gene, which typically result in different phenotypes of the virus. The possible alterations range from the complete removal or inactivation of a gene (null-allele) to the exchange of a single nucleotide only (point mutation). The mutagenesis technique can also be used to introduce additional genes or DNA sequences into CMV genomes (insertions). The mutant allele is transferred into the BAC-carrying *E. coli* host cells with the help of a shuttle vector. Recombination between the mutant and the parental allele makes use of the general recombination machinery of *E. coli*. The RecA protein and exonuclease V of *E. coli* (encoded by *recBCD*) are the minimal requirements for this recombination pathway. In many *E. coli* strains the *recA* gene is inactivated, rendering the strains recombination-incompetent and providing enhanced stability for cloned DNA fragments. Although *recA* expression is required for mutagenesis it increases the risk of instability of the BAC-cloned CMV genomes. Fortunately however, the CMV BACs seem to display remarkable stability in *recA*⁺ strains (Borst et al., 1999; Messerle et al., 1997; Yu et al., 2002). Nevertheless, we recommend keeping the exposure of the CMV BACs to recombination enzymes as short as possible in order to minimize the risk of acquiring adventitious mutations. For this reason, mutagenesis was initially performed in an *E. coli* strain that can be transiently rendered *recA*⁺ (Kempkes et al., 1995; Messerle et al., 1997) and later the *recA* gene was provided on the backbone of the shuttle plasmid (Borst et al., 2004; Hobom et al., 2000).

Homologous recombination between the mutant allele on the shuttle plasmid and the parental allele on the BAC typically leads to the formation of a cointegrate between the two molecules (Figure 4.2). Such recombination events are probably rare, but appropriate selection schemes allow the detection of bacterial clones that contain cointegrates. Two kinds of shuttle vectors have been employed. One is based on the pSC101 replicon and offers a temperature-sensitive mode of replication (Borst et al., 2004; Posfai et al., 1997) and the other one contains the R6K replicon and thus replicates only in *E. coli* strains that express the lambda π gene (Smith and Enquist, 1999; Smith and Enquist, 2000; Yu et al., 2002). Following the transfer of pSC101-based shuttle plasmids, bacterial clones harboring cointegrates can be selected by shifting the bacteria to a temperature that is non-permissive for the replication of the shuttle plasmid and by selecting for the antibiotic resistances that are encoded by the BAC and the shuttle plasmid, respectively (Borst et al., 2004). R6K-based shuttle plasmids can be transferred either by transformation or by conjugation. Since they cannot replicate in BAC-carrying *E. coli*, bacterial clones that survive under selection with antibiotics must have integrated the DNA sequences of the shuttle plasmid into the BAC.

Resolution of the cointegrate occurs spontaneously by homologous recombination but again this is a rather rare event. Accordingly, selection against bacteria that have retained

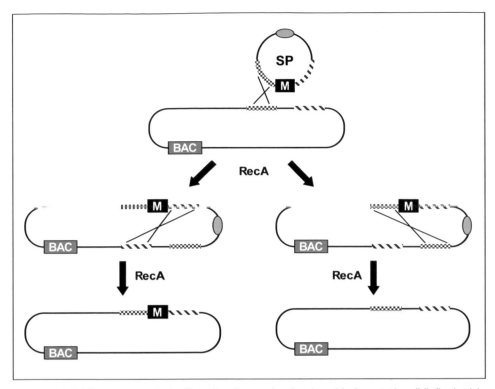

**Figure 4.2** BAC mutagenesis in *E. coli* using a shuttle plasmid. A mutation (M) flanked by viral sequences (hatched and crosshatched) is cloned into a shuttle plasmid (SP). The shuttle plasmid usually contains a selectable and a counterselectable marker (e.g., *kan* and *sac*B). Introduction of the shuttle plasmid into the *E. coli* host containing the CMV BAC leads to homologous recombination and formation of a cointegrate. In a second recombination step, the cointegrate is resolved, leading either to a mutant (left) or an unaltered genome (right). Homologous recombination is dependent on the *E. coli* RecA enzyme and requires homologies of at least 500 bp.

the cointegrates by utilizing the negative selection marker *sacB* proved to be highly useful in identifying those clones that harbor mutated BACs (Borst et al., 1999; Chang and Barry, 2003). Since the *sacB* gene has an intrinsic propensity for spontaneous inactivation, a kan/*lacZ* cassette and blue-white screening of colonies in the presence of X-Gal were also used to identify modified BACs (Yu et al., 2002). Depending on whether resolution occurs via the same homologous sequence that was used during generation of the cointegrate or via the sequence that flanks the mutation on the other side, the outcome is either a BAC with the parental configuration or a BAC carrying the mutation (Figure 4.2). Final conformation of the successful introduction of a mutation has to be obtained by restriction analysis, PCR, and/or sequencing.

This two-step recombination method worked highly reliably in our hands. Admittedly, the construction of the shuttle plasmid may involve several cloning steps and can be time-consuming. The homologous sequences flanking the mutation each have to be 1.5 to 2 kbp in size in order to achieve a reasonable chance of recombination. With respect to the stability of the BAC it is reassuring that recombination only occurs at a low frequency and

and homology arms complementary to regions of the viral genome as much as 119 kbp apart, large deletions covering dozens of genes can be attained. As shown schematically in Figure 4.4, a set of 16 viable MCMV mutants was generated with deletions covering between 4 and 49 genes, confirming that the majority of MCMV genes are non-essential for virus replication in cell culture (M. Wagner, unpublished).

The ability to add a few nucleotides within the oligonucleotide primers was also used to C-terminally epitope tag viral genes in the context of the viral genome. An HA-tagged viral protein can easily be detected by commercially available antibodies specific for the nine amino acid HA peptide, thereby enabling the investigation of its temporal expression pattern and intracellular localization, and of potentially interacting proteins by Western blot, immunofluorescence and co-immunoprecipitation experiments (Atalay et al., 2002; Brune et al., 2003; Bubeck et al., 2004; Kattenhorn et al., 2004; Menard et al., 2003).

For many applications the few nucleotides remaining of the selection marker after FLP or Cre recombinase-mediated excision are acceptable. However, if only one or a few nucleotides are to be exchanged, e.g. in order to change one amino acid of a protein, the concomitant introduction of additional sequences is not tolerable. Two ways have been explored to solve the problem: mutagenesis can either be performed in the absence of a positive selection marker or the selection marker can be replaced in a second round of mutagenesis (Figure 4.3B). $red\alpha\beta$-mediated recombination with linear fragments or single stranded oligonucleotides can occur with such a high frequency that the identification of mutated BACs by PCR screening becomes feasible (Ellis et al., 2001). The construction of a revertant of a MCMV m152 deletion mutant by this strategy proved its applicability for

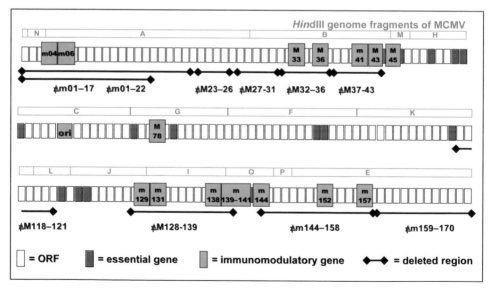

**Figure 4.4** Example for site-directed mutagenesis using linear PCR fragments: defining essential genes of MCMV with a number of large deletions in the viral genome. Deletions of up to 43 kbp were introduced as shown in Figure 4.3A. The deletion mutants shown in this figure were all viable, defining 101 of 169 predicted viral genes (60%) as nonessential for MCMV replication. The 101 deleted genes include 82 genes of unknown function.

CMV BACs (Holtappels et al., 2004). However, a frequency of successful recombination events of about 0.1% meant that screening of several thousand bacterial clones by PCR was required.

In the second approach, a positive and a negative selection marker (or a screening marker) are first inserted into the CMV BAC at the target site. In a second step the markers are replaced by a fragment carrying the desired mutation. Britt et al. used the ampicillin resistance gene in combination with the *lacZ* gene (Britt et al., 2004). Clones in which replacement of the marker genes with a linear fragment occurred could conveniently be detected as white colonies by white-blue screening, although just 0.1 to 0.8% of the clones were positive. Dunn et al. (2003) used the *E. coli rpsL* gene as a negative selection marker for the construction of revertant viruses. *rpsL* confers streptomycin sensitivity when introduced into a streptomycin-resistant *E. coli* host strain. The loss of the negative selection marker by replacement with a fragment providing the respective wild-type ORF of HCMV was detected by selection for streptomycin resistance. Since the loss of the negative selection marker by unwanted rearrangements will also result in streptomycin-resistant clones, it may be necessary to characterize a number of clones until a BAC with an intact genome configuration and the desired mutation is identified.

## Transposon mutagenesis

Transposons are mobile genetic elements that can relocate and insert at any genomic location in a virtually random manner. In bacteria they have been utilized widely as random insertion mutagens both at the genomic level and in the analysis of individual genes (reviewed by Hayes, 2003). With their cloning as BACs, CMV genomes also became accessible to transposon mutagenesis. The power of the technique lies in its simple and rapid application and in particular in its capacity to generate a large number of mutants within a short time.

Several different transposon systems have been tested for their suitability to manipulate CMV genomes (Table 4.2). Because of the potential vulnerability of the large

**Table 4.2** Transposon systems used for BAC mutagenesis

| Transposon | Properties | References |
|---|---|---|
| Tn*1721* (Tn*Max*) | In vivo system, preferential insertion into BAC, high efficiency | Brune et al. (1999, 2001, 2002, 2003), Hobom et al. (2000), Ménard et al. (2003), Yu et al. (2003), Zimmermann et al. (2005) |
| Tn5 | In vivo system, preferential insertion into bacterial genome, requires re-transformation | Smith and Enquist (1999), Yu et al. (2003) |
| Tn5 [a] | In vitro system, results in high percentage of incomplete BACs | McGregor et al. (2004) |
| Tn5 [a] (transposome) | In vivo system, requires transposon-transposase complex (transposome) | McGregor et al. (2004) |
| Tn7 [a] | In vitro system, results in high percentage of incomplete BACs | McGregor et al. (2004) |
| Tn7 | In vivo system, insertion into defined attachment site (non-random) | Hahn et al. (2003a) |

[a]Commercially available through Epicentre and NEB, respectively.

BAC molecules to shearing in vitro, transposon mutagenesis in E. coli was preferred. The first transposon system for the mutagenesis of an entire herpesvirus genome was based on Tn1721 (a member of the Tn3 family), which displays a preference for insertion into negatively supercoiled plasmids (Brune et al., 1999). The transposon is provided on a donor plasmid with a temperature-sensitive replication mode (Figure 4.5A). The transposition control unit with the transposase gene is also located on the donor plasmid but separately from the transposable element and is therefore not translocated. Accordingly, secondary transposition events can be prevented by eliminating the donor plasmid after mutagenesis. Following the introduction of the donor plasmid into E. coli harboring the BAC, the bacteria are kept at the permissive temperature and the transposon can jump into the CMV BAC. Subsequently, the bacteria are shifted to the non-permissive temperature to eliminate the donor plasmid. BAC clones with an insertion are identified by selection with those antibiotics whose resistance is encoded by the transposon and the BAC vector. In this way thousands of BAC mutants can be generated overnight. Transposition into the bacterial chromosome occurred in only 5 to 10% of the clones, obviating a need for enrichment of clones with mutant BACs (Brune et al., 1999; Hobom et al., 2000). Almost all of the BACs

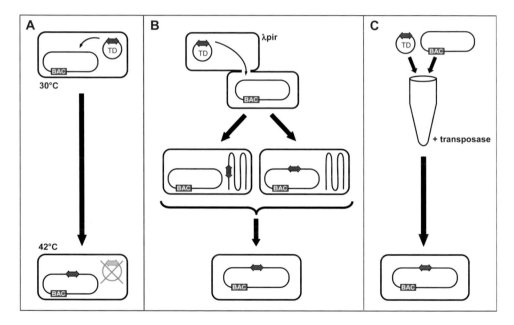

**Figure 4.5** Transposon mutagenesis: different approaches. (A) A temperature-sensitive transposon donor (TD) plasmid containing a selectable marker gene within the transposable element is maintained in E. coli together with the CMV BAC at 30°C. Transpositions occur at a low frequency. Plating the bacteria at 42°C with appropriate selection results in the loss of the TD plasmid. In case of a Tn1721-derived transposon roughly 90% of the BACs contain a transposon insertion as this transposon inserts preferentially into plasmids or BACs. (B) A λ pir-dependent suicide TD plasmid is delivered to the BAC host by conjugation. The Tn5-derived transposon inserts more often into the bacterial genome and less frequently into the BAC. Isolation of BACs from numerous clones and transformation of fresh bacteria facilitates the retrieval of mutant BACs. (C) In vitro mutagenesis. The TD plasmid, the BAC, and transposase enzymes are combined in a test tube, where transposition takes place. The transposition reaction is then transformed into E. coli in order to isolate individual clones.

contained a single insertion, which remained stable in the genome after reconstitution of viruses. Primer binding sites at the respective termini of the transposon allowed the determination of the insertion site by direct sequencing. This method was first used to identify essential and non-essential genes of MCMV (Brune et al., 1999). Later on, mutants of whole families of CMV genes, namely of the HCMV envelope glycoprotein genes and the MCMV US22 family members, were isolated from BAC libraries in order to study their function (Hobom et al., 2000; Menard et al., 2003). Although the distribution of the transposon insertions did not seem to be completely random, the libraries were of sufficient complexity to find mutants of each member of the gene families.

Smith and Enquist (1999) adopted a transposon mutagenesis procedure based on the Tn5 transposon, using a donor plasmid with an R6K origin of replication. Such plasmids need the lambda π protein for propagation and cannot replicate in the recipient bacteria that contain the CMV BAC (Figure 4.5B). Upon transfer of the Tn5 donor plasmid by conjugation or electroporation the transposon is inserted either into the BAC or the bacterial chromosome. Since the latter occurs in a substantial proportion of the transposition events, one needs to isolate the BACs and retransform fresh bacteria in order to enrich for the mutant BACs. Yu et al. (2003) used both previously described transposon systems (i.e. the Tn5- and the Tn1721-derived transposons) to create a comprehensive library of HCMV AD169 mutants that enabled them to define non-essential and essential ORFs of HCMV. In addition to the selection marker (kn), both transposons contained reporter genes (GFP, *lacZ*) for eukaryotic as well as for prokaryotic cells. The *lacZ* marker facilitated the generation of revertant HCMV BACs by allelic exchange and blue-white screening. Another interesting property of one of the transposons is the ability to tag the truncated proteins in four of the six possible ORFs in order to identify proteins that are expressed by an ORF carrying the insertion (Yu et al., 2003).

One could speculate that in vitro transposition systems would be more efficient for the mutagenesis of BACs, because the amount of transposase and donor plasmid can more easily be controlled and optimized if necessary (Figure 4.5C). McGregor et al. (2004) tested two commercially available in vitro mutagenesis systems that are based on Tn5 and Tn7, respectively. In principle both systems were considered to be suitable for the manipulation of a guinea-pig CMV BAC. However, since handling the BAC molecules cannot be avoided during the in vitro reaction and the subsequent electroporation into *E. coli*, the majority of the resulting BACs either displayed deletions or rearrangements. This problem could be elegantly overcome by the use of so-called Tn5 "transposomes". Transposomes are complexes between the transposon molecule and the transposase, which are formed in vitro in the absence of $Mg^{2+}$ ions. Once transferred into bacteria carrying the BAC, the complex is activated by the $Mg^{2+}$ ions inside the cells, resulting in the insertion of the transposon into a target molecule. Since the complex is self-limiting there is no risk for secondary transposition events. Unfortunately, Tn5 is not able to discriminate between the BAC episome and the bacterial chromosome and thus there remains a need to isolate the mutated BACs and retransform them into new *E. coli* bacteria in order to separate them from clones that received an insertion into the bacterial genome (McGregor et al., 2004; McGregor and Schleiss, 2004).

The transposon Tn7 differs from other mobile elements in that it can insert with a high frequency into a single specific site (att Tn7) in the *E. coli* chromosome. Following engineering of attTn7 into a BAC and the removal or sealing of the natural attTn7 site in the *E. coli* genome, Tn7-based transposons can be used with high efficiency to shuttle any desired gene or DNA sequence into a BAC. Hahn et al. (2003a) used the system to demonstrate that the growth of a UL122 (ie2) deleted HCMV can be complemented in cis by the re-introduction of an ie2 cDNA.

In order to perform a comprehensive analysis of a single CMV gene, it would be desirable to generate a panel of mutants with the mutations distributed throughout the ORF. It has indeed been shown that whole series of mutants affecting a single gene can be isolated from CMV BAC libraries (Brune et al., 2001; Hobom et al., 2000; Menard et al., 2003). However, since the bulky transposon insertions usually disrupt the function of the gene, one can gain little insight into the structure-function relation of a protein by this approach. To study the function of individual domains of a protein most of the transposon would have to be removed, leaving behind only a small in-frame insertion within an ORF. This has not yet been achieved in the context of an infectious CMV BAC, because it would require modification of the BAC in vitro with the previously mentioned risk of its destruction. In view of these limitations Bubeck et al. (2004) subcloned the MCMV gene M50, which is involved in capsid egress, into a plasmid vector and subjected it to comprehensive mutagenesis using a modified Tn7-transposon system with a potential for random insertions. The bulk of the transposon was subsequently removed by treatment with a restriction enzyme and religation. The library of plasmid-based mutants was then transferred back into the CMV BAC using a procedure catalyzed by a site-specific recombinase (see below). The resulting series of viral BACs (104 mutants) made it possible to study the M50 gene in the context of the viral replication process and to draw important conclusions regarding the function of different domains of the M50 protein.

Considering the versatility of the transposon-based techniques and their relatively simple, rapid and inexpensive application it becomes obvious that these techniques have a tremendous potential for the analysis of CMV gene functions.

## Site-specific insertions

Construction of a revertant virus by repair of the respective mutation is one of the controls that is mandatory to prove that a specific phenotype is due to the disruption of a candidate gene and does not result from an accidental mutation in another gene. However, as pointed out by Yewdell and Hill (2002), restoration of the original genetic sequence and rescue of a phenotype can be easily misinterpreted, because the repair of a gene may also restore the expression of neighboring or not yet identified overlapping genes. They therefore proposed that the insertion of a candidate gene elsewhere in the genome would be a better control.

The ectopic insertion of genes can be performed with several of the described techniques. The use of Tn7-mediated transposition for the cis-complementation of an HCMV ie2 deletion mutant has already been mentioned (Hahn et al., 2003a). As an alternative, ET recombination was used to insert the gene for the small capsid protein at an ectopic position in the HCMV genome in order to prove that this protein is essential for viral growth (Borst et al., 2001)

Since FLP and Cre recombinases catalyze site-specific recombination in both directions, i.e. excision as well as insertion, it was examined whether the recombinases can be used for rapid insertion of DNA sequences (Menard et al., 2003; Bubeck et al., 2004; Borst and Messerle, 2005). The candidate genes are cloned into a suicide vector that carries an FRT site, the R6K replicon and a small antibiotic resistance gene (zeocin or kanamycin). An FRT target site has to be anchored in the CMV BAC first (e.g. by ET mutagenesis) and after transient expression of FLP recombinase in the bacteria the suicide plasmid can be efficiently and rapidly integrated into the CMV genome. The vector backbone is integrated together with the candidate gene, thus slightly enlarging the viral genome, but the vector sequences encompass 1.5 kbp at most and this is usually tolerated. Bubeck et al. (2004) have successfully used the method to generate hundred MCMV mutants in order to analyze the function of the essential M50 gene.

Despite the success and the justification of the approach one has to add a word of caution: even if a gene is transferred together with its promoter to an ectopic position, one cannot necessarily expect its expression level to be unaffected. For instance, the ectopic re-insertion of the MCMV m157 gene into an m157 deletion genome led to an only partial rescue of the mutant's phenotype, i.e. susceptibility of the resulting virus to NK cell control in C57BL/6 mice (Bubic et al., 2004). Likewise, the transfer of the HCMV lytic origin of replication to another genome position did not restore the full growth capacity (Borst and Messerle, 2005). Obviously, the influence of neighboring genetic elements on the activity of the transferred genetic elements cannot be neglected and ectopic insertion can therefore represent just one of several genetic methods used to investigate viral functions.

## Complementation of essential genes and conditional mutants

Only BAC-based mutagenesis techniques allow the disruption of essential genes, as the construction of mutant genomes becomes independent of their biological properties. Transfection of CMV genomes carrying lethal mutations is characterized by a failure in plaque formation and aside from the information that the gene is essential one does not get any further insight. In order to learn at which step of the replication cycle the gene is involved, it would be desirable initially to complement the missing function and grow up the virus and then to perform infection experiments under non-complementing conditions.

In one of the first analyses of essential CMV genes, transient co-transfection experiments with a glycoprotein B (gB) knock-out genome of HCMV and gB expression plasmids were performed (Strive et al., 2002). Although the complementation was far from satisfactory, leading to mini-plaques and the release of a limited number of infectious virions only, the experiments gave us at least an idea as to which domains of the glycoprotein are important and which are dispensable. In the case of the MCMV ie3 deletion mutant, a complementing cell line based on NIH 3T3 cells could be generated and the mutant grown up (Angulo et al., 2000). Since the titers of the ie3 mutant on the complementing cell line remained one to two orders of magnitude below those of the wild-type virus, it appeared that either the amount of the *trans*-complementing protein or the temporal expression pattern were inadequate for full complementation. A similar observation was made with HCMV UL99 mutants. On complementing cells they produced substantially reduced virus yields in comparison to the parental virus (Silva et al., 2003). Nevertheless, sufficient quantities of the mutant viruses could be obtained to perform infection experiments. The UL99 mutants were not impaired in DNA replication and gene expression but capsids accumulated in the cytoplasm, suggesting that UL99 plays an essential role for the envelopment of HCMV virions.

Although these results were encouraging, it is desirable to turn on and off CMV genes at will. Recently, Rupp and colleagues adopted the widely used tetracycline (tet) regulated expression system to control transcription of essential MCMV genes (Rupp et al., 2005). There have been previous attempts to utilize the tet system in CMV (Kim et al., 1995; McVoy and Mocarski, 1999), but in the absence of effective methods for engineering CMV genomes, it did not find application. Rupp et al. reasoned that the effector molecule, the tet repressor, as well as the regulated gene unit should be included into the viral genome to make the regulation independent of a special cell line. To this end, they constructed a shuttle vector containing the tet repressor gene under the control of the HCMV major immediate-early promoter and an SV40-driven promoter expression cassette containing tet operator sequences downstream of the TATA box. The gene of interest is first deleted from the CMV genome and cloned into a shuttle vector. Following FLP-mediated insertion of the shuttle vector into the viral genome, the virus is reconstituted by transfection into permissive cells and can be grown in the presence of doxycycline. In the absence of doxycycline, the tet repressor binds to the tet-operator sequences and inhibits expression of the gene of interest. Using two different MCMV genes, M50 and a dominant-negative version of the small capsid protein, Rupp et al. (2005) demonstrated that growth of the MCMV mutants can be regulated from nil to levels comparable to the wild-type virus in a tet-dependent manner. Even more importantly, the regulation also worked in vivo, extending the application of the regulator system to the analysis of gene functions involved in pathogenesis. Although it remains to be shown whether the technique can be applied to all temporal classes of CMV genes, it definitely has an enormous potential for deciphering essential CMV functions.

## Reverse genetics

The ability to rapidly mutagenize CMVs made it possible to test specific hypotheses concerning the role of viral genes during the infection cycle. Perhaps unsurprisingly, many colleagues concentrated on those genes that they had previously analyzed using different techniques.

### Immediate-early genes

Given the wealth of data on immediate-early genes, it is understandable that many of the first BAC-based CMV mutants concerned this class of genes. The very first mutation that was introduced into the BAC-cloned MCMV genome led to the disruption of the ie1 gene and an MCMV mutant that expressed a truncated version of the IE1 protein (Messerle et al., 1997). The mutant was viable and displayed only slightly reduced growth kinetics, resembling the phenotype of a similar HCMV ie1 mutant (Mocarski et al., 1996; Greaves and Mocarski, 1998). In the meantime, a mutant with a complete knockout of the MCMV ie1 gene was constructed by directly fusing exon 3 to exon 5 of the ie1-ie3 transcription unit (Ghazal et al., 2005). Surprisingly, the growth of this ie1 mutant in different cell culture systems was indistinguishable from that of WT-MCMV, yet the ie1 mutant was unable to disrupt promyelocytic leukemia (PML) nuclear bodies. Although the mutant displayed reduced virulence in vivo it could nevertheless grow to considerable titers in the absence of the ie1 gene. One cannot rule out that enhanced expression of the IE3 transactivator may somehow compensate for the lack of the IE1 protein. Thus, additional ie1 mutants and further analyses are required to elucidate the roles of the MCMV ie1 gene during productive infection as well as in establishing latency and in reactivation.

The disruption of the MCMV ie3 gene was lethal, confirming the expectation that the IE3 transactivator would be essential for viral growth (Angulo et al., 2000). This replication defect could be rescued when the IE3 protein was provided in *trans* by a complementing cell line. Infection experiments revealed that only ie1 transcripts were synthesized in cells infected with the ie3 mutant, implying a very early block in gene expression. Similar results were obtained with an ie2 deletion mutant of the HCMV Towne strain (Marchini et al., 2001). Whereas mRNA and protein expression of other IE genes were readily detectable in cells transfected with the ie2-deleted HCMV BAC, there was no expression of early genes, suggesting that IE2 transactivates most HCMV promoters. By introducing a point mutation, which led to the exchange of a cysteine to a glycine at position 510 of IE2, Heider et al. (2002a) chose an elegant approach to construct an HCMV variant with a temperature-sensitive IE2 protein. The virus grew at 32.5°C but not at the non-permissive temperature of 39.5°C. Again, RNA analysis confirmed the block at the transactivation step of early promoters.

A number of further mutants with point mutations or subtle alterations in the IE genes allowed other important questions to be addressed. An HCMV ie2 mutant with alanine substitutions at four different positions replacing threonine and serine residues was constructed to test whether phosphorylation modulates the transactivating property of IE2. Surprisingly, the mutant grew with identical kinetics to the parental virus in human fibroblasts and the activation of viral promoters did not seem to be changed (Heider et al., 2002b). An HCMV ie2 mutant with an internal deletion of amino acids 136 to 290 was viable but exhibited altered growth characteristics (Sanchez et al., 2002). Unexpectedly, this mutation seemed to affect the expression of some late proteins. Since the mutation removes the initiator methionine codons for the p40 and p60 forms of the IE2 protein, one may expect that these proteins play an important though not essential role in the expression of a subset of late proteins. Mutants with small deletions at more C-terminal positions in the IE2 protein were nonviable (White et al., 2004), but transfection experiments and RNA analyses allowed conclusions to be drawn regarding the requirement of different protein domains for transactivation or autoregulation. Sumoylation of IE2 was expected to increase its transactivation activity. However, substitution of the major sumoylation sites (Lys175 and Lys180) of IE2 did not impair the growth of the mutant in human fibroblasts (Lee and Ahn, 2004). The HCMV IE1 protein is also sumoylated but the functional significance of this modification was not known. Nevels et al. (2004) constructed an HCMV genome that encodes an IE1 protein lacking the sumoylation site. The virus genome was infectious, although the mutant virus grew more slowly than the wild-type virus and produced a reduced yield. The ability of the mutant to disrupt PML nuclear bodies was not impaired and the authors observed a reduced accumulation of the IE2-86 kDa protein in infected cells, which could explain its reduced replication capacity. A similar mutant based on a BAC of the Towne strain was also infectious, confirming that sumoylation of IE1 is not absolutely necessary for viral growth (Lee et al., 2004).

These examples demonstrate that several hypotheses concerning the role of IE proteins and their subdomains could be verified in the context of viral infection, but, perhaps even more importantly, some hypotheses had to be discarded. Obviously, a lot of questions remain open and a more systematic mutagenesis approach (e.g. linker scanning mutagenesis) may help to dissect the various functions of the IE proteins.

## Mutagenesis of other essential CMV genes

Similarly to the IE genes, several other CMV genes that deserve closer attention were subjected to mutagenesis. The maturational protease of CMVs has an internal cleavage site, which is unique among the assemblins of herpesviruses. The significance of this additional cleavage site was not known. By introducing a single nucleotide change into the HCMV genome, which resulted in an alanine to valine substitution in the assembly protein, Chan et al. (2002) destroyed the cleavage site. The resulting mutant displayed unaltered growth kinetics, leaving open why the CMV proteases have acquired this property.

The major envelope protein of the CMVs, glycoprotein B, carries a recognition motif for proteolytic cleavage by the cellular endoprotease furin and is indeed cleaved during maturation. Since several gB homologs of other herpesviruses lack this motif, it was unclear whether cleavage of gB by furin is essential for the growth of HCMV. Destruction of the furin recognition site abolished cleavage of gB, but the growth of the mutant virus was not impaired, at least not in fibroblasts, thus questioning the significance of the furin cleavage site (Strive et al., 2002). It is possible that the relevance of such motifs can only be revealed by studying comparable mutants of animal CMVs in vivo or at least in various cell culture systems. Glycoprotein B features a number of additional motifs that are worth analyzing. Remarkably, Jarvis and colleagues managed to modify a phosphorylation site (Ser900) in the C-terminal tail of gB by performing homologous recombination in cell culture (Jarvis et al., 2004). The construction of these HCMV mutants probably required a lot of effort as well as additional modifications such as the insertion of reporter genes and promoter elements. Hence, generation of mutant genomes by BAC mutagenesis will save time and moreover will allow the construction of the mutants in an even more precise manner without a need for extraneous alterations.

Such an example is provided by the mutagenesis of the UL99 gene (Silva et al., 2003; Britt et al., 2004). At first, both groups deleted the UL99 ORF from the HCMV genome and observed that the mutated genomes were not infectious, implying that the pp28 protein encoded by UL99 is essential. However, since the UL99 and UL98 ORFs overlap, one could not exclude the possibility that the gross alteration in UL99 interferes with the expression of the neighboring UL98. Therefore, mutants were constructed with only single point mutations that either changed the initiator methionine codon to a leucine codon or interrupted the ORF at position 4 with a stop codon but did not change the overlapping UL98 ORF. Again, both genomes turned out to be non-infectious. By changing the myristylation site at codon two from glycine to alanine, Britt et al. (2004) could also demonstrate that this property of pp28 is needed for virus assembly.

Treatment of patients with antiviral drugs can lead to the emergence of resistant CMV strains that have often acquired multiple mutations in several genes. In order to understand the molecular basis of the resistance as well as of the mode of action of the antiviral substances, it would be of interest to learn which mutations are responsible for the drug-resistant phenotype. By means of BAC mutagenesis, single as well as multiple mutations can now be rapidly inserted into the viral genome and the phenotypes of the resulting viruses tested. Using this approach it was shown that a mutation in UL104 encoding the portal protein of the capsid is not sufficient to mediate resistance against the antiviral substance benzimidazole D-ribonucleoside BDCRB (Komazin et al., 2004).

Whereas all the mutants mentioned so far were developed to answer specific questions, Yu et al. (2003) and Dunn et al. (2003) chose a more systematic approach for reverse genetics of HCMV. Using either redαβ-based recombination or transposon mutagenesis they knocked out each individual ORF of the HCMV genome and investigated the consequences of the mutations for HCMV growth. In this way they recognized 41 and 45 ORFs of the AD169 and the Towne strain, respectively, as essential and 115 and 117 ORFs as nonessential. A few ORFs were found to be augmenting, in that in their presence the virus yields were at least tenfold higher (Yu et al., 2003). Interestingly, several HCMV mutants grew 10- to 500-fold better in different cell types than the parental Towne strain, suggesting that the deleted ORFs encode temperance factors that modulate growth of HCMV in different cell types (Dunn et al., 2003). Although these studies gave us hints of the nature and qualities of HCMV genes, they could, of course, not clarify their functions directly.

## Forward genetics

Forward genetics is a classical genetic procedure that starts out with a physical or measurable characteristic of an organism (the phenotype) and aims at the identification of the gene or genes that are responsible for it. A prerequisite of forward genetics is the generation of a comprehensive library of mutants, i.e. saturating random mutagenesis. Prior to BAC cloning of CMV genomes, this was difficult to achieve and tracking of subtle mutations in the huge genome was hardly conceivable.

Forward genetics for CMVs was pioneered by creating large libraries of transposon BAC clones (Brune et al., 1999; Brune et al., 2001). Literally thousands of mutants can easily be obtained overnight. Yet, there was the problem of effectively converting the mutated genomes into mutant viruses. This obstacle was overcome by endowing the E. coli bacteria with the invasin of Yersinia pseudotuberculosis and the hemolysin O protein of Listeria monocytogenes (Figure 4.6A). These proteins enabled the bacteria to invade fibroblasts, which was followed by the release of the CMV BACs into the cells and the consequent initiation of the viral replication cycle. Thus, the bacterial clones had simply to be inoculated onto fibroblasts in order to produce a collection of viral mutants. The expression of GFP, which was encoded by the transposable element, facilitated the identification of the mutant viruses (Figure 4.6B). Using this procedure a library of approximately 600 mutant MCMV genomes was generated, yielding about 200 viable mutants (Brune et al., 2001). There was no need to characterize the individual mutants prior to screening. The first application was to screen the library for mutants that displayed a growth deficit in endothelial cells. To this end, fibroblasts and endothelial cells were infected in parallel with individual mutants and monitored for the spread of the infection. Whereas all viruses grew on fibroblasts, six of them failed to grow on endothelial cells. By means of the transposon insertions, the mutations could be easily mapped in the viral genomes and were found to reside in one single MCMV gene, M45 (Brune et al., 2001). The protein encoded by M45 is a constituent of the viral tegument and shows homology to the large subunit of the cellular ribonucleotide reductase, but does not represent a functional enzyme subunit (Lembo et al., 2004). Infection experiments with the M45 mutants revealed that M45 is required for protecting endothelial cells from premature apoptosis (Brune et al., 2001). These results

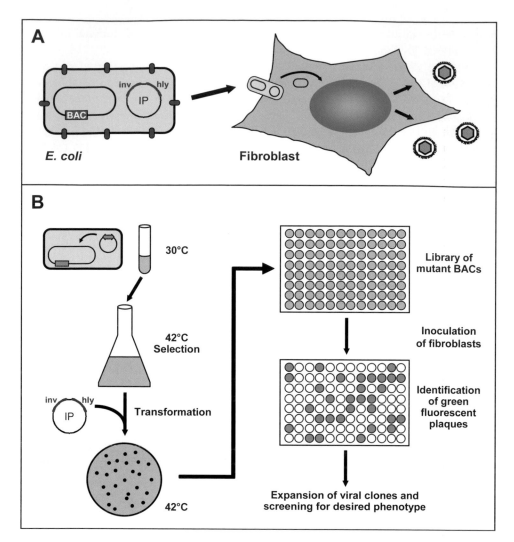

**Figure 4.6** Example for transposon mutagenesis: Identification of viral genes by their function (forward genetics). (A) Direct transfer of BAC DNA from *E. coli* to a mammalian cell with the help of an invasion plasmid (IP). It contains the invasin gene (*inv*, red) of *Yersinia pseudotuberculosis* and the hemolysin O gene (*hly*, blue) of *Listeria monocytogenes*. The Inv protein (red) mediates attachment of *E. coli* to β1 integrins on the cell surface and subsequent internalization. Hly facilitates release of the bacterial contents (including the BAC) from endosomes to the cytoplasm, which ultimately leads to synthesis of infectious progeny virus. (B) Generation of a library of mutant viruses and functional screening. Transposon mutagenesis with a GFP-containing transposon (green) is performed in liquid culture essentially as shown in Figure 4.5A. The invasion plasmid is introduced, and a library of individual bacterial clones is generated using microtiter plates. BACs are transferred to fibroblasts by inoculating the fibroblast cultures with bacterial clones. Two hours after inoculation, the cell culture medium is replaced, and antibiotics are added in order to inhibit bacterial growth. A few days later, mutant viruses can be identified by the green fluorescence of infected cells. The mutant clones are subsequently expanded and used for a functional screen.

indicated that cell tropism of a virus may not only be determined at the entry step but also by influencing the physiological conditions of a cell.

By analyzing the growth behavior of the mutants of the same library another mutant was found that was able to induce apoptosis in several different cell types (Brune et al., 2003). The affected gene m41 encodes a Golgi-localized protein that prevents the early onset of apoptosis in infected cells. Zimmermann and colleagues used the library to find a viral inhibitor of IFN-γ receptor signaling (Zimmermann et al., 2005). Screening was carried out in a cell line carrying an IFN-γ-responsive luciferase reporter construct. WT-MCMV inhibited IFN-γ receptor signaling and thus prevented expression of luciferase upon stimulation with IFN-γ. Accordingly, these authors searched for mutants that had lost the inhibitory activity and were therefore able to induce luciferase expression. Indeed, they found several MCMV mutants with this property, all of which had transposon insertions in the M27 gene. The M27 protein turned out to be an inhibitor of STAT2.

These three examples already highlight the power of the forward genetic approach. Screening a library for a particular phenotype led directly to the identification of the responsible gene.

A similar strategy was used to identify the genes encoding the HCMV Fcγ-receptors (Atalay et al., 2002). As these proteins were already known to be glycosylated, a collection of HCMV mutants was constructed by site-directed mutagenesis of ORFs encoding putative glycoproteins. Screening had to be done by immunoprecipitation of the Fc-receptors from lysates of cells infected with the various mutants. The absence of an immunoprecipitate indicated the disruption of the corresponding gene.

The genetic basis of the growth of clinical isolates of HCMV in endothelial cells and their transfer to leukocytes was not known, but genes in the ULb' region of the HCMV genome were promising candidates for these determinants as laboratory strains lacking this region do not possess these properties. Therefore, Hahn et al. concentrated their analysis on the ULb' region (Hahn et al., 2004). By constructing mutants with small deletions in this region they could rapidly narrow down the candidate genes to UL128, UL130 and UL131. Mutants with single gene knock-outs as well as trans-complementation assays suggested that proteins encoded by each of the three ORFs are required for both endothelial cell tropism and the transfer to leukocytes.

Mutants with large deletions may generally speed up the identification of genes responsible for specific viral properties and reduce the effort needed to screen a complete library. Assuming that immunomodulatory genes, which are not essential for replication in cell culture, are mainly located toward the termini of CMV genomes, Loewendorf and colleagues constructed a set of MCMV mutants with deletions of about 10 to 15 kbp (Loewendorf et al., 2004). Flow-cytometric analysis of macrophages infected with the deletion mutants rapidly led to the definition of a genomic region encoding a protein that interferes with the expression of the co-stimulatory molecule CD86 on antigen presenting cells. By constructing mutants with successively smaller deletions the responsible gene was finally identified. Following this strategy there is, however, a risk of missing a gene if it is located between essential genes within the central region of the CMV genomes. Thus, in the future, it will probably make sense to perform a more systematic screening by combining mutants that carry large deletions and mutants with a disruption of defined genes. The comprehensive libraries of HCMV mutants constructed by Yu et al. (2003) and Dunn et al. (2003) may represent a starting point, and screening for growth determinants in endothelial and epithelial cells with some of these mutants has already begun.

## Potential limitations and perspectives of BAC mutagenesis

After the cloning of the CMV genomes and the establishment of the different mutagenesis techniques one can now introduce any kind of mutation into CMVs, ranging from targeted point mutations to random insertions. Thus, one can now test any hypothesis about the function of a protein by reverse genetics and determine the viral genes responsible for specific properties of CMVs by the powerful techniques of forward genetics.

Some CMVs still await cloning as a BAC. In some cases this may be hampered by a lack of a suitable cell culture system or insufficient experience with the construction of recombinant viruses of a particular CMV. In principle, however, we think that these potential problems can be overcome and that all CMVs will be amenable to BAC cloning and mutagenesis.

The mutagenesis techniques will of course undergo further refinement but the principles appear to be established. These are excellent conditions to see a flourishing of CMV research during the next years and, as outlined in this review, we have already seen a number of brilliant results that were obtained with the new techniques.

So, are the BAC-based mutagenesis techniques definitely superior to the previous techniques or do they have any limitations? Certainly, they are faster and in many cases they do not require the insertion of additional reporter or selection markers. Also, the generation of comprehensive libraries of CMV mutants is hardly conceivable with any other technique. A major concern is the stability of the cloned genomes. As we have seen previously with cell-culture based recombination techniques, viral DNA molecules are always susceptible to mutagenic agents. Of course, this applies also to *E. coli* bacteria – even if they are recombination deficient. Moreover, if we have to allow targeted mutagenesis in order to alter the CMV genome in a specific manner we cannot completely rule out the occurrence of undesired alterations elsewhere in the genome. There is a low but not negligible mutation rate in *E. coli* and there is no selection force – not even for essential genes – to preserve the integrity of the CMV genome. Thus, the longer we propagate the CMV BACs in *E. coli* the higher the risk to acquire unwanted mutations. Fortunately, the risk seems very low. It was very encouraging to see that the BAC-cloned MCMV strain has preserved its wild-type characteristics even after extended rounds of propagation and mutagenesis in *E. coli* (Wagner et al., 1999). Nevertheless, it would be prudent to keep the original CMV BAC clones and return to them whenever possible.

When a multi-copy vector was used to express the *red*αβ recombination functions, instability of the BACs was sometimes observed but this seemed to be a stochastic event. Interestingly, the HCMV BAC appeared to be much more stable than the MCMV BAC. In the meantime, the problem could be largely solved by using either a low-copy vector or a λ-prophage inserted as a single copy into the bacterial genome for expression of the recombination functions and by adjusting the induction times of the functions. The direct repeats in the viral genome remain a special concern, because they provide optimal substrates for recombination. An obligatory precaution is to characterize the mutated genomes thoroughly by treatment of the viral DNA with several different restriction enzymes and to discard clones that show unexpected alterations. In addition, we can now easily construct several mutants affecting the same gene since it is highly unlikely that identical spontaneous mutations will occur in independent clones. When we identify a mutant virus with a specific phenotype we also have to construct a revertant virus in order to prove that the phenotype is really due to the specific mutation.

In which field will we see the most progress? Reverse genetics will of course be important in constructing selected mutants to answer specific questions about individual genes. It will also allow us to define the biological relevance of a CMV gene by performing in vivo experiments with CMV mutants in a relevant animal model. However, without some initial functional information, analysis of a mutant virus with a knock-out of a particular gene will often yield few results. At best, such studies will tell us that a mutant is attenuated and that the respective gene represents a virulence factor, but usually give us little insight into the underlying molecular mechanisms.

In contrast, forward genetics appears to be much more promising. Many characteristics and specific functions of CMVs have been defined and now we have the opportunity to find the responsible viral genes. Thus, we have to consider how we can readily measure viral properties and how to set up manageable read-out systems in order to perform screenings of a virus library. We are convinced that by taking this approach, functions can be assigned to many, if not all, CMV genes.

Finally, the BAC-based mutagenesis techniques will also be instrumental for the development of a CMV vaccine. Still, the rational design of a CMV vaccine may have to await a more comprehensive knowledge of the multifarious CMV gene functions, but the door is now open to tackle this task.

## Acknowledgments

We apologize to all colleagues whose publications have not been included owing to space constraints.

The authors thank Ulrich Koszinowski for giving us the opportunity to develop the BAC techniques and for his encouragement.

Our work has been supported by grants from the Deutsche Forschungsgemeinschaft and the Bundesministerium für Forschung und Bildung (BMBF).

## References

Abbate, J., Lacayo, J.C., Prichard, M., Pari, G., and McVoy, M.A. (2001). Bifunctional protein conferring enhanced green fluorescence and puromycin resistance. Biotechniques 31, 336–340.

Akel, H.M. and Sweet, C. (1993). Isolation and preliminary characterization of twenty-five temperature-sensitive mutants of mouse cytomegalovirus. FEMS Microbiol. Lett. 113, 253–260.

Angulo, A., Ghazal, P., and Messerle, M. (2000). The major immediate-early gene ie3 of mouse cytomegalovirus is essential for viral growth. J. Virol. 74, 11129–11136.

Atalay, R., Zimmermann, A., Wagner, M., Borst, E., Benz, C., Messerle, M., and Hengel, H. (2002). Identification and expression of human cytomegalovirus transcription units coding for two distinct Fcgamma receptor homologs. J. Virol. 76, 8596–8608.

Borst, E.M., Hahn, G., Koszinowski, U.H., and Messerle, M. (1999). Cloning of the human cytomegalovirus (HCMV) genome as an infectious bacterial artificial chromosome in Escherichia coli: a new approach for construction of HCMV mutants. J. Virol. 73, 8320–8329.

Borst, E.M., Mathys, S., Wagner, M., Muranyi, W., and Messerle, M. (2001). Genetic evidence of an essential role for cytomegalovirus small capsid protein in viral growth. J. Virol. 75, 1450–1458.

Borst, E.M., Posfai, G., Pogoda, F., and Messerle, M. (2004). Mutagenesis of herpesvirus BACs by allele replacement. In Bacterial Artificial Chromosomes Volume 2: Functional Studies, S. Zhao and M. Stodolsky, eds. (Totowa, NJ: Humana Press Inc.), pp. 269–280.

Borst, E., and Messerle, M. (2005). Analysis of human cytomegalovirus oriLyt sequence requirements in the context of the viral genome. J. Virol. 79, 2998–3008.

Britt, W.J., Jarvis, M., Seo, J.Y., Drummond, D., and Nelson, J. (2004). Rapid genetic engineering of human cytomegalovirus by using a lambda phage linear recombination system: demonstration that pp28 (UL99) is essential for production of infectious virus. J. Virol. 78, 539–543.

Brune, W. (2002). Random transposon mutagenesis of large DNA molecules in Escherichia coli. Methods Mol. Biol. 182, 165–171.

Brune, W., Menard, C., Heesemann, J., and Koszinowski, U.H. (2001). A ribonucleotide reductase homolog of cytomegalovirus and endothelial cell tropism. Science 291, 303–305.

Brune, W., Menard, C., Hobom, U., Odenbreit, S., Messerle, M., and Koszinowski, U.H. (1999). Rapid identification of essential and nonessential herpesvirus genes by direct transposon mutagenesis. Nature Biotechnol. 17, 360–364.

Brune, W., Nevels, M., and Shenk, T. (2003). Murine cytomegalovirus m41 open reading frame encodes a Golgi-localized antiapoptotic protein. J. Virol. 77, 11633–11643.

Bubeck, A., Wagner, M., Ruzsics, Z., Lotzerich, M., Iglesias, M., Singh, I.R., and Koszinowski, U.H. (2004). Comprehensive mutational analysis of a herpesvirus gene in the viral genome context reveals a region essential for virus replication. J. Virol. 78, 8026–8035.

Bubic, I., Wagner, M., Krmpotic, A., Saulig, T., Kim, S., Yokoyama, W.M., Jonjic, S., and Koszinowski, U.H. (2004). Gain of virulence caused by loss of a gene in murine cytomegalovirus. J. Virol. 78, 7536–7544.

Chan, C.K., Brignole, E.J., and Gibson, W. (2002). Cytomegalovirus assemblin (pUL80a): cleavage at internal site not essential for virus growth; proteinase absent from virions. J. Virol. 76, 8667–8674.

Chang, W.L. and Barry, P.A. (2003). Cloning of the full-length rhesus cytomegalovirus genome as an infectious and self-excisable bacterial artificial chromosome for analysis of viral pathogenesis. J. Virol. 77, 5073–5083.

Chee, M.S., Bankier, A.T., Beck, S., Bohni, R., Brown, C.M., Cerny, R., Horsnell, T., Hutchison, C.A., Kouzarides, T., Martignetti, J.A., Preddie, E., Satchwell, S.C., Tomlinson, P., Weston, K.M., and Barrell, B.G. (1990). Analysis of the protein-coding content of the sequence of human cytomegalovirus strain AD169. Curr. Top. Microbiol. Immunol. 154, 125–169.

Court, D.L., Swaminathan, S., Yu, D., Wilson, H., Baker, T., Bubunenko, M., Sawitzke, J., and Sharan, S.K. (2003). Mini-lambda: a tractable system for chromosome and BAC engineering. Gene 315, 63–69.

Datsenko, K.A. and Wanner, B.L. (2000). One-step inactivation of chromosomal genes in Escherichia coli K-12 using PCR products. Proc. Natl. Acad. Sci. USA 97, 6640–6645.

Davison, A.J., Dolan, A., Akter, P., Addison, C., Dargan, D.J., Alcendor, D.J., McGeoch, D.J., and Hayward, G.S. (2003). The human cytomegalovirus genome revisited: comparison with the chimpanzee cytomegalovirus genome. J. Gen. Virol. 84, 17–28.

Dolan, A., Cunningham, C., Hector, R.D., Hassan-Walker, A.F., Lee, L., Addison, C., Dargan, D.J., McGeoch, D.J., Gatherer, D., Emery, V.C., Griffiths, P.D., Sinzger, C., McSharry, B.P., Wilkinson, G.W., and Davison, A.J. (2004). Genetic content of wild-type human cytomegalovirus. J. Gen. Virol. 85, 1301–1312.

Dunn, W., Chou, C., Li, H., Hai, R., Patterson, D., Stolc, V., Zhu, H., and Liu, F. (2003). Functional profiling of a human cytomegalovirus genome. Proc. Natl. Acad. Sci. USA 100, 14223–14228.

Ehsani, M.E., Abraha, T.W., Netherland-Snell, C., Mueller, N., Taylor, M.M., and Holwerda, B. (2000). Generation of mutant murine cytomegalovirus strains from overlapping cosmid and plasmid clones. J. Virol. 74, 8972–8979.

Ellis, H.M., Yu, D., DiTizio, T., and Court D.L. (2001). High efficiency mutagenesis, repair, and engineering of chromosomal DNA using single-stranded oligonucleotides. Proc. Natl. Acad. Sci. USA 98, 6742–6746.

Ghazal, P., Visser, A.E., Gustems, M., Garcia, R., Borst, E.M., Sullivan, K., Messerle, M., and Angulo, A. (2005). Elimination of ie1 significantly attenuates murine cytomegalovirus virulence but does not alter replicative capacity in cell culture. J. Virol. 79, 7182–7194.

Greaves, R.F., Brown, J.M., Vieira, J., and Mocarski, E.S. (1995). Selectable insertion and deletion mutagenesis of the human cytomegalovirus genome using the Escherichia coli guanosine phosphoribosyl transferase (gpt) gene. J. Gen. Virol. 76, 2151–2160.

Greaves, R.F. and Mocarski, E.S. (1998). Defective growth correlates with reduced accumulation of a viral DNA replication protein after low-multiplicity infection by a human cytomegalovirus ie1 mutant. J. Virol. 72, 366–379.

Hahn, G., Jarosch, M., Wang, J.B., Berbes, C., and McVoy, M.A. (2003a). Tn7-mediated introduction of DNA sequences into bacmid-cloned cytomegalovirus genomes for rapid recombinant virus construction. J. Virol. Methods 107, 185–194.

Hahn, G., Khan, H., Baldanti, F., Koszinowski, U.H., Revello, M.G., and Gerna, G. (2002). The human cytomegalovirus ribonucleotide reductase homolog UL45 is dispensable for growth in endothelial cells, as determined by a BAC-cloned clinical isolate of human cytomegalovirus with preserved wild-type characteristics. J. Virol. 76, 9551–9555.

Hahn, G., Revello, M.G., Patrone, M., Percivalle, E., Campanini, G., Sarasini, A., Wagner, M., Gallina, A., Milanesi, G., Koszinowski, U., Baldanti, F., and Gerna, G. (2004). Human cytomegalovirus UL131–128 genes are indispensable for virus growth in endothelial cells and virus transfer to leukocytes. J. Virol. 78, 10023–10033.

Hahn, G., Rose, D., Wagner, M., Rhiel, S., and McVoy, M.A. (2003b). Cloning of the genomes of human cytomegalovirus strains Toledo, TownevarRIT3, and Towne long as BACs and site-directed mutagenesis using a PCR-based technique. Virology 307, 164–177.

Hansen, S.G., Strelow, L.I., Franchi, D.C., Anders, D.G., and Wong, S.W. (2003). Complete sequence and genomic analysis of rhesus cytomegalovirus. J. Virol. 77, 6620–6636.

Hayes, F. (2003). Transposon-based strategies for microbial functional genomics and proteomics. Annu. Rev. Genet. 37, 3–29.

Heider, J.A., Bresnahan, W.A., and Shenk, T.E. (2002a). Construction of a rationally designed human cytomegalovirus variant encoding a temperature-sensitive immediate-early 2 protein. Proc. Natl. Acad. Sci. USA 99, 3141–3146.

Heider, J.A., Yu, Y., Shenk, T., and Alwine, J.C. (2002b). Characterization of a human cytomegalovirus with phosphorylation site mutations in the immediate-early 2 protein. J. Virol. 76, 928–932.

Hobom, U., Brune, W., Messerle, M., Hahn, G., and Koszinowski, U.H. (2000). Fast screening procedures for random transposon libraries of cloned herpesvirus genomes: mutational analysis of human cytomegalovirus envelope glycoprotein genes. J. Virol. 74, 7720–7729.

Holtappels, R., Podlech, J., Pahl-Seibert, M.F., Julch, M., Thomas, D., Simon, C.O., Wagner, M., and Reddehase, M.J. (2004). Cytomegalovirus misleads its host by priming of CD8 T-cells specific for an epitope not presented in infected tissues. J. Exp. Med. 199, 131–136.

Ihara, S., Hirai, K., and Watanabe, Y. (1978). Temperature-sensitive mutants of human cytomegalovirus: isolation and partial characterization of DNA-minus mutants. Virology 84, 218–221.

Jarvis, M.A., Jones, T.R., Drummond, D.D., Smith, P.P., Britt, W.J., Nelson, J.A., and Baldick, C.J. (2004). Phosphorylation of human cytomegalovirus glycoprotein B (gB) at the acidic cluster casein kinase 2 site (Ser900) is required for localization of gB to the trans-Golgi network and efficient virus replication. J. Virol. 78, 285–293.

Kattenhorn, L., Mills, R., Wagner, M., Lomsadze, A., Vsevolod, M., Borodovsky, M., Ploegh, H., and Kessler, B. (2004). Identification of proteins associated with murine cytomegalovirus virions. J. Virol. 78, 11187–11197.

Kemble, G., Duke, G., Winter, R., and Spaete, R. (1996). Defined large-scale alterations of the human cytomegalovirus genome constructed by cotransfection of overlapping cosmids. J. Virol. 70, 2044–2048.

Kempkes, B., Pich, D., Zeidler, R., Sugden, B., and Hammerschmidt, W. (1995). Immortalization of human B lymphocytes by a plasmid containing 71 kilobase pairs of Epstein–Barr virus DNA. J. Virol. 69, 231–238.

Kim, H.J., Gatz, C., Hillen, W., and Jones, T.R. (1995). Tetracycline repressor-regulated gene repression in recombinant human cytomegalovirus. J. Virol. 69, 2565–2573.

Komazin, G., Townsend, L.B., and Drach, J.C. (2004). Role of a mutation in human cytomegalovirus gene UL104 in resistance to benzimidazole ribonucleosides. J. Virol. 78, 710–715.

Kumura, K., Ibusuki, K., and Minamishima, Y. (1990). Independent existence of mutations responsible for temperature sensitivity and attenuation in a mutant of murine cytomegalovirus. Virology 175, 572–574.

Lee, E.C., Yu, D., Martinez, d., V, Tessarollo, L., Swing, D.A., Court DL, Jenkins, N.A., and Copeland, N.G. (2001). A highly efficient Escherichia coli-based chromosome engineering system adapted for recombinogenic targeting and subcloning of BAC DNA. Genomics 73, 56–65.

Lee, H.R. and Ahn, J.H. (2004). Sumoylation of the major immediate-early IE2 protein of human cytomegalovirus Towne strain is not required for virus growth in cultured human fibroblasts. J. Gen. Virol. 85, 2149–2154.

Lee, H.R., Kim, D.J., Lee, J.M., Choi, C.Y., Ahn, B.Y., Hayward, G.S., and Ahn, J.H. (2004). Ability of the human cytomegalovirus IE1 protein to modulate sumoylation of PML correlates with its functional activities in transcriptional regulation and infectivity in cultured fibroblast cells. J. Virol. 78, 6527–6542.

Lembo, D., Donalisio, M., Hofer, A., Cornaglia, M., Brune, W., Koszinowski, U., Thelander, L., and Landolfo, S. (2004). The ribonucleotide reductase R1 homolog of murine cytomegalovirus is not a functional enzyme subunit but is required for pathogenesis. J. Virol. 78, 4278–4288.

Loewendorf, A., Krueger, C., Borst, E.M., Wagner, M., Just, U., and Messerle, M. (2004). Identification of a mouse cytomegalovirus gene selectively targeting CD86 expression on antigen presenting cells. J. Virol. 78, 13062–13071.

Marchini, A., Liu, H., and Zhu, H. (2001). Human cytomegalovirus with IE-2 (UL122) deleted fails to express early lytic genes. J. Virol. 75, 1870–1878.

McGregor, A., Liu, F., and Schleiss, M.R. (2004). Identification of essential and non-essential genes of the guinea-pig cytomegalovirus (GPCMV) genome via transposome mutagenesis of an infectious BAC clone. Virus Res. 101, 101–108.

McGregor, A. and Schleiss, M.R. (2001). Molecular cloning of the guinea-pig cytomegalovirus (GPCMV) genome as an infectious bacterial artificial chromosome (BAC) in Escherichia coli. Mol. Genet. Metab 72, 15–26.

McGregor, A. and Schleiss, M.R. (2004). Herpesvirus genome mutagenesis by transposon-mediated strategies. Methods Mol. Biol. 256, 281–302.

McVoy, M.A. and Mocarski, E.S. (1999). Tetracycline-mediated regulation of gene expression within the human cytomegalovirus genome. Virology 258, 295–303.

Menard, C., Wagner, M., Ruzsics, Z., Holak, K., Brune, W., Campbell, A.E., and Koszinowski, U.H. (2003). Role of murine cytomegalovirus US22 gene family members in replication in macrophages. J. Virol. 77, 5557–5570.

Messerle, M., Crnkovic, I., Hammerschmidt, W., Ziegler, H., and Koszinowski, U.H. (1997). Cloning and mutagenesis of a herpesvirus genome as an infectious bacterial artificial chromosome. Proc. Natl. Acad. Sci. USA 94, 14759–14763.

Mocarski, E.S. and Kemble, G.W. (1996). Recombinant cytomegaloviruses for study of replication and pathogenesis. Intervirology 39, 320–330.

Mocarski, E.S., Kemble, G.W., Lyle, J.M., and Greaves, R.F. (1996). A deletion mutant in the human cytomegalovirus gene encoding IE1(491aa) is replication defective due to a failure in autoregulation. Proc. Natl. Acad. Sci. USA 93, 11321–11326.

Murphy, E., Rigoutsos, I., Shibuya, T., and Shenk, T.E. (2003a). Reevaluation of human cytomegalovirus coding potential. Proc. Natl. Acad. Sci. USA 100, 13585–13590.

Murphy, E., Yu, D., Grimwood, J., Schmutz, J., Dickson, M., Jarvis, M.A., Hahn, G., Nelson, J.A., Myers, R.M., and Shenk, T.E. (2003b). Coding potential of laboratory and clinical strains of human cytomegalovirus. Proc. Natl. Acad. Sci. USA 100, 14976–14981.

Muyrers, J.P., Zhang, Y., Testa, G., and Stewart, A.F. (1999). Rapid modification of bacterial artificial chromosomes by ET-recombination. Nucleic Acids Res. 27, 1555–1557.

Nevels, M., Brune, W., and Shenk, T. (2004). SUMOylation of the human cytomegalovirus 72-kilodalton IE1 protein facilitates expression of the 86-kilodalton IE2 protein and promotes viral replication. J. Virol. 78, 7803–7812.

Posfai, G., Koob, M.D., Kirkpatrick, H.A., and Blattner, F.R. (1997). Versatile insertion plasmids for targeted genome manipulations in bacteria: isolation, deletion, and rescue of the pathogenicity island LEE of the Escherichia coli O157:H7 genome. J. Bacteriol. 179, 4426–4428.

Rawlinson, W.D., Farrell, H.E., and Barrell, B.G. (1996). Analysis of the complete DNA sequence of murine cytomegalovirus. J. Virol. 70, 8833–8849.

Redwood, A.J., Messerle, M., Harvey, N.L., Hardy, C.M., Koszinowski, U.H., Lawson, M.A., and Shellam, G.R. (2005). Use of a murine cytomegalovirus K181-derived bacterial artificial chromosome as a vaccine vector for immunocontraception. J. Virol. 79, 2998–3008.

Rupp. B., Ruzsics, Z., Sacher, T., and Koszinowski, U.H. (2004). Conditional cytomegalovirus replication in vitro and in vivo. J. Virol. 79, 486–494.

Sanchez, V., Clark, C.L., Yen, J.Y., Dwarakanath, R., and Spector, D.H. (2002). Viable human cytomegalovirus recombinant virus with an internal deletion of the IE2 86 gene affects late stages of viral replication. J. Virol. *76*, 2973–2989.

Shizuya, H., Birren, B., Kim, U.J., Mancino, V., Slepak, T., Tachiiri, Y., and Simon, M. (1992). Cloning and stable maintenance of 300-kilobase-pair fragments of human DNA in Escherichia coli using an F-factor-based vector. Proc. Natl. Acad. Sci. USA *89*, 8794–8797.

Silva, M.C., Yu, Q.C., Enquist, L., and Shenk, T. (2003). Human cytomegalovirus UL99-encoded pp28 is required for the cytoplasmic envelopment of tegument-associated capsids. J. Virol. *77*, 10594–10605.

Smith, G.A. and Enquist, L.W. (1999). Construction and transposon mutagenesis in Escherichia coli of a full-length infectious clone of pseudorabies virus, an alpha-herpesvirus. J. Virol. *73*, 6405–6414.

Smith, G.A. and Enquist, L.W. (2000). A self-recombining bacterial artificial chromosome and its application for analysis of herpesvirus pathogenesis. Proc. Natl. Acad. Sci. USA *97*, 4873–4878.

Spaete, R.R. and Mocarski, E.S. (1987). Insertion and deletion mutagenesis of the human cytomegalovirus genome. Proc. Natl. Acad. Sci. USA *84*, 7213–7217.

Strive, T., Borst, E., Messerle, M., and Radsak, K. (2002). Proteolytic processing of human cytomegalovirus glycoprotein B is dispensable for viral growth in culture. J. Virol. *76*, 1252–1264.

Vink, C., Beuken, E., and Bruggeman, C.A. (2000). Complete DNA sequence of the rat cytomegalovirus genome. J. Virol. *74*, 7656–7665.

Wagner, M., Gutermann, A., Podlech, J., Reddehase, M.J., and Koszinowski, U.H. (2002). Major histocompatibility complex class I allele-specific cooperative and competitive interactions between immune evasion proteins of cytomegalovirus. J. Exp. Med. *196*, 805–816.

Wagner, M., Jonjic, S., Koszinowski, U.H., and Messerle, M. (1999). Systematic excision of vector sequences from the BAC-cloned herpesvirus genome during virus reconstitution. J. Virol. *73*, 7056–7060.

Wagner, M. and Koszinowski, U.H. (2004). Mutagenesis of viral BACs with linear PCR fragments (ET recombination). Methods Mol. Biol. *256*, 257–268.

White, E.A., Clark, C.L., Sanchez, V., and Spector, D.H. (2004). Small internal deletions in the human cytomegalovirus IE2 gene result in nonviable recombinant viruses with differential defects in viral gene expression. J. Virol. *78*, 1817–1830.

Wolff, D., Jahn, G., and Plachter, B. (1993). Generation and effective enrichment of selectable human cytomegalovirus mutants using site-directed insertion of the neo gene. Gene *130*, 167–173.

Yamanishi, K. and Rapp, F. (1977). Temperature-sensitive mutants of human cytomegalovirus. J. Virol. *24*, 416–418.

Yewdell, J.W. and Hill, A.B. (2002). Viral interference with antigen presentation. Nat. Immunol. 3, 1019–1025.

Yu, D., Ellis, H.M., Lee, E.C., Jenkins, N.A., Copeland, N.G., and Court, D.L. (2000). An efficient recombination system for chromosome engineering in Escherichia coli. Proc. Natl. Acad. Sci. USA *97*, 5978–5983.

Yu, D., Silva, M.C., and Shenk, T. (2003). Functional map of human cytomegalovirus AD169 defined by global mutational analysis. Proc. Natl. Acad. Sci. USA *100*, 12396–12401.

Yu, D., Smith, G.A., Enquist, L.W., and Shenk, T. (2002). Construction of a self-excisable bacterial artificial chromosome containing the human cytomegalovirus genome and mutagenesis of the diploid TRL/IRL13 gene. J. Virol. *76*, 2316–2328.

Zhang, Y., Buchholz, F., Muyrers, J.P., and Stewart, A.F. (1998). A new logic for DNA engineering using recombination in Escherichia coli. Nat. Genet. *20*, 123–128.

Zimmermann, A., Trilling, M., Wagner, M., Wilborn, M., Bubic, I., Jonjic, S., Koszinowski, U.H., and Hengel, H. (2004). A cytomegaloviral protein reveals a dual role for STAT2 in IFN-γ signaling and antiviral responses. J. Exp. Med. *201*, 1543–1553.

# A Proteomics Analysis of Human Cytomegalovirus Particles

5

*Daniel N. Streblow, Susan M. Varnum, Richard D. Smith, and Jay A. Nelson*

## Abstract

While the sequence of the AD169 HCMV genome has been known for several years, the viral and cellular proteins that compose the infectious HCMV virion and entry-competent, non-replicating viral particles such as Dense Bodies (DBs) and Non-Infectious Enveloped Particles (NIEPs) are unknown. To approach this problem we have utilized a gel-free 2-D capillary liquid chromatography (LC)-MS/MS and LC-Fourier transform ion cyclotron resonance (FTICR) mass spectrometry to identify and determine the relative abundance of viral and cellular proteins in purified HCMV AD169 particles. This study has identified and quantitated the proteins that compose both HCMV virions and DBs. While a number of previously identified proteins were detected by this method, the number of viral proteins that compose the HCMV virion was doubled in this study suggesting that over a third of the viral open reading frames are part of an infectious virion. This chapter discusses the implications of our findings in relation to what was previously known about HCMV and MCMV virion composition.

## Introduction

The herpesvirus family encompasses a group of large complex enveloped viruses that are 150 nm to 300 nm in size and are ubiquitous in almost every species of animal in nature. The human herpesviruses constitute some of the most important known human viral pathogens. Herpesviruses are grouped together based on virion structure that includes a capsid that contains a large double-stranded linear DNA, a tegument composed of phosphoproteins that surrounds the capsid and an envelope that contains multiple glycoprotein complexes that are the receptors of the virus. Subfamilies of the herpesviruses are based on biological properties and structure of the viral genome and include the $\alpha$, $\beta$, and $\gamma$ viruses. The sequence of almost all of the human herpesviruses is known, however, the viral and cellular proteins that compose an infectious virion are unknown.

Human cytomegalovirus (HCMV) is a prototypic $\beta$-herpesvirus that encodes over 200 predicted open reading frames (ORFs) (Chee et al., 1990; Davison et al., 2003; Dunn et al., 2003; Murphy et al., 2003a and b; Yu et al., 2003). The mature HCMV virion is 150–200nm in diameter, and composed of a 100 nm icosahedral capsid that contains a linear 230-kbp double-stranded DNA genome with attached proteins, a large tegument component in comparison to Herpes Simplex Virus (HSV), surrounded by the envelope that contains a cellular lipid bilayer with viral glycoproteins (see Figure 5.1) (Mocarski and Courcelle, 2001). A number of the viral and cellular proteins that compose an HCMV virion have been identified by biochemical and immunological approaches (Baldick and Shenk, 1996; Gibson, 1996). These methods generally are limited to the identification of single abundant species of protein using immunological methods with specific antibodies. However, other approaches have included the sequencing of peptides from disrupted HCMV particles separated on gels.

HCMV-infected cells generate three different types of particles including infectious mature virions described above, non-infectious enveloped particles (NIEPs), and dense bodies (DBs) (Figure 5.2). NIEPs are composed of the same viral proteins as infectious virions and possess a capsid but lack viral DNA. So, by electron microscopy they can be distinguished from mature virions by their lack of an electron dense DNA core (Mocarski and Courcelle, 2001). DBs are uniquely characteristic of HCMV and simian CMV infections and are non-replicating, fusion-competent enveloped particles composed primarily of the tegument protein pp65 (UL83). The quantities of these different HCMV particles are dependent on the viral strain and the multiplicity of infection. MCMV infected cells do not produce DBs; instead, this virus forms multi-capsid virions, which are not observed in HCMV-infected cells.

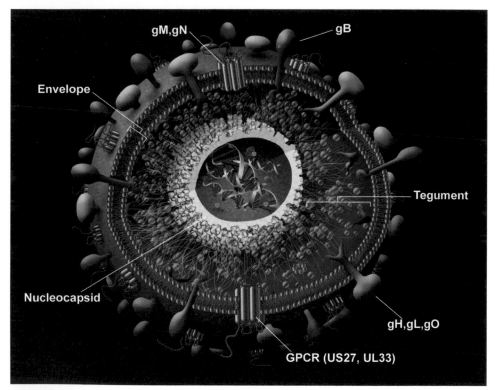

**Figure 5.1** HCMV structure. HCMV virions comprise three major layers. The first layer is the nucleocapsid containing the double-stranded viral DNA genome, which is surrounded by a proteinaceous tegument layer. The tegumented capsids are enveloped by a host-derived lipid bi-layer that is studded with viral glycoproteins.

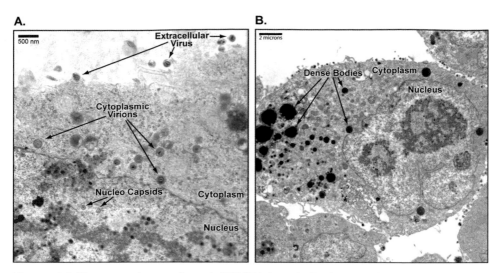

**Figure 5.2** Electron micrographs of HCMV-infected fibroblasts. Transmission electron microscopy of stained ultra-thin sections of HCMV-AD169 infected human dermal fibroblasts. (A) Typical virus production with extracellular virus, enveloped particles in the cytoplasm and nuclear capsids. (B) Normal dense body formation in the infected cell.

The HCMV capsid is assembled in the nucleus from five viral proteins that are encoded by UL86, UL85, UL80, UL48.5, and UL46. The capsid is composed of 162 capsomers made up of hexons and pentons arranged in a T=16 icosahedral lattice structure (Figure 5.3). The capsomers are studded with the smallest capsid protein (UL48.5). The capsid is surrounded by the tegument, which is acquired in both the nucleus and cytoplasm of the infected cell. There are 20–25 virion-associated tegument proteins, many of which are phosphorylated and have unknown functions (Mocarski and Courcelle, 2001). Some of the more prominent tegument proteins include UL83 (pp65), UL82 (pp71), UL99 (pp28), UL32 (pp150), UL48, UL69, UL82, TRS1, and IRS1. Cytoplasmic viral capsids containing tegument acquire the envelope by budding into the Trans Golgi Network (TGN) or a closely apposed cellular compartment. The envelope contains 8 virally encoded glycoproteins, including gB (UL55), gM (UL100), gH (UL75), gL (UL115), gO (UL74), gN (UL73), gp48 (UL4), and UL33 (Mocarski and Courcelle, 2001). In total, approximately 35 virally encoded proteins have been identified in HCMV virions (Baldick and Shenk, 1996; Gibson, 1996). In addition to virally encoded structural proteins, a small number of cellular proteins, including CD13 (aminopeptidase N), $\beta_2$-microglobulin, protein phosphatase I, annexin II and actin-related protein 2/3 (Arp2/3) have also been detected in virion preparations (Baldick and Shenk, 1996; Giugni et al., 1996; Grundy et al., 1987; Michelson et al., 1996; Stannard, 1989; Wright et al., 1995). The numbers of viral and cellular proteins that compose an infectious virion are controversial and depend on the stringency of the isolation procedures. Unfortunately, as viral purification procedures become more stringent, virion preparations tend to lose infectivity. This situation makes identification of essential virion proteins difficult to assess except by abundance.

The development of sophisticated liquid phase separations and mass spectrometry (MS)-based approaches for analyzing complex mixtures of peptides has allowed a largely

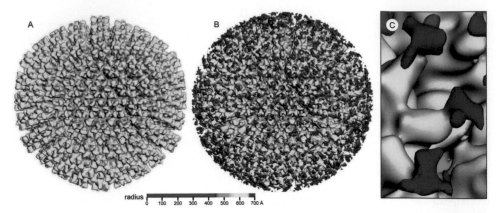

**Figure 5.3** 3D reconstructions of the HCMV capsid (A) Capsid decorated by antibodies against the smallest capsid protein pORF48.5 (B). The structures were determined by electron cryomicroscopy to 22-Å resolution and are colored according to radius (see color bar) so that densities at different radial distances from the particle center are colored differently. The antibody densities are shown in purple and are attached only to major capsid protein (MCP, pORF86) subunits of hexons, not pentons. (C) Close-up of a region clearly showing attachment of antibody to the tips of hexon subunits. (Modified by Z. H. Zhou from Yu et al., 2003, with permission from the publisher.)

unbiased analysis of the proteins in complex organisms such as bacteria and large viruses, as well as cells. The MS-based proteomics approaches have led to the determination of a number of first-generation proteome maps of numerous prokaryotic and eukaryotic species including *Deinococcus radiodurans*, *Plasmodium falciparum*, yeast, and adenovirus (Chelius et al., 2002; Florens et al., 2002; Lasonder et al., 2002; Lipton et al., 2002; Washburn et al., 2001). One MS approach exploits high accuracy mass measurements with enhanced dynamic range using liquid chromatography Fourier transform ion cyclotron resonance mass spectrometry (LC-FTICR) to make very high confidence peptide assignments for an organism, from potential mass and time tags identified using LC-tandem mass spectrometry (LC-MS/MS). This two stage approach provides greater confidence in the identifications, as well as the basis for subsequent higher sensitivity and throughput measurements without the need for routine MS/MS (Smith et al., 2002a; Smith et al., 2002b). We and others have applied the LC-MS based methods to analyze the proteome of HCMV virions and DBs, as well as mouse CMV (MCMV) virions (Kattenhorn et al., 2004; Varnum et al., 2004). Additionally, in our analysis of HCMV particles an algorithm was generated to compare the peptide ion intensities to determine the relative abundances of proteins isolated from virion and DB particles. These analyses resulted in the identification of both new and unknown viral ORFs that encode peptides in the virion and demonstrate the utility of this approach for studying viruses.

## HCMV proteins in virion particles

We determined the protein composition of the HCMV particles purified from the supernatant culture fluids of human fibroblas T-cells infected with HCMV strain AD169 utilizing high-throughput MS (Varnum et al., 2004). This HCMV strain was selected because of the availability of the complete sequence of the viral genome (Chee et al., 1990) that is essential in the analysis of peptides encoded by the viral genome. HCMV particles used for MS analysis were obtained from AD169 infected supernatants containing virions and DBs purified to greater than 96% by sequential sedimentation and density ultracentrifugation gradients as determined by electron microscopy. Electron micrographs of the purified HCMV virion and DB particles are shown in Figure 5.4A as well as SDS-PAGE analysis of the same preparations in Figure 5.4B.

To determine the peptide content of the HCMV particles, purified viral preparations were digested to yield a complex mixture of polypeptides that was analyzed by a two-stage MS approach. The first stage employed both one-dimensional and two-dimensional LC (MudPIT) coupled to MS/MS (Adkins et al., 2002; Link et al., 1999). The results from the LC-MS/MS analysis were verified and extended by employing high-accuracy mass measurements using FTICR combined with chromatographic elution time information. Using this approach we identified 59 proteins, including tentative evidence for 12 proteins encoded by known HCMV ORFs that were not previously shown to reside in virions (Table 5.1). The known virion proteins identified included five capsid proteins (UL46, UL48–49, UL80, UL85, and UL86), fourteen tegument proteins (UL24, UL25, UL26, UL32, UL43, UL47, UL48, UL82, UL83, UL94, UL99, US22, US23, and US24), eleven glycoproteins (RL10, UL22A, UL41A, UL55, UL73, UL74, UL75, UL77, UL100, UL115, and UL119), thirteen proteins involved in DNA replication and transcription (IRS1, TRS1, UL44, UL45, UL54, UL57, UL69, UL72, UL77, UL84, UL89, UL97, and

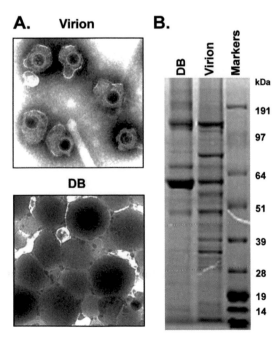

**Figure 5.4** Characterization of HCMV virion and dense body preparations. (A) Transmission electron microscopy of HCMV virion and dense body (DB) preparations (magnification = 84 00×). (B) Analysis of the proteins that constitute the purified HCMV particle preparations. Proteins were separated using NuPAGE MOPS gradient gels and visualized by Coomassie blue staining.

**Table 1. HCMV proteins found in virions.** HCMV proteins isolated from virions that were identified by LC-MS/MS and FTICR.

| Group | HCMV ORF | Comments | LCQ MS/MS # Different Peptides | LCQ MS/MS Max XCorr | FTICR # Different Peptides | FTICR Percent Coverage |
|---|---|---|---|---|---|---|
| Capsid | UL46 | Minor Capsid Binding Protein | 20 | 5.30 | 14 | 44.8 |
| | UL48-49 | Smallest Capsid Protein | 8 | 6.52 | 5 | 54.7 |
| | UL80 | Assemblin Precursor | 37 | 6.36 | 30 | 35.6 |
| | UL85 | Minor Capsid Protein | 21 | 6.73 | 22 | 63.1 |
| | UL86 | Major Capsid Protein | 149 | 3.97 | 123 | 71.0 |
| Tegument | UL24 | Tegument Protein | 8 | 5.06 | 9 | 38.3 |
| | UL25 | Tegument Protein | 60 | 7.04 | 59 | 59.2 |
| | UL26 | | 9 | 4.77 | 10 | 53.7 |
| | UL32 | Tegument Protein, pp150 | 135 | 3.01 | 100 | 70.5 |
| | UL43 | Tegument Protein, US22 Family | 7 | 5.50 | 10 | 28.1 |
| | UL47 | Tegument Protein | 53 | 6.10 | 64 | 57.5 |
| | UL48 | Large Tegument Protein | 111 | 4.29 | 109 | 56.8 |
| | UL82 | Upper matrix phosphoprotein, pp71 | 70 | 6.39 | 47 | 69.3 |
| | UL83 | Lower matrix phosphoprotein, pp83 | 123 | 5.44 | 86 | 92.0 |
| | UL94 | Tegument Protein | 10 | 5.08 | 12 | 26.4 |
| | UL99 | Tegument Protein, pp28 | 8 | 5.87 | 9 | 64.7 |
| | US22 | Tegument Protein | 2 | 3.16 | 2 | 5.4 |
| | US23* | Tegument Protein, US22 Family | 1 | 2.61 | 1 | 4.6 |
| | US24* | Tegument Protein, US22 Family | 1 | 4.83 | 2 | 7.0 |
| Glycoproteins | RL10 | | 5 | 2.36 | 4 | 22.8 |
| | TRL14* | | * | | 1 | 7.5 |
| | UL5* | | * | | 1 | 5.4 |
| | UL22A | | 1 | 5.04 | 1 | 19.4 |
| | UL33 | G-protein Coupled Receptor | 4 | 6.11 | 4 | 14.1 |
| | UL38* | | * | | 1 | 5.7 |
| | UL41A | | 2 | 5.72 | 2 | 25.6 |
| | UL50* | | 1 | 2.82 | 4 | 10.6 |
| | UL55 | gB | 21 | 6.16 | 23 | 24.8 |
| | UL73 | gN | 2 | 3.47 | 2 | 6.5 |
| | UL74 | gO | 4 | 5.07 | 4 | 13.5 |
| | UL75 | gH | 21 | 6.15 | 22 | 35.7 |
| | UL77 | Pyruvoyl decarboxylase | 14 | 5.65 | 12 | 31.2 |
| | UL93 | | 15 | 5.35 | 14 | 31.7 |
| | UL100 | gM | 13 | 5.24 | 7 | 15.9 |
| | UL115 | gL | 11 | 4.73 | 9 | 47.1 |
| | UL119 | | 2 | 2.23 | 1 | 4.6 |
| | UL132* | | 8 | 5.89 | 8 | 47.0 |
| | US27 | G-protein Coupled Receptor | 4 | 4.25 | 2 | 7.7 |
| Transcription/ Replication Machinery | IRS1 | Viral Gene Transactivator | 15 | 6.01 | 17 | 25.8 |
| | TRS1 | Viral Gene Transactivator | 10 | 6.92 | 23 | 34.7 |
| | UL44 | DNA Processivity Factor | 1 | 4.32 | 9 | 31.0 |
| | UL45 | Ribonucleotide Reductase | 43 | 5.85 | 52 | 52.2 |
| | UL54 | DNA polymerase | * | | 1 | 1.6 |
| | UL57 | ssDNA Binding Protein | * | | 1 | 0.4 |
| | UL69 | Tegument Protein, Viral Transactivator | 6 | 4.17 | 7 | 19.0 |
| | UL72 | dUTPase | * | | 1 | 4.6 |
| | UL84 | DNA Replication | 1 | 2.50 | 3 | 12.8 |
| | UL89 | | * | | 1 | 3.1 |
| | UL97 | Phosphotransferase | 13 | 5.95 | 9 | 32.1 |
| | UL122 | Viral And Cellular Gene Transactivator, IE2 | 2 | 4.26 | 4 | 11.7 |
| Uncharacterized | UL35 | UL25 Family | 42 | 6.27 | 40 | 56.1 |
| | UL51 | | * | | 1 | 3.2 |
| | UL71* | | 12 | 6.32 | 11 | 40.4 |
| | UL79* | | * | | 1 | 10.9 |
| | UL88 | | 14 | 6.8 | 17 | 33.6 |
| | UL96 | | 1 | 4.46 | 1 | 19.7 |
| | UL103* | | 8 | 5.18 | 8 | 37.0 |
| | UL104 | | 9 | 4.68 | 9 | 23.0 |
| | UL112 | | 1 | 3.30 | 4 | 4.7 |

* Denotes proteins newly identified associated with HCMV virions.

UL122), and two G-protein-coupled proteins (UL33 and US27) (Table 5.1). This analysis also identified twelve HCMV-encoded polypeptides not previously associated with the virion including UL5, UL38, UL50, UL71, UL79, UL93, UL96, UL103, UL132, US23, US24, and TRL14. Of these, UL71, UL93, and UL103 are high confidence identifications, while the rest are more tentative at this point. Nine of these HCMV-encoded polypeptides

(UL38, UL50, UL71, UL79, UL93, UL96, UL102, US23, US24) are required for efficient virus growth in cultured fibroblasts (Dunn et al., 2003; Yu et al., 2003). Although UL4, UL23, UL53, UL56, UL98a, and US28 had previously been reported to be associated with HCMV particles, none of the peptides detected from these proteins met our criteria for inclusion in our database.

In order to evaluate the presence of small polypeptides in the virion not represented by ORFs in the annotated HCMV genome, a database of stop-to-stop (StoS) protein-coding regions was established from all reading frames of the HCMV genome that were 20 amino acids (aa) or longer in length. The MS/MS spectra were analyzed against the predicted short polypeptides from this StoS database and polypeptides corresponding to 12 short ORFs were identified, including several with very high confidence, that had not been previously characterized (Figure 5.5) Six of these short polypeptides either had very high confidence identifications or were identified from multiple different smaller peptides, and were presumably more abundant. None of these new short polypeptides corresponded to the candidate ORFs described by Murphy et al. (Murphy et al., 2003a), in part because the referenced analysis considered ORFs ≥ 50 aa. BLAST analysis revealed that all of the identified AD169 StoS sequences are present in the HCMV strains TR, PH, FIX, Merlin, Toledo, and Towne and are between 97% and 100% identical to the StoS at the DNA level. The shortest of the small polypeptides detected in this study was 22 aa (StoS-1-0779; 66 bp), and the longest detected was 190 aa (StoS-1-0415, 570 bp) in length, overlapping, but out of frame with UL31. What is unclear from these observations is whether the proteins encoded by the small ORFs are unique or part of larger proteins, which is the subject of future studies.

## Relative quantitation of virion proteins

The relative abundance of each HCMV protein was determined using relative peptide intensities from FTICR spectra and is shown in Table 5.2 as a percentage of the sum of all the virion proteins. These analyses indicated that the virion was composed of 50% tegument proteins, 30% capsid proteins, 13% envelope proteins, and 7% unassigned proteins. The most abundant protein in the virion was the pp65 (UL83) tegument protein which was present in a molar ratio of ~2:1 with UL86 (major capsid protein, MCP). These findings are similar to previous observations using different quantitative techniques (Baldick and Shenk, 1996; Irmiere and Gibson, 1983; Landini et al., 1987; Schmolke et al., 1995). Similarly, we observed that UL85 (minor capsid protein, mCP) was present in a 2:1 molar ratio with UL46 (mCP-binding protein) and that UL82 (pp71) and UL86 (MCP) were in approximately equimolar amounts. Both of these observations are consistent with previous reports (Irmiere and Gibson, 1983).

Surprisingly, ~8% of the total peptide content of the virion preparations was composed of UL80. UL80 encodes a protein that is cleaved into the assembly protein (AP UL80.5) at the carboxyl terminus (UL80-aa336–708) and the protease (assemblin, UL80A) at the amino terminus (UL80-aa1–256) that is auto-catalytically cleaved into two more peptides (Gibson, 1996; Mocarski and Courcelle, 2001). In our study all of the UL80 peptides identified in the virion preparations were derived from AP (UL80 regions aa431–486 and aa566–578). The controversy in this observation is that the AP polypeptide is a major component of immature B capsids that lack viral DNA. This protein is believed to be lost

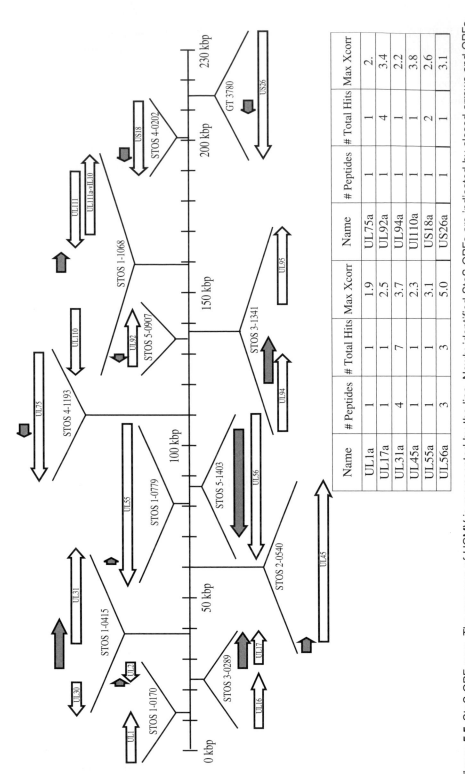

**Figure 5.5** StoS ORF map. The genome of HCMV is represented by the line. Newly identified StoS ORFs are indicated by closed arrows and ORFs that were recognized previously in AD169 are marked by open arrows. Identifications which are based on a single peptide are of lower confidence. The table (bottom right) gives the number of different peptides identified, the number of total identifications from the analyses, and the SEQUEST maximum Xcorr value.

| Name | # Peptides | # Total Hits | Max Xcorr | Name | # Peptides | # Total Hits | Max Xcorr |
|------|-----------|--------------|-----------|------|-----------|--------------|-----------|
| UL1a | 1 | 1 | 1.9 | UL75a | 1 | 1 | 2. |
| UL17a | 1 | 1 | 2.5 | UL92a | 1 | 4 | 3.4 |
| UL31a | 4 | 7 | 3.7 | UL94a | 1 | 1 | 2.2 |
| UL45a | 1 | 1 | 2.3 | Ul110a | 1 | 1 | 3.8 |
| UL55a | 1 | 1 | 3.1 | US18a | 1 | 2 | 2.6 |
| UL56a | 3 | 3 | 5.0 | US26a | 1 | 1 | 3.1 |

**Table 2. Average abundance for HCMV proteins associated with virions and dense bodies**

| HCMV ORF | Virion Average Abundance | Dense Bodies Average Abundance |
|---|---|---|
| UL83 | 15.4 | 60.2 |
| UL48-49 | 12.6 | 4.4 |
| UL100 | 9.2 | 4.4 |
| UL32 | 9.1 | 2 |
| UL82 | 8.9 | 1.7 |
| UL48 | 8.8 | 0.8 |
| UL80 | 7.7 | 1.1 |
| UL86 | 6 | 1.5 |
| UL45 | 4.7 | 0.9 |
| UL85 | 2.8 | 1 |
| UL25 | 2.2 | 12.7 |
| UL46 | 1.5 | 0.3 |
| UL47 | 1.5 | 0.3 |
| UL55 | 1.4 | 0.3 |
| UL94 | 1.2 | 1 |
| IRS1 | 0.8 | 0.3 |
| UL88 | 0.7 | 0.1 |
| UL77 | 0.6 | <0.1 |
| TRS1 | 0.6 | 0.3 |
| UL75 | 0.6 | 0.6 |
| UL35 | 0.5 | 1 |
| UL115 | 0.5 | 0.3 |
| UL119 | 0.5 | <0.1 |
| UL132 | 0.4 | 0.2 |
| UL104 | 0.4 | <0.1 |
| US27 | 0.2 | <0.1 |
| RL10 | 0.2 | 0.1 |
| UL72 | 0.2 | <0.1 |
| UL22A | 0.2 | <0.1 |
| UL97 | 0.1 | 0.5 |
| UL71 | 0.1 | <0.1 |
| UL26 | 0.1 | 2.4 |
| US22 | 0.1 | <0.1 |
| UL103 | 0.1 | 0.1 |
| UL24 | 0.1 | 0.5 |
| UL44 | 0.1 | 0.4 |
| UL33 | 0.1 | <0.1 |
| UL43 | <0.1 | 0.2 |
| UL69 | <0.1 | 0.1 |
| UL93 | <0.1 | 0.1 |
| UL99 | <0.1 | 0.5 |

Abundances are based upon a percentage of the total protein concentration present derived from integrated peptide MS peak intensitites. Detection limit was ≤0.1.

during the maturation of the HCMV B capsid to the C capsid that eventually develops into the infectious virion. However, some B capsids form noninfectious enveloped particles (NIEPs) that may have contaminated our viral preparations. However, when we examined our virus preparations by EM, undetectable amounts of NIEPs were observed in our samples which indicated that AP is a component of the infectious particle. Consistent with this argument, if NIEPs were contributing to the presence of AP, then we would expect that peptides representing assemblin would be present, and this protein was not detected in the virion preparations. This observation was important since AP is a potential antiviral target.

Another interesting observation in our studies was that the most abundant HCMV envelope glycoprotein was UL100 (gM), which constituted 10% of the total virion protein molecules. This observation was surprising since previous reports have suggested that the most predominant envelope glycoprotein was UL55 (gB), reported here to be 10-fold less or 1% of the total virion protein content. The discrepancy in the observations may be accounted for by the highly hydrophobic nature of gM, which contains at least eight trans-membrane domains, making this protein difficult to detect by gel electrophoresis (Lehner et al., 1989). Although gM complexes with UL73 (gN), this glycoprotein represented only 0.1% of the peptide content in our virion preparations. Lastly, UL75 (gH) and UL115 (gL) that exist as a glycoprotein complex (Huber and Compton, 1998; Kaye et al., 1992) were detected at an equal molar ratio. However, UL74 (gO) that is also a member of this complex was detected at lower levels (<0.1% of the total peptides detected). The unexpected lower levels of gN and gO in virions detected by FTICR may be due to the highly glycosylated state of these molecules, which may inhibit complete trypsinization, or due to factors affecting the LC separation or MS detection (Huber and Compton, 1999; Mach et al., 2000).

## HCMV proteins in dense body (DB) particles

Mass spectrometric analysis of DB preparations indicated that these particles were composed of 21 viral proteins including five capsid proteins, nine tegument proteins, four glycoproteins, and four proteins involved in virus transcription and replication (Table 5.3). As previously described, UL83 (pp65) was the predominant protein present in DBs. However, representing 60% of the total protein, this protein was less abundant than indicated by other studies (Baldick and Shenk, 1996; Irmiere and Gibson, 1983). Interestingly, UL25 was significantly more abundant in DBs than in virions (13% in DBs versus 2.2% in virions) (Baldick and Shenk, 1996). Another DB abundant protein was UL26 that represented

**Table 3. HCMV proteins isolated from dense bodies that were identified by LC-MS/MS and FTICR.**

| Viral Protein Group | HCMV ORF | # Unique Peptides | Max Xcorr | FTICR different peptides identified |
|---|---|---|---|---|
| Capsid | UL46 | 1 | 3.6 | 6 |
| | UL48A | 1 | 5.8 | 1 |
| | UL80 | 1 | 6.1 | 2 |
| | UL85 | 4 | 5.0 | 4 |
| | UL86 | 22 | 5.0 | 19 |
| Tegument | UL24 | 1 | 1.9 | 2 |
| | UL25 | 17 | 6.3 | 13 |
| | UL26 | 3 | 3.6 | 3 |
| | UL32 | 11 | 5.4 | 15 |
| | UL35 | 5 | 5.6 | 9 |
| | UL43 | 1 | 3.9 | 1 |
| | UL47 | 2 | 4.3 | 6 |
| | UL48 | 7 | 5.4 | 12 |
| | UL71 | 2 | 5.2 | ND |
| | UL82 | 9 | 5.1 | 6 |
| | UL83 | 40 | 6.3 | 14 |
| | UL94 | 1 | 4.3 | 1 |
| | UL99 | 1 | 2.4 | ND |
| Glycoproteins | UL55 | * | * | 1 |
| | UL74 | 1 | 4.2 | ND |
| | UL75 | 4 | 5.6 | 2 |
| | UL100 | * | * | 3 |
| | UL132 | * | * | 1 |
| | US27 | * | * | 1 |
| | UL69 | 1 | 3.4 | ND |
| Transcription/ Replication Machinery | UL45 | 2 | 4.3 | 6 |
| | UL97 | * | * | 4 |
| | TRS1 | 1 | 4.7 | 5 |
| | IRS1 | 3 | 5.6 | 2 |
| Uncharacterized | UL88 | * | * | 1 |
| | UL103 | * | * | 1 |

\* Identified peptides do not meet minimal criteria for positive identification.
ND: not detected by FTICR.

2.4% of the protein in DBs in comparison to 0.1% of protein in virions. Although the presence of UL26 in DBs was previously reported, this tegument protein was not detected in virions in the earlier work (Baldick and Shenk, 1996), Another HCMV protein detected in DBs was the tegument phosphoprotein UL32 (pp150). In contrast to virions in which pp150 was 9.1% of the particle protein mass, UL32 accounted for 2% of the DB particles. While the amount of UL32 in the virion was lower than previously estimated, our analysis indicates that pp150 is preferentially incorporated into virions rather than DBs (Benko et al., 1988). Additionally, the presence of five nucleocapsid proteins, UL46, UL48–9, UL80, UL85 and UL86, in the DB preparations is perhaps surprising, given that DBs are reported to lack a nucleocapsid (Gibson, 1983; Sarov and Abady, 1975). However, at least two of these proteins, UL85 and UL86, have previously been detected in DBs (Baldick and Shenk, 1996). The five capsid proteins represent a relatively small portion of the total DB protein (~7.8%), whereas these same five proteins represent ~24% of the total protein isolated from virion particles.

With regard to HCMV glycoproteins detected in DBs, UL55, UL100, and UL132 decreased in relative abundance by 4-fold, 2-fold, and 2-fold respectively, while UL75 and UL115 were approximately equivalent in comparison to the abundance of these glycoproteins in virions. The total percentage of viral glycoproteins in the DB preparations was 5.2% compared to 13% in virion preparations. This discrepancy may reflect the variation in size as well as high tegument composition (85.5%) of the DBs. Interestingly, TRL10, UL22A, UL33, UL77, UL119, and US27 that were present in virion particles were not detected in the DB preparations. These observations suggest that these glycoproteins are present in low concentrations in DBs and may not be essential for the functional properties of DBs.

## Host proteins associated with HCMV particles

Another area of controversy in the herpesvirus field centers on the host proteins that are incorporated into HCMV particles. In our studies of HCMV virions a total of 71 host cell proteins were detected with high confidence, significantly increasing the number of previously identified virion proteins (Table 5.4) (Baldick and Shenk, 1996; Giugni et al., 1996; Grundy et al., 1987; Michelson et al., 1996; Mocarski and Courcelle, 2001; Stannard, 1989; Wright et al., 1995). These HCMV cellular virion proteins included cytoskeletal proteins such as $\alpha$- and $\beta$-actin, tubulin, several annexins, $\alpha$-actinin and vimentin, as well as cellular proteins involved in translational control including initiation and elongation factors. Other cellular proteins identified in HCMV virion preparations included clathrin and ADP-ribosylation factor 4. These proteins are involved in vesicular trafficking in the endoplasmic reticulum and Golgi suggesting a role for these proteins in viral envelopment and/or egress. In addition to the above cellular virion proteins four isoforms of the signal transduction protein 14-3-3 were also identified in HCMV preparations together with other signaling proteins such as RasGAP, Casein kinase 2, and $\beta_2$-GTO-binding regulatory protein. Interestingly, although $\beta_2$-microglobulin was previously reported to be a component of the HCMV virion (Grundy et al., 1987; Stannard, 1989), this cellular protein was either present in low amounts or absent in virion preparations and thus eliminated from our database based on the criteria for this study. The presence of cellular proteins in virion preparations is controversial and may be attributed to co-purification of cellular components in virion preparations, or to purified virions sticking to cellular proteins, or that the cellular proteins

**Table 4. Selected host proteins identified by category**

| | |
|---|---|
| ATP binding | DEAD/H (Asp-Glu-Ala-Asp/His) box polypeptide 1, Sodium/Potassium-Transporting ATPase |
| Ca+ Binding | Annexin I, Annexin V, Annexin VI, Annexin A2, Calreticulin, alpha-Actinin 1 |
| Chaperone | Cyclophilin A, Glucose-regulated Protein, Heat shock 70kD protein, Heat-shock protein 90kD, Tumor rejection antigen |
| Cytoskeleton | $\alpha$-Actin, $\beta$-Actin, Cofilin, Filamin, Keratin, Moesin, $\alpha$-Tubulin, $\beta$-Tubulin, Vimentin |
| Enzymes | Aminopeptidase N, Transketolase, Vinculin |
| Glycolysis | Enolase, Glyceraldehyde-3-phosphate dehydrogenase, Lactate dehydrogenase, Phosphoglycerate kinase, Serine/threonine protein phosphatase PP1, Triosephosphate isomerase |
| Protein Transport | ADP-ribosylation factor 4, Clathrin, Polyubiquitin 3, $\beta$- RAB GDP dissociation inhibitor |
| Signal Transduction | $\alpha_1$-Casein kinase 2, 14-3-3 protein (four isoforms) |
| Transcription/Translation | Eukaryotic translation elongation factor 2, Eukaryotic translation elongation factor 1, Enhancer protein, Eukaryotic translation initiation factor 4A, |

are constituents of the viral particles. However, the absence of $\beta_2$-microglobulin and other cellular proteins in HCMV-purified virion preparations suggests that the cellular proteins identified in this study are not contaminants in the virion preparations. In addition, examination of the viral preparations by EM indicated undetectable amounts of cellular organelles and debris suggesting that our purification scheme provided virions lacking these cellular contaminants.

With regard to the analysis of cellular proteins in DBs, only a small number of host cell proteins were identified in these HCMV particles including glyceraldehyde-3-phosphate dehydrogenase, annexin A2, $\beta$-actin and heat shock 70 kDa protein (Table 5.5). Interestingly, the cellular protein composition of DBs differed markedly from the cellular proteins in virion preparations. A possible explanation for this observation is that DB formation and egress is fundamentally different from the envelopment and egress of virion particles.

## Comparison of the HCMV and MCMV proteomes

Concurrent with our analysis of the HCMV virion proteome, a group led by Drs. Kessler and Ploegh at Harvard University identified the proteins that compose the Smith strain of murine CMV (MCMV) virion (Kattenhorn et al., 2004). Since the MCMV genome was not annotated, a putative MCMV ORF database was generated using the gene prediction algorithm GeneMark. Similar to HCMV, the MCMV genome is 230 kbp and contains 170 predicted ORFs of >100 aa. However, at the genome level MCMV and HCMV only share about 42.5% sequence identity, and of the 170 ORFs only 78 have significant amino acid identity with proteins from HCMV. One distinguishing feature of MCMV is the fact that unlike HCMV, MCMV does not produce DBs. Since HCMV mutants lacking pp65 do not form DBs, the inability of MCMV to generate DBs suggests that functional differences exist in pp65 homologs, which is highlighted by their low sequence similarity.

In the Harvard study, purified MCMV preparations were denatured and analyzed using two different sample preparation methods prior to MS including the traditional SDS-PAGE separation followed by in-gel digestion and the tryptical digestion in-solution. The MCMV tryptic peptides were separated by nano-flow liquid chromatography and analyzed by LC-MS/MS. The two methods of peptide preparation yielded highly different levels of detection with the in-solution digestion method identifying 58 viral proteins in contrast to only 19 MCMV proteins detected by in-gel digestions. This study confirms our find-

**Table 5. Host cell proteins identified by MS/MS that are associated with isolated HCMV dense bodies**

| NCBI Accession | Description | LC-IM Trap MS MS | |
|---|---|---|---|
| | | Max Xcorr* | #Unique Peptides |
| 1065361 | Adp-Ribosylation Factor 1 | 4.1 | 1 |
| 4757756 | Annexin A2 | 4.9 | 1 |
| 4501885 | Actin-Beta | 4.2 | 2 |
| 1070613 | Actin-Alpha 2 | 4.1 | 1 |
| 420027 | Collagen alpha 2 chain | 2.4 | 1 |
| 4503295 | Dead/H (Asp-Glu-Ala-Asp/His) Box Polypeptide 3 | 3.4 | 1 |
| 3183005 | Eyes absent homolog | 3.9 | 1 |
| 4503471 | Eukaryotic Translation Elongation Factor 1 alpha 1 | 3.4 | 1 |
| 5031863 | Galectin 6 Binding Protein | 3.9 | 1 |
| 7669492 | Glyceraldehyde-3-Phosphate Dehydrogenase, Liver | 6.1 | 3 |
| 120643 | Glyceraldehyde-3-Phosphate Dehydrogenase, Muscle | 2.4 | 1 |
| 5729877 | Heat Shock 70Kd Protein 10 (Hsc71) | 3.9 | 1 |
| 7512592 | Hypothetical protein DKFZp434G173.1 | 3.1 | 1 |

*Maximal SEQUEST Xcorr value observed.

ings that the sensitivity of detecting proteins associated with viral preparations is far better when using the in-solution digestion approach. A total of 38 MCMV proteins of known function were identified with this approach including capsid, tegument, envelope, replication and immunomodulatory family members. Peptides encoded by 20 other MCMV ORFs without a known function were also identified in this study.

According to this analysis, MCMV contains four capsid proteins (m48.2, the smallest capsid protein; M85, the minor capsid protein; M86, the major capsid protein; and M46, the minor capsid binding protein) in contrast to five HCMV capsid proteins. The MCMV virion proteins detected in this study that have homologs of HCMV virion proteins detected in our studies are listed in Table 5.6. For convenience and consistency with previous publications we have designated the MCMV ORFs that have homologs in HCMV with an upper-case "M", whereas those without HCMV counterparts are designated with a lower-case "m". Similar to our analysis of HCMV, the viral assembly protein/protease M80 was detected in MCMV virions. Also similar to our studies, ten tegument proteins associated with MCMV particles were detected in virion preparations including M25 (UL25), M32 (pp150), M35, M47 (high molecular weight binding protein), M48 (high molecular weight tegument protein), M51, M82 (pp71), M83 (pp65), M94, and M99 (pp28). In MCMV virions, four glycoproteins were detected, including M55 (gB), M74 (gO), M75 (gH), and M100 (gM). Similar to HCMV a number of MCMV replication proteins were also detected in virion preparations including M54 (DNA polymerase), M44 (polymerase accessory protein), M57 (major DNA binding protein), M69 (IE transactivator), M70, M102, and M105 (helicase/primase complex). In addition, other MCMV genes involved in DNA

**Table 6. Homologous viral proteins detected by MS analysis in HCMV and MCMV.**

| Virus Detected | MCMV ORF | HCMV ORF | Comments |
|---|---|---|---|
| MCMV/HCMV | M25 | UL25 | Tegument Protein |
| | m25.2 | US22 | Tegument Protein |
| | M32 | UL32 | Tegument Protein, pp150 |
| | M35 | UL35 | Tegument Protein, UL25 Family |
| | M43 | UL43 | Tegument Protein, US22 Family |
| | M44 | UL44 | DNA Processivity Factor |
| | M45 | UL45 | Ribonucleotide Reductase |
| | M46 | UL46 | Minor Capsid Binding Protein |
| | M47 | UL47 | Tegument Protein |
| | M48 | UL48 | Large Tegument Protein |
| | m48.2 | UL48/49 | Smallest Capsid Protein |
| | M51 | UL51 | |
| | M54 | UL54 | DNA Polymerase |
| | M55 | UL55 | gB |
| | M57 | UL57 | ssDNA Binding Protein |
| | M69 | UL69 | Viral Transactivator |
| | M71 | UL71 | |
| | M72 | UL72 | dUTPase |
| | m74 | UL74 | gO |
| | M75 | UL75 | gH |
| | M77 | UL77 | Pyruvoyl decarboxylase |
| | M80 | UL80 | Assembly/Protease |
| | M82 | UL82 | Upper matrix phosphoprotein, pp71 |
| | M83 | UL83 | Lower matrix phosphoprotein, pp83 |
| | M85 | UL85 | Minor Capsid Protein |
| | M86 | UL86 | Major Capsid Protein |
| | M88 | UL88 | Tegument Protein |
| | M94 | UL94 | Tegument Protein |
| | M97 | UL97 | Phosphotransferase |
| | M99 | UL99 | Tegument Protein, pp28 |
| | M100 | UL100 | gM |
| | M104 | UL104 | Structural Protein |
| MCMV | M28 | | |
| | M31 | | |
| | M56 | | Terminase |
| | M70 | | Helicase/Primase Subunit |
| | M87 | | |
| | m90 | | |
| | M95 | | |
| | M98 | | Alkaline Nuclease |
| | M102 | | Helicase/Primase Subunit |
| | M105 | | Helicase/Primase Subunit |
| | M116 | | |
| | M121 | | |
| HCMV | | UL24 | Tegument Protein |
| | | UL26 | Tegument Protein |
| | | UL33 | G-protein Coupled Receptor |
| | | UL38 | |
| | | UL50 | |
| | | UL73 | gN |
| | | UL79 | |
| | | UL84 | |
| | | UL89 | |
| | | UL93 | |
| | | UL96 | |
| | | UL103 | |
| | | UL112 | |
| | | UL115 | gL |
| | | US23 | |
| | | US24 | |

replication and packaging were detected including the DNA packaging protein M56, the viral dUTPase M72, the viral protein kinase M97, and the viral exonuclease M98. The independent detection of replication enzymes in both HCMV and MCMV virions suggests that CMV may have the potential to replicate the viral DNA template prior to the early phase of replication when these genes are synthesized by the virus. Interestingly, three different MCMV proteins with immunomodulatory activity were observed in MCMV preparations including m138 (Fc-Receptor), M43, and M45 (anti-apoptotic protein). The HCMV homologs of M43 and M45 were also observed in our HCMV virion preparations. In total, thirty-two virion proteins were detected in both MCMV and HCMV preparations of the 43 MCMV proteins detected in virions with homologs in HCMV in the Harvard study (Table 5.6).

A number of annotated MCMV proteins with unknown function were also found associated with virion particles including: m18, m25.2, M28, M31, m39, M71, M87, M88, m90, M95, m107, M121, m150, m151, m163, m165 (Table 5.7). All of these proteins were detected by the in-solution tryptic method and not by the in-gel analysis, which suggests that they are either low copy number or that they are highly insoluble. Only three of these proteins had counterparts that were detected in HCMV particles including m25.2 (US22 family member), M71, and M88. Similar to our studies the Harvard group also detected peptides corresponding to small ORFs previously not described, including m166.5 and ORF 105932–106072 (44 aa). Expression of m166.5 was confirmed in infected cells by Western blot analysis confirming the presence of this novel viral protein. In addition to finding novel ORFs, the 3' proximal ends of m20 and M31 were remapped based on the GeneMark sequence data analysis that was confirmed by the proteomics approach. Previous sequencing errors mapped the 3' end of m20 out of frame that in this study was correctly extended by an extra 226 bp. The combined use of new sequencing methods with proteomics demonstrates the utility of these techniques to correctly define ORFs.

A number of MCMV virion proteins that have homologs to HCMV ORFs were not found in HCMV virions. These unique MCMV virion proteins include M28, M31, M39, M70, M87, M95, M98, M102, M105, M116, and M121 (Table 5.6). Only a few of these unique MCMV virion proteins have putative functions including M70 (helicase/primase subunit), M98 (alkaline nuclease), and M105 (helicase/primase subunit). Comparison of the HCMV and MCMV proteomes also indicated that HCMV contained a number of unique virion proteins with MCMV homologs. These unique HCMV proteins include UL24 (tegument; an endothelial cell tropism determinant), UL26, UL33 (vGPCR), UL38, UL50, UL73 (gN), UL79, UL84, UL89, UL93, UL96, UL103, UL112, UL115 (gL), US23, and US24 (Table 5.6). Interestingly, the HCMV homolog (UL115) of M115 (gL) forms a complex with gH and gO that was detected in HCMV virions at levels similar to that of gH. Similarly, HCMV gN (UL73), a chaperone for the most abundant glycoprotein gM, was detected in HCMV but not MCMV preparations. Lastly, although both MCMV and HCMV contain virally encoded GPCRs, only HCMV virions contained these viral proteins that included UL33 and US27. Neither the virally encoded GPCR UL78 nor the MCMV homolog M78 was found in either virion preparation, although M78 has been reported in MCMV virions in previous studies (Oliveira and Shenk, 2001).

A number of HCMV-associated proteins that do not have homologs in MCMV were found in HCMV virions in our studies including UL5, UL22A, UL41A, UL119, UL122,

**Table 7. Viral proteins detected by MS analysis in MCMV or HCMV with unknown homology.**

| Virus Detected | MCMV ORF | HCMV ORF | Comments |
|---|---|---|---|
| MCMV | m02 | | |
| | m18 | | |
| | m20 | | |
| | m39 | | |
| | m90 | | |
| | m107 | | |
| | m117.1 | | |
| | m138 | | Fc Receptor |
| | m147 | | |
| | m150 | | m145 Family |
| | m151 | | m145 Family |
| | m163 | | |
| | m165 | | |
| | m165.5 | | Novel ORF |
| HCMV | | TRS1 | Viral Gene Transactivator |
| | | IRS1 | Viral Gene Transactivator |
| | | RL10 | |
| | | TRL14 | |
| | | UL5 | |
| | | UL22A | |
| | | UL41A | |
| | | UL119 | |
| | | UL122 | |
| | | UL132 | |
| | | US27 | G-protein Coupled Receptor |

UL132, US27 (vGPCR), IRS1, TRS1, RL10, and TRL14 (Table 5.7). This observation may reflect the sequence divergence between the two genomes and thus we would predict that there would be a number of MCMV-associated proteins without proper positional homologs in HCMV including; m02, m18, m20, m39, m107, m117.1, m147, m150, m151, m163, m165, and m165.5. Whether these proteins are functional homologs to the HCMV-associated proteins that lack MCMV positional homologs is yet to be determined; studies in this area may prove interesting.

The Harvard group also identified a number of cellular proteins in MCMV virion preparations including: actin-γ, annexin I/IV, cofilin, histone H2A, the translation factor EF1α, glyceraldehyde 3-phosphate dehydrogenase, cadherin, and the RhoGDP dissociation factor. The majority of these cellular proteins were also found in HCMV AD169 virion preparations, independently confirming our studies. Whether these proteins have a role in virus replication is yet to be determined but a subject of intense study.

## Conclusions

In this chapter we described the detailed analysis of the viral and cellular proteins that compose HCMV particles using high-throughput LC-MS analysis. Although a significant amount of information has been accumulated over the years concerning the protein content of HCMV particles, the data presented in this review indicate that HCMV virions and DBs are much more complex than previously predicted from other studies. Previous work using conventional approaches have suggested that 30–40 ORFs encode proteins necessary for the formation of an infectious HCMV virion (Baldick and Shenk, 1996). However, using LC-MS analysis of purified virion preparations indicates that at least 71 ORFs encode proteins that compose the virion. In addition, a significantly higher number of cellular proteins were also present in virion preparations with at least 75 proteins meeting our criteria for significance in this analysis. Some of the salient observations are discussed below.

One of the more surprising findings from our studies was the presence of twelve small peptides from previously unidentified ORFs in the viral genome. Similar findings were also observed in the analysis of the MCMV proteome (Kattenhorn et al., 2004). Some of these small ORFs overlapped with previously identified ORFs or occurred in areas of the viral genome overlooked because of the size of the predicted proteins. An analysis of the five other strains of HCMV indicated that the small ORFs are conserved between strains, suggesting that these proteins are important for the virus (Murphy et al., 2003a). Whether the peptides are part of other proteins or are unique viral proteins is the subject of current studies.

Several interesting observations were derived from our quantitative evaluation of the viral proteins present in the virion. The first surprise was that gM rather than gB was the predominant glycoprotein in the virion. We observed that gM was prevalent in at least a 10-fold excess to the other viral glycoproteins. The unexplained underestimation of this glycoprotein in the virion is most likely due to the insoluble nature of this multi-membrane spanning protein in combination with the lack of adequate antibodies to analyze the protein (Mach et al., 2000). The prevalence of this glycoprotein receptor on the virion suggests the importance of this protein in viral entry processes. Second, while US28 has been shown by ligand binding assays to be present in HCMV virion preparations (Penfold et al., 2003), our analysis failed to detect this viral GPCR. However, two other virally encoded GPCRs previously reported to be incorporated into HCMV virions, US27 and UL33, were present in our analysis. The third interesting observation from our quantitative studies was the high prevalence of AP in our virion preparations. This protein was believed to only occur in immature B capsids and NIEPs. However, our studies indicated that a portion of AP was a major part of the infectious virion similar to what has been characterized in HSV-1. MCMV virion preparations also contained AP (M80). These observations are important and need to be followed-up in subsequent studies since this protein is a target for interventional therapy.

Another unexpected observation was the presence of enzymes associated with replication of the viral DNA template in the virion. This observation was also noted in the analysis of the MCMV proteome. These findings are intriguing with respect to the hypothesis that HCMV may undergo an initial round of DNA template replication prior to the synthesis of the early genes that encode the replication enzymes. The role of these enzymes in the virion is unknown, and they may represent remnants of the encapsidation of viral templates undergoing DNA synthesis. However, the enzymes may be important for amplifying the viral genome upon initial infection.

An analysis of the cellular proteins detected in purified virion preparations indicated the presence of over 75 cellular proteins. Cellular proteins were also detected in the MCMV proteome studies. The major concern in these observations is whether the proteins that we detected are actually part of the virion or a co-purification contaminant with the particle.

An interesting finding that would argue against the co-contaminant theory is the fact that different cellular proteins were observed in the DB preparations that were prepared by similar purification procedures. Whether or not the cellular proteins are an integral constituent of the purified particle, the virion with these proteins forms an infectious unit and these cellular proteins must be considered part of this complex.

In conclusion, while this study has addressed a very straightforward question as to the make-up of a complex enveloped virus, the observations in this study have also created more questions than answers. This study is the first to demonstrate the utility of this technique to approach a very complex problem and will certainly pave the way for future studies with other herpesviruses, as well as other families of viruses. Comparison of the features of these viral proteomes will be important in our understanding of the viral and cellular proteins that compose infectious virions and lead to new areas of future investigations.

## Acknowledgments
Portions of this research were supported in part by the U.S. Department of Energy (DOE), Office of Biological and Environmental Research, LDRD under the Biomolecular Systems Initiative at the Pacific Northwest National Laboratory (PNNL), NIH National Center for Research Resources (RR18522), and the Environmental Molecular Science Laboratory (a DOE user facility located at the PNNL in Richland, WA). PNNL is operated by Battelle Memorial Institute for the DOE under contract DE-AC05-76RLO-1830 Pacific Northwest National Laboratory. The work at OHSU was supported in part by a Public Service Grant from the National Institutes of Health (AI 21640) (J.A.N). D.N.S. is supported by an AHA Scientist Development Grant. We would also like to thank Andrew Townsend for his contributions to these graphics.

## References
Adkins, J.N., Varnum, S.M., Auberry, K.J., Moore, R.J., Angell, N.H., Smith, R.D., Springer, D.L., and Pounds, J.G. (2002). Toward a human blood serum proteome: analysis by multidimensional separation coupled with mass spectrometry. Mol. Cell. Proteomics 1, 947–955.

Baldick, C.J., Jr., and Shenk, T. (1996). Proteins associated with purified human cytomegalovirus particles. J. Virol. 70, 6097–6105.

Benko, D.M., Haltiwanger, R.S., Hart, G.W., and Gibson, W. (1988). Virion basic phosphoprotein from human cytomegalovirus contains O-linked N-acetylglucosamine. Proc. Natl. Acad. Sci. USA 85, 2573–2577.

Chee, M.S., Bankier, A.T., Beck, S., Bohni, R., Brown, C.M., Cerny, R., Horsnell, T., Hutchison, C.A., 3rd, Kouzarides, T., Martignetti, J.A., and et al. (1990). Analysis of the protein-coding content of the sequence of human cytomegalovirus strain AD169. Curr. Top. Microbiol. Immunol. 154, 125–169.

Chelius, D., Huhmer, A.F., Shieh, C.H., Lehmberg, E., Traina, J.A., Slattery, T.K., and Pungor, E., Jr. (2002). Analysis of the adenovirus type 5 proteome by liquid chromatography and tandem mass spectrometry methods. J. Proteome Res. 1, 501–513.

Davison, A.J., Dolan, A., Akter, P., Addison, C., Dargan, D.J., Alcendor, D.J., McGeoch, D.J., and Hayward, G.S. (2003). The human cytomegalovirus genome revisited: comparison with the chimpanzee cytomegalovirus genome. J. Gen.Virol. 84, 17–28.

Dunn, W., Chou, C., Li, H., Hai, R., Patterson, D., Stolc, V., Zhu, H., and Liu, F. (2003). Functional profiling of a human cytomegalovirus genome. Proc. Natl. Acad. Sci. USA 100, 14223–14228.

Florens, L., Washburn, M.P., Raine, J.D., Anthony, R.M., Grainger, M., Haynes, J.D., Moch, J.K., Muster, N., Sacci, J.B., Tabb, D.L., et al. (2002). A proteomic view of the Plasmodium falciparum life cycle. Nature 419, 520–526.

Gibson, W. (1983). Protein counterparts of human and simian cytomegaloviruses. Virology 128, 391–406.

Gibson, W. (1996). Structure and assembly of the virion. Intervirology 39, 389–400.

Giugni, T.D., Soderberg, C., Ham, D.J., Bautista, R.M., Hedlund, K.O., Moller, E., and Zaia, J.A. (1996). Neutralization of human cytomegalovirus by human CD13-specific antibodies. J. Infect. Dis. 173, 1062–1071.

Grundy, J.E., McKeating, J.A., and Griffiths, P.D. (1987). Cytomegalovirus strain AD169 binds beta 2 microglobulin in vitro after release from cells. J. Gen.Virol. 68 (Pt 3), 777–784.

Huber, M.T., and Compton, T. (1998). The human cytomegalovirus UL74 gene encodes the third component of the glycoprotein H-glycoprotein L-containing envelope complex. J. Virol. 72, 8191–8197.

Huber, M.T., and Compton, T. (1999). Intracellular formation and processing of the heterotrimeric gH–gL–gO (gCIII) glycoprotein envelope complex of human cytomegalovirus. J. Virol. 73, 3886–3892.

Irmiere, A., and Gibson, W. (1983). Isolation and characterization of a noninfectious virion-like particle released from cells infected with human strains of cytomegalovirus. Virology 130, 118–133.

Kattenhorn, L.M., Mills, R., Wagner, M., Lomsadze, A., Makeev, V., Borodovsky, M., Ploegh, H.L., and Kessler, B.M. (2004). Identification of proteins associated with murine cytomegalovirus virions. J. Virol. 78, 11187–11197.

Kaye, J.F., Gompels, U.A., and Minson, A.C. (1992). Glycoprotein H of human cytomegalovirus (HCMV) forms a stable complex with the HCMV UL115 gene product. J. Gen.Virol. 73, 2693–2698.

Landini, M.P., Severi, B., Furlini, G., and Badiali De Giorgi, L. (1987). Human cytomegalovirus structural components: intracellular and intraviral localization of p28 and p65–69 by immunoelectron microscopy. Virus Res. 8, 15–23.

Lasonder, E., Ishihama, Y., Andersen, J.S., Vermunt, A.M., Pain, A., Sauerwein, R.W., Eling, W. M., Hall, N., Waters, A.P., Stunnenberg, H.G., and Mann, M. (2002). Analysis of the Plasmodium falciparum proteome by high-accuracy mass spectrometry. Nature 419, 537–542.

Lehner, R., Meyer, H., and Mach, M. (1989). Identification and characterization of a human cytomegalovirus gene coding for a membrane protein that is conserved among human herpesviruses. J. Virol. 63, 3792–3800.

Link, A.J., Eng, J., Schieltz, D.M., Carmack, E., Mize, G.J., Morris, D.R., Garvik, B.M., and Yates, J.R., 3rd (1999). Direct analysis of protein complexes using mass spectrometry. Nature Biotechnol. 17, 676–682.

Lipton, M.S., Pasa-Tolic, L., Anderson, G.A., Anderson, D.J., Auberry, D.L., Battista, J.R., Daly, M.J., Fredrickson, J., Hixson, K.K., Kostandarithes, H., et al. (2002). Global analysis of the Deinococcus radiodurans proteome by using accurate mass tags. Proc. Natl. Acad. Sci. USA 99, 11049–11054.

Mach, M., Kropff, B., Dal Monte, P., and Britt, W. (2000). Complex formation by human cytomegalovirus glycoproteins M (gpUL100) and N (gpUL73). J. Virol. 74, 11881–11892.

Michelson, S., Turowski, P., Picard, L., Goris, J., Landini, M.P., Topilko, A., Hemmings, B., Bessia, C., Garcia, A., and Virelizier, J.L. (1996). Human cytomegalovirus carries serine/threonine protein phosphatases PP1 and a host cell derived PP2A. J. Virol. 70, 1415–1423.

Mocarski, E.S., and Courcelle, C.T. (2001). Cytomegalovirus and Their Replication. In Fields Virology, S. E. Straus, ed. (Philadelphia, Lippincott-Raven), pp. 2629–2673.

Murphy E., Yu D., Grimwood J., Schmutz J., Dickson M., Jarvis M.A., Hahn G, Nelson J.A., Myers R.M., and Shenk, T.E. (2003a). Coding potential of laboratory and clinical strains of human cytomegalovirus. Proc. Natl. Acad. Sci. USA 100, 14976–14981.

Murphy, E., Rigoutsos, I., Shibuya, T., and Shenk, T.E. (2003b). Reevaluation of human cytomegalovirus coding potential. Proc. Natl. Acad. Sci. USA.

Oliveira, S.A., and Shenk, T.E. (2001). Murine cytomegalovirus M78 protein, a G protein-coupled receptor homologue, is a constituent of the virion and facilitates accumulation of immediate-early viral mRNA. Proc. Natl. Acad. Sci. USA 98, 3237–3242.

Penfold, M.E., Schmidt, T.L., Dairaghi, D.J., Barry, P.A., and Schall, T.J. (2003). Characterization of the rhesus cytomegalovirus US28 locus. J. Virol. 77, 10404–10413.

Sarov, I., and Abady, I. (1975). The morphogenesis of human cytomegalovirus. Isolation and polypeptide characterization of cytomegalovirions and dense bodies. Virology 66, 464–473.

Schmolke, S., Kern, H., Drescher, P., Jahn, G., and Plachter, B. (1995). The dominant phosphoprotein pp65 (UL83) of Human cytomegalovirus is dispensable for growth in cell culture. J. Virol. 69, 5959–5968.

Smith, R.D., Anderson, G.A., Lipton, M.S., Masselon, C., Pasa-Tolic, L., Shen, Y., and Udseth, H.R. (2002a). The use of accurate mass tags for high-throughput microbial proteomics. Omics 6, 61–90.

Smith, R.D., Anderson, G.A., Lipton, M.S., Pasa-Tolic, L., Shen, Y., Conrads, T.P., Veenstra, T.D., and Udseth, H.R. (2002b). An accurate mass tag strategy for quantitative and high-throughput proteome measurements. Proteomics 2, 513–523.

Stannard, L.M. (1989). Beta 2 microglobulin binds to the tegument of cytomegalovirus: an immunogold study. J. Gen.Virol. *70 (Pt 8)*, 2179–2184.

Varnum, S.M., Streblow, D.N., Monroe, M.E., Smith, P., Auberry, K.J., Pasa-Tolic, L., Wang, D., Camp, D.G., 2nd, Rodland, K., Wiley, S., *et al.* (2004). Identification of Proteins in Human Cytomegalovirus (HCMV) Particles: the HCMV Proteome. J. Virol. *78*, 10960–10966.

Washburn, M.P., Wolters, D., and Yates, J.R., 3rd (2001). Large-scale analysis of the yeast proteome by multidimensional protein identification technology. Nature Biotechnol. *19*, 242–247.

Wright, J.F., Kurosky, A., Pryzdial, E.L., and Wasi, S. (1995). Host cellular annexin II is associated with cytomegalovirus particles isolated from cultured human fibroblasts. J. Virol. *69*, 4784–4791.

Yu, D., Silva, M.C., and Shenk, T. (2003). Functional map of human cytomegalovirus AD169 defined by global mutational analysis. Proc. Natl. Acad. Sci. USA *100*, 12396–12401.

# Virus Entry and Activation of Innate Immunity

6

*Karl W. Boehme and Teresa Compton*

## Abstract

All viruses must deliver their genomes to host cells to initiate infection. The cell plasma membrane serves as an initial barrier that must be crossed if an infection is going to take place. This chapter will summarize what is known about the entry pathway of human cytomegalovirus (HCMV) with certain parallels and commonalities noted between HCMV and other beta-herpesviruses. The roles of HCMV envelope glycoproteins in mediating critical virus entry events such as attachment and fusion as well as the current knowledge of the identification of cellular receptors that serve as entry mediators will also be described. This chapter will also discuss entry-associated innate immune activation and the emerging role of signaling pathways in the early events in infection. Lastly, we will examine how virus entry and innate immune activation may be coordinated.

## Introduction

In the simplest context, entry requires that all enveloped viruses, including HCMV, adhere to the cell surface and trigger fusion between the virus envelope and a cellular membrane that results in the deposition of virion components into the cytoplasm (Figure 6.1). Following content delivery to the cytoplasm, certain tegument proteins and genome-containing capsids translocate to the nucleus in a process collectively known as uncoating. For these structurally complex viruses whose envelopes contain as many as 20 proteins and glycoproteins, attachment is a multi-step process typically involving more than one envelope glycoprotein interacting with a series of cell-surface molecules that serve as receptors and co-receptors. A likely consequence of these virus-cell interactions is receptor-activated conformational changes in the envelope glycoproteins that play roles in membrane fusion. Here too, multiple envelope glycoproteins are required to fuse membranes. Another consequence of these initial virus–host interactions may be the formation and/or delivery of bound virions to specialized membrane domains or compartments that are optimal for fusion and activation of signal transduction cascades. We have also recently learned that HCMV entry is accompanied by innate immune activation. This considerably heightens the complexity of the molecular events occurring during the early events in HCMV infection.

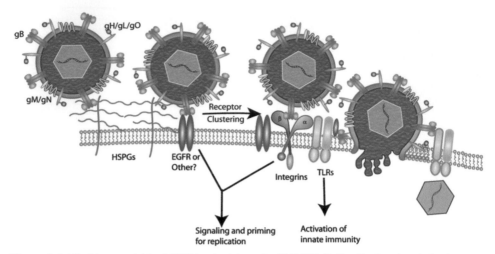

**Figure 6.1** Working model for HCMV entry into cells. HCMV initially attaches in a tethering step to HSPGs via gM/gN and/or gB. The gB protein interacts in a stable docking step to EGFR on many HCMV permissive cell types or to as yet unidentified receptors in hematopoietic cell types. Other interactions between HCMV envelope glycoproteins and cellular integrins promote receptor clustering. At least one of these interactions triggers fusion that leads to internalization of virion components. Signal transduction events are initiated via EGFR and/or integrins and these events are hypothesized to prime and facilitate downstream steps in the virus life cycle such as nuclear translocation of the capsid and efficient viral gene expression. Toll-like receptors detect an HCMV-displayed pathogen-associated molecular pattern during virus entry leading to distinct signaling events and activation of innate immunity.

## HCMV entry

To begin a discussion of virus entry at the cellular level, one must first consider the basis of cellular tropism since receptors involved in entry are expressed on permissive cells. In the human host, HCMV causes systemic infection and exhibits a tropism for fibroblasts, endothelial cells, epithelial cells, monocytes/macrophages, smooth muscle cells, stromal cells, neuronal cells, neutrophils, and hepatocytes (Ibanez et al., 1991; Myerson et al., 1984; Sinzger et al., 2000). This exceptionally broad cellular tropism is the root of HCMV disease manifestation of most organ systems and tissue types in the immunocompromised host. Although HCMV is considered to have a very restricted cell tropism in vitro, entry into target cells is very promiscuous, as HCMV is able to bind, fuse and initiate replication in all tested vertebrate cell types. Productive in vitro replication is only supported by primary fibroblasts, endothelial and differentiated myeloid cells as well as certain astrocyte cell lines (Ibanez et al., 1991; Nowlin et al., 1991). The ability of HCMV to enter such a wide range of cells is highly indicative of multiple cell specific receptors, broadly expressed receptors or a complex entry pathway in which a combination of both cell specific and broadly expressed cellular receptors are utilized.

### History of cellular receptors for HCMV
It has been known for some time that HCMV initiates infection via a tethering interaction of virions and cell-surface heparan sulfate proteoglycans (HSPGs) (Compton et al., 1993) (Figure 6.1). HSPG engagement is a relatively conserved feature of herpesvirus entry path-

ways and is thought to also play a role in HHV-6 and HHV-7 interactions with lymphoid cells (Conti et al., 2000). At least in cell culture systems, HCMV engagement of HSPGs is thought to play a crucial role in recruiting virions to the cell surface and enhancing the engagement of other receptors (Compton et al., 1993). This hypothesis is further supported by biochemical analysis of HCMV binding, which indicates biphasic binding properties with multiple distinct affinities (Boyle and Compton, 1998).

The broad ability of HCMV to enter cells in culture has hampered efforts to identify cellular receptors using modern molecular approaches such as expression cloning. Over the past 20 years, numerous receptor candidates have been put forward. These candidate receptor molecules were selected on the basis of sound logic and solid initial criteria but none of these molecules have held up to commonly accepted parameters for virus receptors. HCMV virions were initially shown to bind $\beta_2$ microglobulin ($\beta_2$m) in urine samples (Grundy et al., 1988; McKeating et al., 1986; McKeating et al., 1987). This observation led to numerous binding studies concluding that the HCMV tegument binds $\beta_2$m as it is released from cells (Grundy et al., 1987b; McKeating et al., 1987; Stannard, 1989). This $\beta_2$m–HCMV complex was then thought to associate with the alpha chain of HLA class I antigens (Beersma et al., 1990; Beersma et al., 1991; Browne et al., 1990; Grundy et al., 1987a). These data led to a model in which $\beta_2$m-coated HCMV bound MHC class I molecules and displaced $\beta_2$m from the MHC class I $\alpha$ chain. However, it was later determined that $\beta_2$m expression had no correlation with in vitro entry or in vivo spread of infectivity (Beersma et al., 1991; Polic et al., 1996; Wu et al., 1994).

Virus-cell overlay blots were used to globally analyze cellular proteins that could bind to HCMV virions. These studies identified a cell-surface protein of approximately 30 kDa, whose expression correlated with cells permissive for entry (Nowlin et al., 1991; Taylor and Cooper, 1990). This protein was later identified as annexin II, a phospholipid and calcium binding protein that has membrane bridging function (Wright et al., 1995). Annexin II was also shown to bind HCMV virions (Wright et al., 1994; Wright et al., 1995). Upon further study it was found that gB was able to directly interact with annexin II and that this protein was able to enhance HCMV binding and fusion to phospholipid membranes (Pietropaolo and Compton, 1997; Raynor et al., 1999). However, cells devoid of annexin II are fully permissive for entry and initiation of infection (Pietropaolo and Compton, 1999). What role, if any, annexin II plays in the life cycle of HCMV is unknown but, given its membrane-bridging activity, it remains formally possible that this enhances entry, cell–cell spread and/or maturation and egress.

CD13, or human aminopeptidase N, has also been implicated as a HCMV receptor. This hypothesis was based on that fact that only human peripheral blood mononuclear cells (PBMCs) that were CD13 positive supported productive infection (Larsson et al., 1998; Soderberg et al., 1993a; Soderberg et al., 1993b). This led to a more thorough study of this possibility in which CD13-specific antibodies and chemical inhibitors of CD13 activity were both shown to inhibit HCMV binding and entry (Soderberg et al., 1993a). Excitement from this report was dampened by later reports that CD13 antibodies could bind and neutralize virus before contact with cells, coupled with normal entry of HCMV into CD13-depleted cells (Giugni et al., 1996). More recently, an interaction between HCMV and CD13 was shown to be important in inhibition of differentiation of monocytes into macrophages suggesting this may be a strategy for interference with cellular differentiation pathways (Gredmark et al., 2004).

A consideration of HCMV-induced signaling cascades led Wang et al. (2003) to hypothesize a role for epidermal growth factor receptor (EGFR) as an HCMV receptor. It was found that EGFR became phosphorylated in response to HCMV and this phosphorylation event correlated with the activation of phosphatidylinositol 3-kinase (PI-3 kinase) and Akt, as well as the mobilization of intracellular $Ca^{++}$. Signaling events were blocked by EGFR antibodies. In addition, chemically crosslinked virus provided evidence for a gB-EGFR interaction. A limitation of the study, however, was that there was no experimental evidence that EGFR functioned in an entry event per se. Nor was it shown if EGFR was required for the delivery of virion components across the plasma membrane. Also, conflicting results exist in the literature. Fairley and colleagues demonstrated that HCMV promoted inactivation of EGFR phosphorylation and signaling (Fairley et al., 2002). In an ironic twist, studies investigating CD13 as an HCMV receptor used antibodies to EGFR as a "negative control". In these experiments, EGFR polyclonal antibodies had no effect on HCMV entry (Soderberg et al., 1993a). Lastly, it is important to note that EGFR is not expressed on all HCMV permissive cells, including those of hematopoietic lineage. Combined, these observations and the history of HCMV entry receptors engender caution. There is a need for careful duplication of the data and verification of this molecule as a receptor.

## Cellular integrins may serve as co-receptors for beta-herpesviruses

The first and foremost observation about HCMV biology was its namesake characteristic, cytomegaly, or cell enlargement. In vitro studies initially demonstrated a unique cytopathogenic effect (CPE) of infected cells, with HCMV-infected cells appearing round and enlarged with intracellular viral inclusion bodies (Albrecht and Weller, 1980). Infection proceeded with two waves of cell rounding, the first beginning as early as 1 h post infection, corresponding to the entry event, and another at approximately 24 h post infection. The cause of this phenomenon was widely speculated upon; theories for HCMV-induced cell rounding included cation influx, suppression of fibronectin synthesis and integrin down-regulation (Albrecht et al., 1983; Albrecht and Weller, 1980; Ihara et al., 1982; Warren et al., 1994). Cellular integrins are ubiquitously expressed cell-surface receptors that, when activated, lead to major reorganization of the cytoskeleton. Integrins exist on the plasma membrane as a noncovalently linked heterodimer consisting of an α and β subunit, which conveys specificity in cell–cell and cell–ECM (extracellular matrix) attachment, immune cell recruitment, extravasation, and signaling (Berman and Kozlova, 2000; Berman et al., 2003; Cary et al., 1999). In addition, integrins have emerged as receptors for a broad range of pathogens including pathogenic plant spores, bacteria and viruses. Feire et al. (2004) tested the hypothesis that integrins were involved in the HCMV entry pathway. Analysis of the effects of various neutralizing antibodies showed α2β1, α6β1, and αVβ3 were involved in HCMV entry (Feire et al., 2004). Furthermore, cells devoid of beta 1 integrins exhibited little to no infection by HCMV or mouse cytomegalovirus while entry and spread were restored when the expression of beta 1 integrin was re-introduced into the cells. The integrin antibodies had no effect on virus attachment but specifically blocked the delivery of a virion component, pp65 (pUL83), suggesting that integrins function at a post-attachment stage of infection, possibly at the level of membrane fusion. The fact that multiple integrin heterodimers are utilized is consistent with integrin biology in that many natural integrin ligands, such as extracellular matrix proteins, bind to multiple heterodimers. Indeed, most integrin-binding viruses frequently interact with one to three individual integrin molecules. Integrins are capable of engaging ligands through a number of identified ECM protein motifs, the most common of which contain the amino acid sequence RGD. However, there are a number of RGD-independent integrin binding motifs, including the disintegrin domain proteins of the ADAM (a disintegrin and a metalloprotease) family of proteins. After inspection of all HCMV structural glycoproteins, the only homology to an integrin-binding domain was a disintegrin-like consensus sequence ($RX_{5-7}DLXXF/L$) (Eto et al., 2002; Stone et al., 1999; Wolfsberg et al., 1995) on the amino-terminus of gB. Sequence alignments confirmed that the gB disintegrin loop was more than 98% identical among 44 clinical isolates analyzed. The role of this sequence in entry was confirmed since synthetic peptides to this sequence inhibited both HCMV and MCMV entry, but not the disintegrin-loop lacking HSV. Furthermore, the HCMV gB disintegrin-loop was conserved throughout much of the gamma and all of the beta-herpesvirus subfamilies, but not in the alpha-herpesvirus subfamily where there are previously identified RGD sequences. The presence of integrin binding sequences among conserved *Herpesviridae* glycoproteins strongly suggests that integrins may be important for entry and signaling throughout this virus family. EGFR has also been shown to become phosphorylated and signal indirectly, as a result of integrin activation through src family kinases or focal adhesion kinase (FAK) (Jones et al., 1997; Miyamoto et al., 1996; Moro et al., 1998). Future work will no doubt be aimed at an analysis of the integrin-triggered signaling events and defining their roles in entry and infection.

### Entry-activated cell signaling

It has been apparent for many years that cells respond to HCMV virions by activation of numerous cell signaling pathways including changes in $Ca^{2+}$ homeostasis, activation of phospholipases C and A2, as well as increased release of arachidonic acid and its metabolites (for review see, Fortunato et al., 2000). All of these changes can be triggered by UV-inactivated virions, suggesting that structural components of the virus are responsible for activation during virus-cell contact and/or virus entry. Virus-cell contact also results in the activation of transcription factors such as c-fos/jun, myc, NF-κB, SP-1, as well as mitogen-activated protein (MAP) kinase ERK1/2 and p38 (Boldogh et al., 1991; Boyle et al., 1999; Kowalik et al., 1993; Yurochko et al., 1995). These virally induced cellular physiological changes are associated with a profound effect on host cell gene expression. Thousands of transcripts are altered in their expression and most transcriptional changes do not require viral gene expression (Browne et al., 2001; Simmen et al., 2001; Zhu et al., 1998). These data are consistent with the interpretation that HCMV engages a cellular receptor(s) that activate(s) signal transduction pathways culminating in reprogramming of cellular transcription.

## Activation of innate immunity by cytomegaloviruses

Cytomegaloviruses trigger strong anti-pathogen responses upon infection. Effective control of HCMV in vivo requires cooperation between many facets of both the innate and adaptive arms of the host immune system. A comprehensive discussion of the immune response to CMVs is a daunting task and well beyond the scope of a single chapter. The discussion here is limited to activation of innate responses triggered during entry of the virus into host cells, focusing on recent advances in the induction of interferon-α/β responses by HCMV and the emerging role of Toll-like receptors in detection of HCMV virions. Other aspects of the innate immune response will be explored in Chapter 15.

## Activation of the interferon-α/β response by HCMV

Like most viruses, HCMV elicits an extremely potent interferon-α/β response from its host. Interferon-α/β confers an antiviral state on cells, causing them to become refractory to viral infection and limiting virus replication at the site of infection (Stark et al., 1998). Interferon-α/β is also a critical component of the cytokine cocktail that activates T-cell responses. The cardinal role of interferon-α/β in the immune response to CMVs is evidenced by the observation that mice lacking the interferon-α receptor (both interferon-α and interferon-β utilize the same receptor complex) are highly susceptible to MCMV infection as compared to wild-type mice (Salazar-Mather et al., 2002). CMVs have long been known to elicit interferon-α/β; however the mechanism(s) by which these responses are induced have only recently begun to be defined.

A growing body of evidence indicates that HCMV activates innate immunity during binding and entry of virions into the host cell. Transcriptional profiling studies demonstrated that HCMV is a potent inducer of many indicators of innate immunity, including inflammatory cytokines and interferon-α/β (Browne et al., 2001; Simmen et al., 2001; Zhu et al., 1997; Zhu et al., 1998). The induction is rapid with transcriptional up-regulation of innate markers observed within 2 hours of infection (Browne et al., 2001). In addition, a variety of cell signaling pathways are activated upon infection. NF-κB, a central player in a variety of host cell defenses, is activated as early as 15 minutes post-infection, and IRF3, the key transcriptional regulator of interferon-α/β responses, is also activated upon HCMV infection (Navarro et al., 1998; Preston et al., 2001; Yurochko and Huang, 1999; Yurochko et al., 1995; Yurochko et al., 1997a; Yurochko et al., 1997b). Furthermore, activation of innate effectors by HCMV is unaffected by the presence of the protein synthesis inhibitor cycloheximide, indicating that the production of additional secreted factors such as interferon-α/β or inflammatory cytokines is not required to elicit these responses (Boehme et al., 2004; Browne et al., 2001; Navarro et al., 1998; Preston et al., 2001; Zhu et al., 1997; Zhu et al., 1998). In conclusion, the rapid and direct induction of innate responses by HCMV suggests that they are triggered during virus binding and entry into its target cell.

## Overview of the interferon pathway

The interferon-α/β response to viral infection can be divided into two phases: an activation phase and an amplification phase (Taniguchi and Takaoka, 2002; Taniguchi et al., 2001). The activation phase begins with the initial detection of the virus and propagation of intracellular signals that culminate in the secretion of interferon-α/β (Stark et al., 1998). The mechanisms by which cells detect viruses and initiate the interferon response have not been completely defined, however one means by which cells can activate interferon-α/β responses is through specific members of the Toll-like receptor (TLR) family (TLRs 3, 4, 7, 8, and 9). An increasing amount of evidence indicates that these TLRs play a critical role in the interferon response to herpesviral infection (Alexopoulou et al., 2001; Diebold et al., 2004; Heil et al., 2004; Krug et al., 2004a; Krug et al., 2004b; Lund et al., 2003; Lund et al., 2004) and the specific roles of TLRs in the interferon response to CMVs will be discussed below. Other mechanisms for induction of interferon-α/β responses to viruses exist, but these processes remain poorly understood.

Viral detection leads to activation of the interferon regulatory factor 3 (IRF3) transcription factor via phosphorylation of specific serine and threonine residues in its carboxyterminal domain by the cellular kinases IKKε and TBK1 (Fitzgerald et al., 2003; Servant et al., 2001; Sharma et al., 2003). IRF3 normally resides in the cytoplasm, however upon phosphorylation it translocates to the nucleus where it complexes with p300/CBP to drive expression of a subset of interferon stimulated genes (ISGs), as well as interferon-β (Suhara et al., 2002; Yoneyama et al., 1998). Secreted interferon-β acts in an paracrine manner through the IFN- α/β receptor and the well defined Janus kinases and signal transducers and activators of transcription (JAK-STAT) pathway to promote expression of the full complement of ISGs (Stark et al., 1998). Among these ISGs is IRF7 which, like IRF3, is activated by phosphorylation (presumably by IKKε and TBK1) and complexes with IRF3 to drive the expression the IFN-α genes (Sato et al., 1998; Sato et al., 2000). IFN-α, like IFN-β, acts in an autocrine/paracrine manner to induce ISG expression. This amplification loop provides a rapid and effective mechanism by which host cells can respond to viral infection.

Interestingly, the specialized cell type primarily responsible for secreting type I interferon in response to a wide variety of viruses, plasmacytoid DCs (pDCs), selectively produce interferon-α as opposed to interferon-β (Asselin-Paturel et al., 2001; Cella et al., 1999; Dalod et al., 2002; Dalod et al., 2003; Feldman et al., 1994; Fonteneau et al., 2003; Perussia et al., 1985; Poltorak et al., 1998b; Siegal et al., 1999). New evidence indicates that IRF7 is initially activated in pDCs, as opposed to IRF3 (Dai et al., 2004; Honda et al., 2004; Kawai et al., 2004). The specificity of IRF7 for the interferon-α genes provides the mechanism by which pDCs induce secretion of interferon-α, but not interferon-β, which is dependent upon IRF3. Functional differences exist between the different interferon-α subtypes and interferon-β and it is likely that these differences are the reason for the preference of interferon-α exhibited by pDCs (Evinger et al., 1981; Foster et al., 1996; Weck et al., 1981).

## Role of Toll-like receptors in innate immune activation

As described above, TLRs are one way in which cells can detect viruses and initiate interferon-α/β responses. TLRs are not limited to induction of interferon responses, but rather serve a larger function as general pathogen recognition receptors (PRRs) that detect and initiate immune responses to a myriad of bacteria, fungi, and viruses (Takeda and Akira, 2003). The primary consequences of TLR activation are inflammatory cytokine secretion, expression of immune co-stimulatory molecules, dendritic cell maturation, and for certain TLRs, the production of interferon-α/β (Takeuchi and Akira, 2001). Together these factors limit viral replication at the site of infection, elicit infiltration of immune cells to the site of infection, and initiate and modulate adaptive immune responses by B and T-cells.

TLRs are type I transmembrane proteins composed of leucine-rich repeat (LRR) ectodomains and cytoplasmic regions with Toll-interleukin-1 receptor (TIR) domains (Rock et al., 1998). To date 11 TLRs have been identified in humans and, although all cells express at least a subset of TLRs, they are expressed predominantly by immune sentinel cells such as macrophages and dendritic cells (DCs) (Chuang and Ulevitch, 2000; Chuang and Ulevitch, 2001; Du et al., 2000; Hornung et al., 2002; Medzhitov et al., 1997; Rock et al., 1998; Takeuchi et al., 1999; Zarember and Godowski, 2002; Zhang et al., 2004). TLRs recognize a wide variety of microbial pathogens on the basis of pathogen associated molecular patterns (PAMPs). PAMPs are structural elements found uniquely on microbes, but not normally present in the host organism. Some examples of TLR ligands are lipopolysaccharide (LPS) and peptidoglycan from bacteria, dsRNA from viruses, and unmethylated CpG DNA motifs from bacteria and viruses (Alexopoulou et al., 2001; Bauer et al., 2001; Hemmi et al., 2000; Poltorak et al., 1998a; Qureshi et al., 1999). An expanded list of TLR PAMPs can be found in Table 6.1. Ligand-induced TLR clustering brings the cytoplasmic tails of the receptors into close proximity, thereby creating a docking site for intracellular signaling intermediates. TLRs activate NF-κB and mitogen-activated protein kinases, however a subset of TLRs (TLRs 3, 7, 8, and 9) concomitantly activate interferon-α/β secretion by activating members of the interferon regulator factor family (Wagner, 2004). A comprehensive discussion of TLR signaling is beyond the scope of this chapter, however a number of reviews are available detailing various aspects of TLR biology (Akira, 2003; Akira and Hemmi, 2003; Boehme and Compton, 2004; O'Neill et al., 2003; Takeda and Akira, 2003).

## Roles of HCMV envelope glycoproteins in virus entry

### Composition of the HCMV envelope

The HCMV envelope is exceedingly complex and currently incompletely defined. The HCMV genome encodes ORFs to at least 57 putative glycoproteins; far more than other herpesviruses, however, the extent of transcription, translation and function of the major-

**Table 6.1** Viral ligands of Toll-like receptors

| Toll-like receptor | Viral ligand |
| --- | --- |
| TLR2 | Peptidoglycan |
| | Zymosan |
| | Measles virus (hemagglutinin) |
| | Human cytomegalovirus (gB) |
| | HSV-1 |
| TLR3 | dsRNA |
| | MCMV |
| TLR4 | LPS |
| | RSV (F protein) |
| | MMTV (envelope protein) |
| TLR7 and TLR8 | Influenza A (genomic RNA) |
| | HIV-1 (synthetic RNA oligonucleotide from U5 region of genome) |
| | VSV |
| TLR9 | HSV-1 |
| | HSV-2 (genomic DNA) |
| | MCMV |

ity of these glycoproteins remains unknown. Biochemical studies of HCMV virions have revealed that fourteen glycoproteins are structural; eight of these experimentally shown to reside in the envelope (Britt et al., 2004). Among structural glycoproteins, a group homologous to other herpesviruses and thus thought to serve conserved functions in entry exists. These include glycoproteins B (gB), H (gH), L (gL), O (gO), M (gM), and N (gN), while others (gpTRL10, gpTRL11, gpTRL12 and gpUL132) are HCMV-specific (Table 6.2).

For many years, the large genome and complicated reverse genetics system have made the creation of HCMV knockout and mutant viruses difficult. Recently, a system capable of such mutations was developed whereby HCMV is maintained as an infectious bacterial artificial chromosome (BAC) within *Escherichia coli* (Borst et al., 1999). This development has greatly hastened the process of mutating individual ORFs (see also Chapter 4) and will generate much information regarding both the structure and function of many envelope glycoproteins. In fact, the BAC system has demonstrated the requirement for several glycoprotein genes in the production of replication-competent virus (Dunn et al., 2003; Hobom et al., 2000; Yu et al., 2002). The HCMV glycoprotein homologs gB, gM, gN, gH, gL, have been shown to be essential for growth, while gO knockout virus remained viable with a small plaque phenotype (Hobom et al., 2000). Genes for all the currently identified HCMV-specific envelope glycoproteins, including UL4 (gp48), TRL10 (gpTRL10), TRL11 (gpTRL11), TRL12 (gpTRL12), US27, UL33, UL132, have been shown to be nonessential and play no known roles in the entry pathway (Dunn et al., 2003). The HCMV-encoded chemokine receptor US28 is present in the virion envelope and has been shown to promote cell–cell fusion mediated by HIV and VSV viral proteins, however the gene has been shown to be nonessential and there is no evidence for a role for gpUS28 in either HCMV-cell or cell–cell fusion (Dunn et al., 2003; Pleskoff et al., 1997; Pleskoff et al., 1998).

**Table 6.2** Envelope proteins of HCMV

| ORF | Protein name | Essential | Complex partner | Role in entry |
|-----|-----|-----|-----|-----|
| UL4 | gpUL4; gp48 | No | None known | None known |
| UL33 | UL33 | No | None known | None known |
| UL55 | gB | Yes | Forms homodimers | Receptor binding, fusion, signal transduction, innate immune activation |
| UL73 | gN | Not determined | UL100; gM | None known |
| UL74 | gO | No; defect in cell-to-cell spread | UL75;gH,, UL115;gL | Enhancer of cell-to-cell spread |
| UL75 | gH | Yes | UL74;gO, UL115;gL | Fusion, receptor binding (?), innate immune activation |
| UL100 | gM | Yes | UL73;gN | HSPG binding |
| UL115 | gL | Yes | UL75;gH, UL74;gO | Required for gH activity |
| TRL10 | gpTRL10 | Not determined | None known | None known |
| TLR12 | gpTRL12 | Not determined | None known | None known |
| US27 | US27 | No | None known | None known |
| US28 | US28 | No | None known | None known |

The essential and abundant HCMV envelope glycoproteins conserved throughout the *Herpesviridae* (gB, gM, gN, gH, and gL) were originally classified as three distinct disulfide-linked high molecular weight complexes (gCI–gCIII) (Gretch et al., 1988). The gCI complex is composed of homodimers of glycoprotein B (gB) (Britt, 1984; Britt and Auger, 1986). The gCII heterodimeric complex is composed of glycoprotein M (gM) and glycoprotein N (gN) (Mach et al., 2000). Finally, the gCIII complex is a heterotrimeric complex composed of glycoprotein H (gH), glycoprotein L (gL), and glycoprotein O (gO) (Huber and Compton, 1997; Huber and Compton, 1998; Li et al., 1997).

### Envelope glycoproteins that bind to receptors

At least two glycoprotein complexes have heparin binding ability. Both gB and the gM component of gCII can bind to soluble heparin suggesting a critical role in the initial tethering event (Carlson et al., 1997; Kari and Gehrz, 1992; Kari and Gehrz, 1993). Functional redundancy, in particular with respect to HSPG binding, is also a shared property with other herpesviruses. The gB protein appears to be the primary receptor binding protein. Soluble forms of gB exhibit biphasic cell binding properties, and cells treated with gB are refractory to infection suggesting that gB is tying up critical receptor sites used by the virus (Boyle and Compton, 1998). One of the binding sites for gB is HSPGs in that cells lacking HSPGs had a single component Scatchard plot as compared to a biphasic plot for HSPG-bearing cells. As noted above, it now seems clear that a second binding partner is an integrin (Feire et al., 2004), but much work remains to formally prove the disintegrin hypothesis and confirm the role of this domain in receptor engagement. The gB protein may also engage EGFR at least in certain cell types; however, it is not yet known if this interaction is a consequence of initial integrin binding. The gH complex may also have a distinct receptor. Syngeneic monoclonal anti-idiotypic antibodies were created that bear the "image" of this glycoprotein complex (Keay et al., 1988). This reagent led to a putative gH receptor; a phosphorylated 92.5 kDa cell-surface glycoprotein that appears to mediate virus-cell fusion, but not attachment and $Ca^{2+}$ influxes at the plasma membrane (Keay and Baldwin, 1991; Keay and Baldwin, 1992; Keay and Baldwin, 1996; Keay et al., 1989; Keay et al., 1995). The data set relies heavily on a single reagent however (anti-idiotypic antibodies) and has led to only a partially sequenced receptor clone, with no identified homology to any known human protein (Baldwin, 2000; Keay and Baldwin, 1996). Since HCMV gH is essential, infectivity can be neutralized with anti-gH antibodies. Anti-idiotypic antibodies can also neutralize infectivity, presumably by occupying the receptor for gH. The closest relative of HCMV, HHV-6, contains a homologous complex gH/gL/gQ with an identified receptor (Santoro et al., 1999), however the identity of an HCMV gH/gL/gO receptor remains unknown, yet quite possible.

### Roles of envelope glycoproteins in membrane fusion

Membrane fusion remains a big black box in herpesvirology. Unlike orthomyxoviruses, paramyxoviruses, filoviruses and retroviruses that use a single envelope glycoprotein for membrane fusion, herpesviruses typically employ multi-component fusion machines frequently consisting of gB, gH and gL (Spear and Longnecker, 2003). Both the HCMV gB- and gH-containing complex trigger neutralizing antibodies that block infection at a post-attachment stage of entry, presumably at the level of fusion (Bold et al., 1996; Britt,

1984; Keay and Baldwin, 1991; Tugizov et al., 1994; Utz et al., 1989). One limitation of these conclusions, however, is the lack of a direct fusion assay and thus a role for these glycoproteins in fusion is inferred. Despite the complexity of multi-component fusion machines, it is very likely that there are strong parallels to single-component fusion proteins. Alpha-helical coiled-coils are considered critical structural domains involved in fusion that function to drive the energetic folding of membranes together. Conformational changes in fusogenic proteins bearing these coiled-coils are also a defining paradigm. Using an algorithm to detect potential coiled coils, Lopper and Compton (2004) identified heptad repeat regions in gB and gH that were predicted to form coiled coils. Synthetic peptides to these motifs substantially inhibited HCMV entry including virion content delivery suggesting that these motifs play a fundamental role in membrane fusion. Genetic analysis of these motifs in the context of HCMV virions will be required to further analyze the importance of alpha-helical coiled coils in HCMV entry. Another fundamental question will be to determine if the gB–integrin interaction is a trigger of conformational change that leads to exposure of membrane fusion domains. Intriguingly, disintegrin-bearing cellular proteins in the ADAM family are known to trigger fusion via integrin interaction in a variety of processes including sperm–egg fusion and myoblast fusion (White, 2003). Development of a reliable fusion assay is also greatly needed to begin a dissection of the biophysical properties of HCMV fusion glycoproteins.

## Role of the envelope glycoproteins in activation of innate immunity

Transcriptionally incompetent virions are equally efficient, if not more robust, in their ability to induce expression of genes involved in innate immunity (Browne et al., 2001; Navarro et al., 1998; Preston et al., 2001; Simmen et al., 2001; Zhu et al., 1997; Zhu et al., 1998). These observations suggest that one or more virion structural components are responsible for initiating innate responses. The rapidity of the response coupled with the ability of a virion structural component to elicit the responses pointed to the possibility that glycoproteins displayed on the surface of the virion were activating innate immunity by interacting with host cell receptors.

## Glycoprotein B is a component capable of activating the interferon pathway

The ability of the envelope glycoprotein B (gB), the primary viral ligand, to directly bind cells make it a candidate for direct activation of innate receptors (Boyle and Compton, 1998). A soluble form of gB comprising the ectodomain of gB and retaining cell binding properties consistent with virion-associated gB was utilized to test the ability of gB to activate innate immunity independent of other virion components (Boyle et al., 1999). Microarray analysis revealed remarkable similarities with respect to ISG induction between HCMV-infected cells and cells treated with soluble gB (Simmen et al., 2001). In addition to direct induction of ISG transcription, soluble gB activates IRF3, NF-κB, and elicits the production of IFN-β (Boehme et al., 2004; Boyle and Compton, 1998; Chin and Cresswell, 2001; Simmen et al., 2001; Yurochko et al., 1997a). Together these data suggest that virus entry is not strictly required for the induction of interferon-α/β responses to CMV, but rather that interaction of gB with a cell surface receptor during virus binding is sufficient to elicit antiviral responses. Recently we tested the ability of HCMV to activate interferon responses in cells expressing a dominant negative form of TLR2 (DN-TLR2), a receptor known to trigger inflammatory cytokine responses to HCMV (Compton et al., 2003). Our results showed that the interferon pathway was intact in cells expressing DN-TLR2 suggesting that it is not the receptor involved in activation of the interferon response (K. Boehme and T. Compton, unpublished results).

Although the specific receptor responsible for activation of the interferon pathway by HCMV has not been identified, other members of the herpesvirus family may provide clues into mechanisms by which HCMV can activate the interferon-α/β pathway. MCMV, HSV-1, and HSV-2 elicit interferon-α secretion via TLR9 (Krug et al., 2004a; Krug et al., 2004b; Lund et al., 2003; Tabeta et al., 2004). Each of these viruses has a CpG motif-rich genome that constitutes the molecular basis for recognition of these viruses by TLR9 (Krug et al., 2004a; Lund et al., 2003). HCMV also contains a CpG-rich genome and is therefore also likely subject to innate sensing by TLR9, however this has not been formally established. Recognition of viral genomic DNA by TLR9, which resides in intracellular compartments, would likely require degradation of the virion within an appropriate compartment in order to release the viral genome from its housing and make it available for detection by TLR9 (Ahmad-Nejad et al., 2002; Takeshita et al., 2001). In addition, TLR3 is critical for MCMV-induced interferon-α production (Alexopoulou et al., 2001; Tabeta et al., 2004). dsRNA species resulting from bidirectional transcription of the viral genome during the late stages of MCMV infection are hypothesized to activate TLR3 (Tabeta et al., 2004). Such a mechanism might also be important for the interferon-α/β response to HCMV. Finally, RNAs in the tegument of the HCMV virion may also render it subject to sensing by TLR7 and/or TLR8, which detect ssRNA from several different viruses (Table 6.1) (Diebold et al., 2004; Heil et al., 2004; Lund et al., 2004). A growing possibility is that HCMV activates innate responses by several different means. Dissection of the multiple mechanisms by which HCMV may induce interferon responses will undoubtedly be the focus of intense future study.

## Envelope glycoproteins activate TLR-2-mediated inflammatory cytokine induction

Envelope glycoproteins including gB are also likely molecular triggers for the TLR2-mediated inflammatory cytokine induction. Similar to the interferon-α/β response, UV-inactivated virions elicit TLR2-dependent responses indicating that gene expression is not required for TLR activation, but rather a structural component(s) of the virion is responsible for triggering TLR2 (Compton et al., 2003). Envelope glycoprotein B appears to be the trigger for TLR2 as the two physically associate with one another and can be co-immunoprecipitated from cells. Furthermore, recombinant gB has intrinsic ability to induce IL-6 and IL-8 in a TLR2-dependent manner (M. Guerrero and T. Compton, unpublished results). The gH-containing complex may also be a trigger. Neutralizing antibodies to gH impair the ability of HCMV to activate cytokine production (M. Guerrero and T. Compton, unpublished results). The ability of glycoproteins to activate TLRs is intriguing because TLRs are theorized to recognize structures not normally found on in the host, such as LPS or CpG DNA. However, viral glycoproteins are produced by the host cell's own protein synthesis and glycosylation machinery and therefore do not intuitively harbor modifications or moieties that differ significantly from the host. It is possible that these proteins may possess unique structural conformations that are not assumed by host cell proteins. Similar to HCMV, respiratory syncytial virus activates TLR4 through its envelope protein, Env (Rassa, 2002). A common property of these two viral glycoproteins is that they play critical roles in the binding and entry of their respective viruses. This observation suggests that viruses are detected at the earliest stages of infection, simply via contact between glycoproteins on the viral envelope and TLRs on the cell surface. The ability of TLRs to recognize viral proteins that mediate cell entry is highly advantageous to the host because it allows the cell to begin mounting the appropriate response before the virus has actually invaded.

Additionally, innate activation may occur as a result of the fusion process itself. Virus-cell fusion is typically driven by helical bundle formation by viral glycoproteins. These structures could be subject to recognition by TLRs, or other PRRs. Both HCMV gB and gH are important for HCMV fusion (Kinzler and Compton, 2005) and contain putative coiled-coil domains which are the basis of helical bundle formation (Lopper and Compton, 2004). A soluble gB comprising the amino-terminal segment of the molecule (residues 1–460), but lacking the carboxy-terminal domain where the predicted coiled-coil domain resides is unable to activate IRF3 or induce ISG transcription (Boehme et al., 2004). These data support a role for fusion-generated structures as important determinants for innate immunity. Intriguingly, a small molecule inhibitor of HCMV entry that allows cell binding but prevents entry inhibits the induction of interferon-$\alpha/\beta$ responses (Netterwald et al., 2004). However, the inhibitor targets gB and may disrupt critical interactions between gB and receptor(s) that initiate the interferon-$\alpha/\beta$ response (Jones et al., 2004). Further study will be required to fully elucidate the variety of mechanisms by which HCMV activates innate immunity.

## Coordination of entry and innate immune activation

We are left with an apparent dichotomy. As HCMV enters cells to establish infection, the host recognizes the virions and activates innate immune responses. How are the two processes coordinated or are they at all? At this time, there is no apparent role for TLRs in driving an actual entry event. Rather at some point in the entry process, host immune sensors detect a pathogen-associated molecular pattern displayed on HCMV envelope glycoproteins and activate host innate defenses (Figure 6.1). One possibility is that entry receptors (HSPGs, integrins and signaling accessory molecules) and innate immunity machinery (TLR2, its membrane-associated partners, cytoplasmic adaptors and signaling machinery) coalesce into specialized membrane microdomains with integrins playing central ligating role. Concentration of all of these cell-surface receptors into a defined platform likely facilitates cell signaling events, some of which are optimal for replication and others of which are clearly hostile to the virus. Intriguingly integrins have also been shown to associate with TLR2 and to partition into cholesterol-rich lipid rafts (Ogawa et al., 2002; Triantafilou et al., 2002). The complexity of events at the cell surface during the initial encounter of HCMV and cells represents an exciting opportunity to better understand the molecular underpinnings of the early virus–host interactions. The recent identification of cell-surface molecules involved in the early steps of infection has greatly enhanced our knowledge of entry events in infection. Yet much remains to be done to elucidate aspects of mechanism of entry events and the corresponding innate immune activation.

References

Ahmad-Nejad, P., Hacker, H., Rutz, M., Bauer, S., Vabulas, R. M., Wagner, H. (2002). Bacterial CpG-DNA and lipopolysaccharides activate Toll-like receptors at distincT-cellular compartments. Eur. J. Immunol. 32, 1958–1968.

Akira, S. (2003). Toll-like receptor signaling. J. Biol. Chem. 278, 38105–38108.

Akira, S., and Hemmi, H. (2003). Recognition of pathogen-associated molecular patterns by TLR family. Immunol. Lett. 85, 85–95.

Albrecht, T., Speelman, D.J., and Steinsland, O.S. (1983). Similarities between cytomegalovirus-induced cell rounding and contraction of smooth muscle cells. Life. Sci. 32, 2273–2278.

Albrecht, T., and Weller, T.H. (1980). Heterogeneous morphologic features of plaques induced by five strains of human cytomegalovirus. Am. J. Clin. Pathol. 73, 648–654.

Alexopoulou, L., Holt, A.C., Medzhitov, R., and Flavell, R.A. (2001). Recognition of double-stranded RNA and activation of NF-kappaB by Toll-like receptor 3. Nature 413, 732–738.

Asselin-Paturel, C., Boonstra, A., Dalod, M., Durand, I., Yessaad, N., Dezutter-Dambuyant, C., Vicari, A., O'Garra, A., Biron, C., Briere, F., and Trinchieri, G. (2001). Mouse type I IFN-producing cells are immature APCs with plasmacytoid morphology. Nature Immunol. 2, 1144–1150.

Baldwin, B.R., Zhang, C., Keay, S. (2000). Cloning and epitope mapping of a functional partial fusion receptor for human cytomegalovirus gH. J. Gen. Virol. 81, 27–35.

Bauer, S., Kirschning, C.J., Hacker, H., Redecke, V., Hausmann, S., Akira, S., Wagner, H., and Lipford, G.B. (2001). Human TLR9 confers responsiveness to bacterial DNA via species-specific CpG motif recognition. Proc. Natl. Acad. Sci. USA 98, 9237–9242.

Beersma, M.F., Wertheim, van, D.P., and Feltkamp, T.E. (1990). The influence of HLA-B27 on the infectivity of cytomegalovirus for mouse fibroblasts. Scand. J. Rheumatol. Suppl 87, 102–103.

Beersma, M.F., Wertheim, van, D.P., Geelen, J.L., and Feltkamp, T.E. (1991). Expression of HLA class I heavy chains and beta 2-microglobulin does not affect human cytomegalovirus infectivity. J. Gen. Virol. 72, 2757–2764.

Berman, A.E., and Kozlova, N.I. (2000). Integrins: structure and functions. Membr. Cell. Biol. 13, 207–244.

Berman, A.E., Kozlova, N.I., and Morozevich, G.E. (2003). Integrins: structure and signaling. Biochemistry (Mosc.) 68, 1284–1299.

Boehme, K.W., and Compton, T. (2004). Innate sensing of viruses by Toll-like receptors. J. Virol. 78, 7867–7873.

Boehme, K.W., Singh, J., Perry, S.T., and Compton, T. (2004). Human cytomegalovirus elicits a coordinated cellular antiviral response via envelope glycoprotein B. J. Virol. 78, 1202–1211.

Bold, S., Ohlin, M., Garten, W., and Radsak, K. (1996). Structural domains involved in human cytomegalovirus glycoprotein B-7mediated cell–cell fusion. J. Gen. Virol. 77, 2297–2302.

Boldogh, I., AbuBakar, S., Deng, C.Z., and Albrecht, T. (1991). Transcriptional activation of cellular oncogenes fos, jun, and myc by human cytomegalovirus. J. Virol. 65, 1568–1571.

Borst, E.M., Hahn, G., Koszinowski, U.H., and Messerle, M. (1999). Cloning of the human cytomegalovirus (HCMV) genome as an infectious bacterial artificial chromosome in Escherichia coli: a new approach for construction of HCMV mutants. J. Virol. 73, 8320–8329.

Boyle, K.A., and Compton, T. (1998). Receptor-binding properties of a soluble form of human cytomegalovirus glycoprotein B. J. Virol. 72, 1826–1833.

Boyle, K.A., Pietropaolo, R.L., and Compton, T. (1999). Engagement of the cellular receptor for glycoprotein B of human cytomegalovirus activates the interferon-responsive pathway. Mol. Cell. Biol. 19, 3607–3613.

Britt, W.J. (1984). Neutralizing antibodies detect a disulfide-linked glycoprotein complex within the envelope of human cytomegalovirus. Virology 135, 369–378.

Britt, W.J., and Auger, D. (1986). Human cytomegalovirus virion-associated protein with kinase activity. J. Virol. 59, 185–188.

Britt, W.J., Jarvis, M., Seo, J.Y., Drummond, D., and Nelson, J. (2004). Rapid genetic engineering of human cytomegalovirus by using a lambda phage linear recombination system: demonstration that pp28 (UL99) is essential for production of infectious virus. J. Virol. 78, 539–543.

Browne, E.P., Wing, B., Coleman, D., and Shenk, T. (2001). Altered cellular mRNA levels in human cytomegalovirus-infected fibroblasts: viral block to the accumulation of antiviral mRNAs. J. Virol. 75, 12319–12330.

Browne, H., Smith, G., Beck, S., and Minson, T. (1990). A complex between the MHC class I homologue encoded by human cytomegalovirus and beta 2 microglobulin. Nature 347, 770–772.

Carlson, C., Britt, W.J., and Compton, T. (1997). Expression, purification and characterization of a soluble form of human cytomegalovirus glycoprotein B. Virology 239, 198–205.

Cary, L.A., Han, D.C., and Guan, J.L. (1999). Integrin-mediated signal transduction pathways. Histol. Histopathol. 14, 1001–1009.

Cella, M., Jarrossay, D., Facchetti, F., Alebardi, O., Nakajima, H., Lanzavecchia, A., and Colonna, M. (1999). Plasmacytoid monocytes migrate to inflamed lymph nodes and produce large amounts of type I interferon. Nature Med. 5, 919–923.

Chin, K.C., and Cresswell, P. (2001). Viperin (cig5), an IFN-inducible antiviral protein directly induced by human cytomegalovirus. Proc. Natl. Acad. Sci. USA 98, 15125–15130.

Chuang, T.H., and Ulevitch, R.J. (2000). Cloning and characterization of a sub-family of human Toll-like receptors: hTLR7, hTLR8 and hTLR9. Eur. Cytokine Netw. 11, 372–378.

Chuang, T.H., and Ulevitch, R.J. (2001). Identification of hTLR10: a novel human Toll-like receptor preferentially expressed in immune cells. Biochim. Biophys. Acta 1518, 157–161.

Compton, T., Kurt-Jones, E.A., Boehme, K.W., Belko, J., Latz, E., Golenbock, D.T., and Finberg, R.W. (2003). Human cytomegalovirus activates inflammatory cytokine responses via CD14 and Toll-like receptor 2. J. Virol. 77, 4588–4596.

Compton, T., Nowlin, D.M., and Cooper, N.R. (1993). Initiation of human cytomegalovirus infection requires initial interaction with cell surface heparan sulfate. Virology 193, 834–841.

Conti, C., Cirone, M., Sgro, R., Altieri, F., Zompetta, C., and Faggioni, A. (2000). Early interactions of human herpesvirus 6 with lymphoid cells: role of membrane protein components and glycosaminoglycans in virus binding. J. Med. Virol. 62, 487–497.

Dai, J., Megjugorac, N.J., Amrute, S.B., and Fitzgerald-Bocarsly, P. (2004). Regulation of IFN regulatory factor-7 and IFN-alpha production by enveloped virus and lipopolysaccharide in human plasmacytoid dendritic cells. J. Immunol. 173, 1535–1548.

Dalod, M., Hamilton, T., Salomon, R., Salazar-Mather, T.P., Henry, S.C., Hamilton, J.D., and Biron, C.A. (2003). Dendritic cell responses to early murine cytomegalovirus infection: subset functional specialization and differential regulation by interferon alpha/beta. J. Exp. Med. 197, 885–898.

Dalod, M., Salazar-Mather, T.P., Malmgaard, L., Lewis, C., Asselin-Paturel, C., Briere, F., Trinchieri, G., and Biron, C.A. (2002). Interferon alpha/beta and interleukin 12 responses to viral infections: pathways regulating dendritic cell cytokine expression in vivo. J. Exp. Med. 195, 517–528.

Diebold, S.S., Kaisho, T., Hemmi, H., Akira, S., and Reis, E.S.C. (2004). Innate antiviral responses by means of TLR7-mediated recognition of single-stranded RNA. Science 303, 1529–1531.

Du, X., Poltorak, A., Wei, Y., and Beutler, B. (2000). Three novel mammalian Toll-like receptors: gene structure, expression, and evolution. Eur. Cytokine Netw. 11, 362–371.

Dunn, W., Chou, C., Li, H., Hai, R., Patterson, D., Stolc, V., Zhu, H., and Liu, F. (2003). Functional profiling of a human cytomegalovirus genome. Proc. Natl. Acad. Sci. USA 100, 14223–14228.

Eto, K., Huet, C., Tarui, T., Kupriyanov, S., Liu, H.Z., Puzon-McLaughlin, W., Zhang, X.P., Sheppard, D., Engvall, E., and Takada, Y. (2002). Functional classification of ADAMs based on a conserved motif for binding to integrin alpha 9beta 1: implications for sperm-egg binding and other cell interactions. J. Biol. Chem. 277, 17804–17810.

Evinger, M., Rubinstein, M., and Pestka, S. (1981). Antiproliferative and antiviral activities of human leukocyte interferons. Arch. Biochem. Biophys. 210, 319–329.

Fairley, J.A., Baillie, J., Bain, M., and Sinclair, J.H. (2002). Human cytomegalovirus infection inhibits epidermal growth factor (EGF) signalling by targeting EGF receptors. J. Gen. Virol. 83, 2803–2810.

Feire, A.L., Koss, H., and Compton, T. (2004). Cellular integrins function as entry receptors for human cytomegalovirus via a highly conserved disintegrin-like domain. Proc. Natl. Acad. Sci. USA 101, 15470–15475.

Feldman, S.B., Ferraro, M., Zheng, H.M., Patel, N., Gould-Fogerite, S., and Fitzgerald-Bocarsly, P. (1994). Viral induction of low frequency interferon-alpha producing cells. Virology 204, 1–7.

Fitzgerald, K.A., McWhirter, S.M., Faia, K.L., Rowe, D.C., Latz, E., Golenbock, D.T., Coyle, A.J., Liao, S.M., and Maniatis, T. (2003). IKKepsilon and TBK1 are essential components of the IRF3 signaling pathway. Nature Immunol. 4, 491–496.

Fonteneau, J.F., Gilliet, M., Larsson, M., Dasilva, I., Munz, C., Liu, Y.J., and Bhardwaj, N. (2003). Activation of influenza virus-specific CD4$^+$ and CD8$^+$ T-cells: a new role for plasmacytoid dendritic cells in adaptive immunity. Blood 101, 3520–3526.

Fortunato, E.A., McElroy, A.K., Sanchez, I., and Spector, D.H. (2000). Exploitation of cellular signaling and regulatory pathways by human cytomegalovirus. Trends Microbiol. 8, 111–119.

Foster, G.R., Rodrigues, O., Ghouze, F., Schulte-Frohlinde, E., Testa, D., Liao, M.J., Stark, G.R., Leadbeater, L., and Thomas, H.C. (1996). Different relative activities of human cell-derived interferon-alpha subtypes: IFN-alpha 8 has very high antiviral potency. J. Interferon Cytokine Res. 16, 1027–1033.

Giugni, T.D., Soderberg, C., Ham, D.J., Bautista, R.M., Hedlund, K.O., Moller, E., and Zaia, J.A. (1996). Neutralization of human cytomegalovirus by human CD13-specific antibodies. J. Infect. Dis. 173, 1062–1071.

Gredmark, S., Britt, W.B., Xie, X., Lindbom, L., and Soderberg-Naucler, C. (2004). Human cytomegalovirus induces inhibition of macrophage differentiation by binding to human aminopeptidase N/CD13. J. Immunol. 173, 4897–4907.

Gretch, D.R., Kari, B., Rasmussen, L., Gehrz, R.C., and Stinski, M.F. (1988). Identification and characterization of three distinct families of glycoprotein complexes in the envelopes of human cytomegalovirus. J. Virol. 62, 875–881.

Grundy, J.E., McKeating, J.A., and Griffiths, P.D. (1987a). Cytomegalovirus strain AD169 binds beta 2 microglobulin in vitro after release from cells. J. Gen. Virol. 68, 777–784.

Grundy, J.E., McKeating, J.A., Sanderson, A.R., and Griffiths, P.D. (1988). Cytomegalovirus and beta 2 microglobulin in urine specimens. Reciprocal interference in their detection is responsible for artifactually high levels of urinary beta 2 microglobulin in infected transplant recipients. Transplantation 45, 1075–1079.

Grundy, J.E., McKeating, J.A., Ward, P.J., Sanderson, A.R., and Griffiths, P.D. (1987b). Beta 2 microglobulin enhances the infectivity of cytomegalovirus and when bound to the virus enables class I HLA molecules to be used as a virus receptor. J. Gen. Virol. 68, 793–803.

Heil, F., Hemmi, H., Hochrein, H., Ampenberger, F., Kirschning, C., Akira, S., Lipford, G., Wagner, H., and Bauer, S. (2004). Species-specific recognition of single-stranded RNA via Toll-like receptor 7 and 8. Science 303, 1526–1529.

Hemmi, H., Takeuchi, O., Kawai, T., Kaisho, T., Sato, S., Sanjo, H., Matsumoto, M., Hoshino, K., Wagner, H., Takeda, K., and Akira, S. (2000). A Toll-like receptor recognizes bacterial DNA. Nature 408, 740–745.

Hobom, U., Brune, W., Messerle, M., Hahn., G., Koszinowski, U. (2000). Fast screening procedures for random transposon libraries of cloned herpesvirus genomes: mutational analysis of human cytomegalovirus envelope glycoprotein genes. J. Virol. 74, 7720–7729.

Honda, K., Yanai, H., Mizutani, T., Negishi, H., Shimada, N., Suzuki, N., Ohba, Y., Takaoka, A., Yeh, W.C., and Taniguchi, T. (2004). Role of a transductional-transcriptional processor complex involving MyD88 and IRF-7 in Toll-like receptor signaling. Proc. Natl. Acad. Sci. USA 101, 15416–15421.

Hornung, V., Rothenfusser, S., Britsch, S., Krug, A., Jahrsdorfer, B., Giese, T., Endres, S., and Hartmann, G. (2002). Quantitative expression of Toll-like receptor 1–10 mRNA in cellular subsets of human peripheral blood mononuclear cells and sensitivity to CpG oligodeoxynucleotides. J. Immunol. 168, 4531–4537.

Huber, M.T., and Compton, T. (1997). Characterization of a novel third member of the human cytomegalovirus glycoprotein H-glycoprotein L complex. J. Virol. 71, 5391–5398.

Huber, M.T., and Compton, T. (1998). The human cytomegalovirus UL74 gene encodes the third component of the glycoprotein H-glycoprotein L-containing envelope complex. J. Virol. 72, 8191–8197.

Ibanez, C.E., Schrier, R., Ghazal, P., Wiley, C., and Nelson, J.A. (1991). Human cytomegalovirus productively infects primary differentiated macrophages. J. Virol. 65, 6581–6588.

Ihara, S., Saito, S., and Watanabe, Y. (1982). Suppression of fibronectin synthesis by an early function(s) of human cytomegalovirus. J. Gen. Virol. 59, 409–413.

Jones, P.L., Crack, J., and Rabinovitch, M. (1997). Regulation of tenascin-C, a vascular smooth muscle cell survival factor that interacts with the alpha v beta 3 integrin to promote epidermal growth factor receptor phosphorylation and growth. J. Cell. Biol. 139, 279–293.

Jones, T.R., Lee, S.W., Johann, S.V., Razinkov, V., Visalli, R.J., Feld, B., Bloom, J.D., and O'Connell, J. (2004). Specific inhibition of human cytomegalovirus glycoprotein B-mediated fusion by a novel thiourea small molecule. J. Virol. 78, 1289–1300.

Kari, B., and Gehrz, R. (1992). A human cytomegalovirus glycoprotein complex designated gC-II is a major heparin-binding component of the envelope. J. Virol. 66, 1761–1764.

Kari, B., and Gehrz, R. (1993). Structure, composition and heparin binding properties of a human cytomegalovirus glycoprotein complex designated gC-II. J. Gen. Virol. 74, 255–264.

Kawai, T., Sato, S., Ishii, K.J., Coban, C., Hemmi, H., Yamamoto, M., Terai, K., Matsuda, M., Inoue, J., Uematsu, S., Takeuchi, O.,and Akira, S. (2004). Interferon-alpha induction through Toll-like receptors involves a direct interaction of IRF7 with MyD88 and TRAF6. Nature Immunol. 5, 1061–1068.

Keay, S., and Baldwin, B. (1991). Anti-idiotype antibodies that mimic gp86 of human cytomegalovirus inhibit viral fusion but not attachment. J. Virol. 65, 5124–5128.

Keay, S., and Baldwin, B. (1992). The human fibroblast receptor for gp86 of human cytomegalovirus is a phosphorylated glycoprotein. J. Virol. 66, 4834–4838.

Keay, S., and Baldwin, B.R. (1996). Evidence for the role of cell protein phosphorylation in human cytomegalovirus/host cell fusion. J. Gen. Virol. 77, 2597–2604.

Keay, S., Baldwin, B.R., Smith, M.W., Wasserman, S.S., and Goldman, W.F. (1995). Increases in [$Ca^{2+}$]i mediated by the 92.5-kDa putative cell membrane receptor for HCMV gp86. Am. J. Physiol. 269, 11–21.

Keay, S., Merigan, T.C., and Rasmussen, L. (1989). Identification of cell surface receptors for the 86-kilodalton glycoprotein of human cytomegalovirus. Proc. Natl. Acad. Sci. USA 86, 10100–10103.

Keay, S., Rasmussen, L., and Merigan, T.C. (1988). Syngeneic monoclonal anti-idiotype antibodies that bear the internal image of a human cytomegalovirus neutralization epitope. J. Immunol. 140, 944–948.

Kinzler, E.R, and Compton, T. (2005). Characterization of human cytomegalovirus glycoprotein-induced cell–cell fusion. J. Virol. 79, 7827–7837.

Kowalik, T.F., Wing, B., Haskill, J.S., Azizkhan, J.C., Baldwin, A.S., Jr., and Huang, E.S. (1993). Multiple mechanisms are implicated in the regulation of NF-kappa B activity during human cytomegalovirus infection. Proc. Natl. Acad. Sci. USA 90, 1107–1111.

Krug, A., French, A.R., Barchet, W., Fischer, J.A., Dzionek, A., Pingel, J.T., Orihuela, M.M., Akira, S., Yokoyama, W.M., and Colonna, M. (2004a). TLR9-dependent recognition of MCMV by IPC and DC generates coordinated cytokine responses that activate antiviral NK cell function. Immunity 21, 107–119.

Krug, A., Luker, G.D., Barchet, W., Leib, D.A., Akira, S., and Colonna, M. (2004b). Herpes simplex virus type 1 activates murine natural interferon-producing cells through Toll-like receptor 9. Blood 103, 1433–1437.

Larsson, S., Soderberg-Naucler, C., Wang, F.Z., and Moller, E. (1998). Cytomegalovirus DNA can be detected in peripheral blood mononuclear cells from all seropositive and most seronegative healthy blood donors over time. Transfusion 38, 271–278.

Li, L., Nelson, J.A., and Britt, W.J. (1997). Glycoprotein H-related complexes of human cytomegalovirus: Identification of a third protein in the gCIII complex. J. Virol. 71, 3090–3097.

Lopper, M., and Compton, T. (2004). Coiled-coli domains in glycorpteoins B and H are involved in human cytomegalovirus membrane fusion. J. Virol. 78, 8333–8341.

Lund, J.M., Alexopoulou, L., Sato, A., Karow, M., Adams, N.C., Gale, N.W., Iwasaki, A., and Flavell, R.A. (2004). Recognition of single-stranded RNA viruses by Toll-like receptor 7. Proc. Natl. Acad. Sci. USA 101, 5598–5603.

Lund, J.M., Sato, A., Akira, S., Medzhitov, R., and Iwasaki, A. (2003). Toll-like receptor 9-mediated recognition of herpes simplex virus-2 by plasmacytoid dendritic cells. J. Exp. Med. 198, 513–520.

Lusso, P., Secchiero, P., Crowley, R.W., Garzino-Demo, A., Berneman, Z.N., and Gallo, R.C. (1994). CD4 is a critical component of the receptor for human herpesvirus 7: interference with human immunodeficiency virus. Proc. Natl. Acad. Sci. USA 91, 3872–3876.

Mach, M., Kropff, B., Dal Monte, P., and Britt, W. (2000). Complex formation by human cytomegalovirus glycoproteins M (gpUL100) and N (gpUL73). J. Virol. 74, 11881–11892.

McKeating, J.A., Griffiths, P.D., and Grundy, J.E. (1987). Cytomegalovirus in urine specimens has host beta 2 microglobulin bound to the viral envelope: a mechanism of evading the host immune response? J. Gen. Virol. 68, 785–792.

McKeating, J.A., Grundy, J.E., Varghese, Z., and Griffiths, P.D. (1986). Detection of cytomegalovirus by ELISA in urine samples is inhibited by beta 2 microglobulin. J. Med. Virol. 18, 341–348.

Medzhitov, R., Preston-Hurlburt, P., and Janeway, C.A., Jr. (1997). A human homologue of the Drosophila Toll protein signals activation of adaptive immunity. Nature 388, 394–397.

Miyamoto, S., Teramoto, H., Gutkind, J.S., and Yamada, K.M. (1996). Integrins can collaborate with growth factors for phosphorylation of receptor tyrosine kinases and MAP kinase activation: roles of integrin aggregation and occupancy of receptors. J. Cell. Biol. 135, 1633–1642.

Moro, L., Venturino, M., Bozzo, C., Silengo, L., Altruda, F., Beguinot, L., Tarone, G., and Defilippi, P. (1998). Integrins induce activation of EGF receptor: role in MAP kinase induction and adhesion-dependenT-cell survival. EMBO J. *17*, 6622–6632.

Myerson, D., Hackman, R.C., Nelson, J.A., Ward, D.C., and McDougall, J.K. (1984). Widespread presence of histologically occult cytomegalovirus. Hum. Pathol. *15*, 430–439.

Navarro, L., Mowen, K., Rodems, S., Weaver, B., Reich, N., Spector, D., and David, M. (1998). Cytomegalovirus activates interferon immediate-early response gene expression and an interferon regulatory factor 3-containing interferon-stimulated response element-binding complex. Mol. Cell Biol. *18*, 3796–3802.

Netterwald, J.R., Jones, T.R., Britt, W.J., Yang, S.J., McCrone, I.P., and Zhu, H. (2004). Postattachment events associated with viral entry are necessary for induction of interferon-stimulated genes by human cytomegalovirus. J. Virol. *78*, 6688–6691.

Nowlin, D.M., Cooper, N.R., and Compton, T. (1991). Expression of a human cytomegalovirus receptor correlates with infectibility of cells. J. Virol. *65*, 3114–3121.

Ogawa, T., Asai, Y., Hashimoto, M., and Uchida, H. (2002). Bacterial fimbriae activate human peripheral blood monocytes utilizing TLR2, CD14 and CD11a/CD18 as cellular receptors. Eur. J. Immunol. *32*, 2543–2550.

O'Neill, L.A., Fitzgerald, K.A., and Bowie, A.G. (2003). The Toll-IL-1 receptor adaptor family grows to five members. Trends Immunol. *24*, 286–290.

Perussia, B., Fanning, V., and Trinchieri, G. (1985). A leukocyte subset bearing HLA-DR antigens is responsible for in vitro alpha interferon production in response to viruses. Nat. Immun. Cell Growth Regul. *4*, 120–137.

Pietropaolo, R., and Compton, T. (1997). Direct interaction between human cytomegalovirus glycoprotein B and cellular annexin II. J. Virol. *71*, 9803–9807.

Pietropaolo, R., and Compton, T. (1999). Interference with annexin II has no effect on entry of human cytomegalovirus into fibroblasT-cells. J. Gen. Virol. *80*, 1807–1816.

Pleskoff, O., Treboute, C., and Alizon, M. (1998). The cytomegalovirus-encoded chemokine receptor US28 can enhance cell–cell fusion mediated by different viral proteins. J. Virol. *72*, 6389–6397.

Pleskoff, O., Treboute, C., Brelot, A., Heveker, N., Seman, M., and Alizon, M. (1997). Identification of a chemokine receptor encoded by human cytomegalovirus as a cofactor for HIV-1 entry. Science *276*, 1874–1878.

Polic, B., Jonjic, S., Pavic, I., Crnkovic, I., Zorica, I., Hengel, H., Lucin, P., and Koszinowski, U.H. (1996). Lack of MHC class I complex expression has no effect on spread and control of cytomegalovirus infection in vivo. J. Gen. Virol. *77*, 217–225.

Poltorak, A., He, X., Smirnova, I., Liu, M.Y., Huffel, C.V., Du, X., Birdwell, D., Alejos, E., Silva, M., Galanos, C., Freudenberg, M., Ricciardi-Castagnoli, P., Layton, B., and Beutler, B. (1998a). Defective LPS signaling in C3H/HeJ and C57BL/10ScCr mice: mutations in Tlr4 gene. Science *282*, 2085–2088.

Poltorak, A., Smirnova, I., He, X., Liu, M.Y., Van Huffel, C., McNally, O., Birdwell, D., Alejos, E., Silva, M., Du, X., Thompson, P., Chan, E.K., Ledesma, J., Roe, B., Clifton, S., Vogel, S.N.,and Beutler, B. (1998b). Genetic and physical mapping of the Lps locus: identification of the toll-4 receptor as a candidate gene in the critical region. Blood Cells Mol. Dis. *24*, 340–355.

Preston, C.M., Harman, A.N., and Nicholl, M.J. (2001). Activation of interferon response factor-3 in human cells infected with herpes simplex virus type 1 or human cytomegalovirus. J. Virol. *75*, 8909–8916.

Qureshi, S.T., Lariviere, L., Leveque, G., Clermont, S., Moore, K.J., Gros, P., and Malo, D. (1999). Endotoxin-tolerant mice have mutations in Toll-like receptor 4 (Tlr4). J. Exp. Med. *189*, 615–625.

Rassa, J.C., Meyers, J.L., Zhang, Y., Kudaravalli, R., and Ross, S.R. (2002). Murine retroviruses activate B-cells via interaction with Toll-like receptor 4. Proc. Natl. Acad. Sci. USA *99*, 2281–2286.

Raynor, C.M., Wright, J.F., Waisman, D.M., and Pryzdial, E.L. (1999). Annexin II enhances cytomegalovirus binding and fusion to phospholipid membranes. Biochemistry *38*, 5089–5095.

Rock, F.L., Hardiman, G., Timans, J.C., Kastelein, R.A., and Bazan, J.F. (1998). A family of human receptors structurally related to Drosophila Toll. Proc. Natl. Acad. Sci. USA *95*, 588–593.

Salazar-Mather, T.P., Lewis, C.A., and Biron, C.A. (2002). Type I interferons regulate inflammatory cell trafficking and macrophage inflammatory protein 1alpha delivery to the liver. J. Clin. Invest. *110*, 321–330.

Santoro, F., Kennedy, P.E., Locatelli, G., Malnati, M.S., Berger, E.A., and Lusso, P. (1999). CD46 is a cellular receptor for human herpesvirus 6. Cell 99, 817–827.

Sato, M., Hata, N., Asagiri, M., Nakaya, T., Taniguchi, T., and Tanaka, N. (1998). Positive feedback regulation of type I IFN genes by the IFN-inducible transcription factor IRF-7. FEBS Lett. 441, 106–110.

Sato, M., Suemori, H., Hata, N., Asagiri, M., Ogasawara, K., Nakao, K., Nakaya, T., Katsuki, M., Noguchi, S., Tanaka, N., and Taniguchi, T. (2000). Distinct and essential roles of transcription factors IRF-3 and IRF-7 in response to viruses for IFN-alpha/beta gene induction. Immunity 13, 539–548.

Servant, M.J., ten Oever, B., LePage, C., Conti, L., Gessani, S., Julkunen, I., Lin, R., and Hiscott, J. (2001). Identification of distinct signaling pathways leading to the phosphorylation of interferon regulatory factor 3. J. Biol. Chem. 276, 355–363.

Sharma, S., TenOever, B.R., Grandvaux, N., Zhou, G.P., Lin, R., and Hiscott, J. (2003). Triggering the interferon antiviral response through an IKK-related pathway. Science. 300, 1148–1151

Siegal, F.P., Kadowaki, N., Shodell, M., Fitzgerald-Bocarsly, P.A., Shah, K., Ho, S., Antonenko, S., and Liu, Y.J. (1999). The nature of the principal type 1 interferon-producing cells in human blood. Science 284, 1835–1837.

Simmen, K.A., Singh, J., Luukkonen, B.G., Lopper, M., Bittner, A., Miller, N.E., Jackson, M.R., Compton, T., and Fruh, K. (2001). Global modulation of cellular transcription by human cytomegalovirus is initiated by viral glycoprotein B. Proc. Natl. Acad. Sci. USA 98, 7140–7145.

Sinzger, C., Kahl, M., Laib, K., Klingel, K., Rieger, P., Plachter, B., and Jahn, G. (2000). Tropism of human cytomegalovirus for endothelial cells is determined by a post-entry step dependent on efficient translocation to the nucleus. J. Gen. Virol. 81, 3021–3035.

Soderberg, C., Giugni, T.D., Zaia, J.A., Larsson, S., Wahlberg, J.M., and Moller, E. (1993a). CD13 (Human Aminopeptidase-N) mediates human cytomegalovirus infection. J. Virol. 67, 6576–6585.

Soderberg, C., Larsson, S., Bergstedtlindqvist, S., and Moller, E. (1993b). Definition of a subset of human peripheral blood mononuclear cells that are permissive to human cytomegalovirus infection. J. Virol. 67, 3166–3175.

Spear, P.G., and Longnecker, R. (2003). Herpesvirus entry: an update. J. Virol. 77, 10179–10185.

Stannard, L.M. (1989). Beta 2 microglobulin binds to the tegument of cytomegalovirus: an immunogold study. J. Gen. Virol. 70, 2179–2184.

Stark, G.R., Kerr, I.M., Williams, B.R., Silverman, R.H., and Schreiber, R.D. (1998). How cells respond to interferons. Annu. Rev. Biochem. 67, 227.

Stone, A.L., Kroeger, M., and Sang, Q.X. (1999). Structure-function analysis of the ADAM family of disintegrin-like and metalloproteinase-containing proteins (review). J. Protein Chem. 18, 447–465.

Suhara, W., Yoneyama, M., Kitabayashi, I., and Fujita, T. (2002). Direct involvement of CREB-binding protein/p300 in sequence-specific DNA binding of virus-activated interferon regulatory factor-3 holocomplex. J. Biol. Chem. 277, 22304–22313.

Tabeta, K., Georgel, P., Janssen, E., Du, X., Hoebe, K., Crozat, K., Mudd, S., Shamel, L., Sovath, S., Goode, J., Alexopoulou, L., Flavell, R.A., and Beutler, B. (2004). Toll-like receptors 9 and 3 as essential components of innate immune defense against mouse cytomegalovirus infection. Proc. Natl. Acad. Sci. USA 101, 3516–3521.

Takeshita, F., Leifer, C. A., Gursel, I., Ishii, K. J., Takeshita, S., Gursel, M., Klinman, D. M. (2001). Cutting edge: Role of Toll-like receptor 9 in CpG DNA-induced activation of human cells. J. Immunol. 167, 3555–3558.

Takeda, K., and Akira, S. (2003). Toll receptors and pathogen resistance. Cell. Microbiol. 5, 143–153.

Takeuchi, O., and Akira, S. (2001). Toll-like receptors; their physiological role and signal transduction system. Int. Immunopharmacol. 1, 625–635.

Takeuchi, O., Hoshino, K., Kawai, T., Sanjo, H., Takada, H., Ogawa, T., Takeda, K., and Akira, S. (1999). Differential roles of TLR2 and TLR4 in recognition of gram-negative and gram-positive bacterial cell wall components. Immunity 11, 443–451.

Taniguchi, T., Ogasawara, K., Takaoka, A., and Tanaka, N. (2001). IRF family of transcription factors as regulators of host defense. Annu. Rev. Immunol. 19, 623–655.

Taniguchi, T., and Takaoka, A. (2002). The interferon-alpha/beta system in antiviral responses: a multimodal machinery of gene regulation by the IRF family of transcription factors. Curr. Opin. Immunol. 14, 111–116.

Taylor, H.P., and Cooper, N.R. (1990). The human cytomegalovirus receptor on fibroblasts is a 30-kilodalton membrane protein. J. Virol. 64, 2484–2490.

activation as well as their affect on host cells is required. This is particularly critical for fast growing hosts such as embryos or newborns. A substantial amount of work has already been published on the function of these IE proteins, including work showing how host cells can interfere with immediate-early protein function and how viruses respond to circumvent this host cell defense mechanism.

This review will summarize findings that relate to the selective localization of actively transcribing viral genomes in the host nucleus and to the formation of a privileged space where replication takes place. Within this context, it is clear that a balance exists between the host's defense and the virus' virulence that can be shifted to successful suppression or successful productive viral replication, influenced by the host's ability to silence the viral genome and the viral countermeasure to control its own transcription at the major immediate-early promoter (MIEP), and might be described as feed back loops whose outcomes depend on many factors and cellular backgrounds. A shift in balance, even slightly in favor of the host, may result in disproportionate repressive effects as measured by the number of infected cells where replication can take place or in the amount of viral particles produced in those cells where replication does take place.

## Nuclear positioning of the transcriptionally active and replicating viral genome

The last physical hurdle for the infecting viral genome appears to be the nuclear pore complex. How viral DNA enters the host nucleus is not entirely clear, nor is it clear how it is protected from degradation by endonucleases. One has to assume that neutralization of the viral DNA must first take place before it can enter the host nucleus. Such neutralization can be mediated either by the host's histones (resulting in the formation of a chromatin package) or by positively charged spermidine (resulting in the neutralization of the virus' negative charge). Late during infection, spermidine and spermine are up-regulated (Clarke and Tyms, 1991) and may function in neutralizing and compacting the replicated viral genome, the latter of which is essential for packaging viral genomes into capsids. Thus, spermidine may still be associated with the viral genome when it leaves the capsid and enters the host nucleus. Because the nucleus is not a homogenous "bag" of solutes but instead is a highly structured environment, viral genomes are excluded from certain domains most easily imagined for the tightly packed nucleolus. However, diffusion of macromolecules, such as mRNA, occurs nearly as fast as could be expected by free diffusion within this environment (Politz et al., 1999).

We do not know the size of the large viral genome of CMV in the nucleus. At one extreme, a completely extended circular genome may reach ~78 μm, which clearly would have difficulties diffusing much beyond the nuclear pore complex. At the other extreme, a tightly coiled, chromatinized genome may be just slightly larger than an encapsidated genome. Whether the viral genome moves through the nucleus by passive diffusion or by active transport is presently unknown. To date it has proven difficult to examine an active mechanism because tests for active transport use energy deprivation techniques. Such approaches change so many other functions that the results are likely to "prove transport" but not what prevented the respective read-out for transport such as arriving at a specific site seen by the appearance of transcripts.

Identification and localization of CMV genomes by in situ hybridization is aided by their large size, which enables individual genomes to be visualized. The size of these signals is close to that obtained by imaging tegument proteins like pp71 (Figure 7.1A and B). This is known as the diffraction point size, which is limited by the wavelength of light. Signals can only become weaker when originating from smaller sources. Comparison of encapsidated viruses and in situ hybridization signals representing CMV genomes suggests that the infected host nucleus contains genomes that are highly condensed. One caveat is that completely extended genomes would not be visible using this technique, because signals from any point along the genomes would be too weak to register with our present techniques. The maximum detectable size therefore, as estimated from what is visible, is equal to or greater than the wavelength of light (~300 nm).

The opposite technical problem is the much larger appearance of a structure due to the dispersion of light from a high-intensity source. In this instance, a structure can appear much larger than it is in reality. One such source are ND10 (nuclear domain 10); most often identified by the promyelocytic leukemia protein (PML) Figure 7.1B–D), which figure prominently in the early stages after DNA-virus infections and appear to function like nuclear depots (Ishov and Maul, 1996; Maul, 1998). Most ND10 appear to be substantially larger than 300 nm when examined by fluorescent microscopy but smaller than 300 nm when examined by electron microscopy. This disparity is an important issue to consider when interpreting images that show physical association between viral genomes and ND10. When human fibroblasts are examined at 3 h post infection (p.i.) (Figure 7.1B), HCMV genomes appear throughout the nucleus as diffraction-spot-sized signals. Only a few of these signals localize beside ND10; none localizes *in* these domains. Localization of genomes next to ND10 could be due to random events, particularly when one excludes the large volume of the nucleolus and the apparent location of various DNA viruses in only the interchromosomal space, that is, not within the chromosomal territories but the space occupied where ND10 are positioned (Bell et al., 2001; and unpublished results).

When IE transcripts are imaged, they appear to emanate only from a few ND10 (Figure 7.1C and D) (Ishov et al., 1997). The conclusion drawn from these images was that most of the viral genomes that had reached the nucleus were incompetent to transcribe and only those that reached ND10 found a congenial space where transcription was possible ("honeypot" hypothesis). The fly in that pot, however, is that most of the major ND10-associated proteins, such as PML and Sp100, are interferon-up-regulated proteins (Maul, 1998). More ominous, all, including the PML-interacting Daxx, are transcriptional repressors involved in the formation of heterochromatin (Ishov et al., 2004; Seeler et al., 1998; Xu et al., 2001). ND10, therefore, appear more like sites for nuclear defense. Also, CMV and other DNA viruses possess genes whose products can eliminate or disperse these nuclear domains, as first identified for herpes simplex virus (Maul et al., 1993). For CMV the dispersing protein is IE1, which is first accumulated at ND10 for some time before ND10 are dispersed (Ahn et al., 1998; Ishov et al., 1997). Surprisingly, the highest concentration of the MIEP repressor IE2 is located at the site of the highest concentration of viral IE transcripts (Figure 7.1C). These transcripts seem to pass from the IE2-covered site into the splicing-factor-containing domain (Figure 7.1D). The general arrangement of ND10, the region of accumulated transcripts, and the IE2-containing site was described as the immediate transcript environment (Ishov et al., 1997). At first glance, these observa-

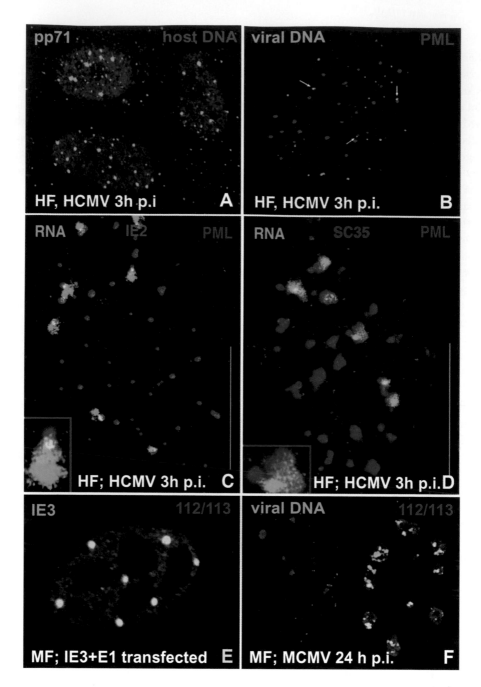

**Figure 7.1** Distribution of viral and certain host proteins after infection. (A) An HCMV-infected fibroblast nucleus probed for the tegument protein pp71 (green) and DAPI (blue) to label nuclei. Green small dots represent single viral particles. Larger whitish dots in nuclei represent pp71 accumulation at ND10. (B) An HCMV-infected fibroblast nucleus at 3 h post-infection (p.i.) containing individual in situ hybridization signals (green) that correspond to viral genomes, only three of which are in the immediate vicinity of ND10 (red). (C) Same as B but stained for IE2 (blue) and IE transcripts *(green)*. *IE2* aggregates surround the IE transcripts (green) emanating

tions indicate that the virus dispersed a site inhibitory to its replicative success. As will become evident, however, the immediate-early transcript environment also includes 112/113 gene products (Figure 7.1E) (i.e., the expanding prereplication domain) and apparently becomes the replication compartment many hours later (Figure 7.1F). More than a decade after the first observation that viruses transcribe predominantly at ND10, it is still unclear what advantage is afforded the virus by dispersing ND10-associated proteins. Is it possible that the questions asked are not relevant and that transcription at ND10, and ultimately replication where ND10 was present, is just a consequence of various layers of interactions, none of them simply representing either advantages for the virus or defense mechanisms of the host?

Because of the limited resolving power of light microscopy, we still do not know the precise location of transcribing viral genomes relative to ND10 to better than the huge molecular space gap of 300 nm. This is important since it could mean the virus has become part of ND10, is localizing at the ND10 interface within the interchromosomal space or does not associate with ND10 at all, but randomly localizes in the limited interchromosomal space where ND10 also resides. However, the observations that viral transcription occurs at ND10, IE1 disperses ND10-associated proteins, and interferon induces ND10-associated proteins suggest that the relationship between viral transcription and ND10 is not a casual but a causal relationship. To understand this relationship better we may have to tear the layers of this onion down to its tearful center to reconstruct the evolutionary balance achieved with a multitude of interactions each modifying others.

## IE1 counteracts the host's silencing mechanisms

The major immediate-early transcript is differentially spliced to produce a number of proteins. The two major and best-investigated proteins, IE1 and IE2, have in common exons 2 and 3 but differ in the larger exon 4 (IE1) and exon 5 (HCMV IE2 and its MCMV homolog, IE3). These proteins seem to act synergistically to activate early proteins, but antagonistically to auto-regulate MIEP (Keil et al., 1987; Stenberg et al., 1990). Both are considered to be promiscuous because they activate or augment both viral and host gene transcription (Hagemeier et al., 1992b). Most transactivators become part of the basal transcription machinery when they bind to DNA in the promoter region. Apparently, IE1 and IE2 are no exception. IE2 binds to specific sequences on early promoters (Cherrington and Mocarski 1989; Meier and Stinski, 1997). In addition, early in infection, IE2 interacts with TAFIID, TBP, TFIIB, and TAF250 (Caswell et al., 1993; Hagemeier et al., 1992a; Jupp et al., 1993;

---

from ND10 (red). Insert shows an immediate transcript environment at higher magnification (white line represents 10μm). (D) An HCMV-infected fibroblast nucleus triple labeled for ND10 (red), IE transcripts (green), and the interchromosomal splicing component containing SC35 (blue). Transcripts emanate from ND10 and spread into the SC35 domain. Insert shows an immediate transcript environment at higher magnification (white line represents 10μm). This image is represented in the center of Figure 7.5 in diagrammatic form. (E) Mouse fibroblast nucleus transfected with IE3 (green) and M112/113 (red) demonstrating that these two proteins colocalize. (F). Two nuclei infected at the same time with MCMV. Different stages of viral replication are present. Strong in situ hybridization (green) signals are present in a few replication compartments in one cell (right) but none in the other cell (left). Unit-strength signals at the surface of 112/113 domains in the nucleus (left) are interpreted as input genomes represented at the lower right of Figure 7.5 in diagrammatic form.

Lukac et al., 1997) proteins of the basal transcription machinery. Some, but not all, transcription factors (TBP, TFIIB) accumulate in the domain adjacent to ND10 together with IE2 (Ishov et al., 1997). Interestingly, IE2 is found in complexes containing other transcription factors (Lukac et al., 1997), suggesting that it functions like a TBP-associated factor (TAF). Acting as a TAF, IE2 can apparently substitute for TAFII250, since it rescued the transcriptional defect displayed by a ts mutant of TAFII250 (Lukac et al., 1997).

The mechanisms of IE1 functions are not understood as well as those of IE2. Apparently, IE1 is not essential to produce viral progeny but is necessary for the more natural mode of low particle infection involving low levels of particles. Fibroblasts infected with IE1-deletion mutants of HCMV require a much larger number of mutant viral particles to achieve the same degree of replicative success as that of wild-type viruses, indicating the necessity of multiple hits of IE1 mutants (Mocarski et al., 1996). Mocarski and colleagues also found that viral transactivators, such as tegument proteins, can compensate for IE1 at high multiplicities of infection. At low moi infected cells produce no replication compartments despite the near equal amount of IE2 synthesis (Greaves and Mocarski; 1998). IE1-deletion mutants of MCMV produce the same kind of results (Tang and Maul, 2003). This evidence does not support the idea that IE1 is a necessary component of the transcription machinery, which is in contrast with IE2. IE1, however, may interact with various TAFs (Table 7.1). Perhaps by using TAFII130 as a bridge to connect to IE2, IE1 can cause some augmentation of early gene transcription (Lukac et al., 1997). The temporal localization of IE1 in specific nuclear compartments, along with the potential interactions of IE1 with nuclear proteins in these compartments, points to additional functions of IE1 that are in line with the often noted augmentation of transcription observed in transfection experiments.

Shortly after infection, IE1 is first observed colocalizing with all ND10 with increasing concentration. Figure 7.2 quantitates the temporal appearance of IE1 at ND10 when cells are infected at 3 moi resulting in near 100% of cells producing IE1 by 6 h p.i. Also evident from this analysis is that IE2 appears simultaneously with IE1, albeit at only a few ND10 (i.e., the ones that show IE transcripts). Two other effects are evident in the plot: the disappearance of ND10 and the number of IE1 and IE2 domains observed. The cross-over of IE1 appearance and ND10 disappearance is at 60% instead of 50% expected from an immediate dispersion. From these data, derived from single-cell observations during the first 6 h p.i., we deduced that IE1 is first segregated and remains highly accumulated at ND10 for approximately 1 h (Ishov et al., 1997). ND10 dispersion, therefore, does not appear to be mediated enzymatically, but instead, may come about via sequestration; specifically, through PML binding of IE1 throughout the nucleus (Ahn et al., 1998). This explanation, however, is undercut by the recent finding that a 7 aa deletion mutant of MCMV IE1 can bind ND10 but cannot disperse it (Tang et al., submitted). We also found that IE2 remains associated with the immediate transcript environment after ND10 dispersal. It is not until 5 h p.i., a time when transcription of IE proteins diminishes, that these accumulations begin to disappear and the mRNA within them disperses (Ishov et al., 1997).

Intuitively, one might suppose that the nuclear site of the highest concentration of a protein is the same site where it functions. IE1 should therefore function in all ND10, and IE2 should function beside a few ND10. However, because not all ND10 have transcribing viral genomes, it follows that IE1 would not act on viral genomes nor does this

**Table 7.1** IE1 interactions and suspected effects

| IE1 interaction | Effect(s) | Reference |
| --- | --- | --- |
| IE1 | Autoregulation through 18 bp repeat | Cherrington and Mocarski (1989) |
| CTF-1/NF1 | Activates the DNAα pol promoter | Hayhurst et al. (1995) |
| E2F1 | Activates the DFHR promoter | Margolis et al. (1995) |
| E2F | Kinase activity | Pajovic et al. (1997) |
| p107 | Relieves transcriptional repression | Poma et al. (1996), Zhang et al. (2003) |
| Sp1 | | Yurochko et al. (1997) |
| dTAFII40 | Augments transcriptional activity | Lukac et al. (1997) |
| dTAFII110 | | Lukac et al. (1997) |
| hTAF130 | aa 215–378 | Lukac et al. (1997) |
| hTAFII130/IE2 | Activates; augmentation of transcription | Lukac et al. (1997) |
| PML | Reduces repressive effect of PML; desumoylates PML | Xu et al. (2001), Tang and Maul (2003), Lee et al. (2004) |
| Daxx | Reduces repressive effect of Daxx? | Tang and Maul (2003) |
| HDAC | Reduces deacetylation of MIEP | Tang and Maul (2003) |

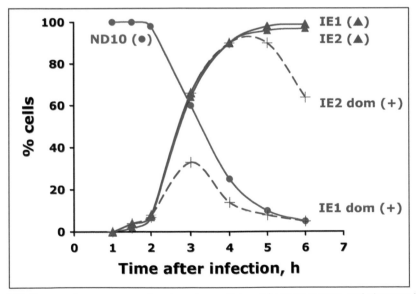

**Figure 7.2** Temporal appearance of various immediate-early components and changes in host structures. IE1 and IE2 appear simultaneously and within the same time frame as UL112/113 products (the latter not shown). The appearance and disappearance of ND10-IE1 and ND10-IE2 aggregates is also plotted, showing that IE1 appearance in ND10 precedes ND10 dispersion by about one hour.

IE1 act on host genes as no host genes have been found *in* ND10. Functions other than transactivation with the basal transcription machinery must therefore exist. Identifying proteins that interact with IE1 is one way to determine its other functions. IE colocalization with ND10 proteins has been used, therefore, to identify IE1 interaction partners. The interaction between IE1 and PML in HCMV (Ahn et al., 1998) has been confirmed also in MCMV (Tang and Maul, 2003). In immunoprecipitation analyses, both of the ND10-associated proteins PML and Daxx co-immunoprecipitate with MCMV IE1, suggesting that all three proteins form a complex. At present, no functional assay exists for examining the interaction between IE1 and Daxx. However, MCMV replicates successfully in both PML$^{-/-}$ and Daxx$^{-/-}$ cells, indicating that IE1-PML and IE1-Daxx interactions are not essential although no careful analysis as to the influence of these proteins on the overall replicative success has been performed in cells where these proteins and another ND10 protein, Sp100, have been eliminated or strongly down regulated by siRNA.

For PML in vitro evidence has been presented to show that IE1 can reduce the repressive effect of this protein (Lee et al., 2004). For Daxx a reduction in viral production was reported presumably due to the lack of viral genome deposition at ND10. The tegument transactivator pp71 and Daxx interact in the N-terminal half of the molecule (Hofmann et al., 2002; Ishov et al., 2002), i.e. away from the PML interacting C-terminal end, and pp71 is therefore highly enriched in all ND10 after infection. The viral genome deposition at ND10 is thought to come about by interaction of the viral DNA binding to pp71, which in turn is complexed to Daxx and thus deposited at highly increased frequency to the high concentration of PML at ND10. If Daxx is a repressor and binds pp71 to produce an inactive transactivator complex, the viral genome should be suppressed at ND10, a nuclear defense. Here IE1 may counter this defense by its binding of Daxx. How these separate interactions are balanced should be instructive in the choreography of sequential, temporal and spatial interactions that set the stage for the progress or suppression of the lytic cycle.

Our experiments also show that IE1 appears to interact with PML or Daxx independently. Because Daxx interacts with the histone deacetylases (HDAC) (Li et al., 2000), we tested the possibility that IE1 binds indirectly to HDAC. Identifying an IE1-HDAC interaction, however, may have been fortuitous, since HDAC does not normally localize to ND10, but does so in the presence of IE1. HDAC may be recruited to ND10 by IE1 during the early stages of infection when IE1 accumulates at ND10. The IE1-dependent segregation of HDAC to ND10 may not be significant, because there is little segregated, relative to the amount present in the nucleus. On the other hand, the large amounts of IE1 expressed, especially after ND10 dispersal, may be more significant, because IE1 might flood the nucleus sufficiently to reduce free HDAC to relieve the HDAC-associated suppression of chromatinized viral genomes. This is consistent with results from HDAC-activity assays showing that IE1 binding to HDAC inhibits HDAC deacetylation (Tang and Maul, 2003). IE1, therefore, may not exert its primary effects at sites where it is most concentrated; rather, it appears that IE1 functions throughout the nucleus as an HDAC scavenger, and possibly, as a scavenger for other host proteins.

When entering the host nucleus, the viral genome may become chromatinized to reduce its size; this would facilitate diffusion through the nucleus and aid repression by the host's deacetylating agents. Such a silencing mechanism would be an effective host defense. Indeed, evidence for silenced viral genomes has been found by precipitating deacetylated chromatinized MIEP with antibodies against deacetylated histones (Tang and Maul, 2003; Reeves, personal communication) shortly after infection of permissive cells. Consistent with such a host defense mechanism, as well as with the viral counteracting mechanism (i.e., IE1-mediated inhibition of the deacetylation of chromatinized viral DNA), is the finding that the deacetylation inhibitor trichostatin A (TSA) rescues an MCMV IE1 deletion mutant (Tang and Maul, 2003) and also the HCMV deletion mutant (unpublished results). Somewhat unexpectedly, evidence also shows that TSA substantially enhances the viral productivity of permissive cells infected with wild-type virus and significantly increases the number of cells exhibiting signs of productive infection. The latter observation suggests that even permissive cells, in the absence of an immune system or the cytokine-based innate immune response, can suppress viral replication after infection by a competent virus and can limit the initial production of IE1 and IE2 (or IE3, in the case of MCMV) (Tang and Maul, 2003). The suppression of many individual viral genomes may take place in the same nucleus where some other genomes are actively transcribing (those at ND10). The ability of the host cell to completely suppress the initiation of the viral replication cycle without complete inhibition of IE transcription, may rely on several factors: (1) the number of viral genomes entering the nucleus; (2) the increased amount of tegument-associated transcription factors internalized as a result of fusing dense particles; (3) the cell cycle stage of the cell; and (4) the amounts of silencing factors, such as HDAC, and repressors, such as Daxx and PML. An implication of these observations (i.e., that host cells can win) is that these mechanisms might be exploited to further shift the defense-virulence balance in favor of defense. The reverse has been shown by shifting the balance in favor of the virus. TSA treatment either in permissive cells or in cells that contained the virus over long times in a non-replicating state can be reactivated by the shift in the acetylation status (Murphy et al., 2002; Tang and Maul, 2003).

To shift the balance in favor of a nuclear defense, the precise function and structural properties of IE1 must be understood clearly to develop strategies that circumvent the pro virus activities of this immediate-early protein. One directly observable action of IE1 is that it regulates the dispersion of ND10-associated proteins (Ishov et al., 1997). Herpes simplex 1 destructs ND10-associated proteins and other host proteins by tagging with ubiquitin using the IE protein ICP0 as a conjugating enzyme (Boutell et al., 2002). IE1 is in essence the CMV equivalent, ultimately causing the dispersal of ND10-associated proteins. In contrast to Herpes simplex, CMV counteracts host repressors by inactivating them via IE1 binding, rather than by shunting them to the proteosomal pathway for degradation. After finding several repressors binding to IE1 including p107 (Poma et al., 1996; Zhang et al., 2003) it may be asked whether IE1 acts as a scavenger for specific host repressor proteins detrimental to the progression of viral productive infection. Whether IE1 inactivates host repressors by tightly binding to individual repressors or by binding to repressors that are components of larger complexes should be determined using column chromatography. The domain structure of IE1 might also reveal how such binding takes place and whether multiple proteins might bind individual IE1 domains.

## Structural and functional aspects of IE1

IE1 has several functional properties that have been used to probe its structure through mutational analysis. These include augmentation of viral and host gene transcription,

ND10 dispersal, and repressor protein binding. Table 7.1 lists proteins that interact with IE1. These interactions are related either to the indirect augmentation of transcription, possibly through the alleviation of repression of IE1 binding to p107 (Poma et al,. 1996) or of IE1 binding to HDAC (Tang et al., 2003), or to the direct augmentation of transcription through IE1 binding of transcription factors (Lukac et al., 1997). Determining how these proteins bind to IE1, and to which interface, is important for developing effective interference strategies. Because IE1 plays an important role during low-particle infections (i.e., the normal infection mode), a strategy aimed at IE1 inactivation might be successful in blocking the HDAC-binding capacity of IE1 and allow the host cell to silence competent viral genomes.

Deletion analysis has revealed that the HDAC binding site in MCMV IE1 is located between amino acid residues 100 and 310. A peptide composed of these amino acids retains HDAC-binding capability and the potential to augment transcription from MIEP. A deletion in a similar helical region of HCMV IE1 also eliminates the augmenting effect (Stenberg et al., 1990) and abolishes IE1-mediated dispersal of ND10 (Ishov et al., 1997; Lee et al., 2004). More detailed deletion analysis revealed that removal of amino acid sequences surrounding the HDAC binding domain eliminates the ability of IE1 to disperse ND10 (Tang et al., submitted). At the three-dimensional level this suggests that IE1 possesses a bipartite ND10 binding domain that is different from the HDAC binding domain. A bipartite p107 binding domain has also been reported for HCMV IE1 (Poma et al., 1996).

HCMV IE1 and MCMV IE1 share only 12% amino acid homology, mostly in the highly acidic C-terminal region. Moreover, HCMV IE1 is 20% shorter than MCMV IE1. From these figures, these two proteins appear to have very little in common. However, they share the same genetic structure, as well as the ability to disperse ND10 and augment viral transcription. Additional comparative and functional analyses are necessary, particularly if the mouse system is to be used as an experimental model to examine what cannot be done with the HCMV-host interactions during various phases of the natural infection cycle. Although the primary structures of HCMV IE1 and MCMV IE1 differ significantly, their secondary structures are surprisingly similar (Figure 7.3). This similarity is evident in the high degree of helicity in the first 400 amino acids of each protein and in the spacing of shorter and longer helical sections. In addition, a large proportion of the C-terminal region of each protein forms a random coil. In the absence of crystallographic data, in silico structural predictions will aid in the identification of putative binding interfaces and in the experimental analysis of IE1.

Secondary structure predictions indicate that both MCMV and HCMV IE1 proteins have the same folding or helical pattern, and thus, are likely to have a similar tertiary structure. Because neither MCMV IE1 nor HCMV IE1 share significant sequence homology with any other known proteins, threading calculations (Jones, 1998) were performed to fit an overlapping series of domain-sized sequence segments (100–200 aa) of IE1 onto all known three-dimensional (3D) protein structures. The threading calculations suggest that the first two-thirds of the IE1 sequence folds to form two structural domains. A model of the N-terminal domain of MCMV IE1 (residues 36–195) was based on the 3D structure of influenza M1 (PDB: 1aa7A2) (Sha and Luo, 1997). These N-terminal residues form a pair of anti-parallel bundles comprising four helices. A model of the central domain of

**MCMV IE1**

**HCMV IE1**

**Figure 7.3** Schematic representation of MCMV IE1 and HCMV IE1 depicting a similar distribution of helical elements.

MCMV IE1 (residues 215–345) was based on the 3D structure of the tetrapeptide repeat (TPR) domain of protein phosphatase 5 (PDB: 1a17) (Das *et al.*, 1998). The regions associated with ND10 binding, ND10 dispersal, and HDAC binding, are color-coded in Figure 7.4.

From this model (Figure 7.4), one can see that the two regions responsible for loss of ND10 binding (red and yellow) are located on the same face of the MCMV IE1 molecule and thus could be part of a single interaction interface. This indicates that MCMV IE1 possesses distinct sites that carry out these two binding functions. Deletion analysis indicates that the HDAC binding capacity of IE1 fails to be disrupted by removal of all residues located outside of the HDAC binding region. On the other hand, the region responsible for HDAC binding is located on a different part of the IE1 molecule. It is positioned opposite to that containing the two ND10 binding domains. This also indicates that maintenance

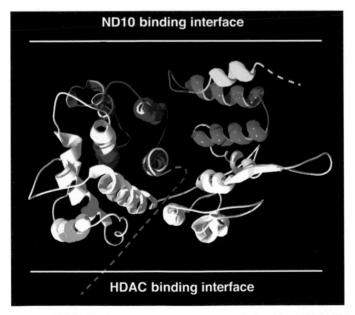

**Figure 7.4** Model of MCMV IE1 (a very similar model was derived for HCMV IE1) showing the helical regions of the 400 amino acids at the N-terminal. Dashed lines represent amino acids that could not be modeled. Color codes correspond to sections essential for HDAC binding (white) and ND10 binding (red as deleted from the N-terminal and yellow as deleted from the C-terminal). The green segment denotes two turns of a helix. Deletion of this helical region abolishes ND10 dispersion but not binding.

of the HDAC binding structure is independent of the surrounding residues. The HDAC binding site may also be bipartite in that it has the potential to form a binding cleft between N-terminal and central structural domains. Nonetheless, the effective or minimum HDAC binding site could involve only a subset of these residues. Indeed, recent observations in yeast identify tetratricopeptide domains as potential candidates for the minimum HDAC binding site of IE1. Davie et al. (2003) demonstrated that Tup1-Ssn6 interactions with multiple class I HDACs in vivo occur via distinct tetratricopeptide domains within Ssn6. The C-terminal half of the HDAC binding region of IE1 also has a pair of tetratricopeptide repeats that could also bind HDAC. These tetratricopeptide repeats represent, therefore, the likely candidate for the minimum HDAC binding site.

Residues 135–141 (Figure 7.4, green) of IE1 form half of a helix that corresponds, according to threading calculations, to the influenza M1 dimer interface. Deletion of these residues are predicted to disrupt the putative dimer interface of IE1, but not to destabilize its ND10 binding site (red). The site important for ND10 dispersal is located within the HDAC binding domain (white), and thus is located in an area distinct from the ND10 binding site. Deletion analysis has yet to define the full functional aspects of HCMV IE1. However, if the structures of HCMV IE1 and MCMV IE1 are as similar as the threading calculation models suggest, it is clear why deletion of the putative HDAC binding domain abolishes ND10 binding or more specifically PML binding (Ishov et al 1997; Lee et al., 2004). The entire 3D structure is disrupted when the helical region between residues 132 and 274 is deleted from HCMV IE1. Although these observations may suggest that IE1 possesses ND10 and HDAC binding interfaces that overlap, this is unlikely, because wild-type IE1 can bind at ND10 and sequester HDAC to this domain (Tang and Maul, 2003). Deletion of either aa 45–52 or aa 312–317 of HCMV IE1 abolishes IE1 binding to p107 (Zhang et al., 2003). These sites are located in the same region containing the bipartite ND10 binding site; thus, removal of these residues may also interrupt a homologous domain. Comparative mutational analysis should clarify this issue. Three-dimensional structural modeling of IE1 is consistent with findings from deletion analyses and provides a more rational approach for identifying interaction domains and mapping them with higher resolution. This approach may prove to be very useful when searching for small molecule docking sites that can prevent HDAC binding.

One unresolved aspect of the apparent structural symmetry of modeled HCMV IE1 and MCMV IE1 is the position of the small ubiquitin modifier (SUMO). In HCMV IE1, SUMO is positioned at aa 450 (Xu et al., 2001) and in MCMV IE1 we find the SUMO consensus sequence at aa 223. These rather large covalent modifications could have a strong differential influence on the 3D structure of IE1, and thus could influence the functional properties of the two different IE1. Since SUMO modifies very small amounts of protein at any given time (Johnson, 2004), these SUMO subsets may have additional functions. In HCMV IE1, deletion of the SUMO modification site reduces the levels of IE2 transcript and their translation products (Nevels et al, 2004) and PML desumoylation (Lee et al., 2004). In MCMV IE1, the putative SUMO modification site is located within the HDAC binding site; its precise functions, however, are yet to be determined.

Despite the non-essential aspect of IE1 for productive infection, IE1 appears very important for replicative success. Isolating the respective functions of this molecule and assigning them to its different interfaces may provide a rational basis for the search for small interfering molecules. Such molecules may induce an IE1 minus phenotype with a substantial lowering of productive infection.

## Segregation of host and viral proteins into a privileged virus-generated nuclear space

IE3, the major transactivator of MCMV early proteins, and its HCMV homolog IE2 are essential for replicative success (Angulo et al., 2000; White et al., 2004). As with MCMV IE1 and HCMV IE1 sequence similarity, MCMV IE3 and HCMV IE2 lack much sequence similarity, except in the C-terminal portion of each molecule. These proteins interact with several proteins involved in transcription, such as the TATA-binding protein TBP (Casswell et al., 1993) and the transcription-associated factor TAFIID (Hagemeier et al., 1992 a), and function like TAFs (Lukac et al., 1997). Moreover, MCMV IE3 and HCMV IE2 are essential for transcriptional activation of early proteins, such as those expressed from the 112/113 locus (Malone et al., 1990; Messerle et al., 1992), even though these early proteins become recognizable at the same time as IE1 and IE2/IE3 at about 2 h p.i. (Ahn et al 1999; Bühler et al., 1990). This suggests that very little IE3 is needed to activate the early promoter of M112/113. Although MCMV IE3 and HCMV IE2 are auto-repressors of MIEP, they do not repress at the same apparent concentration effective in activation of the 112/113 promoter. The highest concentration of IE2 in the region of the actively transcribing viral genome (the immediate transcript environment) should ensure repression despite the low concentration of HCMV IE2 or MCMV IE3 in the nucleus. Additional controls must exist to block these repressors from suppressing MIEP.

For HCMV, early proteins transcribed from UL112/113 bind single- and double-stranded DNA and localize to nuclear inclusions (Iwayama et al., 1994) that reside with IE2 proteins and are situated adjacent to ND10 (Ahn et al., 1999). These domains expand and represent the prereplication domains. When replication starts within these domains, they become the replication compartment (Tang et al., 2005). It is unknown why these prereplication domains expand and exist for an extended time before viral DNA synthesis begins. One possible explanation is that UL112/113 must be present prior to viral DNA synthesis, because they enhance transcription of proteins essential for the replication complexes (Iwayama et al., 1994). Indeed, UL112/113 proteins have been shown to enhance the IE-mediated activation of the DNA polymerase (UL54) promoter (Kerry et al., 1996). Although no other functions have been described for the products of the 112/113 locus, this should be expected, because the expression kinetics of different 112/113 splice products are quite different. Surprisingly, individual 112/113 proteins follow a complicated and quantitatively dynamic accumulation pattern (Wright and Spector, 1989). Various methods have identified four gene products each for UL112/113 (34, 43, 50, and 84 kDa) and for M112/113 (33, 36, 38, and 87 kDa) (Bühler et al., 1990; Ciocco-Schmitt et al., 2002; Wright and Spector, 1989). Additional variants may represent secondary modifications of these products, most likely phosphorylation, which may modify the 112/113 isoforms to function in different contexts.

As with early proteins transcribed from UL112/113, those transcribed from M112/113 also form nuclear inclusions that localize adjacent to a limited number of ND10. Later on during the viral replication cycle, these nuclear inclusions house replicated viral progeny, and exclude host DNA (Tang et al., 2005). This expanding and PML and Daxx recruiting domain can be mimicked by transfecting cells with M112/113 alone or with both M112/113 and IE3. In cells transfected with M112/113 alone, M112/113 forms a space where cellular and viral proteins are segregated whereas IE3 segregates only in the presence of M112/113 (Figure 7.1E) and accumulates to a higher level than when expressed alone suggesting that IE3 is protected by M112/113 from degradation. Immunoprecipitation shows that in cells transfected with M112/113 and IE3, M112/113 binds IE3, and MIEP-driven luciferase reporter assays demonstrate that despite the M112/113-dependent higher accumulation of IE3 its repressive effect is substantially diminished. This may account for the paradoxical finding demonstrated by in situ hybridization that high concentrations of the auto-repressor IE3 accumulate at the site of viral immediate-early transcription. Thus, despite the high concentrations of IE3 present early during infection around MIEP, MIEP is not repressed because IE3 may not be functionally available due to binding with M112/113 products.

This observation reveals a novel function for M112/113 proteins: their ability to block the interaction between an immediate-early protein with its own promoter. We have constructed three models to describe this newly discovered function. In the first model, binding of IE3 and 112/113 proteins can occur via an essential C-terminal IE3 promoter-binding site, thus rendering IE3 inactive. Such binding also results in the deposition of IE3 at the 112/113 protein-induced prereplication domain. In this case, this deposition would be an irrelevant side-effect. In the second model, after nucleation of the viral genome at ND10, M112/113 accumulates to form a dense matrix (similar to the PML-based ND10 matrix) that can trap incoming IE3. To act as an IE3 sink, this matrix would need sufficient binding sites for IE3. Once bound by this "sink," IE3 is prevented from reaching MIEP and interacting with it. The determination of IE3 resident times in the prereplication domain may prove instructive in the alternatives.

These two models are not mutually exclusive. IE3 and its HCMV homolog IE2 have binding sites for a number of cellular factors and probably more than one interface (Lukac et al., 1997). M112/113 proteins are postulated to possess at least two interfaces, one in the common N-terminal region, which may be responsible for aggregating all four 112/113 translation products within the prereplication domain, and one in the C-terminal end of the 87 kDa component, which appears to contain the binding site for IE3 (Tang et al., 2005). These multiple binding sites may contribute to the formation of a dense matrix around the viral genome. The viral genome may serve as the site of nucleation, because it has many IE3 binding sites as well as binding sites for early promoters. Although these two models adequately represent the simple removal of IE3 from the soluble pool, they fail to account for how such removal affects transactivation of the early protein promoters. These promoters appear to be activated synergistically by all four 112/113 products (Kerry et al., 1996). The various 112/113 splice products, as well as their secondary phospho-modified products, may modulate the repression of IE3 and the activation of early protein transcription. Alternatively, the N-terminal region of the 112/113 proteins, necessary for activation of early promoters (Malone et al., 1990; Pizzorno and Hayward 1990), may remain functional, even though the repressive region (i.e., IE3 binding site) in the C-terminal domain remains occupied by the 112/113 proteins. Since viral replication starts much later than the appearance of 112/113 early proteins, important proteins integral to the replicative complement may not be present. An accurate timing of their appearance needs to be considered to clarify this issue. Some of these proteins may appear substantially later than 112/113 proteins (see time sequence of IE1 and IE2 appearance; Figure 7.2). There seems to be a need to dissect how the different components of the 112/113 locus inhibit IE3 repression of MIEP under the assumption that it would be desirable to enhance this repression with a virus-specific method.

In the third model, upon entry into the host nucleus, most viral genomes briefly undergo transcription. Shortly thereafter, however, HDAC quickly silences these viral genomes. Through this brief window of transcription, many of these genomes can contribute to the production of IE1 and IE3, thereby providing the activation potential for early proteins, particularly those directed by the 112/113 promoter. With IE1 segregation to all ND10 and the resulting inhibition of HDAC-mediated deacetylation, only genomes reaching ND10 continue to be transcribed. Although ND10 do not segregate HDAC despite the presence of Daxx, HDAC levels may decrease in the immediate vicinity of ND10 through HDAC binding that occurs via IE1. Sites containing viral DNA then attract 112/113 products that form the prereplication domain and that inhibit the repressive properties of IE3. This enables the prereplication domain to extend further and to generate a privileged space. Instead of silencing MIEP, further transcription would be a consequence of the formation of prereplication sites where transcription complexes are preformed and repressive complexes are kept inactive. All other viral genomes outside of the immediate vicinity of this privileged space are silenced either by HDAC or by IE3. Such a scenario is consistent with the finding that a large number of competent viruses in the nucleus become silenced when they are not at ND10. The formation of a privileged space by either UL112/113 or M112/113 resolves a number of difficult-to-explain findings, including the presence of active viral genomes at sites having the highest levels of interferon-activated repressive proteins and having the highest concentrations of auto-repressive immediate-early proteins. Figure 7.5 illustrates a few of the positional stages possible with a question mark at the viral genome transcribing at sites other than ND10, an essential feature of this model which will be difficult to prove.

## Summary

The following is a simplified sequence envisioned for the temporal sequence of the several counteracting activations and repressions that occur very early during CMV infection of host cells. After entering the host nucleus, viral genomes encounter histones and are likely chromatinized before being activated by tegument-associated transactivators like pp71. At this time, transcription of these genomes begins and continues throughout the nucleus until the genomes are silenced by HDAC, IE2, hydrolysis or loss of activation by the segregation of pp71 by Daxx to ND10. Genomes that reach ND10 are preferentially protected from degradation. These viral genomes are bound to ND10 possibly via foreign DNA/protein complexes formed by the binding of pp71 to the genome, caught by ND10 via binding of pp71 to Daxx interaction with PML. Whether Daxx with its extreme short residency rate at ND10 inactivates or promotes activation by pp71 needs to be established. After the initial, but minimal, translation of the IE proteins, the very early 112/113 promoter becomes activated and 112/113 proteins are produced. At the same time, IE1 and IE2/IE3 attempt to maintain MIEP open and shut it down, respectively. However, the 112/113 products, being produced at the 'same' time as IE2/IE3, scavenge these MIEP repressors and early protein activators thus protecting the MIEP from suppression, and holding off replication

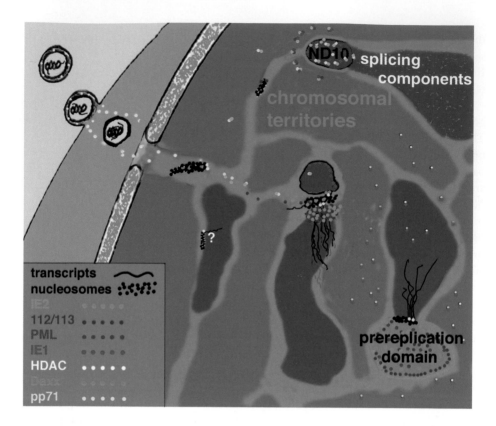

**Figure 7.5** Diagrammatic representation of temporal and spatial aspects of early CMV infection. The interacting proteins are coded as double dots with the respective colors at the left lower corner.

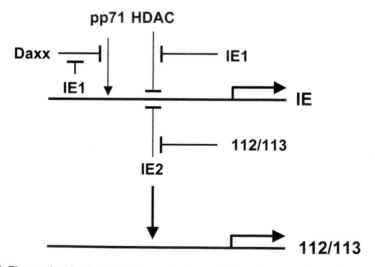

**Figure 7.6** Players for dominance at the major immediate-early promoter–enhancer.

protein transcription and thus replication. How this suppression is relieved is unclear and needs to be investigated with the intent to discern additional layers of control mechanisms. It is clear, however, that this suppression originates from the host (via HDAC) and from the virus (via IE2/IE3) and is modified by viral IE1 and 112/113 proteins, respectively shown diagrammatically (Figure 7.6). Understanding the regulation of the temporal balance of these very early events which may either lead to lytic progression or latency gains heightened interest because it can lead to strategies aimed at shifting the balance in favor of the host, strategies that do not involve nucleotide analogs and target only viral proteins.

## Acknowledgments
This study was supported by funds from NIH AI 41136 and NIH GM 57599, the G. Harold and Leila Y. Mathers Charitable Foundation, and the Commonwealth Universal Research Enhancement Program, Pennsylvania Department of Health.

## References
Ahn, J.H., Brignole, E.J., 3rd, and Hayward, G.S. (1998). Disruption of PML subnuclear domains by the acidic IE1 protein of human cytomegalovirus is mediated through interaction with PML and may modulate a RING finger-dependent cryptic transactivator function of PML. Mol. Cell Biol. *18*, 4899–4913.

Ahn, J.H., Jang, W.J., and Hayward, G.S. (1999). The human cytomegalovirus IE2 and UL112–113 proteins accumulate in viral DNA replication compartments that initiate from the periphery of promyelocytic leukemia protein-associated nuclear bodies (PODs or ND10). J. Virol. *73*, 10458–10471.

Angulo, A., Ghazal, P., and Messerle, M. (2000). The major immediate-early gene ie3 of mouse cytomegalovirus is essential for viral growth. J. Virol. *74*, 11129–11136.

Bell, P., Montaner, L.J., and Maul, G.G. (2001). Accumulation and intranuclear distribution of unintegrated human immunodeficiency virus type 1 DNA. J. Virol. *75*, 7683–7691.

Boutell, C., Sadis, S., and Everett, R.D. (2002). Herpes simplex virus type 1 immediate-early protein ICP0 and its isolated RING finger domain act as ubiquitin E3 ligases in vitro. J. Virol. *76*, 841–850.

Bühler, B., Keil, G.M., Weiland, F., and Koszinowski, U. H. (1990). Characterization of the murine cytomegalovirus early transcription unit e1 that is induced by immediate-early proteins. J. Virol. *64*, 1907–1919.

Caswell, R., Hagemeier, C., Chiou, C.J., Hayward, G., Kouzarides, T., and Sinclair, J. (1993). The human cytomegalovirus 86K immediate-early (IE) 2 protein requires the basic region of the TATA-box binding protein (TBP) for binding, and interacts with TBP and transcription factor TFIIB via regions of IE2 required for transcriptional regulation. J. Gen. Virol. *74*, 2691–2698.

Cherrington, J.M., and Mocarski, E.S. (1989). Human cytomegalovirus ie1 transactivates the alpha promoter-enhancer via an 18-base-pair repeat element. J. Virol. *63*, 1435–1440.

Clarke, J.R., and Tyms, A.S. (1991). Polyamine biosynthesis in cells infected with different clinical isolates of human cytomegalovirus. J. Med. Virol. *34*, 212–216.

Das, A.K., Cohen, P.W., and Barford, D. (1998). The structure of the tetratricopeptide repeats of protein phosphatase 5: implications for TPR-mediated protein–protein interactions. Embo J. *17*, 1192–1199.

Davie, J.K., Edmondson, D.G., Coco, C.B., and Dent, S.Y. (2003). Tup1-Ssn6 interacts with multiple class I histone deacetylases in vivo. J. Biol. Chem. *278*, 50158–50162.

Greaves, R.F., and Mocarski, E.S. (1998). Defective growth correlates with reduced accumulation of a viral DNA replication protein after low-multiplicity infection by a human cytomegalovirus ie1 mutant. J. Virol. *72*, 366–379.

Hagemeier, C., Walker, S., Caswell, R., Kouzarides, T., and Sinclair, J. (1992a). The human cytomegalovirus 80-kilodalton but not the 72-kilodalton immediate-early protein transactivates heterologous promoters in a TATA box-dependent mechanism and interacts directly with TFIID. J. Virol. *66*, 4452–4456.

Hagemeier, C., Walker, S.M., Sissons, P.J., and Sinclair, J.H. (1992b). The 72K IE1 and 80K IE2 proteins of human cytomegalovirus independently trans-activate the c-fos, c-myc and hsp70 promoters via basal promoter elements. J. Gen. Virol. 73, 2385–2393.

Hayhurst, G.P., Bryant, L.A., Caswell, R.C., Walker, S.M., and Sinclair, J.H. (1995). CCAAT box-dependent activation of the TATA-less human DNA polymerase alpha promoter by the human cytomegalovirus 72-kilodalton major immediate-early protein. J. Virol. 69, 182–188.

Hofmann, H., Sindre, H., and Stamminger, T. (2002). Functional interaction between the pp71 protein of human cytomegalovirus and the PML-interacting protein human Daxx. J. Virol. 76, 5769–5783.

Ishov, A.M., and Maul, G.G. (1996). The periphery of nuclear domain 10 (ND10) as site of DNA virus deposition. J. Cell Biol. 134, 815–826.

Ishov, A.M., Stenberg, R.M., and Maul, G.G. (1997). Human cytomegalovirus immediate-early interaction with host nuclear structures: definition of an immediate transcript environment. J. Cell Biol. 138, 5–16.

Ishov, A.M., Vladimirova, O.V. and Maul. G.G. (2002). Daxx-Mediated Accumulation of Human Cytomegalovirus Tegument Protein pp71 at ND10 Facilitates Initiation of Viral Infection at These Nuclear Domains. J Virol 76:7705–12.

Ishov, A.M., Vladimirova, O.V., and Maul. G.G. (2004). Heterochromatin and ND10 are cell cycle regulated and phosphorylation-dependent alternate nuclear sites of the transcription repressor Daxx and SWI/SNF protein ATRX. J Cell Sci 117:3807–20.

Iwayama, S., Yamamoto, T., Furuya, T., Kobayashi, R., Ikuta, K., and Hirai, K. (1994). Intracellular localization and DNA-binding activity of a class of viral early phosphoproteins in human fibroblasts infected with human cytomegalovirus (Towne strain). J. Gen. Virol. 75, 3309–3318.

Jones, D.T. (1998) Protein Sequence Threading by Double Dynamic Programming. In Computational Methods in Molecular Biology. Steven Salzberg, David Searls, and Simon Kasif, eds. (Elsevier Science), chapter 13.

Johnson, E.S. (2004). Protein modification by SUMO. Annu. Rev. Biochem. 73, 355–382.

Jupp, R., Hoffmann, S., Stenberg, R.M., Nelson, J.A., and Ghazal, P. (1993). Human cytomegalovirus IE86 protein interacts with promoter-bound TATA- binding protein via a specific region distinct from the autorepression domain. J. Virol. 67, 7539–7546.

Keil, G.M., Ebeling-Keil, A., and Koszinowski, U.H. (1987). Immediate-early genes of murine cytomegalovirus: location, transcripts, and translation products. J. Virol. 61, 526–533.

Kerry, J.A., Priddy, M.A., Jervey, T.Y., Kohler, C.P., Staley, T.L., Vanson, C.D., Jones, T.R., Iskenderian, A.C., Anders, D.G., and Stenberg, R.M. (1996). Multiple regulatory events influence human cytomegalovirus DNA polymerase (UL54) expression during viral infection. J. Virol. 70, 373–382.

Lee, H.R., Kim, D.J., Lee, J.M., Choi, C.Y., Ahn, B.Y., Hayward, G.S., and Ahn, J.H. (2004). Ability of the human cytomegalovirus IE1 protein to modulate SUMOylation of PML correlates with its functional activities in transcriptional regulation and infectivity in cultured fibroblasT-cells. J. Virol. 78, 6527–6542.

Li, H., Leo, C., Zhu, J., Wu, X., O'Neil, J., Park, E.J., and Chen, J.D. (2000). Sequestration and inhibition of Daxx-mediated transcriptional repression by PML. Mol. Cell Biol. 20, 1784–1796.

Lukac, D.M., Harel, N.Y., Tanese, N., and Alwine, J.C. (1997). TAF-like functions of human cytomegalovirus immediate-early proteins. J. Virol. 71, 7227–7239.

Lukac, D.M., Manuppello, J.R., and Alwine, J.C. (1994). Transcriptional activation by the human cytomegalovirus immediate-early proteins: requirements for simple promoter structures and interactions with multiple components of the transcription complex. J. Virol. 68, 5184–5193.

Malone, C.L., Vesole, D.H., and Stinski, M.F. (1990). Transactivation of a human cytomegalovirus early promoter by gene products from the immediate-early gene IE2 and augmentation by IE1: mutational analysis of the viral proteins. J. Virol. 64, 1498–1506.

Margolis, M.J., Pajovic, S., Wong, E.L., Wade, M., Jupp, R., Nelson, J.A., and Azizkhan, J. C. (1995). Interaction of the 72-kilodalton human cytomegalovirus IE1 gene product with E2F1 coincides with E2F-dependent activation of dihydrofolate reductase transcription. J. Virol. 69, 7759–7767.

Maul, G.G., Guldner, H.H. and Spivack J.G. 1993. Modification of discrete nuclear domains induced by herpes simplex virus type 1 immediate-early gene 1 product (ICP0). J Gen Virol 74:2679–90.

Maul, G.G. (1998). Nuclear domain 10, the site of DNA virus transcription and replication. Bioessays 20, 660–667.

Meier, J.L., and Stinski, M.F. (1997). Effect of a modulator deletion on transcription of the human cytomegalovirus major immediate-early genes in infected undifferentiated and differentiated cells. J. Virol. 71, 1246–1255.

Messerle, M., Buhler, B., Keil, G.M., and Koszinowski, U.H. (1992). Structural organization, expression, and functional characterization of the murine cytomegalovirus immediate-early gene 3. J. Virol. 66, 27–36.

Mocarski, E.S., Kemble, G.W., Lyle, J.M., and Greaves, R.F. (1996). A deletion mutant in the human cytomegalovirus gene encoding IE1(491aa) is replication defective due to a failure in autoregulation. Proc. Natl. Acad. Sci. U.S.A 93, 11321–11326.

Murphy, J.C., Fischle, W., Verdin, E., and Sinclair, J.H. (2002). Control of cytomegalovirus lytic gene expression by histone acetylation. EMBO J. 21, 1112–1120.

Nevels, M., Brune, W., and Shenk, T. (2004). SUMOylation of the human cytomegalovirus 72-kilodalton IE1 protein facilitates expression of the 86-kilodalton IE2 protein and promotes viral replication. J. Virol. 78, 7803–7812.

Pajovic, S., Wong, E.L., Black, A.R., and Azizkhan, J.C. (1997). Identification of a viral kinase that phosphorylates specific E2Fs and pocket proteins. Mol. Cell Biol. 17, 6459–6464.

Pizzorno, M.C., and Hayward, G.S. (1990). The IE2 gene products of human cytomegalovirus specifically down- regulate expression from the major immediate-early promoter through a target sequence located near the cap site. J. Virol. 64, 6154–6165.

Politz, J.C., Tuft, R.A., Pederson, T., and Singer, R.H. (1999). Movement of nuclear poly(A) RNA throughout the interchromatin space in living cells. Curr. Biol. 9, 285–291.

Poma, E.E., Kowalik, T.F., Zhu, L., Sinclair, J.H., and Huang, E.S. (1996). The human cytomegalovirus IE1-72 protein interacts with the cellular p107 protein and relieves p107-mediated transcriptional repression of an E2F-responsive promoter. J. Virol. 70, 7867–7877.

Seeler, J.S., Marchio, A., Sitterlin, D., Transy, C., and Dejean, A. (1998). Interaction of SP100 with HP1 proteins: a link between the promyelocytic leukemia-associated nuclear bodies and the chromatin compartment. Proc. Natl. Acad. Sci. U.S.A 95, 7316–7321.

Sha, B., and Luo, M. (1997). Structure of a bifunctional membrane-RNA binding protein, influenza virus matrix protein M1. Nat. Struct. Biol. 4, 239–244.

Stenberg, R.M., Fortney, J., Barlow, S.W., Magrane, B.P., Nelson, J.A., and Ghazal, P. (1990). Promoter-specific trans activation and repression by human cytomegalovirus immediate-early proteins involves common and unique protein domains. J. Virol. 64, 1556–1565.

Stenberg, R.M., Thomsen, D.R., and Stinski, M.F. (1984). Structural analysis of the major immediate-early gene of human cytomegalovirus. J. Virol. 49, 190–199.

Tang, Q., and Maul, G.G. (2005). Mouse cytomegalovirus immediate-early protein 1 binds with host cell repressors to relieve suppressive effects on viral transcription and replication during lytic infection. J. Virol. 77, 1357–1367.

Tang, Q., Li, L., Negorev, D., Rux, J.J. and Maul, G.G. (2005). Separate molecular sites of mouse cytomegalovirus immediate-early protein 1 function in ND10 binding and inactivation of histone deacetylase. Submitted.

Tang, Q., Li, L. and Maul, G.G. 2005. Mouse cytomegalovirus early 112/113 proteins control the repressive effect of IE3 on the major immediate-early promoter. J. Virol. 79, 257–263.

White, E.A., Clark, C.L., Sanchez, V., and Spector, D.H. (2004). Small internal deletions in the human cytomegalovirus IE2 gene result in nonviable recombinant viruses with differential defects in viral gene expression. J. Virol. 78, 1817–1830.

# Major Immediate-early Enhancer and its Gene Products

8

*Jeffery L. Meier and Mark F. Stinski*

## Abstract

The major immediate-early (MIE) regulatory region plays a key role in the control of lytic and latent infections. Transcription from the MIE promoter is rate-limiting for the lytic cycle, whereas transcriptional quiescence is linked to viral latency. The MIE enhancer governs these outcomes by integrating a diverse array of input provided by the cell, the virus, and external surroundings. Its complex structure also affords the regulatory means for partly determining cell tropism. The degree of enhancer-dependent transcriptional activation determines the level of expression of the IE1 p72 and IE2 p86 proteins that are vital for viral replication. These proteins function to not only activate other essential viral genes, but to also vastly change host cell physiology and behavior. Hence, the mechanisms underlying MIE gene regulation and function contribute importantly to the genesis of disease caused by CMV. A better understanding of the MIE enhancer and its gene products is hoped to spawn novel strategies for preventing CMV-related illness.

## Introduction

CMV replicates in a wide range of well-differentiated cell types of endoderm, mesoderm, and ectoderm origin (Sinzger and Jahn, 1996). Failure to replicate in poorly differentiated cells results in latent infection in select cell types until reactivated by cellular differentiation and/or stimulation (Sinclair and Sissons, 1996; Soderberg-Naucler and Nelson, 1999). Different infection outcomes partly arise from the regulatory control of transcription from the CMV major immediate-early (MIE) promoter, whose gene products are rate-limiting for the lytic cycle and are rarely expressed during viral latency (Meier and Stinski, 1996). The enhancer of the MIE regulatory region is pivotal in the coordinate regulation of MIE promoter activity, viral replication, and pathogenesis. Engineered enhancer alterations in the CMV genome have recently produced several findings to support this view.

## Enhancer requirement in viral replication and pathogenesis

Removal of murine CMV's enhancer *in toto* nearly abolishes viral replication in cultured cells (Angulo et al., 1998) and eliminates disease and viral spread in the mouse (Ghazal et al., 2003). The replication defect *ex vivo* is corrected by replacing the missing enhancer with the comparable human CMV enhancer (Angulo et al., 1998). A murine CMV recombinant having a human MIE enhancer/promoter in place of its own exhibits normal replication kinetics in cultured fibroblasts (Grzimek et al., 1999). However, after intravenous in-oculation into mice, the virus falters in spread or initiation of infection in lung, spleen, and adrenal glands, unlike the infection in liver (Grzimek et al., 1999). Replacing the enhancers in rat (Sanford et al., 2001) and human (Isomura and Stinski, 2003) CMVs with that of murine CMV reduces viral replication in cultured rat and human fibroblasts, respectively. The crippled rat CMV also falls short at establishing infection in rat salivary gland after intraperitoneal inoculation, despite replicating well in the spleen (Sanford et al., 2001). Thus, enhancer swaps between different CMV species result in deviations in dissemination efficacy and host cell tropism.

The human CMV enhancer of ~540 bp is composed of distal and proximal halves that differ in structural makeup yet function jointly to provide efficient MIE promoter activation and viral replication (Isomura and Stinski, 2003; Meier and Pruessner, 2000). Removal of the distal enhancer significantly decreases MIE promoter activity, viral replication, and viral plaquing efficiency (number of plaques and rate of their growth) at levels commensurate with decreasing multiplicity of infection (MOI) (Meier and Pruessner, 2000). These phe-notypic defects are not observed at MOI of $\geq 1$ PFU per cell or when inactivated purified CMV particles are added in abundance (Meier and Pruessner, 2000; Meier et al., 2002). This suggests that one or more virion components enhance transcription from the MIE promoter. Multiple cis-acting elements in the distal enhancer generate the MIE promoter activity, whereas the short hypothetical open readings frames in the region are noncon-tributory (Meier et al., 2002). The proximal enhancer is also composed of multiple positive cis-acting elements, as viruses with successively larger 5′-truncation deletions of this region are progressively more defective in MIE promoter activity and viral replication (Isomura et al., 2004). A human CMV with only the first 28-bp of proximal enhancer containing a putative SP1 binding site exhibits only vestiges of replication and MIE promoter activity, whereas virus devoid of all enhancer parts is unable to replicate (Isomura et al., 2004). The functional abnormalities created by taking out either the distal or proximal enhancer are not remedied by inserting the murine CMV enhancer equivalent (Isomura and Stinski, 2003). Hence, the distal and proximal enhancers both contribute multiple cis-acting ele-ments that govern the initiation and magnitude of MIE promoter-dependent transcription and viral replication, but such regulatory units are not always functionally interchangeable between CMV species.

## Enhancer mechanics

The precise regulatory mechanisms underlying enhancer function are incompletely under-stood. Some insight into this issue comes from studies focusing on MIE enhancer/promot-er segments placed in cell-free, cell culture, or transgenic animal settings. In these models, transcriptional activity of the enhancer/promoter varies in accord with cell type, stage of cellular differentiation, and activity of particular signaling pathways (Meier and Stinski, 1996). This conditional activity results from dynamic interplay among a diverse array of cellular and viral factors that act on the enhancer/promoter in either a positive or negative way. Multiple types of specific cellular transcription factors bind to cognate sites in the enhancer to regulate transcription. Several of them bind to multiple sites, potentially al-lowing for cooperative or synergistic interaction among the same or different transcription factor families. Enhancers vary among different CMV species in binding site composition and arrangement, suggesting a mechanistic basis for the functional differences among the enhancers. As exemplified in Figure 8.1, human and chimpanzee CMV enhancers have four or more copies of binding sites for both NF-κB and CREB/ATF, the murine CMV enhancer has five or more bindings sites for NF-κB and AP-1, while the rat CMV enhancer (Maastricht strain) has several bindings sites for AP-1 but none for NF-κB. Differences in binding sites for other types of transcription factors not depicted also likely impart individuality to enhancer function. These structural variations presumably reflect evolutionary adaptations to different hosts and viral needs.

The enhancer's complexity is further reflected in findings with isolated fragments of the human CMV MIE regulatory region, which contains the core promoter, enhancer, unique region, and modulator (Figure 8.2). These data indicate that cellular factors NF-κB, CREB/ATF, AP-1, SP1, serum response factor, ELK-1, and retinoic acid/retinoid X receptor bind to cognate cis-acting elements in the enhancer to stimulate transcription from the core promoter (Meier and Stinski, 1996). Their quantities and activities are modulated by scores of conditions specified by the host cell, tissue type, surrounding milieu, as well as the virus. The binding of specific factors is thought to strengthen basal transcriptional complex formation and recruit co-activators with ability of modifying the transcription apparatus and chromatin to enhance transcriptional initiation and elongation. The binding of other cellular transcription factors to recognition sites located upstream in the unique region (e.g., PDX1) and modulator may augment transcriptional activity from either the MIE promoter (Chao et al., 2004; Nelson et al., 1987) or the transcriptionally divergent UL127 promoter residing at the unique region-modulator junction (Lashmit et al., 2004; Angulo et al., 2000b; Lundquist et al., 1999).

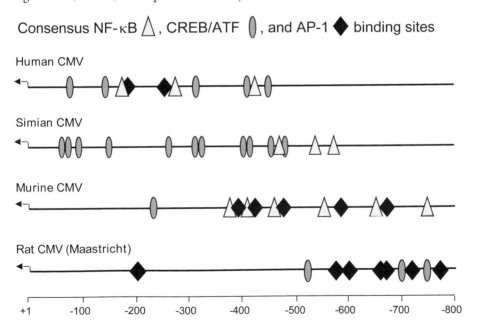

Consensus NF-κB △, CREB/ATF (|), and AP-1 ◆ binding sites

Human CMV

Simian CMV

Murine CMV

Rat CMV (Maastricht)

+1  -100  -200  -300  -400  -500  -600  -700  -800

**Figure 8.1** Arrangement and organization of selected *cis*-acting elements in the MIE enhancers of human, simian, murine, and rat CMVs. Locations of consensus recognition sequences for NF-κB (off-white triangle), CREB/ATF (gray oval), and AP-1 (black diamond) are depicted in reference to the transcriptional start-site (arrow, +1) in the viral genome.

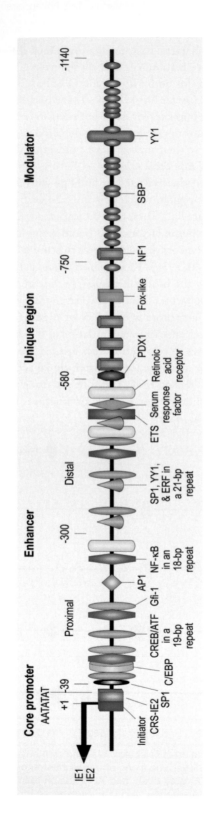

**Figure 8.2** The human CMV MIE regulatory region is composed of core promoter, enhancer, unique region, and modulator. The enhancer is divided into proximal (-39 to -299) and distal (-300 to -579) portions. Positions of putative binding sites for initiator complex, C/EBP, CREB/ATF, Gfi-1, AP1, NF-κB, SP1, YY1, ERF, ETS, serum response factor, retinoic acid receptor, PDX1, NF1, and SBP are shown relative to the +1 RNA start-site (leftward arrow) of the core promoter. Binding sites for NF-κB, CREB/ATF, and the combination of SP1, YY1, ERF are nested in the 18-, 19-, and 21-bp repeats, respectively. IE2 p86 binds to the *cis*-repression signal (CRS-IE2).

Studies of human CMV genomes containing targeted mutations indicate that MIE enhancer/promoter activity is unaltered by elimination of the unique region and/or modulator (Meier, 2001; Meier and Pruessner, 2000; Meier and Stinski, 1997) and is minimally to negligibly affected by mutation of the enhancer's cyclic AMP response elements (CRE) (Keller et al., 2003) or NF-κB binding sites (Benedict et al., 2004). Thus, other regulatory mechanisms may dominate or compensate for such mutations in some acutely infected cell types. These mechanisms are likely coupled to one or more of the numerous cellular signaling pathways that are promptly activated by viral particles as they bind to and enter the cell to dramatically change the cellular gene expression and physiological control programs (Compton, 2004; Fortunato et al., 2000). Indeed, a preliminary report suggests that triggering of the interferon signaling pathway by incoming virus stimulates MIE enhancer/promoter activity via gamma interferon-activating sequence elements embedded in the enhancer (Zhu et al., 2004). While human CMV glycoprotein B independently induces the interferon-responsive pathway (Boyle et al., 1999), several different virion components, i.e., pp71 (ppUL82), UL69, TRS1/IRS1, and ppUL35 proteins (Bresnahan and Shenk, 2000; Homer et al., 1999; Liu and Stinski, 1992; Romanowski et al., 1997; Schierling et al., 2004; Winkler and Stamminger, 1996), are known to stimulate MIE enhancer/promoter activity. Enhancer/promoter activity also increases in response to positive feedback provided by the viral MIE gene product, IE1 p72 (Cherrington and Mocarski, 1989).

Evidence suggests that the CMV genome in the nucleus of a lytically infected cell is not organized into an ordered nucleosomal array akin to cellular chromatin, but may partially acquire structural characteristics reminiscent of chromatin (Kierszenbaum and Huang, 1978; St. Jeor et al., 1982). This paradigm parallels that for herpes simplex virus, in which cellular histones assemble on viral genomes during lytic infection to form nucleosome-like structures having variable distance in spacing (Herrara and Trienzenberg, 2004; Kent et al., 2004). Histone modifications signifying transcriptionally permissive chromatin (e.g., acetylation of histone H3 Lys9 and Lys14 and methylation of H3 Lys4) are preferentially located on viral promoters of actively expressed genes and map to locations in the herpes simplex genome in accord with the temporal order of viral gene expression (Kent et al., 2004). A preliminary report suggests that human CMV genomes are likewise associated with histones that are similarly modified in relationship to lytic cycle gene expression (Reeves et al., 2004). The combined action of the multiple transcription factors attracted to the enhancer likely alters chromatin in a manner favorable to transcription. Enhancer performance is also influenced by the outcome of targeting incoming viral genomes to nuclear domains 10 (ND10 domains; also named PML-associated nuclear bodies or oncogenic domains) that are rich in chromatin modifying enzymes and other transcriptional regulators.

## Enhancer quiescence

Studies performed in conditionally permissive cells of monocytic and neuronal lineages have long shown that the human CMV MIE enhancer/promoter and subsequent lytic cycle events are silenced in undifferentiated cells, but activated in terminally differentiated cellular counterparts. These findings mirror human CMV latency in macrophage-dendritic cell precursors, in which both MIE promoter and viral replication are quiescent until reactivated by cellular stimulation and differentiation (Hahn et al., 1998; Kondo et al., 1994; Soderberg-Naucler et al., 1997; Soderberg-Naucler et al., 2001; Taylor-Wiedeman et al.,

1991; Taylor-Wiedeman et al., 1994). Murine CMV's MIE enhancer/promoter is also inactive during viral latency in vivo, aside from occasional transcriptional reactivation events (Grzimek et al., 2001; Kurz et al., 1999; Kurz and Reddehase, 1999; Simon et al., 2005). The mechanisms by which the MIE enhancer/promoter is silenced remain poorly understood. Three copies of the 21-bp-repeat in the enhancer, as well as the upstream modulator, decrease MIE enhancer/promoter activity in reporter plasmids transfected into undifferentiated nonpermissive cells (Kothari et al., 1991; Nelson et al., 1987; Sinclair et al., 1992). Cellular transcriptional repressors YY1, ETS2-repressor factor (ERF), and/or silencing binding protein (SBP) bind to these negative cis-acting elements in vitro (Figure 8.2) (Bain et al., 2003; Huang et al., 1996; Liu et al., 1994; Thrower et al., 1996). However, removal of the 21-bp repeats, modulator, or both 21-bp repeats and modulator from the viral genome does not alleviate MIE enhancer/promoter silencing in quiescently infected cells (Meier, 2001; Meier and Stinski, 1997). Transient assays also implicate cellular growth factor independence-1 (Gfi-1) (Zweidler-McKay et al., 1996) and CCAAT/enhancer binding proteins (Prösch et al., 2001) in enhancer/promoter silencing. However, mutation of the two Gfi-1 binding sites in the human CMV genome fails to relieve silencing in undifferentiated monocytic cells (J. Schnetzer and M.F. Stinski., unpublished data). Whether enhancer quiescence in the intact viral genome is brought about by extensive redundancy in previously recognized negative cis-acting sites or from undiscovered silencing mechanisms remains to be resolved. Heterochromatin likely participates in the silencing, as its constituents of hypoacetylated histones, heterochromatin protein 1, and histone deacetylases (HDAC) are observed to condense on the inactive MIE enhancer/promoter in viral genomes in quiescently infected cell models (Murphy et al., 2002) and in latent viral genomes in endogenously infected dendritic cell precursors (Reeves et al., 2003).

Messenger RNAs mapping to both DNA strands of the human CMV MIE locus are detected in a small subset of latently infected monocytic-dendritic cell precursors, but not in peripheral blood monocytes (Hahn et al., 1998; Lofgren-White et al., 2000; Kondo et al., 1994; Kondo et al., 1996; Slobedman and Mocarski, 1999). Messages expressed from the same DNA strand as the MIE genes originate from within the enhancer at nucleotide positions −292 and −356, with respect to the +1 start-site for lytic cycle transcription. The largest of the open reading frames in these mRNAs corresponds to UL126 and codes for the ORF94 protein (Kondo et al., 1994; Kondo et al., 1996). UL126 function is dispensable for latent and lytic infections in cultured cells (Lofgren-White et al., 2000). Therefore, the purpose for these viral mRNAs and the mechanism(s) by which they are regulated remain enigmatic.

## Enhancer's role in reactivation

The frequency of reactivation of latent CMV increases as host cellular immunity wanes. Allogeneic stimulation, cellular differentiation, and exposure to select pro-inflammatory cytokines can spark the reactivation process. The MIE promoter regulation is first among multiple check-points in place to control viral reactivation (Kurz and Reddehase, 1999; Simon et al., 2005; Taylor-Wiedeman et al., 1994). The triggering of murine CMV MIE promoter reactivation by TNF-α (Hummel et al., 2001; Simon et al., 2005) is linked to the signaling-mediated activation of NF-κB and AP-1 that bind to and stimulate the enhancer (Hummel and Abecassis, 2002). In quiescently infected embryonal NTera2 cells, the human CMV enhancer is strongly repressed by mechanisms that block retinoic acid-induced activation via the retinoic acid response elements, despite retinoic acid's activation of trans-fected enhancer/promoter segments under comparable cellular conditions (Meier, 2001). Treatment with an HDAC inhibitor overcomes the repression and permits viral replication (Meier, 2001). This result corresponds to disruption of heterochromatin nucleation at the MIE enhancer/promoter (Murphy et al., 2002), a finding that is also linked to human CMV reactivation in endogenously infected dendritic cell precursors (Reeves et al., 2003). In the embryonal cell model, MIE enhancer/promoter silencing is also partially relieved by stimulating the enhancer's CRE repetition via cyclic AMP signaling, and combining this stimulation with an HDAC inhibitor results in synergistic levels of reactivation (M.J. Keller and J.L Meier, unpublished data). Hence, MIE enhancer/promoter reactivation like-ly results from the concerted efforts of multiple regulatory mechanisms that stimulate and de-repress transcription.

## Major IE proteins

The human CMV MIE gene locus comprises the MIE regulatory region, five exons, and two polyadenylation signals. Its promoter generates a primary transcript that undergoes differential splicing and polyadenylation to produce multiple mRNA species (Figure 8.3). Two of these messages are abundantly made to code for the IE1 p72 (72 kDa) and IE2 p86 (86 kDa) proteins, both of which are major pleiotropic regulators of the viral lytic cycle. Although they share some characteristics, IE1 p72 and IE2 p86 are distinctly different in overall structure and function. Minor splice variants are also made though their biological roles in the infected cell are unclear.

Human CMV IE1 p72 (also called IE1$_{491aa}$ and ppUL123) has 491 amino acids encoded by exons 2, 3, and 4 and is modified by covalent attachment of phosphate and small ubiquitin-like modifiers (SUMO) (Pajovic et al., 1997; Spengler et al., 2002; Xu et al., 2001). The amino-terminal 85 amino acids corresponding to exons 2 and 3 are also contained in IE2 p86. IE1 p72 localizes to the nucleus of productively infected cells where it disperses the subnuclear ND10 domains (Ahn and Hayward, 1997; Ahn et al., 1998b;

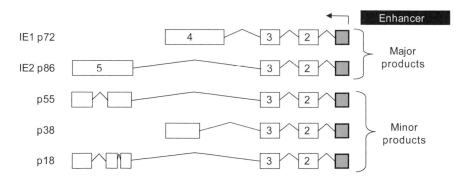

**Figure 8.3** Differential splicing of the primary transcript in lytically infected cells results in multiple MIE gene products. In human CMV infected cells, IE1 p72 and IE2 p86 predominate. These major products share exons 2 and 3 (open boxes), but only IE1 p72 has exon 4 and IE2 p86 has exon 5. The minor RNA splice variants produce the p55, p38, and p18 proteins. All of the spliced RNAs contain the same 5′-untranslated region (gray box) and are regulated by the enhancer (black box).

external surroundings. Its complex structure also affords the regulatory means for partly determining cell tropism. The degree of enhancer-dependent transcriptional activation determines the level of expression of the IE1 p72 and IE2 p86 proteins that are vital for viral replication. These proteins function to not only activate other essential viral genes, but to also vastly change host cell physiology and behavior in support of viral propagation. Hence, the mechanisms underlying MIE gene regulation and function also contribute importantly to the genesis of disease caused by CMV. While our knowledge about such mechanisms has advanced greatly, it is far from complete. A better understanding of the enhancer and its gene products is hoped to spawn novel strategies for combating CMV infection and disease.

## Acknowledgments

We regret that other publications and contributions have not been included because of space constraints.

Our work is supported by grants from the National Institutes of Health (AI-13562 to M.F.S. and AI-27661 to J.L.M.), Department of Veterans Affairs (Merit to J.L.M.), and American Heart Association (to J.L.M.).

## References

Ahn, J.H., and Hayward, G.S. (1997). The major immediate-early proteins IE1 and IE2 of human cytomegalovirus colocalize with and disrupt PML-associated nuclear bodies at very early times in infected permissive cells. J. Virol. 71, 4599–4613.

Ahn, J.H., Chiou, C.J., and Hayward, G.S. (1998a). Evaluation and mapping of the DNA binding and oligomerization domains of the IE2 regulatory protein of human cytomegalovirus using yeast one and two hybrid interaction assays. Gene 210, 25–36.

Ahn, J.H., Xu, Y., Jang, W.J., Matunis, M., and Hayward, G.S. (1998b). Disruption of PML subnuclear domains by the acidic IE1 protein of human cytomegalovirus is mediated through interaction with PML and may modulate a RING finger-dependent cryptic transactivator function of PML. Mol. Cell Biol. 18, 4899–4913.

Ahn, J.H., Xu, Y., Jang, W.J., Matunis, M.J., and Hayward, G.S. (2001). Evaluation of interactions of human cytomegalovirus immediate-early IE2 regulatory protein with small ubiquitin-like modifiers and their conjugation enzyme Ubc9. J. Virol. 75, 3859–3872.

Angulo, A., Ghazal, P., and Messerle, M. (2000a). The major immediate-early gene IE3 of mouse cytomegalovirus is essential for viral growth. J. Virol. 74, 11129–11136.

Angulo, A., Kerry, D., Haung, H., Borst, E.M., Razinsky, A., Wu, J., Hobom, U., Messerle, M., and Ghazal, P. (2000b). Identification of a boundry domain adjacent to the potent human cytomegalovirus enhancer that represses transcription of the divergent UL127 promoter. J. Virol. 74, 2826–2839.

Angulo, A., Messerle, M., Koszinowski, U.H., and Ghazal, P. (1998). Enhancer requirement for murine cytomegalovirus growth and genetic complementation by the human cytomegalovirus enhancer. J. Virol. 72, 8502–8509.

Bain, M., Mendelson, M., and Sinclair, J. (2003). Ets-2 Repressor Factor (ERF) mediates repression of the human cytomegalovirus major immediate-early promoter in undifferentiated non-permissive cells. J. Gen. Virol. 84, 41–49.

Barrasa, M.I., Harel, N.Y., and Alwine, J.C. (2005). The phosphorylation status of the serine rich region of the human cytomegalovirus 86-kilodalton major immediate-early protein HCMV IE2/IEP86 protein affects temporal viral gene expression. J. Virol. 79, 1428–1437

Barrasa, M.I., Harel, N., Yu, Y., and Alwine, J.C. (2003). Strain variations in single amino acids of the 86-kilodalton human cytomegalovirus major immediate-early protein (IE2) affect its functional and biochemical properties: implications of dynamic protein confirmation. J. Virol. 77, 4760–4772.

Benedict, C.A., Angulo, A., Patterson, G., Ha, S., Huang, H., Messerle, M., Ware, C.F., and Ghazal, P. (2004). Neutrality of the canonical NF-κB-dependent pathway for human and murine cytomegalovirus transcription and replication in vitro. J. Virol. 78, 741–750.

Bonin, L.R., and McDougall, J.K. (1997). Human cytomegalovirus IE2 86-kilodalton protein binds p53 but does not abrogate G1 checkpoint function. J. Virol. 71, 5861–5870.

Boyle, K.A., Pietropaolo, R.L., and Compton, T. (1999). Engagement of the cellular receptor for glyco-protein B of human cytomegalovirus activates the interferon-responsive pathway. Mol. Cell Biol. 19, 3607–3613.

Bresnahan, W.A., and Shenk, T.E. (2000). UL82 virion protein activates expression of immediate-early genes in human cytomegalovirus infected cells. Proc. Natl. Acad. Sci. USA 97, 14506–14511.

Bryant, L.A., Mixon, P., Davidson, M., Bannister, A.J., Kouzarides, T., and Sinclair, J.H. (2000). The hu-man cytomegalovirus 86-kilodalton major immediate-early protein interacts physically and function-ally with histone acetyltransferase P/CAF. J. Virol. 74, 7230–7237.

Castillo, J.P., Yurochko, A.D., and Kowalik, F. (2000). Human cytomegalvirus immediate-early proteins and cell growth. J. Virol. 74, 8028–8037.

Caswell, R.C., Hagemeier, C., Chiou, C.J., Hayward, G.S., Kouzarides, T., and Sinclair, J. (1993). The hu-man cytomegalovirus 86K immediate-early (IE2) protein requires the basic region of the TATA-box binding protein for binding and interacts with TBP and transcription factor TFIIB via regions IE2 required for transcriptional regulation. J. Gen. Virol. 74, 2691–2698.

Chao, S.H., Harada, J.N., Hyndman, F., Gao, X., Nelson, C.G., Chanda, S.K., and Caldwell, J.S. (2004). PDX-1, a cellular homeoprotein, binds to and regulates the activity of human cytomegalovirus imme-diate-early promoter. J. Biol. Chem. 279, 16111–16120.

Cherrington, J.M., and Mocarski, E.S. (1989). Human cytomegalovirus ie1 transactivates the α promoter-enhancer via an 18-base-pair repeat element. J. Virol. 63, 1435–1440.

Compton, T. (2004). Receptors and immune sensors: the complex entry path of human cytomegalovirus. Trends Cell Biol. 14, 5–8.

Fortunato, E.A., McElroy, A.K., Sanchez, V., and Spector, D.H. (2000). Exploitation of cellular signaling and regulatory pathways by human cytomegalovirus. Trends Mircobiol. 8, 111–119.

Fortunato, E.A., Sommer, M.H., Yoder, K., and Spector, D.H. (1997). Identification of domains within the human cytomegalovirus major immediate-early 86-kilodalton protein and the retinoblastoma pro-tein required for physical and functional interactions with each other. J. Virol. 71, 8176–8185.

Furnari, B.A., Poma, E.T., Kowalik, F., Huong, S.M., and Huang, E.S. (1993). Human cytomegalovirus immediate-early gene 2 protein interacts with itself and with several novel cellular proteins. J. Virol. 67, 4981–4991.

Gawn, J.M., and Greaves, R.F. (2002). Absence of IE1 p72 protein function during low-multiplicity of infection by human cytomegalovirus results in a broad block to viral delayed-early gene expression. J. Virol. 76, 4441–4455.

Gebert, S., Schmolke, S., Sorg, S., Floss, S., Plachter, B., and Stamminger, T. (1997). The UL84 protein of human cytomegalovirus acts as a transdominant inhibitor of immediate-early-mediated transactiva-tion that is able to prevent viral replication. J. Virol. 71, 7048–7060.

Ghazal, P., Messerle, M., Osborn, K., and Angulo, A. (2003). An essential role for the enhancer for murine cytomegalovirus in vivo growth and pathogenesis. J. Virol. 77, 3217–3228.

Greaves, R.F., and Mocarski, E.S. (1998). Defective growth correlates with reduced accumulation of a viral DNA replication protein after low-multiplicity infection by a human cytomegalovirus ie1 mutant. J. Virol. 72, 366–379.

Grzimek, N.K.A., Dreis, D., Schmalz, S., and Reddehase, M.J. (2001). Random, asynchronous, and asym-metric transcriptional activity of enhancer-flanking major immediate-early genes IE1/3 and IE2 dur-ing murine cytomegalovirus latency in the lungs. J. Virol. 75, 2692–2705.

Grzimek, N.K.A., Podlech, J., Steffens, H.P., Holtappels, R., Schmalz, S., and Reddehase, M.J. (1999). In vivo replication of recombinant murine cytomegalovirus driven by the paralogous major immediate-early promoter-enhancer of human cytomegalovirus. J. Virol. 73, 5043–5055.

Hagemeier, C., Caswell, R., Hayhurst, G., Sinclair, J., and Kouzarides, T. (1994). Functional interaction between the HCMV IE2 transactivator and the retinoblastoma protein. EMBO J. 13, 2897–2903.

Hagemeier, C., Walker, S., Caswell, R., Kouzarides, T., and Sinclair, J. (1992). The human cytomegalovirus 80-kilodalton but not the 72-kilodalton immediate-early protein transactivates heterologous promoters in a TATA box dependent mechanism and interacts directly with TFIID. J. Virol. 66, 4452–4456.

Hahn, G., Jores, R., and Mocarski, E.S. (1998). Cytomegalovirus remains latent in a common precursor of dendritic and myeloid cells. Proc. Natl. Acad. Sci. USA 95, 3937–3942.

Harel, N.Y., and Alwine, J.C. (1998). Phosphorylation of the human cytomegalovirus 86-kilodalton im-mediate-early protein IE2. J. Virol. 72, 5481–5492.

Mocarski, E.S., Kemble, G., Lyle, J., and Greaves, R.F. (1996). A deletion mutant in the human cytomegalovirus gene encoding IE1 (491aa) is replication defective due to a failure in autoregulation. Proc. Natl. Acad. Sci. USA 93, 11321–11326.

Muller, S., and Dejean, A. (1999). Viral immediate-early proteins abrogate the modification by SUMO-1 of PML and Sp100 proteins, correlating with nuclear body disruption. J. Virol. 73, 5173–5143.

Murphy, E.A., Streblow, D.N., Nelson, J.A., and Stinski, M.F. (2000). The human cytomegalovirus IE86 protein can block cell cycle progression after inducing transition into the S phase of permissive cells. J. Virol. 74, 7108–7118.

Murphy, J., Fischle, W., Verdin, E., and Sinclair, J. (2002). Control of cytomegalovirus lytic gene expression by histone acetylation. EMBO J. 21, 1112–1120.

Nelson, J.A., Reynolds-Kohler, C., and Smith, B. (1987). Negative and positive regulation by a short segment in the 5′-flanking region of the human cytomegalovirus major immediate-early gene. Mol. Cell Biol. 7, 4125–4129.

Nevels, M., Brune, W., and Shenk, T. (2004). SUMOylation of the human cytomegalovirus 72-kilodalton protein facilitates expression of the 86-kilodalton IE2 protein and promotes viral replication. J. Virol. 78, 7803–7812.

Noris, E., Zannetti, C., Demurtas, A., Sinclair, J., De Andrea, M., Garigilio, M., and Landolfo, S. (2002). Cell cycle arrest by human cytomegalovirus 86-kDa IE2 protein resembles premature senescence. J. Virol. 76, 12135–12148.

Pajovic, S., Wong, E.L., Black, A.R., and Azizkhan, J.C. (1997). Identification of a viral kinase that phosphorylates specific E2Fs and pocket proteins. Mol. Cell Biol. 17, 6459–6464.

Pizzorno, M.C., and Hayward, G.S. (1990). The IE2 gene products of human cytomegalovirus specifically down-regulate expression from the major immediate-early promoter through a target sequence located near the cap site. J. Virol. 64, 6154–6165.

Poma, E.E., Kowalik, T.F., Zhu, L., Sinclair, J.H., and Huang, E.S. (1996). The human cytomegalovirus IE1-72 protein interacts with cellular p107 protein and relieves p107-mediated transcriptional repression of an E2F-responsive promoter. J. Virol. 70, 7876–7877.

Prösch, S., Heine, A.K., Volk, H.D., and Krüger, D.H. (2001). CAAT/enhancer-binding proteins α and β negatively influence capacity of tumor necrosis factor α to up-regulate the human cytomegalovirus IE1/2 enhancer/promoter by nuclear factor κB during monocyte differentiation. J. Biol. Chem. 276, 40712–40720.

Reeves, M., Baillie, J., Greaves, R., and Sinclair, J. (2004). Major changes in chromatization of the HCMV major immediate-early promoter occur during all phases of productive infection. 29th International Herpesvirus Workshop Reno, Nevada U.S.A. Abstract 1.09.

Reeves, M.B., Sissons, J.G.P., Lehner, P.J., and Sinclair, J.H. (2003). Analysis of the regulation of endogenous HCMV latency and reactivation in myeloid dendritic cells. 28th International Herpesvirus Workshop Madison, Wisconsin U.S.A. Abstract 8.12.

Reinhardt, J., Smith, G.B., Himmelheber, C.T., Azizkhan, J.C., and Mocarski, E.S. (2005). The carboxyl-terminal region of human cytomegalovirus IE1491aa contains an acidic domain that plays a regulatory role and a chromatin-tethering domain that is dispensible during viral replication. J. Virol. 79, 225–233.

Rodems, S.M., Clark, C.L., and Spector, D.H. (1998). Separate DNA elements containing ATF/CREB and IE86 binding sites differentially regulate the human cytomegalovirus UL112–113 promoter at early and late times in the infection. J. Virol. 72, 2697–2707.

Romanowski, M.J., Garrido-Guerrero, E., and Shenk, T. (1997). pIRS1 and pTRS1 are present in human cytomegalovirus virions. J. Virol. 71, 5703–5705.

Sanford, G.R., Brock, L.E., Voight, S., Forester, C.M., and Burns, W.H. (2001). Rat cytomegalovirus major immediate-early enhancer switching results in altered growth characteristics. J. Virol. 75, 5076–5083.

Schierling, K., Stamminger, T., Mertens, T., and Winkler, M. (2004). Human cytomegalovirus tegument proteins ppUL82 (pp71) and ppUL35 interact and cooperatively activate the major immediate-early enhancer. J. Virol. 78, 9512–9523.

Schwartz, R., Helmich, B., and Spector, D.H. (1996). CREB and CREB-binding proteins play an important role in the IE2 86-kilodalton protein-mediated transactivation of the human cytomegalovirus 2.2-kilobase RNA promoter. J. Virol. 70, 6955–6966.

Scully, A.R., Sommer, M.H., Schwartz, R., and Spector, D.H. (1995). The human cytomegalovirus IE2 86-kilodalton protein interacts with an early gene promoter via site-specific DNA binding and protein–protein associations. J. Virol. *69*, 6533–6540.

Simon, C.O., Seckert, C.K., Dreis, D., Reddehase, M.J., and Grzimek, N.K.A. (2005). Role of tumor necrosis factor alpha in murine cytomegalovirus transcriptional reactivation in latently infected lungs. J. Virol. *79*, 326–340.

Sinclair, J., and Sissons, P. (1996). Cytomegalovirus: Latent and persistent infection of monocytes and macrophages. Intervirology *39*, 293–301.

Sinclair, J.H., Baillie, J., Bryant, L.A., Taylor-Wiedeman, J. A., and Sissons, J. G. (1992). Repression of human cytomegalovirus major immediate-early gene expression in a monocytic cell line. J. Gen. Virol. *73*, 433–435.

Singh, J., and Compton, T. (2004). The IE1 protein of human cytomegalovirus is a key component of host innate immunity. 29th International Herpesvirus Workshop Reno, Nevada U.S.A. *Abstract 11.02.*

Sinzger, C., and Jahn, G. (1996). Human cytomegalovirus cell tropism and pathogenesis. Intervirology *39*, 302–319.

Slobedman, B., and Mocarski, E. (1999). Quantitative analysis of latent human cytomegalovirus. J. Virol. *73*, 4806–4812.

Soderberg-Naucler, C., Fish, K.N., and Nelson, J.A. (1997). Reactivation of latent human cytomegalovirus by allogenic stimulation of blood cells from healthy donors. Cell *91*, 119–126.

Soderberg-Naucler, C., and Nelson, J.A. (1999). Human cytomegalovirus latency and reactivation – a delicate balance between the virus and its host's immune system. Intervirology *42*, 314–321.

Soderberg-Naucler, C., Streblow, D.N., Fish, K.N., Allan-Yorke, J., Smith, P.P., and Nelson, J.A. (2001). Reactivation of latent human cytomegalovirus in CD14+ monocytes is differentiation dependent. J. Virol. *75*, 7543–7554.

Sommer, M.H., Scully, A.L., and Spector, D.H. (1994). Transactivation by the human cytomegalovirus IE2 86-kilodalton protein requires a domain that binds to both the TATA box-binding protein and the retinoblastoma protein. J. Virol. *68*, 6232–6242.

Song, Y.J., and Stinski, M.F. (2002). Effect of the human cytomegalovirus IE86 protein on expression of E2F-responsive genes: a DNA microarray analysis. Proc. Natl. Acad. Sci. USA *99*, 2836–2841.

Song, Y.J., and Stinski, M.F. (2005). Inhibition of cell division by the human cytomegalovirus IE86 protein: role of the p53 pathway or cyclin-dependent kinase 1/cyclin B1. J. Virol. *79*, 2597–2603.

Spector, D.J., and Tevethia, M.J. (1994). Protein–protein interactions between human cytomegalovirus IE2–580aa and pUL84 in lytically infected cells. J. Virol. *68*, 7549–7533.

Speir, E., Modali, R., Huang, E., Leon, M.B., Shawl, F., Finkel, T., and Epstein, S.E. (1994). Potential role of human cytomegalovirus and p53 interaction in coronary restenosis. Science *265*, 391–394.

Spengler, M.L., Kurapatwinski, K., Black, A.R., and Azizkhan-Clifford, J. (2002). SUMO-1 modification of human cytomegalovirus IE1/IE72. J. Virol. *76*, 2990–2996.

St. Jeor, S., Hall, C., McGraw, C., and Hall, M. (1982). Analysis of human cytomegalovirus nucleoprotein complexes. J. Virol. *41*, 309–312.

Tang, Q., Li, L., and Maul, G.G. (2005). Mouse cytomegalovirus early M112/113 proteins control the repressive effect of IE3 on the major immediate-early promoter. J. Virol. *79*, 257–263.

Tang, Q., and Maul, G.G. (2003). Mouse cytomegalovirus immediate-early protein 1 binds with host cell repressors to relieve suppressive effects on viral transcription and replication during lytic infection. J. Virol. *77*, 1357–1367.

Taylor-Wiedeman, J., Sissons, J.G., Borysiewicz, L.K., and Sinclair, J.H. (1991). Monocytes are a major site of persistence of human cytomegalovirus in peripheral blood mononuclear cells. J. Gen. Virol. *72*, 2059–2064.

Taylor-Wiedeman, J.A., Sissons, J.G.P., and Sinclair, J.H. (1994). Induction of endogenous human cytomegalovirus gene expression after differentiation of monocytes from healthy carriers. J. Virol. *68*, 1597–1604.

Thrower, A.R., Bullock, G.C., Bissell, J.E., and Stinski, M.F. (1996). Regulation of a human cytomegalovirus immediate-early gene (US3) by a silencer/enhancer combination. J. Virol. *70*, 91–100.

Wang, X., and Sonenshein, G.E. (2005). Induction of the RelB NF-κB subunit by the cytomegalovirus IE1 protein is mediated via Jun kinase and c-Jun/Fra-2 AP-1 complexes. J. Virol. *79*, 95–105.

White, E.A., Clark, C.L., Sanchez, V., and Spector, D.H. (2004). Small internal deletions in the human cytomegalovirus IE2 gene result in nonviable recombinant viruses with differential defects in viral gene expression. J. Virol. *78*, 1817–1830.

Wiebusch, L., and Hagemeier, C. (1999). Human cytomegalovirus 86-kilodalton IE2 protein blocks cell cycle progression in G1. J. Virol. 73, 9274–9283.

Wiebusch, L., and Hagemeier, C. (2001). The human cytomegalovirus immediate-early 2 protein dissociates cellular DNA synthesis from cyclin-dependent kinase activation. EMBO J. 20, 1086–1098.

Wilkinson, G.W., Kelly, C., Sinclair, J.H., and Richards, C. (1998). Disruption of PML-associated nuclear bodies mediated by the human cytomegalovirus major immediate-early gene product. J. Gen. Virol. 79, 1233–1245.

Winkler, M., and Stamminger, T. (1996). A specific subform of the human cytomegalovirus transactivator protein pUL69 is contained within the tegument of virus particles. J. Virol. 70, 8984–8987.

Xu, Y., Ahn, J.H., Cheng, M., Apryhs, C.J., Chiou, C.-J., Zong, J., Matunis, M.J., and Hayward, G.S. (2001). Proteosome-independent disruption of PML oncogenic domains (PODS), but not covalent modification by SUMO1, is required for human cytomegalovirus immediate-early protein IE1 to inhibit PML-mediated transcriptional repression. J. Virol. 75, 10683–10695.

Xu, Y., Cei, S.A., Huete, A.R., Colletti, K.S., and Pari, G.S. (2004). Human cytomegalovirus DNA replication requires transcriptional activation via an IE2- and UL84-responsive bidirectional promoter element within oriLyt. J. Virol. 78, 11664–11677.

Yu, Y., and Alwine, J.C. (2002). Human cytomegalovirus major immediate-early proteins and simian virus 40 large T antigen can inhibit apoptosis through activation of the phosphatidylinositide 3'-OH kinase pathway and the cellular kinase Akt. J. Virol. 76, 3731–3738.

Yurochko, A.D., Mayo, M.W., Poma, E.E., Baldwin, A.S., and Huang, E.S. (1997). Induction of the transcription factor Sp1 during human cytomegalovirus infection mediates up-regulation of the p65 and p105/p50 NF-kB promoters. J. Virol. 71, 4638–4648.

Zhang, Z., Huong, S.M., Wang, X., Haung, D.Y., and Huang, E.S. (2003). Interactions between human cytomegalovirus IE-72 and cellular p107: functional domains and mechanisms of up-regulation of cyclin E/cdk2 kinase activity. J. Virol. 77, 12660–12670.

Zhu, H., Netterwald, J., and Yang, S. (2004). An HCMV-induced signal transduction pathway which activates interferon-stimulated genes is important for major immediate-early gene expression. 29th International Herpesvirus Workshop Reno, Nevada, USA. *Abstract 1.61*.

Zhu, H., Shen, Y., and Shenk, T. (1995). Human cytomegalovirus IE1 and IE2 proteins block apoptosis. J. Virol. 69, 7960–7970.

Zweidler-McKay, P.A., Grimes, H.L., Flubacher, M.M., and Tsichlis, P.N. (1996). Gfi-1 encodes a nuclear zinc finger protein that binds DNA and functions as a transcripitonal repressor. Mol. Cell Biol. 16, 4024–4034.

# Regulation of Human Cytomegalovirus Gene Expression by Chromatin Remodeling

*Mark Bain, Matthew Reeves, and John Sinclair*

## Abstract

Although the myeloid lineage is important for the carriage of latent HCMV genomes, the mechanisms underlying how the latent state is maintained and how latent virus reactivates, are unclear. In this review we discuss how initial findings using model cell lines, mobility shift assays and transient transfection experiments, together with more recent developments such as chromatin immunoprecipitation assays, have led to an understanding that the higher-order chromatin structure around the viral major immediate-early promoter region, has profound effects on the control of viral latency and reactivation.

## Introduction

It has been recognized for many years that transcription regulation is not merely the formation or inhibition of the pre-initiation complex (PIC; comprising TFIIA, TFIIB, TBP and its associated factors, and RNA Polymerase II) on naked DNA templates. Whilst such recruitment of basal transcription factors to promoters is of crucial importance, the regulation of gene expression was, for some time, also known to involve higher-order chromatin structure. However, the mechanism by which chromatin regulated transcription was largely left unaddressed until specific reagents, which could distinguish between modified chromatin components, became available. As our understanding of chromatin structure and its role in the control of cellular gene expression increased, it became possible to directly assess chromatin's influence not only on cellular gene expression but viral gene expression also.

## The packaging of cellular DNA

The structure of the nucleosome, the basic unit of chromatin structure, was first elucidated over thirty years ago as a complex of chromosomal DNA and histone proteins (Kornberg, 1974; Oudet et al., 1975). These nucleosomes, with their characteristic "beads on a string" appearance under the electron microscope (Oudet et al., 1975), typically protect a ~200 base pair fragment of DNA from the action of nucleases (Hewish and Burgoyne, 1973). Further data indicated that distinct arrangements of histone proteins (H1, H2A, H2B, H3, and H4) were required for the association with DNA (Figure 9.1A; Chung et al., 1978). Although this association between basic cellular proteins and nucleic acid is clearly required for spatially condensing the large amounts of DNA contained in the eukaryotic cell, such DNA compaction does present problems in terms of accessibility of the DNA to,

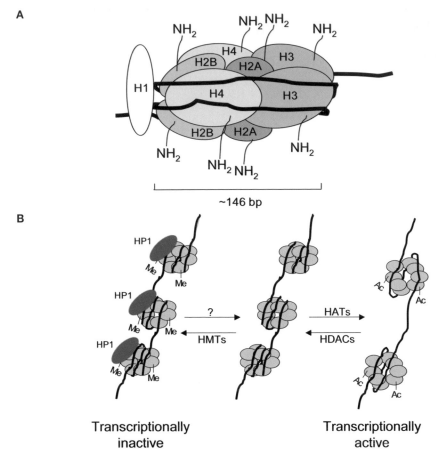

A

B

Transcriptionally
inactive

Transcriptionally
active

**Figure 9.1** (A) Structure of the nucleosome. Shown is the characteristic octamer comprising two molecules each of histones H2A, H2B, H3 and H4. Each nucleosome has approximately 140 base pairs of DNA wrapped around it with histone H1 acting as a "linker" holding the structure intact. (B) Simplified model showing the interplay of histone acetyltransferases (HATs), histone deacetylases (HDACs) and histone methyltransferases (HMTs) in modulating chromatin structure. "?" indicates a proposed "demethylase" activity which removes methyl groups. HP1 is the heterochromatin protein-1, Me are methyl groups and Ac are acetyl groups. See main text for further details.

for example, the DNA replication and transcription apparatus. Indeed, it was initially proposed that DNA associated with histones was transcriptionally inactive and that only the free DNA observed between nucleosomes encoded for genes (Oudet et al., 1975). Therefore, activated transcription obviously was believed to require some form of remodeling of chromatin structure and it soon emerged that the structure of chromatin at specific regions on DNA could play a role in the regulation of cellular gene transcription, as well as its more established role in the packaging of DNA (Lilley and Pardon, 1979; Weintraub and Groudine, 1976).

Whilst it has become clear that transcriptionally active DNA is not necessarily free of histone association (Kuo et al., 1998), it is now accepted that biochemical modifications of these histones on specific amino acid residues in their amino-terminal domains are responsible for a more open, or a more closed, chromatin structure leading to gene expression or gene silencing, respectively. These histone modifications include: methylation (lysine and arginine residues; Zhang and Reinberg, 2001), acetylation (lysine residues; Doenecke and Gallwitz, 1982), deacetylation (lysine residues; Lopez-Rodas et al., 1993), sumoylation (lysine residues; Shiio and Eisenman, 2003), ubiquitination and phosphorylation (lysine and serine and threonine residues respectively; Lusser, 2002). This diverse array of modifications, which determines the overall structure of chromatin, is the basis for the "histone code hypothesis" (Strahl and Allis, 2000). This hypothesis suggests that patterns, and not just levels, of enzyme-mediated modifications of histones results in the generation of a code that determines the binding of regulatory proteins. This, in turn, leads to downstream effects, based specifically on these specific histone modifications, determining levels of gene expression.

## Histone acetylation

Probably the most well-characterized histone modification is that of acetylation, with which there is a strong correlation with transcriptionally active regions of the genome (Kuo and Allis, 1998; Lusser, 2002). The acetylation of histone tails is catalyzed by a family of enzymes known as histone acetyltransferases or "HATs" (Figure 9.1B; Eberharter and Becker, 2002) and these include p300, CREB-binding protein (CBP) and p300/CBP-associated factor (P/CAF). The removal of acetyl groups from histones is catalyzed by histone deacetylases ("HDACs"; Figure 9.1B), an evolutionarily conserved family of enzymes. The HDACs, structurally and functionally divided into distinct classes, are involved in transcriptional repression of gene expression (Khochbin et al., 2001).

It is probable that amino-terminal positively charged lysine residues within histones form multiple stable interactions with negatively charged DNA. This results in a close association between the histones and DNA, compaction of the chromatin, and a transcriptionally inactive state. Acetylation of these lysine residues, particularly those of histones H3 and H4, is likely to disrupt the histone–DNA interaction leading to a looser association and thereby facilitating access of transcription factors and the PIC to the DNA template (Figure 9.1B; Cary et al., 1982; Bode et al., 1983; Morgan et al., 1987). In addition, these acetyl-lysines may also provide binding or recognition sites for the so-called bromodomains of other proteins. These bromodomains are acetyl-lysine binding motifs (Zeng and Zhou, 2002) found in a diverse array of chromatin-associated proteins such as HATs (Dhalluin et al., 1999), and TAFII250, a component of the TFIID transcription complex (Jacobson et al., 2000). Conversely, the removal of acetyl groups from lysine residues, catalyzed by HDACs, re-creates the basic charge on the lysine leading to a strong association between DNA and histones and hence the basis of a transcriptionally inactive state (Figure 9.1B; Lusser, 2002).

However, it should be noted that acetylation of histones does not always correlate with transcriptional activation. Some promoters become active in the absence of significant acetylation changes (Dudley et al., 1999), while other genes are activated despite no change in the hypersensitivity of the promoter (by definition, acetylation of histones results in a looser association with DNA and hence the DNA is more sensitive to nuclease activity) (Urnov et al., 2000; Wong et al., 1998).

## Histone methylation

Methylation of histones is historically associated with transcriptionally silent regions of the genome (Razin, 1998) and is also linked with epigenetic inheritance (Lachner et al., 2003). Addition of methyl groups to lysine residues on histones H3 and H4 has been associated with transcriptional repression (Figure 9.1B; Cao et al., 2002; Schotta et al., 2002) and X chromosome inactivation (Mermoud et al., 2002; Plath et al., 2003). Lysine residues can be multiply methylated (up to three) by histone methyltransferases (HMTs). Trimethylation of lysine residue 9 is performed by the SUvar family of histone methylases, whereas mono- and dimethylation is carried out by the G9a family of histone methyltransferases (Rice et al., 2003). Interestingly, trimethylation appears to occur at heterochromatic regions (that is, densely staining chromatin regions believed generally to be transcriptionally inactive) of the genome, while mono- and dimethylation occurs at euchromatic regions of the genome (that is, regions believed generally to be transcriptionally active; Rice et al., 2003).

The marking of transcriptionally silent regions of chromatin is also associated with the binding of specific transcriptional silencing proteins such as the heterochromatin protein-1 (HP1). As the name suggests, HP1 is found at heterochromatic sites (James et al., 1989) and is recruited to these regions by methylation events on lysine residues. It appears that there is a two-step process: (i) methylation of histone H3 on lysine residue 9 or 27 results in (ii) the creation of a target for the chromatin binding domain of HP1. This binding by HP1 increases with the simultaneous methylation of both lysine 9 and 27 (Khorasanizadeh, 2004). Thus, heterochromatic regions can be identified by both methylated histone residues as well as binding by HP1 (Bannister et al., 2001).

Interestingly, recent data has suggested a link between histone methylation and another mechanism of transcriptional silencing – that of DNA methylation. Fuks and co-workers have demonstrated a physical and functional interaction between components of the lysine methylation complex (such as HMTs) and methyl-CpG-binding protein (Fuks et al., 2003a; 2003b). Such cross-talk between repression mechanisms would serve to further enforce transcriptional silencing.

One obvious, but so far unanswered, question refers to the ability of transcriptionally silenced chromatin regions to respond to gene activation signals. That is, how is silenced, methylated chromatin demethylated? Although several HMTs have been cloned and characterized, only one histone demethylase has been characterized so far, the amine oxidase homolog LSD1, although it is presumed that other demethylases do exist (Shi et al., 2004). Consequently, although there is a definite lack in our knowledge as to the mechanisms by which such methyl groups may be removed to allow "de-repression" of chromatin regions, several models, including histone replacement or proteolytic removal of methylated tails, have been proposed and have not been ruled out (reviewed in Bannister et al., 2002).

However, whilst methylation of lysine residues on histones H3 and H4 invariably is a marker of transcriptional repression, methylation per se is not always an indicator of inactive chromatin. Methylation of arginine residue 3 for example, catalyzed by enzymes containing the protein R methyltransferase motif, supports histone H4 acetylation and transcriptional activation (Wang et al., 2001).

## Chromatin and the analysis of virus latency

Clearly, chromatin structure plays a profound role in the control of eukaryotic gene expression. This understanding, and the recent advances in molecular tools to analyze chromatin-mediated regulation of gene expression, has allowed a more molecular analysis of the control of gene expression in vitro which, in turn, has led to an understanding of the possible mechanisms regulating viral latency and reactivation in vivo.

## HCMV latency and the myeloid lineage

It has been clear for some time that the myeloid lineage is an important site for the carriage of latent viral genomes (reviewed in Sinclair and Sissons, 1996; Sissons et al., 2002). Early experiments showed that HCMV could be transmitted in peripheral blood and that this transmission was reduced upon leukocyte depletion (Gilbert et al., 1989). Whilst this implicated leukocytes in the carriage of HCMV in vivo, infectious HCMV cannot be directly isolated from the blood of healthy seropositive individuals. Clearly, peripheral blood leukocytes contain reactivatable, but not necessarily infectious virus. Further experiments, utilizing highly sensitive DNA PCR for example, identified viral genomes in the myeloid lineage from primitive CD34+ progenitor cells through to monocytes (Bevan et al., 1993; Mendelson et al., 1996; Minton et al., 1994; Taylor-Wiedeman et al., 1991), but sensitive RT-PCR protocols showed that no viral lytic gene expression occurred during the carriage of virus in these cells (Mendelson et al., 1996; Taylor-Wiedeman et al., 1994). Intriguingly, terminal differentiation of monocytes to macrophages was shown to reactivate IE gene expression (Figure 9.2). However, under these conditions, infectious virus could not be recovered, as determined by co-culture with fully permissive fibroblasts (Taylor-Wiedeman et al., 1994). One subsequent report showed that allogeneic stimulation of monocytes, from healthy seropositive carriers, by co-culture with activated T-cells, could result in the reactivation of latent virus and the release of infectious virus (Soderberg-Naucler et al., 1997). However, the precise phenotype of the cell that contained reactivating virus in this study was unclear.

What was clear from these studies though, was that there is a very strong link between reactivation of latent virus and the state of differentiation of the cell. Interestingly, this apparent dependence of myeloid differentiation for reactivation of viral lytic gene expression also correlates well with the known levels of permissiveness of myeloid cells for IE expression after exogenous infection with HCMV in vitro; many workers have shown that whilst monocytes are non-permissive, macrophages are fully permissive for infection (reviewed in Sinclair and Sissons, 1996).

## The study of latency: model cell systems

Many studies, seeking to understand the molecular mechanisms regulating HCMV reactivation from latency, have used this differentiation-dependent permissiveness of myeloid cells for HCMV infection to carry out experiments that rely on infection of non-permissive cell types and the subsequent investigation of the effects of differentiation on viral IE gene expression and productive infection. These systems include CD33+ fetal liver stem cells (Kondo et al., 1994), myeloid lineage-committed progenitor cells which display both myeloid and DC markers (Hahn et al., 1998) and bone marrow progenitor cells (includ-

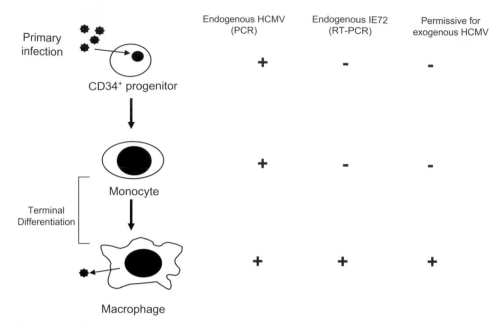

**Figure 9.2** HCMV and the myeloid lineage.

ing CD34$^+$ cells; Goodrum et al., 2002; Goodrum et al., 2004; Maciejewski et al., 1992; Minton et al., 1994).

Similarly, because of the difficulties of obtaining and working with primary material, such as monocytes and CD34$^+$ bone marrow progenitors, many studies have also utilized cell lines that, in vitro, show differentiation state-dependent control of IE gene expression. The myelo-monocytic cell line THP-1 and the NTera2 (T2) embryonal carcinoma cell line are non-permissive for HCMV infection due to a block in IE gene expression. Following terminal differentiation of THP-1 cells to macrophage-like cells with phorbol esters or differentiation of T2 cells with retinoic acid, these cells become permissive for IE gene expression and productive infection (Gonczol et al., 1984; Weinshenker et al., 1988). Thus, these cell lines are good models for the differentiation state-dependence of IE gene expression and should give insight into the mechanisms regulating reactivation of viral IE gene expression upon differentiation of monocytes to macrophages.

## The control of HCMV latency in the myeloid lineage

Model systems, based on exogenous infection of myeloid cells, have reproducibly confirmed the differentiation-dependent regulation of viral IE gene expression after infection, but they have also been used to attempt to determine whether viral transcripts are associated with latent carriage of HCMV. Undifferentiated myeloid cells generally carry the latent genome in the absence of appreciable viral lytic gene expression; HCMV does not appear to express viral gene products equivalent to the levels of Epstein–Barr virus nuclear antigens (EBNAs) or latent membrane proteins (LMPs), proteins associated with the establishment and maintenance of the latent state in EBV (Young et al., 2000). However, Kondo and colleagues, using infection of granulocyte–macrophage progenitor cells (GMPs) as a model for latent infection, have identified both sense and antisense transcripts originating from the major IE region (Kondo et al., 1994; Kondo et al., 1996). These transcripts, also found in some healthy seropositive carriers (Kondo et al., 1996), initially suggested that HCMV resembles HSV by encoding a transcript(s) that could, perhaps, provide antisense–mediated inhibition of critical lytic genes (Preston, 2000). Subsequent studies, however, have not identified a precise role for these latency-associated transcripts. Only a small proportion of the infected GMPs express these transcripts (Slobedman and Mocarski, 1999) and viral deletion mutants unable to express these RNAs establish latency normally (White et al., 2000). Similarly, these transcripts are also found during productive virus infection (Lunetta and Wiedeman, 2000) and thus their function remains unclear.

Interestingly, using the same model system Jenkins et al. (2004) have recently identified an additional, apparently latent, viral transcript encoding a protein with homology to the potent immunosuppressor IL-10. The presence of this RNA was confirmed in mononuclear cells of healthy seropositive carriers but also in some seronegative individuals. It has been suggested that expression of this protein during latency may aid the virus in avoiding host immune system recognition by creating an immunosuppressed micro-environment around the latently infected cell.

A number of putative latency-associated transcripts (including those from the major IE region) have also been reported in other HCMV latency models which used exogenous infection of myeloid cells (Goodrum et al., 2002; Goodrum et al., 2004). However, the marker for viral latency/reactivation in this system was the presence of GFP driven by an SV40 promoter. Consequently, it is difficult to know if viral transcripts that were detected are truly latency-associated or represent residual viral gene expression that occurs following infection at high MOI as viral genomes expressing GFP, driven by an heterologous promoter, become quiescent after infection.

Therefore, whether the RNAs detected in these systems are bona fide latent transcripts with defined functions for the latent phenotype remains to be resolved. However, it is worth noting that, although some transcripts identified initially using exogenously infected latent model systems have also been identified in bone marrow aspirates and bone marrow mononuclear cells during natural infection, none have yet been shown to be present in latently infected monocytes from healthy carriers in vivo (Jenkins et al., 2004; Kondo et al., 1996).

Whether or not specific viral gene products are required for latent infection in vivo is, therefore, still a crucial question that is under intense investigation by several laboratories. What is evident, however, is that undifferentiated cells carrying viral genomes are generally silent with respect to viral IE gene expression. This suggests that the cell carrying the viral genome is not conducive to viral lytic gene expression. Specifically, the HCMV major IE promoter which drives expression of the viral major IE genes appears to be transcriptionally repressed in these undifferentiated cells.

Interestingly, Slobedman et al. (2004) have shown that there is a dramatic change in cellular gene expression following experimental infection of non-permissive myeloid progenitor cells, and propose that such changes may alter the cellular environment to promote the latent state by, for instance, repression of viral IE gene expression.

## HCMV latency and chromatin remodeling

The physical association between cellular DNA and histones is a prerequisite for compaction of the eukaryotic genome and crucial for the architecture of the nucleus. It is, perhaps, not surprising that genomes of DNA viruses have been investigated as to whether they are themselves chromatinized; early studies suggested that they were. Oudet et al (1975) showed that lambda or adenovirus DNA, when incubated with histone proteins, reconstituted bona fide nucleosome structures in vitro. Later experiments, using the SV40 promoter-enhancer element for example (Cereghini and Yaniv, 1984; Cremisi et al., 1975;), showed a similar association between viral DNA sequences and cellular histones. Although it is still unclear to what extent chromatinization of viral genomes exactly mimics the structure of cellular chromatin, it has emerged that a number of herpesvirus genomes are associated with histones and that changes in chromatin structure appear to specifically regulate viral gene expression (for only a few examples see Krithivas et al., 2000; Kubat et al., 2004; Radkov et al., 1999).

One obvious mechanism for regulating HCMV IE gene expression and, therefore, reactivation from latency, is chromatin remodeling. Initial experiments have indicated that chromatin remodelers have a critical role in regulating major IE promoter (MIEP) activity and virus infection. Treatment of ordinarily non-permissive T2 cells with the broad-spectrum HDAC inhibitor Trichostatin A (TSA) renders these cells permissive for viral IE gene expression (Meier, 2001; Murphy et al., 2002). Moreover, overexpression of HDACs in normally permissive, differentiated, T2 cells (T2RA cells) inhibited viral IE gene expression following infection (Murphy et al., 2002). Similar results were observed in transient transfection assays of these cells with MIEP reporter constructs; TSA treatment of T2 cells rendered the cells permissive for MIEP activity, whereas super-expression of HDACs in T2RA cells repressed MIEP activity (Murphy et al., 2002).

The apparent importance of HDACs in regulating the MIEP strongly suggested that the MIEP might be subject to chromatinization. The advent of chromatin immunoprecipitation (ChIP) assays, normally used extensively to analyze the chromatin structure of cellular genes, allowed detailed direct analyses of the chromatin structure around viral genomes. In addition, because of the sensitivity of these assays, only small samples are required and this makes them ideal not only for analysis of exogenously infected cells in vitro, but also for latently infected material obtained from healthy seropositive individuals in vivo.

ChIP assays are used to analyze proteins (transcription factors, chromatin components etc.) that are bound to DNA. DNA-binding proteins are cross-linked in vivo to DNA using formaldehyde. Chromatin is prepared and sheared into smaller fragments and then antibody is added that is specific for the protein molecule of interest (this may be an anti-acetylated histone antibody for example). Using standard immunoprecipitation techniques antibody–antigen complexes are isolated. Cross-linking is reversed to release the DNA, proteins are digested and then the purified DNA is amplified using PCR to assess whether certain sequences were precipitated with the antibody. With such an approach it is possible to characterize transcription factors bound to promoter elements or, as described below, analyze the histone modifications around such promoters.

Using such ChIP assays, it has been shown that in exogenously infected undifferentiated non-permissive cells, in which the MIEP is transcriptionally inactive, the MIEP is preferentially associated with deacetylated histones (Figure 9.3; Murphy et al., 2002;

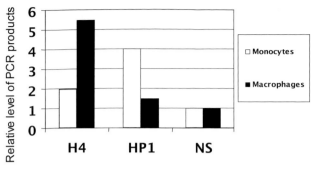

**Figure 9.3** Chromatin immunoprecipitation assays were performed on exogenously infected monocytes and monocyte-derived macrophages as described (Murphy et al., 2002). H4 indicates immunoprecipitation using an anti-acetylated histone H4 antiserum, HP1 indicates immunoprecipitation using an anti-HP1 antiserum, and NS is a non-specific control antiserum. Figure adapted from Murphy et al. (2003).

Murphy et al., 2003). This hypo-acetylation of histones associated with the MIEP was observed after infection of both non-permissive T2 cells as well as non-permissive peripheral blood monocytes. Similarly, in monocytes equivalent analyses showed that not only was the MIEP associated with hypo-acetylated histones but it was also associated with the silencing protein HP1 (Figure 9.3; Murphy et al., 2002; Murphy et al., 2003). These data are entirely consistent with transcriptional silencing of the MIEP in these undifferentiated cells being due to a closed chromatin conformation. Conversely, infection of cells after they had been differentiated to a permissive phenotype showed an association of the MIEP with acetylated histones and a concomitant loss of association with HP1; all entirely consistent with transcriptional activation of the MIEP, expression of the IE proteins and a productive lytic infection (Murphy et al., 2002; Murphy et al., 2003). It appears, therefore, that the differentiation–dependent state of chromatin around the MIEP plays a pivotal role in regulating lytic gene expression and hence the outcome of infection.

Clearly, the viral chromatin studies detailed above show a direct correlation between an open chromatin conformation and the transcriptional activity of the MIEP, after exogenous infection of undifferentiated/differentiated model cell lines or monocytes/monocyte-derived macrophages. Other cell types of the myeloid lineage have also been used to analyze differentiation-dependent regulation of viral IE gene expression upon infection. A number of workers have shown that terminally differentiated myeloid DCs are fully permissive for HCMV infection whilst their undifferentiated progenitors show no such permissiveness (Hertel et al., 2003; Riegler et al., 2000). Interestingly, these cell types are likely to be of real biological relevance as one report, detailing reactivation of endogenous latent HCMV after differentiation of monocytes, suggested the cell type that reactivated virus in vivo carries some DC markers (Soderberg-Naucler et al., 1997).

Consistent with this, chromatin studies using exogenous infection of CD34[+] progenitor cells have shown that, upon infection, the MIEP is preferentially associated with hypo-acetylated histones and HP1 (M.R. and J.S, unpublished data). In contrast, after exogenous infection of mature CD34[+] -derived DCs and mature monocyte-derived DCs, the MIEP becomes associated with acetylated histones with no evidence of HP1 association (M.R. and J.S, unpublished data). This is consistent with the differentiation-dependent permissiveness of these cells for IE gene expression and productive infection, following infection in vitro (Hertel et al., 2003; Riegler et al., 2000). Thus, a mature dendritic phenotype, supporting IE gene expression, results from differentiation-mediated changes in the cellular milieu resulting in chromatin remodeling events around the MIEP (such as hyper-acetylation and the loss of an association with HP1), similar to the differentiation-dependent chromatin remodeling observed in monocyte/macrophage cells.

These studies have been taken further in an attempt to model reactivation associated with differentiation of CD34+ cells. Non-permissive CD34+ cells, infected and then cultured for seven days in the absence of differentiation promoting signals such as cytokines, maintain the viral genome with no IE expression and without detectable infectious virus. Consistent with the transcriptional silencing of major IE expression, the MIEP is associated with hypo-acetylated histones and HP1 protein. After long-term differentiation however, resulting in a mature DC phenotype and the release of infectious virus, the MIEP loses its association with HP1 and becomes associated with hyper-acetylated histones (Reeves et al., 2005). Thus, the processes of reactivation from latency in this exogenously infected model system is also linked to chromatin remodeling events.

One caveat with all experiments using exogenous infection, however, is that care must be taken in interpreting observations from in vitro infection of non-permissive cells and applying them to bona fide latency in vivo. However, the use of highly sensitive PCR-based ChIP assays allows this type of analysis to go one step further: it can be used to analyze endogenous latent virus from normal healthy seropositive carriers.

Recent work using ChIP assays on monocytes and CD34+ cells from normal seropositive donors has substantiated the results from exogenous infection studies. That is, the MIEP of endogenous latent virus in these cells during natural infection is also associated with HP1 and hypo-acetylated histones – entirely consistent with the silencing of the MIEP. Ex vivo differentiation of monocytes to monocyte-derived DCs or CD34+ cells to mature CD34+ -derived DCs results in association of the MIEP with hyper-acetylated histones, a decreased association with HP1 and the release of infectious virus (M.R. and J.S, unpublished data).

## Factors which direct chromatin remodeling of the HCMV MIEP

Clearly the chromatin structure around the MIEP is likely to play an important role in regulating its activity during latency and reactivation, but this begs the question as to what regulates the acetylation/deacetylation state of MIEP-associated histones in cells carrying viral genomes?

A number of studies have shown that, using transient transfection assays with truncated versions of the MIEP as reporter constructs, distinct regions of the MIEP either positively or negatively regulate MIEP activity in non-permissive/permissive cell types. For example, the 21 base pair repeats and the dyad symmetry element (modulator) are negative regulatory elements for MIEP activity in non-permissive cell types (Kothari et al., 1991; Lubon et al., 1989; Nelson et al., 1987). Further analyses of such negative regulatory sequences, using in vitro and in vivo assays, has allowed the identification of binding sites for cellular transcription factors. Analyses of these factors confirm their binding to, and repression of, the MIEP. Factors so far identified include yin yang-1 (YY1; Liu et al., 1994), Ets2 repressor factor (ERF; Bain et al., 2003), growth factor independence-1 (Gfi-1; Zweidler-Mckay et al., 1996) and modulator recognition factor (MRF; Huang et al., 1996).

Current models suggest that some of these factors – YY1 for example – are preferentially expressed in non-permissive cell types where they repress MIEP activity. Following differentiation to the permissive phenotype, levels of expression of these factors decrease (Liu et al., 1994). In contrast, no such change is observed in the absolute levels of expression of the other repressor factors, for example ERF (Wright et al., 2005), suggesting that changes in the levels of, so far, unidentified, co-factors are necessary for these factors to mediate repression. This relative shift in the balance of specific activators or repressors of the MIEP is likely to be crucial for regulation of its activity in non-permissive and permissive cells.

Interestingly, several of the factors identified as binding to, and repressing, the MIEP are thought to mediate their effects through interactions with chromatin remodeling co-factors. For example, it is known that YY1 interacts with multiple HDAC family members (HDAC-1, -2 and -3; Yang et al., 1996; Yang et al., 1997). Similarly, ERF interacts with HDAC1 and this interaction is important for ERF to mediate repression of the MIEP and may also serve to recruit the histone methyltransferase SUvar to the MIEP (Wright et al., 2005). This suggests a multi-step process of initial binding of repressor factors that sequentially recruit chromatin remodeling enzymes, eventually leading to transcriptional repression. Following differentiation to permissive phenotypes, however, the level of YY1, ERF and/or HDAC-1, -2 and -3 decreases, an observation that is common to many of the model systems discussed here, including T2 and THP-1 cells, monocytes and CD34+ cells (Liu et al., 1994; Murphy et al., 2002; Wright et al., 2005). These observations, showing de-repression of the MIEP, support a model in which the MIEP in non-permissive or latent cell types is inhibited by the action of numerous cellular transcriptional repressors. These act through multiple sites within the MIEP and function by recruiting HDAC and methylase activities, leading to a more compact chromatin structure and transcriptional silencing. In contrast, in permissive cells or upon virus reactivation, levels of repressor complexes are reduced resulting in chromatin remodeling to a more open chromatin configuration and subsequent transcriptional activation of the MIEP (Figure 9.4).

## HCMV and interactions with histone acetyltransferases

As outlined above, a current model for the differentiation state-dependent control of the MIEP relies on a delicate interplay between transcriptional repressors and transcriptional activators. These activators, including NF-κB and CREB, have numerous binding sites within the MIEP (see the accompanying review by Meier and Stinski, Chapter 8). In addition, it is likely that at least some of these activating factors mediate their effects by the recruitment of HATs. For example, the p65 component of NF-κB interacts with CBP that has intrinsic HAT activity (Sheppard et al., 1998; Wadgaonkar et al., 1999). Thus, HAT activity is probably as important in directing high-level MIEP activity as the action of HDACs are in causing MIEP repression. However, the role of HATs in mediating MIEP activity, though probable, has not been as extensively investigated as the repressive effects of the HDACs.

Although chromatin remodeling is believed to play a major role in the establishment and maintenance of the latent HCMV genome, as well as in reactivation from latency, such

Undifferentiated (non-permissive)          Differentiated (permissive)

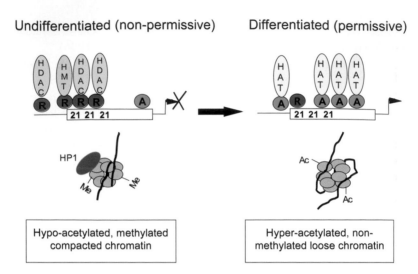

| Hypo-acetylated, methylated compacted chromatin | Hyper-acetylated, non-methylated loose chromatin |

**Figure 9.4** Model for the differentiation state-dependent control of the MIEP. The MIEP in non-permissive undifferentiated cells is silenced due to the action of numerous transcriptional repressors upon the MIEP. Several of these repressors function by the recruitment of HDAC chromatin remodelers. Following differentiation, the activity of the MIEP is induced due to a relative shift in the balance of repressors and activators of the MIEP. Repressor levels decrease accompanied with a decrease in HDAC co-factor levels. Also shown are activators of the MIEP, some of which function by the recruitment of histone acetyltransferase activities to the promoter. **R** are repressors such as YY1 and ERF, **A** are activators such as NF-κB. 21 indicates the 21 base pair repeats, HMT is histone methyltransferase activity, HDAC is histone deacetylase activity, HAT is histone acetyltransferase activity. HP1 is the heterochromatin protein-1, Me and Ac represent methylated and acetylated histones, respectively.

changes in chromatin structure of both viral and cellular gene promoters are also likely to be important in the course of a productive lytic infection. It is known that HCMV infection has profound effects on normal host gene expression. Precisely how all these cellular genes are affected is not yet known, but it is possible that at least some of them are activated/repressed in response to chromatin remodeling around their promoters. Such effects are likely to be mediated by viral factors in concert with cellular transcription factors. One such candidate viral factor is the potent and promiscuous transcriptional activator immediate-early factor IE2-p86. This protein is known to activate a wide range of viral genes, ensuring E and L viral gene expression, but is also known to promiscuously activate cellular gene expression (reviewed in Stenberg, 1996). The mechanisms by which IE2-p86 mediates such effects are not fully understood but it is probable that they also involve chromatin remodeling. For instance, it has been shown that IE2-p86 interacts with the cellular protein P/CAF, a protein with HAT activity (Bryant et al., 2000). IE2-p86 and P/CAF physically and functionally interact and this interaction enhances IE2-p86-mediated activation of target cellular promoters.

Conversely, it has also been shown that IE2-p86 can also interact with the acetyltransferase domains of proteins such as p300 and CBP. These interactions block the intrinsic HAT activity of these proteins and, as a consequence, can affect the acetylation status, and therefore the function of, target proteins such as p53 (Hsu et al., 2004). It is therefore likely that the acetylation status of promoters, and indeed transcription factors themselves, of many HCMV genes activated by IE2-p86 are also altered during the course of infection.

## Concluding remarks

This review has detailed a number of studies that indicate a role for chromatin remodeling in the control of HCMV gene expression. It is clear that in many in vitro model systems, silencing of the MIEP in undifferentiated non-permissive cells appears to result from the association of this viral promoter with non-acetylated histones and silencing proteins such as HP1; heterochromatic proteins indicative of compacted transcriptionally inactive chromatin. Differentiation-induced activation of the viral MIEP, on the other hand, correlates with an open chromatin conformation. Similar changes in the chromatin structure of the viral MIEP of endogenous virus during natural latent infection are also seen in myeloid progenitors upon their differentiation to more mature DCs, resulting in reactivation of endogenous latent virus. Clearly, latency of HCMV in DCs during natural persistence, and this critical linkage of DCs with chromatin remodeling of the MIEP and subsequent reactivation, is likely to play an important role in the pathogenesis of HCMV infection.

It is perhaps inevitable that the models described above will need to be revised or modified in the near future as our understanding of the histone code hypothesis develops. For example, no mention is made here of histone ubiquitination, phosphorylation or sumoylation and how they may influence HCMV biology. In addition, given the complexity of the HCMV MIEP and its multiplicity of binding sites for both cellular and viral factors, it may emerge that other, as yet uncharacterized, factors may use these processes for regulating HCMV gene expression. Nevertheless, the analysis of the molecular mechanisms regulating the viral MIEP with respect to chromatin structure has begun to shed light on the mechanisms of viral latency and reactivation, which are a defining characteristic of the herpesvirus family, and will be crucial to our understanding of the biology of HCMV infection.

## Acknowledgments

The authors thank past and present members of the laboratory whose work has contributed to the studies described here. The work from this laboratory was funded by the Medical Research Council and The Wellcome Trust.

## References

Bain, M., Mendelson, M., and Sinclair, J. (2003). Ets-2 Repressor Factor (ERF) mediates repression of the human cytomegalovirus major immediate-early promoter in undifferentiated non-permissive cells. J. Gen. Virol. 84, 41 19.

Bannister, A.J., Schneider, R., and Kouzarides, T. (2002). Histone methylation: dynamic or static? Cell 109, 801–806.

Bannister, A.J., Zegerman, P., Partridge, J.F., Miska, E.A., Thomas, J.O., Allshire, R.C., and Kouzarides, T. (2001). Selective recognition of methylated lysine 9 on histone H3 by the HP1 chromo domain. Nature 410, 120–124.

Bevan, I.S., Walker, M.R., and Daw, R.A. (1993). Detection of human cytomegalovirus DNA in peripheral blood leukocytes by the polymerase chain reaction. Transfusion 33, 783–784.

Bode, J., Gomez-Lira, M.M., and Schroter, H. (1983). Nucleosomal particles open as the histone core becomes hyperacetylated. Eur. J. Biochem. 130, 437–445.

Bryant, L.A., Mixon, P., Davidson, M., Bannister, A.J., Kouzarides, T., and Sinclair, J.H. (2000). The human cytomegalovirus 86-kilodalton major immediate-early protein interacts physically and functionally with histone acetyltransferase P/CAF. J. Virol. 74, 7230–7237.

Cao, R., Wang, L., Wang, H., Xia, L., Erdjument-Bromage, H., Tempst, P., Jones, R.S., and Zhang, Y. (2002). Role of histone H3 lysine 27 methylation in Polycomb-group silencing. Science 298, 1039–1043.

Cary, P.D., Crane-Robinson, C., Bradbury, E.M., and Dixon, G.H. (1982). Effect of acetylation on the binding of N-terminal peptides of histone H4 to DNA. Eur. J. Biochem. 127, 137–143.

Cereghini, S., and Yaniv, M. (1984). Assembly of transfected DNA into chromatin: structural changes in the origin-promoter-enhancer region upon replication. EMBO J. 3, 1243–1253.

Chung, S.Y., Hill, W.E., and Doty, P. (1978). Characterization of the histone core complex. Proc. Natl. Acad. Sci. USA 75, 1680–1684.

Cremisi, C., Pignatti, P.F., Croissant, O., and Yaniv, M. (1975). Chromatin-like structures in polyoma virus and simian virus 10 lytic cycle. J. Virol. 17, 204–211.

Dhalluin, C., Carlson, J.E., Zeng, L., He, C., Aggarwal, A.K., and Zhou, M.M. (1999). Structure and ligand of a histone acetyltransferase bromodomain. Nature 399, 491–496.

Doenecke, D., and Gallwitz, D. (1982). Acetylation of histones in nucleosomes. Mol. Cell. Biochem. 44, 113–128.

Dudley, A.M., Rougeulle, C., and Winston, F. (1999). The Spt components of SAGA facilitate TBP binding to a promoter at a post-activator-binding step in vivo. Genes Dev. 13, 2940–2945.

Eberharter, A., and Becker, P.B. (2002). Histone acetylation: a switch between repressive and permissive chromatin. Second in review series on chromatin dynamics. EMBO Rep 3, 224–229.

Fuks, F., Hurd, P.J., Deplus, R., and Kouzarides, T. (2003a). The DNA methyltransferases associate with HP1 and the SUV39H1 histone methyltransferase. Nucleic Acids Res. 31, 2305–2312.

Fuks, F., Hurd, P.J., Wolf, D., Nan, X., Bird, A.P., and Kouzarides, T. (2003b). The methyl-CpG-binding protein MeCP2 links DNA methylation to histone methylation. J. Biol. Chem. 278, 4035–4040.

Gilbert, G.L., Hayes, K., Hudson, I.L., and James, J. (1989). Prevention of transfusion-acquired cytomegalovirus infection in infants by blood filtration to remove leucocytes. Neonatal Cytomegalovirus Infection Study Group. Lancet 1, 1228–1231.

Gonczol, E., Andrews, P.W., and Plotkin, S.A. (1984). Cytomegalovirus replicates in differentiated but not in undifferentiated human embryonal carcinoma cells. Science 224, 159–161.

Goodrum, F., Jordan, C.T., Terhune, S.S., High, K., and Shenk, T. (2004). Differential outcomes of human cytomegalovirus infection in primitive hematopoietic cell subpopulations. Blood 104, 687–695.

Goodrum, F.D., Jordan, C.T., High, K., and Shenk, T. (2002). Human cytomegalovirus gene expression during infection of primary hematopoietic progenitor cells: a model for latency. Proc. Natl. Acad. Sci. USA 99, 16255–16260.

Hahn, G., Jores, R., and Mocarski, E.S. (1998). Cytomegalovirus remains latent in a common precursor of dendritic and myeloid cells. Proc. Natl. Acad. Sci. USA 95, 3937–3942.

Hertel, L., Lacaille, V.G., Strobl, H., Mellins, E.D., and Mocarski, E.S. (2003). Susceptibility of immature and mature Langerhans cell-type dendritic cells to infection and immunomodulation by human cytomegalovirus. J. Virol. 77, 7563–7574.

Hewish, D.R., and Burgoyne, L.A. (1973). Chromatin sub-structure. The digestion of chromatin DNA at regularly spaced sites by a nuclear deoxyribonuclease. Biochem. Biophys. Res. Commun. 52, 504–510.

Hsu, C.H., Chang, M.D., Tai, K.Y., Yang, Y.T., Wang, P.S., Chen, C.J., Wang, Y.H., Lee, S.C., Wu, C.W., and Juan, L.J. (2004). HCMV IE2-mediated inhibition of HAT activity down-regulates p53 function. EMBO J. 23, 2269–2280.

Huang, T.H., Oka, T., Asai, T., Okada, T., Merrills, B.W., Gertson, P.N., Whitson, R.H., and Itakura, K. (1996). Repression by a differentiation-specific factor of the human cytomegalovirus enhancer. Nucleic Acids Res. 24, 1695–1701.

Jacobson, R.H., Ladurner, A.G., King, D.S., and Tjian, R. (2000). Structure and function of a human TAFII250 double bromodomain module. Science 288, 1422–1425.

James, T.C., Eissenberg, J.C., Craig, C., Dietrich, V., Hobson, A., and Elgin, S.C. (1989). Distribution patterns of HP1, a heterochromatin-associated nonhistone chromosomal protein of Drosophila. Eur. J. Cell Biol. 50, 170–180.

Jenkins, C., Abendroth, A., and Slobedman, B. (2004). A novel viral transcript with homology to human interleukin-10 is expressed during latent human cytomegalovirus infection. J. Virol. 78, 1440–1447.

Khochbin, S., Verdel, A., Lemercier, C., and Seigneurin-Berny, D. (2001). Functional significance of histone deacetylase diversity. Curr. Opin. Genet. Dev. *11*, 162–166.

Khorasanizadeh, S. (2004). The nucleosome: from genomic organization to genomic regulation. Cell *116*, 259–272.

Kondo, K., Kaneshima, H., and Mocarski, E. S. (1994). Human cytomegalovirus latent infection of granulocyte–macrophage progenitors. Proc. Natl. Acad. Sci. USA *91*, 11879–11883.

Kondo, K., Xu, J., and Mocarski, E.S. (1996). Human cytomegalovirus latent gene expression in granulocyte–macrophage progenitors in culture and in seropositive individuals. Proc. Natl. Acad. Sci. USA *93*, 11137–11142.

Kornberg, R.D. (1974). Chromatin structure: a repeating unit of histones and DNA. Science *184*, 868–871.

Kothari, S., Baillie, J., Sissons, J.G., and Sinclair, J.H. (1991). The 21bp repeat element of the human cytomegalovirus major immediate-early enhancer is a negative regulator of gene expression in undifferentiated cells. Nucleic Acids Res. *19*, 1767–1771.

Krithivas, A., Young, D.B., Liao, G., Greene, D., and Hayward, S.D. (2000). Human herpesvirus 8 LANA interacts with proteins of the mSin3 corepressor complex and negatively regulates Epstein–Barr virus gene expression in dually infected PEL cells. J. Virol. *74*, 9637–9645.

Kubat, N.J., Tran, R.K., McAnany, P., and Bloom, D.C. (2004). Specific histone tail modification and not DNA methylation is a determinant of herpes simplex virus type 1 latent gene expression. J. Virol. *78*, 1139–1149.

Kuo, M.H., and Allis, C.D. (1998). Roles of histone acetyltransferases and deacetylases in gene regulation. BioEssays *20*, 615–626.

Kuo, M.H., Zhou, J., Jambeck, P., Churchill, M.E., and Allis, C.D. (1998). Histone acetyltransferase activity of yeast Gcn5p is required for the activation of target genes in vivo. Genes Dev. *12*, 627–639.

Lachner, M., O'Sullivan, R.J., and Jenuwein, T. (2003). An epigenetic road map for histone lysine methylation. J. Cell Sci. *116*, 2117–2124.

Lilley, D.M., and Pardon, J.F. (1979). Structure and function of chromatin. Annu. Rev. Genet. *13*, 197–233.

Liu, R., Baillie, J., Sissons, J.G., and Sinclair, J.H. (1994). The transcription factor YY1 binds to negative regulatory elements in the human cytomegalovirus major immediate-early enhancer/promoter and mediates repression in non-permissive cells. Nucleic Acids Res. *22*, 2453–2459.

Lopez-Rodas, G., Brosch, G., Georgieva, E.I., Sendra, R., Franco, L., and Loidl, P. (1993). Histone deacetylase. A key enzyme for the binding of regulatory proteins to chromatin. FEBS Lett. *317*, 175–180.

Lubon, H., Ghazal, P., Hennighausen, L., Reynolds-Kohler, C., Lockshin, C., and Nelson, J. (1989). Cell-specific activity of the modulator region in the human cytomegalovirus major immediate-early gene. Mol. Cell. Biol. *9*, 1342–1345.

Lunetta, J.M., and Wiedeman, J.A. (2000). Latency-associated sense transcripts are expressed during in vitro human cytomegalovirus productive infection. Virology *278*, 467–476.

Lusser, A. (2002). Acetylated, methylated, remodeled: chromatin states for gene regulation. Curr. Opin. Plant Biol. *5*, 437–443.

Maciejewski, J.P., Bruening, E.E., Donahue, R.E., Mocarski, E.S., Young, N.S., and St Jeor, S. C. (1992). Infection of hematopoietic progenitor cells by human cytomegalovirus. Blood *80*, 170–178.

Meier, J.L. (2001). Reactivation of the human cytomegalovirus major immediate-early regulatory region and viral replication in embryonal NTera2 cells: role of trichostatin A, retinoic acid, and deletion of the 21-base-pair repeats and modulator. J. Virol. *75*, 1581–1593.

Mendelson, M., Monard, S., Sissons, P., and Sinclair, J. (1996). Detection of endogenous human cytomegalovirus in CD34$^+$ bone marrow progenitors. J. Gen. Virol. *77*, 3099–3102.

Mermoud, J.E., Popova, B., Peters, A.H., Jenuwein, T., and Brockdorff, N. (2002). Histone H3 lysine 9 methylation occurs rapidly at the onset of random X chromosome inactivation. Curr. Biol. *12*, 247–251.

Minton, E.J., Tysoe, C., Sinclair, J.H., and Sissons, J.G. (1994). Human cytomegalovirus infection of the monocyte/macrophage lineage in bone marrow. J. Virol. *68*, 4017–4021.

Morgan, J.E., Blankenship, J.W., and Matthews, H.R. (1987). Polyamines and acetylpolyamines increase the stability and alter the conformation of nucleosome core particles. Biochemistry *26*, 3643–3649.

Murphy, J.C., Bain, M., Mendelson, M., Liu, R., and Sinclair, J.H. (2003). HCMV latency and lytic gene regulation: a question of chromatin structure? In New Aspects of CMV-Related Immunopathology. Monogr. Virol., S. Prosch, J. Cinatl, and M. Scholz, (Basel: Karger) pp. 96–104.

Murphy, J.C., Fischle, W., Verdin, E., and Sinclair, J.H. (2002). Control of cytomegalovirus lytic gene expression by histone acetylation. EMBO J. *21*, 1112–1120.

Nelson, J.A., Reynolds-Kohler, C., and Smith, B.A. (1987). Negative and positive regulation by a short segment in the 5′-flanking region of the human cytomegalovirus major immediate-early gene. Mol. Cell. Biol. *7*, 4125–4129.

Oudet, P., Gross-Bellard, M., and Chambon, P. (1975). Electron microscopic and biochemical evidence that chromatin structure is a repeating unit. Cell *4*, 281–300.

Plath, K., Fang, J., Mlynarczyk-Evans, S.K., Cao, R., Worringer, K.A., Wang, H., de la Cruz, C.C., Otte, A.P., Panning, B., and Zhang, Y. (2003). Role of histone H3 lysine 27 methylation in X inactivation. Science *300*, 131–135.

Preston, C.M. (2000). Repression of viral transcription during herpes simplex virus latency. J. Gen. Virol. *81*, 1–19.

Radkov, S.A., Touitou, R., Brehm, A., Rowe, M., West, M., Kouzarides, T., and Allday, M.J. (1999). Epstein–Barr virus nuclear antigen 3C interacts with histone deacetylase to repress transcription. J. Virol. *73*, 5688–5697.

Razin, A. (1998). CpG methylation, chromatin structure and gene silencing-a three-way connection. EMBO J. *17*, 4905–4908.

Rice, J.C., Briggs, S.D., Ueberheide, B., Barber, C.M., Shabanowitz, J., Hunt, D.F., Shinkai, Y., and Allis, C.D. (2003). Histone methyltransferases direct different degrees of methylation to define distinct chromatin domains. Mol. Cell *12*, 1591–1598.

Reeves, M.B., MacAry, P.A., Lehner, P.J., Sissons, J.G., and Sinclair, J. H. (2005). Latency, chromatin remodeling and reactivation of human cytomegalovirus in the dendritic cells of healthy carriers. Proc. Natl. Acad. Sci. USA *102*, 4140–4145.

Riegler, S., Hebart, H., Einsele, H., Brossart, P., Jahn, G., and Sinzger, C. (2000). Monocyte-derived dendritic cells are permissive to the complete replicative cycle of human cytomegalovirus. J. Gen. Virol. *81*, 393–399.

Schotta, G., Ebert, A., Krauss, V., Fischer, A., Hoffmann, J., Rea, S., Jenuwein, T., Dorn, R., and Reuter, G. (2002). Central role of Drosophila SU(VAR)3–9 in histone H3-K9 methylation and heterochromatic gene silencing. EMBO J. *21*, 1121–1131.

Sheppard, K.A., Phelps, K.M., Williams, A.J., Thanos, D., Glass, C.K., Rosenfeld, M.G., Gerritsen, M.E., and Collins, T. (1998). Nuclear integration of glucocorticoid receptor and nuclear factor-kappaB signaling by CREB-binding protein and steroid receptor coactivator-1. J. Biol. Chem. *273*, 29291–29294.

Shi, Y., Matson, C., Mulligan, P., Whetstine, J.R., Cole, P.A., Casero, R.A., and Shi, Y. (2004). Histone demethylation mediated by the nuclear amine oxidase homolog LSD1. Cell *119*, 941–953.

Shiio, Y., and Eisenman, R.N. (2003). Histone sumoylation is associated with transcriptional repression. Proc. Natl. Acad. Sci. USA *100*, 13225–13230.

Sinclair, J., and Sissons, P. (1996). Latent and persistent infections of monocytes and macrophages. Intervirology *39*, 293–301.

Sissons, J.G.P., Bain, M., and Wills, M.R. (2002). Latency and reactivation of human cytomegalovirus. J. Infect. *44*, 73–77.

Slobedman, B., and Mocarski, E.S. (1999). Quantitative analysis of latent human cytomegalovirus. J. Virol. *73*, 4806–4812.

Slobedman, B., Stern, J.L., Cunningham, A.L., Abendroth, A., Abate, D.A., and Mocarski, E.S. (2004). Impact of human cytomegalovirus latent infection on myeloid progenitor cell gene expression. J. Virol. *78*, 4054–4062.

Soderberg-Naucler, C., Fish, K.N., and Nelson, J.A. (1997). Reactivation of latent human cytomegalovirus by allogeneic stimulation of blood cells from healthy donors. Cell *91*, 119–126.

Stenberg, R.M. (1996). The human cytomegalovirus major immediate-early gene. Intervirology *39*, 343–349.

Strahl, B.D., and Allis, C.D. (2000). The language of covalent histone modifications. Nature *403*, 41–45.

Taylor-Wiedeman, J., Sissons, J.G., Borysiewicz, L.K., and Sinclair, J.H. (1991). Monocytes are a major site of persistence of human cytomegalovirus in peripheral blood mononuclear cells. J. Gen. Virol. *72*, 2059–2064.

Taylor-Wiedeman, J., Sissons, P., and Sinclair, J. (1994). Induction of endogenous human cytomegalovirus gene expression after differentiation of monocytes from healthy carriers. J. Virol. *68*, 1597–1604.

Urnov, F.D., Yee, J., Sachs, L., Collingwood, T.N., Bauer, A., Beug, H., Shi, Y.B., and Wolffe, A.P. (2000). Targeting of N-CoR and histone deacetylase 3 by the oncoprotein v-erbA yields a chromatin infrastructure-dependent transcriptional repression pathway. EMBO J. *19*, 4074–4090.

Wadgaonkar, R., Phelps, K.M., Haque, Z., Williams, A.J., Silverman, E.S., and Collins, T. (1999). CREB-binding protein is a nuclear integrator of nuclear factor-kappaB and p53 signaling. J. Biol. Chem. *274*, 1879–1882.

Wang, H., Huang, Z.Q., Xia, L., Feng, Q., Erdjument-Bromage, H., Strahl, B.D., Briggs, S.D., Allis, C.D., Wong, J., Tempst, P., and Zhang, Y. (2001). Methylation of histone H4 at arginine 3 facilitating transcriptional activation by nuclear hormone receptor. Science *293*, 853–857.

Weinshenker, B.G., Wilton, S., and Rice, G.P. (1988). Phorbol ester-induced differentiation permits productive human cytomegalovirus infection in a monocytic cell line. J. Immunol. *140*, 1625–1631.

Weintraub, H., and Groudine, M. (1976). Chromosomal subunits in active genes have an altered conformation. Science *193*, 848–856.

White, K.L., Slobedman, B., and Mocarski, E.S. (2000). Human cytomegalovirus latency-associated protein pORF94 is dispensable for productive and latent infection. J. Virol. *74*, 9333–9337.

Wong, J., Patterton, D., Imhof, A., Guschin, D., Shi, Y.B., and Wolffe, A.P. (1998). Distinct requirements for chromatin assembly in transcriptional repression by thyroid hormone receptor and histone deacetylase. EMBO J. *17*, 520–534.

Wright, E., Bain, M., Teague, L., Murphy, J., and Sinclair, J. (2005). Ets-2 repressor factor recruits histone deacetylase to silence human cytomegalovirus immediate-early gene expression in non-permissive cells. J. Gen. Virol. *86*, 535–544.

Yang, W.M., Inouye, C., Zeng, Y., Bearss, D., and Seto, E. (1996). Transcriptional repression by YY1 is mediated by interaction with a mammalian homolog of the yeast global regulator RPD3. Proc. Natl. Acad. Sci. USA *93*, 12845–12850.

Yang, W.M., Yao, Y.L., Sun, J.M., Davie, J.R., and Seto, E. (1997). Isolation and characterization of cDNAs corresponding to an additional member of the human histone deacetylase gene family. J. Biol. Chem. *272*, 28001–28007.

Young, L.S., Dawson, C.W., and Eliopoulos, A.G. (2000). The expression and function of Epstein–Barr virus encoded latent genes. Mol. Pathol. *53*, 238–247.

Zeng, L., and Zhou, M.M. (2002). Bromodomain: an acetyl-lysine binding domain. FEBS Lett. *513*, 124–128.

Zhang, Y., and Reinberg, D. (2001). Transcription regulation by histone methylation: interplay between different covalent modifications of the core histone tails. Genes Dev. *15*, 2343–2360.

Zweidler-McKay, P.A., Grimes, H.L., Flubacher, M.M., and Tsichlis, P.N. (1996). Gfi-1 encodes a nuclear zinc finger protein that binds DNA and functions as a transcriptional repressor. Mol. Cell. Biol. *16*, 4024–4034.

# Regulation of Viral mRNA Export from the Nucleus

10

*Peter Lischka and Thomas Stamminger*

## Abstract

Nuclear export of mRNA is a central step in eukaryotic gene expression. Thus, viruses that replicate within the nucleus have evolved regulatory proteins that are believed to up-regulate replication by facilitating the selective nuclear export of viral mRNA transcripts. Complex retroviruses such as human immunodeficiency virus type 1 encode sequence-specific RNA binding proteins that recruit the cellular nuclear export receptor CRM1 to incompletely spliced viral mRNAs. In contrast, herpesviruses encode nucleocytoplasmic shuttling proteins that direct their intronless mRNAs to the cellular mRNA export pathway whose key components are the heterodimeric mRNA export receptor TAP–p15, the adapter protein REF and the DEAD-box RNA helicase UAP56. Although these viral shuttling proteins are conserved among all herpesviruses, it appears that individual members have shared and separate features. Thus, these proteins share the ability (1) to shuttle between the nucleus and the cytoplasm; (2) to bind RNA; and (3) to export RNAs from the nucleus. In other aspects, these viral shuttling proteins are different. Whereas the HSV-1 mRNA export factor ICP27 and its EBV counterpart EB2 facilitate RNA export by accessing the TAP pathway via direct protein interaction with REF, the HCMV UL69 protein targets this pathway by binding to the RNA helicase UAP56 that acts upstream of REF. In addition, a functional interaction between pUL69 and the transcription elongation factor hSPT6 was described. Since UAP56 also affects transcription elongation, we hypothesize that pUL69, by interacting with both hSPT6 and UAP56, is a factor which optimizes the coupling of transcription elongation to mRNA export during viral replication.

## Introduction

A defining feature of eukaryotic cells is their division into nucleoplasm and cytoplasm. This segregation requires specific mechanisms for the continuous transport of large numbers of macromolecules between both compartments (Vinciguerra and Stutz, 2004). Viruses that replicate in the nucleus have to make extensive use of, or even modify these transport mechanisms in order to optimize the cellular environment for efficient viral multiplication. One essential class of macromolecules that are produced in the nucleus, yet are primarily used in the cytoplasm, are mRNAs. Eukaryotic pre-mRNAs require extensive nuclear processing after synthesis in the nucleus, including addition of the 5′ cap, splicing and polyadenylation. Following these processing events mRNAs are efficiently transported to the cytoplasm where they direct protein synthesis. These reactions are highly interdependent and occur co-transcriptionally (Erkmann and Kutay, 2004; Vinciguerra and Stutz, 2004). At all stages of their biogenesis mRNAs are in dynamic association with many proteins and are exported to the cytoplasm as large ribonucleoprotein complexes (mRNPs). Some of these mRNP proteins are important for RNA export from the nucleus to the cytoplasm. A striking characteristic of these export factors is that they shuttle continuously between the nucleus and the cytoplasm. Nucleocytoplasmic shuttling is based on the fact that these proteins contain a signal sequence mediating their import into (nuclear localization signal; NLS), as well as a sequence that signals their export from the nucleus (nuclear export signal; NES). The transport of macromolecules occurs through the nuclear pore complex (NPC) and is mediated by specific transport receptor(s) (Cullen, 2003b; Dimaano and Ullman, 2004).

While most cellular mRNAs are derived from spliced genes, viruses often contain intronless genes (e.g. herpesviruses) or are dependent on the nuclear export and translation of unspliced messages (e.g. retroviruses). Since early observations suggested that splicing is a prerequisite for efficient nuclear export of mRNAs, it was thought that viruses had to develop specific mechanisms in order to enhance the cytoplasmic accumulation of their intronless or unspliced mRNAs. In this regard the analysis of nuclear export of viral mRNAs and virus-encoded mRNA export factors has provided several key insights into the mechanisms of nuclear export in general. In particular, studies of nuclear RNA export pathways accessed by different primate retroviruses led to the identification of CRM1, as an important member of the karyopherin/exportin family of nucleocytoplasmic transport receptors and of TAP as a key nuclear mRNA export factor in metazoan cells. More recent research on herpesviral mRNA export suggested that these viruses use and modify the recently discovered export pathway for cellular mRNAs. This review will summarize our current knowledge on various viral strategies for nuclear mRNA export with emphasis on herpesviruses. In particular, we will focus on the UL69 protein of HCMV, presenting a model that describes the HCMV UL69 protein as a factor which links viral transcription to mRNA export.

## Nuclear export of incompletely spliced retroviral RNA

Although the retrovirus genome is a single-stranded mRNA, viral replication proceeds via a double stranded DNA intermediate (provirus) which is integrated into the host genome. This proviral DNA is transcribed as an initial, genome-length viral transcript by RNA polymerase II and undergoes processing, including alternative splicing or polyadenylation like other metazoan pre-mRNAs. Importantly, the retroviral replication cycle requires the nuclear export and translation of both fully spliced and incompletely spliced variants of the same initial mRNA transcript. In the cytoplasm, incompletely spliced RNAs serve both as templates for the expression of essential viral proteins and as genomes for progeny viral particles. This requirement presents a problem for retroviruses, since eukaryotic cells have developed a stringent, splicing-coupled proofreading mechanism to ensure that incompletely spliced pre-mRNAs are retained in the nucleus. This should avoid the cytoplasmic translation of cellular pre-mRNAs that would be likely to encode useless or deleterious protein products (Legrain and Rosbash, 1989). Consequently, retroviruses have evolved at least two distinct mechanisms to circumvent this nuclear retention by promoting the recruitment of cellular nuclear export receptors to incompletely spliced viral mRNAs.

## CRM1-dependent mRNA export of HIV-1 transcripts

Human immunodeficiency virus type 1 (HIV-1), which belongs to the complex retroviruses, encodes a regulatory protein, termed Rev, that binds directly to a structured cis-acting RNA target sequence (Rev-response element, RRE) found on all incompletely spliced HIV-1 mRNAs (Figure 10.1A) (Malim et al., 1990). Rev is a nucleocytoplasmic shuttling protein that contains a potent leucine-rich NES (Fischer et al., 1995). Functionally related export sequences of this type have since then been detected in various proteins of diverse functions and have been shown to bind CRM1, a nuclear export factor belonging to the karyopherin or importin/exportin family of nuclear transport receptors (Fornerod et al., 1997). CRM1, like other transport receptors of the exportin family binds its NES substrates in the nucleus in the presence of the GTP-bound form of the GTPase Ran. After nuclear export, conversion of RanGTP to RanGDP causes a conformational shift that induces cytoplasmic substrate release, thus providing the directionality of this export pathway (Fried and Kutay, 2003). Together, these data identified Rev as the first sequence-specific viral mRNA export factor which acts as an adaptor between RRE-containing mRNA and the cellular export receptor CRM1. In addition to HIV-1 several other retroviruses encode adaptor proteins that recruit CRM1 to incompletely spliced viral mRNAs. However, it is now clear that CRM1 not only mediates the nuclear export of retrovirus RNAs but is also responsible for the nuclear export of the vast majority of nucleocytoplasmic shuttling proteins containing a leucine rich NES and of some noncoding cellular RNAs (e.g. U snRNAs or rRNAs). Additionally, although nuclear export of bulk cellular poly(A)+ RNA does not depend on CRM1 function, a small subset of cellular mRNAs might also rely on the CRM1 pathway (Cullen, 2003b). CRM1 function can be specifically blocked by the antibiotic leptomycin B (LMB), resulting in a block of the nuclear export of all CRM1 dependent cargos (Wolff et al., 1997).

## TAP-dependent mRNA export of MPMV mRNA

Whereas complex retroviruses encode their own Rev-like mRNA export factor, simple retroviruses lack such a regulatory protein (Cullen, 2003a). Yet, these retroviruses, exemplified by the Mason-Pfizer monkey virus (MPMV), face the same difficulty of having to export both fully spliced and incompletely spliced forms of one initial transcript. To avoid nuclear retention of intron-containing genomic RNA, some simple retroviruses contain cis-acting elements that interact directly with the cellular mRNA-export machinery. MPMV has been shown to express a cis-acting RNA sequence, the constitutive transport element (CTE), that is sufficient to induce nuclear RNA export in the absence of any viral gene product (see Figure 10.1A) (Ernst et al., 1997). TAP (later also termed NXF1) was first identified as the cellular cofactor interacting with the CTE (Gruter et al., 1998) and TAP was shown to promote the nuclear export of CTE containing RNAs (Kang and Cullen, 1999). Later on it was demonstrated that in addition to its role for MPMV mRNA export, TAP/NXF1 functions as the major mRNA export receptor in metazoan cells (see below). However, unlike CTE-RNA, which binds directly to TAP, binding of cellular mRNAs to TAP is mediated through export adaptor proteins. Thus, simple retroviruses circumvented the need for adaptor proteins by the development of an RNA secondary structure, the CTE, with high binding affinity for the cellular export receptor TAP.

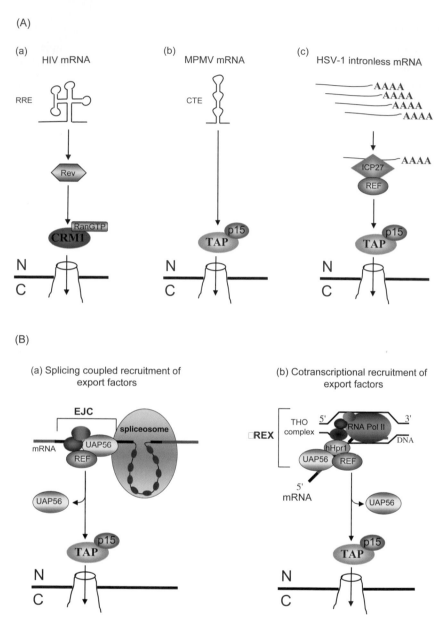

**Figure 10.1** Nuclear mRNA export pathways. Shown is a schematic overview of several nuclear mRNA export pathways used by retro- and herpesviruses or by metazoan cells. (A) Viral mRNA export. (a) Unspliced mRNA encoded by HIV-1 is exported through the viral adapter protein Rev and the cellular karyopherin transport receptor CRM1. Rev binds to the highly structured RRE in unspliced HIV-1 mRNA and interacts with CRM1 through its leucine-rich NES. This interaction can be blocked by LMB. CRM1-mediated export of HIV-1 mRNA requires the formation of a trimeric complex consisting of RNA bound Rev, CRM1 and the small GTPase Ran in its GTP-bound form. (b and c) Unspliced mRNAs encoded by the simple retrovirus MPMV and intronless mRNAs encoded by HSV-1 are exported through the cellular mRNA export receptor TAP–p15. MPMV promotes export of unspliced viral mRNAs by direct binding of TAP–p15 to the constitutive transport element (CTE) RNA target found in unspliced mRNAs. HSV-1 utilizes ICP27

## Nuclear export of cellular mRNA

### mRNA export receptors and adaptor proteins

After identification of the cellular export receptor TAP, all major constituents of the meta-zoan nuclear mRNA export pathway have been identified successively and appear to be conserved from yeast to humans (see Figure 10.1B). The functional mRNA export receptor consists of a large and a small subunit. TAP (alternatively termed NXF1), the larger subunit, forms a heterodimer with the small cofactor p15 (alternatively termed NXT1). In yeast, these proteins are termed Mex67p and Mtr2p, respectively. This heterodimerization is essential for the direct, high-affinity interaction of TAP with the nuclear pore to facilitate mRNP transport to the cytoplasm (Segref et al., 1997; Strasser et al., 2000). Importantly, binding to RanGTP is not required for TAP export activity (Erkmann and Kutay, 2004). Thus, TAP–p15 heterodimers are distinct from the well-characterized family of nuclear transport receptors known as karyopherins or importins/exportins. This protein family has instead been implicated in the nuclear export of non-coding RNAs (tRNAs, UsnRNAs, rRNAs) and of unspliced lentivirus mRNA (Cullen, 2003b).

Although TAP is able to interact directly with cellular RNA in vitro in a sequence-nonspecific fashion, several lines of evidence suggest that additional factors are needed to bridge the interaction between TAP and metazoan mRNA (Braun et al., 2001; Liker et al., 2000). Using both genetic and biochemical approaches, proteins that might act as adaptors between mRNPs and TAP have been identified. Among these are several RNA-binding proteins, including members of the evolutionarily conserved REF protein family (alter-natively termed Aly, in yeast termed Yra1). REF proteins bind directly to both RNA and TAP, thereby facilitating the association of the export receptor TAP with cellular mRNAs (Strasser and Hurt, 2000; Stutz et al., 2000; Zenklusen et al., 2001). A role for metazoan REFs as mRNA export adaptors is supported by the observations that REF shuttles be-tween the nucleus and the cytoplasm, associates with vertebrate mRNAs during splicing and stimulates the export of spliced mRNAs. Subsequently, REF was shown to stimulate the export of intronless mRNAs (Rodrigues et al., 2001). Interestingly, recent RNAi-based knock-down experiments in Drosophila and C. *elegans* revealed that REF proteins are dis-pensable for mRNA export (Gatfield and Izaurralde, 2002; Longman et al., 2003), whereas TAP is required for the nuclear export of most transcripts. These data suggest that TAP association with mRNA may occur through other adaptor proteins in addition to REF. Consistently, additional TAP-binding partners that include several shuttling SR proteins have been identified (Huang et al., 2004).

---

to target intronless viral mRNAs to the TAP–p15 pathway via its interaction with REF proteins. (B) Cellular mRNA export. (a) Splicing-coupled model for the recruitment of mRNA export factors. During splicing of vertebrate mRNA, a complex of proteins (the exon–exon junction complex; EJC) that contains UAP56 and REF is deposited on spliced mRNAs upstream of exon–exon junctions. REF molecules present in EJCs subsequently recruit the export receptor TAP–p15. (b) Cotranscriptional model for the recruitment of mRNA export factors. The THO complex, involved in transcription elongation, recruits the mRNA export factors UAP56 and REF (together called the transcription-export complex; TREX) on nascent mRNAs cotranscriptionally. UAP56 and REF recruit TAP–p15 which targets mRNPs through the NPC. It has been proposed that the association of TAP–p15 with the mRNPs displaces UAP56.

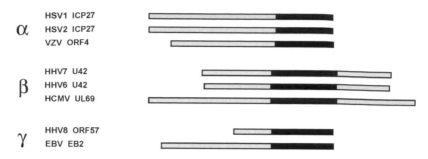

**Figure 10.2** Alignment of proteins encoded by human alpha -, beta-, and gamma-herpesviruses that show homology to ICP27 of HSV-1. The region of highest homology is indicated by a black bar. The beta-herpesviral members of the protein family are characterized by a C-terminal extension that is not present within the alpha- and gamma-herpesviral proteins.

residues (Sandri-Goldin and Mendoza, 1992). This highly conserved amino acid sequence corresponds to a central domain within the HCMV protein UL69, since this beta-herpesvirus polypeptide differs from its homologous proteins by a unique C-terminal extension that is not contained within the alpha- or gamma-herpesvirus members of this protein family (Figure 10.2).

## ICP27-mediated nuclear export of intronless HSV-1 RNA

ICP27 is an essential viral protein expressed from an immediate-early (IE) gene during the lytic HSV-1 life cycle. The protein is a critical regulatory factor, mediating the efficient expression of certain early and most late HSV-1 genes (Sandri-Goldin, 2001). ICP27 can also function to repress expression from host genes by inhibiting splicing and, hence, blocking the nuclear export of cellular mRNAs (Bryant et al., 2001; Sciabica et al., 2003). It was shown that ICP27 functions mainly at the post-transcriptional level. The protein has been demonstrated to shuttle between the nucleus and the cytoplasm, suggesting a role for it in the nuclear export of viral transcripts. Consistent with these observations, ICP27 facilitates HSV-1 RNA export by direct binding to intronless viral RNAs (Phelan and Clements, 1997; Sandri-Goldin, 1998; Soliman et al., 1997). Stuctural domains of ICP27 implicated in RNA binding are an arginine- and glycine-rich region (RGG box) (Mears and Rice, 1996; Sandri-Goldin, 1998) and three putative KH-like RNA interaction motifs (Soliman and Silverstein, 2000b). Although ICP27 interacts with various RNAs in vitro and in vivo, no ICP27 specific RNA motifs on potential ICP27 target RNAs have been identified so far (Mears and Rice, 1996; Sandri-Goldin, 1998; Sokolowski et al., 2003; Soliman and Silverstein, 2000a). Thus, it remains unclear how ICP27 is able to specifically recognize the approximately 80 different viral mRNAs expressed in infected cells. Although ICP27 was first proposed to be exported out of the nucleus via a CRM1-dependent leucine-rich NES (Sandri-Goldin, 1998; Soliman and Silverstein, 2000b), recent reports demonstrated that ICP27 neither shuttles nor exports viral mRNA via the CRM1 pathway (Koffa et al., 2001). These results therefore suggested that the protein accesses a different nuclear RNA-export pathway. It has now been demonstrated that ICP27 directly interacts with the cellular mRNA export adaptor REF, and hence indirectly with the TAP–p15 heterodimer (Figure 10.1A, c). The region of ICP27 required for REF interaction overlaps the NLS and is adjacent to the RGG motif (Chen et al., 2002). Therefore, ICP27 acts to stimulate nuclear export of intronless viral mRNAs by accessing the cellular mRNA export pathway.

## ICP27 homologs in human gamma-herpesviruses

The ICP27 homologs EB2 of EBV and ORF57 of HHV-8 were described as proteins whose inactivation abolishes the production of viral particles, demonstrating that the proteins are essential for viral replication (Gruffat et al., 2002; Sun et al., 1999). Evidence has been presented to suggest that EB2 and ORF57 also act at the post-transcriptional level and share properties with mRNA export factors. In fact, it has been demonstrated that both, EB2 and ORF57 (1) shuttle between the nucleus and the cytoplasm, (2) export RNAs out of the nucleus and (3) interact with REF and TAP to activate mRNA export (Bollo et al., 1999, Farjot et al., 2000; Hiriart et al., 2003b; Malik et al., 2004; Semmes et al., 1998). Furthermore, EB2 was shown to bind directly to RNA via an arginine-rich RNA binding motif that, however, shares no homology with the ICP27 RGG box. Similar to ICP27, no specific RNA sequences have been identified on EB2 RNA targets up to now (Hiriart et al., 2003a). To date, it has not been determined whether ORF57 directly binds to RNA.

Taken together, these findings predict that all members of the ICP27 protein family share conserved functions and that the selective recruitment of cellular mRNA export factors might be the key to nuclear export of intronless herpesviral mRNAs. However, there is also evidence for differences between the various herpesviruses concerning their gene expression as well as the function of the respective ICP27 homologs. The vast majority of HSV-1 and EBV genes are unspliced. Thus, it was assumed that the expression of a viral adaptor protein is required to access cellular export pathways efficiently. By contrast, although transcript mapping is not yet complete, the HHV-8 genome appears to contain a higher number of spliced genes, including the gene encoding the ICP27 homolog ORF57 (Dourmishev et al., 2003; Zheng, 2003). Whereas ICP27 has been reported to repress gene expression of intron-containing transcripts, HHV-8 ORF57 exerted no negative regulation depending on the presence of introns (Gupta et al., 2000; Kirshner et al., 2000). Furthermore, there are conflicting reports on whether EB2 inhibits expression of intron-containing genes (Farjot et al., 2000; Ruvolo et al., 1998). Consistent with this, EBV EB2 protein only partially complemented ICP27 when inserted into an ICP27-deleted viral genome (Boyer et al., 2002). In summary, it appears that the ICP27 herpesvirus homologs may have shared and separate features.

## The UL69 protein of human cytomegalovirus

The ICP27 homolog of HCMV is the 744 amino acid protein pUL69. The relationship between these two proteins is based on an amino acid identity of about 24% and on their location at identical positions within a gene block that is conserved between the various subclasses of herpesviruses and codes for several proteins involved in DNA replication (e.g. dUTPase, components of the helicase–primase complex) (Davison et al., 2003). It is interesting that HCMV pUL69 differs from its counterparts in alpha- or gamma-herpesviruses in having a unique carboxy-terminal amino acid sequence, indicating that pUL69 is the most distantly related homolog of this protein family (Cook et al., 1994; Winkler et al., 2000). Consistent with this, it has been demonstrated that, in contrast to the IE protein

conserved regulator of transcription elongation by RNA polymerase II. This is supported by several observations: (1) in yeast, ySPT6 was shown to be important for the transcription of a distinct subset of genes (Compagnone-Post and Osley, 1996); (2) in *Drosophila*, SPT6 is associated with several transcriptionally active chromosomal loci, including hsp70 loci, after heat shock induction (Kaplan et al., 2000); (3) yeast SPT6 interacts genetically with SPT5 and recent experiments also suggest a physical interaction between human SPT6 and SPT5 (Endoh et al., 2004; Hartzog et al., 1998). Human SPT5 and SPT4 form the biochemically defined transcription elongation factor complex DSIF which associates with RNA polymerase II and is able to both repress and activate transcription elongation in vitro; (4) a mutation in the yeast SPT6 gene causes a 6-azauracil-sensitive phenotype, indicating a defect in transcription elongation (Hartzog et al., 1998); (5) finally, it was recently demonstrated that human SPT6 stimulates transcription elongation by RNA polymerase II in an in vitro transcription assay using naked DNA templates suggesting a direct influence of this protein on the transcription machinery (Endoh et al., 2004). Thus, although no experimental evidence for a modulation of distinct hSPT6 functions by pUL69 is available yet, this interaction with a cellular transcription elongation factor might be one explanation for the pleiotropic transactivation mediated by pUL69.

## Nucleocytoplasmic shuttling activity of pUL69

Based on the findings that have defined some of the pUL69 homologous proteins as virally encoded mRNA export factors, studies on pUL69 nuclear trafficking and mRNA export activity were initiated. By the use of the interspecies heterokaryon analysis, in which pUL69 expressing cells were fused with murine cells, it was shown that pUL69 can shuttle efficiently between the nucleus and the cytoplasm of infected cells (Figure 10.3). Furthermore, pUL69 nucleocytoplasmic shuttling activity could be demonstrated likewise in transfected cells indicating that nuclear export of pUL69 does not depend on viral infection or the expression of any other HCMV-encoded cofactors (Lischka et al., 2001). The vast majority of shuttling proteins described so far possesses a NES similar to the leucine-rich NES of the HIV-1 Rev protein which has been shown to interact with the CRM1 export receptor. Previous studies demonstrated that the antibiotic LMB inhibits export of leucine-rich NES-containing proteins by a covalent modification of CRM1, thereby preventing its association with the NES substrate without directly affecting the other known nuclear transport pathways (Kudo et al., 1999). Interestingly, the predicted amino acid sequence of pUL69 does not contain a leucine-rich, classical NES, suggesting that the pUL69 export is not mediated by CRM1. Consistent with this observation, it was shown that nucleocytoplasmic shuttling of pUL69 is completely unaffected by blocking of the CRM-1 mediated export pathway using LMB. Furthermore, an LMB-insensitive NES was identified that maps to the unique C-terminal extension of pUL69 which is characteristic of the beta-herpesviral members of the ICP27 protein family (Figure 10.4). The export activity of this signal could be confirmed by microinjection experiments and was transferable to heterologous proteins, thus fulfilling the criteria for a novel type of nuclear export signal. The pUL69 NES comprises 28 amino acids and consists of an N-terminal cluster of proline and glutamine residues and a C-terminal domain containing mainly acidic amino acids (Figure 10.4) (Lischka et al., 2001). In this context it is of note that, in contrast to earlier studies, a CRM1-independent shuttling of both HSV-1 ICP27 and EBV EB2 has been

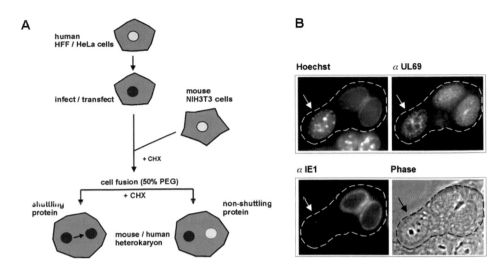

**Figure 10.3** Nucleocytoplasmic shuttling of pUL69 in HCMV-infected fibroblasts. (A) Schematic diagram showing the interspecies heterokaryon assay. The appearance of a viral protein in the mouse nucleus of an interspecies heterokaryon suggests that the protein has shuttled from the infected human nucleus to the mouse nucleus. (B) Interspecies heterokaryons were generated by fusion of HCMV-infected human fibroblasts and NIH 3T3 mouse cells. Two hours after fusion, the cells were fixed and the localization of pUL69 ($\alpha$ UL69) and IE1 ($\alpha$ IE1) was assessed by indirect immunofluorescence. Hoechst, staining of the DNA with Hoechst dye 33258 was used to differentiate between human and murine nuclei within the heterokaryon. Murine nuclei display a characteristic punctate pattern, whereas human nuclei are diffusely stained with this reagent; murine nuclei are indicated by arrows. The panel marked "Phase" shows the phase contrast image of heterokaryons; the cytoplasmic edge is highlighted by a broken line. The UL69 protein could be detected in murine nuclei that are part of heterokaryons indicating that the protein migrated from infected fibroblast human nuclei to the cytoplasm and subsequently entered the murine nuclei. In contrast, IE1, which is an important nuclear regulatory protein of HCMV, was not detected in murine nuclei.

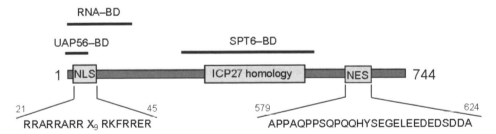

**Figure 10.4** Domain organization of the HCMV UL69 protein showing the positions of the known functional regions. The sequence of the bipartite NLS from amino acids 21–45 is depicted, as is the sequence of the LMB-insensitive NES, from residues 597–624. ICP27 homology: domain of pUL69 with high homology to ICP27; UAP56-BD: domain required for binding to the cellular mRNA export factor UAP56; RNA-BD: RNA-binding region; SPT6-BD: domain required for the interaction to the cellular transcription elongation factor hSPT6.

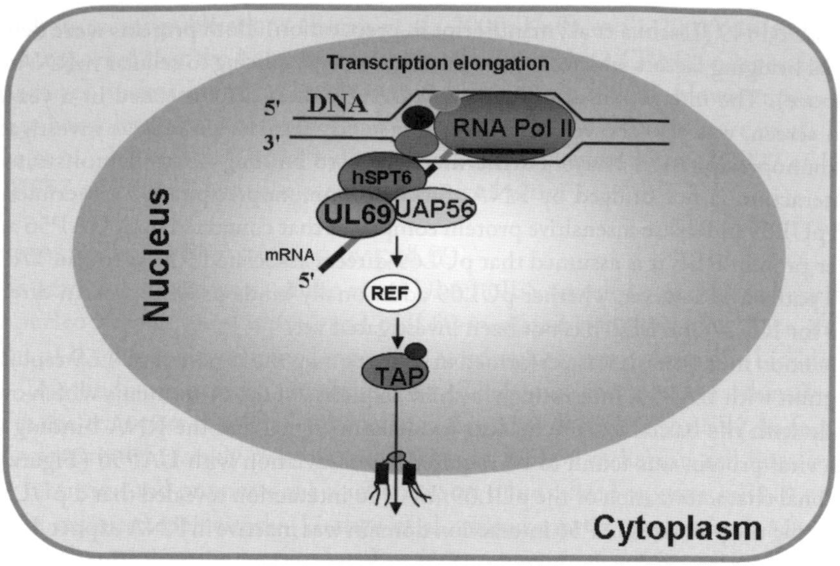

**Figure 10.5** Model of pUL69-mediated stimulation of gene expression. The UL69 protein interacts with both the cellular factors hSPT6 and UAP56. These have been described as being regulators of transcription elongation which travel with RNA polymerase II. In addition, UAP56 acts as an essential mRNA export factor which couples transcription to mRNA export due to recruitment of the adaptor protein REF. Thus, pUL69 may facilitate mRNA export by optimizing the co-transcriptional loading of RNA export factors to nascent mRNA.

## Acknowledgments

We are grateful to Manfred Marschall and Zsolt Toth for critical reading of the manuscript and we thank Bernhard Fleckenstein for continuous support. This work was supported by the Deutsche Forschungsgemeinschaft (SFB473), the BMBF (IZKF Erlangen), and the Wilhelm Sander Stiftung.

## References

Abruzzi, K.C., Lacadie, S., and Rosbash, M. (2004). Biochemical analysis of TREX complex recruitment to intronless and intron-containing yeast genes. EMBO J. 23, 2620–2631.

Bello, L.J., Davison, A.J., Glenn, M.A., Whitehouse, A., Rethmeier, N., Schulz, T.F., and Barklie, C.J. (1999). The human herpesvirus-8 ORF 57 gene and its properties. J. Gen. Virol. 80, 3207–3215.

Bortvin, A. and Winston, F. (1996). Evidence that Spt6p controls chromatin structure by a direct interaction with histones. Science 272, 1473–1476.

Boyer, J.L., Swaminathan, S., and Silverstein, S.J. (2002). The Epstein–Barr virus SM protein is functionally similar to ICP27 from herpes simplex virus in viral infections. J. Virol. 76, 9420–9433.

Braun, I.C., Herold, A., Rode, M., Conti, E., and Izaurralde, E. (2001). Overexpression of TAP/p15 heterodimers bypasses nuclear retention and stimulates nuclear mRNA export. J. Biol. Chem. 276, 20536–20543.

Bryant, H.E., Wadd, S.E., Lamond, A.I., Silverstein, S.J., and Clements, J.B. (2001). Herpes simplex virus IE63 (ICP27) protein interacts with spliceosome- associated protein 145 and inhibits splicing prior to the first catalytic step. J. Virol. 75, 4376–4385.

Chen, I.H., Sciabica, K.S., and Sandri-Goldin, R.M. (2002). ICP27 Interacts with the RNA export factor Aly/REF to direct herpes simplex virus type 1 intronless mRNAs to the TAP export pathway. J. Virol. 76, 12877–12889.

Clark-Adams, C.D. and Winston, F. (1987). The SPT6 gene is essential for growth and is required for delta-mediated transcription in Saccharomyces cerevisiae. Mol. Cell Biol. 7, 679–686.

Compagnone-Post, P.A. and Osley, M.A. (1996). Mutations in the SPT4, SPT5, and SPT6 genes alter transcription of a subset of histone genes in Saccharomyces cerevisiae. Genetics 143, 1543–1554.

Cook, I.D., Shanahan, F., and Farrell, P.J. (1994). Epstein–Barr virus SM protein. Virology 205, 217–227.

Cullen, B.R. (2003a). Nuclear mRNA export: insights from virology. Trends Biochem. Sci. 28, 419–424.

Cullen, B.R. (2003b). Nuclear RNA export. J. Cell Sci. 116, 587–597.

Davison, A.J., Dolan, A., Akter, P., Addison, C., Dargan, D.J., Alcendor, D.J., McGeoch, D.J., and Hayward, G.S. (2003). The human cytomegalovirus genome revisited: comparison with the chimpanzee cytomegalovirus genome. J. Gen. Virol. 84, 17–28.

Dimaano, C. and Ullman, K.S. (2004). Nucleocytoplasmic transport: integrating mRNA production and turnover with export through the nuclear pore. Mol. Cell Biol. 24, 3069–3076.

Dourmishev, L.A., Dourmishev, A.L., Palmeri, D., Schwartz, R.A., and Lukac, D.M. (2003). Molecular genetics of Kaposi's sarcoma-associated herpesvirus (human herpesvirus-8) epidemiology and pathogenesis. Microbiol. Mol. Biol. Rev. 67, 175–212.

Endoh, M., Zhu, W., Hasegawa, J., Watanabe, H., Kim, D.K., Aida, M., Inukai, N., Narita, T., Yamada, T., Furuya, A., Sato, H., Yamaguchi, Y., Mandal, S.S., Reinberg, D., Wada, T., and Handa, H. (2004). Human Spt6 stimulates transcription elongation by RNA polymerase II in vitro. Mol. Cell Biol. 24, 3324–3336.

Erkmann, J.A. and Kutay, U. (2004). Nuclear export of mRNA: from the site of transcription to the cytoplasm. Exp. Cell Res. 296, 12–20.

Ernst, R.K., Bray, M., Rekosh, D., and Hammarskjold, M.L. (1997). A structured retroviral RNA element that mediates nucleocytoplasmic export of intron-containing RNA. Mol. Cell Biol. 17, 135–144.

Farjot, G., Buisson, M., Duc, D.M., Gazzolo, L., Sergeant, A., and Mikaelian, I. (2000). Epstein–Barr virus EB2 protein exports unspliced RNA via a crm-1- independent pathway. J. Virol. 74, 6068–6076.

Fischer, U., Huber, J., Boelens, W.C., Mattaj, I.W., and Luhrmann, R. (1995). The HIV-1 Rev activation domain is a nuclear export signal that accesses an export pathway used by specific cellular RNAs. Cell 82, 475–483.

Fornerod, M., Ohno, M., Yoshida, M., and Mattaj, I.W. (1997). CRM1 is an export receptor for leucine-rich nuclear export signals. Cell 90, 1051–1060.

Fried, H. and Kutay, U. (2003). Nucleocytoplasmic transport: taking an inventory. Cell Mol. Life Sci. 60, 1659–1688.

Gatfield, D. and Izaurralde, E. (2002). REF1/Aly and the additional exon junction complex proteins are dispensable for nuclear mRNA export. J. Cell Biol. 159, 579–588.

Gatfield, D., Le Hir, H., Schmitt, C., Braun, I.C., Kocher, T., Wilm, M., and Izaurralde, E. (2001). The DExH/D box protein HEL/UAP56 is essential for mRNA nuclear export in Drosophila. Curr. Biol. 11, 1716–1721.

Gruffat, H., Batisse, J., Pich, D., Neuhierl, B., Manet, E., Hammerschmidt, W., and Sergeant, A. (2002). Epstein–Barr virus mRNA export factor EB2 is essential for production of infectious virus. J. Virol. 76, 9635–9644.

Gruter, P., Tabernero, C., von Kobbe, C., Schmitt, C., Saavedra, C., Bachi, A., Wilm, M., Felber, B.K., and Izaurralde, E. (1998). TAP, the human homolog of Mex67p, mediates CTE-dependent RNA export from the nucleus. Mol. Cell 1, 649–659.

Gupta, A.K., Ruvolo, V., Patterson, C., and Swaminathan, S. (2000). The human herpesvirus 8 homolog of Epstein–Barr virus SM protein (KS-SM) is a post-transcriptional activator of gene expression. J. Virol. 74, 1038–1044.

Hardy, W.R. and Sandri-Goldin, R.M. (1994). Herpes simplex virus inhibits host cell splicing, and regulatory protein ICP27 is required for this effect. J. Virol. 68, 7790–7799.

Hartzog, G.A., Wada, T., Handa, H., and Winston, F. (1998). Evidence that Spt4, Spt5, and Spt6 control transcription elongation by RNA polymerase II in Saccharomyces cerevisiae. Genes Dev. 12, 357–369.

Hayashi, M.L., Blankenship, C., and Shenk, T. (2000). Human cytomegalovirus UL69 protein is required for efficient accumulation of infected cells in the G1 phase of the cell cycle. Proc. Natl. Acad. Sci. USA 97, 2692–2696.

Herold, A., Teixeira, L., and Izaurralde, E. (2003). Genome-wide analysis of nuclear mRNA export pathways in Drosophila. EMBO J. 22, 2472–2483.

Strasser, K., Bassler, J., and Hurt, E. (2000). Binding of the Mex67p/Mtr2p heterodimer to FXFG, GLFG, and FG repeat nucleoporins is essential for nuclear mRNA export. J. Cell Biol. *150*, 695–706.

Strasser, K. and Hurt, E. (2000). Yra1p, a conserved nuclear RNA-binding protein, interacts directly with Mex67p and is required for mRNA export. EMBO J. *19*, 410–420.

Strasser, K. and Hurt, E. (2001). Splicing factor Sub2p is required for nuclear mRNA export through its interaction with Yra1p. Nature *413*, 648–652.

Strasser, K., Masuda, S., Mason, P., Pfannstiel, J., Oppizzi, M., Rodriguez-Navarro, S., Rondon, A.G., Aguilera, A., Struhl, K., Reed, R., and Hurt, E. (2002). TREX is a conserved complex coupling transcription with messenger RNA export. Nature *417*, 304–308.

Stutz, F., Bachi, A., Doerks, T., Braun, I.C., Seraphin, B., Wilm, M., Bork, P., and Izaurralde, E. (2000). REF, an evolutionary conserved family of hnRNP-like proteins, interacts with TAP/Mex67p and participates in mRNA nuclear export. RNA. *6*, 638–650.

Sun, R., Lin, S.F., Staskus, K., Gradoville, L., Grogan, E., Haase, A., and Miller, G. (1999). Kinetics of Kaposi's sarcoma-associated herpesvirus gene expression. J. Virol. *73*, 2232–2242.

Swanson, M.S. and Winston, F. (1992). SPT4, SPT5 and SPT6 interactions: effects on transcription and viability in Saccharomyces cerevisiae. Genetics *132*, 325–336.

Vinciguerra, P. and Stutz, F. (2004). mRNA export: an assembly line from genes to nuclear pores. Curr. Opin. Cell Biol. *16*, 285–292.

Wiegand, H.L., Lu, S., and Cullen, B.R. (2003). Exon junction complexes mediate the enhancing effect of splicing on mRNA expression. Proc. Natl. Acad. Sci. USA *100*, 11327–11332.

Winkler, M., aus dem Siepen, T., and Stamminger, T. (2000). Functional interaction between pleiotropic transactivator pUL69 of human cytomegalovirus and the human homolog of yeast chromatin regulatory protein SPT6. J. Virol. *74*, 8053–8064.

Winkler, M., Rice, S.A., and Stamminger, T. (1994). UL69 of human cytomegalovirus, an open reading frame with homology to ICP27 of herpes simplex virus, encodes a transactivator of gene expression. J. Virol. *68*, 3943–3954.

Winkler, M. and Stamminger, T. (1996). A specific subform of the human cytomegalovirus transactivator protein pUL69 is contained within the tegument of virus particles. J. Virol. *70*, 8984–8987.

Wolff, B., Sanglier, J.J., and Wang, Y. (1997). Leptomycin B is an inhibitor of nuclear export: inhibition of nucleo- cytoplasmic translocation of the human immunodeficiency virus type 1 (HIV-1) Rev protein and Rev-dependent mRNA. Chem. Biol. *4*, 139–147.

Zenklusen, D., Vinciguerra, P., Strahm, Y., and Stutz, F. (2001). The yeast hnRNP-Like proteins Yra1p and Yra2p participate in mRNA export through interaction with Mex67p. Mol. Cell Biol. *21*, 4219–4232.

Zenklusen, D., Vinciguerra, P., Wyss, J.C., and Stutz, F. (2002). Stable mRNP formation and export require cotranscriptional recruitment of the mRNA export factors Yra1p and Sub2p by Hpr1p. Mol. Cell Biol. *22*, 8241–8253.

Zheng, Z.M. (2003). Split genes and their expression in Kaposi's sarcoma-associated herpesvirus. Rev. Med. Virol. *13*, 173–184.

Zhi, Y. and Sandri-Goldin, R.M. (1999). Analysis of the phosphorylation sites of herpes simplex virus type 1 regulatory protein ICP27. J. Virol. *73*, 3246–3257.

# Exploitation of Host Cell Cycle Regulatory Pathways by HCMV 11

*Veronica Sanchez and Deborah H. Spector*

## Abstract

Human cytomegalovirus has evolved multiple mechanisms to manipulate the host cell's metabolic and regulatory systems for the purposes of creating an environment favorable for productive infection. From the very early phases of the infection, virus-encoded proteins target key components of the cell cycle machinery to effect arrest and prevent cellular DNA replication while maintaining an active state that provides the intermediates required for the virus to replicate its own DNA. The prolonged replicative cycle of the virus necessitates that it attenuates the cell's response to the stresses imposed by viral infection. To this end, the virus has also developed methods to circumvent the apoptotic program and extend cell survival. In this chapter we will discuss the changes induced by viral infection with specific focus on the viral proteins that contribute to these effects and the possible consequences that these modifications may have in cells that do not become productively infected.

## Introduction

The serious clinical problems associated with HCMV infection provide a strong impetus to understand the complex interactions between the virus and the host cell that contribute to viral pathogenesis. A number of groups have described the multiple effects of HCMV on signaling pathways, the cell cycle, immune evasion, and apoptosis (for review see Castillo and Kowalik, 2002; Cinatl Jr. et al., 2004; Fortunato et al., 2000b), but much of the work is beyond the scope of this chapter. The overall theme is that the virus induces metabolic and structural changes in the host cell that are favorable for viral gene expression, replication, and virion morphogenesis as well as cell survival (Albrecht et al., 1991; Bresnahan et al., 1996; Jault et al., 1995; Johnson et al., 2001; McElroy et al., 2000; Muranyi et al., 2002). In this chapter, we focus on the interplay between the viral and cellular factors that seem to impact directly on the cell cycle and cell survival. Space limitations prevent an exhaustive review of the literature. Moreover, we have concentrated on the most current studies that by their very nature raise more questions than they answer.

To understand the effects of HCMV infection on the host cell cycle, it is necessary to briefly describe the various phases and the effector proteins that regulate the many events involved in cell cycle progression (Figure 11.1). The cell cycle is the highly coordinated process by which the cell prepares for division (for review see Sherr, 1996; 2000). Proliferative signals including growth factors and serum stimulation induce quiescent ($G_0$) cells to enter

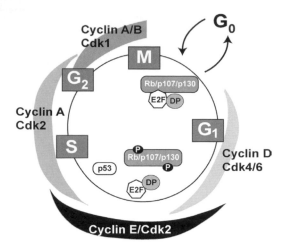

**Figure 11.1** The cell cycle. The cell cycle is a tightly regulated process through which the cell replicates its DNA and divides into two daughter cells. $G_0$ is the resting or quiescent phase. Stimulation by growth signals results in the cells entering $G_1$ phase and stimulation of the D-type cyclins that form a complex with Cdk4 or Cdk6. $G_1$ phase prepares the cell to enter S phase and replicate its DNA. The passage into S phase is governed by Cdk2/cyclin E. Replication of the cellular DNA requires the activity of Cdk2/cyclin A. $G_2$ phase follows S phase and marks the transition prior to cell division in M phase. Cdk1/cyclin B and Cdk1/cyclin A are required during the $G_2$/M period. The major guardians of the cell cycle are p53 and the Rb family of proteins (Rb, p107, and p130). At the beginning of $G_1$, the proteins in the Rb family are in a hypophosphorylated state and form a complex with E2F/DP transcription factors. This is associated with repression of transcription from E2F-responsive promoters, which direct the synthesis of many gene products required for DNA synthesis. In response to growth signals, the Rb-related proteins are phosphorylated, resulting in the release of E2F and activation of the E2F-responsive promoters. The p53 protein prevents the cell from entering S phase in the case of DNA damage or other insults to the cell.

the cycle. These mitogenic factors stimulate the expression of D-type cyclins, which form kinase complexes with cyclin-dependent kinase 4 or 6. Phosphorylation of substrates by Cdk4 and Cdk6 kinase complexes relieves growth suppression and allows cells to enter the $G_1$ phase of the cell cycle. During $G_1$, or gap period, there is a commitment to cell division once the necessary requirements are met. Past this restriction point, the progression through the cell cycle no longer depends on the continued presence of growth stimuli. The shift into S phase is regulated by the cyclin-dependent kinase complex Cdk2/cyclin E. In S phase, the cell's replication machinery is activated and regulated by Cdk2/cyclin A and Cdc7/Dbf4 complexes (for review see Nishitani and Lygerou, 2002). Together these kinase complexes mediate the firing of DNA origins of replication and promote DNA replication. After the genome has been duplicated in S phase, cells enter $G_2$, the transition period before cytokinesis. Cell division, or M phase, is mediated by the activity of Cdk1/cyclin A and Cdk1/cyclin B complexes. These complexes phosphorylate many substrates that help modify the cell's architecture to ensure that cellular material is segregated properly between daughter cells. At each of these steps, checkpoints or sensor modules have evolved that halt cell cycle progression in cases of DNA damage, nutrient deprivation, and improperly assembled spindles (for review see Lukas et al., 2004). The most extensively studied "guardians" of the cell cycle are the tumor suppressor proteins p53, pRb, p107, and p130; the last three proteins are referred to as the Rb family of pocket proteins (PP). The cell cycle continues if the insult that induced arrest is overcome. Apoptosis, or programmed cell death, is an alternative pathway that may be triggered if the integrity of cell division is challenged for delayed periods. This prevents the generation of defective daughter cells. Consequently, the faithful duplication of cells results from maintenance of a fine balance between survival and death signals that ensure fidelity of the genome.

## Effect of the cell cycle on HCMV replication

Although the majority of this review will focus on the effect of HCMV on the cell cycle, it is clear that the phase of the cell cycle at the time of infection impacts significantly on the replication of the virus. Early work by Salvant et al. (1998) showed that when cells are infected in $G_0$ phase, or just after they have been released from $G_0$ into $G_1$, viral gene expression initiates very quickly and the infection progresses without delay. There is also a corresponding block to host cell DNA replication (Bresnahan et al., 1996; Dittmer and Mocarski, 1997; Lu and Shenk, 1996; Salvant et al., 1998; Wiebusch and Hagemeier, 1999). In contrast, when cells are infected near or during S phase, many cells are able to pass through S phase and undergo mitosis prior to cell cycle arrest. Infection during S phase also produces a delay in the appearance of the virus-induced cytopathic effect and in the synthesis of IE and early proteins (Salvant et al., 1998). Labeling of cells with bromodeoxyuridine immediately prior to HCMV infection in S phase revealed that viral protein expression occurs primarily in cells that are not engaged in DNA synthesis at the time of infection (Salvant et al., 1998).

In a subsequent study, Fortunato et al. (2002) showed that most cells that are actively synthesizing their DNA at the beginning of the infection do not express the IE proteins until the cells undergo mitosis and enter the following $G_1$ phase. A minor population, however, is able to express IE proteins in the $G_2/M$ phase of the cell cycle. This blockade to activation of IE gene expression occurs very quickly after the initiation of DNA replication. Some of the cells close to the $G_1/S$ boundary at the time of infection are able to express IE antigens and traverse through S phase, but there is a very narrow window of time during which cells can both initiate viral gene expression and carry out cellular DNA synthesis. Thus, there is not an absolute block to IE gene expression in either the S or $G_2/M$ phases for cells that are not actively replicating their DNA at the beginning of the infection.

Of particular interest is the finding by Fortunato et al. (2002) that treatment of cells with the proteasome inhibitor MG132 partially relieves the block to IE gene expression in S phase. The authors hypothesized that IE gene expression might require some factor that has a short half-life in cells replicating their DNA. This factor could be a direct activator of IE gene expression or could counter the effects of some inhibitor. However, since proteasome inhibition did not fully restore the expression of IE genes during S phase, additional factors appear to be involved.

Although it might seem that the virus is inactive when infection occurs during S phase, this is not the case. Infection of cells with HCMV during S phase results in two specific breaks in chromosome 1, at positions 1q42 and 1q21 (Fortunato et al., 2000a). The induction of these breaks is not specific to a particular strain of HCMV or to the source of the cells. Interestingly, although incubation of the virus with neutralizing antibody prior to infection prevents the induction of breaks, UV-inactivated virus is as efficient as untreated virus in inducing specific damage to chromosome 1. Thus, this event requires viral adsorption and penetration but not new viral gene expression. The identification of genes affected by these two specific breaks and the determination of the mechanisms underlying the generation of this damage may provide important clues regarding the role of HCMV in the development of neurological defects in the congenitally infected infant and its potential role as a cofactor during cancer progression.

## HCMV and the cell cycle

The principal objectives of the virus are to induce metabolic changes that create a cellular milieu optimal for gene expression and DNA replication and to inhibit select host cell functions ensuring that viral replication is favored over that of the host. The effects of HCMV on the cell cycle begin with interactions between virion proteins and components of the cell cycle machinery immediately following viral entry. The viral DNA is transported to the nucleus, and if the cell is in $G_0/G_1$ phase, transcription of IE genes begins without delay. The products of the IE genes interact with some cell cycle proteins and play an important role in promoting cell survival and inhibiting apoptosis. The IE proteins also induce the expression of the early class of viral genes. Early proteins are required for viral DNA synthesis, cleavage and packaging of the viral genome, and assembly of viral particles. In addition, the early proteins play a role in maintaining a favorable intra- and extracellular environment for viral gene expression and DNA replication through subversion of the host's cell cycle machinery (McElroy et al., 2000; Sanchez et al., 2003).

Early work from our laboratory and others described the disruption of cell cycle progression in cells infected with HCMV (Figure 11.2) (Bresnahan et al., 1996; Dittmer and Mocarski, 1997; Jault et al., 1995; Lu and Shenk, 1996; Salvant et al., 1998; Wiebusch and Hagemeier, 1999; 2001). Cells infected in the $G_0/G_1$ phase of the cycle do not replicate their DNA and arrest in a pseudo-$G_1$ state (Bresnahan et al., 1996; Dittmer and Mocarski, 1997; Salvant et al., 1998; Wiebusch and Hagemeier, 2001). The tumor suppressors pRb, p130, and p107, which inhibit the expression of E2F-responsive genes and the activity of Cdk2 kinase complexes, are maintained in a phosphorylated state in infected cells (Jault et al., 1995; McElroy et al., 2000). In addition, the tumor suppressor protein p53 accumulates, but its effects on target promoters are not observed (Fortunato and Spector, 1998; Muganda et al., 1994). Infection also stimulates expression of the proto-oncogenes fos, jun, and myc, as well as expression of enzymes important for DNA replication (Boldogh et al., 1991; Isom, 1979). Cyclin-dependent kinase activity is also dysregulated. The $G_1/S$ and $G_2/M$ cyclins E and B1, respectively, accumulate in infected cells with a concomitant increase in associated kinase activity (Jault et al., 1995; Sanchez et al., 2003; Wiebusch and Hagemeier, 2001). In contrast, the steady-state levels of the S phase cyclin A and $G_1$ phase cyclin D1 are reduced (Bresnahan et al., 1996; Jault et al., 1995; Salvant et al., 1998; Wiebusch and Hagemeier, 2001). Infection also inhibits the licensing of host cell DNA origins of replication (Biswas et al., 2003; Wiebusch et al., 2003b).

Several studies have used DNA arrays to examine the effects of HCMV infection on the accumulation of cellular RNAs. Early work by Zhu et al. (1998) demonstrated that infection of fibroblasts led to significant changes in the levels of 258 cellular mRNAs prior to the onset of viral DNA replication. Included in these genes were those encoding several cell cycle proteins such as pRb, Wee1, and Cdc25B, which all increased in expression at the

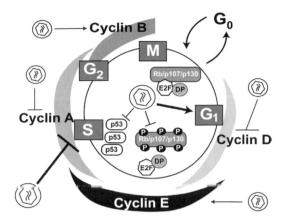

**Figure 11.2** Effects of the HCMV infection on the cell cycle. Infection of the cell with HCMV during $G_0/G_1$ induces progression through $G_1$ and activation of many genes required for entry into S phase and synthesis of cellular DNA. However, the normal expression of the cyclin-dependent kinases is disrupted, and the cell cycle is blocked before replication of the cellular DNA. Cyclins E and B1 accumulate in their active kinase form, but the expression of the cyclins D1 and A is inhibited. The tumor suppressors Rb, p130, and p107 accumulate in a phosphorylated state and do not inhibit the activity of the E2F transcription factors. Infection also results in accumulation of p53, but it cannot activate its target promoters.

RNA level. More recent studies have shown that the up-regulation of cellular genes could be clustered into two groups based on whether the increase occurred before or after the IE/early phase of the infection (Challacombe et al., 2004). This work demonstrated increases in the RNAs encoding PCNA, Cdk2, cyclins D2, D3, and E1, Cdc7, E2F4, and MCM3, all of which have previously been examined in HCMV-infected cells in the context of their roles in cell cycle progression (Challacombe et al., 2004). This work also showed that Cdk6 and cyclin A2 mRNA levels increased later in the infection. A comprehensive profiling of cellular gene expression at the late phases of infection was recently reported by Hertel and Mocarski (2004). Among the many changes observed was a dysregulation of RNAs encoding proteins that function during mitosis. Of particular note were the effects of the infection on the steady state levels of RNAs encoding several kinases and cyclins, transcription factors, and regulators of proteasome activity, all of which influence the cell cycle. There are several caveats in the interpretation of the array data. One is that the relative increase or decrease in specific RNAs in the infected cells depends on the reference uninfected cell RNA used for comparison. For example, in the study by Hertel and Mocarski (2004), the reference uninfected cell RNA was from cells that were maintained in a $G_0$ state. Under these conditions, the RNAs from many genes involved in cell cycle regulation decline to very low levels and only begin to accumulate once the cell is released from $G_0$ to $G_1$. In addition, it should be noted that levels of RNA do not necessarily reflect changes in protein accumulation. This highlights a point that will be discussed later regarding the differential effects of the infection on protein stability. Nevertheless, the gene array analyses do provide a good starting point for examining the infection-associated changes in transcription, and there is clearly much gold to be mined from these studies. The challenge is to separate the "sand from the gold nuggets" that will lead to unraveling the cellular pathways that are most important for viral replication and pathogenesis.

## Immediate effects of HCMV on cellular metabolism – cell cycle arrest and inhibition of apoptosis

### Cell cycle arrest

*Viral entry*
Two virion proteins, ppUL69 and pp71 (ppUL82), have been implicated in mediating cell cycle arrest (Figure 11.3). As components of the virion tegument, they can exert their effects upon entry of the virus into the cell. The protein encoded by the UL69 gene was initially described as a transactivator of transcription (Winkler et al., 1994; Winkler and Stamminger, 1996). Subsequent work by Lu and Shenk (1999) implicated ppUL69 as an initiator of the cell cycle arrest observed in HCMV-infected cells. Their studies showed that uninfected human fibroblasts or U2OS cells transduced with a retrovirus containing the HCMV UL69 gene accumulated in the $G_1$ phase of the cell cycle. More recently, Hayashi

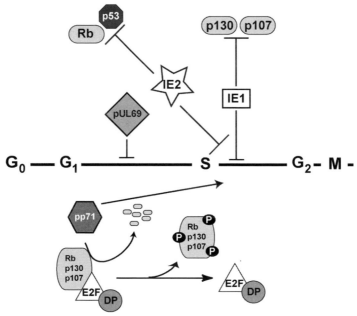

**Figure 11.3** HCMV proteins modulate cell cycle progression. The tegument proteins ppUL69 and pp71 can exert their effects on the cell cycle upon viral entry. ppUL69 causes the accumulation of cells in $G_1$ phase by an unknown mechanism. pp71, the product of the UL82 gene, targets the hypophosphorylated forms of the pocket proteins pRb, p130, and p107, for proteasome-mediated destruction and promotes entry into S-phase. This degradation could serve to induce E2F-driven transcription in a manner similar to Cdk-dependent phosphorylation of the pocket proteins that causes the release of E2F/DP complexes and transcriptional derepression during the cell cycle. The IE proteins IE1-72 and IE2-86 both induce the accumulation of cells in S phase. IE1-72 inhibits the activity of the pocket proteins p107 and p130 while IE2-86 can interfere with Rb and p53 functions.

and colleagues (2000) demonstrated that HCMV virions that were genotypically and phe-notypically negative for UL69 did not effectively induce the cell cycle block after infection of primary fibroblasts. The mutant virus also did not replicate to the same level as the par-ent virus; however, by providing ppUL69 in trans, the growth defect was repaired. This requirement for ppUL69 in the lytic cycle of HCMV was reiterated by the work of Dunn et al. (2003), which showed that an HCMV BAC containing a deletion of the UL69 ORF has a severe growth defect in fibroblasts. Taken together, these data suggest that ppUL69 may play a very early role in initiating cell cycle arrest in HCMV-infected cells.

The pp71 protein also was initially identified as a transactivator that was necessary for efficient activation of the major IE promoter at the beginning of the infection (Baldick et al., 1997; Bresnahan and Shenk, 2000; Liu and Stinski, 1992). Like UL69, deletion of the UL82 gene from the virus causes a severe growth defect that can be complemented in trans (Bresnahan and Shenk, 2000; Dunn et al., 2003). The protein also contains an LXCXD motif found in cellular and viral proteins that can interact with the cell growth suppres-sors pRb, p107, and p130 (Kalejta et al., 2003). One of the ways that these suppressor proteins exert their effects on the cell cycle is by forming a complex with E2F transcription factors bound to DNA, resulting in the repression of transcription (for review see Cam and Dynlacht, 2003; Stevens and La Thangue, 2003). The E2F–pocket protein interac-tions occur in a cell cycle-dependent manner and require that pRb, p107, and p130 are in a hyposphosphorylated state. In response to growth signals, the phosphorylation of the pocket proteins by cyclin-dependent kinases results in the release of the E2Fs thus allow-ing the recruitment of the transcriptional machinery to the DNA and the expression of genes involved in cell cycle progression. A role for pp71 in disruption of this pathway was suggested by the finding that pp71 expressed transiently in uninfected cells was able to target the hypophosphorylated forms of the pocket proteins for degradation by the pro-teasome independent of ubiquitin (Kalejta et al., 2003; Kalejta and Shenk, 2003b). These data, coupled with the observation that expression of pp71 in uninfected cells accelerates progression through $G_1$ phase but does not change the overall doubling time of these cells (Kalejta and Shenk, 2003a), indicate that input pp71 may play a role in stimulating the cell cycle at the beginning of the infection prior to viral gene expression. It should be noted, however, that the above studies were performed in uninfected cells expressing pp71 in the absence of other viral proteins. The question of whether pp71 can modulate the cell cycle in the context of infection remains to be addressed and may prove difficult to answer because of the protein's other roles in activation of viral gene transcription.

*Immediate-early proteins*
In addition to the virion proteins described above, HCMV encodes two IE proteins that have been shown to disrupt normal cell cycle progression in uninfected cells. The major IE region of the viral genome encodes the IE1-72 and IE2-86 proteins. As reviewed elsewhere (Fortunato and Spector, 1999), both of these proteins play important roles in the activation of viral early gene expression. Independently, they also have the potential to modulate the cell cycle by virtue of their transcriptional activities and by binding to key cell cycle proteins (Figure 11.3).

Transient expression of IE1-72 in asynchronously cycling cells alters the distribution of the cells so that greater percentages reside in the S and $G_2/M$ phases (Castillo et al.,

2000). In the absence of p53, IE1-72 also induces quiescent cells to enter S phase and delays cell cycle exit (Castillo et al., 2000). Two properties of IE1-72 may be related to these observations. First, it alleviates p107-mediated repression of E2F-responsive promoters in transient transfection assays, and second, it can reverse the inhibitory effects of p107 on Cdk2/cyclin E activity (Castano et al., 1998; Hansen et al., 2001; Margolis et al., 1995; Poma et al., 1996; Woo et al., 1997; Zhang et al., 2003). The molecular mechanisms underlying these activities of IE1-72 have yet to be defined. One possibility is that it is simply due to the ability of IE1-72 to form a complex with p107 (Johnson et al., 1999; Poma et al., 1996; Zhang et al., 2003). The finding that IE1-72 can phosphorylate p107, p130, and some members of the E2F family (Pajovic et al., 1997), also evokes the hypothesis that some of the effects of IE1-72 on transcription and on the cell cycle may be a result of its reported kinase activity. An apparent contradiction to this hypothesis is that active Cdk2/cyclin E is present in complexes with p107 and p130 in cells infected with a virus that has a deletion in IE1-72 (McElroy and Spector, unpublished data). This mutant virus also replicates normally at a high multiplicity of infection (Greaves and Mocarski, 1998). However, as we learn more about the interaction of the virus with the host, it becomes clear that redundancy is an intrinsic feature of the viral infection.

One of the most extensively studied HCMV proteins is IE2-86, encoded by the UL122 gene (for review see Fortunato and Spector, 1999). This IE protein is a potent transactivator of transcription from HCMV and heterologous promoters and is essential for progression of the HCMV lytic cycle. Microarray analysis of RNAs from human fibroblasts infected with an adenovirus expressing IE2-86 from an inducible promoter has revealed that the protein induces an increase in the steady state levels of RNA from multiple genes important in the synthesis of DNA precursors as well as from known E2F-responsive genes, including E2F-1, cyclin E, Cdk2, and mini-chromosome maintenance (MCM) proteins 3 and 7 (Song and Stinski, 2002).

The functions of IE2-86 are believed to be mediated via protein–protein and protein–DNA interactions. IE2-86 interacts with itself and the product of the viral UL84 gene as well as with an expanding list of cellular proteins. These host factors include: components of the basal transcription complex TBP, TFIIB, and TBP-associated factors (TAFs); the tumor suppressor proteins Rb and p53; multiple transcription factors such as Sp1, Tef-1, c-Jun, JunB, ATF-2, NF-κB, protein kinase A-phosphorylated delta CREB, Nil-2A, CHD-1, Egr-1, and UBF; and the histone acetylases p300, CBP, and P/CAF (see White et al., 2004 and Sanchez et al., 2002 and references therein; Hsu et al., 2004). Based on the observations that IE2-86 can interact with proteins regulating the cell cycle, a number of investigators subsequently examined the effect of IE2-86 expression on cell cycle progression. An early paper by Wiebusch and Hagemeier (1999) reported that U373 cells transfected with a plasmid expressing IE2-86 failed to enter S phase after serum stimulation while cells transfected with control vector entered the cell cycle. This effect was observed in other cell lines as well. In addition, HCMV-infected U373 cells appeared to arrest in $G_1$ as had previously been described for infected primary fibroblasts. The arrest was shown to require some viral gene expression as UV-inactivated HCMV failed to induce the cell cycle block (Wiebusch and Hagemeier, 1999). IE2-86 expressed in trans restored the block to cell cycle progression in U373 cells infected with UV-inactivated virus, suggesting that IE2-86 is responsible for this effect, at least in this experimental system (Wiebusch and

Hagemeier, 1999). These results also indicated that in U373 cells, the input virion proteins ppUL69 and pp71 do not induce the cell cycle block. Later work by this group showed that IE2-86 expressed transiently could promote the $G_1/S$ transition and that the transfected U373 cells arrested after entering S phase with high levels of cyclin E activity (Wiebusch et al., 2003a; Wiebusch and Hagemeier, 2001).

Using a different experimental design, Murphy et al. (2000) also addressed IE2-86 associated cell cycle arrest. The use of adenoviral vectors expressing IE2-86 allowed this group to achieve >95% transduction of U373 cells. Similar to the other reports, their results suggested that IE2-86 blocks cells in S phase. In contrast, a recent study by Noris and colleagues (2002) provides conflicting data regarding the phase at which IE2-86 halts the cell cycle in transient assay systems. In this case, the authors reported that the cell cycle block induced by HCMV infection of fibroblasts resembles cellular senescence by virtue of the elevated expression of p16$^{Ink4}$ and increased senescence-associated beta-galactosidase activity. These effects were reproduced by transducing fibroblasts with a baculovirus expressing IE2-86, and the expression of IE2-86 also prevented quiescent cells from entering S phase (Noris et al., 2002). The differences between these studies may be attributed to the cell types used and the methods used for their synchronization. Regardless of the exact point of the cell cycle block, however, consideration of all of the data leads to the conclusion that there are likely bona fide interactions between IE2-86 and key cell cycle proteins that contribute to some of the dysregulation of the cell cycle in the infected cells.

The critical nature of IE2-86 in the viral replicative cycle makes it difficult to address its direct role in cell cycle modulation during the infection in untransformed cells since efforts to make permissive, primary cell lines that fully complement IE2-86 mutants have not yet been successful. Most mutations in IE2-86 in the context of the viral genome result in non-viable recombinants that do not proceed past IE gene expression, but a few mutants have been generated that exhibit tolerable growth defects and produce infectious progeny (Marchini et al., 2001; Sanchez et al., 2002; White et al., 2004; White, Clark, and Spector, unpublished data). In general, the inhibitory effects on the infection can be attributed to the inability of the protein to activate target viral promoters; however, it seems likely that a more extensive cellular role for IE2-86 derives from its interaction with host cell proteins. Unfortunately, we still do not know which of the many potential complexes between IE2-86 and cellular proteins actually form in vivo during the infection. This information is a necessary prerequisite for determining the effect of mutations on the binding of IE2-86 to specific cellular proteins and for correlating these effects with alterations in cell cycle progression or resistance to apoptosis.

## Inhibition of apoptosis

The effects described above all have the potential to induce programmed cell death, or apoptosis, in infected cells; thus it follows that HCMV should have the means to inhibit early destruction of its host cell. In general, apoptosis is either caspase-mediated or caspase-independent. Our discussion will focus on the pathways controlled by caspases, as the others are not well described in the context of HCMV infection (for review see Cinatl Jr. et al., 2004). The processes designated caspase-dependent are further subdivided on the basis of the stimuli that elicit the death response (Figure 11.4). Apoptosis can be triggered by events within a cell that challenge homeostasis including genotoxic stress, ER stress

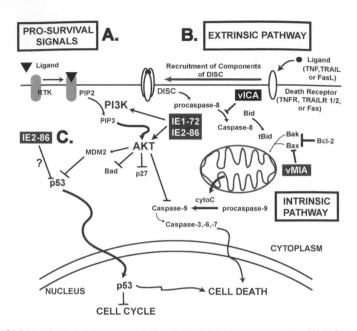

**Figure 11.4** HCMV inhibits premature cell death by inhibiting apoptosis. (A) HCMV activates the PI3K/Akt signaling pathway. Binding of ligands to receptor tyrosine kinases (RTKs) leads to activation of lipid kinase PI3K. PIP3 molecules induce AKT activity and recruitment of this kinase to the plasma membrane. Several downstream substrates that promote cell survival or inhibit apoptosis are regulated by AKT-mediated phosphorylation. The expression of the HCMV IE proteins IE1-72 and IE2-86 is sufficient for PI3K/AKT activation. (B) Intracellular stresses (intrinsic pathway) and death receptor signaling (extrinsic pathway) trigger the apoptotic program. Binding of soluble or membrane-bound ligands to death receptors causes aggregation of these receptors at the plasma membrane, followed by recruitment of the death-inducing signaling complex (DISC). Activation of DISC results in the cleavage of procaspase-8 into its active form. Caspase-8 mediates the processing of Bid into tBid, a protein that facilitates the insertion of Bak/Bax into the mitochondrial membrane. These proteins participate in the permeabilization of the membrane. Cytochrome C (cytoC) release leads to the activation of caspase-9, which further activates downstream caspases, thus amplifying the death signal. HCMV encodes two proteins, vMIA and vICA, that interfere with these pathways at the indicated stages. (C) HCMV IE2-86 binds to p53 and inhibits transcription from p53-dependent promoters and potentially interferes with p53 apoptotic functions.

reponses, or the withdrawal of growth stimuli. In contrast to this 'intrinsic' pathway, the signals for cell death can also come from the plasma membrane as a result of the external activation of death receptors, such as Fas/CD95 and the TNF receptor. These events trigger the 'extrinsic' pathway of apoptosis.

*p53*

p53 coordinates multiple cellular processes through its activity as a transcriptional activator and repressor in response to stress and growth factors, and it has a fundamental role in the decision of whether the cell will grow or undergo apoptosis (for review see Slee et al., 2004; Vousden and Lu, 2002). The activation of p53 is modulated by phosphorylation, and it is a substrate for numerous cellular kinases. Phosphorylation controls the association of p53 with MDM2, a protein that targets p53 for destruction by the proteasome. In response to

DNA damage, heat shock, nutrient deprivation, or other insults, p53 levels are stabilized resulting in the expression of p53-responsive genes such as the Cdk inhibitor p21; however, the regulation of apoptosis by p53 is independent of the cell cycle arrest that is induced by p21 expression (Rowan et al., 1996). Several pro-apoptotic genes are induced by p53 including Bax, PUMA, and Noxa. In contrast, the expression of the pro-survival gene Bcl-2 is repressed by activated p53. It has also been suggested that p53 plays a direct role in apoptosis by promoting cytochrome c release at the mitochondria (Mihara et al., 2003).

In cells infected with HCMV, p53 levels are stabilized but the expression of its down-stream target gene p21 is not observed, suggesting that the virus encodes proteins that impede p53 function (Bresnahan et al., 1996; Chen et al., 2001; Fortunato and Spector, 1998; Muganda et al., 1994). As described above, HCMV IE2-86 protein has been shown to physically interact with p53 (Bonin and McDougall, 1997; Speir et al., 1994). In fact, expression of IE2-86 protects some cells from p53-mediated apoptosis under conditions that would otherwise induce the pathway (Tanaka et al., 1999). Recent work by Hsu and colleagues (2004) indicates that in transient expression systems, IE2-86 inhibits the binding of p53 to target promoters, such as the p21 promoter, and thus represses p53-driven transcription as was previously reported (Muganda et al., 1998; Speir et al., 1994; Tsai et al., 1996). IE2-86 expression also inhibited the acetylation of p53 and of histones in proximity to p53-dependent promoters. These effects appear to be due to the down-regulation of the histone acetyl transferase (HAT) activity of p300/CBP, which was detected in complex with p53 and IE2-86 (Hsu et al., 2004). While the IE2-86 mediated repression of p300/CBP HAT activity in the infected cells remains to be investigated, these observations suggest that very early in the infection, IE2-86 inhibits p53 transcriptional activity and contributes to the inhibition of apoptosis observed in infected cells. Later in the infection, the sequestration of p53 in viral replication centers (Fortunato and Spector, 1998) may play a role in maintaining p53 in an inactive state. The localization of IE2-86 to viral replication centers supports the theory that this viral protein mediates the sequestration of p53 during productive infection, but it is possible that other proteins are involved or that p53 might specifically target to promoters on the viral DNA to activate transcription. In addition, a role for p53 in virus-specific DNA repair has been suggested, strengthening the hypothesis that p53 does not colocalize with viral DNA during the infection merely to prevent apoptosis. It must be noted, however, that if the sequestered p53 in HCMV-infected cells cannot induce expression of pro-apoptotic genes, it follows that it cannot repress the expression of pro-survival genes either. In this regard, it has been reported that HCMV infection of tumor cells results in the up-regulation of Bcl-2 and that the cells are resistant to death induced by measures that trigger apoptosis (Cinatl Jr. et al., 1998; Harkins et al., 2002); however, the activity of p53 in these experimental systems was not specifically examined. Nonetheless, the results from numerous studies indicate that the targeting of p53 is one of the means that the virus may use to block premature death of infected cells.

## HCMV UL37x1 and UL36

Two viral IE proteins that directly inhibit steps in the apoptotic pathway upstream of p53 are encoded by the UL36–38 locus (Kouzarides et al., 1988; Tenney and Colberg-Poley, 1991a, b). This region includes three promoters that give rise to at least five transcripts and is required for HCMV DNA replication in a transient complementation assay (Colberg-

Poley et al., 1992; Iskenderian et al., 1996; Pari and Anders, 1993; Pari et al., 1993; Pari and Anders, 1993). Only the UL37 gene, however, is essential in the context of infection (Dunn et al., 2003). The most abundant transcript expressed from the UL37 promoter is a 1.7-kb unspliced RNA. This RNA is present throughout the infection and encodes a glycoprotein, UL37x1, that targets to mitochondria along with the related proteins UL37 and $UL37_M$ that are specified from alternatively spliced longer transcripts (Goldmacher et al., 1999; Mavinakere and Colberg-Poley, 2004; Tenney and Colberg-Poley, 1991a, b). It is at this site that UL37x1 helps to suppress apoptosis induced in response to viral infection by preventing the permeabilization of mitochondria and release of cytochrome c, which activates procaspase-9 and contributes to cell death (Goldmacher et al., 1999). UL37x1 was therefore named vMIA, viral mitochondria-localized inhibitor of apoptosis. Early work showed that there was an interaction between vMIA and the cellular adenine nucleotide translocator, a resident mitochondrial protein important in maintaining membrane integrity (Goldmacher et al., 1999). More recently, however, the true mechanism by which vMIA ensures cell survival has been elucidated. Arnoult and coworkers have reported that vMIA sequesters Bax in the mitochondria and hinders the ability of Bax to promote cytochrome c release (Arnoult et al., 2004). Therefore, UL37x1 is effective in preventing death in cells that have triggered the mitochondrial apoptotic signaling pathway predominantly through Bax (Arnoult et al., 2004).

The functions described for UL37x1 place its position in the apoptotic pathway after the activation of procaspase-8 but before the activation of procaspase-9. Another member of the UL36–38 family of proteins has been shown to act upstream of UL37x1 and to interfere with the activation of procaspase-8. The protein encoded by UL36 was named vICA, viral inhibitor of caspase activation (Figure 11.4). Unlike UL37x1, UL36 is not essential for viral growth in culture (Dunn et al., 2003; Skaletskaya et al., 2001). vICA is important in protecting cells from apoptosis induced by the activation of death receptors on the cell surface. It does so by short-circuiting the signaling pathway that results in cytochrome c release. By directly preventing the cleavage of procaspase-8 into the active caspase-8, vICA delays the processing of Bid to tBid, the protein that leads to the insertion of Bax and Bak into the mitochondrial membrane. Because vICA functions upstream of the activation of caspase-8, it is likely that it protects a wide variety of infected cells from apoptosis induced by the host immune response.

*IE1-72 and IE2-86*
Early work by Zhu et al. (1995) demonstrated that expression of IE1 and IE2 in HeLa cells protected against apoptosis induced by TNF-alpha or by infection with a mutant adenovirus lacking the E1B 19-kDa protein. While HCMV does not establish a lytic infection in HeLa cells, it is possible that the inhibition of apoptosis by the IE proteins may be a mechanism that prolongs survival of transformed cells. These pro-survival events may also contribute to the disease process in vivo in abortively infected primary cell types that only express IE proteins, or in cells that are persistently infected and yield low levels of virus after a prolonged replication cycle. Relevant to this is the observation that in primary coronary artery smooth muscle cells, which do support productive infection, expression of IE2, but not IE1, was shown to inhibit apoptosis induced by doxorubicin treatment (Tanaka et al., 1999).

*Virus-induced changes in death receptor signaling*

One mechanism by which the virus eludes an immune response to infection and the death of the infected cell may be by up-regulating expression of death receptor ligands, such as FasL and TRAIL, on the surface of infected cells. In healthy hosts, the expression of FasL is restricted to lymphocytes, NK cells, and some non-lymphoid cells including macrophages (for review see Curtin and Cotter, 2003; Dockrell, 2003). In contrast, the receptor for FasL, CD95/Fas, is widely expressed. During an immune response, activated CTLs seek out and destroy cells expressing foreign antigens on their surfaces. The membrane bound FasL (mFasL) on the infiltrating lymphocytes causes aggregation of Fas receptor on the target cell. This leads to the recruitment of FADD and other proteins that comprise the death-inducing signaling complex (DISC), a protease complex that activates caspase-8. This cascade of events induces apoptosis in the target cell and is a primary mechanism by which CTLs function in the immune response. The HCMV IE2-86 protein was shown to induce expression of mFasL in human retinal pigment epithelial cells, a model system for CMV retinitis (Chiou et al., 2001) and of TRAIL in human fibroblasts (Sedger et al., 1999). Since it is believed that the recognition and binding of the membrane-bound ligands by death receptors on immune cells stimulates the apoptotic pathways in the T-cells themselves resulting in death, the above effects of IE2-86 may be another mechanism for promoting survival of the infected cell in vivo.

*Pro-survival signals*

In addition to the pro-apoptotic pathways disrupted or exploited by HCMV, the virus also stimulates pathways involved in the transduction of growth signals from the cell surface to the nucleus. A major cellular regulatory protein is Akt, a kinase that is involved in translational control, glucose metabolism, cell cycle progression, and inhibition of apoptosis (for review see Cooray, 2004; Vara et al., 2004). In response to the activation of receptor tyrosine kinases by growth factors, the lipid kinase PI3K phosphorylates the inositol ring of PIP2 (Figure 11.4). The product, PIP3, leads to the recruitment and activation of Akt at the plasma membrane. Active Akt can then phosphorylate many substrates including Bad, caspase-9, and FKHR transcription factors. The modification of these proteins inhibits their activities. Phosphorylated Bad is sequestered by 14-3-3 proteins, and as a result, the binding of Bad to anti-apoptotic members of the Bcl-2 family is inhibited. The dimerization of Bcl-2 or Bcl-xl with Bad neutralizes their anti-apoptotic effects; thus phosphorylation of Bad by Akt promotes cell survival. Similarly, phosphorylation of caspase-9, a protease critical for the amplification of death signals, blocks its activity and inhibits apoptosis. Akt-mediated phosphorylation also stimulates cell cycle progression. The Cdk inhibitors p21 and p27 are substrates for Akt and their translocation into the nucleus is inhibited by Akt modification, thus preventing them from inactivating Cdk complexes. Akt also phosphorylates MDM2, which promotes binding and degradation of p53.

Studies by Johnson and coworkers (2001) have shown that the activation of PI3K and Akt occurs in cells infected with HCMV. Two phases of activation were observed. Binding of the virus to the plasma membrane induced a transient activation of PI3K that declined after 2 h p.i. Viral gene expression was necessary for the sustained activity of PI3K and Akt, which began after 4 h p.i. Inhibitors of the PI3K/Akt pathway inhibited viral replication, suggesting that the survival signals mediated by this pathway were important. Later work

al., 1995). The $G_1$ and $G_2$/M cyclins E and B1, respectively, accumulate in an active kinase form, while the expression of the S phase cyclin A is inhibited. These preliminary results led to the hypothesis that these perturbations of the cell cycle machinery result in cell cycle arrest. Later work determined that the accumulation of cyclin E is due to an increase in the cyclin E mRNA (Salvant et al., 1998). Cyclin A mRNA is not induced in infected cells and cyclin B1 mRNA shows only a modest increase. These data were supported by the work of Bresnahan and coworkers (1996) who additionally found that the levels of cyclin D1 protein decrease in virus-infected cells while the levels of Cdk4 are unchanged. More recent work has explored the Cdk activities that regulate transcription and has examined the pathways that regulate the individual Cdk-cyclin complexes found in HCMV-infected cells (Sanchez et al., 2003; Tamrakar, Kapasi, Clark, and Spector, unpublished data). In addition, compounds that inhibit Cdk activity have been utilized in order to understand what impact activation of these kinases has during the infection (Bresnahan et al., 1997; Sanchez et al., 2004).

*Cdk2/cyclin E*
The accumulation of cyclin E in HCMV-infected cells has been documented by several laboratories, but the precise mechanisms leading to induction of the cyclin E mRNA and accumulation of the cyclin E protein in the infected cell are not fully understood (Bresnahan et al., 1996; Jault et al., 1995; McElroy et al., 2000; Salvant et al., 1998). Several studies have suggested a role for IE2-86 in accumulation of cyclin E, but the contribution of IE2-86 has primarily been examined in an isolated context (Song and Stinski, 2002; Wiebusch and Hagemeier, 2001; Wiebusch et al., 2003a). The majority of the experiments have focused on transcriptional regulation of the cyclin E promoter and involved the use of transient expression systems. For example, Bresnahan et al. (1998) showed that IE2-86 could bind to sequences in the cyclin E promoter in vitro and could activate expression of a cyclin E promoter-driven reporter construct in transient assays. In other work, it was found that expression of IE2-86 from an adenovirus vector induces synthesis of endogenous cyclin E mRNA, again suggesting that IE2-86 plays a role in cyclin E accumulation in HCMV-infected cells (Song and Stinski, 2002). A different result was reported by McElroy et al. (2000) who found that cyclin E protein did not accumulate in human fibroblasts expressing IE2-86 from a recombinant baculovirus. Their work also demonstrated that viral early gene expression is required for accumulation of the very high levels of cyclin E observed in infected cells. These conflicting results suggest that several factors might affect the accumulation of cyclin E protein in infected cells and that the increase in cyclin E might be regulated at both the mRNA and protein level.

The activity of Cdk2/cyclin E complexes is modulated at several levels in uninfected cells. Cyclin E mRNA is induced and translated into protein in $G_1$, and the protein is then degraded after the transition to S phase. The associated kinase Cdk2 is activated through phosphorylation by the Cdk-activating kinase (CAK) Cdk7/cyclin H. Cdk2/cyclin E complexes can also be inhibited by the binding of p21 and the pocket proteins p107 and p130. In infected cells, there are changes in each of these pathways that result in the maintenance of high levels of active Cdk2/cyclin E. One of the most striking effects of the virus is that cyclin E mRNA accumulates at early times in the infection to a level that is at least 10-fold higher than the peak level found in the uninfected cells. In addition, there is a decrease in p21 (Bresnahan et al., 1996; Chen et al., 2001), and high levels of active Cdk2/cyclin E are found in complexes with p107 and p130 (McElroy, Biswas, and Spector; unpublished data). It has been suggested that the binding of IE1-72 to p107 might relieve the repression of the Cdk2/cyclin E kinase (Johnson et al., 1999; Zhang et al., 2003). However, because high levels of Cdk2/cyclin E activity are also observed in cells infected with an HCMV recombinant that does not encode a functional IE1-72 (McElroy and Spector, unpublished data), it seems likely that other viral functions are also involved in overriding the inhibition imposed by the pocket proteins. It is possible that the utilization of multiple mechanisms to achieve the same goal assures that the virus can replicate efficiently in different cellular environments.

Taken together, the above studies suggest that up-regulation of Cdk2/cyclin E is important for viral replication, but the role of Cdk2/cyclin E in the infection is still uncertain. Work by Bresnahan et al. (1997) suggested that the activity of Cdk2/cyclin E complexes is essential for viral replication based on the observation that introduction of a dominant-negative form of Cdk2 by transient transfection or treatment of infected cells with the cyclin-dependent kinase inhibitor Roscovitine causes a block to viral replication and viral gene expression (see below). There are several concerns regarding the results. First, the efficiency of the transient transfection was very low, with less than 2% of the infected cells expressing the dominant negative form of Cdk2. In addition, only one late viral protein was analyzed by immunostaining. Finally, as will be discussed below, recent studies have revealed that the effects of Roscovitine in the infected cell are very complex (Sanchez et al., 2004).

## Cdk1/cyclin B

Our recent studies have investigated the modification of Cdk1 regulatory pathways in infected fibroblasts that leads to sustained Cdk1/cyclin B activity (Sanchez et al., 2003). Because it plays such a crucial role in cell division, Cdk1 activity is subject to several levels of regulation (Figure 11.6) (O'Farrell, 2001). First, the activity is modulated by phosphorylation of the Cdk1 catalytic subunit. Inhibitory phosphates are added to Thr14 and Tyr15 of Cdk1 by two kinases, Wee1 and Myt1, whose expression and activity are also cell cycle regulated (Booher et al., 1997; Heald et al., 1993; Liu et al., 1997; 1999; Mueller et al., 1995; Watanabe et al., 1995). The inhibition is relieved by dephosphorylation of these sites by members of the Cdc25 family of dual-specificity protein phosphatases. Cdc25B is expressed during $S/G_2$ phases of the cell cycle and is thought to initially activate Cdk1 (Hoffmann et al., 1993; Karlsson et al., 1999; Lammer et al., 1998). Cdk1/cyclin B can then phosphorylate and activate Cdc25C, beginning a feedback loop that amplifies the Cdk-dependent kinase activity and the signal for cell division. Cdk1 is also phosphorylated at Thr161 by the Cdk-activating kinase CAK (Cdk7/cyclin H) (Harper and Elledge, 1998).

Because the levels of cyclin B fluctuate during the cell cycle, the availability of the cyclin subunit adds another layer of complexity to regulation of Cdk1/cyclin B kinase activity. Cyclin B mRNA begins to accumulate in $G_1$ phase, but the protein levels do not increase until the cells enter S phase and they peak during $G_2/M$. Cyclin B is then ubiquitinated and targeted for destruction by the anaphase-promoting complex (APC) during mitosis, and this degradation of newly synthesized cyclin B continues during early $G_0/G_1$ (for review see Harper et al., 2002). There are two forms of cyclin B that mediate the substrate specificity

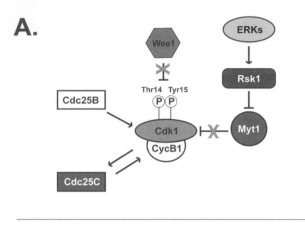

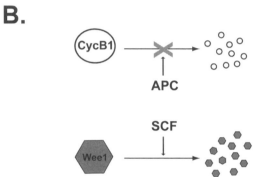

**Figure 11.6** Model for activation of Cdk1/cyclin B1 complexes in HCMV-infected cells. (A) The addition of inhibitory phosphates to the catalytic subunit, Cdk1, is mediated by Myt1 and Wee1 kinases. Myt1 is inhibited by phosphorylation mediated by p90Rsk1, which itself is activated by the ERK kinases. The removal of the Cdk1 inhibitory phosphates is catalyzed by Cdc25B, an S/$G_2$ phase phosphatase, and Cdc25C, a $G_2$/M phosphatase. Cdc25B initially activates Cdk1/cyclin B1 complexes, which in turn activate Cdc25C. Cdc25C amplifies the activation of Cdk1/cyclin B1 complexes during mitosis. In HCMV-infected cells, Myt1 and Wee1 expression is reduced while the Cdc25 phosphatases accumulate. This results in the accumulation of the nonphosphorylated (Thr14/Tyr15), active form of Cdk1 in HCMV-infected cells. (B) The E3 ubiquitin ligase APC regulates the degradation of cyclin B1 and the E3 ubiquitin ligase SCF promotes degradation of Wee1. In HCMV-infected cells, there is accumulation of cyclin B1 and degradation of Wee1. These results are consistent with down-regulation of the APC in virus-infected cells.

of Cdk1-containing complexes (Draviam et al., 2001). Cyclin B1 has been shown to shuttle between the nucleus and cytoplasm during interphase, but it accumulates in the nucleus during mitosis. In contrast, cyclin B2 is cytoplasmic and Golgi-associated throughout the cell cycle. Thus, cyclin B-associated kinase activity is controlled by the availability of the cyclin subunit in terms of expression and localization.

Recent studies in our laboratory have shown that during the infection, the accumulation of Cdk1/cyclin B1 in its active form is regulated at multiple levels. In the infected cells, there is down-regulation of the Myt1 and Wee1 inhibitory pathways in terms of reduced expression and activity (Sanchez et al., 2003). In contrast, the Cdc25 protein phosphatases accumulate. These alterations in the proteins that regulate Cdk1 phosphorylation are consistent with the very low level of inactive, Tyr15-phosphorylated Cdk1 found in infected cells. In addition, we have found that accumulation of cyclin B1 in virus-infected cells requires early viral gene expression. Interestingly, the levels of cyclin B1 in infected cells do not change upon treatment with the proteasome inhibitor MG132. These data indicate that accumulation of cyclin B1 in virus-infected cells results from an increased abundance of mRNA coupled with an inhibition of cyclin B1 protein degradation (Salvant et al., 1998). The stabilization of cyclin B1 is significant because along with the accumulation of Cdc6 and geminin in infected cells, it provides additional evidence that some event leading to proteasome-mediated degradation of these substrates is inhibited (Biswas et al., 2003; Wiebusch et al., 2003b). Interestingly, all of these proteins are targeted for degradation by the anaphase-promoting complex, an E3 ubiquitin ligase, leading to the hypothesis that the activity of the APC is inefficient in virus-infected cells. It does not appear that there is a global inhibition of the proteasome because Wee1 does not accumulate at the same time as cyclin B1. The lack of accumulation of Wee1 is due to proteasome degradation since treatment of the infected cells with the proteasome inhibitor MG132 prevents its destruction. The differential degradation of substrates ubiquitinated by specific E3 ligases, SCF for Wee1 and APC for cyclin B1, respectively, suggests that the virus has evolved mechanisms for accumulating specific proteins and down-regulating others. It is possible, however, that a process downstream of ubiquitination, such as transport to the proteasome or some substrate-specific activity of the proteasome itself, is affected by virus infection. Work is currently under way to explore these possibilities.

The localization of cyclin B1 is also altered in infected cells. By immunofluorescence, it was found that the level of cyclin B1 detected in individual infected cells is low but that it is evenly distributed between the nucleus and cytoplasm (Sanchez et al., 2003). Cdk1 immunostaining was evenly distributed throughout the cell with the exception of concentration at the centrosome. By biochemical fractionation of HCMV-infected cells, we found that the active form of Cdk1 is localized primarily to the cytoplasm. These data suggest that substrates for activated Cdk1/cyclin B1 in infected cells reside primarily in the cytoplasm.

*Role of Cdks in HCMV-infected cells*
In order to establish a role for the Cdk activity induced by the infection, we and others have used chemical inhibitors that have high specificity for individual Cdks (Bresnahan et al., 1997; Sanchez et al., 2004). The drug Roscovitine is a purine analog that competes with ATP for binding to Cdk1, 2, 5, 7, and 9 and reversibly inhibits their activity (De Azevedo et al., 1997; Meijer et al., 1997). As noted above, early work indicated that Roscovitine is effective in blocking viral replication and the production of infectious virus in HCMV-infected cells (Bresnahan et al., 1997).

Recent studies in our lab have looked more closely at the effect of this inhibitor at various times during the infection. We have found that Cdk activity is required during two intervals in the infection; one between 0–6 h p.i., and the second after the onset of viral DNA replication (Sanchez et al., 2004). Roscovitine imposes a concentration-dependent inhibitory effect on virus production if the drug is added at the time of infection. This inhibition is characterized by a significant decrease in the expression of early viral genes and a reduction in the replication of the viral DNA. In addition, the processing of the RNAs encoding the IE proteins UL122/123 and UL37x1/UL37 is affected. The effects on processing are negated if the addition of the drug is delayed until 6 h p.i. Similarly, the effects of Roscovitine on gene expression and viral DNA replication do not occur if addition of the drug is delayed; however, the drop in virus titer is observed even if Roscovitine is not added until 48 h p.i. These data suggest that a step after replication, perhaps assembly of infectious virus, is also inhibited by Roscovitine. Several observations support this hypothesis. First, a delay in the accumulation of pp65 in the cytoplasm late in infection is observed if the drug is first added after 24 h p.i. (Sanchez, Mahr, McElroy, and Spector, unpublished data). This protein is a major component of the virion particle (see Chapter 5), and its transition from the nucleus to the juxtanuclear assembly compartment characterizes a shift to the late phase of infection. In addition, the formation of the assembly compartment itself is disrupted in the presence of the drug. Taken together, these results point to a role for active Cdk in assembly of the virus.

Of particular interest are the specific effects on IE gene expression when Roscovitine is added at the beginning of the infection. During the course of the normal infection, IE1-72 RNA initially accumulates to a higher level than the IE2-86 RNA. Over time, more IE2-86 transcripts accumulate while there is only a slight increase in IE1-72 transcripts. When Roscovitine is added at the beginning of the infection, the ratio of IE1-72 RNA to IE2-86 RNA is reversed. The level of the IE2-86 transcript increases and there is a decrease in IE1-72 RNA. It is possible that these findings reflect altered stability of the IE1-72 mRNA. However, given the proximity of the polyadenylation and splicing signals, the most plausible explanation for these results is that the differential splicing and polyadenylation of the UL122–123 transcript is altered by treatment with the Cdk inhibitor, leading to suppression of the first cleavage/polyadenylation site that generates IE1-72 and enhanced utilization of the adjacent downstream 3′ splice acceptor site that generates IE2-86. In support of this hypothesis, the differential processing of the IE UL37 RNAs, which also have signals for polyadenylation and splicing juxtaposed, is similarly affected by the presence of Roscovitine.

What is the mechanism governing these effects on the processing of the RNA and how are the Cdks involved? These questions remain to be answered, but there are some clues from our current knowledge of RNA processing in normal cells. The consensus is that the steps of RNA processing, involving mRNA capping, splicing, and cleavage/polyadenylation, occur as the RNA is synthesized, and that these processes are interdependent (for review see Bentley, 2002; Zorio and Bentley, 2004). Since the polyadenylation site and the splice acceptor site in the IE RNAs are in close proximity, it seems likely that factors involved in both processes compete for sites on the RNA. Multiple factors mediate splicing and cleavage/polyadenylation, and thus a change in the abundance, activity, or localization of any of these proteins could shift the equilibrium to favor splicing over cleavage/polyadenylation. The phosphorylation of proteins involved in RNA processing determines their activity, and Cdk activity may directly or indirectly influence this modification. It is therefore possible that inhibition of the Cdks by Roscovitine suppresses the cleavage/polyadenylation of IE1-72 and UL37X1 by the loss of activity of a positive factor or the gain of activity of a negative factor involved in RNA processing.

Roscovitine may also affect the phosphorylation of the C-terminal domain (CTD) of the large subunit of RNA polymerase II. The CTD consists of 52 repeats of the consensus heptapeptide sequence Tyr-Ser-Pro-Thr-Ser-Pro-Ser and serves as a platform for the recruitment and transport of factors involved in RNA initiation, elongation, 5' capping, splicing and cleavage/polyadenylation (for review, see Bentley, 2002; Orphanides and Reinberg, 2002; Proudfoot et al., 2002). The serines at positions 2 and 5 within the heptapeptide repeats of the CTD are differentially phosphorylated during the initiation, elongation, and termination phases of transcription (for review, see Majello and Napolitano, 2001; Prelich, 2002). After the hypophosphorylated form of RNAPII is recruited to the initiation complex, the CTD becomes phosphorylated at serine 5 by Cdk7/cyclin H (a component of the basal transcription factor complex TFIIH). Subsequently, serine 2 is phosphorylated by Cdk9/cyclin T (P-TEFb). These modifications, particularly the phosphorylation at serine 2, commit RNAPII to the elongation phase of transcription. As already mentioned, proteins involved in 5' capping, elongation, and processing of the RNA are recruited to the CTD. The binding of these factors is determined in part by the differential phosphorylation of serine 2 and serine 5 within the 52 repeats. This is an active field of research and many of the mechanisms regulating association of RNA processing factors with RNAPII have yet to be elucidated. There is little doubt that infection of cells with viruses will also introduce new layers to the complexity of these processes. In this regard, Baek et al. (2004) have recently reported that HCMV induces an intermediate form of phosphorylated RNAP II, and although preliminary, evidence from our lab indicates that there is a change in the phosphorylation of serine 2 and serine 5 within the repeats of CTD during the infection. This would suggest that the effects of Roscovitine on the processing of IE RNAs may be due to inhibition of Cdk7 or Cdk9, as both of these kinases phosphorylate the CTD of RNAPII. The CTD has also been shown to be a substrate for Cdk1, and although both Cdk1 and cyclin B1 are low or below the level of detection between 0 and 6 h p.i., we cannot exclude the possibility that changes in RNAPII phosphorylation in infected cells are due to Cdk1-associated kinase activity. The use of dominant-negative Cdks in combination with siRNA technology to knock-down expression of cyclin subunits will hopefully elucidate the roles of individual Cdks during productive HCMV infection.

## Conclusions

The studies described here demonstrate that HCMV has evolved multiple mechanisms to undermine the cell's regulatory machinery for the purposes of promoting viral gene expression and production of progeny virions. Beginning with the primary attachment of the virus particles to the cell, signaling cascades are induced that prime the cell's metabolism and create an atmosphere conducive for productive infection by the virus. Furthermore, the virus encodes gene products that influence cell cycle progression and apoptotic pathways prolonging the cell's survival until it meets its inevitable doom. The roles of many viral proteins in these pathways remain to be elucidated and thus these unanswered questions provide an impetus to continue investigating the mechanisms that underlie the changes observed in the infected host cell. These ongoing studies will add to our current understanding of HCMV disease processes and contribute to the development of antiviral therapies.

## Acknowledgments

Research in the authors' laboratory was supported by NIH grants to D.H.S. (CA34729 and CA073490) and V.S. (CA102094).

Iskenderian, A.C., Huang, L., Reilly, A., Stenberg, R.M., and Anders, D.G. (1996). Four of eleven loci required for transient complementation of human cytomegalovirus DNA replication cooperate to activate expression of replication genes. J. Virol. 70, 383–392.

Isom, H.C. (1979). Stimulation of ornithine carboxylase by human cytomegalovirus. J. Gen. Virol. 42, 265–278.

Jault, F.M., Jault, J.M., Ruchti, F., Fortunato, E.A., Clark, C., Corbeil, J., Richman, D.D., and Spector, D.H. (1995). Cytomegalovirus infection induces high levels of cyclins, phosphorylated Rb, and p53, leading to cell cycle arrest. J. Virol. 69, 6697–6704.

Johnson, R.A., Wang, X., Ma, X.L., Huong, S.M., and Huang, E.S. (2001). Human cytomegalovirus up-regulates the phosphatidylinositol 3-kinase (PI3-K) pathway: inhibition of PI3-K activity inhibits viral replication and virus-induced signaling. J. Virol. 75, 6022–6032.

Johnson, R.A., Yurochko, A.D., Poma, E.E., Zhu, L., and Huang, E.S. (1999). Domain mapping of the human cytomegalovirus IE1-72 and cellular p107 protein–protein interaction and the possible functional consequences. J. Gen. Virol. 80, 1293–1303.

Kalejta, R.F., Bechtel, J.T., and Shenk, T. (2003). Human cytomegalovirus pp71 stimulates cell cycle progression by inducing the proteasome-dependent degradation of the retinoblastoma family of tumor suppressors. Mol. Cell. Biol. 23, 1885–1895.

Kalejta, R.F., and Shenk, T. (2003a). The human cytomegalovirus UL82 gene product (pp71) accelerates progression through the G1 phase of the cell cycle. J. Virol. 77, 3451–3459.

Kalejta, R.F., and Shenk, T. (2003b). Proteasome-dependent, ubiquitin-independent degradation of the Rb family of tumor suppressors by the human cytomegalovirus pp71 protein. Proc. Natl. Acad. Sci. USA 100, 3263–3268.

Karlsson, C., Katich, S., Hagting, A., Hoffmann, I., and Pines, J. (1999). Cdc25B and cdc25C differ markedly in their properties as initiators of mitosis. J. Cell Biol. 146, 573–584.

Kouzarides, T., Bankier, A.T., Satchwell, A.C., Preddy, E., and Barrell, B.G. (1988). An immediate-early gene of human cytomegalovirus encodes a potential membrane glycoprotein. Virology 165, 151–164.

Lammer, C., Wagerer, S., Saffrich, R., Mertens, D., Ansorge, W., and Hoffmann, I. (1998). The cdc25B phosphatase is essential for the G2/M phase transition in human cells. J. Cell Sci. 111, 2445–2453.

Liu, F., Rothblum-Oviatt, C., Ryan, C.E., and Piwnica-Worms, H. (1999). Overproduction of human Myt1 kinase Induces a G2 cell cycle delay by interfering with the intracellular trafficking of cdc2-cyclin B1 complexes. Mol. Cell. Biol. 19, 5113–5123.

Liu, F., Stanton, J.J., Wu, Z., and Piwnica-Worms, H. (1997). The human Myt1 kinase preferentially phosphorylates cdc2 on threonine 14 and localizes to the endoplasmic reticulum and Golgi complex. Mol. Cell. Biol. 17, 571–583.

Liu, B., and Stinski, M.F. (1992). Human cytomegalovirus contains a tegument protein that enhances transcription from promoters with upstream ATF and AP-1 cis-acting elements. J. Virol. 66, 4434–4444.

Lu, M., and Shenk, T. (1996). Human cytomegalovirus infection inhibits cell cycle progression at multiple points, including the transition from G1 to S. J. Virol. 70, 8850–8857.

Lu, M., and Shenk, T. (1999). Human cytomegalovirus UL69 protein induces cells to accumulate in G1 phase of the cell cycle. J. Virol. 73, 676–683.

Lukas, J., Lukas, C., and Bartek, J. (2004). Mammalian cell cycle checkpoints: signalling pathways and their organization in space and time. DNA Repair 3, 997–1007.

Majello, B., and Napolitano, G. (2001). Control of RNA polymerase II activity by dedicated CTD kinases and phosphatases. Front. Biosci. 6, D1358–1368.

Marchini, A., Liu, H., and Zhu, H. (2001). Human cytomegalovirus with IE-2 (UL122) deleted fails to express early lytic genes. J. Virol. 75, 1870–1878.

Margolis, M.J., Panjovic, S., Wong, E.L., Wade, M., Jupp, R., Nelson, J.A., and Azizkhan, J.C. (1995). Interaction of the 72-kilodalton human cytomegalovirus IE1 gene product with E2F1 coincides with E2F-dependent activation of dihydrofolate reductase transcription. J. Virol. 69, 7759–7767.

Mavinakere, M.S., and Colberg-Poley, A.M. (2004). Dual targeting of the human cytomegalovirus UL37 exon 1 protein during permissive infection. J. Gen. Virol. 85, 323–329.

McElroy, A.K., Dwarakanath, R.S., and Spector, D.H. (2000). Dysregulation of cyclin E gene expression in human cytomegalovirus-infected cells requires viral early gene expression and is associated with changes in the Rb-related protein p130. J. Virol. 74, 4192–4206.

Meijer, L., Borgne, A., Mulner, O., Chong, J.P.J., Blow, J.J., Inagaki, N., Inagaki, M., Delcros, J.G., and Molinoux, J.P. (1997). Biochemical and cellular effects of roscovitine, a potent and selective inhibitor of the cyclin-dependent kinases cdc2, cdk2 and cdk5. Eur. J. Biochem. 243, 527–536.

Mihara, M., Erster, S., Zaika, A., Petrenko, O., Chittenden, T., Pancoska, P., and Moll, U.M. (2003). p53 has a direct apoptogenic role at the mitochondria. Mol. Cell *11*, 577–590.

Mueller, P.R., Coleman, T.R., Kumagai, A., and Dunphy, W.G. (1995). Myt1: a membrane-associated inhibitory kinase that phosphorylates cdc2 on both threonine-14 and tyrosine-15. Science *270*, 86–90.

Muganda, P., Carrasco, R., and Qian, Q. (1998). The human cytomegalovirus IE2 86 kDa protein elevates p53 levels and transactivates the p53 promoter in human fibroblasts. Cell. Mol. Biol. *44*, 321–331.

Muganda, P., Mendoza, O., Hernandez, J., and Qian, Q. (1994). Human cytomegalovirus elevates levels of the cellular protein p53 in infected fibroblasts. J. Virol. *68*, 8028–8034.

Muranyi, W., Haas, J., Wagner, M., Krohne, G., and Koszinowski, U. (2002). Cytomegalovirus recruitment of cellular kinases to dissolve the nuclear lamina. Science *297*, 854–857.

Murphy, E.A., Streblow, D.N., Nelson, J.A., and Stinski, M.F. (2000). The human cytomegalovirus IE86 protein can block cell cycle progression after inducing transition into the S phase of permissive cells. J. Virol. *74*, 7108–7118.

Nishitani, H., and Lygerou, Z. (2002). Control of DNA Replication. Genes Cells *7*, 523–534.

Noris, E., Zannetti, C., Demurtas, A., Sinclair, J., De Andrea M., Gariglio, M., and Landolfo, S. (2002). Cell cycle arrest by human cytomegalovirus 86-kDa IE2 protein resembles premature senescence. J. Virol. *76*, 12135–12148.

O'Farrell, P.H. (2001). Triggering the all-or-nothing switch into mitosis. Trends Cell Biol. *11*, 512–519.

Orphanides, G., and Reinberg, D. (2002). A unified theory of gene expression. Cell *108*, 439–451.

Pajovic, S., Wong, E.L., Black, A.R., and Azizkhan, J.C. (1997). Identification of a viral kinase that phosphorylates specific E2Fs and pocket proteins. Mol. Cell. Biol. *17*, 6459–6464.

Pari, G.S., and Anders, D.G. (1993). Eleven loci encoding trans-acting factors are required for transient complementation of human cytomegalovirus oriLyt-dependent DNA replication. J. Virol. *67*, 6979–6988.

Pari, G.S., Kacica, M.A., and Anders, D.G. (1993). Open reading frames UL44, IRS1/TRS1, and UL36–38 are required for transient complementation of human cytomegalovirus oriLyt-dependent DNA synthesis. J. Virol. *67*, 2575–2582.

Poma, E.E., Kowalik, T.M., Zhu, L., Sinclair, J.H., and Huang, E.S. (1996). The human cytomegalovirus IE1-72 protein interacts with the cellular p107 protein and relieves p107-mediated transcriptional repression of an E2F-responsive promoter. J. Virol. *70*, 7867–7877.

Prelich, G. (2002). RNA polymerase II carboxy-terminal domain kinases: emerging clues to their function. Eukaryot. Cell *1*, 153–162.

Proudfoot, N.J., Furger, A., and Dye, M.J. (2002). Integrating mRNA processing with transcription. Cell *108*, 501–512.

Rowan, S., Ludwig, R.L., Haupt, Y., Bates, S., Lu, X., Oren, M., and Vousden, K.H. (1996). Specific loss of apoptotic but not T-cell cycle arrest function in a human tumor derived p53 mutant. EMBO J. *15*, 827–838.

Salvant, B.S., Fortunato, E.A., and Spector, D.H. (1998). Cell cycle dysregulation by human cytomegalovirus: influence of the cell cycle at the time of infection and effects on cyclin transcription. J. Virol. *72*, 3729–3741.

Sanchez, V., Clark, C.L., Yen, J.Y., Dwarakanath, R., and Spector, D.H. (2002). Viable human cytomegalovirus recombinant virus with an internal deletion of the IE2 86 gene affects late stages of viral replication. J. Virol. *76*, 2973–2989.

Sanchez, V., McElroy, A.K., and Spector, D.H. (2003). Mechanisms governing maintenance of cdk1/cyclin B1 kinase activity in cells infected with human cytomegalovirus. J. Virol. *77*, 13124–13224.

Sanchez, V., McElroy, A.K., Yen, J., Tamrakar, S., Clark, C.L., Schwartz, R.A., and Spector, D.H. (2004). Cyclin-dependent kinase activity is required at early times for accurate processing and accumulation of the human cytomegalovirus UL122–123 and UL37 immediate-early transcripts and at later times for virus production. J. Virol. *78*, 11219–11232.

Sedger, L.M., Shows, D.M., Blanton, R.A., Peschon, J.J., Goodwin, R.G., Cosman, D., and Wiley, S.R. (1999). IFN-gamma mediates a novel antiviral activity through dynamic modualtion of TRAIL and TRAIL receptor expression. J. Immunol. *163*.

Sherr, C.J. (1996). Cancer cell cycles. Science *274*, 1672–1677.

Sherr, C.J. (2000). The Pezcoller Lecture: Cancer cell cycles revisited. Cancer Res. *60*, 3689–3695.

Skaletskaya, A., Bartle, L.M., Chittenden, T., McCormick, A.L., Mocarski, E.S., and Goldmacher, V.S. (2001). A cytomegalovirus-encoded inhibitor of apoptosis that suppresses caspase-8 activation. Proc. Natl. Acad. Sci. USA *98*, 7829–7834.

ated with essentially unchanged enzymatic activity (Holwerda et al., 1994). C-site cleavage is thought to provide an alternate processing pathway (Jones et al., 1994), and recent findings with mutant viruses blocked at one or both sites are consistent with that interpretation (Chan et al., 2002). When just one of the sites is blocked for cleavage, replication is only moderately reduced, but when both the sites are blocked production of infectious virus is decreased by $\approx$ 90% – demonstrating their biological relevance (unpublished results).

The proteolytic portion of pPR, assemblin, has been purified and characterized structurally and enzymatically. Although a serine protease (DiIanni et al., 1994; Welch et al., 1993), it has a Ser-His-His (instead of Ser-His-Asp) catalytic triad that distinguishes it from other members of its family (Chen et al., 1996; Qiu et al., 1996; Shieh et al., 1996; Tong et al., 1996). The active enzyme is a homodimer (Cole, 1996; Darke et al., 1996; Margosiak et al., 1996); each subunit has a separate catalytic site; and its substrate- and buffer-specific cleavage rate against peptidyl substrate mimics is calculated to be $\approx$ 2/min with a kcat of $\approx$ 10 $\mu$M (Bonneau et al., 1998; Khayat et al., 2003; Khayat et al., 2004; Waxman and Darke, 2000), making it one of the slowest viral maturational proteases.

The crystal structure of the protease precursor has not been determined, but there is evidence that its enzymatic characteristics differ from those of assemblin. The first indication of this came from in vitro comparisons of purified native assemblin with purified denatured/renatured precursor, and from in-cell comparisons of the two proteins expressed from plasmids (Wittwer et al., 2002). In both assays, assemblin was more active than its precursor. More recently, assemblin was determined to be 3- to 10-fold more active than its precursor in cleaving a fluorogenic substrate in vitro, after both proteins were expressed, denatured, purified, and assayed in parallel and under identical conditions (Gibson et al., 2004). In addition, assemblin inactivated by mutating its catalytically critical His63 (i.e., H63G) can be chemically rescued with imidazole to restore I-site cleavage in living cells and in vitro, whereas pPR with the same mutation and under the same conditions cannot be rescued (Gibson et al., 2004; McCartney et al., 2003a; McCartney et al., 2003b). This difference between assemblin and its precursor may have significance for antiviral drug discovery.

### Early stages: organization and translocation of the capsid proteins

Self-interaction of pAP is one of the earliest steps in the capsid assembly pathway, as deduced from results of yeast GAL4 two-hybrid assays (Desai and Person, 1996; Pelletier et al., 1997; Wood et al., 1997). This interaction is mediated by the amino conserved domain (ACD, Figure 12.1B) in CMV and is abolished by deleting the ACD or by substituting Ala for Leu47 within it (Wood et al., 1997). The structural and biological importance of this interaction is indicated by the findings that (i) the L47A mutant pPR protein sediments as a monomer, rather than a tetramer, and (ii) a mutant virus carrying the L47A point mutation replicates poorly and produces aberrant intranuclear capsids (Gibson et al., 2004).

A second important early cytoplasmic interaction of pAP is with the major capsid protein (MCP, pUL86) (Beaudet-Miller et al., 1996; Desai and Person, 1996; Hong et al., 1996; Wood et al., 1997). This interaction is mediated by the carboxyl conserved domain of pAP (CCD, Figure 12.1B) and is potentiated or stabilized by pAP multimerization

(Pelletier et al., 1997; Wood et al., 1997). The pAP–MCP interaction is also required to translocate MCP into the nucleus. MCP is too large to enter by diffusion and has no nuclear localization signals (NLS) of its own. NLS present in pAP provide for nuclear translocation of the pAP-MCP complex (Plafker and Gibson, 1998). Conspicuously, only the betaherpesvirus pAP homologs contain two NLS; those of other herpesviruses have just one (Plafker and Gibson, 1998). The molecular composition of the cytoplasmic pAP–MCP complex is unknown. Minimally or initially it might be $pAP_2$–MCP, but if higher order or complete capsomer precursors (protocapsomers) form in the cytoplasm, the structure may be more complex (e.g., $pAP_{12}$–$MCP_6$). The time required to assemble such a large structure could account for the comparatively long delay observed for movement of both pAP and MCP from the cytoplasmic into the nuclear fraction of infected cells (Gibson et al., 1990; Manabu et al., 1985).

Triplexes are the other integral subunits of the capsid shell (Newcomb et al., 1993), and are composed of two copies of the minor capsid protein (mCP, pUL85) and one copy of the mCP-binding protein (mC-BP, pUL46). mCP monomers are small enough to diffuse into the nucleus, but they self-associate to form nucleus-excluded ≈ 60-kDa dimers lacking NLS (Baxter and Gibson, 1997; Gibson et al., 1996a). Interaction of mC-BP with the mCP dimer constitutes the triplex and enables it to move into the nucleus (Figure 12.2), indicating that mC-BP contains an NLS (Baxter and Gibson, 1997).

Less is known about early interactions and movement into the nucleus of the recently identified portal protein (Newcomb et al., 2001). By extension of information obtained with the HSV protein pUL6 (Lamberti and Weller, 1996; Newcomb et al., 2001; Newcomb and Brown, 2002; Newcomb et al., 2003; Patel and MacLean, 1995; Patel et al., 1996; Trus et al., 2004), the CMV homolog (pUL104) is expected to form a ring structure composed of 12 pUL104 monomers, inserted at a single vertex of the capsid shell; forming a channel for the viral DNA to enter and leave the capsid. It has been found that the HSV portal protein interacts with the HSV pAP homolog (preVP22a, pUL26.5) through a sequence containing the highly conserved amino acid triplet, ProGlyGlu/Asp (Singer et al., 2005), also present in the CMV pAP (Figure 12.1B herein; and Figure 11 in Plafker and Gibson, 1998). It is unknown, however, whether this interaction occurs in the cytoplasm, or to what extent the 12-subunit complex may be assembled there. Expressed alone, the HSV portal protein localizes to the nucleus (Patel et al., 1996), despite its large size, suggesting that it contains NLS and in that regard has no need to interact with pAP in the cytoplasm.

In addition to these proteins, an abundant fifth species called the smallest capsid protein (SCP, pUL48/49) (Gibson et al., 1996b; Lai and Britt, 2003) is bound tightly to the outer surface of the hexons (Booy et al., 1994; Butcher et al., 1998; Gibson et al., 1996b; Lai and Britt, 2003; Zhou et al., 1994). Although the HSV homolog, VP26, is not essential for assembly and infectivity (Desai et al., 1998), the CMV SCP is essential for infectivity (Borst et al., 2001). If there are functional similarities between the CMV and HSV counterparts of this protein, as their location on the capsid surface implies, then the CMV SCP may likewise enter the nucleus as part of the pAP–MCP complex and cycle on and off nascent capsid intermediates (Chi and Wilson, 2000).

The mechanics of DNA packaging are not fully defined, even in bacteriophage systems, but the process requires ATP hydrolysis and movement of the portal complex to generate force (Grimes et al., 2002; Smith et al., 2001). Packaging incorporates a single, full-length genome into the capsid and ends when terminase cleaves the DNA as the next *pac* site arrives at the portal. The large volume and high density of electronegativity associated with the DNA could help displace the cleaved remnants of pAP and pPR from the capsid volumetrically and/or by charge repulsion. The packing density of the DNA is high within the capsid and the genome resides there in a liquid crystalline form (Booy et al., 1991; Bhella et al., 2000). Polyamines present in herpes virus particles may serve as biological counterions to neutralize some of the DNA electronegativity and facilitate packaging (Gibson and Roizman, 1971; 1973; Gibson et al., 1984). Packaged DNA requires stabilization for retention within the capsid and this may be a function of CMV pUL77, whose HSV homolog (pUL25), when mutated, gives rise to capsids that lose mature DNA (McNab et al., 1998; Sheaffer et al., 2001).

## Does pAP/pPR phosphorylation influence capsid formation?

As a final note, and returning to the original focus on involvements of the assembly protein and protease in capsid formation, attention is given to the role that pAP phosphorylation may have in these processes. Experiments done with the SCMV pAP homolog have established that it is phosphorylated at a mitogen-activated protein (MAP) kinase consensus site, and at a glycogen synthase kinase 3 (GSK-3) site four residues upstream (Casaday et al., 2004; Plafker et al., 1999). Both phosphorylations are interpreted to cause conformational changes in the protein, and both sites have apparent counterparts in the HCMV pAP. More interestingly, phosphorylation of these sites was found to diminish pAP self-interaction and to enhance interaction of pAP with MCP (Casaday et al., 2004). One in-

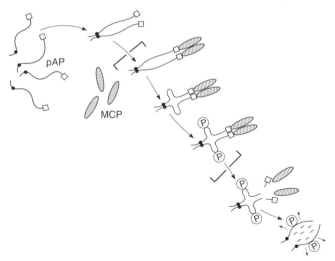

**Figure 12.3** Hypothetical involvement of assembly protein phosphorylation in nucleocapsid formation. The amino conserved domain is represented by a filled circle at the left-hand end of pAP; the carboxyl conserved domain is represented at the opposite end by an empty box. Brackets include possible conformational changes resulting from pAP interactions. Phosphorylation is represented by encircled letter P; negative charges are indicated by dash marks at lower right. See text for additional description.

terpretation of how these findings may relate to capsid assembly is illustrated in Figure 12.3 and can be summarized as follows. Newly synthesized pAP is not phosphorylated, favoring its dimerization which may help expose the phosphorylation sites. Phosphorylation of the MAP and GSK-3 sites would enhance pAP–MCP interaction and help stabilize the complexes during nuclear translocation, oligomerization into protocapsomers, and incorporation into procapsids. Conformation-induced changes in pAP resulting from interaction with MCP could also enable timely exposure of the phosphorylation sites. The same phosphorylations predicted to stabilize pAP–MCP complexes during procapsid formation would then destabilize the AP complexes released by M-site cleavage of pAP → AP during the procapsid → angular capsid transition. Weakened self-interaction of phosphorylated AP would promote its dissociation into monomers and facilitate elimination through the capsid shell, perhaps enhanced by charge repulsion from the incoming DNA.

## Late maturational events

Once DNA packaging is completed the capsid undergoes primary envelopment at the inner nuclear membrane, tegumentation, and secondary or final envelopment in the cytoplasm to become an infectious virion, as discussed in Chapter 13. Other particles formed as byproducts of the virion assembly pathway during HCMV replication (e.g., noninfectious enveloped particles, NIEPs, and dense bodies, DB) are discussed elsewhere (Gibson and Irmiere, 1984; Gibson, 1996; Irmiere and Gibson, 1983; see also Chapter 5).

## Acknowledgments
I thank Ed Brignole, Steve McCartney, Amy Loveland, and Keli Kolegraff for their valuable contributions to yet unpublished studies of the protease precursor, virus mutants (L47A, I⁻, C⁻, I⁻/C⁻), and chemical rescue. I also thank Amy Loveland for help in preparing the manuscript, and Joe Dieter in the Department of Art as Applied to Medicine, JHMI, for computer renderings of figures. Work referenced from our lab was aided by Public Health Service research grants AI13719 and AI032957.

## References
Baines, J., Poon, A., Rovnak, J., and Roizman, B. (1994). The herpes virus 1 UL15 gene encodes two proteins and is required for cleavage of genomic viral DNA. J. Virol. *68*, 8118–8124.

Baxter, M.K., and Gibson, W. (1997). The putative human cytomegalovirus triplex proteins, minor capsid protein (mCP) and mCP-binding protein (mC-BP), form a heterotrimeric complex that localizes to the cell nucleus in the absence of other viral proteins. Paper presented at: 22nd International Herpesvirus Workshop, La Jolla, CA.

Beaudet-Miller, M., Zhang, R., Durkin, J., Gibson, W., Kwong, A.D., and Hong, Z. (1996). Viral specific interaction between the human cytomegalovirus major capsid protein and the C-terminus of the precursor assembly protein. J. Virol. *70*, 8081–8088.

Becker, A., and Murialdo, H. (1990). Bacteriophage lambda DNA: The beginning of the end. J. Bacteriol. *172*, 2819–2824.

Bhella, D., Rixon, F.J., and Dargan, D.J. (2000). Cryomicroscopy of human cytomegalovirus virions reveals more densely packed genomic DNA than in herpes simplex virus type 1. J. Mol. Biol. *295*, 155–161.

Black, L.W. (1988). DNA packaging in dsDNA bacteriophages., In The Bacteriophages Vol. 2, R. Calendar, ed. (New York: Plenum Press), pp. 321–373.

Bogner, E., Radsak, K., and Stinski, M.F. (1995). The HCMV UL56 gene product is a DNA binding protein. 5th International Cytomegalovirus Conference *Abstract P051*, 85.

Bogner, E., Radsak, K., and Stinski, M.F. (1998). The gene product of human cytomegalovirus open reading frame UL56 binds the pac motif and has specific nuclease activity. J. Virol. *72*, 2259–2264.

Bogner, E., Reschke, M., Reis, B., Richter, A., Mockenhaupt, T., and Radsak, K. (1993). Identification of the gene product encoded by ORF UL56 of the human cytomegalovirus. Virology 196, 290–293.

Bonneau, P.R., Plouffe, C., Pelletier, A., Wernic, D., and Poupart, M.A. (1998). Design of fluorogenic peptide substrates for human cytomegalovirus protease based on structure-activity relationship studies. Anal. Biochem. 255, 59–65.

Booy, F.P., Newcomb, W.W., Trus, B.L., Brown, J.C., Baker, T.S., and Steven, A.C. (1991). Liquid-crystalline, phage-like packing of encapsidated DNA in herpes simplex virus. Cell 64, 1007–1015.

Booy, F.P., Trus, B.L., Newcomb, W.W., Brown, J.C., Conway, J.F., and Steven, A.C. (1994). Finding a needle in a haystack: detection of a small protein (the 12-kDa VP26) in a large complex (the 200-MDa capsid of herpes simplex virus). Proc. Natl. Acad. Sci. USA 91, 5652–5656.

Borst, E.M., Mathys, S., Wagner, M., Muranyi, W., and Messerle, M. (2001). Genetic evidence of an essential role for cytomegalovirus small capsid protein in viral growth. J. Virol. 75, 1450–1458.

Brown, J.C., McVoy, M.A., and Homa, F.L. (2002). Packaging DNA into herpesvirus capsids, In Structure-Function Relationships of Human Pathogenic Viruses, A. Holzenburg, E. Bogner, ed. (New York: Kluwer Academic/Plenum Publishers), pp. 111–153.

Butcher, S.J., Aitken, J., Mitchell, J., Gowen, B., and Dargan, D.J. (1998). Structure of the human cytomegalovirus B capsid by electron cryomicroscopy and image reconstruction. J. Struct. Biol. 124, 70–76.

Casaday, R.J., Bailey, J.R., Kalb, S.R., Brignole, E.J., Loveland, A.N., Cotter, R.J., and Gibson, W. (2004). Assembly protein precursor (pUL80.5 homolog) of simian cytomegalovirus is phosphorylated at a glycogen synthase kinase 3 site and its downstream "priming" site: phosphorylation affects interactions of protein with itself and with major capsid protein. J. Virol. 78, 13501–13511.

Chan, C.K., Brignole, E.J., and Gibson, W. (2002). Cytomegalovirus assemblin (pUL80a): cleavage at internal site not essential for virus growth; proteinase absent from virions. J. Virol. 76, 8667–8674.

Chen, P., Tsuge, H., Almassy, R.J., Gribskov, C.L., Katoh, S., Vanderpool, D.L., Margosiak, S.A., Pinko, C., Matthews, D.A., and Kan, C.-C. (1996). Structure of the human cytomegalovirus protease catalytic domain reveals a novel serine protease fold and catalytic triad. Cell 86, 835–843.

Chi, J.H., and Wilson, D.W. (2000). ATP-Dependent localization of the herpes simplex virus capsid protein VP26 to sites of procapsid maturation. J. Virol. 74, 1468–1476.

Cole, J.L. (1996). Characterization of human cytomegalovirus protease dimerization by analytical centrifugation. Biochemistry 35, 15601–15610.

Darke, P.L., Cole, J.L., Waxman, L., Hall, D.L., Sardana, M.K., and Kuo, L.C. (1996). Active human cytomegalovirus protease is a dimer. J. Biol. Chem. 271, 7445–7449.

Desai, P., DeLuca, N.A., and Person, S. (1998). Herpes simplex virus type 1 VP26 is not essential for replication in cell culture but influences production of infectious virus in the nervous system of infected mice. Virology 247, 115–124.

Desai, P., and Person, S. (1996). Molecular interactions between the HSV-1 capsid proteins as measured by the yeast two-hybrid system. Virology 220, 516–521.

Desai, P., Watkins, S.C., and Person, S. (1994). The size and symmetry of B capsids of herpes simplex virus type 1 are determined by the gene products of the UL26 open reading frame. J. Virol. 68, 5365–5374.

DiIanni, C.L., Stevens, J.T., Bolgar, M., O'Boyle, D.R., Weinheimer, S.P., and Colonno, R.J. (1994). Identification of the serine residue at the active site of the herpes simplex virus type I protease. J. Biol. Chem. 269, 12672–12676.

Dolan, A., Arbuckle, M., and McGeoch, D.J. (1991). Sequence analysis of the splice junction in the transcript of herpes simplex virus type 1 gene UL15. Virus Res. 20, 97–104.

Feiss, M., and Becker, A. (1983). DNA packaging and cutting., In Lambda II, R.W. Hendrix, J.W. Roberts, F.W. Stahl, and R.A. Weisberg, eds. (Cold Spring Harbor: Cold Spring Harbor Press), pp. 305–330.

Gibson, W. (1996). Structure and assembly of the virion. Intervirology 39, 389–400.

Gibson, W., Baxter, M. K., and Clopper, K. S. (1996a). Cytomegalovirus "missing" capsid protein identified as heat-aggregable product of human cytomegalovirus UL46. J. Virol. 70, 7454–7461.

Gibson, W., Brignole, E.J., Casaday, R.J., and McCartney, S.A. (2004). Self-interaction, phosporylation, cleavage: scaffolding features with potential to modulate herpesvirus capsid formation. Paper presented at: FASEB Conference on Virus Assembly (Saxton River, VT).

Gibson, W., Clopper, K.S., Britt, W.J., and Baxter, M.K. (1996b). Human cytomegalovirus smallest capsid protein identified as product of short open reading frame located between HCMV UL48 and UL49. J. Virol. 70, 5680–5683.

Gibson, W., and Irmiere, A. (1984). Selection of particles and proteins for use as human cytomegalovirus subunit vaccines. Birth Defects 20, 305–324.

Gibson, W., Marcy, A.I., Comolli, J.C., and Lee, J. (1990). Identification of precursor to cytomegalovirus capsid assembly protein and evidence that processing results in loss of its carboxy-terminal end. J. Virol. 64, 1241–1249.

Gibson, W., and Roizman, B. (1971). Compartmentalization of spermine and spermidine in the herpes simplex virion. Proc. Natl. Acad. Sci. USA 68, 2818–2821.

Gibson, W., and Roizman, B. (1973). The structural and metabolic involvement of polyamines with herpes simplex virus, In Polyamines in normal and neoplastic growth, D.H. Russell, ed. (New York: Raven), pp. 123–135.

Gibson, W., VanBreemen, R., Fields, A., LaFemina, R., and Irmiere, A. (1984). D,L-alpha-difluoromethylornithine inhibits human cytomegalovirus replication. J. Virol. 50, 145–154.

Gibson, W., Welch, A.R., and Hall, M.R.T. (1995). Assemblin, a herpes virus serine maturational proteinase and new molecular target for antivirals. Perspectives in Drug Discovery and Design 2, 413–426.

Grimes, S., Jardine, P.J., and Anderson, D. (2002). Bacteriophage phi 29 DNA packaging. Adv. Virus Res. 58, 255–294.

Higgins, R.R., and Becker, A. (1994). The lambda terminase enzyme measures the point of its endonucleolytic attack 47 +/- 2 bp away from its site of specific DNA binding, the R site. EMBO J. 13, 6162–6171.

Holwerda, B.C., Wittwer, A.J., Duffin, K.L., Smith, C., Toth, M.V., Carr, L.S., Wiegand, R.C., and Bryant, M.L. (1994). Activity of two-chain recombinant human cytomegalovirus protease. J. Biol. Chem. 269, 25911–25915.

Holzenburg, A., and Bogner, E. (2002). From concatemeric DNA into unit-length genomes – a miracle or clever genes? In Structure-Function Relationships of Human pathogenic Viruses, A. Holzenburg, and E. Bogner, eds. (New York: Plenum Publishers).

Hong, Z., Beaudet-Miller, M., Burkin, J., Zhang, R., and Kwong, A.D. (1996). Identification of a minimal hydrophobic domain in the herpes simplex virus type 1 scaffolding protein which is required for interaction with the major capsid protein. J. Virol. 70, 533–540.

Irmiere, A., and Gibson, W. (1983). Isolation and characterization of a noninfectious virion-like particle released from cells infected with human strains of cytomegalovirus. Virology 130, 118–133.

Jones, T.R., Sun, L., Bebernitz, G.A., Muzithras, V.P., Kim, H.-J., Johnston, S.H., and Baum, E.Z. (1994). Proteolytic activity of human cytomegalovirus UL80 proteinase cleavage site mutants. J. Virol. 68, 3742–3752.

Khayat, R., Batra, R., Bebernitz, G.A., Olson, M.W., and Tong, L. (2004). Characterization of the monomer-dimer equilibrium of human cytomegalovirus protease by kinetic methods. Biochemistry 43, 316–322.

Khayat, R., Batra, R., Qian, C., Halmos, T., Bailey, M., and Tong, L. (2003). Structural and biochemical studies of inhibitor binding to human cytomegalovirus protease. Biochemistry 42, 885–891.

Lai, L., and Britt, W.J. (2003). The interaction between the major capsid protein and the smallest capsid protein of human cytomegalovirus is dependent on two linear sequences in the smallest capsid protein. J. Virol. 77, 2730–2735.

Lamberti, C., and Weller, S.K. (1996). The herpes simplex virus type 1 UL6 protein is essential for cleavage and packaging but not for genomic inversion. Virology 226, 403–407.

Liu, F., and Roizman, B. (1991). The promoter, transcriptional unit, and coding sequence of herpes simplex virus 1 family 35 proteins are contained within and in frame with the UL26 open reading frame. J. Virol. 65, 206–212.

Manabu, Y., Nishiyama, Y., Hisashi, F., Isomura, S., and Maeno, K. (1985). On the intracellular transport and the nuclear association of human cytomegalovirus structural proteins. J. Gen. Virol. 66, 675–684.

Margosiak, S.A., Vanderpool, D.L., Sisson, W., Pinko, C., and Kan, C.C. (1996). Dimerization of the human cytomegalovirus protease – kinetic and biochemical characterization of the catalytic homodimer. Biochemistry 35, 5300–5307.

McCartney, S.A., Brignole, E.J., Kolegraff, K.N., and Gibson, W. (2003a). Imidazole rescues histidine mutants of cytomegalovirus protease, assemblin (pUL80a), in living cells and in vitro. Paper presented at: Meeting of the American Society for Virology (Davis, CA).

# Glycoprotein Trafficking in Virion Morphogenesis

# 13

*Markus Eickmann, Dorothee Gicklhorn, and Klaus Radsak*

## Abstract

Assembled cytomegalovirus nucleocapsids are exported from the nuclear compartment by sequential events including (i) *primary envelopment* at the inner nuclear membrane and (ii) de-envelopment at the outer nuclear membrane prior to (iii) *secondary envelopment* of naked cytoplasmic nucleocapsids at cytoplasmic cisternae, and (iv) release of mature enveloped particles by exocytosis. In permissive cells viral envelope glycoproteins are subject to retrograde transport from the site of biosynthesis, namely the rough endoplasmic reticulum, into the inner nuclear membrane presumably by lateral diffusion, as well as to anterograde transport along the cellular exocytosis pathway to the plasma membrane from where they are retrieved to the cytoplasmic compartment of final viral maturation. Recent investigations have unraveled some of the molecular mechanisms involved in cellular trafficking of viral gene products and their interaction with host cell products during virion morphogenesis. Continued efforts along this line will contribute to elucidate pivotal questions of cell biology.

## Introduction

### General ultrastructural aspects of virion morphogenesis

Studies of the ultrastructure of herpesvirus-infected cells performed by several groups as long ago as the 1950s (Morgan 1954, 1959; Stoker 1958) agreed in that viral morphogenesis starts with the assembly of nucleocapsids in the infected nucleus. It was initially, however, less clear how viral egress from the nucleus was achieved. The first investigators who raised the possibility that herpes simplex virus nuclear egress involves an envelopmental event at the inner nuclear membrane (INM) were Falke et al. (1959). It became generally accepted that this budding process at the inner leaflet of the nuclear envelope leads to the presence of enveloped capsids in the enlarged perinuclear cisterna (Roizman and Furlong, 1974). The route by which the herpesvirus particles ultimately reach the extracellular space, however, remained for some time a matter of debate. The favored view for quite some time was that intracisternal enveloped particles bud at the outer nuclear membrane (ONM) thereby receiving an additional envelope and traverse the cytoplasm singly or in groups in vesicular structures which finally fuse with the plasma membrane for release of mature virions. Maturational processing of viral envelope glycoproteins that are acquired at the

tural component which represents a proteinaceous meshwork of 10-nm type V intermediate filaments (Fawcett, 1966; Gerace et al., 1978; Krohne et al., 1978) termed lamins A/C and B (Moir et al., 1995; Stuurman et al., 1998). B-type lamins are constitutively expressed in all somatic cells, while A-type lamins are expressed only in differentiated cells.

The INM has a unique composition (Figure 13.2) and contains a distinct set of integral proteins, including e.g. the *lamin B receptor* (LBR), *lamina associated polypeptides* 1 and 2 (LAP1, 2), *emerin, MAN1, otefin* and *nurim*. Recent reports emphasize the functional interaction of the cellular INM proteins with constituents of the lamina and the nucleoskeleton, e.g. nuclear actin (reviewed in Foisner, 2001; Shumaker et al., 2004). These aspects appear to open novel views and therefore meet with increasing interest. Nuclear lamina and nucleoskeleton are required for maintaining nuclear size and shape during the interphase of the cell cycle and may also play a role in chromatin organization, regulation of transcription and DNA replication as well as positioning of the nuclear pore complex. The NL undergoes dynamic structural changes of disassembly and reassembly during the cell cycle. Disassembly prior to mitosis is thought to be induced by hyperphosphorylation of lamins. Both, in vitro and in vivo assays suggest that the mitotic CDC 2 kinase, protein kinase C (PKC) and cyclic AMP-dependent kinase (PKA) phosphorylation sites are important in lamin assembly as well as disassembly (Hutchison, 2002).

## The nuclear envelope in infected cells

Morphological changes observed in infected cells include focal NL and INM thickening and formation of blebs during the late phase of the infectious cycle. The alterations are presumably induced by specific viral gene products that are translocated into this cellular compartment and are possibly involved in interaction and/or modification of resident (endogenous) host cell products.

## Translocation of viral gene products into the inner nuclear membrane

### Viral glycoproteins

It is generally accepted that viral transmembrane glycoproteins in herpesvirus-infected cells are transported from the site of their synthesis in the RER not only along the exocytic pathway but are also subject to retrograde translocation into the INM (Gilbert et al., 1994; Gong and Kieff, 1990; Heinemann et al., 2000). In the case of HCMV, presence in the INM has been demonstrated for the conserved glycoproteins B and H (gB and gH; Bogner et al., 1992; Radsak et al., 1990). More detailed information is available for gB (ORF UL55) which represents a type I transmembrane protein whose amino-terminal ectodomain is co-translationally dimerized and modified by mannose-rich sugar residues (Britt and Vugler, 1992). Studies with *brefeldin A*, which interferes with the exocytic cellular transport, have shown that gB remained uncleaved with its export being interrupted but that it was still translocated into the nuclear compartment under these conditions (Eggers et al. 1992).

For the cellular INM-resident type II transmembrane proteins, e.g. LBR, LAP2, MAN1 and emerin it has been described that their nuclear targeting depends on specific sequence motifs that are located in the nucleoplasmic amino-terminal domain (Lin et al., 2000; Östlund et al., 1999; Smith and Blobel, 1993; Soullam and Worman, 1993; 1995). Accordingly, regarding nuclear translocation of the viral gB which is a type I transmem-

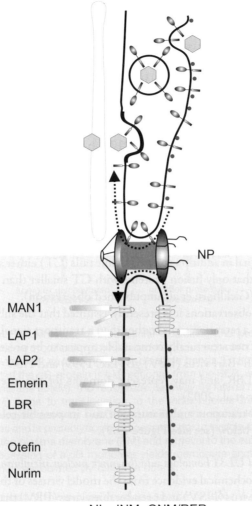

**Figure 13.2** Retrograde transport of viral transmembrane glycoproteins. The scheme depicts the structural components of the nuclear envelope which consists of the nuclear lamina (NL), inner (INM) and outer nuclear membrane (ONM). *Upper part:* Following biosynthesis of HCMV gB at the RER/ONM (see legend of Figure 13.1) viral transmembrane glycoproteins are translocated to the inner nuclear membrane (INM) by lateral diffusion via the nuclear pore (NP) membrane. The retrograde transport route is indicated by the dotted arrow. In the *lower shaded part* biosynthesis, retrograde translocation and positioning of cellular integral constituents of the INM are illustrated (see text).

brane protein, initial experiments suggested an essential role of a terminal domain of the nucleo-/cytoplasmic tail (Bogner et al., 1997). Extension of this work using CD8 as a reporter protein revealed that the hexameric amino acid motif DRLRHR at the extreme nucleo-/cytoplasmic carboxy terminus of gB (aa 885–890) was sufficient for INM localization. Furthermore, directed mutations of single residues of the DRLRHR motif indicated

the nuclear periphery in HCMV-infected cells (DalMonte et al., 2002). From observations of the phenotype of a viable UL97 deletion mutant of HCMV the conclusion was drawn that this nonessential protein kinase may also play a role in nucleocapsid egress from the infected cell nucleus (Krosky et al., 2003).

## Mechanisms involved in primary envelopment at the inner nuclear membrane

### Modification of the nuclear lamina

Early studies on permissive fibroblasts have suggested HCMV-induced alterations of nuclear lamins A/C late after infection (Radsak et al., 1989) and have indicated changes of the state of lamin phosphorylation (Radsak et al. 1991). Stability, dissolution and reassembly of the NL during the cell cycle has been shown to be regulated by the state of lamin phosphorylation (Gant and Wilson, 2000). The described herpesvirus-induced focal NL alterations implicated circumscribed loosening of the tight lamina meshwork to allow the viral 100-nm nucleocapsids, which are too large for passage through the nuclear pores, to interact with the INM for *primary envelopment*. This view prompted experiments by Muranyi et al. (2002) to look for M50-induced changes at the INM that would possibly influence the state of lamin phosphorylation. They demonstrated for the MCMV model that M50/p35 recruited host cell protein kinase C to the INM for lamin hyperphosphorylation and thus possibly for focal lamina dissolution (Muranyi et al., 2002). The authors proposed that foci of the NL decorated by the complex formed of M50/p35 and M53/p38 may represent the docking sites for viral nucleocapsids. It remains to be examined whether this scenario also holds true for HCMV and whether in the case of α-herpesviruses other mechanisms are involved as was suggested by Reynolds et al. (2004).

### Role of viral glycoproteins

The presence of herpesvirus glycoproteins, of gB in particular, in the INM provokes the question of their need for *primary envelopment*. By use of envelope glycoprotein-deficient viral mutants of HSV (Cai et al., 1988; Desai et al., 1988) and PRV (Granzow et al., 2001) it has been shown previously that virions – albeit non-infectious – are released from infected cells. This observation implies that *primary* and *secondary envelopment* apparently proceed in the absence of viral glycoproteins in the case of α-herpesviruses. Regarding the β-herpesvirus HCMV a viable Ad169-derived mutant was generated harboring a gB protein that lacks the carboxy-terminal signal for retention in the INM (Strive et al., 2004). Although growth of this mutant in culture was impaired, infectious progeny was released from infected cells suggesting again that *primary envelopment* was apparently not dependent on the gB INM retention motif (Meyer et al., 2002; Strive et al., 2004). Interestingly, growth impairment of this mutant virus correlated well with significantly reduced incorporation of mutagenized gB into the envelope of extracellular viral particles (Strive et al., 2004) (Figure 13.4). This intriguing observation, though calling for closer investigation, rather suggests a role of the carboxy-terminal gB signal in *secondary envelopment*.

Regarding presence of gB in the INM of infected fibroblasts, as shown by immunogold labeling (Radsak et al., 1990), there is presently no direct proof for viral glycoproteins of cytomegaloviruses being an integral component of the temporary envelope of intracisternal

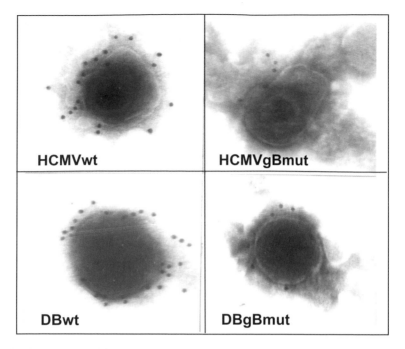

**Figure 13.4** Immunogold labeling with gB-specific monoclonal antibody of extracellular virions and dense bodies released from human fibroblasts infected with wild-type HCMV (HCMVwt and DBwt) or mutant HCMV harboring a gB protein in which the carboxy-terminal nuclear localization signal was functionally deleted (HCMVgBmut and DBgBmut; Strive et al., 2004).

virions. Whilst this is apparently not the case in the PRV model system, β-herpesviruses having fewer envelope glycoproteins may differ in this respect from α-herpesviruses (Table 13.1).

From recent experiments there is evidence that HCMV gB might interact in infected cells with cellular INM-resident proteins, e.g. emerin and lamins A/C, and that emerin is associated with extracellular virions (Gicklhorn et al., unpublished observation). However, further work is needed to substantiate these observations and to elucidate or exclude a role of viral glycoproteins in *primary envelopment* or egress from the nuclear cisterna (see below).

## Exit of nucleocapsids from the perinuclear cisterna

While the process of *primary envelopment* at the INM has been addressed by a number of recent investigations, little information is available for cytomegaloviruses regarding the subsequent step of exit from the perinuclear cisterna, i.e. the loss of the temporary envelope of intracisternal virions by de-envelopment at the ONM followed by release of naked nucleocapsids into the cytoplasm of infected cells. It seems clear for the PRV model (Granzow et al., 2001; Klupp et al., 2001) that gB, gH and gC are needed neither for *primary envelopment* nor for fusion of the temporary envelope with the ONM to promote egress of viral particles from the perinuclear space. This observation implicated that membrane fusion processes occurring during egress are essentially different from those during entry which

**Table 13.1** Conserved herpesvirus envelope glycoproteins

| HCMV | | MCMV | HSV/PRV |
|---|---|---|---|
| gc I | gB (gp UL55) | M55 | gB |
| | | | gC |
| | | | gD |
| | | | gE |
| | | | gG |
| gc III | gH (gp UL75) | M75 | gH |
| | | | gI |
| | | | gK |
| gc III | gL (gp UL115) | M115 | gL |
| gc II | gM (gp UL100) | M100 | gM |
| gc II | gN (gp UL73) | M73 | gN |
| gc III | gO (gp UL74) | M74 | |

Abbreviations: HSV, herpes simplex virus; PRV, pseudorabies virus; MCMV, murine cytomegalovirus; HCMV, human cytomegalovirus; gc, glycoprotein complex.

depend on the presence of at least gB in all herpesvirus systems examined so far (Granzow et al., 2001). The group of Mettenleiter (Klupp et al., 2001) could attribute a possible role in egress to the protein kinase encoded by US3 which is conserved in the α-herpesviruses but not in the β-herpesvirus subfamily (Granzow et al., 2004). In cells infected with PRV mutants deficient in the non-essential US3 gene, enveloped virions accumulated in the nuclear cisterna. Functional homologs of US3 are unknown in cytomegalovirus systems.

## Cytoplasmic phase of viral maturation

### Cytoplasmic tegumentation

For cytomegaloviruses a number of tegument proteins have been described which functionally may be considered equivalent to the matrix proteins of other enveloped viruses (Adair et al., 2002; Spaete et al., 1994). According to current view they are sequentially added to the nucleocapsids during the nuclear and more so during the cytoplasmic phase of morphogenesis (Sanchez et al., 1998; 2000a; Trus et al., 1999). Tegument protein ppUL69 exhibits a nuclear localization, whereas pp65 (ppUL83) has been identified in the nucleus as well as in the cytoplasm, and pp28 (UL99) and pp71 (UL82) were detected exclusively in the cytoplasm late post infection (Sanchez et al., 2000b). Somewhat controversial observations were reported for the gene product of UL32 of HCMV, pp150, a tegument protein with homologs in other β-herpesviruses found predominantly in the cytoplasm (Homman-Loudiyi et al., 2003; Sanchez et al., 2000a), while Hensel et al. (1995) had localized it in the nuclear compartment (Hensel et al., 1995). Evidence has been provided that cytoplasmic tegumentation takes place in a stable cytoplasmic compartment where virion tegument and envelope proteins accumulate during the late infectious cycle (Homman-Loudiyi et al., 2003; Sanchez et al., 2000a). For some time the herpesvirus tegument had been considered an amorphous proteinaceous matrix but there is increasing evidence for α- as well as β-herpesviruses for an ordered, possibly symmetrical structure (Chen et al., 1999; reviewed by Mettenleiter, 2002; Trus et al., 1999). The distinct differences between nuclear and cytoplasmic nucleocapsids, the latter showing a thickened electron-dense coat at the electron microscopic level (Eggers et al., 1992; reviewed by Mettenleiter, 2002; Silva et al., 2003) may well be due to the varying protein composition.

Regarding the role of tegument proteins that are acquired in the cytoplasm, an essential role has recently been attributed to HCMV UL99-encoded pp28 which colocalizes in infected cells with other cytoplasmic tegument and viral envelope proteins (Sanchez et al., 2000a; Silva et al., 2003). Viral mutants with a functional deletion of pp28 failed to produce infectious progeny and large numbers of tegumented but unenveloped nucleocapsids accumulated in the cytoplasm suggesting that pp28 is essential for *secondary envelopment*. Apart from colocalization studies there is limited information at the molecular level regarding possible direct interactions between tegument proteins and envelope glycoproteins of cytomegaloviruses. It is intriguing that most of the tegument proteins are highly phosphorylated, an aspect that remains to be elucidated (Mettenleiter, 2002; Roby and Gibson, 1986).

In addition to their role in virion morphogenesis, tegument proteins being introduced into the host cell during viral entry evidently exhibit defined functions during the early infectious cycle (see Chapter 6).

## Anterograde envelope glycoprotein transport

In extracellular HCMV, three major membrane-anchored glycoprotein complexes (designated gc I to III; see Table 13.1) have been identified consisting of gene products that are highly conserved among herpesviruses and are essential for viral replication in cell culture (Gehrz and Kari, 1988; Hobom et al., 2000). gc I represents a disulfide-linked homodimer of glycoprotein B (gB or gpUL55) (Britt 1984; Britt and Auger, 1986; Eickmann et al., 1998) with multimerization potential (Scheffczik et al., 2001). gc II has been reported to comprise gM (gpUL100) associated with gN (gpUL73) through disulfide bonds (Kari et al., 1990; Mach et al., 2000). gH (gpUL75), gL (gpUL115), and gO (gpUL74) together represent the non-covalently associated gc III complex (Bogner et al., 1992; Huber et al., 1997; Li et al., 1997).

The constituents of these protein complexes are type I glycoproteins, with the exception of gM, which is a type III glycoprotein, and are formed and processed during transport along the constitutive exocytic pathway from the RER via Golgi complex and trans-Golgi network (TGN) to the plasma membrane. Regarding gc I, binding to the molecular chaperone calnexin is required for correct complex formation (Yamashita et al., 1996). In the case of gc II and gc III, complexation of the heterologous proteins is required for the export from the RER to more distal parts of the secretory pathway (Huber and Compton, 1999; Li et al., 1997; Mach et al., 2000). For the dominant envelope glycoprotein gB it has been shown by constitutive expression of mutagenized products that a single membrane spanning domain (aa 751–771) is essential as well as sufficient for membrane anchoring (Reschke et al., 1995). During transport along the exocytic pathway via Golgi apparatus and TGN the dimerized gB precursor is further modified by complex glycosylation and by proteolytic processing. The latter maturation event yields for each precursor constituent (160 kDa) of the dimer products of about 110 and 55 kDa that remain associated by disulfide bonds (Figure 13.4) (Britt et al., 1992; Eickmann et al., 1998). It has been shown recently by construction of a respective viral mutant that proteolytic cleavage of gB is not needed for viral infectivity and replication in cell culture (Strive et al., 2002).

Processed HCMV gB was shown to follow the secretory pathway to the plasma membrane prior to retrieval by endocytosis to be available at cytoplasmic cisternae for *secondary envelopment* (Figure 13.5 for an EM image and Figure 13.6 for a model) (Jarvis et al., 2002; Radsak et al., 1996). The extended nucleo-/cytoplasmic gB tail of 134 aa exhibits tyrosine and dileucine aa motifs for protein sorting that presumably regulate the transport in the distal secretory pathway and recycling via an endosomal recycling compartment (Figure 13.6) (Fish et al., 1998). Internalization of gB from the endosomal compartment was shown to depend on phosphorylation at a single site (Ser900) by a tautomycin-sensitive phosphatase (Fish et al., 1998; Jarvis et al., 2004). In addition, phosphofurin acidic cluster sorting protein 1 (PACS-1) appears to subsequently serve in retrieval of gB to the TGN (Figure 13.6) (Molloy et al., 1999; Crump et al., 2003).

Except for gH, whose cellular transport is PACS-1 independent, data comparable to those for gB are not available for the other constituents of the glycoprotein complexes of HCMV.

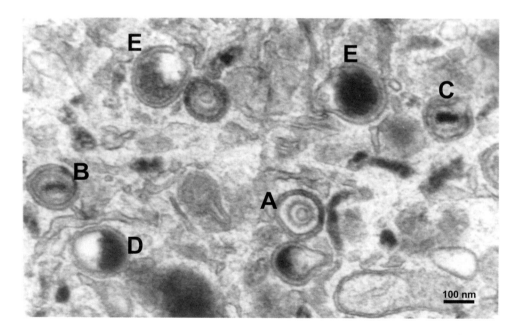

**Figure 13.5** Secondary envelopment of viral nucleocapsids and dense bodies at cytoplasmic cisternae. Human fibroblasts were incubated late during HCMV infection with the fluid-phase tracer horseradish peroxidase. In order to visualize the early endocytic compartment involved in wrapping of viral particles, infected cells were subsequently processed for fixation, staining with hydrogen peroxide and diaminobenzidine, and transmission electron microscopy (Tooze et al., 1993). A, nucleocapsid with lucent core (wrapping by endocytic cisterna completed); B, nucleocapsid with dense core (wrapping by endocytic cisterna completed); C, nucleocapsid with dense core (in the process of wrapping by endocytic cisterna); D, dense body (wrapping by endocytic cisterna completed); E, dense body (in the process of wrapping by endocytic cisterna). The black bar corresponds to 100 nm.

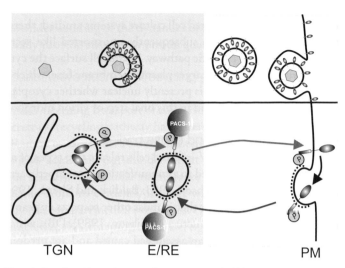

**TGN**  **E/RE**  **PM**

**Figure 13.6** Signals involved in anterograde transport of HCMV gB along the exocytic pathway from trans-Golgi network (TGN) to the plasma membrane (PM), and in internalization and retrieval to the early/recycling endosome (E/RE) and TGN. *Lower part:* Internalization and retrieval of gB appears to depend on carboxy-terminal phosphorylation and interaction with phosphofurin acidic cluster sorting protein 1 (PACS-1). *In the upper part* the morphogenetic events of final viral maturation are schematically depicted. The shaded area indicates the endocytic compartment of *secondary envelopment*.

## Secondary envelopment and exocytosis

Initial experiments demonstrated that the cellular compartment of envelopment of tegumented cytoplasmic nucleocapsids in HCMV-infected fibroblasts can be labeled with the fluid phase endocytic tracer horseradish peroxidase (Tooze et al., 1993) (Figure 13.5). Furthermore, wrapped viral particles were not directed to the late endosome but were destined directly to the plasma membrane and released from the cells. It therefore was concluded that the cytoplasmic compartment of HCMV *secondary envelopment* represented cisternae of the tubular early/recycling endosome (E/RE). A consecutive study, however, indicated that the recycling pathway from E/RE to the TGN (Figure 13.6) is greatly enhanced in virus-infected cells leading to the appearance of significant amounts of fluid-phase marker in the lumen of the TGN and Golgi complex (Schmelz et al., 1994). Homman-Loudiyi et al. (2003) recently confirmed by immunogold labeling and electron microscopic analysis that the cisternae involved in wrapping of cytoplasmic nucleocapsids in HCMV-infected cells exhibited specific staining for the dominant envelope protein gB and the TGN markers TGN 46 and mannosidase II as well as for Rab 3, a tracer of secretory vesicles. Taking into account that these authors also observed some colocalization with Rab 4, which is a tracer for the E/RE, it may well be that a new stable intermediate compartment is formed in infected cells for secondary envelopment (Homman-Loudiyi et al., 2003) comprising portions of E/RE and TGN-derived secretory vesicles as well (Figure 13.6). The role of multivesicular endosomes (Fraile-Ramos et al., 2002) in the context of HCMV maturation remains to be elucidated.

Britt, W.J., and Auger, D. (1986). Synthesis and processing of the envelope gp55–116 complex of human cytomegalovirus. J. Virol. *58*, 185–191.

Britt, W.J., and Vugler, L.G. (1992). Oligomerization of the human cytomegalovirus major envelope glycoprotein complex gB (gp55–116). J. Virol. *66*, 6747–6754.

Bubeck, A., Reusch, U., Wagner, M., Ruppert, T., Muranyi, W., Kloetzel, P.M., and Koszinowski, U.H. (2002). The glycoprotein gp48 of murine cytomegalovirus: proteasome-dependent cytosolic dislocation and degradation. J. Biol. Chem. *277*, 2216–2224.

Cai, W., Gu, B., and Person, S. (1988). Role of glycoprotein B of herpes simplex virus type I in viral entry and cell fusion. J. Virol. *62*, 2596–2604.

Campadelli-Fiume, G., Farabegoli, F., Di Gaeta, S., and Roizman, B. (1991). Origin of unenveloped capsids in the cytoplasm of cells infected with herpes simplex virus 1. J. Virol. *65*, 1589–1595.

Cha, T.A., Tom, E., Kemble, G.W., Duke, G.M., Mocarski, E.S. and. Spaete, R.R. (1996). Human cytomegalovirus clinical isolates carry at least 19 genes not found in laboratory strains. J. Virol. *70*, 78–83.

Chang Y.E., Van Sant, C., Krug, P.W., Sears, A.E., and Roizman, B. (1997). The null mutant of the U(L)31 gene of herpes simplex virus 1: construction and phenotype in infected cells. J. Virol. *71*, 8307–8315.

Chee, M.S., Bankier, A.T., Beck, S., Bohni, R., Brown, C.M., Cerny, R., Horsnell, T., Hutchison, C.A., Kouzarides, T., Martignetti, J.A., Preddie, E., Satchwell, S. Tomloson, P., Weston, K., and Barrell B. (1990). Analysis of the protein-coding content of the sequence of human cytomegalovirus strain AD169. Curr. Top. Microbiol. Immunol. *154*, 125–169.

Chen, D.H., Jiang, H., Lee, M., Liu, F. and Zhou, Z.H. (1999). Three-dimensional visualization of tegument/capsid interactions in the intact human cytomegalovirus. Virology *260*, 10–24.

Compton, T., and Courtney, R.J. (1984). Virus specific glycoproteins associated with the nuclear fraction of herpes simplex virus type 1-infected cells. J. Virol. *49*, 594–597.

Crump C.M., Hung, C.H., Thomas, L., Wan, L., and Thomas, G. (2003). Role of PACS-1 in trafficking of human cytomegalovirus glycoprotein B and virus production. J. Virol. *77*, 11105–11113.

DalMonte P, Pignatelli, S., Zini, N., Maraldi, N.M., Perret, E., Prevost, M.C., and Landini, M.P. (2002). Analysis of intracellular and intraviral localization of the human cytomegalovirus UL53 protein. J. Gen. Virol. *83*, 1005–1012.

Desai, P., Schaffer, M., and Minson, A. (1988). Excretion of non-infectious virus particles lacking glycoprotein H by a temperatur-sensitive mutant of herpes simplex virus type I: evidence that gH is essential for virion infectivity. J. Gen. Virol. *69*, 1147–1156.

Eggers, M., Bogner, E., Agricola, B.,. Kern, H.F, and Radsak, K. (1992). Inhibition of human cytomegalovirus maturation by brefeldin A. J. Gen. Virol. *73*, 2679–2692.

Eickmann, M., Lange, R., Ohlin, M., Reschke, M., and Radsak, K. (1998). Effect of cysteine substitutions on dimerization and interfragment linkage of human cytomegalovirus glycoprotein B (gp UL55). Arch. Virol. *143*, 1865–1880

Ellenberg, J., Siggia, E.D., Moreira, J.E., Smith, C., Presley, J.F., Worman, H.J., and Lippincott-Schwartz, J. (1997). Nuclear membrane dynamics and reassembly in living cells: targeting of an inner nuclear membrane protein in interphase and mitosis. J. Cell Biol. *138*, 1193–1206.

Falke, D., Siegert, R., and Vogell, W. (1959). Electron microscopic findings on the problem of double membrane formation in herpes simplex virus. Arch. Gesamte Virusforsch. *9*, 484–496.

Fawcett, D.W. (1966). On the occurrence of a fibrous lamina on the inner aspect of the nuclear envelope in certain cells of vertebrates. Am. J. Anat. *119*, 129–145.

Fish, K.N., Soderberg-Naucler, C., and Nelson, J.A. (1998). Steady-state plasma membrane expression of human cytomegalovirus gB is determined by the phosphorylation state of Ser900. J. Virol. *72*, 6657–6664.

Foisner, R. (2001). Inner nuclear membrane proteins and the nuclear lamina. J. Cell Sci. *114*, 3791–3792.

Fraile-Ramos, A., Kledal, T.N., Pelchen-Matthews, A., Bowers, K., Schwartz, T.W., and Marsh, M. (2001). The human cytomegalovirus US28 protein is located in endocytic vesicles and undergoes constitutive endocytosis and recycling. Mol. Biol. Cell. *12*, 1737–1749.

Fraile-Ramos, A., Pelchen-Matthews, A.,. Kledal, T.N., Browne, H., Schwartz, T.W., and Marsh M. (2002). Localization of HCMV UL33 and US27 in endocytic compartments and viral membranes. Traffic *3*, 218–232.

Fuchs W., Klupp, B.G., Granzow, H., Osterrieder, N., and Mettenleiter, T.C. (2002). The interacting UL31 and UL34 gene products of pseudorabies virus are involved in egress from the host cell nucleus and represent components of primary enveloped but not mature virions. J. Virol *76*, 364–378.

Furukawa, K., Panté, N., Aebi, U., and Gerace, L. (1995). Cloning of a cDNA for lamina-associated polypeptide 2 (LAP2) and identification of regions that specify targeting to the nuclear envelope. EMBO J. *14*, 1626–1636.

Gallina, A., Percivalle, E., Simoncini, L., Revello, M.G., Gerna, G., and Milanesi, G. (1996). Human cytomegalovirus pp65 lower matrix phosphoprotein harbours two transplantable nuclear localization signals. J. Gen. Virol. *77*, 1151–1157.

Gant, T.M., and Wilson, K.L. (1997). Nuclear assembly. Annu. Rev. Cell Dev. Biol. *13*, 669–695.

Gerace, L., Blum, A., and Blobel, G. (1978). Immunocytochemical localization of the major polypeptides of the nuclear pore complex-lamina fraction. Interphase and mitotic distribution. J. Cell Biol. *79*, 546–566.

Gibson, W., and Irmiere, A. (1984). Selection of particles and proteins for use as human cytomegalovirus subunit vaccines. In: March of Dimes Birth Defects Foundation. Birth Defects: Original Article Series, Vol. 20, pp. 305–324, Alan R. Liss, Inc., New York.

Gibson, W., Marcy, A.I., Comolli, J.C., and Lee, J. (1990). Identification of precursor to cytomegalovirus capsid assembly protein and evidence that processing results in loss of its carboxy-terminal end. J. Virol. *64*, 1241–1249.

Gilbert, R., Gosh, K., Rasile, L., and Gosh, H.P. (1994). Membrane anchoring domain of herpes simplex virus glycoprotein B is sufficient for nuclear envelope localization. J. Virol. *68*, 2272–2285.

Gong, M., and Kieff, E. (1990). Intracellular trafficking of two major Epstein–Barr virus glycoproteins, gp350/220 and gp110. J. Virol. *64*, 1507–1516.

Granzow, H., Klupp, B.G., Fuchs W., Veits, J., Osterrieder, N., and Mettenleiter, T.C. (2001). Egress of alpha-herpesviruses: comparative ultrastructural study. J. Virol. *75*, 3675–3684.

Granzow, H., Klupp, B.G., and Mettenleiter, T.C. (2004). The pseudorabies virus US3 protein is a component of primary and of mature virions. J. Virol. *78*, 1314–1323.

Heineman, T.C., Krudwig, N., and Hall, S.L. (2000). Cytoplasmic domain signal sequences that mediate transport of varicella-zoster virus gB from the endoplasmic reticulum to the Golgi. J. Virol. *74*, 9421–9430.

Hensel, G., Meyer, H., Gärtner S., Brand, G., and. Kern, H.F. (1995). Nuclear localization of the human cytomegalovirus tegument protein pp150 (UL32). J. Gen. Virol. *76*, 1591–1601.

Hobom, U., Brune, W., Messerle, M., Hahn, G., and Koszinowski, U.H. (2000). Fast screening procedures for random transposon libraries of cloned herpesvirus genomes: mutational analysis of human cytomegalovirus envelope glycoprotein genes. J. Virol. *74*, 7720–7729.

Homman-Loudiyi, M., Hultenby, K., Britt, W., and Soderberg-Naucler, C. (2003). Envelopment of human cytomegalovirus occurs by budding into Golgi-derived vacuole compartments positive for gB, Rab 3, trans-golgi network 46, and mannosidase II. J. Virol. *77*, 3191–3203.

Huber, M.T., and Compton, T. (1997). Characterization of a novel third member of the human cytomegalovirus glycoprotein H-glycoprotein L complex. J. Virol. *71*, 5391–5398.

Huber, M.T., and Compton, T. (1999). Intracellular formation and processing of the heterotrimeric gH-gL-gO (gCIII) glycoprotein envelope complex of human cytomegalovirus. J Virol. *73*, 3886–3892.

Hutchison, C. (2002). Lamins: building blocks or regulators of gene expression. Nature Rev. Mol. Cell Biol. *3*, 848–858.

Irmiere, A., and Gibson, W. (1984). Isolation and characterization of a noninfectious virion-like particle released from cells infected with human strains of cytomegalovirus. Virology *130*, 118–133.

Iwasaki, Y., Furukawa, T., Plotkin, S., and Koprowski, H. (1973). Ultrastructural study on the sequence of human cytomegalovirus infection in human diploid cells. Arch. Gesamte Virusforsch. *40*, 311–324.

Jarvis, M.A., Fish, K.N., Soderberg-Naucler, C., Streblow, D.N., Meyers, H.L., Thomas, G., and Nelson, J.A. (2002). Retrieval of human cytomegalovirus glycoprotein B from cell surface is not required for virus envelopment in astrocytoma cells. J. Virol. *76*, 5147–5155.

Jarvis, M.A., Jones, T.R., Drummond, D.D., Smith, P.P., Britt, W.J., Nelson, J.A., and Baldick, C.J. (2004). Phosphorylation of human cytomegalovirus glycoprotein B (gB) at the acidic cluster casein kinase 2 site (Ser900) is required for localization of gB to the trans-Golgi network and efficient virus replication. J. Virol. *78*, 285–293.

Johnson, D., and Ligas, M. (1988). Herpes simplex virus lacking glycoprotein D are unable to inhibit virus penetration: Quantitative evidence for virus specific cell surface receptors. J. Virol. *62*, 4605–4612.

Kari, B., Goertz, R., and Gehrz, R. (1990). Characterization of cytomegalovirus glycoproteins in a family of complexes designated gC-II with murine monoclonal antibodies. Arch. Virol. *112*, 55–65.

Severi, B., Landini, M.P., Musiani, M., and Zerbini, M. (1979). A study of the passage of human cytomegalovirus from the nucleus to the cytoplasm. Microbiologica 2, 265–273.

Shumaker, D.K., Kuczmarski, E.R., and Goldman. R.D. (2003). The nucleoskeleton: lamins and actin are major players in essential nuclear functions. Curr. Opin. Cell Biol. 15, 358–366.

Silva, M.C., Yu, Q.C., Enquist, L., and Shenk, T. (2003). Human cytomegalovirus UL99-encoded pp28 is required for the cytoplasmic envelopment of tegument-associated capsids. J. Virol. 77, 10594–10605.

Smith, S., and Blobel, G. (1993). The first membrane spanning region of the lamin B receptor is sufficient for sorting to the inner nuclear membrane. J. Cell Biol. 120, 631–637.

Soullam, B., and Worman, H.J. (1993). The amino-terminal domain of the lamin B receptor is a nuclear envelope targeting signal. J. Cell Biol. 120, 1093–1100.

Soullam, B., and Worman, H.J. (1995). Signals and structural features involved in integral membrane protein targeting to the inner nuclear membrane. J. Cell Biol. 130, 15–27.

Spaete, R.R., Gehrz, R.C., and Landini, M.P. (1994). Human cytomegalovirus structural proteins. J. Gen. Virol. 75, 3287–3308.

Stoker, M.G. (1958). Mode of intercellular transfer of herpes virus. Nature 182, 1525–1526.

Strive, T., Borst, E., Messerle, M., and Radsak, K. (2002). Proteolytic processing of human cytomegalovirus glycoprotein B is dispensable for viral growth in culture. J. Virol. 76, 1252–1264.

Strive, T., Gicklhorn, D., Wohlfahrt, M., Kolesnikowa, L., Eickmann, M., Borst, E., Messerle, M. and Radsak, K. (2004). Site directed mutagenesis of the carboxylterminus of human cytomegalovirus glycoprotein B leads to attenuation of viral growth in cell culture. Arch. Virol. 150, 585–593.

Stuurman, N., Heins, S., and Aebi, U. (1998). Nuclear lamins: their structure, assembly, and interactions. J. Struct. Biol. 122, 42–66.

Tooze, J., Hollinshead, M., Reis, B., Radsak, K., and Kern, H. (1993). Progeny vaccinia and human cytomegalovirus particles utilize early endosomal cisternae for their envelopes. Eur. J. Cell Biol. 60, 163–178.

Trus, B.L., Gibson. W., Cheng, N., and Steven, A.C. (1999). Capsid structure of simian cytomegalovirus from cryoelectron microscopy: evidence for tegument attachment sites. J Virol. 73, 2181–2192.

Weiland F, Keil, G.M., Reddehase, M.J., and Koszinowski, U.H. (1986). Studies on the morphogenesis of murine cytomegalovirus. Intervirology 26, 192–201.

Wiebusch, L., Asmar, J., Uecker, R., and Hagemeier, C. (2003). Human cytomegalovirus immediate-early protein 2 (IE2)-mediated activation of cyclin E is cell cycle-independent and forces S-phase entry in IE2-arrested cells. J. Gen. Virol. 84, 51–60.

Wiebusch, L., Uecker, R., and Hagemeier, C. (2003). Human cytomegalovirus prevents replication licensing by inhibiting MCM loading onto chromatin. EMBO Rep. 4, 42–46.

Wilson, K.L., Zastrow, M.S., and Lee, K.K. (2001). Lamins and disease: insights into nuclear infrastructure. Cell 9, 647–650.

Yamashita, Y., Shimokata, K., Mizuno, S., Daikoku, T., Tsurumi, T., and Nishiyama, Y. (1996). Calnexin acts as a molecular chaperone during the folding of glycoprotein B of human cytomegalovirus. J Virol. 70, 2237–2246.

Yamauchi Y, Shiba, C., Goshima, F., Nawa, A., Murata, T., and Nishiyama, Y. (2001). Herpes simplex virus type 2 UL34 protein requires UL31 protein for its relocation to the internal nuclear membrane in transfected cells. J. Gen. Virol. 82,1423–1428.

Ye, G.J., and Roizman, B. (2000). The essential protein encoded by the UL31 gene of herpes simplex virus 1 depends for its stability on the presence of UL34 protein. Proc. Natl. Acad. Sci. USA 97, 11002–11007.

Ye, G.J., Vaughan, K.T., Vallee, R.B., and Roizman, B. (2000). The herpes simplex virus 1 UL34 protein interacts with a cytoplasmic dynein intermediate chain and targets nuclear membrane. J. Virol. 74,1355–1363.

# Antibody-mediated Neutralization of Infectivity

# 14

*Michael Mach*

## Abstract

Neutralization of virus infectivity by antibodies represents a powerful tool of the immune system in the battle against viral infections. Targets for CMV-neutralizing antibodies are glycoproteins in the viral envelope. However, our knowledge about overall composition of the viral envelope and the importance of individual proteins with respect to virus neutralization is still rudimentary and more detailed information is only available for glycoproteins B and H. We need to extend this knowledge considerably with respect to CMV envelope composition as well as the antibody response against it. As a persistent virus CMV has also developed mechanisms to evade efficient neutralization by antibodies and antigenic variation has emerged as an important factor in CMV biology. The contribution of this variability and additional factors that influence the neutralizing antibody response need to be established. Only if we considerably increase our understanding about the delicate balance between the persistence of this virus and the presence of neutralizing antibodies we will be able to device strategies to strengthen the humoral immune system in the control of CMV infection. Results from the animal CMVs are encouraging since they provide clear evidence that antibodies are protective.

## Introduction

The clinical significance of disease arising from HCMV infections is well accepted. In the healthy host the vast majority of clinical illnesses are the result of a primary infection rather than endogenous reactivation or reinfection with antigenically different strains indicating that preexisting immunity is beneficial for the course of the infection. In patients with a defective immune response reactivation and reinfection can also lead to significant clinical problems (see Chapter 1).

The potential clinical consequences of a primary infection could most effectively be prevented by the development of a vaccine (see Chapter 26). Based on the experience with other vaccines the success of this vaccine will entirely be dependent on the induction of potent virus-neutralizing antibodies in the vaccinee (Zinkernagel, 2003). Which immune effector functions will be beneficial in combating endogenous virus reactivation or reinfection is more difficult to predict since these conditions occur in the face of a fully functioning immune system. Most likely neutralizing antiviral antibodies as well as specific cellular immunity will contribute to the control of viral replication and thus limit clinical symptoms.

In any case, providing the host with a sufficient level of potent neutralizing antibodies will play a central role in our efforts to prevent the clinically relevant consequences of the infection. These antibody levels could either be elicited following vaccination or passively transferred in form of polyclonal IgG preparations or mAb or both. In either case, it is essential to have knowledge on the antigens that result in induction of neutralizing antibodies following immunization and on antibody mediated mechanisms that result in efficient virus neutralization in vivo.

Knowledge about HCMV antigens that could be useful for immunoprophylaxis and/or therapy is far from being complete. The virus codes for > 150 proteins, and most of them would be expected to be immunogenic with respect to antibody induction during natural infection. Some of these antigens represent excellent markers for detecting HCMV-specific humoral response for diagnostic purposes but in most cases the induction of these antibodies might be without consequences for the course of the infection and hence useless for neutralization of infectious virions. Examples are antibodies against non-structural proteins which are never expressed on the surface of infected cells such as the immediate-early proteins. In this respect, HCMV represents an interesting case since overall anti-HCMV antibody titers and neutralizing antibody titers do not correlate following natural infection indicating that a significant fraction of antibodies is generated against HCMV antigens that do not contribute to virus neutralization (Leogrande et al., 1992). Antibodies with the highest impact on virus replication will be directed at viral envelope components. However, the composition of the HCMV-envelope with respect to number of individual proteins, abundance, function or relevance for induction of neutralizing antibodies is far from clear.

An additional major hurdle for the assessment of strategies for active or passive immunization is the lack of an animal system in which efficacy could be directly determined using HCMV as the pathogen. There are, however, a number of animal systems available using related cytomegaloviruses such as MCMV, RhCMV or GPCMV with comparable, yet not identical, biology of the respective viruses. These systems have provided valuable information on in vivo neutralization of infectivity.

Lacking a relevant animal model for HCMV, investigators have been forced to rely on in vitro assays and hope that the results of these assays bear some relevance to the activity of antiviral antibodies in vivo. In vitro tests usually determine reduction of input infectivity. Whether this reflects true "neutralization" of viruses by binding of antibodies to critical sites on virion surface or is mediated by other factors such as aggregation of input virus by antibodies that do not neutralize is usually not determined. What exactly determines neutralization mechanistically is still a matter of intense discussion. In the past, neutralization has been viewed as a complex process that could be mediated by a number of different mechanisms, such as aggregation, inhibition of viral entry by inhibition of viral attachment and/or fusion with the target cell as well as post entry mechanisms such as uncoating (Dimmock, 1993). It has been suggested that some antibodies may neutralize by acting via several mechanisms simultaneously or sequentially. More recently, a more simple model was put forward which argues that neutralization is a function of occupancy of binding sites on the virion and may be mostly mediated by the bulk of the antibody molecule (Burton et al., 2001; Klasse and Sattentau, 2002). Experimentally the results of neutralization tests can depend on a number of variables such as cell type or source of virus (Dimmock, 1993).

For the following discussion it should be kept in mind that in general neutralization of HCMV is assayed in vitro by using a single cell type, human fibroblasts, as the target cell and viruses which have been isolated from and passaged on fibroblasts, and which are able to infect as cell free virions. The relevance of this assay system for the in vivo situation is not clear since the virus infects a variety of different cells in vivo and cell to cell spread may be a major mode of viral transmission throughout the body.

## The envelope glycoproteins and the antibody response against them

HCMV codes for more than 50 glycoproteins (see Chapters 3 and 5). How many of these proteins will be incorporated in the viral envelope is unknown but can be expected to exceed the number found in alpha herpesviruses which contain at least eleven glycoproteins. With respect to virus neutralization we have more detailed information only on two proteins: gB and gH.

### Glycoprotein B

Glycoprotein B is perhaps the most highly conserved envelope component of all members of the herpesvirus family. Studies on deletion mutants have also demonstrated that gB is an essential protein for virus replication in fibroblasts (Hobom et al., 2000). The function of gB in the replication cycle of HCMV is not completely understood, but the protein participates in both the attachment to target cells as well as fusion of the virion with host cell membranes (see Chapter 6). In this respect, it resembles the gB molecules of alphaherpesviruses.

The serologic response to gB as a result of HCMV infection is universal. Assays using recombinant gB, expressed in mammalian cells, have detected gB-specific antibodies in sera from naturally infected persons without exception (Britt et al., 1990; Marshall et al., 1992). The immunoglobulin isotype that dominates this response is IgG1 (Urban et al., 1994). Using E. coli-derived bacterial fusion proteins, anti-gB antibodies can also be detected with high frequency, indicating that a fraction of the anti-gB specific antibodies does not depend on the native structure (most probably oligomeric) and/or modification of the protein in order to bind. Using this approach, three antigenic domains (AD) have been defined on gB: AD-1 to AD-3 (reviewed in Britt and Mach, 1996) (Figure 14.1). gB molecules from related cytomegaloviruses such as MCMV and RhCMV seem to have a similar distribution of antigenic regions within the molecule (Kropff and Mach, 1997; Xu et al., 1996;).

AD-1 was the first antibody binding site defined on HCMV gB. This "epitope" consists of approximately 80 amino acids (aa) between positions 552 and 635 of gB (Britt and Mach, 1996). Seropositivity rates for AD-1 in sera from convalescent persons approach 100% (Schoppel et al., 1997). In studies using AD-1-specific mAbs as competitor, Kniess et al. (1991) demonstrated that approx. 50% of the gB-specific response in human sera was directed against AD-1. Binding of antibodies to AD-1 requires formation of a disulfide bond between residues 573 and 610 of gB (Speckner et al., 2000). AD-2 is located at the extreme amino terminus of gB between aa 50 and 77 (Meyer et al., 1990). With seropositivity rates of approximately 50% in human HCMV-convalescent sera, the antigenicity of this domain appears to be considerably lower (Schoppel et al., 1997). The relative contribution of AD-2-specific antibodies to the overall serologic response against gB is unknown. Very

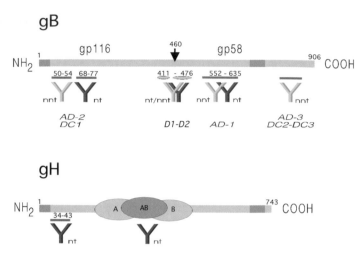

**Figure 14.1** Antibody binding sites on gB and gH. The gB and gH proteins are depicted as linear molecules. The signal sequence and the membrane anchor regions are shown as dark gray areas. Cleavage at aa 460 of gB generates the surface (gp116) and transmembrane (gp58) components. Antibody binding sites that were defined by using unmodified and denatured protein are depicted as black bars. Binding sites that depend on conformation are shown as blue ovals. The actual conformational binding sites on gH are not known. Neutralizing (nt) antibodies and non-neutralizing (nnt) antibodies are shown in green and red, respectively. The antibody binding site at the amino terminus of gH differs between AD169-like and Towne-like strains. Thus, antibodies binding to this site are strain specific.

little is known about the serologic response against AD-3 (located between aa 783 and 906) except that in a small number of samples from healthy blood donors a 100% reactivity was observed (Kniess et al., 1991). AD-3, however, is located at the intraluminal (intraviral) part of gB, which most probably does not give rise to antibodies which are capable of neutralizing infectious virus. During primary infection antibodies reacting with AD-1 or AD-2 develop with a considerable delay of 50–100 days compared to antibodies against other structural or non-structural HCMV proteins (Schoppel et al., 1997). A delay in appearance of serum neutralizing antibodies has been reported in a number of studies and seems to be a general phenomenon associated with primary HCMV infection (Alberola et al., 2000; Eggers et al., 1998; Marshall et al., 1994). This delay may offer the virus a chance to disseminate more effectively in the host. In fact, an inverse correlation between severity of symptoms and levels of neutralizing antibodies has been observed in immunocompetent persons undergoing primary infection (Alberola et al., 2000).

It is highly likely that in addition to those antigenic regions which were defined by procaryotically expressed parts of gB a number of additional conformational epitopes do exist on gB although their exact nature has not been defined. However, the existence of human mAbs that recognize exclusively non-denatured gB strongly argues for the existence of conformational epitopes on gB with antibody-forming potential during the natural infection (Masuho et al., 1987).

Antibodies specific for gB are not only induced with high frequency following infection, they also contribute significantly to the virus-neutralizing capacity in human sera. The evidence for this comes from several studies:

1    Titers of anti-gB antibodies and overall neutralizing capacity of sera correlate (Marshall et al., 1992). This is in contrast to antibody titers measured with antigen produced from HCMV-infected cell lysates (Leogrande et al., 1992). However, this correlation is strong but not perfect thus emphasizing that there exist neutralizing epitopes within other virion proteins. Similar observations have also been made in the closely related RhCMV system (Yue et al., 2003).

2    Adsorption of gB-specific antibodies from human sera shows that in most cases between 40% and 70% of the neutralizing capacity can be removed. However, in some cases adsorption had no effect on the neutralizing activity of the respective sera, again pointing to the fact that additional antigens do exist on the virus which are important for the induction of neutralizing antibodies (Britt et al., 1990; Gonczol et al., 1991; Marshall et al., 1992).

3    A number of gB-specific human mAbs with virus neutralizing capacity have been isolated (Ehrlich et al., 1988; Masuho et al., 1990; Ohlin et al., 1993).

A characteristic feature of gB is the induction of not only those antibodies which have potent neutralizing capacity but also those which do not reduce infectivity when tested in vitro (Qadri et al., 1992; Utz et al., 1989). This became apparent during studies using AD-1 and AD-2 specific murine and human mAbs as well as human polyclonal antibodies. For some antibodies binding to gB is competitive (Ohlin et al., 1993). The fine specificity of AD-1 binding mAbs (human and murine) as well as of affinity-purified human polyclonal antibodies was analyzed by using recombinant proteins containing single amino acid substitutions spanning the entire AD-1 domain. All mAbs had individual binding patterns to the mutant proteins indicating the presence of a considerable number of distinct antibody binding sites on AD-1. At least six of those could be identified by using this relatively crude assay (Schoppel et al., 1996). Neutralization capacity of antibodies could not be predicted from their binding pattern to AD-1 mutant proteins. Polyclonal human antibodies purified from different convalescent sera showed identical binding patterns to the mutant proteins suggesting that the sum of antibody specificities present in human sera is comparable between individuals. However, neutralization capacities of polyclonal human AD-1-specific antibodies do not exceed 50% (Speckner et al., 1999). Thus, although individual antibody molecules targeting AD-1 were found to completely block virus neutralization, polyclonal preparations specific for AD-1 are only able to partly prevent infection suggesting that such preparations contain different types of antibodies mediating and preventing virus neutralization. The mechanisms by which AD-1-specific antibodies block virus infectivity were also investigated. Dependent on the recognized substructure within AD-1, an influence of avidity as well as effects mediated by the Fc-portion have been noted (Schoppel et al., 1996).

AD-2 consists of at least two sites located between aa 50 and 77 of gB (Figure 14.1). Site I is common to all strains and induces neutralizing antibodies. The amino acid sequence of site II differs between strains and is recognized by strain-specific antibodies (Meyer et al., 1992). Antibodies recognizing site II, either purified from human sera or murine mAbs are non-neutralizing. Binding of neutralizing and non-neutralizing antibodies to AD-2 does not seem to be competitive (Meyer et al., 1992). In addition to AD-1 and AD-2, a number of conformation-dependent neutralizing antibodies have been mapped to the amino-terminal half as well as to the mid-region of gB using murine mAbs (Qadri et al., 1992).

gB probably has multiple roles in attachment to and penetration into cells (see Chapter 6). Hence it is not surprising that gB-specific neutralizing antibodies can interfere with multiple steps of infection. The available data suggest that the neutralizing activities of anti-gB mAbs involve different stages of the virus adsorption process to cells. The reports on potential mechanisms by which anti-gB antibodies mediate neutralization have included attachment (Ohizumi et al., 1992), penetration of virions into cells (Navarro et al., 1993) and fusion (Gicklhorn et al., 2003). Efficient inhibition of virion-mediated fusion was consistently observed for an AD-2-specific human mAb as determined by a reporter gene activation assay based on permissive astrocytoma cells. In contrast, antibodies directed against AD-1 reduced fusion only by 20–60% (Gicklhorn et al., 2003). The effects mediated by anti-gB antibodies are most probably dependent on the intact IgG molecule. Monomeric single chain Fv-fragments (scFv) reactive with AD-2 were incapable of neutralizing virus (Lantto et al., 2002). The dissociation rate of scFv compared to whole IgG from the protein seemed not to be a factor. Instead, it was suggested that the scFv is too small to exert a blocking effect, while the dimeric form may mediate neutralization through steric hindrance, an effect that would be in agreement with the occupancy model of neutralization. An intact IgG molecule is also required for efficient neutralization via AD-1-specific antibodies since scFv fragments exert neutralization only after dimerization (S. Knör, M. Mach, M. Ohlin; unpublished observation)

## The gH/gL/gO complex

Homologs of gH and gL are found in all herpesviruses studied to date. In some herpesviruses an additional protein is complexed with gH/gL forming a tripartite complex. For HCMV, this protein has been termed gO. For most herpesviruses gH and gL are part of the fusion machinery and HCMV is no exception (see Chapter 6). Whereas gH and gL seem to be essential for viral replication, gO negative recombinant viruses can replicate in fibroblasts in vitro albeit at a reduced rate (Hobom et al., 2000).

Following natural infection with HCMV, gH-specific antibodies have been detected with frequencies between 95% and 100% (Boppana et al., 1995; Urban et al., 1996). In some sera, anti-gH antibodies constitute the majority of the neutralizing activity (Urban et al., 1996). The protein regions relevant to induction of antibodies are not well defined. A linear antibody binding site has been localized to the amino terminus of the protein between residues 34–43 (Urban et al., 1992). Antibodies recognizing this site are neutralizing. In addition, at least one conformational epitope exists on gH which induces neutralizing antibodies during natural infection as is indicated by the isolation of a human mAb reactive with native gH but not with denatured protein (Drobyski et al., 1991). Using a set of seven neutralizing murine mAbs, a topological map based on binding to gH has been constructed (Simpson et al., 1993). Two unique binding sites that are bridged by a third site have been defined (Simpson et al., 1993). Three antigenically variable and three conserved epitopes were identified within the three antigenic regions. Whether antibodies induced during natural infection recognize these sites on gH has not been defined. The proteins gL and gO have not been investigated with respect to antibody reactivity in human sera or neutralization via mAbs.

In a "fusion-from-without" system, antibodies reactive with gH have been shown to inhibit fusion as well as cell to cell spread in monolayers of both embryonic fibroblasts and continuous astrocytoma cells (Milne et al., 1998).

## The gM/gN complex

The gM/gN complex is the third major constituent of the viral envelope (Mach et al., 2000). The presence of both proteins is required for proper modification and transport of the constituents to the virus assembly compartment within infected cells. Using 293T-cells transiently expressing the gM/gN complex, seropositivity rates of 62% have been reported (Mach et al., 2000). Antibody reactivity against the individual components was considerably lower indicating that some antigenic determinants are formed only after complex formation of gM and gN. The fact that a murine mAb neutralizes infectious virus very efficiently in the presence of complement suggests that antibodies reactive with gM/gN complexes and present in human sera could also have virus-neutralizing activity (Britt and Auger, 1985). Indeed, anti-gM/gN antibodies affinity purified from pooled HCMV immune serum have been shown to neutralize HCMV at concentrations of approximately 0.1 µg/ml, levels similar to the neutralizing activity of anti-gH antibodies found in human immune serum (W. Britt, personal communication).

## Additional HCMV envelope glycoproteins

A number of additional envelope glycoproteins have been identified but the antibody response against them either following natural infection or after immunization with the respective protein has not been studied (Table 14.1).

## Escape mechanisms from the neutralizing antibody response

CMVs have developed a plethora of mechanisms to avoid elimination from the host by the immune response. For cellular immunity this has been well documented. It would be more than naive to assume that the virus did not also develop mechanisms to evade the humoral

**Table 14.1.** Envelope glycoproteins of HCMV and induction of neutralizing antibodies

| Glycoprotein (ORF) | Neutralizing antibodies after natural infection | Neutralizing antibodies after immunization |
|---|---|---|
| gB (UL55) | Yes | Yes |
| gH (UL75) | Yes | Yes |
| gL (UL115) | ? | ? |
| gO (UL74) | ? | ? |
| gM (UL100) | No | No |
| gN (UL73) | ? | Yes |
| gp48 (UL4)[a] | ? | ? |
| gpTRL10 (TRL10)[a] | No | No |
| gpTRL11 (TRL11)[a] | ? | ? |
| gpUL118-119[a] | ? | ? |
| gpUL132 (UL132)[a] | No | No |

[a]For review see Britt and Boppana (2004).

immune response, most importantly the action of neutralizing antibodies, and the accumulated data strongly suggest that this is indeed the case.

Evasion of cellular and humoral immune responses follow fundamentally different mechanisms. The evasion of the cellular immune response is mediated by viral proteins which interfere with the function of proteins essential for this process e.g. MHC class I and MHC class II molecules. This mechanism is species-specific and reflects the adaptation of a CMV species to its host species. In contrast, the evasion of an efficient neutralizing antibody response can be mediated by minor alterations in different virally encoded envelope glycoproteins with variable impact on binding or induction of (neutralizing) antibodies. The formation of antibodies, however, is a process that is specific for the individual. It is conceivable then that mutations in individual proteins might lead to escape from a neutralizing humoral immune response in one case but not in others.

## Induction of neutralizing and competing non-neutralizing antibodies

As mentioned above, during natural infection AD-1 of gB gives rise to antibodies with widely differing neutralization capacity and some of these antibodies compete for binding. As a result, virus infectivity cannot be completely neutralized by polyclonal anti-AD-1 antibodies purified from human sera, even at high IgG concentrations (Speckner et al., 2000). Thus, the induction of competing non-neutralizing antibodies may represent a mechanism to evade a protective antibody response in vivo, especially if non-neutralizing antibodies are produced in excess (Figure 14.2B).

## Antigenic variation

Antigenic variability is a frequently used mechanism for evasion of a potent neutralizing antibody response. For herpesviruses such a mechanism has long been ignored because of the conservation of the DNA sequence between strains. And indeed, when first sequence comparisons between HCMV isolates were available it became clear that the major antigenic glycoproteins were highly conserved between strains.

In sharp contrast to the conservation of the major structural glycoproteins have been the results of analysis of cross-reactivity of sera raised against different HCMV isolates. In work published in 1978, Waner and Weller already noted antigenic diversity among HCMV isolates (Waner and Weller, 1978). They used sera produced in rabbits against lysates from HCMV-infected cells and noted significant differences in the complement-dependent neutralization capacity. Klein et al. (1999) confirmed strain-specific neutralization with human sera. They used nine recent HCMV-isolates and their corresponding sera in cross-neutralization assays in the presence and absence of complement. They observed differences, independent of variance in the absolute neutralization capacity, between the sera ranging from 6-fold to more than 60-fold when tested against the individual HCMV isolates. For one of the isolates complete resistance to neutralization by two human sera was seen. Thus, antigenic variation can render an existing pool of antibodies ineffective for neutralization of heterologous virus isolates (Figure 14.2D).

The antigens that are responsible for these differences have not been identified. However, if the reduced capacity of human polyclonal sera to neutralize heterologous HCMV isolates in vitro represents a mechanism to escape efficient neutralization in vivo, gB and gH would be prime candidates since they contribute significantly to the overall neutralization

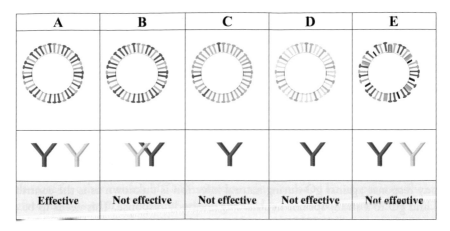

**Figure 14.2** Factors potentially influencing neutralization of CMV by antibodies. (A) Effective neutralization of virus through binding of antibodies to critical sites on the viral envelope. Antibody and target antigen are shown in identical color. (B) Efficiency of neutralization is reduced through competition of antibodies with different neutralization capacity for the same binding site on the surface of virions. (C) Reduced incorporation of proteins in the viral envelope that are targets for neutralizing antibodies results in inefficient neutralization. (D) Antigenic variation of envelope glycoproteins will result in inefficient neutralization. (E) The presence of Fc-binding (black boxes) and complement-regulating proteins (dark green squares) in the envelope of CMV virions reduces the neutralizing activity of antibodies.

capacity in human sera (see above). Whether this is indeed the case has not been tested. An indirect piece of evidence that gH and gB could be involved in the induction of strain-specific antibodies comes from studies using mAbs. In general, human or murine mAbs reacting with gB or gH show virus strain-specific neutralization capacity (e.g., Baboonian et al., 1989; Rasmussen et al., 1985; Simpson et al., 1993). The differences can be dramatic resulting in lack of binding to some HCMV strains. Even human mAbs recognizing AD-1 show dramatically different binding and neutralization capacity despite the fact that AD-1 is almost perfectly conserved between HCMV strains (Ohlin et al., 1993). The only exception from strain-specific neutralization seems to be the human mAb C23 that reacts with AD-2 of gB (Meyer et al., 1990). This antibody neutralized all tested HCMV strains with equal capacity (M. Mach and M. Klein, unpublished results). During natural infection, however, AD-2-specific antibodies are induced with low frequency (Schoppel et al., 1997). Altogether, the available data justify the conclusion that the minimal changes in the primary amino acid sequence of gB and gH between different HCMV strains have a significant impact on the antigenic structure of the respective molecule and could thus contribute to a strain-specific antibody response.

If the relatively minor differences between HCMV strains found in gB and gH have already an impact on strain-specific neutralization, what could be expected from proteins such as gN and gO? Both of these proteins are derived from hypervariable genes in HCMV. Protein sequences from gN can be grouped into four major subtypes with divergence between the subtypes of ~15% to 24% (Pignatelli et al., 2003). Interestingly, the sequence variations are not randomly distributed over the entire molecule but are rather clustered in the N-terminal region of the protein, where differences between strains can reach 40%. This

complex. The lack of correlation between genotypes and seroptypes has also been noted for other viral systems, e.g. HIV (Moore et al., 1996). Addressing this question with a virus as complex as CMV will be extremely difficult. However, the generation of recombinant viruses with defined alterations in glycoprotein genes on a homogeneous genetic background might allow us to tackle this problem in the future.

## Neutralization of infectivity in vivo

If neutralizing antibodies play a role in the course of an infection it should be possible to demonstrate this in vivo. The classical approach to this question has been to passively transfer antibodies from immune donors or mAbs to naive recipients and determine the protection provided. In this respect it is important to distinguish phenomenologically between two types of antiviral antibody activity: the ability of an antibody to protect against infection when it is present before the infection and the ability of an antibody to interfere with an established infection, reactivation or reinfection.

There is good evidence that passive transfer of polyclonal antibodies prior to infection can protect against the consequences of this infection. This has convincingly been shown by using MCMV. Transfer of polyclonal antibodies either directed against whole virus or single envelope proteins expressed from a recombinant vaccinia virus were able to protect the animals from a lethal challenge dose (Rapp et al., 1993; Shanley et al., 1981). Monoclonal antibodies have also been tested by passive immunization. Specifically, Farrell and Shellam (1991) compared eight mAbs specific for different structural proteins of MCMV for their capacity to protect mice against MCMV infection. Administration of mAb resulted in a significantly lower mortality rate upon challenge with a lethal virus dose. Yet, no direct correlation was observed between the level of protection against lethal CMV disease and the neutralization titers in vitro. mAbs which were incapable of neutralizing MCMV in vitro afforded similar protection in vivo compared to the potently neutralizing antibodies. In contrast, the reduction in MCMV titers in the livers of BALB/c and C57BL/10 mice by the mAbs closely correlated with their neutralization titer in vitro. Unlike in the liver, almost all of the mAbs tested were ineffective in reducing MCMV replication in the spleen of BALB/c mice, an organ in which MCMV replicates to high titer. Thus, mAbs specific for single proteins of MCMV are protective in vivo. However, the extent to which the mAb protected could not be predicted from the neutralization titer in vitro or by their effect on splenic MCMV replication in vivo. Moreover, mAbs that required complement in vitro were as effective in vivo in C5-deficient mice as in normal mice indicating that the mechanism(s) of neutralization of MCMV in vitro and in vivo are different.

These data highlight two important points:

1   In vitro neutralizing capacity and in vivo protection do not correlate.
2   There exists an organ-specific effect of antibody.

The guinea-pig model of CMV was used to evaluate the effects of passively transferred antibody given to pregnant dams on pup survival. When antibody was administered before CMV challenge, pup survival significantly increased even if the challenge inoculum was high (Bratcher et al., 1995). Transfer of antibody started after viral challenge increased pup survival only at the lowest challenge dose. Infection of the pups was not prevented by either treatment.

Effect of antibody during CMV reactivation was investigated in mouse models of latency. Jonjic et al. (1994) used latently infected B-cell deficient mice to show that after reactivation of virus following immunosuppression the viral titers in salivary glands, lungs and spleens were 100- to 1000-fold higher than in B-cell-positive heterozygous or C57BL/6 wild-type controls. Adoptive transfer of immune serum to the B-cell deficient mice reduced titers to levels seen in the controls, thus demonstrating that the effect was indeed based on antibody. In a latency model comparing the risk of reactivation in BALB/c mice that had experienced primary infection as either neonates or adults, Reddehase et al. (1994) observed a higher incidence of virus recurrence in the neonatally infected mice due to a higher latent genome load in spite of a 4-fold higher titer of neutralizing antibody in the serum, thus demonstrating that neutralizing antibodies do not prevent virus reactivation and recurrence. However, recurrent infection remained confined to the organs in which reactivation had occurred and serum transfer comparing serum from wild-type or B-cell-deficient donor mice identified antiviral antibody as the factor that prevented virus spread to unaffected organs. A role for antibody was further supported by the finding that latently infected mice in which cellular immunity was abolished could not be superinfected even if very high doses of challenge virus were given intravenously. Collectively, these two reports concurred in the conclusion that serum antibodies cannot prevent virus reactivation within the organs but efficiently limit subsequent dissemination of recurrent virus.

A common effect observed in all studies was the reduced viral titer in most organs after administration of neutralizing antibodies. Since the extent of viral replication and dissemination during primary infection determines the risk for reactivation and disease, neutralizing antibodies will have a beneficial effect on the clinical outcome of the infection despite being unable to prevent infection (Reddehase et al., 1994; Emery et al., 2000).

A somewhat more indirect evidence for the efficacy of antibodies to combat a CMV infection comes from vaccination studies using recombinant antigens capable of inducing neutralizing antibodies in different animal systems. In general, protection was observed (see Chapter 26). However, T-cell-mediated responses might have contributed to the observed effects since most of the envelope glycoproteins are also targets for the cellular immune response.

Results on protection by passive antibody treatment in humans are less clear. Polyclonal intravenous immunoglobulin preparations have been extensively used in patients at risk of CMV disease. However, even after almost 20 years of using this treatment either prophylactically or therapeutically, uncertainty about benefits of immunoprophylaxis for the prevention of CMV infection and disease is evident in the literature (reviewed in Sokos et al., 2002). Monoclonal antibodies have also been used for CMV therapy/prophylaxis in different patient populations. Boeckh et al. (2001) reported on the use of the gH-specific human mAb MSL-109 in trials of preventing HCMV infection after allogeneic hematopoietic stem cell transplantation. Recipients with positive donor and/or recipient serology for HCMV prior to transplantation were treated with the antibody at 60 mg/kg. No statistically significant difference in CMV antigenemia or viremia and overall survival rates were seen. Adjuvant therapy with MSL-109 was also ineffective in clearing CMV DNA and CMV antigen from plasma in AIDS patients (Jabs et al., 2002).

The question arises why a protective effect of antibody is so difficult to demonstrate in humans while it is so easily observed in animal models. A number of explanations can be postulated:

1   HCMV infection and the immune response against it differ from the animal systems such that neutralizing antibodies are ineffective. However, it could be argued that the role of the protective antibody responses in the biology of HCMV infection should not differ fundamentally from the role of antibody in CMV infections in other species such as the mouse or the guinea-pig.

2   In general, in the animal studies the antigen that was used for the production of neutralizing antibodies came from the same CMV strain as the challenge virus. In real life, however, this might be the exception rather than the rule, even for the animal CMVs (Booth et al., 1993). In this case, antigenically different viruses could be involved and protection might therefore be incomplete. In addition, in the animal systems the challenge inoculum usually consists of high doses of cell free virus administered by the intraperitoneal or intravenous route. Whether this route of infection is comparable to infection in humans can be questioned. Moreover, infection with cell-associated virus might be a common situation found in transplant recipients.

3   A major problem in the interpretation of the human studies has been the use of different unstandardized intravenous immunoglobulin preparations, all of which could be expected to vary in their titer of HCMV-specific neutralizing antibodies (Schmitz and Essuman, 1986). In addition, the titer of neutralizing antibodies in these preparations is usually not significantly increased as compared to individual human sera. As a consequence, only low titers of neutralizing antibodies (50% neutralization titer between 1:14 and 1:43) were recovered after passive immunization of bone marrow transplant recipients with four different intravenous IgG products (Filipovich et al., 1992). During an active infection, however, even in immunocompromised patients significantly higher neutralization titers (50% neutralization titer between 1:500 and 1:2000) can be generated by the host immune system. These titers were found to correlate with absence of infectious virus in plasma and protection from HCMV disease in bone marrow transplant recipients (Schoppel et al., 1998). Induction of high titers of neutralizing antibodies has also been found to be associated with protection from secondary infections among women of childbearing age (Adler et al., 1995). In animal CMV systems, on the other hand, sera with high titer of neutralizing antibodies are usually generated resulting in more effective protection after passive immunization. The correlation between levels of serum antibody neutralizing activity and in vivo protection has been shown for many viral systems (reviewed in Parren and Burton, 2001). Therefore, it would seem logical to develop immunoglobulin preparations for human use with significantly higher titers of neutralizing antibodies for future trials. This could be achieved by pretesting serum donors either directly for virus neutralizing antibodies or by using surrogate tests such as ELISAs specific for HCMV glycoprotein antibodies (Kropff et al., 1993).

4    Therapy using a single mAb is necessarily limited due to the virus strain specificity of recognition as discussed above.

## Conclusion

The identification of viral glycoproteins and the characterization of antigenic structures capable of binding neutralizing antibodies had induced considerable interest in strategies for preventing or treating HCMV infection and/or disease in the eighties and nineties of the last century. However, attempts to correlate neutralizing antibody activity with the course of infection in humans or to induce protective immunity by vaccination with recombinant gB were largely disappointing. As a result, many researchers came to the conclusion that neutralizing antibodies have only limited benefit for attenuating the course of the infection, and therefore activities in the field of humoral immunity have considerably dropped. On the other hand, there is no real reason to believe that antibodies are ineffective in combating HCMV infection. Results from the animal systems clearly show that antibodies do what they are supposed to do: they neutralize virus, limit dissemination, and ultimately prevent disease.

Various reasons are conceivable why this could not be demonstrated unequivocally in the case of HCMV infections:

1    The antigenic variability of HCMV has been underestimated. There are more and more data showing that HCMV isolates differ significantly from each other. How many different subtypes exist and what consequences this has for the neutralizing antibody response is basically unknown. We should analyze in animal models to what extent the neutralizing antibody response provides protection against antigenically different virus isolates. At least for MCMV genetic and antigenic differences as well as coinfections with multiple strains have been demonstrated (Booth et al., 1993). So the tools are there.

2    We have so far used fibroblasts as the only cell source for in vitro assays to evaluate antiviral activity of antibodies. Whether this assay provides a correlate to in vivo protection can be questioned. Data from the animal systems tell us that this is most probably not the case. We should try to establish assays which more closely correlate with in vivo protection. This may not be easy but using non-fibroblast based cell systems for isolation of HCMV strains from patients as well as for evaluation of virus-neutralizing capacity are conceivable as an alternative or as a complement to the existing assay. In addition, antibodies could be explored for their activity to prevent cell to cell transmission between different cell types, e.g. from monocytes to endothelial cells.

3    Studies in which vaccination with recombinant protein was used to induce a protective antibody response in humans may have suffered from using a sub-optimal antigen. We know from studies on HIV that a vaccine antigen as close to the native structure as possible provides the best antigen for induction of potently neutralizing antibodies (see Chapter 26).

In summary, virus neutralization by antibodies should still be considered as a major mechanism to limit the consequences of an HCMV infection, and efforts should be made to improve the antibody response in situations where active infection could cause clinical problems. This will certainly be to the benefit of those who are at risk of developing CMV disease.

## Acknowledgments

I would like to thank W. Britt for helpful discussions and critical reading of the manuscript. Work from the author's laboratory was supported by the Deutsche Forschungsgemeinschaft.

## References

Adler, S.P., Starr, S.E., Plotkin, S.A., Hempfling, S.H., Buis, J., Manning, M.L., and Best, A.M. (1995). Immunity induced by primary human cytomegalovirus infection protects against secondary infection among women of childbearing age. J. Infect. Dis. *171*, 26–32.

Alberola, J., Tamarit, A., Igual, R., and Navarro, D. (2000). Early neutralizing and glycoprotein B (gB)-specific antibody responses to human cytomegalovirus (HCMV) in immunocompetent individuals with distinct clinical presentations of primary HCMV infection. J. Clin. Virol. *16*, 113–122.

Atalay, R., Zimmermann, A., Wagner, M., Borst, E., Benz, C., Messerle, M., and Hengel, H. (2002). Identification and expression of human cytomegalovirus transcription units coding for two distinct Fcgamma receptor homologs. J. Virol. *76*, 8596–8608.

Baboonian, C., Blake, K., Booth, J.C., and Wiblin, C.N. (1989). Complement-independent neutralising monoclonal antibody with differential reactivity for strains of human cytomegalovirus. J. Med. Virol. *29*, 139–145.

Baldanti, F., Sarasini, A., Furione, M., Gatti, M., Comolli, G., Revello, M.G., and Gerna, G. (1998). Coinfection of the immunocompromised but not the immunocompetent host by multiple human cytomegalovirus strains. Arch. Virol. *143*, 1701–1709.

Boeckh, M., Bowden, R.A., Storer, B., Chao, N.J., Spielberger, R., Tierney, D.K., Gallez-Hawkins, G., Cunningham, T., Blume, K.G., Levitt, D., and Zaia, J.A. (2001). Randomized, placebo-controlled, double-blind study of a cytomegalovirus-specific monoclonal antibody (MSL-109) for prevention of cytomegalovirus infection after allogeneic hematopoietic stem cell transplantation. Biol. Blood Marrow Transplant. *7*, 343–351.

Booth, T.W., Scalzo, A.A., Carrello, C., Lyons, P.A., Farrell, H.E., Singleton, G.R., and Shellam, G.R. (1993). Molecular and biological characterization of new strains of murine cytomegalovirus isolated from wild mice. Arch. Virol. *132*, 209–220.

Boppana, S.B., Polis, M.A., Kramer, A.A., Britt, W.J., and Koenig, S. (1995). Virus-specific antibody responses to human cytomegalovirus (HCMV) in human immunodeficiency virus type 1-infected persons with HCMV retinitis. J. Infect. Dis. *171*, 182–185.

Boppana, S.B., Rivera, L.B., Fowler, K.B., Mach, M., and Britt, W.J. (2001). Intrauterine transmission of cytomegalovirus to infants of women with preconceptional immunity. N. Engl. J. Med. *344*, 1366–1371.

Borza, C.M. and Hutt-Fletcher, L.M. (2002). Alternate replication in B-cells and epithelial cells switches tropism of Epstein–Barr virus. Nat. Med. *8*, 594–599.

Bratcher, D.F., Bourne, N., Bravo, F.J., Schleiss, M.R., Slaoui, M., Myers, M.G., and Bernstein, D.I. (1995). Effect of passive antibody on congenital cytomegalovirus infection in guinea-pigs. J. Infect. Dis. *172*, 944–950.

Britt, W.J. and Auger, D. (1985). Identification of a 65000 Dalton virion envelope protein of human cytomegalovirus. Virus Res. *4*, 31–36.

Britt, W.J., and Boppana, S. (2004). Human cytomegalovirus virion proteins. Hum. Immunol. *65*, 395–402.

Britt, W.J. and Mach, M. (1996). Human cytomegalovirus glycoproteins. Intervirology *39*, 401–412.

Britt, W.J., Vugler, L., Butfiloski, E.J., and Stephens, E.B. (1990). Cell surface expression of human cytomegalovirus (HCMV) gp55–116 (gB): use of HCMV-recombinant vaccinia virus-infected cells in analysis of the human neutralizing antibody response. J. Virol. *64*, 1079–1085.

Burton, D.R., Saphire, E.O., and Parren, P.W. (2001). A model for neutralization of viruses based on antibody coating of the virion surface. Curr. Top. Microbiol. Immunol. *260*, 109–143.

Chou, S.W. (1989). Neutralizing antibody responses to reinfecting strains of cytomegalovirus in transplant recipients. J. Infect. Dis. *160*, 16–21.

Crnkovic-Mertens, I., Messerle, M., Milotic, I., Szepan, U., Kucic, N., Krmpotic, A., Jonjic, S., and Koszinowski, U.H. (1998). Virus attenuation after deletion of the cytomegalovirus Fc receptor gene is not due to antibody control. J. Virol. *72*, 1377–1382.

Dimmock, N.J. (1993). Neutralization of animal viruses. Curr. Top. Microbiol. Immunol. *183*, 1–149.

Drobyski, W.R., Gottlieb, M., Carrigan, D., Ostberg, L., Grebenau, M., Schran, H., Magid, P., Ehrlich, P., Nadler, P.I., and Ash, R.C. (1991). Phase I study of safety and pharmacokinetics of a human anticytomegalovirus monoclonal antibody in allogeneic bone marrow transplant recipients. Transplantation *51*, 1190–1196.

Eggers, M., Metzger, C., and Enders, G. (1998). Differentiation between acute primary and recurrent human cytomegalovirus infection in pregnancy, using a microneutralization assay. J. Med. Virol. *56*, 351–358.

Ehrlich, P.H., Moustafa, Z.A., Justice, J.C., Harfeldt, K.E., and Ostberg, L. (1988). Further characterization of the fate of human monoclonal antibodies in rhesus monkeys. Hybridoma 7, 385–395.

Emery, V.C., Sabin, C.A., Cope, A.V., Gor, D., Hassan-Walker, A.F., and Griffiths, P.D. (2000). Application of viral-load kinetics to identify patients who develop cytomegalovirus disease after transplantation. Lancet 355, 2032–2036.

Farrell, H.E. and Shellam, G.R. (1991). Protection against murine cytomegalovirus infection by passive transfer of neutralizing and non-neutralizing monoclonal antibodies. J. Gen. Virol. 72, 149–156.

Filipovich, A.H., Peltier, M.H., Bechtel, M.K., Dirksen, C.L., Strauss, S.A., and Englund, J.A. (1992). Circulating cytomegalovirus (CMV) neutralizing activity in bone marrow transplant recipients: comparison of passive immunity in a randomized study of four intravenous IgG products administered to CMV-seronegative patients. Blood *80*, 2656–2660.

Gicklhorn, D., Eickmann, M., Meyer, G., Ohlin, M., and Radsak, K. (2003). Differential effects of glycoprotein B epitope-specific antibodies on human cytomegalovirus-induced cell–cell fusion. J. Gen. Virol. 84, 1859–1862.

Gonczol, E., deTaisne, C., Hirka, G., Berencsi, K., Lin, W.C., Paoletti, E., and Plotkin, S. (1991). High expression of human cytomegalovirus (HCMV)-gB protein in cells infected with a vaccinia-gB recombinant: the importance of the gB protein in HCMV immunity. Vaccine 9, 631–637.

Grundy, J.E., Lui, S.F., Super, M., Berry, N.J., Sweny, P., Fernando, O.N., Moorhead, J., and Griffiths, P.D. (1988). Symptomatic cytomegalovirus infection in seropositive kidney recipients: reinfection with donor virus rather than reactivation of recipient virus. Lancet 2, 132–135.

Hobom, U., Brune, W., Messerle, M., Hahn, G., and Koszinowski, U. (2000). Fast screening procedures for random transposon libraries of cloned herpesvirus genomes: Mutational analysis of human cytomegalovirus envelope glycoprotein genes. J. Virol. 74, 7720–7729.

Jabs, D.A., Gilpin, A.M., Min, Y.I., Erice, A., Kempen, J.H., and Quinn, T.C. (2002). HIV and cytomegalovirus viral load and clinical outcomes in AIDS and cytomegalovirus retinitis patients: Monoclonal Antibody Cytomegalovirus Retinitis Trial. AIDS 16, 877–887.

Jonjic, S., Pavic, I., Polic, B., Crnkovic, I., Lucin, P., and Koszinowski, U.H. (1994). Antibodies are not essential for the resolution of primary cytomegalovirus infection but limit dissemination of recurrent virus. J. Exp. Med. *179*, 1713–1717.

Klasse, P.J. and Sattentau, Q.J. (2002). Occupancy and mechanism in antibody-mediated neutralization of animal viruses. J. Gen. Virol. *83*, 2091–2108.

Klein, M., Schoppel, K., Amvrossiadis, N., and Mach, M. (1999). Strain-specific neutralization of human cytomegalovirus isolates by human sera. J. Virol. 73, 878–886.

Kniess, N., Mach, M., Fay, J., and Britt, W.J. (1991). Distribution of linear antigenic sites on glycoprotein gp55 of human cytomegalovirus. J. Virol. *65*, 138–146.

Kropff, B., Landini, M.P., and Mach, M. (1993). An ELISA using recombinant proteins for the detection of neutralizing antibodies against human cytomegalovirus. J. Med. Virol. 39, 187–195.

Kropff, B. and Mach, M. (1997). Identification of the gene coding for rhesus cytomegalovirus glycoprotein B and immunological analysis of the protein. J. Gen. Virol. 78, 1999–2007.

Lantto, J., Fletcher, J.M., and Ohlin, M. (2002). A divalent antibody format is required for neutralization of human cytomegalovirus via antigenic domain 2 on glycoprotein B. J. Gen. Virol. 83, 2001–2005.

Leogrande, G., Merchionne, F., Lazzarotto, T., and Landini, M.P. (1992). Large-scale testing of human serum to determine cytomegalovirus neutralising antibody. J. Infect. 24, 289–299.

# Innate Immunity to Cytomegaloviruses

# 15

*Stipan Jonjic, Ivan Bubic, and Astrid Krmpotic*

## Abstract

Cells of the innate immune system, including macrophages, DCs and NK cells play an important role in the control of CMV infection during the time that precedes the induction of a specific immune response. NK cells are considered the most important effector cells in early CMV surveillance. An evolutionary struggle between NK cells and CMV can be inferred from the existence of a broad range of viral mechanisms designed to compromise NK cell function. Infection of mice with MCMV is the most developed model for studying the role of NK cells during CMV infection. Despite a tremendous increase in knowledge about cellular receptors and their ligands involved in regulation of NK cell activity during CMV infection we are only beginning to understand the complexity of these interactions and their significance during primary as well as latent CMV infection. Recent studies emphasize the importance of the NKG2D receptor and its ligands in the regulation of the innate and adaptive immune response. The data highlight the importance of cross-talk between DCs and NK cells in instructing the innate and specific immune response against CMV.

## Introduction

Innate immunity is devoted to sensing and controlling microbial replication during the time that precedes the induction of the adaptive immune response (Figure 15.1). Unlike T and B lymphocytes, which undergo somatic recombination and clonal specification to generate antigen-specific receptors, the cells of the innate immune system utilize receptors that recognize structurally conserved molecular patterns on a broad spectrum of microorganisms and infected cells. Major receptors by which cells of the innate immune response sense the invasion of microorganisms are Toll-like receptors (TLRs) that recognize pathogen-associated molecular patterns (PAMPs) as well as endogenous ligands such as heat shock proteins (Hsp) (Akira and Takeda, 2004). In addition to their role in the initiation and regulation of the innate and adaptive immune response, macrophages are characterized by their remarkable capacity of phagocytosis that makes them important players in the clearance and inactivation of most microbial pathogens (Burke and Lewis, 2002). DCs play a predominant role not only in induction of adaptive immunity but also in activation of NK cells that exert their functions against infected cells (Reis e Sousa, 2004).

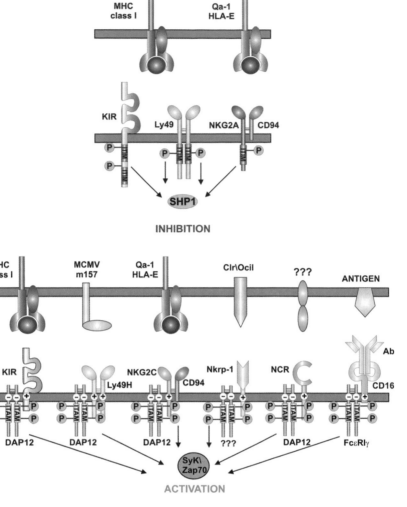

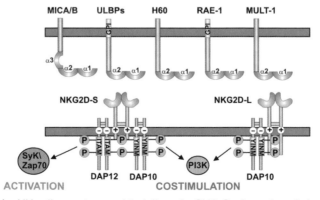

**Figure 15.2** Major NK cell receptors and their ligands. SH2, Src homology 2 domain; SHP1, Src homology phosphatase; Syk, spleen tyrosine kinase; ZAP70, 70 kDa zeta-associated protein; PI3K, phosphoinositide-3 kinase; Clr, C-type lectin-related protein; Ocil, osteoclast inhibitory lectin.

either inhibitory (CD94/NKG2A) or activating function (CD94/NKG2C and CD94/NKG2E) to the heterodimeric receptor. In contrast to KIRs and Ly49 receptors, which directly monitor expression of MHC class Ia molecules, the CD94/NKG2 heterodimers sense the expression of these molecules indirectly. Specifically, they recognize the conserved nine-amino acid peptide sequence contained in the signal peptide domain of MHC class Ia molecules that is presented by non-classical MHC class Ib molecules (HLA-E in humans and Qa-1 in mice).

The activating NK cell receptor NKG2D is classified as a member of the NKG2 family, although it has a different structure and does not pair with CD94 (Raulet, 2003b). Mouse NKG2D exists in two alternatively spliced isoforms (Carayannopoulos and Yokoyama, 2004; Raulet, 2003b). The shorter isotype, NKG2D-S binds not only to DAP10, but also to DAP12, whereas NKG2D-L binds exclusively to DAP10 (Figure 15.2). In contrast to its role in NK cells, NKG2D operates in T-lymphocytes only as a costimulatory receptor. NKG2D-mediated target cell killing is relatively resistant to inhibition through engagement of classical MHC class I molecules by inhibitory NK cell receptors (Cerwenka et al., 2001). NKG2D receptor recognizes a variety of cell-surface ligands that are distantly related to MHC class I molecules and that are poorly expressed on healthy cells but up-regulated on stressed, infected or transformed cells (Cerwenka and Lanier, 2003). Known NKG2D ligands on human cells are the MHC class I chain-related molecules (MICA and MICB) (Bauer et al., 1999; Steinle et al., 2001) and the UL-16 binding proteins (ULBP-1, ULBP-2, ULBP-3 and ULBP-4) (Chalupny et al., 2003; Cosman et al., 2001), whereas mouse NKG2D ligands are H60 (Diefenbach et al., 2000; Malarkannan et al., 1998), retinoic acid early inducible genes 1 (RAE-1$\alpha$, $\beta$, $\gamma$, $\delta$ and $\varepsilon$ isoforms) (Cerwenka et al., 2000) and murine UL-16 binding protein-like transcript 1 (MULT-1) (Carayannopoulos et al., 2002; Diefenbach et al., 2003). The MICA and MICB proteins consist of $\alpha_1$, $\alpha_2$ and $\alpha_3$ domains with MHC class I folds and, unlike classical MHC class I molecules, do not bind $\beta_2$-microglobulin and peptide (Cerwenka and Lanier, 2003). The ULBPs, with the exception of ULBP-4, are glycosylphosphatidylinositol (GPI)-anchored proteins and possess only $\alpha_1$ and $\alpha_2$ domains (Cosman et al., 2001). Like ULBPs, mouse NKG2D ligands have only MHC class I-like $\alpha_1$ and $\alpha_2$ domains (Raulet, 2003b). H60 is a type I transmembrane protein that was originally described as a minor histocompatibility antigen recognized by T-cells from C57BL/6 mice in reaction to BALB.B splenocytes (Malarkannan et al., 1998). RAE-1 is a GPI-linked membrane protein strongly induced on embryonal carcinoma F9 mouse cells after treatment with retinoic acid and has a role during embryonic development (Zou et al., 1996). MULT-1 is a type I transmembrane protein with a large cytoplasmic domain. In contrast to H60 and RAE-1, its mRNA is expressed in a wide variety of tissues (Carayannopoulos et al., 2002; Diefenbach et al., 2003).

The NK cell receptors listed above do not account for all NK cell specificities. The additional lectin-like receptors encoded within the NK gene complex include the NK receptor protein-1 (Nkrp-1) family (Figure 15.2) (Giorda et al., 1990; Lanier et al., 1994; Plougastel et al., 2001). This family is conserved between rodents and humans and includes mouse Nkrp-1c (NK1.1) (Karlhofer and Yokoyama, 1991), an activating receptor also known as a marker for NK cells in some mouse strains including the C57BL/6 strain. Interestingly, the ligands for these receptors are encoded by the same region of the NK gene complex, and this receptor–ligand interaction may thus represent an ancient mechanism for self-

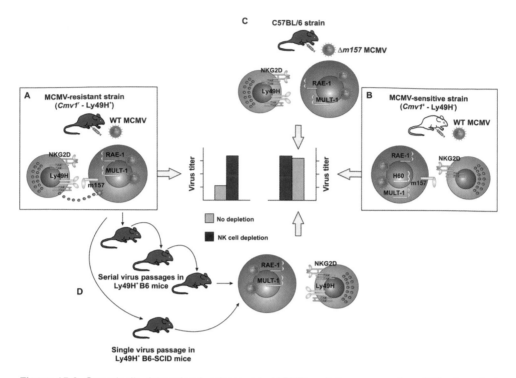

**Figure 15.3** Genetically determined resistance to MCMV and virus escape from NK cell control by mutation of viral ligands for activating NK cell receptors. (A) MCMV-resistance of Ly49H+ (Cmv1$^r$) mouse strains is mediated by activation of NK cells through interaction between Ly49H and virally encoded m157 protein. Depletion of NK cells abolishes the resistance which results in high virus titers. (B) In MCMV-sensitive Ly49H− (Cmv1$^s$) mouse strains, MCMV is NK resistant and depletion of NK cells has no significant influence on virus titers. (C) Deletion of gene m157 results in MCMV resistance to NK cells even in Ly49H+ (Cmv1$^r$) mice. (D) Under the selective pressure by Ly49H+ NK cells (e.g. by serial virus passages through C57BL/6 mice or by single passage through C57BL/6-SCID mice), MCMV escape variants are selected in which gene m157 carries mutations that reduce or prevent the binding of the viral ligand m157 to the activating NK cell receptor Ly49H. As a consequence, these escape variants are NK resistant.

Ly49H+ NK cells fail to control MCMV infection, and, as a consequence, the deletion of the m157 gene has no longer an influence (Bubic et al., 2004). Conversely, transgenic complementation of Ly49H in otherwise MCMV-susceptible mice results in an NK cell-mediated MCMV-resistant phenotype (Lee et al., 2003).

The study of Δm157 virus in vivo also highlighted the complexity of NK cell control in different organs (Orange et al., 1995; Tay and Welsh, 1997). Although the Cmv1 locus has been described as a host resistance locus that regulates the NK cell response to MCMV infection primarily in the spleen (Scalzo et al., 1990), we have recently shown that Ly49H NK receptor is also important for virus control in the lungs (Bubic et al., 2004). Our results also confirmed previous observations that MCMV control in the liver is largely independent of m157/Ly49H interaction (Brown et al., 2001). It has been speculated that Ly49H-independent virus control in the liver may be associated with an early activation of NK cells. Specifically, it has been proposed that in Cmv1$^r$ mice NK cells pass through two different stages of activation after primary MCMV infection (Dokun et al., 2001). During the early, nonspecific phase that lasts for the first two days of MCMV infection, IFN-$\gamma$ production and NK cell proliferation occur independently of Ly49H expression, whereas the second, specific phase is characterized by selective proliferation of Ly49H$^+$ NK cells (Dokun et al., 2001).

An interesting question is whether human NK cell receptors function in a manner similar to Ly49H. In this context it is notable that a delayed progression of AIDS has been observed in HIV patients who co-express the activating KIR receptor KIR3DS1 and HLA-B (Martin et al., 2002). Furthermore, it has been recently shown that inhibitory NK cell receptor KIR2DL3 and HLA-C1 directly influence the resolution of hepatitis C virus infection (Khakoo et al., 2004). These findings suggest that not only activating but also inhibitory NK cell interactions can be decisive in antiviral immunity in humans and that diminished inhibition can confer protection against virus infection.

It would be of interest to investigate whether additional MCMV-resistant mouse strains, apart from Ly49H$^+$ mice, exist in nature. If so, one would like to know whether such resistance is linked to viral ligands for activating NK cell receptors, inability of MCMV to down-modulate cellular ligands for activating receptors or to strong down-regulation of cellular ligands for inhibitory NK cell receptors.

## Are NK cells indeed essential for successful control of CMV infection?

The notion that NK cells are essential for surviving a CMV infection is often found in the literature. Is there sufficient evidence to support this view? A frequently used model for studying the role of NK cells in CMV infection utilizes infection of mice with highly virulent salivary gland-derived MCMV (SGV). In MCMV sensitive mouse strains, inocula of $10^5$ PFU and higher doses of SGV result in high mortality due to multi-organ disease (Trgovcich et al., 2000). In addition, a dramatic loss of spleen cells and induction of a surplus of cytokines such as TNF-$\alpha$ and IFN-$\gamma$ occurs. The epigenetic phenomena of higher virulence of SGV as compared to the virus derived from other tissues and cell culture-grown virus can be reversed by a single passage of SGV through cell culture. This particular pathogenesis of infection with SGV MCMV is caused, at least in part, by the tissue components within the preparation of the virus, rather than by virus itself. Similar pathophysiological and pathohistological changes can be observed even in resistant mouse strains (e.g. C57BL/6) by using higher doses ($2 \times 10^5$ PFU and higher) of SGV. Therefore, direct extrapolation of results obtained in such experimental settings to draw general conclusions on the role of NK cells in the control of CMV infections should be seen with some caution. To the best of our knowledge, there exists no clinical situation of HCMV infection as a counterpart to the SGV experimental model (Podlech et al., 2002).

Whilst there is no doubt that NK cells are important for early control of CMV infection, there exists also evidence arguing against their role being an essential one. For instance, most laboratory mouse strains mount hardly any NK cell response to CMV; yet they are able to control primary CMV infection and establish latency by generating a specific immune response. Furthermore, with the exception of NK cells from *Cmv1$^r$* mice, adoptively transferred non-primed spleen cells are unable to protect an immunodeficient syngeneic host from CMV disease unless transferred in very high numbers. In contrast, strikingly low numbers of virus-specific CD8$^+$ T-cells are sufficient to protect an MCMV infected host. Finally, mice selectively deficient in the adaptive immune response, with the NK cells

Ly49H⁺ mice, they select *m157* escape variants that no longer activate Ly49H (Figure 15.3D) (Voigt et al., 2003). Furthermore, the strong selective pressure by Ly49H⁺ NK cells during the primary infection of B6-SCID mice is sufficient for rapid selection of a large number of escape variants with no need for serial passages (Figure 15.3D) (French et al., 2004). The escape variants isolated from these mice are as resistant to Ly49H⁺ NK cells as the virus bearing the deliberate mutation of *m157* gene (Bubic et al., 2004; French et al., 2004). By contrast, escape variants with mutations in *m157* are not observed after MCMV passage in *Cmv1ˢ* mice indicating that the selective pressure by Ly49H⁺ NK cells is essential for the occurrence of escape variants (French et al., 2004).

The rapid emergence of *m157* escape variants indicates that exposure of MCMV to NK cell control represents a strong pressure for selecting viral variants carrying mutations in the *m157* gene. Therefore, it is unclear why the *m157* gene was nevertheless preserved during evolution. In this context it is important to note that most MCMV strains isolated from wild mice have m157 proteins that do not activate NK cells via Ly49H in spite of the fact that Ly49H is uncommon in wild mice and can therefore not account for selective pressure enforcing a loss of binding (Scalzo, 2002; Voigt et al., 2003). An interesting question is whether other members of the m145 protein family, in addition to m157, can serve as ligands for activating NK cell receptors. It is unknown whether other MCMV immunomodulatory proteins of the m145 family exhibit a degree of sequence variability comparable to m157. If so, this would indicate that MCMV continuously evolves to acquire additional immunomodulatory functions. MCMV escape mutants generated by selective immune pressure are not restricted to *m157*. High level of variability has been reported for the gB-encoding *M55* gene, perhaps as a consequence of selective pressure from neutralizing antibodies (Xu et al., 1996); see also Chapter 14 by M. Mach).

## Modulation of cellular ligands for activating NK cell receptors

In order to escape CD8⁺ CTL control, CMV proteins efficiently down-regulate MHC class I expression (Alcami and Koszinowski, 2000). The altered MHC class I expression might predispose infected cells to lysis by NK cells, as suggested by the "missing self" hypothesis (Ljunggren and Karre, 1990). However, there is evidence to suggest that sensitivity to NK cells is not only, and perhaps not even primarily, determined by the level of MHC class I expression. A prominent example is the activating NK cell receptor NKG2D, whose binding to its ligands on target cells triggers NK cell activation, even if the cells express normal levels of MHC class I molecules (Cerwenka and Lanier, 2003; Raulet, 2003b). Therefore, activation of NK cells via NKG2D is predestined to represent a dominant mechanism to cope with viruses and other intracellular pathogens. NKG2D is expressed on all NK cells, and the inducible expression of its ligands provides efficient activation of NK cells (Cerwenka and Lanier, 2003; Raulet, 2003b). Like with MHC class I down-modulation to compromise CD8⁺ T-cell function, the cellular ligands for the NKG2D receptor are targets for CMV escape from NK cell control (Cosman et al., 2001; Krmpotic et al., 2002).

The first CMV protein described to be involved in regulation of cellular expression of NKG2D ligands is the HCMV-encoded UL16 protein (Cosman et al., 2001). UL16 is an early viral non-structural glycoprotein that binds to ULBP-1 and -2 and MICB, but not to ULBP-3 and MICA (Cosman et al., 2001). The soluble form of UL16 blocks the interaction between NKG2D and its cognate activating ligand but fails to inhibit target killing via NKG2D–ligand interaction (Cosman et al., 2001; Kubin et al., 2001). Although the precise mechanisms by which UL16 down-modulates NKG2D ligands are not defined, UL16 is known to accumulate and colocalize with MICB at the ER/*cis*-Golgi. UL16 is proposed to form stable intracellular complexes with the NKG2D ligands in the ER/*cis*-Golgi compartment, thereby preventing their translocation to the cell surface (Dunn et al., 2003; Wu et al., 2003). In the absence of UL16, NKG2D ligands move rapidly through the Golgi apparatus to the cell surface. Accordingly, deletion of the *UL16* gene was found to enhance the susceptibility of HCMV-infected cells to NK cell lysis (Vales-Gomez et al., 2003). However, one should bear in mind that not all human NKG2D ligands are affected by UL16. One must therefore propose the existence of additional, as yet unidentified, HCMV proteins that target NKG2D ligands which are not affected by UL16. When considering the in vivo relevance of UL16-mediated NK evasion one has to bear in mind that HCMV induces expression of MICs to serve as co-stimulatory ligands for T-cell activation (Groh et al., 2001). Therefore, the relative significance of these two phenomena needs to be determined. Although the genes encoding NKG2D ligands in humans and mouse are different, their protein products have functional homology, and results obtained in the mouse model can be used to predict the significance of NKG2D ligands in humans.

MCMV encodes several proteins that inhibit NK cells by down-regulation of NKG2D ligands (Figure 15.4). With the exception of the C57BL/6 strain in which NK cell activation occurs via m157/Ly49H interaction (Arase et al., 2002; Bubic et al., 2004; Smith et al., 2002), most laboratory mouse strains, as well as wild mice, fail to generate a significant NK cell response to MCMV (Scalzo, 2002). Furthermore, Ly49H$^+$ mice behave like Ly49H$^-$ mice when infected with a virus mutant lacking the *m157* gene (Bubic et al., 2004). We were puzzled over the fact that Ly49H$^-$ mice, although capable of mounting an efficient NK cell response against other pathogens (Warfield et al., 2004), are unable to create an effective NK cell control of MCMV. MCMV *m152* was originally described as a gene encoding protein gp40 that is responsible for retention of MHC class I complexes in the ERGIC/*cis*-Golgi compartment, thereby preventing recognition of infected cells by CD8$^+$ T-cells (Ziegler et al., 1997). Thanks to the observation that an MCMV deletion mutant lacking *m152* is attenuated as early as on day 3 after infection, long before CD8$^+$ T-cells take over virus control, we came to the idea that *m152*/gp40 might be actively involved in regulation of NK cell function as well. The fact that this early attenuation of the Δ*m152* virus was abolished by depletion of NK cells supported our hypothesis. To discriminate between different possibilities that might have explained inhibition of NK cell control by *m152*/gp40 we tested the expression of cellular ligands for NKG2D on uninfected and MCMV-infected cells. Our own work provided the first evidence to conclude that MCMV indeed strongly down-modulates expression of NKG2D ligands (Krmpotic et al., 2002). Specifically, cells infected with the *m152* deletion mutant remained positive for the cell-surface expression of NKG2D ligands indicating that *m152*/gp40 is responsible for their down-regulation in WT virus. In addition, in contrast to WT virus, Δ*m152* mutant virus was sensitive to NK cell control both in vitro and in vivo. Deletion of the *m152* gene resulted in conversion of NK cell-resistant virus to NK cell-sensitive virus (Figure 15.4). Subsequent studies by Lewis Lanier's group identified RAE-1α, β and γ as the NKG2D ligands that are regulated by *m152*/gp40 (Lodoen et al., 2003).

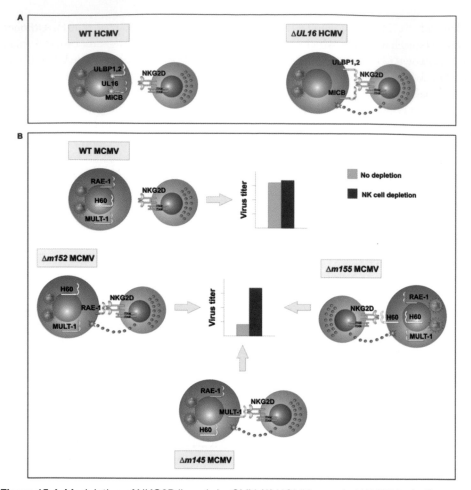

**Figure 15.4** Modulation of NKG2D ligands by CMV. (A) HCMV-encoded UL16 blocks the export of ULBP1, 2 and MICB from the ER/*cis*-Golgi compartment, thereby preventing their expression on the surface of infected cells and, as a consequence, also the activation of NK cells via the NKG2D receptor. In cells infected with Δ*UL16* HCMV, the block of NKG2D ligands is lifted leading to NK cell activation via NKG2D. (B) MCMV encodes at least three proteins that down-modulate NKG2D ligands and prevent activation of NK cells in MCMV-infected *Cmv1*[s] mice (e.g. BALB/c). Viruses with a deletion of either *m152*, *m145* or *m155*, which are responsible for down-regulation of RAE-1, MULT-1 and H60 respectively, become NK sensitive.

    To escape NK cell control, it is not sufficient that MCMV down-regulates only one of at least three different NKG2D ligands, since the remaining ligands might still be sufficient to trigger NK cell activation. We therefore postulated that in addition to *m152*/gp40 there must exist other MCMV proteins for controlling the expression of NKG2D ligands other than RAE-1. Considering the fact that staining of MCMV-infected cells with NKG2D tetramers shows a complete absence of NKG2D ligands on the plasma membrane and that *m152*/gp40 down-regulates only RAE-1 (Lodoen et al., 2003), we predicted MCMV proteins that down-regulate H60 and MULT-1. Since *m152* belongs to the *m145* gene family, we supposed that other members of this gene family might be responsible for modulation of other NKG2D ligands. By using MCMV deletion mutants lacking genes of the *m145* gene family we and others were able to identify two additional genes that down-regulate NKG2D ligands. Staining with monoclonal antibodies specific for RAE-1 excluded that two other MCMV proteins regulate RAE-1 family members. By using monoclonal antibodies to H60 and MULT-1 it could be demonstrated that the proteins encoded by genes *m155* (Hasan et al., 2005; Lodoen et al., 2004) and *m145* (Krmpotic et al., 2005) are responsible for the down-modulation of cell-surface H60 and MULT-1 proteins, respectively. Recombinant vaccinia viruses expressing *m145* or *m155* were finally used to show that these two MCMV proteins do not require a cooperation with other MCMV functions in order to down-modulate their respective target proteins. Moreover, deletion mutants for either *m145* or *m155* were attenuated in vivo in an NK cell-dependent manner (Hasan et al., 2005; Krmpotic et al., 2005; Lodoen et al., 2004). This finding confirmed a role for these viral proteins in suppressing the NK cell response, similar to the role that *m152*/gp40 exerts through down-modulation of RAE-1 (Figure 15.4). Since MCMV infection strongly induces the expression of MULT-1 mRNA and protein, the observed in vivo phenotype was predicted for Δ*m145* virus, but the phenotype of Δ*m155* virus came as a surprise because H60 is poorly expressed in normal tissue and is not induced upon MCMV infection. Furthermore, it appears from the data that *m155* is not the only MCMV gene causing down-regulation of H60, although the additional gene(s) involved remained elusive. Perhaps, MCMV infection does induce H60 expression in vivo in cell types that have not been tested so far in vitro. In this context it is worth noting that H60 mRNA is expressed in resting spleen cells and DCs (Malarkannan et al., 1998) that are in vivo targets for MCMV infection. At this point it is also worth mentioning that MCMV NK immunoevasins are operative in MCMV-sensitive as well as in MCMV-resistant mouse strains, but obviously the in vivo inhibitory forces are overridden in *Cmv1*[r] mice by the superior activating m157/Ly49H interaction (Bubic et al., 2004). When a double deletion mutant lacking *m155* and *m157* was tested in vivo, the impact of NKG2D ligand down-regulation became apparent. Because the NKG2D receptor is also expressed on T-cells, it is likely that interference with the function of NKG2D ligands represents an efficient and perhaps essential requirement for virus-mediated attenuation of both innate and specific immune responses relevant for the establishment of life-long infection and/or successful virus transmission.

What could be the rational explanation for the existence of more than one ligand for the NKG2D receptor? One possibility is a differential expression of NKG2D ligands in different cells and tissues (Cerwenka and Lanier, 2003). Hamerman et al. (2004) have recently shown that signaling through TLR in macrophages induces transcription of RAE-1 but not of H60 or MULT-1 genes, and we have shown that MCMV induces MULT-1 in infected cells (Krmpotic et al., 2005). Furthermore, different ligands may be differentially expressed during tissue development and cell differentiation and have distinguishable biological functions. For instance, RAE-1 can be detected in the embryonic central nervous system (CNS) but appears to be down-regulated in cells of the adult CNS and other adult organs (Nomura et al., 1996). The role of viral immune evasion genes in modulating expression of NK cell ligands such as RAE-1, H60 or MULT-1 in developing tissues like the embryonic CNS is still unknown but may be relevant for our understanding of CMV-mediated embryopathies. Promotion of virus persistence by viral immune evasion proteins is likely to represent a viral virulence factor, particularly in an organ with limited capacity of self-renewal. It also remains to be determined whether CMVs use immunoevasins to specifically limit immunopathology in certain tissues, in order to preserve vital functions of their host.

The molecular mechanisms by which MCMV genes down-regulate the NKG2D ligands are not yet completely understood. The $m152/gp40$ arrests MHC class I molecules in the ERGIC/$cis$-Golgi compartment and alters the half-life of the arrested proteins (Ziegler et al., 1997). The mechanism of $m152/gp40$-mediated down-modulation of surface RAE-1, however, remains elusive. We have recently found that the regulation of NKG2D ligands by m145 and m155 takes place in a compartment beyond ERGIC/$cis$-Golgi. Specifically, immunoprecipitation of NKG2D ligands from metabolically labeled cells that were co-infected with recombinant vaccinia viruses (VAC) VAC-m155 and VAC-H60 or with VAC-m145 and VAC-MULT-1 revealed no alteration in the glycosylation pattern, the rate of intracellular transport as indicated by acquisition of EndoH resistance and the half-life of MULT-1 (Krmpotic et al., 2005) and H60 (Hasan et al., 2005) proteins. Therefore, the lack of MULT-1 and H60 surface expression in infected cells may be the consequence of an interference with their ability to reach the cell membrane after leaving the ERGIC/$cis$-Golgi compartment or an alteration in the kinetics of their internalization/turn-over after having reached the cell surface. Although highly speculative, one or both of these molecules may accumulate at the cell surface in an altered conformation or in association with another protein that may mask their function and/or detection by specific mAbs and NKG2D tetramers (Hasan et al., 2005). This would not be without precedent as $m04/gp34$ binds to MHC class I and their complex is expressed at the cell surface putatively for serving as a decoy for NK cell activation (Kavanagh et al., 2001b; Kleijnen et al., 1997). Clearly, the remarkable reduction of membrane expression in spite of the apparent maturation of the glycoproteins requires further experimentation aiming at the identification of the compartment to which MULT-1 and H60 are targeted in the presence of m145 and m155, respectively.

The fact that the virus engages several specific genes for the regulation of NKG2D ligands points to the importance of these mechanisms in the immunobiology of the virus. Four genes known to be involved in evading NK cell control ($m145$, $m152$, $m155$, and $m157$) belong to the $m145$ gene family and one may speculate that the whole gene family may be involved in counteracting NK cell function, particularly if one bears in mind that the m157 also serves as a ligand for an inhibitory NK cell receptor, namely Ly49I (Arase et al., 2002). The possible role of other members of this gene family in regulation of the innate immune response is yet to be discovered.

*Possible importance of NKG2D evasion mechanisms in vivo*
Although the MCMV-mediated evasion of NK cells has been well documented both in vitro and in vivo, many aspects of virally encoded evasion functions in virus-mediated histopathology remain unknown. Are these genes equally effective in various cell types and tissues and what is the significance of the fact that some ligands are induced by the virus and others are not? Is there any tissue-specific pathology due to NK cell activation? Do viral immunoevasion genes play a role in limiting a putative autoimmune NK cell response against healthy bystander cells in which NKG2D ligands are induced through stimulation

by cytokines and other factors induced by the virus? At present it is unknown whether the activation of NK cells via the NKG2D receptor could have differential roles in virus control in different tissues. In our work we have found that the significance of viral genes that down-modulate NKG2D ligands is less pronounced in the liver than in the spleen and lungs. Thus, it is well conceivable that MCMV-encoded immune evasion functions exhibit organ-specific phenotypes. At present it is unknown whether the tissue-specific differences observed so far are related to differential recruitment of NK cells expressing different NK cell receptors. Chemokines such as IP10 and MIP-1 can recruit NK cells to sites of inflammation (Salazar-Mather et al., 2000; Trifilo et al., 2004) but it is unknown whether these and other chemokines/cytokines select different subsets of NK cells. In the case of the Ly49H$^+$ NK cell subset this could play a role because Ly49H is expressed on only 50% of the NK cells in C57BL/6 mice, whereas NKG2D is expressed on all NK cells. It is also unknown whether in certain tissues the expression of ligands is increased in response to cellular stress and infection, as it has been shown for MULT-1 (Krmpotic et al., 2005), and whether viral immunoevasion genes could play a beneficial role in reducing the collateral damage of tissue by invading NK cells.

CMVs are easily controlled by immunocompetent hosts which raise the question of why the virus needs multiple evasion genes to subvert innate and adaptive immune response. One could speculate that such "smart" mechanisms enable the virus to avoid immune recognition not only during the primary infection but also, and perhaps dominantly, during the reactivation from latency, which occurs in spite of a fully primed immune response. On the other hand, the virus might need these mechanisms in order to gain some advantage over the ongoing adaptive immune response for avoiding its elimination.

Human herpesviruses are implicated in the pathogenesis of several human malignancies (Howley et al., 2001). A high percentage of malignant gliomas, as well as human colorectal polyps and adenocarcinomas are infected by HCMV (Cobbs et al., 2002; Harkins et al., 2002). Multiple HCMV gene products are expressed in these tumors suggesting a possible association between HCMV and malignancies (Cobbs et al., 2002; Harkins et al., 2002). HCMV immunomodulatory genes, if expressed in the tumors, could play an active role in tumor pathogenesis by protecting the tumor from NK cells. Some human tumors do secrete soluble forms of NKG2D ligands that act to suppress the NK and T-cell responses by down-modulating NKG2D receptors on these cells (Groh et al., 2002). It is unknown whether HCMV-induced NKG2D ligands can also be secreted to mediate the immunosuppression that frequently accompanies infection.

## CMV MHC class I homologs as ligands for inhibitory NK cell receptors

HCMV encodes an MHC class I homolog, UL18, that was originally described as a decoy ligand for inhibitory NK cell receptors (Figure 15.5A) (Reyburn et al., 1997). Like MHC class I, UL18 possesses a peptide-binding groove and associates with β2-microglobulin (Beck and Barrell, 1988; Browne et al., 1990). It has been reported that HCMV UL18 expressed in HLA-A, B and C-deficient human cells reduces their susceptibility to NK cell lysis (Reyburn et al., 1997). However, this finding is controversial because the known cellular receptor for UL18, LIR-1 (leukocyte immunoglobulin-like receptor 1), is only marginally expressed on NK cells but heavily expressed on monocytic and B lymphoid cell types (Cosman et al., 1997). Furthermore, the UL18 protein does not contain a peptide sequence

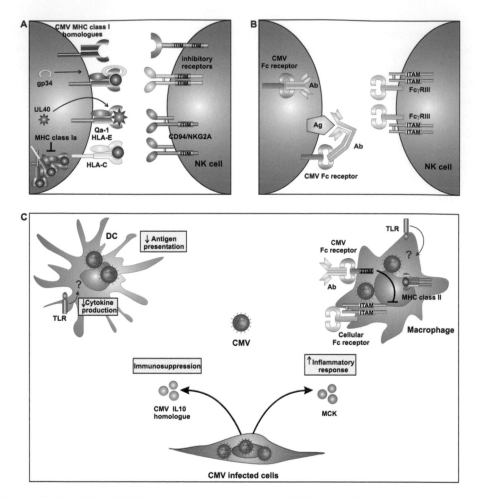

**Figure 15.5** Additional CMV mechanisms for evasion of NK cell-mediated control. (A) CMVs encode MHC class I homologs (UL18 and m144) to serve as ligands for inhibitory NK cell receptors. By differential modulation of MHC class I molecules, preserving MHC class Ib (HLA-E and Qa-1) to serve as ligands for the inhibitory receptor CD94/NKG2A while simultaneously down-modulating MHC class Ia molecules, CMVs can evade lysis by both NK and CD8+ T-cells. An HCMV UL40-derived peptide binds to HLA-E and up-regulates its expression. (B) In order to prevent ADCC against infected cells, CMVs encode Fc receptors which bind the Fc portion of antibody and thereby prevent ligation of the cellular Fc receptor on NK cells. (C) By targeting DCs and macrophages, CMVs can modulate their surface receptors and their ability to secrete cytokines. As a consequence, the NK cell response is modulated as well.

that would form a complex with HLA-E, the cellular ligand for the inhibitory NK cell receptor CD94/NKG2A (Tomasec et al., 2000). UL18 binds with high affinity to LIR-1 and inhibits cellular functions such as cytotoxicity, cytokine production, and activation of APC. Moreover, LIR-1 binds with a thousand-fold higher affinity to UL18 than to classical MHC class I proteins (Chapman et al., 1999). Interestingly, in lung-transplant patients who later developed CMV disease, the proportion of cells expressing LIR-1 was found to have increased several weeks before viral DNA became detectable by PCR, suggesting that LIR-1 expression may be a very early prognostic marker in transplant patients (Berg et al., 2003). On the other hand, because myeloid cells are considered to be a cellular site of CMV latency, the UL18/LIR-1 interaction might be important for protecting persistently infected myeloid cells.

MCMV also encodes an MHC class I homolog, m144, which can inhibit NK cell activation (Figure 15.5A) (Farrell et al., 1997). A recombinant virus lacking the *m144* gene shows reduced pathogenesis that can be reversed by NK cell depletion. In addition, the expression of m144 on tumor cells has an inhibitory effect on NK cells (Cretney et al., 1999) and transfection of *m144* inhibits ADCC (Kubota et al., 1999). The mechanism of m144-mediated NK cell evasion is not understood because its cellular receptor remained elusive. A homolog of *m144* was identified in RCMV, and the respective deletion mutant was found to be attenuated in vivo (Kloover et al., 2002). Although the expression of viral MHC class I homologs is regarded as an attractive strategy for immunoevasion of NK cells, firm evidence for direct inhibition of NK cells by interaction of these viral proteins with inhibitory receptors is lacking.

## Differential modulation of MHC class I expression

Both HCMV and MCMV down-regulate MHC class I molecules by using several proteins that interfere with distinct stages of the MHC class I processing and presentation pathway (Alcami and Koszinowski, 2000; Reddehase, 2002). By selective down-modulation of MHC class Ia molecules and simultaneously preserved expression of MHC class Ib molecules (HLA-E in humans and Qa-1 in mice), which are ligands for the inhibitory NK cell receptor CD94/NKG2A, CMV can evade both CD8+ T-cell and NK cell control (Figure 15.5A). The MHC class Ib molecules are loaded with a peptide derived from a conserved signal sequence of MHC class Ia molecules (Llano et al., 1998; Kraft et al., 2000). Thus, only cells that have normal MHC class Ia synthesis will be able to express MHC class Ib to serve as a ligand for CD94/NKG2A. Since the presentation of MHC class Ia molecules is TAP dependent (Hengel and Koszinowski, 1997), cells that either lack TAP or express viral functions that compromise TAP should fail to express MHC class Ib molecules. Therefore, while KIRs and Ly49 NK receptors directly recognize surface MHC class Ia molecules, the HLA-E-CD94/NKG2A ligand-receptor interaction is an indirect monitoring system for MHC class Ia expression. Thus, viral functions that spare the expression of HLA-E in human cells or Qa-1 in mouse cells, while simultaneously down-modulating MHC class Ia molecules, should be an optimal concept to evade lysis by both NK and CD8+ T-cells. This concept is actually used by HCMV. A peptide identical to a nonameric human HLA-C leader peptide is derived from the signal sequence of the HCMV glycoprotein UL40, binds to HLA-E and up-regulates its surface expression (Figure 15.5A) (Tomasec et al., 2000; Ulbrecht et al., 2000). The UL40-derived peptide is loaded onto HLA-E independently of the peptide transporter TAP and the complex is recognized by CD94/NKG2A. Through this pathway, NK cell activation can be prevented even in cells in which TAP is nonfunctional (Wang et al., 2002). Yet, the in vivo significance of UL40 function is questionable since a peptide derived from the signal sequence of the stress-induced chaperone Hsp60 can efficiently replace MHC class Ia-derived or UL40-derived peptide in HLA-E and thereby prevent binding of HLA-E to CD94/NKG2A and, as a consequence, prevent the inhibition of NK cells (Michaelsson et al., 2002). Likewise, peptide replacement in Qa-1 may provide a mechanism for the activation/de-repression of mouse NK cells (Kraft et al., 2000) to oppose viral evasion. Binding of stress peptides to HLA-E should not affect the inhibition of NK cells through KIRs but may nevertheless result in the activation of NK cells as a result of reduced CD94/NKG2A-mediated inhibition. Although the concept of evasion of NK cells mediated through the UL40-derived peptide is attractive, one should keep in mind that UL40-mediated protection from NK cell attack was revealed under overexpressing conditions in which HLA-E was loaded with UL40 leader sequence peptide. Its role and significance in the context of viral infection has been questioned (Falk et al., 2002). Generally, it is mandatory to study the immunomodulatory viral genes in the context of viral infection in order to establish their true significance.

Notably, HLA-E also serves as a ligand for an activating NK cell receptor, namely CD94/NKG2C (Figure 15.2) (Llano et al., 1998). Furthermore, when HLA-E is loaded with an HLA-G-derived nonapeptide, this interaction has a higher avidity than the interaction with the inhibitory receptor CD94/NKG2A, thus resulting in a net activation of NK cells (Llano et al., 1998). By analysis of peripheral blood lymphocytes from healthy HCMV seropositive humans it has been demonstrated that HCMV shapes the NK cell receptor repertoire by expanding CD94/NKG2C$^+$ NK cells (Guma et al., 2004). One may speculate that the increase in CD94/NKG2C$^+$ NK cells relates to the rate of reactivation episodes. CD94/NKG2C may serve as a receptor for HLA-E-bound viral peptide (e.g. UL40) or Hsp60-derived peptide and this ligand recognition may drive the expansion of the CD94/NKG2C$^+$ subsets of NK as well as T-cells. Clearly, these ideas beg for further investigation.

US2 and US11 HCMV proteins differentially down-modulate MHC class I molecules in a way that reroutes certain MHC class I molecules for proteasomal degradation while sparing HLA-C and HLA-E, which are ligands for inhibitory NK cell receptors (Figure 15.5A) (Llano et al., 2003; Schust et al., 1998). However, it is unknown whether this discrimination in down-regulating HLA molecules is beneficial to the virus. Namely, selective down-regulation of MHC class Ia molecules in HCMV-infected cells will remain unrecognized by CD94/NKG2A, which is specific for MHC class Ib molecules, whereas NK cells regulated by a KIR with specificity for the down-regulated MHC class Ia molecules may become activated. Other proteins of HCMV, US3 and US6, are less selective and inhibit the cell-surface expression of HLA-C and HLA-G (Ahn et al., 1997; Jun et al., 2000). Although the down-modulating function of US3 and US6 could make infected cells generally susceptible to NK cell lysis, there is currently no evidence for this. The nonclassical MHC class I molecule HLA-G has been described as an inhibitor of the NK cell response (Le Bouteiller, 1997). HLA-G is a ligand for inhibitory receptors of the *LIR* gene family expressed on subsets of lymphocytes and on monocytes and DCs (Colonna et al., 1997; 1998; Rajagopalan and Long, 1999). Recognition of membrane-bound HLA-G by the receptors of *LIR* gene family may have inhibitory effects toward NK cells, T-cells, and macrophages (Le Gal et al., 1999). Therefore, by up-regulating HLA-G or by encoding a ligand for LIR, both of which applies to HCMV, viruses can inhibit both the innate and the adaptive immune response (Onno et al., 2000).

Three MCMV immune evasion genes, namely *m04*, *m06*, and *m152*, modulate MHC class I expression in order to avoid CD8$^+$ T-cell recognition (Wagner et al., 2002). However, the product of the *m04* gene, gp34, appears to interfere with MHC class I in such a way as to simultaneously avoid NK cell activation. gp34 is a type I-anchored integral membrane glycoprotein that tightly associates with mouse MHC class I molecules (Kleijnen et al., 1997). gp34 binding to MHC class I is $\beta_2$-microglobulin dependent and differentially affects different mouse MHC class I alleles (Kavanagh et al., 2001a; Wagner et al., 2002). After their formation in the ER, the gp34-MHC class I complexes are maintained throughout the secretory pathway to reach the cell surface where they are stably expressed. It has been speculated that these molecules can serve as ligands for inhibitory NK cell receptors (Figure 15.5A). However, firm functional evidence for gp34 serving as an NK cell decoy is missing. A virus lacking the entire *m02* gene family (*m02-m16*) is clearly attenuated in an NK cell-dependent manner (Oliveira et al., 2002). Interestingly, this attenuation was observed both in BALB/c and C57BL/6 mice, which is remarkable because it indicates that NK cell inhibition mediated by *m02* gene family members can balance the NK cell activation that is caused by m157/Ly49H interaction. It remains to be determined whether this attenuation is due to deletion of the *m04* gene or if another gene within the *m02* family is responsible for NK cell inhibition. In our own study with the virus that lacks the *m02* gene family we did not observe any effect of the deletion on the expression of NKG2D ligands (S.J., unpublished) indicating that the mechanism of NK inhibition might be based on ligation of inhibitory receptors rather than on preventing the expression of ligands for an activating NK cell receptor.

## Viral Fc receptor

Cellular Fc receptors are important for the antimicrobial activity of antigen-bound IgG antibodies. Different pathogens use the same principle of binding the IgG Fc fragment in order to oppose Fc-controlled effector functions. For example, in HSV-1, human IgG is bound by a complex formed by glycoproteins gE and gI (Johnson et al., 1988). gE and gI homologs are also found in some other viruses including HCMV (Budt et al., 2004). The gE–gI complex forms a high-affinity receptor which binds monomeric IgG, thereby preventing HSV-infected cells from ADCC (Dubin et al., 1991). The effect is explained by "bipolar bridging" which reflects IgG binding to an antigenic epitope via the Fab domain, and binding of the Fc domain to the microbial Fc receptor (Frank and Friedman, 1989). Thus, the engagement of the IgG Fc domain prevents it from binding to the Fc receptor on NK cells (Figure 15.5B).

Three viral genes encoding two different Fc binding glycoproteins, gp68 and gp34, were identified within the HCMV genome (Atalay et al., 2002; Lilley et al., 2001). gp68 is encoded by *UL119–118* genes and gp34 is encoded by two identical gene copies, *TRL11* and *IRL11*. In MCMV a single gene, *m138*, encodes a viral FcγR (Thäle et al., 1994). Interestingly, the deletion mutant for MCMV *m138* is clearly attenuated in vivo but irrespective of the presence or absence of antibodies suggesting that the observed phenotype is not dependent on the Ig binding property (Crnkovic-Mertens et al., 1998). There are a few speculations about the possible role of CMV-encoded Fc receptors. Sequence analysis of HCMV gp68 identified a potential ITIM sequence in its cytoplasmic tail (Atalay et al., 2002) suggesting that this viral Fc receptor might serve as a competitive antagonist of a cellular Fc receptor such that engagement leads to inhibitory signaling (Figure 15.5C). Furthermore, the cytoplasmic tail of HCMV gp34 contains a di-leucine consensus motif indicating that this viral Fc receptor may have a role in displacing antiviral antibodies from the cell surface to the endolysosome for destruction (Figure 15.5B) (Atalay et al., 2002).

## Macrophages and dendritic cells in CMV infection

Mononuclear phagocytes (blood monocytes and tissue macrophages) are widely distributed throughout the body and play an important role in innate antiviral immunity. In response to inflammatory stimuli, monocytes are recruited to the site of inflammation from bone marrow and blood, and by differentiating into macrophages they acquire potent effector functions (Gordon, 1998). Macrophages are highly phagocytic cells that detect, ingest, and destroy infectious agents. The phagocytic function is mediated through their opsonic and nonopsonic pattern-recognition receptors. Ligation of antibody-coated microorganisms to cellular Fc receptors results in their neutralization and destruction. In contrast, a broad spectrum of nonopsonic receptors, including TLRs, serves to directly recognize ligands on microorganisms or infected cells. In addition to phagocytosis, macrophages can also process and present antigens to T-cells and produce chemokines and cytokines (Gordon, 1998).

As a known site of CMV replication and latency, monocytes and macrophages play an important role in viral pathogenesis and virus dissemination (Mocarski and Courcelle, 2001; Reddehase et al., 2002). The fact that their susceptibility to CMV infection is differentiation-dependent (see also Chapters 9 by Bain et al. and Chapter 20 by Hanson and Campbell) led to the conclusion that monocytes disseminate infection and differentiate into permissive mature cells as they enter tissues (Hanson et al., 1999; Jarvis and Nelson, 2002; Pollock et al., 1997). MCMV dissemination is potentiated by the expression of a viral proinflammatory chemokine, referred to as MCMV chemokine (MCK), which amplifies the host's inflammatory response (Figure 15.5C) (MacDonald et al., 1997; Mocarski, 2002; Saederup et al., 2001). By recruiting leukocytes to initial sites of infection, MCK facilitates their infection and subsequent virus dissemination (Mocarski, 2002). MCMV lacking the MCK-encoding genes *m131-m129* is attenuated in vivo at early times post infection, but the mechanism by which MCK affects NK cell function is unknown (Fleming et al., 1999; Saederup et al., 2001). HCMV encodes a CC chemokine homolog (UL128) (Akter et al., 2003) as well as a CXC chemokine (gpUL146/vCXCL1) that signals via CXCR2 (Penfold et al., 1999), but it is currently unknown whether either plays any role in HCMV pathogenesis.

MCMV infected macrophages are less sensitive to MHC class I down-regulation (Hengel et al., 2000). Retaining MHC class I molecules on the cell surface may be important for macrophages to resist early destruction by NK cells. In another study, however, virus-induced down-regulation of MHC class I molecules on macrophages was observed (LoPiccolo et al., 2003), which may reflect allelic differences in the interaction between immune evasion molecules and class I molecules. It remains to be tested whether MCMV infected macrophages down-modulate MHC class I molecules in vivo. Down-modulation of MHC class II molecules has also been described in MCMV-infected macrophages and could not be reversed by IFN-γ (Figure 15.5C) (Heise et al., 1998). It may well be that host IL-10, which is strongly induced by MCMV and HCMV infection, is involved in down-regulating MHC class II (Nordoy et al., 2003; Redpath et al., 1999). HCMV UL111a encodes an IL-10 homolog with immunosuppressive properties that may interfere with the host immune response (Figure 15.5C) (Chang et al., 2004; Kotenko et al., 2000).

DCs are the major antigen-presenting cells and play a central role in translating innate into adaptive immune response (Cooper et al., 2004). Through secretion of cytokines (IFN-α/β, IL-2, IL12, IL-15, and IL-18) and activation of NK cells, DCs act as a part of the innate immune system. Like other cells of the innate immune system they express pattern-recognition receptors able to recognize microbial PAMPs. Immature DCs are dispersed throughout peripheral tissues where they take up antigens, degrade them into peptides, and load these peptides onto MHC class I and II molecules. After that, DCs undergo maturation that results in a marked decrease in internalization efficacy and in up-regulation of surface expression of MHC molecules and costimulatory molecules. DC maturation also induces cytokine production and modifies surface expression of adhesion molecules and chemokine receptors (Reis e Sousa, 2004). Upon maturation, DCs migrate to secondary lymphoid organs. DC-NK cell interaction has been shown to require direct cell-to-cell contact independently of IFN-α/β and IL-12 production (Cooper et al., 2004). This interaction takes place at sites of inflammation and in lymph nodes (Ferlazzo and Munz, 2004). DCs are also able to produce IL-2 early after stimulation with microbial products (Granucci et al., 2001; 2003). The timing of IL-2 secretion by DCs corresponds with the earliest production of IFN-γ by NK cells (Lertmemongkolchai et al., 2001). Although IL-2 is necessary for efficient production of IFN-γ by NK cells, direct interaction between DCs and NK cells is needed as well (Granucci et al., 2004). Therefore, bearing in mind the key role of DCs in induction and maintenance of both the innate and the adaptive immune response it is very attractive to propose that the virus counteracts the functions of DCs in order to replicate to high titers and survive in its host.

CMVs can infect DCs in vitro and in vivo (Andrews et al., 2001; Dalod et al., 2003; Raftery et al., 2001; Riegler et al., 2000). Key questions are how CMV infection of DCs influences the ongoing innate and adaptive immune response and what the role of immune control mechanisms is in protecting DCs from virus infection. MCMV infection in vivo compromises DC function resulting in impaired NK cell activity, reduced capacity to prime T-cells, and general immunosuppression (Figure 15.5C) (Andrews et al., 2001; 2003). The mechanisms by which CMVs affect DC and macrophage functions include the block of IFN-γ-induced MHC class II antigen presentation (Heise et al., 1998), the inhibition of monocyte differentiation (Gredmark and Soderberg-Naucler, 2003), and an interference with maturation and migration of immature DCs (Moutaftsi et al., 2002; 2004). In addition, altered production of cytokines and altered expression of surface MHC class I and II as well as costimulatory molecules on mature DCs have been described (Andrews et al., 2001; Hertel et al., 2003; Mathys et al., 2003; Raftery et al., 2001). Induction of apoptosis-inducing ligands CD95L (FasL) and TRAIL after HCMV infection of DCs may result in deletion of activated T-lymphocytes (Raftery et al., 2001).

In mice that possess Ly49H receptor, Ly49H+ NK cells are essential for retaining CD8α+ DCs in the spleen and, in turn, DCs are important for the expansion of Ly49H+ NK cells (Andrews et al., 2003). In contrast to Ly49H+ mice which maintain DCs during the course of infection, dramatic loss of DCs was observed in BALB/c mice (Andrews et al., 2003). Similar results were obtained in C57BL/6 mice depleted of Ly49H+ NK cells, thus confirming that Ly49H+ NK cells are responsible for protection of DCs (Andrews et al., 2003). The finding that DCs are so dramatically affected by MCMV in Ly49H− mice raised the question of why the ongoing specific immune response is not significantly

## Future perspectives

In recent years, efforts to understand the significance of the different immune response mechanisms have profited a great deal from studies of CMV infections, in particular of the mouse model of CMV infection. The roles of NK cell activation receptors and their ligands have been identified. The work on the Ly49H NK cell receptor and its virally encoded ligand has led to a better understanding of the physiology of the NK cell response, and has highlighted the strong selective pressure imposed on the virus to generate variants in order to avoid NK cells. A deeper insight was gained by studying CMV mechanisms aimed at modulating the expression of the cellular ligand for activating NK cell receptor, NKG2D. The growing evidence for a role of TLRs in DC and NK cell responses to CMV infection leads us to speculate that these receptors are likely also a target of CMV evasion mechanisms. Future studies will have to answer several pending questions. First, and perhaps most obvious, why do CMVs need so many evasion mechanisms though being fairly harmless to immunocompetent hosts? We need to learn how important NK cells and other components of the innate immune response are in limiting horizontal and vertical virus spread and whether viral immunoevasins provide some advantage to the virus in terms of efficient dissemination and transmission prior to full activation of immune effector functions that eventually resolve the productive phase of the infection. A long-standing question that remained unanswered is how important NK cells are for the maintenance of CMV latency and the ability of the virus to reactivate in face of a fully primed immune memory. It is also not clear why so many NK cell receptors are needed and why some receptors have more than one cellular ligand.

Testing of MCMV mutants in mice (see also Chapter 4) has been instrumental in understanding the role of different components of the innate immune system in the control of CMV infections and the role of viral immunoevasins in modulating the immune response. Because of the strict species specificity of CMVs, MCMV as well as other animal CMVs (see Chapters 23, 24, and 27) will remain valuable model systems to our understanding of CMV pathogenesis. Only continuing investigation of these many aspects of CMV interaction with the host immune system will finally reveal the biological framework that might hopefully lead to more efficient treatment of HCMV infection and disease.

## Acknowledgments

We would like to thank U.H. Koszinowski for permanent support of our work. We would also like to thank H. Hengel, M. Messerle, M.J. Reddehase and W.J. Britt for help and collaboration in our work. We apologize to all colleagues whose original work could not be included owing to space limitations.

Our work has been funded by grants from the Croatian Ministry of Science and Technology (projects 62004 and 62007).

## References

Ahn, K., Gruhler, A., Galocha, B., Jones, T.R., Wiertz, E.J., Ploegh, H.L., Peterson, P.A., Yang, Y., and Fruh, K. (1997). The ER-luminal domain of the HCMV glycoprotein US6 inhibits peptide translocation by TAP. Immunity 6, 613–621.

Akira, S., and Takeda, K. (2004). Toll-like receptor signalling. Nature Rev. Immunol. 4, 499–511.

Akter, P., Cunningham, C., McSharry, B.P., Dolan, A., Addison, C., Dargan, D.J., Hassan-Walker, A.F., Emery, V.C., Griffiths, P.D., Wilkinson, G.W., and Davison, A.J. (2003). Two novel spliced genes in human cytomegalovirus. J. Gen. Virol. *84*, 1117–1122.

Alcami, A., and Koszinowski, U.H. (2000). Viral mechanisms of immune evasion. Immunol. Today *21*, 447–455.

Andrews, D.M., Andoniou, C.E., Granucci, F., Ricciardi-Castagnoli, P., and Degli-Esposti, M.A. (2001). Infection of dendritic cells by murine cytomegalovirus induces functional paralysis. Nature Immunol. *2*, 1077–1084.

Andrews, D.M., Scalzo, A.A., Yokoyama, W.M., Smyth, M.J., and Degli-Esposti, M.A. (2003). Functional interactions between dendritic cells and NK cells during viral infection. Nature Immunol. *4*, 175–181.

Arase, H., Mocarski, E.S., Campbell, A.E., Hill, A.B., and Lanier, L.L. (2002). Direct recognition of cytomegalovirus by activating and inhibitory NK cell receptors. Science *296*, 1323–1326.

Atalay, R., Zimmermann, A., Wagner, M., Borst, E., Benz, C., Messerle, M., and Hengel, H. (2002). Identification and expression of human cytomegalovirus transcription units coding for two distinct Fcgamma receptor homologs. J. Virol. *76*, 8596–8608.

Bauer, S., Groh, V., Wu, J., Steinle, A., Phillips, J.H., Lanier, L.L., and Spies, T. (1999). Activation of NK cells and T-cells by NKG2D, a receptor for stress-inducible MICA. Science *285*, 727–729.

Beck, S., and Barrell, B.G. (1988). Human cytomegalovirus encodes a glycoprotein homologous to MHC class I antigens. Nature *331*, 269–272.

Berg, L., Riise, G.C., Cosman, D., Bergstrom, T., Olofsson, S., Karre, K., and Carbone, E. (2003). LIR-1 expression on lymphocytes, and cytomegalovirus disease in lung-transplant recipients. Lancet *361*, 1099–1101.

Biron, C.A. (1997). Natural killer cell regulation during viral infection. Biochem. Soc. Trans. *25*, 687–690.

Biron, C.A., Byron, K.S., and Sullivan, J.L. (1989). Severe herpesvirus infections in an adolescent without natural killer cells. N. Engl. J. Med. *320*, 1731–1735.

Biron, C.A., Nguyen, K.B., Pien, G.C., Cousens, L.P., and Salazar-Mather, T.P. (1999). Natural killer cells in antiviral defense: function and regulation by innate cytokines. Annu. Rev. Immunol. *17*, 189–220.

Brown, M.G., Dokun, A.O., Heusel, J.W., Smith, H.R., Beckman, D.L., Blattenberger, E.A., Dubbelde, C.E., Stone, L.R., Scalzo, A.A., and Yokoyama, W.M. (2001). Vital involvement of a natural killer cell activation receptor in resistance to viral infection. Science *292*, 934–937.

Browne, H., Smith, G., Beck, S., and Minson, T. (1990). A complex between the MHC class I homologue encoded by human cytomegalovirus and beta 2 microglobulin. Nature *347*, 770–772.

Bubic, I., Wagner, M., Krmpotic, A., Saulig, T., Kim, S., Yokoyama, W.M., Jonjic, S., and Koszinowski, U.H. (2004). Gain of Virulence Caused by Loss of a Gene in Murine Cytomegalovirus. J. Virol. *78*, 7536–7544.

Budt, M., Reinhard, H., Bigl, A., and Hengel, H. (2004). Herpesviral Fcgamma receptors: culprits attenuating antiviral IgG? Int. Immunopharmacol. *4*, 1135–1148.

Bukowski, J.F., Warner, J.F., Dennert, G., and Welsh, R.M. (1985). Adoptive transfer studies demonstrating the antiviral effect of natural killer cells in vivo. J. Exp. Med. *161*, 40–52.

Bukowski, J.F., Woda, B.A., and Welsh, R.M. (1984). Pathogenesis of murine cytomegalovirus infection in natural killer cell-depleted mice. J. Virol. *52*, 119–128.

Bukowski, J.F., Yang, H., and Welsh, R.M. (1988). Antiviral effect of lymphokine-activated killer cells: characterization of effector cells mediating prophylaxis. J. Virol. *62*, 3642–3648.

Burke, B., and Lewis, C.E. (2002). The macrophage., 2nd Ed. (New York:, Oxford University Press).

Carayannopoulos, L.N., Naidenko, O.V., Fremont, D.H., and Yokoyama, W.M. (2002). Cutting edge: murine UL16-binding protein-like transcript 1: a newly described transcript encoding a high-affinity ligand for murine NKG2D. J. Immunol. *169*, 4079–4083.

Carayannopoulos, L.N., and Yokoyama, W.M. (2004). Recognition of infected cells by natural killer cells. Curr. Opin. Immunol. *16*, 26–33.

Cavanaugh, V.J., Deng, Y., Birkenbach, M.P., Slater, J.S., and Campbell, A.E. (2003). Vigorous innate and virus-specific cytotoxic T-lymphocyte responses to murine cytomegalovirus in the submaxillary salivary gland. J. Virol. *77*, 1703–1717.

Cerwenka, A., Bakker, A.B., McClanahan, T., Wagner, J., Wu, J., Phillips, J.H., and Lanier, L.L. (2000). Retinoic acid early inducible genes define a ligand family for the activating NKG2D receptor in mice. Immunity *12*, 721–727.

Cerwenka, A., Baron, J.L., and Lanier, L.L. (2001). Ectopic expression of retinoic acid early inducible-1 gene (RAE-1) permits natural killer cell-mediated rejection of a MHC class I-bearing tumor in vivo. Proc. Natl. Acad. Sci. USA 98, 11521–11526.

Cerwenka, A., and Lanier, L.L. (2001). Natural killer cells, viruses and cancer. Nature Rev. Immunol. 1, 41–49.

Cerwenka, A., and Lanier, L.L. (2003). NKG2D ligands: unconventional MHC class I-like molecules exploited by viruses and cancer. Tissue Antigens 61, 335–343.

Chalupny, N.J., Sutherland, C.L., Lawrence, W.A., Rein-Weston, A., and Cosman, D. (2003). ULBP4 is a novel ligand for human NKG2D. Biochem. Biophys. Res. Commun. 305, 129–135.

Chang, W.L., Baumgarth, N., Yu, D., and Barry, P.A. (2004). Human cytomegalovirus-encoded interleukin-10 homolog inhibits maturation of dendritic cells and alters their functionality. J. Virol. 78, 8720–8731.

Chapman, T.L., Heikeman, A.P., and Bjorkman, P.J. (1999). The inhibitory receptor LIR-1 uses a common binding interaction to recognize class I MHC molecules and the viral homolog UL18. Immunity 11, 603–613.

Cobbs, C.S., Harkins, L., Samanta, M., Gillespie, G.Y., Bharara, S., King, P.H., Nabors, L.B., Cobbs, C.G., and Britt, W.J. (2002). Human cytomegalovirus infection and expression in human malignant glioma. Cancer Res. 62, 3347–3350.

Colonna, M., Navarro, F., Bellon, T., Llano, M., Garcia, P., Samaridis, J., Angman, L., Cella, M., and Lopez-Botet, M. (1997). A common inhibitory receptor for major histocompatibility complex class I molecules on human lymphoid and myelomonocytic cells. J. Exp. Med. 186, 1809–1818.

Colonna, M., Samaridis, J., Cella, M., Angman, L., Allen, R.L., O'Callaghan, C.A., Dunbar, R., Ogg, G.S., Cerundolo, V., and Rolink, A. (1998). Human myelomonocytic cells express an inhibitory receptor for classical and nonclassical MHC class I molecules. J. Immunol. 160, 3096–3100.

Colucci, F., Di Santo, J.P., and Leibson, P.J. (2002). Natural killer cell activation in mice and men: different triggers for similar weapons? Nature Immunol. 3, 807–813.

Compton, T. (2004). Receptors and immune sensors: the complex entry path of human cytomegalovirus. Trends Cell Biol. 14, 5–8.

Compton, T., Kurt-Jones, E.A., Boehme, K.W., Belko, J., Latz, E., Golenbock, D.T., and Finberg, R.W. (2003). Human cytomegalovirus activates inflammatory cytokine responses via CD14 and Toll-like receptor 2. J. Virol. 77, 4588–4596.

Cooper, M.A., Fehniger, T.A., Fuchs, A., Colonna, M., and Caligiuri, M.A. (2004). NK cell and DC interactions. Trends Immunol. 25, 47–52.

Cosman, D., Fanger, N., Borges, L., Kubin, M., Chin, W., Peterson, L., and Hsu, M.L. (1997). A novel immunoglobulin superfamily receptor for cellular and viral MHC class I molecules. Immunity 7, 273–282.

Cosman, D., Mullberg, J., Sutherland, C.L., Chin, W., Armitage, R., Fanslow, W., Kubin, M., and Chalupny, N.J. (2001). ULBPs, novel MHC class I-related molecules, bind to CMV glycoprotein UL16 and stimulate NK cytotoxicity through the NKG2D receptor. Immunity 14, 123–133.

Cretney, E., Degli-Esposti, M.A., Densley, E.H., Farrell, H.E., Davis-Poynter, N.J., and Smyth, M.J. (1999). m144, a murine cytomegalovirus (MCMV)-encoded major histocompatibility complex class I homologue, confers tumor resistance to natural killer cell-mediated rejection. J. Exp. Med. 190, 435–444.

Crnkovic-Mertens, I., Messerle, M., Milotic, I., Szepan, U., Kucic, N., Krmpotic, A., Jonjic, S., and Koszinowski, U.H. (1998). Virus attenuation after deletion of the cytomegalovirus Fc receptor gene is not due to antibody control. J. Virol. 72, 1377–1382.

Dalod, M., Hamilton, T., Salomon, R., Salazar-Mather, T.P., Henry, S.C., Hamilton, J.D., and Biron, C.A. (2003). Dendritic cell responses to early murine cytomegalovirus infection: subset functional specialization and differential regulation by interferon alpha/beta. J. Exp. Med. 197, 885–898.

Dalod, M., Salazar-Mather, T.P., Malmgaard, L., Lewis, C., Asselin-Paturel, C., Briere, F., Trinchieri, G., and Biron, C.A. (2002). Interferon alpha/beta and interleukin 12 responses to viral infections: pathways regulating dendritic cell cytokine expression in vivo. J. Exp. Med. 195, 517–528.

Daniels, K.A., Devora, G., Lai, W.C., O'Donnell, C.L., Bennett, M., and Welsh, R.M. (2001). Murine cytomegalovirus is regulated by a discrete subset of natural killer cells reactive with monoclonal antibody to Ly49H. J. Exp. Med. 194, 29–44.

Depatie, C., Muise, E., Lepage, P., Gros, P., and Vidal, S.M. (1997). High-resolution linkage map in the proximity of the host resistance locus Cmv1. Genomics 39, 154–163.

Diefenbach, A., Hsia, J.K., Hsiung, M.Y., and Raulet, D.H. (2003). A novel ligand for the NKG2D receptor activates NK cells and macrophages and induces tumor immunity. Eur. J. Immunol. 33, 381–391.

Diefenbach, A., Jamieson, A.M., Liu, S. D., Shastri, N., and Raulet, D.H. (2000). Ligands for the murine NKG2D receptor: expression by tumor cells and activation of NK cells and macrophages. Nature Immunol. 1, 119–126.

Dimasi, N., Moretta, L., and Biassoni, R. (2004). Structure of the Ly49 family of natural killer (NK) cell receptors and their interaction with MHC class I molecules. Immunol. Res. 30, 95–104.

Dokun, A.O., Kim, S., Smith, H.R., Kang, H.S., Chu, D.T., and Yokoyama, W.M. (2001). Specific and nonspecific NK cell activation during virus infection. Nature Immunol. 2, 951–956.

Dubin, G., Socolof, E., Frank, I., and Friedman, H.M. (1991). Herpes simplex virus type 1 Fc receptor protects infected cells from antibody-dependenT-cellular cytotoxicity. J. Virol. 65, 7046–7050.

Dunn, C., Chalupny, N.J., Sutherland, C.L., Dosch, S., Sivakumar, P.V., Johnson, D.C., and Cosman, D. (2003). Human cytomegalovirus glycoprotein UL16 causes intracellular sequestration of NKG2D ligands, protecting against natural killer cell cytotoxicity. J. Exp. Med. 197, 1427–1439.

Falk, C.S., Mach, M., Schendel, D.J., Weiss, E.H., Hilgert, I., and Hahn, G. (2002) NK cell activity during human cytomegalovirus infection is dominated by US2–11-mediated HLA class I down-regulation. J. Immunol. 169, 3257–3266.

Farrell, H.E., Vally, H., Lynch, D.M., Fleming, P., Shellam, G.R., Scalzo, A.A., and Davis-Poynter, N.J. (1997). Inhibition of natural killer cells by a cytomegalovirus MHC class I homologue in vivo. Nature 386, 510–514.

Ferlazzo, G., and Munz, C. (2004). NK cell compartments and their activation by dendritic cells. J. Immunol. 172, 1333–1339.

Fleming, P., Davis-Poynter, N., Degli-Esposti, M., Densley, E., Papadimitriou, J., Shellam, G., and Farrell, H. (1999). The murine cytomegalovirus chemokine homolog, m131/129, is a determinant of viral pathogenicity. J. Virol. 73, 6800–6809.

Forbes, C.A., Brown, M.G., Cho, R., Shellam, G.R., Yokoyama, W.M., and Scalzo, A.A. (1997). The Cmv1 host resistance locus is closely linked to the Ly49 multigene family within the natural killer cell gene complex on mouse chromosome 6. Genomics 41, 406–413.

Frank, I., and Friedman, H.M. (1989). A novel function of the herpes simplex virus type 1 Fc receptor: participation in bipolar bridging of antiviral immunoglobulin G. J. Virol. 63, 4479–4488.

French, A.R., Pingel, J.T., Wagner, M., Bubic, I., Yang, L., Kim, S., Koszinowski, U., Jonjic, S., and Yokoyama, W.M. (2004). Escape of mutant double-stranded DNA virus from innate immune control. Immunity 20, 747–756.

French, A.R., and Yokoyama, W.M. (2003). Natural killer cells and viral infections. Curr. Opin. Immunol. 15, 45–51.

Gessner, J.E., Heiken, H., Tamm, A., and Schmidt, R.E. (1998). The IgG Fc receptor family. Ann. Hematol. 76, 231–248.

Giorda, R., Rudert, W.A., Vavassori, C., Chambers, W.H., Hiserodt, J.C., and Trucco, M. (1990). NKR-P1, a signal transduction molecule on natural killer cells. Science 249, 1298–1300.

Gordon, S. (1998). The role of the macrophage in immune regulation. Res. Immunol. 149, 685–688.

Granucci, F., Feau, S., Angeli, V., Trottein, F., and Ricciardi-Castagnoli, P. (2003). Early IL-2 production by mouse dendritic cells is the result of microbial-induced priming. J. Immunol. 170, 5075–5081.

Granucci, F., Vizzardelli, C., Pavelka, N., Feau, S., Persico, M., Virzi, E., Rescigno, M., Moro, G., and Ricciardi-Castagnoli, P. (2001). Inducible IL-2 production by dendritic cells revealed by global gene expression analysis. Nature Immunol. 2, 882–888.

Granucci, F., Zanoni, I., Pavelka, N., Van Dommelen, S.L., Andoniou, C.E., Belardelli, F., Degli Esposti, M.A., and Ricciardi-Castagnoli, P. (2004). A contribution of mouse dendritic cell-derived IL-2 for NK cell activation. J. Exp. Med. 200, 287–295.

Gredmark, S., and Soderberg-Naucler, C. (2003). Human cytomegalovirus inhibits differentiation of monocytes into dendritic cells with the consequence of depressed immunological functions. J. Virol. 77, 10943–10956.

Groh, V., Rhinehart, R., Randolph-Habecker, J., Topp, M.S., Riddell, S.R., and Spies, T. (2001). Costimulation of CD8alphabeta T-cells by NKG2D via engagement by MIC induced on virus-infected cells. Nature Immunol. 2, 255–260.

Groh, V., Wu, J., Yee, C., and Spies, T. (2002). Tumour-derived soluble MIC ligands impair expression of NKG2D and T-cell activation. Nature 419, 734–738.

Grundy, J.E., Mackenzie, J.S., and Stanley, N.F. (1981). Influence of H-2 and non-H-2 genes on resistance to murine cytomegalovirus infection. Infect. Immunol. *32*, 277–286.

Guma, M., Angulo, A., Vilches, C., Gomez-Lozano, N., Malats, N., and Lopez-Botet, M. (2004). Imprint of human cytomegalovirus infection on the NK cell receptor repertoire. Blood.

Gunturi, A., Berg, R.E., and Forman, J. (2004). The role of CD94/NKG2 in innate and adaptive immunity. Immunol. Res. *30*, 29–34.

Hamerman, J.A., Ogasawara, K., and Lanier, L.L. (2004). Cutting edge: Toll-like receptor signaling in macrophages induces ligands for the NKG2D receptor. J. Immunol. *172*, 2001–2005.

Hanson, L.K., Slater, J.S., Karabekian, Z., Virgin, H.W.t., Biron, C.A., Ruzek, M.C., van Rooijen, N., Ciavarra, R.P., Stenberg, R.M., and Campbell, A.E. (1999). Replication of murine cytomegalovirus in differentiated macrophages as a determinant of viral pathogenesis. J. Virol. *73*, 5970–5980.

Harkins, L., Volk, A.L., Samanta, M., Mikolaenko, I., Britt, W.J., Bland, K.I., and Cobbs, C.S. (2002). Specific localisation of human cytomegalovirus nucleic acids and proteins in human colorectal cancer. Lancet *360*, 1557–1563.

Harte, M.T., Haga, I.R., Maloney, G., Gray, P., Reading, P.C., Bartlett, N.W., Smith, G.L., Bowie, A., and O'Neill, L.A. (2003). The poxvirus protein A52R targets Toll-like receptor signaling complexes to suppress host defense. J. Exp. Med. *197*, 343–351.

Hasan, M., Krmpotic, A., Ruzsics, Z., Bubic, I., Lenac, T., Halenius, A., Loewendorf, A., Messerle, M., Hengel, H., Jonjic, S., and Koszinowski, U.H. (2005). Selective down-regulation of the NKG2D ligand H60 by mouse cytomegalovirus m155 glycoprotein. J. Virol. *79*, 2920–2930.

Heath, W.R., and Carbone, F.R. (2001). Cross-presentation in viral immunity and self-tolerance. Nature Rev. Immunol. *1*, 126–134.

Heise, M.T., Connick, M., and Virgin, H.W.t. (1998). Murine cytomegalovirus inhibits interferon gamma-induced antigen presentation to CD4 T-cells by macrophages via regulation of expression of major histocompatibility complex class II-associated genes. J. Exp. Med. *187*, 1037–1046.

Hengel, H., and Koszinowski, U.H. (1997). Interference with antigen processing by viruses. Curr. Opin. Immunol. *9*, 470–476.

Hengel, H., Reusch, U., Geginat, G., Holtappels, R., Ruppert, T., Hellebrand, E., and Koszinowski, U.H. (2000). Macrophages escape inhibition of major histocompatibility complex class I-dependent antigen presentation by cytomegalovirus. J. Virol. *74*, 7861–7868.

Hertel, L., Lacaille, V.G., Strobl, H., Mellins, E.D., and Mocarski, E.S. (2003). Susceptibility of immature and mature Langerhans cell-type dendritic cells to infection and immunomodulation by human cytomegalovirus. J. Virol. *77*, 7563–7574.

Hornung, V., Rothenfusser, S., Britsch, S., Krug, A., Jahrsdorfer, B., Giese, T., Endres, S., and Hartmann, G. (2002). Quantitative expression of Toll-like receptor 1–10 mRNA in cellular subsets of human peripheral blood mononuclear cells and sensitivity to CpG oligodeoxynucleotides. J. Immunol. *168*, 4531–4537.

Howley, P.M., Ganem, D., and Kieff, E. (2001). Cancer: Principles and Practice of Oncology. In DNA viruses, V. T. De Vita, S. Jr. Hellman, and S. A. Rosenberg, eds. (Philadelphia, Lippincott Williams and Wilkins), pp. 168–173.

Iizuka, K., Naidenko, O.V., Plougastel, B.F., Fremont, D.H., and Yokoyama, W.M. (2003). Genetically linked C-type lectin-related ligands for the NKRP1 family of natural killer cell receptors. Nature Immunol. *4*, 801–807.

Jarvis, M.A., and Nelson, J.A. (2002). Mechanisms of human cytomegalovirus persistence and latency. Front Biosci. *7*, d1575–1582.

Johnson, D.C., Frame, M.C., Ligas, M.W., Cross, A.M., and Stow, N.D. (1988). Herpes simplex virus immunoglobulin G Fc receptor activity depends on a complex of two viral glycoproteins, gE and gI. J. Virol. *62*, 1347–1354.

Jun, Y., Kim, E., Jin, M., Sung, H.C., Han, H., Geraghty, D.E., and Ahn, K. (2000). Human cytomegalovirus gene products US3 and US6 down-regulate trophoblast class I MHC molecules. J. Immunol. *164*, 805–811.

Karlhofer, F.M., and Yokoyama, W.M. (1991). Stimulation of murine natural killer (NK) cells by a monoclonal antibody specific for the NK1.1 antigen. IL-2-activated NK cells possess additional specific stimulation pathways. J. Immunol. *146*, 3662–3673.

Kavanagh, D.G., Gold, M.C., Wagner, M., Koszinowski, U.H., and Hill, A.B. (2001a). The multiple immune-evasion genes of murine cytomegalovirus are not redundant: m4 and m152 inhibit antigen presentation in a complementary and cooperative fashion. J. Exp. Med. *194*, 967–978.

Kavanagh, D.G., Koszinowski, U.H., and Hill, A.B. (2001b). The murine cytomegalovirus immune evasion protein m4/gp34 forms biochemically distinct complexes with class I MHC at the cell surface and in a pre-Golgi compartment. J. Immunol. 167, 3894–3902.

Khakoo, S.I., Thio, C.L., Martin, M.P., Brooks, C.R., Gao, X., Astemborski, J., Cheng, J., Goedert, J.J., Vlahov, D., Hilgartner, M., Cox S, Little, A.M., Alexander, G.J., Cramp, M.E., O'Brien, S.J., Rosenberg, W.M., Thomas D.L., and Carrington, M. (2004). HLA and NK cell inhibitory receptor genes in resolving hepatitis C virus infection. Science 305, 872–874.

Kleijnen, M.F., Huppa, J.B., Lucin, P., Mukherjee, S., Farrell, H., Campbell, A.E., Koszinowski, U.H., Hill, A.B., and Ploegh, H.L. (1997). A mouse cytomegalovirus glycoprotein, gp34, forms a complex with folded class I MHC molecules in the ER which is not retained but is transported to the cell surface. EMBO J. 16, 685–694.

Kloover, J.S., Grauls, G.E., Blok, M.J., Vink, C., and Bruggeman, C.A. (2002). A rat cytomegalovirus strain with a disruption of the r144 MHC class I-like gene is attenuated in the acute phase of infection in neonatal rats. Arch. Virol. 147, 813–824.

Kotenko, S.V., Saccani, S., Izotova, L.S., Mirochnitchenko, O.V., and Pestka, S. (2000). Human cytomegalovirus harbors its own unique IL-10 homolog (cmvIL-10). Proc. Natl. Acad. Sci. USA 97, 1695–1700.

Kraft, J.R., Vance, R.E., Pohl, J., Martin, A.M., Raulet, D.H., and Jensen, P.E. (2000). Analysis of Qa-1(b) peptide binding specificity and the capacity of CD94/NKG2A to discriminate between Qa-1-peptide complexes. J. Exp. Med. 192, 613–624.

Krmpotic, A., Bubic, I., Polic, B., Lucin, P., and Jonjic, S. (2003). Pathogenesis of murine cytomegalovirus infection. Microbes Infect 5, 1263–1277.

Krmpotic, A., Busch, D.H., Bubic, I., Gebhardt, F., Hengel, H., Hasan, M., Scalzo, A.A., Koszinowski, U.H., and Jonjic, S. (2002). MCMV glycoprotein gp40 confers virus resistance to CD8+ T-cells and NK cells in vivo. Nature Immunol. 3, 529–535.

Krmpotic, A.; Hasan, M.; Loewendorf, A.; Saulig, T.; Halenius, A.; Lenac, T.; Polic, B.; Bubic, I.; Kriegeskorte, A.; Pernjak-Pugel, E.; Messerle, M.; Hengel, H.; Busch, D.H.; Koszinowski, U.H.; and Jonjic, S.(2005). NK cell activation through the NKG2D ligand MULT-1 is selectively prevented by the glycoprotein encoded by mouse cytomegalovirus gene m145. J. Exp. Med. 201, 211–220.

Krug, A., French, A.R., Barchet, W., Fischer, J.A., Dzionek, A., Pingel, J.T., Orihuela, M.M., Akira, S., Yokoyama, W.M., and Colonna, M. (2004). TLR9-dependent recognition of MCMV by IPC and DC generates coordinated cytokine responses that activate antiviral NK cell function. Immunity 21, 107–119.

Kubin, M., Cassiano, L., Chalupny, J., Chin, W., Cosman, D., Fanslow, W., Mullberg, J., Rousseau, A.M., Ulrich, D., and Armitage, R. (2001). ULBP1, 2, 3: novel MHC class I-related molecules that bind to human cytomegalovirus glycoprotein UL16, activate NK cells. Eur. J. Immunol. 31, 1428–1437.

Kubota, A., Kubota, S., Farrell, H.E., Davis-Poynter, N., and Takei, F. (1999). Inhibition of NK cells by murine CMV-encoded class I MHC homologue m144. Cell Immunol. 191, 145–151.

Lanier, L.L. (2003). Natural killer cell receptor signaling. Curr. Opin. Immunol. 15, 308–314.

Lanier, L.L., Chang, C., and Phillips, J.H. (1994). Human NKR-P1A. A disulfide-linked homodimer of the C-type lectin superfamily expressed by a subset of NK and T-lymphocytes. J. Immunol. 153, 2417–2428.

Le Bouteiller, P. (1997). HLA-G: on the track of immunological functions. Eur. J. Immunogenet. 24, 397–408.

Le Gal, F.A., Riteau, B., Sedlik, C., Khalil-Daher, I., Menier, C., Dausset, J., Guillet, J.G., Carosella, E.D., and Rouas-Freiss, N. (1999). HLA-G-mediated inhibition of antigen-specific cytotoxic T-lymphocytes. Int. Immunol. 11, 1351–1356.

Lee, S.H., Girard, S., Macina, D., Busa, M., Zafer, A., Belouchi, A., Gros, P., and Vidal, S.M. (2001). Susceptibility to mouse cytomegalovirus is associated with deletion of an activating natural killer cell receptor of the C-type lectin superfamily. Nature Genet. 28, 42–45.

Lee, S.H., Zafer, A., de Repentigny, Y., Kothary, R., Tremblay, M.L., Gros, P., Duplay, P., Webb, J.R., and Vidal, S.M. (2003). Transgenic expression of the activating natural killer receptor Ly49H confers resistance to cytomegalovirus in genetically susceptible mice. J. Exp. Med. 197, 515–526.

Lertmemongkolchai, G., Cai, G., Hunter, C.A., and Bancroft, G.J. (2001). Bystander activation of CD8+ T-cells contributes to the rapid production of IFN-gamma in response to bacterial pathogens. J. Immunol. 166, 1097–1105.

Lilley, B.N., Ploegh, H.L., and Tirabassi, R.S. (2001). Human cytomegalovirus open reading frame TRL11/IRL11 encodes an immunoglobulin G Fc-binding protein. J. Virol. 75, 11218–11221.

Llano, M., Guma, M., Ortega, M., Angulo, A., and Lopez-Botet, M. (2003). Differential effects of US2, US6 and US11 human cytomegalovirus proteins on HLA class Ia and HLA-E expression: impact on target susceptibility to NK cell subsets. Eur. J. Immunol. 33, 2744–2754.

Llano, M., Lee, N., Navarro, F., Garcia, P., Albar, J.P., Geraghty, D.E., and Lopez-Botet, M. (1998). HLA-E-bound peptides influence recognition by inhibitory and triggering CD94/NKG2 receptors: preferential response to an HLA-G-derived nonamer. Eur. J. Immunol. 28, 2854–2863.

Lodoen, M., Ogasawara, K., Hamerman, J.A., Arase, H., Houchins, J.P., Mocarski, E.S., and Lanier, L.L. (2003). NKG2D-mediated natural killer cell protection against cytomegalovirus is impaired by viral gp40 modulation of retinoic acid early inducible 1 gene molecules. J. Exp. Med. 197, 1245–1253.

Lodoen, M.B., Abenes, G., Umamoto, S., Houchins, J.P., Liu, F., and Lanier, L.L. (2004). The Cytomegalovirus m155 Gene Product Subverts Natural Killer Cell Antiviral Protection by Disruption of H60-NKG2D Interactions. J. Exp. Med..

LoPiccolo, D.M., Gold, M.C., Kavanagh, D.G., Wagner, M., Koszinowski, U.H., and Hill, A.B. (2003). Effective inhibition of K(b)- and D(b)-restricted antigen presentation in primary macrophages by murine cytomegalovirus. J. Virol. 77, 301–308.

Ljunggren, H.G., and Karre, K. (1990). In search of the 'missing self': MHC molecules and NK cell recognition. Immunol. Today 11, 237–244.

MacDonald, M.R., Li, X.Y., and Virgin, H.W.t. (1997). Late expression of a beta chemokine homolog by murine cytomegalovirus. J. Virol. 71, 1671–1678.

Malarkannan, S., Shih, P.P., Eden, P.A., Horng, T., Zuberi, A.R., Christianson, G., Roopenian, D., and Shastri, N. (1998). The molecular and functional characterization of a dominant minor H antigen, H60. J. Immunol. 161, 3501–3509.

Martin, M.P., Gao, X., Lee, J.H., Nelson, G.W., Detels, R., Goedert, J.J., Buchbinder, S., Hoots, K., Vlahov, D., Trowsdale, J., Wilson, M., O'Brien, S.J., and Carrington, M.(2002). Epistatic interaction between KIR3DS1 and HLA-B delays the progression to AIDS. Nature Genet. 31, 429–434.

Mathys, S., Schroeder, T., Ellwart, J., Koszinowski, U.H., Messerle, M., and Just, U. (2003). Dendritic cells under influence of mouse cytomegalovirus have a physiologic dual role: to initiate and to restrict T-cell activation. J. Infect. Dis. 187, 988–999.

Medzhitov, R., Preston-Hurlburt, P., and Janeway, C.A., Jr. (1997). A human homologue of the Drosophila Toll protein signals activation of adaptive immunity. Nature 388, 394–397.

Medzhitov, R., Preston-Hurlburt, P., Kopp, E., Stadlen, A., Chen, C., Ghosh, S., and Janeway, C.A., Jr. (1998). MyD88 is an adaptor protein in the hToll/IL-1 receptor family signaling pathways. Mol. Cell 2, 253–258.

Merrill, J.E., Ullberg, M., and Jondal, M. (1981). Influence of IgG and IgM receptor triggering on human natural killer cell cytotoxicity measured on the level of the single effector cell. Eur. J. Immunol. 11, 536–541.

Michaelsson, J., Teixeira de Matos, C., Achour, A., Lanier, L.L., Karre, K., and Soderstrom, K. (2002). A signal peptide derived from hsp60 binds HLA-E and interferes with CD94/NKG2A recognition. J. Exp. Med. 196, 1403–1414.

Mocarski, E.S., and Courcelle, C.T. (2001). Cytomegaloviruses and Their Replication. In Fields Virology, D. M. Knipe, and P. M. Howley, eds. (Philadelphia, Lippincott Williams&Wilkins), pp. 2629–2674.

Mocarski, E.S., Jr. (2002). Immunomodulation by cytomegaloviruses: manipulative strategies beyond evasion. Trends Microbiol. 10, 332–339.

Moretta, A., Bottino, C., Vitale, M., Pende, D., Cantoni, C., Mingari, M.C., Biassoni, R., and Moretta, L. (2001). Activating receptors and coreceptors involved in human natural killer cell-mediated cytolysis. Annu. Rev. Immunol. 19, 197–223.

Moutaftsi, M., Brennan, P., Spector, S.A., and Tabi, Z. (2004). Impaired lymphoid chemokine-mediated migration due to a block on the chemokine receptor switch in human cytomegalovirus-infected dendritic cells. J. Virol. 78, 3046–3054.

Moutaftsi, M., Mehl, A.M., Borysiewicz, L.K., and Tabi, Z. (2002). Human cytomegalovirus inhibits maturation and impairs function of monocyte-derived dendritic cells. Blood 99, 2913–2921.

Nomura, M., Zou, Z., Joh, T., Takihara, Y., Matsuda, Y., and Shimada, K. (1996). Genomic structures. and characterization of Rae1 family members encoding GPI-anchored cell surface proteins and expressed predominantly in embryonic mouse brain. J. Biochem. (Tokyo) 120, 987–995.

Nordoy, I., Rollag, H., Lien, E., Sindre, H., Degre, M., Aukrust, P., Froland, S.S., and Muller, F. (2003). Cytomegalovirus infection induces production of human interleukin-10 in macrophages. Eur. J. Clin. Microbiol. Infect. Dis. 22, 737–741.

Oliveira, S.A., Park, S.H., Lee, P., Bendelac, A., and Shenk, T.E. (2002). Murine cytomegalovirus m02 gene family protects against natural killer cell-mediated immune surveillance. J. Virol. 76, 885–894.

Onno, M., Pangault, C., Le Friec, G., Guilloux, V., Andre, P., and Fauchet, R. (2000). Modulation of HLA-G antigens expression by human cytomegalovirus: specific induction in activated macrophages harboring human cytomegalovirus infection. J. Immunol. 164, 6426–6434.

Orange, J.S., Wang, B., Terhorst, C., and Biron, C.A. (1995). Requirement for natural killer cell-produced interferon gamma in defense against murine cytomegalovirus infection and enhancement of this defense pathway by interleukin 12 administration. J. Exp. Med. 182, 1045–1056.

Penfold, M.E., Dairaghi, D.J., Duke, G.M., Saederup, N., Mocarski, E.S., Kemble, G.W., and Schall, T.J. (1999). Cytomegalovirus encodes a potent alpha chemokine. Proc. Natl. Acad. Sci. USA 96, 9839–9844.

Plougastel, B., Matsumoto, K., Dubbelde, C., and Yokoyama, W.M. (2001). Analysis of a 1-Mb BAC contig overlapping the mouse Nkrp1 cluster of genes: cloning of three new Nkrp1 members, Nkrp1d, Nkrp1e, and Nkrp1f. Immunogenetics 53, 592–598.

Podlech, J., Holtappels, R., Grzimek, N.K., and Reddehase, M.J. (2002). Animal models: murine cytomegalovirus. In Methods in Microbiology, S.H.E. Kaufmann, and D. Kabelitz, eds. (London, UK and San Diego, CA, Academic Press), pp. 493–525.

Pollock, J.L., Presti, R.M., Paetzold, S., and Virgin, H.W.t. (1997). Latent murine cytomegalovirus infection in macrophages. Virology 227, 168–179.

Raftery, M.J., Schwab, M., Eibert, S.M., Samstag, Y., Walczak, H., and Schonrich, G. (2001). Targeting the function of mature dendritic cells by human cytomegalovirus: a multilayered viral defense strategy. Immunity 15, 997–1009.

Rajagopalan, S., and Long, E.O. (1999). A human histocompatibility leukocyte antigen (HLA)-G-specific receptor expressed on all natural killer cells. J. Exp. Med. 189, 1093–1100.

Raulet, D.H. (2003a). Natural Killer Cells. In Fundamental Immunology, W.E. Paul, ed. (Philadelphia, Lippincott Williams & Wilkins), pp. 365–391.

Raulet, D.H. (2003b). Roles of the NKG2D immunoreceptor and its ligands. Nature Rev. Immunol. 3, 781–790.

Raulet, D.H., Vance, R.E., and McMahon, C.W. (2001). Regulation of the natural killer cell receptor repertoire. Annu. Rev. Immunol. 19, 291–330.

Reddehase, M.J. (2002). Antigens and immunoevasins: opponents in cytomegalovirus immune surveillance. Nature Rev. Immunol. 2, 831–844.

Reddehase, M.J., Podlech, J., and Grzimek, N.K. (2002). Mouse models of cytomegalovirus latency: overview. J. Clin. Virol. 25 Suppl 2, S23–36.

Redpath, S., Angulo, A., Gascoigne, N.R., and Ghazal, P. (1999). Murine cytomegalovirus infection down-regulates MHC class II expression on macrophages by induction of IL-10. J. Immunol. 162, 6701–6707.

Reis e Sousa, C. (2004). Activation of dendritic cells: translating innate into adaptive immunity. Curr. Opin. Immunol. 16, 21–25.

Reyburn, H.T., Mandelboim, O., Vales-Gomez, M., Davis, D.M., Pazmany, L., and Strominger, J.L. (1997). The class I MHC homologue of human cytomegalovirus inhibits attack by natural killer cells. Nature 386, 514–517.

Riegler, S., Hebart, H., Einsele, H., Brossart, P., Jahn, G., and Sinzger, C. (2000). Monocyte-derived dendritic cells are permissive to the complete replicative cycle of human cytomegalovirus. J. Gen. Virol. 81, 393–399.

Saederup, N., Aguirre, S.A., Sparer, T.E., Bouley, D.M., and Mocarski, E.S. (2001). Murine cytomegalovirus CC chemokine homolog MCK-2 (m131–129) is a determinant of dissemination that increases inflammation at initial sites of infection. J. Virol. 75, 9966–9976.

Salazar-Mather, T.P., Hamilton, T.A., and Biron, C.A. (2000). A chemokine-to-cytokine-to-chemokine cascade critical in antiviral defense. J. Clin. Invest. 105, 985–993.

Scalzo, A.A. (2002). Successful control of viruses by NK cells–a balance of opposing forces? Trends Microbiol. 10, 470–474.

042

||||||||||||||||||||||||||||||||||||||||||||||
\*P01OBE2D7\*

**LIVERPOOL JOHN MOORES UNIVERSITY**

7

| | | | |
|---|---|---|---|
| | | UK 18589001 F | |
| **Ship To:** | | | |
| LIVERPOOL JOHN MOORES UNi | | | |
| GROUND FLOOR,RM RF07, ACQ | | | |
| MARYLAND STREET | | | |
| LIVERPOOL | | | |
| MERSEYSIDE | | | |
| L1 9DE | | | |
| **Volume:** | | | |
| **Edition:** | | | |
| **Year:** | | 2005 | |
| **Pagination:** | | 650 p. : | |
| **Size:** | | 24 cm. | |

| | |
|---|---|
| **Routing** | 1 |
| **Sorting** | |
| **Y14A05X** | |
| **Inpro** | |
| **RFID** | |
| **Covering — BXAXX** | |
| **Despatch** | |

042384073 ukrwlg10 RC1

| ISBN | Qty | Sales Order |
|---|---|---|
| 9781904455028 | 1 | F 16995196 1 |
| **Customer P/O No** | | **Cust P/O List** |
| 272085 | | 140.00 GBP |
| **Title:** Cytomegaloviruses : | | |

**Format:** C (Cloth/HB)
**Author:**
**Publisher:** Caister Academic Press
**Fund:** BML1–2013
**Location:** NONE
**Loan Type:** NONE
**Coutts CN:** 4662562

**Order Specific Instructions**
LIBRARY NOTE: WEB — # DC,
1X21DM # DUPLICATE

tion, expressing the luciferase gene under control of a promoter containing five consecutive ISRE elements and thus responding to IFN in a dose-dependent manner (Zimmermann et al., 2005). Following MCMV infection, the ISRE-dependent luciferase expression in response to IFN-α continuously declined and was completely abolished within 36 hours. This effect required MCMV E gene expression. To identify gene products responsible for the inhibition of IFN-α in infected cells, a forward-genetic procedure based on BACs of the complete MCMV genome (see Chapter 4) was used. A library of BAC-derived MCMV mutants obtained by random single TnMax16 transposon insertion mutagenesis of the MCMV genome (Brune et al., 2001) was phenotypically screened using the ISRE reporter gene (Zimmermann et al., 2005). This procedure identified three independent TnMax16 clones whose insertion site in all three mutants was mapped to the M27 ORF. Interestingly, the M27 ORF has previously been reported to be dispensable for MCMV replication in vitro but to play an important role in growth and virulence in mice (Abenes et al., 2001).

In order to confirm the effect of the *M27* gene on IFN-dependent gene expression, a targeted MCMV mutant lacking the complete M27 ORF and a revertant virus in which the M27 ORF was reinserted into the ΔM27 genome (M27rev) were constructed. As expected, the BAC-derived MCMV WT and M27rev inhibited the IFN-α induced ISRE-dependent luciferase activity, whereas it was largely preserved after infection with the mutant ΔM27 (Zimmermann et al., 2005). An M27-HA epitope-tagged mutant was used to determine the kinetics of M27 protein expression in the course of productive infection. The 79 kDa M27 protein (pM27) exhibited a typical E-L pattern of gene expression compatible with the time course of the MCMV-induced inhibition of ISRE-dependent gene expression. A multistep replication analysis with MCMV WT and ΔM27 in the presence of IFN yielded 20-fold lower titers in the presence of IFN-α at 96 hours p.i., thus confirming an enhanced antiviral effect of IFN-α in the absence of *M27* (Zimmermann et al., 2005).

To identify the elements of the Jak-STAT pathway targeted by pM27, the levels of STAT proteins were compared in cells infected with MCMV WT, ΔM27 and M27rev by immunoblot analysis. STAT2 was found to be strongly down-regulated at 24 hours p.i. and almost undetectable at 48 hours p.i in cells infected with WT virus and M27-HA, but not in cells infected with ΔM27. The data indicated that MCMV pM27 controls the response of infected cells to IFNs by down-regulating the expression levels of STAT2. Expression of pM27-FLAG by a recombinant vaccinia virus proved that pM27 alone is sufficient to mediate the down-regulation of STAT2. The pM27 mechanism of action was determined to involve ubiquitination of STAT2, resulting in subsequent proteasomal degradation (Zimmermann et al., 2005).

In contrast to the situation in HCMV, STAT1 and IRF9/p48 were both not affected by MCMV infection. No interference with STAT1 phosphorylation and binding to DNA was seen at any time point after MCMV infection indicating different targets being attacked by both CMVs (Zimmermann et al., 2005). Strikingly, in spite of the fact that STAT2 is the principal target of pM27 a much more dramatic effect of *M27* expression upon MCMV replication was observed after pretreatment with IFN-γ compared with IFN-α. The replication of ΔM27 was almost completely inhibited resulting in a reduction of the viral progeny by three to four orders of magnitude, indicating that pM27 acts antagonistically to both IFN-γ and IFN-α (Zimmermann et al., 2005). This finding underlined the importance of type I and type II synergism to combat MCMV infection.

Collectively, control of the Jak-STAT pathway is an important condition for MCMV spread and pathogenicity in vivo. HCMV and MCMV employ different molecular mechanisms to achieve this goal by blocking Jak-STAT during the early and late phase of virus replication.

## Interference of CMV with IFN-induced effector proteins and gene transcription

In simplified terms, the IFN-inducible gene products can be subdivided into two major groups. One group of factors exhibits a direct antiviral function by interfering with viral gene expression or viral protein synthesis. The other group regulates immune responses by acting on antigen presentation and immune cell activation, thereby connecting the innate and the adaptive immune system (Figure 16.5). Several type I IFN-induced proteins are established inhibitors of RNA virus replication. Among these are PKR (protein kinase R), OAS (2′,5′-oligoadenylate synthetases) and RNase L, the ADAR (RNA-specific adenosine desaminase), and the Mx GTPases (Samuel, 2001). IDO and iNOS are also enzymes with direct antiviral potency, but they are mainly induced by IFN-γ.

PKR is a multifunctional cellular protein kinase, which becomes activated upon binding of dsRNA generated during viral infection leading to activation by autophosphorylation and subsequent dimerization. Activated PKR is capable of phosphorylating eIF-2α (eukaryotic translation initiation factor 2α). Phosphorylation results in inactivation of eIF-2α and a general shut-off of cellular protein synthesis, thereby blocking viral gene expression and preventing replication. OAS catalyze the synthesis of oligoadenylate. Binding of these unusual nucleic acids triggers dimerization and activation of RNase L. Activated RNase L cleaves both rRNA and mRNA. Again, inhibition of viral gene expression is achieved by a global inhibition of cellular protein synthesis. ADAR performs RNA editing and modifies RNA by deamination of adenosine to inosine, resulting in hypermutation of viral but also cellular transcripts. Due to their global effects on cellular transcription and protein synthesis, PKR and OAS require additional activation via their dsRNA-binding motifs for complete activation.

Mx proteins are IFN-induced GTPases belonging to the dynamin superfamily of large GTPases and exhibit strong antiviral activity against a wide range of RNA viruses. Mx GTPases appear to detect viral infection by sensing nucleocapsid-like structures. As a consequence, these viral components are trapped and sorted to locations where they become unavailable for the generation of new virus particles (Haller et al., 2002). Further GTP-binding proteins (GBP) exhibiting potential antiviral activity, namely the p65 GBPs and the p47 GTPases, have been described (reviewed by Taylor et al., 2004).

IDO induces the host cells to degrade tryptophan to kynurenine and exerts its antiviral effect by limitation of the cellular tryptophan pool (Bodaghi et al., 1999). iNOS synthesizes nitric oxide (NO) from arginine. Besides its immunomodulatory properties NO may provide a direct antiviral potency by inhibiting cellular and viral iron-dependent ribonucleotide reductases, which results in a limited supply of deoxynucleotides (Croen, 1993).

Despite of the CMV-mediated inhibition of IFN induction and IFNR signaling described above there is still a need for CMVs to deal with IFN-induced effector proteins. For replicating in an IFN-producing environment, viruses have to develop mechanisms

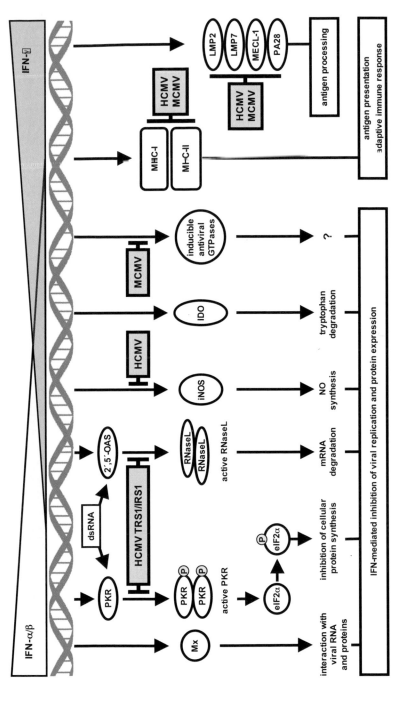

**Figure 16.5** Antiviral gene expression induced by IFNs. Proteins, which suppress virus replication by global interference with transcription or protein synthesis (left), are potentially harmful to the host cell and vigorously controlled. CMV targets PKR and OAS via their requirement for activation by dsRNA (mediated by HCMV TRS1/IRS1), iNOS and possibly antiviral GTPases. Host proteins engaged in antigen presentation and adaptive immune response (right) become affected either by subversion of the IFN signaling pathways or by CMV proteins directly interacting with the MHC pathway of antigen presentation (Benz et al., 2000).

allowing them to enter and to disarm the host, which has already reached the IFN-induced antiviral state. The most prominent mechanism to circumvent dsRNA-triggered antiviral activities is the expression of viral dsRNA binding proteins. A striking example provided by a DNA virus is the vaccinia-encoded interferon antagonist E3L, which has been described to be a dsRNA-binding protein capable of inhibiting both the PKR and OAS-mediated antiviral response (Smith et al., 1997). The first evidence for HCMV counteracting PKR and RNase L came from the observation that HCMV infection provides functional complementation of E3L function in the context of a vaccinia virus deletion mutant (VVΔE3L) lacking E3L (Child et al., 2002). The responsible genes were identified from an HCMV genome library expressed in VVΔE3L. The gene product TRS1 and the closely related IRS1 proved to rescue replication of VVΔE3L in human fibroblasts, to prevent PKR-mediated phosphorylation of eIF-2α, to block activation of RNase L, and to maintain active cellular protein synthesis (Child et al., 2004).

The role of IFN-induced antiviral proteins in the antiviral defense against MCMV infection in vivo has been studied in mouse gene knockout models lacking the p47 GTPAses IGTP, IRG-47, and LRG-47. Infection studies revealed a complete resistance of MCMV against inducible GTPases by determination of virus replication in liver and spleen (Collazo et al., 2001; Taylor et al., 2000). The data might indicate that p47-GTPases do not target MCMV. Alternatively, resistance of MCMV against these factors could be due to the presence of specific viral inhibitors of these GTPases.

Besides their function in the induction of the antiviral state, the IFNs are potent stimulators of MHC gene transcription and antigen presentation. In fact, a multitude of genes is transcriptionally regulated by IFNs, among them MHC class I and MHC class II as well as essential components of the antigen processing machinery including TAP (transporter associated with antigen processing) and the inducible subunits of the immunoproteasome, LMP2, LMP7, and MECL-1. This gives the IFNs an outstanding role in antigen presentation and immune regulation.

Besides its effect on IFN receptor signal transduction, CMVs have developed mechanisms enabling them to modulate IFN-inducible genes directly at the transcriptional level. HCMV is capable of repressing the expression of the IFN-γ induced class II transactivator (CIITA) within the first hours of infection. Notably, in contrast to the pp65-mediated pro-viral effect, CIITA repression depends on viral protein synthesis, but appears to take place independent of HCMV inhibition of the Jak-STAT pathway (Le Roy et al., 1999). Similarly, MCMV interferes with IFN-γ inducible MHC class II expression in macrophages at a stage subsequent to STAT1 activation and nuclear translocation. The proposed mechanism for the MCMV interference with CIITA gene transcription is a specific inhibition of IFN-γ-induced chromosomal promoter assembly (Popkin et al., 2003).

## Future perspectives

CMVs are well known as immune escape artists evading from antigen specific immune responses (Hengel et al., 1998; Mocarski, 2004). Recent research has demonstrated that CMVs are also able to occlude the innate IFN response at multiple sites. While fast replicating viruses with a smaller coding capacity like RNA viruses appear to be specialized to handle distinct steps of the complex IFN response such as IFN induction (e.g. influenza viruses), IFN signaling (e.g. parainfluenza viruses), or IFN effector proteins like PKR (e.g.

polioviruses), CMVs handle all of them. CMVs have been shown to interfere with the primary induction of type I interferons, to block signal transduction pathways of both type I and type II IFNs, and last but not least to inhibit IFN-induced gene products at the same time. In such an scenario, the extremely complex IFN system of the host appears to be checked and balanced by an adapted and similarly complex anti-IFN system of the virus. The CMV inhibitors known to date represent probably just the tip of an iceberg, and it is worthwhile to hunt for further CMV genes that counteract IFNs. Phenotypes still awaiting a genetic basis are the HCMV functions dealing with Jak1 and IRF9 inactivation as well as the unknown MCMV genes preventing IFN induction. A promising tool supporting the identification of these genes is the BAC technology (Wagner et al., 2002; see Chapter 4), which allows forward genetic approaches (Zimmermann et al., 2005), the construction of targeted CMV knockout viruses as well as introduction of point mutations into CMV genomes. The phenotypic screening of CMV mutants should provide a more complete understanding of the CMV genes involved in blocking IFN responses. Once a candidate gene is identified, the next step is the elucidation of the precise molecular mechanisms by which this CMV inhibitor corrupts the IFN response.

Recent work has shown that a single IFN antagonist, pM27, is strictly required for efficient CMV replication in vivo. Accordingly, lack of pM27 resulted in a drastically attenuated phenotype in both immunocompetent as well as in immunocompromised mice deficient in specific components of the IFN signal transduction pathway (Abenes et al., 2001; Zimmermann et al., 2005). From such a perspective, IFN antagonists appear to be attractive new drug targets, the inactivation of which could allow effective antiviral chemotherapy.

Besides their direct activity in promoting virus replication in the presence of IFNs, IFN antagonists are likely to have immunomodulatory pro-viral effects as well. In the mouse system, the antiviral effector function of both CD8$^+$ as well as CD4$^+$ T-cells depends critically on IFN-$\gamma$ enhancing the antigen processing and presentation function of target cells subsequently infected with CMV (Geginat et al., 1997; Hengel et al., 1994; Lucin et al., 1992). As outlined above, once infected with CMV, cells become resistant to IFN-$\gamma$ induced MHC class II expression as well as to IFN-$\gamma$ enhanced antigen presentation to CD8$^+$ T-cells (Cebulla et al., 2002; Heise et al., 1998; Hengel et al., 1994; Hengel et al., 1995; Miller et al., 1998; Miller et al., 2000; Popkin et al., 2003). It is therefore conceivable that, with regard to antigen presentation, substantial differences must exist between MCMV-infected cells expressing or lacking IFN-antagonistic proteins. In fact, expression of IFN-$\gamma$ inducible proteasome subunits, which is prevented in WT-MCMV infected cells, is preserved in the absence of *M27* (Khan et al., 2004). Given the aforementioned multiple immunomodulatory effects one may presume that CMVs lacking distinct IFN-antagonistic proteins might be optimal inducers of the innate as well to the adaptive immune response due to enhanced antigen presentation. In consequence, a live attenuated vaccine, which is still safe and effective, may be developed from CMV mutants exhibiting reduced ability to subvert IFN responses of the host.

## Acknowledgments
Our work was supported by grants from the Deutsche Forschungsgemeinschaft and EU grant QLRT-2001-01112.

References

Aaronson, D.S. and Horvath, C.M. (2002). A road map for those who know JAK-STAT. Science 296, 1653–1655.

Abate, D.A., Watanabe, S., and Mocarski, E.S. (2004). Major human cytomegalovirus structural protein pp65 (ppUL83) prevents interferon response factor 3 activation in the interferon response. J. Virol. 78, 10995–11006.

Abenes, G., Lee, M., Haghjoo, E., Tong, T., Zhan, X., and Liu, F. (2001). Murine cytomegalovirus open reading frame M27 plays an important role in growth and virulence in mice. J. Virol. 75, 1697–1707.

Akira, S. and Takeda, K. (2004). Toll-like receptor signalling. Nature Rev. Immunol. 4, 499–511.

Barchet, W., Cella, M., Odermatt, B., Asselin-Paturel, C., Colonna, M., and Kalinke, U. (2002). Virus-induced interferon alpha production by a dendritic cell subset in the absence of feedback signaling in vivo. J. Exp. Med. 195, 507–516.

Benz, C. and Hengel, H. (2000). MHC class I-subversive gene functions of cytomegalovirus and their regulation by interferons-an intricate balance. Virus Genes 21, 39–47.

Bodaghi, B., Goureau, O., Zipeto, D., Laurent, L., Virelizier, J.L., and Michelson, S. (1999). Role of IFN-gamma-induced indoleamine 2,3 dioxygenase and inducible nitric oxide synthase in the replication of human cytomegalovirus in retinal pigment epithelial cells. J. Immunol. 162, 957–964.

Boehm, U., Klamp, T., Groot, M., and Howard, J.C. (1997). Cellular responses to interferon-gamma. Annu. Rev. Immunol. 15, 749–795.

Boehme, K.W. and Compton, T. (2004). Innate sensing of viruses by Toll-like receptors. J. Virol. 78, 7867–7873.

Boehme, K.W., Singh, J., Perry, S.T., and Compton, T. (2004). Human cytomegalovirus elicits a coordinated cellular antiviral response via envelope glycoprotein B. J. Virol. 78, 1202–1211.

Browne, E.P. and Shenk, T. (2003). Human cytomegalovirus UL83-coded pp65 virion protein inhibits antiviral gene expression in infected cells. Proc. Natl. Acad. Sci. USA 100, 11439–11444.

Browne, E.P., Wing, B., Coleman, D., and Shenk, T. (2001). Altered cellular mRNA levels in human cytomegalovirus-infected fibroblasts: viral block to the accumulation of antiviral mRNAs. J. Virol. 75, 12319–12330.

Brune, W., Menard, C., Heesemann, J., and Koszinowski, U.H. (2001). A ribonucleotide reductase homolog of cytomegalovirus and endothelial cell tropism. Science 291, 303–305.

Cebulla, C.M., Miller, D.M., Zhang, Y., Rahill, B.M., Zimmerman, P., Robinson, J.M., and Sedmak, D.D. (2002). Human cytomegalovirus disrupts constitutive MHC class II expression. J. Immunol. 169, 167–176.

Child, S.J., Hakki, M., De Niro, K.L., and Geballe, A.P. (2004). Evasion of cellular antiviral responses by human cytomegalovirus TRS1 and IRS1. J. Virol. 78, 197–205.

Child, S.J., Jarrahian, S., Harper, V.M., and Geballe, A.P. (2002). Complementation of vaccinia virus lacking the double-stranded RNA-binding protein gene E3L by human cytomegalovirus. J. Virol. 76, 4912–4918.

Collazo, C.M., Yap, G.S., Sempowski, G.D., Lusby, K.C., Tessarollo, L., Woude, G.F., Sher, A., and Taylor, G.A. (2001). Inactivation of LRG-47 and IRG-47 reveals a family of interferon gamma-inducible genes with essential, pathogen-specific roles in resistance to infection. J. Exp. Med. 194, 181–188.

Compton, T., Kurt-Jones, E.A., Boehme, K.W., Belko, J., Latz, E., Golenbock, D.T., and Finberg, R.W. (2003). Human cytomegalovirus activates inflammatory cytokine responses via CD14 and Toll-like receptor 2. J. Virol. 77, 4588–4596.

Croen, K.D. (1993). Evidence for antiviral effect of nitric oxide. Inhibition of herpes simplex virus type 1 replication. J. Clin. Invest. 91, 2446–2452.

Cull, V.S., Bartlett, E.J., and James, C.M. (2002). Type I interferon gene therapy protects against cytomegalovirus-induced myocarditis. Immunology 106, 428–437.

Darnell, J.E., Jr. (1997). STATs and gene regulation. Science 277, 1630–1635.

Davignon, J.L., Castanie, P., Yorke, J.A., Gautier, N., Clement, D., and Davrinche, C. (1996). Anti-human cytomegalovirus activity of cytokines produced by CD4+ T-cell clones specifically activated by IE1 peptides in vitro. J. Virol. 70, 2162–2169.

Dorner, B.G., Smith, H.R., French, A.R., Kim, S., Poursine-Laurent, J., Beckman, D.L., Pingel, J.T., Kroczek, R.A., and Yokoyama, W.M. (2004). Coordinate expression of cytokines and chemokines by NK cells during murine cytomegalovirus infection. J. Immunol. 172, 3119–3131.

Erlandsson, L., Blumenthal, R., Eloranta, M.L., Engel, H., Alm, G., Weiss, S., and Leanderson, T. (1998). Interferon-beta is required for interferon-alpha production in mouse fibroblasts. Curr. Biol. 8, 223–226.

Fennie, E.H., Lie, Y.S., Low, M.A., Gribling, P., and Anderson, K.P. (1988). Reduced mortality in murine cytomegalovirus infected mice following prophylactic murine interferon-gamma treatment. Antiviral. Res. 10, 27–39.

Garcia-Sastre, A. (2001). Inhibition of interferon-mediated antiviral responses by influenza A viruses and other negative-strand RNA viruses. Virology 279, 375–384.

Geginat, G., Ruppert, T., Hengel, H., Holtappels, R., and Koszinowski, U.H. (1997). IFN-gamma is a prerequisite for optimal antigen processing of viral peptides in vivo. J. Immunol. 158, 3303–3310.

Gil, M.P., Bohn, E., O'Guin, A.K., Ramana, C.V., Levine, B., Stark, G.R., Virgin, H.W., and Schreiber, R.D. (2001). Biologic consequences of Stat1-independent IFN signaling. Proc. Natl. Acad. Sci. USA 98, 6680–6685.

Goodbourn, S., Didcock, L., and Randall, R.E. (2000). Interferons: cell signalling, immune modulation, antiviral response and virus countermeasures. J. Gen. Virol. 81, 2311–2364.

Gotoh, B., Komatsu, T., Takeuchi, K., and Yokoo, J. (2002). Paramyxovirus strategies for evading the interferon response. Rev. Med. Virol. 12, 337–357.

Gribaudo, G., Ravaglia, S., Caliendo, A., Cavallo, R., Gariglio, M., Martinotti, M.G., and Landolfo, S. (1993). Interferons inhibit onset of murine cytomegalovirus immediate-early gene transcription. Virology 197, 303–311.

Hakki, M. and Geballe, A.P. (2004). The dsRNA binding activity of the HCMV TRS1 gene is necessary but not sufficient for rescue of vaccinia virus ΔE3L. 29th International Herpesvirus Workshop, Reno, NV, USA, Abstract 11.16

Haller, O. and Kochs, G. (2002). Interferon-induced mx proteins: dynamin-like GTPases with antiviral activity. Traffic 3, 710–717.

Heise, M.T., Connick, M., and Virgin, H.W.t. (1998). Murine cytomegalovirus inhibits interferon gamma-induced antigen presentation to CD4 T-cells by macrophages via regulation of expression of major histocompatibility complex class II-associated genes. J. Exp. Med. 187, 1037–1046.

Hengel, H., Brune, W., and Koszinowski, U.H. (1998). Immune evasion by cytomegalovirus—survival strategies of a highly adapted opportunist. Trends Microbiol. 6, 190–197.

Hengel, H., Esslinger, C., Pool, J., Goulmy, E., and Koszinowski, U.H. (1995). Cytokines restore MHC class I complex formation and control antigen presentation in human cytomegalovirus-infected cells. J. Gen. Virol. 76, 2987–2997.

Hengel, H., Lucin, P., Jonjic, S., Ruppert, T., and Koszinowski, U.H. (1994). Restoration of cytomegalovirus antigen presentation by gamma interferon combats viral escape. J. Virol. 68, 289–297.

Hirsch, M.S., Schooley, R.T., Cosimi, A.B., Russell, P.S., Delmonico, F.L., Tolkoff-Rubin, N.E., Herrin, J.T., Cantell, K., Farrell, M.L., Rota, T.R., and Rubin, R.H. (1983). Effects of interferon-alpha on cytomegalovirus reactivation syndromes in renal-transplant recipients. N. Engl. J. Med. 308, 1489–1493.

Khan, S., Zimmermann, A., Basler, M., Groettrup, M., and Hengel, H. (2004). A cytomegalovirus inhibitor of gamma interferon signaling controls immunoproteasome induction. J. Virol. 78, 1831–1842.

Kisseleva, T., Bhattacharya, S., Braunstein, J., and Schindler, C.W. (2002). Signaling through the JAK/STAT pathway, recent advances and future challenges. Gene 285, 1–24.

Knight, D.A., Waldman, W.J., and Sedmak, D.D. (1997). Human cytomegalovirus does not induce human leukocyte antigen class II expression on arterial endothelial cells. Transplantation 63, 1366–1369.

Kotenko, S.V., Gallagher, G., Baurin, V.V., Lewis-Antes, A., Shen, M., Shah, N.K., Langer, J.A., Sheikh, F., Dickensheets, H., and Donnelly, R.P. (2003). IFN-lambdas mediate antiviral protection through a distinct class II cytokine receptor complex. Nature Immunol. 4, 69–77.

Krug, A., French, A.R., Barchet, W., Fischer, J.A., Dzionek, A., Pingel, J.T., Orihuela, M.M., Akira, S., Yokoyama, W.M., and Colonna, M. (2004). TLR9-dependent recognition of MCMV by IPC and DC generates coordinated cytokine responses that activate antiviral NK cell function. Immunity 21, 107–119.

Le Roy, E., Muhlethaler-Mottet, A., Davrinche, C., Mach, B., and Davignon, J.L. (1999). Escape of human cytomegalovirus from HLA-DR-restricted CD4(+) T-cell response is mediated by repression of gamma interferon-induced class II transactivator expression. J. Virol. 73, 6582–6589.

Levy, D.E. and Garcia-Sastre, A. (2001). The virus battles: IFN induction of the antiviral state and mechanisms of viral evasion. Cytokine Growth Factor Rev. 12, 143–156.

# Adaptive Cellular Immunity to Human Cytomegalovirus

17

*Mark R. Wills, Andrew J. Carmichael, and J. G. Patrick Sissons*

## Abstract

Primary human cytomegalovirus (HCMV) infection induces robust CD8[+] cytotoxic, and CD4[+] helper, T-cell mediated immune responses, which are associated with the resolution of acute primary infection: these responses are maintained at high frequency in long-term memory as the virus establishes persistent infection, with latency and periodic reactivation. Many of these T-cells are specific for epitopes in the pp65 and IE1 HCMV proteins, but it is becoming apparent that many other viral proteins can also be T-cell targets, and in some individuals pp65 and IE1 responses are not immunodominant. During long-term carriage of the virus a balance is established between the T-cell-mediated immune response and viral reactivation: the T-cell response controls viral spread following reactivation, but the virus encodes multiple genes that interfere with class I MHC processing (US2, 3, 6, and 11), and with class II MHC processing and NK cell killing, allowing limited viral evasion of the response. Loss of this balance is most evident in the immunocompromised host where reactivation of latent virus or primary infection can lead to unchecked viral replication, with consequent disease and mortality. This chapter describes current understanding of CD8[+] and CD4[+] T-cell responses to HCMV, and of how these responses reconstitute after bone marrow transplantation and might be used as therapies to protect against HCMV disease in immunocompromised subjects.

## Introduction

It is clear that primary HCMV infection elicits strong virus-specific CD8[+] T-cell responses to numerous viral proteins: evidence suggesting that these responses are protective is inferential, derived from the increased rates of HCMV disease in subjects with impaired T-cell immunity and from data obtained from patients undergoing reconstitution of their immune systems following bone marrow transplantation or stem cell transplantation (BMT/ SCT). In murine models of CMV, mice are protected from lethal murine cytomegalovirus (MCMV) challenge by CD8[+] T-cells specific for immediate-early antigens (Reddehase et al., 1987). In murine models of MCMV reactivation which are deficient in B-cell responses it has been demonstrated that CD4[+] and NK cells can substitute for CD8[+] T-cells in controlling reactivation (Polic et al., 1998). However in murine models of BMT, removal of reconstituted CD8[+] cells leads to lethal disease and reconstituted CD8[+] cells transferred to immunocompromised mice could prevent disease (Podlech et al., 2000). Following BMT

**Table 17.1** Class I MHC peptide epitopes recognized by human CD8[+] T-cells derived from a number of HCMV proteins

| ORF | Protein | Peptide | MHC Class I molecule | Tetramer* | Reference |
|-----|---------|---------|---------------------|-----------|-----------|
| UL32 | pp150 | 945-TTVYPPSSTAK-955 | HLA-A3 | | Longmate et al. (2001) |
| UL44 | pp50 | 792-QTVTSTPVQGR-802 | HLA-A68 | | Longmate et al. (2001) |
| | | 245-VTEHDTLLY-253 | HLA-A1 | | Elkington et al. (2003) |
| UL55 | gB | 618-(F)IAGNSAYEYV-628 | HLA-A2 | | Parker et al. (1992), Utz and Biddison (1992) |
| | | 731-AVGGAVASV-739 | HLA-A2 | | Elkington et al. (2003) |
| UL83 | pp65 | 363-YSEHPTFTSQY-373 | HLA-A1 | | Hebart et al. (2002), Longmate et al. (2001) |
| | | 14-VLGPISGHV-22 | HLA-A2 | | Solache et al. (1999) |
| | | 120-MLNIPSINV-128 | HLA-A2 | | Solache et al. (1999) |
| | | 490-ILARNLVPM-498 | HLA-A2 | | Elkington et al. (2003) |
| | | 495-NLVPMVATV-503 | HLA-A2 | Gillespie et al. (2000) | Diamond et al. (1997), Wills et al. (1996) |
| | | 522-RIFAELEGV-530 | HLA-A0207 | | Kondo et al. (2004) |
| | | 16-GPISGHVLK-24 | HLA-A11 | | Hebart et al. (2002; Longmate et al. (2001) |
| | | 501-ATVQGQNLK-509 | HLA-A11 | | Kondo et al. (2004) |
| | | 113-VYALPLKML-121 | HLA-A24 | a | Masuoka et al. (2001) |
| | | 341-QYDPVAALF-349 | HLA-A24 | Kuzushima et al. (2001) | Kuzushima et al. (2001) |
| | | 369-FTSQYRIQGKL-37 | HLA-A24 | | Longmate et al. (2001) |
| | | 186-FVFPTKDVALR-196 | HLA-A68 | | Longmate et al. (2001) |
| | | 265-RPHERNGFTV-274 | HLA-B7 | | Weekes et al. (1999b) |
| | | 417-(T)PRVTGGGAM-426 | HLA-B7 | Gillespie et al. (2000) | Kern et al. (1998), Wills et al. (1996) |
| | | 215-KMQVIGDQY-223 | HLA-B15 | a | Kondo et al. (2004) |
| | | 123-IPSINVHHY-131 | HLA-B35 | Hassan-Walker et al. (2001) | Gavin et al. (1993) |
| | | 187-VFPTKDVAL-195 | HLA-B35 | | Wills et al. (1996) |
| | | 174-NQWKEPDVY-182 | HLA-B35 | | Kern et al. (2002) |
| | | 367-PTFTSQYRIQGKL-379 | HLA-B38 | | Longmate et al. (2001) |

| Gene | Protein | Epitope | HLA restriction | | Reference |
|---|---|---|---|---|---|
| | | 232-CEDVPSGKL-240 | HLA-B40 | | Kondo et al. (2004) |
| | | 267-HERNGFTVL-275 | HLA-B40 | | Kondo et al. (2004) |
| | | 525-AELEGVWQPA-534 | HLA-B40 | | Kondo et al. (2004) |
| | | 364-SEHPTFTSQY-373 | HLA-B44 | | Kondo et al. (2004) |
| | | 512-EFFWDANDIY-521 | HLA-B44 | | Wills et al. (1996) |
| | | 545-DALPGPCI-552 | HLA-B51 | | Kondo et al. (2004) |
| | | 155-QMWQARLTV-163 | HLA-B52 | | Kern et al. (2002) |
| | | 7-RCPEMISVL-15 | HLA-Cw1 | | Kondo et al. (2004) |
| | | 341-QYDPVAALF-349 | HLA-Cw4 | | Kondo et al. (2004) |
| | | 198-VVCAHELVC-206 | HLA-Cw8 | | Kern et al. (2002), Kondo et al. (2004) |
| | | 294-VAFTSHEHF-302 | HLA-Cw12 | | Kondo et al. (2004) |
| | | 198-VVCAHELVC-206 | HLA-Cw15 | | Kondo et al. (2004) |
| UL98 | pp28 | 277-ARVYEIKCR-285 | HIA-B27 | | Elkington et al. (2003) |
| UL123 | IE1, pp72 | 81-VLAELVKQI-89 | HLA-A2 | | Elkington et al. (2003) |
| | | 315-Y(V/I)LEETSVM-323 | HLA-A2 | | Khan et al. (2002a), Retiere et al. (2000) |
| | | 316-VLEETSVML-324 | HLA-A2 | Khan et al. (2002a) | Khan et al. (2002a) |
| | | 354-YILGADPLRV-363 | HLA-A2 | | Frankenberg et al. (2002) |
| | | 309-CRVLCCYVL-317 | HLA-B7 | | Kern et al. (1999), Wills et al. (2002) |
| | | 88-QIKVRVDMV-96 | HLA-B8 | | Elkington et al. (2003) |
| | | 198-(D)ELRRKMMYM-207 | HLA-B8 | Wills et al. (2002) | Kern et al. (1999), Wills et al. (2002) |
| | | 199-ELKRKMIYM-207 | HLA-B18 | | Retiere et al. (2000) |
| | | 279-CVETMCNEY-287 | HLA-B18 | | Retiere et al. (2000) |
| | | 379-DEEDAIAAY-387 | HLA-B18 | | Retiere et al. (2000) |
| US2 | | 190-SMMWMRFFV-198 | HLA-A2 | | Elkington et al. (2003) |

All the epitopes listed elicit both a functional IFNγ response and have been established as functional cytotoxic T cell lines or clones. Peptides that have been used to generate tetramers are also listed with a reference to the first paper to generate the tetramer.

a Listed as commercially available from ProImmune, UK.

Taken together this evidence suggests that a very large number of the 200 or more HCMV ORFs are potentially CD8$^+$ T-cell targets. As discussed, considerable efforts have gone into identifying viral proteins that are recognized by CD8$^+$ T-cells, as well as mapping fine specificity for peptides and attempting to define ORFs and peptides in terms of their immunodominance and MHC class I restriction. However, to date such analyses only cover some 14 ORFs, leaving a vast portion of the CMV genome for which we have no information. This situation has now changed radically: work from Picker's laboratory aimed to identify all ORFs from HCMV that can be recognized by either CD4$^+$ or CD8$^+$ T-cells derived from a cohort of 33 MHC disparate HCMV positive donors. Approximately 14,000 consecutive 15-mer peptides overlapping each other by 10 amino acids and comprising 213 potential ORFs were synthesized. Initial analysis utilized the peptides organized into ORF pools to stimulate PBMC from donors, followed by intracellular detection of IFN$\gamma$ and quantitation by flow cytometry (Picker et al., manuscript in preparation). The results show that 47% of HCMV ORFs are recognized by CD8 T-cells from at least one of the 33 donors, and on average, each donor exhibits CD8$^+$ T-cell responses to seven different ORFs. Traditionally studied ORFs like pp65 (UL83) and IE1 (UL123) are among the most common recognized (the most frequent specificity in about 18% of donors each, but recognized with lower frequencies in many more donors): however, many other ORFs are also commonly recognized and pp65 and IE1 are by no means universally immunodominant.

An important methodological advance in the last 10 years is the technique of constructing soluble class I MHC "tetramers". These are composed of four class I MHC heavy chains loaded with their cognate viral peptide, and biotinylated so that they can be linked with avidin bearing a fluorescent tag. This provides a reagent which binds directly to the T-cell receptor (TCR) of virus-specific CD8$^+$ T-cells, and allows the visualization of peptide-specific T-cells by flow cytometry: the use of such tetramers is making an important contribution to the analysis of HCMV-specific T-cell responses, as well as in many other human virus infections. It should be noted that tetramers bind to all TCRs which recognize a specific combination of class I allele and cognate peptide, and do not distinguish between different clonotypic TCRs recognizing the same peptide. Antigenic peptides for which MHC class I tetramers have been derived and published are indicated in Table 17.1: at present they are all pp65 and IE1 peptides, but their number is likely to increase significantly in the light of the recent work above, identifying many more class I allele-specific peptides from HCMV ORFs.

## HCMV-encoded immune evasion mechanisms

It has been recognized since the early 1990s that class I MHC expression on the surface of HCMV-infected cells progressively diminishes with increasing time after infection. The disruption of components of the class I MHC processing pathway is a property that many viruses display in an attempt to evade virus-specific CD8$^+$ T-cells (Petersen et al., 2003). However, one may question whether this evasion is at all effective given that it is clear that upon primary HCMV infection the host mounts a strong T-cell response composed of both CD8$^+$ CTL and CD4$^+$ helper T-cells which produce antiviral cytokines, as well as the emerging evidence of the wide extent of antigens recognized by this response. It is also clear from clinical observation that the outcome of acute primary infection in normal healthy individuals is control of viral replication, although HCMV is not cleared from the host but becomes latent with periodic reactivation and production of new virions.

Thus the real functional significance in vivo of the viral immune evasion genes which target normal class I MHC antigen processing and presentation remains to be determined. Following primary infection HCMV persists for life in the host and will periodically reactivate from latency to produce new virions which may potentially be passed to new hosts. It seems plausible that the immune evasion mechanisms give the virus a 'window of protection' during reactivation from latency in the face of an expanded population of antigen-experienced T-cells, enabling the virus to complete its life cycle to produce new virions. Following reactivation the IE proteins are abundant at a time when the virus is retaining class I MHC complexes in the ER, blocking antigen presentation, thus preventing T-cell surveillance and allowing the cell to release progeny virus. However, subsequent infection of other cells will deliver preformed viral structural proteins (such as pp65) to the cytosol, whence they can enter the class I antigen processing pathway prior to expression of the viral immune evasion genes and be presented in association with class I MHC molecules at the infected cell surface, resulting in local containment of the reactivation episode.

HCMV can disrupt the normal class I MHC antigen processing pathways by a number of mechanisms, principally mediated by genes in the US2–11 region of the genome. These include the degradation of newly synthesized class I heavy chains mediated by US2 (Wiertz et al., 1996b) and US11 (Jones et al., 1995; Wiertz et al., 1996a), which dislocate the nascent heavy chains from the ER into the cytoplasm where they undergo proteosomal degradation. The US3 gene product causes retention of MHC class I peptide complexes in the ER (Jones et al., 1996). The US6 gene product blocks peptide translocation into the ER by binding to the cytosolic face of the TAP (transporter of antigenic peptides) heterodimer (Ahn et al., 1997; Lehner et al., 1997). The subject has recently been reviewed by Reddehase (2002) and is summarized updated in Figure 17.1. The number of viral genes apparently dedicated to interfering with normal peptide presentation seems surprising. However the human immune system and HCMV have co-evolved, and it is possible that this diversity of immune evasion mechanisms reflects this co-evolution, and that some of the immune evasion mechanisms may be redundant. It may also be the case that different mechanisms working together may be a very efficient way of down-regulating class I MHC; HCMV infects and becomes latent in a number of different cell types in vivo and it is possible that some of these viral gene products may be more efficient in some cell types than others.

These decreased levels of surface class I MHC on HCMV-infected cells might be predicted to render the cells more susceptible to lysis by NK cells: according to the 'missing self' hypothesis NK cells are normally inhibited by surface class I MHC molecules, but would be less inhibited by HCMV-infected cells and would consequently lyse the cells (Karre et al., 1986). However, it has been demonstrated in a number of experimental systems that HCMV-infected cells are in fact relatively resistant to NK-mediated lysis (Cerboni et al., 2000; Fletcher et al., 1998; Reyburn et al., 1997; Vales-Gomez et al., 2003; Wang et al., 2002). This suggests that in addition to genes to evade T-cell surveillance, HCMV may also have evolved viral encoded functions to evade NK surveillance.

Three distinct virus encoded proteins and mechanisms have been proposed to prevent NK activation and lysis of HCMV-infected cells (see also Chapter 15). A viral MHC class I homolog gpUL18 has been reported to inhibit NK lysis (Reyburn et al., 1997). The viral

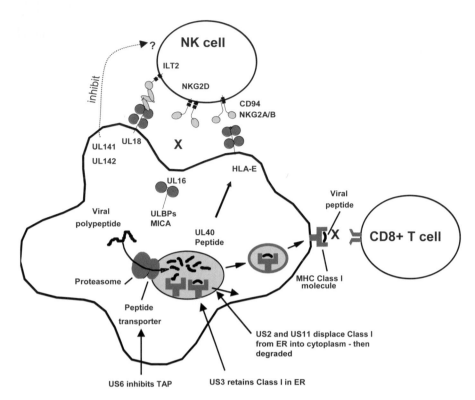

**Figure 17.1** CD8[+] T-cell and NK cell evasion by HCMV-infected cells. Disruption of class I MHC processing and presentation to CD8[+] T-cells by HCMV proteins US2, 3, 6 and 11 causes the surface level of class I MHC to progressively diminish, which would fail to inhibit NK cells. Expression of viral proteins UL16, UL18, UL40, UL141 and UL142 in order to inhibit NK cell activation.

protein encoded by HCMV UL40 also includes a signal sequence peptide similar to those of class I MHC molecules which normally bind to and stabilize the non-classical class I molecule HLA-E. Expression of UL40 has likewise been reported to stabilize HLA-E, and the maintenance of surface HLA-E expression inhibits NK cell lysis via engagement of the CD94/NKG2A heterodimer (Tomasec et al., 2000; Wang et al., 2002). Virus infected cells often express self molecules that indicate that the cell is stressed such as MICA and MICB and the retinoic acid inducible early transcripts also named the UL binding proteins (ULBPs), which ligate the activating receptor NKG2D. A third postulated mechanism of HCMV mediated NK cell evasion is via the expression of the viral membrane glycoprotein UL16, which is able to bind to MICB, ULBP-1 and ULBP-2 and sequester them in the ER/cis-Golgi, preventing them interacting with NK cells bearing NKG2D and thus presenting an activating signal to the NK cell (Cosman et al., 2001; Dunn et al., 2003; Sutherland et al., 2001) (summarized in Figure 17.1).

In addition, a striking difference in their ability to resist NK-mediated lysis has been demonstrated between the laboratory adapted strain AD169, and strains of HCMV (such as Toledo) that are more closely related to clinical isolates (Cerboni et al., 2000; Fletcher et al., 1998). It has been recognized since 1996 that clinical isolates of the virus have a larger genome, and comparison of AD169 with Toledo virus shows that Toledo has a 13.5 kbp insert which encodes 20 predicted ORFs (Davison et al., 2003) that are absent from AD169. It seems likely that genes encoded in this ULb' region are responsible for rendering infected fibroblasts resistant to NK lysis: this region does contain an MHC class I like molecule, UL142, and work from our laboratory suggests that this is able to inhibit some NK cell clones and that other viral proteins encoded in this region are also likely to inhibit NK cell lysis. Recent data (CMV workshop, Virginia, USA) suggest that the protein encoded by UL141 is also able to inhibit NK-mediated lysis by a novel mechanism. Their data also suggests that this inhibition is clonally distributed (Tomasec et al., 2005): this may account for the clones that are not inhibited by UL142 but still leaves the possibility that other HCMV genes are also involved in NK evasion.

## HCMV-specific CD8+ T-cell phenotype and memory

The availability of HCMV-specific class I MHC tetramers, monoclonal antibodies against cell-surface molecules and multi-parameter flow cytometry, as well as techniques for T-cell receptor analysis, have made possible the detailed characterization of antigen-experienced HCMV-specific CD8+ T-cells. The expression of different isoforms of the leukocyte common antigen CD45 was previously thought to distinguish naïve CD45RA(high) from memory CD45RO(high) T-cells. However in healthy carriers it has now been comprehensively shown that HCMV (pp65 or IE1) specific T-cells are present in both the CD45RO(high) and CD45RA(high) subpopulations (Gillespie et al., 2000; Khan et al., 2002a; Wills et al., 1999; Wills et al., 2002). Using a molecular clonotype probing technique it has also been shown that a particular HCMV-specific clone (as identified by the sequence of its hypervariable TCR) can segregate into both the CD45RO(high) and CD45RA(high) subpopulations (Wills et al., 1999). A more detailed phenotypic analysis of the CD45RA(high) HCMV-specific T-cells using markers to cell-surface adhesion, co-stimulation and chemokine receptor molecules has shown that they lack the costimulatory molecule CD28 and the chemokine receptor CCR7.

## CD8+ T-cell clonality

Analysis of the clonal composition of the memory CD8+ T-cells specific for defined pp65 and IE1 epitopes, by sequencing of the TCRs of multiple independently derived epitope-specific CTL clones, reveals a high degree of clonal focusing. In many donors, all of the CTL clones specific to a defined peptide use only one or two different TCRs as defined by their nucleotide sequence (Khan et al., 2002a; Weekes et al., 1999b). This implies that in many donors the circulating HCMV peptide-specific CD8+ T-cells are composed of only a few individual CD8+ clones that have undergone extensive clonal expansion in vivo.

As the extensive number of new HCMV proteins that can be recognized by a diverse population are defined at the level of individual peptides and class I MHC restrictions, it will be of interest to determine whether these T-cells show similar clonal TCR focusing and phenotype to those observed with pp65 and IE1 specific T-cells. These individual HCMV-specific T-cell clones persist for years, as the same clones can be repeatedly isolated over time. HCMV pp65-specific CD8+ T-cell clones also show the property of containing public epitopes, which have been described previously in EBV specific T-cells: clones obtained from unrelated subjects that recognize the same defined peptide–MHC complex often use the same TCR Vβ segment, and have similar amino acid sequences within the hypervariable VDJ region of the TCR that binds to the viral peptide (Argaet et al., 1994; Weekes et al., 1999b). Furthermore, these oligoclonal responses can increase over time to form a very high proportion of the CD8⁺ T-cells (up to 25% of all CD8⁺ T-cells in an individual) as demonstrated by an analysis of elderly donors (Khan et al., 2002b).

### CD8⁺ T-cell response in primary infection and selection into memory

The functional and surface marker phenotypes of HCMV-specific T-cells described so far have been derived from normal healthy donors who are latently infected with HCMV, and thus do not inform us of how these phenotypes arose following primary infection, nor if intermediate phenotypes are present. The kinetics of pp65-specific CD8⁺ T-cell clonal focusing following primary HCMV infection is currently under investigation: does the clonal focusing occur over a long period, perhaps secondary to periodic HCMV reactivation from latency with consequent antigenic rechallenge of the CD8⁺ memory T-cell pool, or is it established early after primary infection? Initial results suggest that the clonal focusing occurs surprisingly rapidly with multiple HCMV-specific clones with particular Vβ usage being lost between 2 and 4 weeks after infection and early selection of clones that are still dominating the response 5 years later (Figure 17.2). Initial analyses of the phenotypic markers on HCMV-specific CD8 T-cells (it must be stressed that dominant pp65 and IE1 peptide specificities only have been studied) show that during primary HCMV infection, all the highly activated HCMV-specific effector T-cells express CD45RO. These activated CD45RO(high) effector CTL are already CD28 and CCR7 negative and remain negative as they enter the memory pool, but many accumulate in the CD45RA(high) subpopulation from 5–8 weeks following acute infection (Wills et al., 1999) (Figure 17.3).

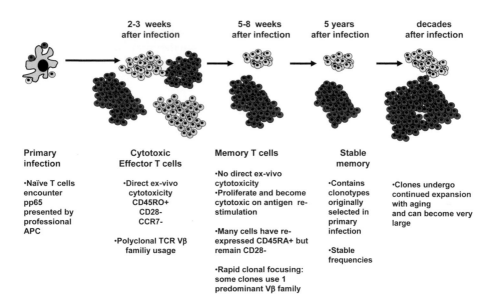

**Figure 17.2** Naïve T-cells expressing multiple Vβ TCR (colored cells) respond to a given HCMV pp65 peptide during primary infection. Clonal focusing occurs rapidly during and after resolution of primary infection, with T-cells bearing particular Vβ TCR being deleted from the repertoire. The pp65-specific clones established in primary infection persist for years, maybe for life.

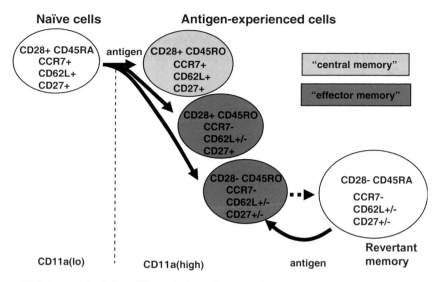

**Figure 17.3** A model of the differentiation of human CD8⁺ T-cells. Following activation of a CD28+ CD45RA(high) naïve cell by antigen, CD11a is up-regulated permanently and CD45RO is initially. Following resolution of primary infection some CD28– CD45RO+ cells revert to CD28- CD45RA(high). In vitro restimulation of these CD28-CD45RA(high) CCR7– cells induces them to proliferate and give rise to activated and cytotoxic CD28– CD45RO+ cells.

T-cells in the CD28–, CD45RA(high), CCR7– subpopulation have previously been described as "terminally differentiated" (Champagne et al., 2001). However, following stimulation in vitro with specific HCMV peptide these cells undergo sustained clonal proliferation, up-regulate CD45RO and CCR5, and show strong peptide-specific cytotoxic activity (Wills et al., 2002) (Figure 17.3). Thus CD8⁺ T-cell memory to HCMV is maintained by cells of expanded HCMV-specific clones that show heterogeneity of activation state and co-stimulation molecule expression, within both CD45RO(high) and CD28-, CD45RA(high) T-cell pools. Similar techniques have been used by a number of groups to examine the memory T-cells generated in response to EBV: it is interesting to note that T-cells specific for lytic EBV antigens display very similar phenotypic profiles to HCMV pp65-specific T-cells, but also of interest that this does not hold for all persistent virus infections (Appay et al., 2002; Faint et al., 2001). A recent review by van Lier and colleagues attempts to unify these complex phenotypic differences in a variety of viral systems including HCMV (van Lier et al., 2003).

## The human CD4 T-cell response to HCMV

The analysis of HCMV-specific CD4⁺ T-cells has lagged behind that of the CD8⁺ responses, although advances in the production of synthetic peptides (speed and cost) to predicted viral ORFs coupled with rapid assay techniques such as intracellular cytokine expression by flow cytometry (Waldrop et al., 1997; Waldrop et al., 1998) have started to change this situation. In particular, the whole genome approach recently adopted by Picker and colleagues and described in greater detail in the CD8 section of this chapter has identified many ORFs which elicit CD4⁺ responses: in due course these will be no doubt be refined to minimal peptides and mapped to restricting MHC class II alleles. The definition

in signal transduction) and CIITA (class II transactivator) are critical components of the IFNγ signaling pathway which leads to class II MHC expression. HCMV can disrupt IFNγ-induced class II MHC up-regulation by affecting Jak1 expression and by repressing CIITA mRNA expression. In addition, the viral protein US2, which has previously been shown to affect class I MHC processing, can also target the class II MHC molecules HLA-DRα and HLA-DMα and redirect them to the cytoplasm for degradation (reviewed in (Miller et al., 2001).

## HCMV-specific T-cell reconstitution following hematopoietic stem cell transplantation

The use of bone marrow transplantation (BMT) and more recently peripheral stem cell transplantation (SCT) as part of the treatment for hematological malignancies has steadily grown over the last few decades. It is well documented that primary infection or reactivation of latent HCMV can lead to serious morbidity and mortality in these patients. It has also been recognized for a considerable period of time that there is a direct correlation between effective HCMV-specific cytotoxicity and recovery from CMV infection or reactivation (Quinnan et al., 1982). The basic research reviewed earlier in this chapter on the function, specificity, frequency and phenotype of HCMV-specific CD4+ and CD8+ T-cells in normal immunocompetent individuals provides the essential tools and baselines in order to understand the reconstitution of HCMV-specific T-cell responses following engraftment. It could be hoped that this knowledge would better inform clinicians as to when antiviral or adoptive immunotherapy interventions would be most appropriate.

Understanding of the generation or reconstitution of HCMV-specific CD4+ and CD8+ T-cells following BMT or SCT is complicated by a number of important factors. The donor (D) and recipient (R) may be matched siblings, but this is in a minority of transplants and the donor/recipient pairs are more usually matched but unrelated (allo-SCT). In addition, the HCMV serostatus of the donor and recipient needs to be considered with there being four possible combinations: D+/R+, D−/R−, D+/R−, and D−/R+. Other variables to be considered include (1) the source of the stem cells, either bone marrow or peripheral stem cells, and in the case of peripheral stem cells whether the graft has been manipulated to remove mature T-cells, (2) the use of antiviral drugs such as ganciclovir to suppress HCMV replication, and (3) the use of steroid therapy to combat graft versus host disease (GVHD) which will of course act to globally impair CD4+ and CD8+ T-cells including those specific for CMV.

The techniques used to assess reconstitution inevitably mirror those used in research on normal HCMV-specific T-cell responses and initially employed proliferation assays and the generation of CTL. More recently, pp65-specific MHC class I tetramers have been used to study the kinetics and frequency of CD8+ T-cell reconstitution (Aubert et al., 2001; Cwynarski et al., 2001; Gratama et al., 2001) in conjunction with functional assays, such as IFNγ production, which are appropriate for assessing a wider range of HCMV epitopes and MHC class I alleles, applied to CD8+ T-cell function, as well as Th1 CD4+ T-cells (Hebart et al., 2002).

## Donor and recipient HCMV serostatus

In solid organ transplantation (liver, kidney, heart or lung), transfer of HCMV in the graft from a seropositive donor into a seronegative recipient who also receives immunosuppressive treatment frequently leads to serious HCMV disease. In contrast, in the case of hemo-poietic transplantation from a seropositive donor, HCMV is infrequently transmitted in the graft (less than 20%) (Goodrich et al., 1994).

In hemopoietic transplantation, it should be noted that the serostatus of the graft do-nor indicates whether there are antigen experienced HCMV-specific donor T-cells that are transferred within the allograft (in addition to whether latent virus is present in the donor); the serostatus of the recipient indicates whether there is latent virus in the recipient. The serostatus of both donor and recipient therefore affect the reconstitution of donor-derived HCMV-specific CD8$^+$ T-cells in the recipient (whose own CD8$^+$ T-cells have of course been deliberately ablated prior to allografting). In the absence of HCMV infection, there is no antigen to stimulate a primary T-cell response; in two studies which included 10 D$^-$/R$^-$ transplantations, there was no detectable pp65 tetramer specific CD8$^+$ T-cell reconsti-tution. In contrast, studies consistently report that almost all D$^+$/R$^+$ transplantations show early and sustained reconstitution of pp65 tetramer-positive T-cells (Aubert et al., 2001; Cwynarski et al., 2001; Gandhi et al., 2003; Gratama et al., 2001).

The greatest risk of HCMV disease in hemopoietic transplantation thus occurs in D$^-$/R$^+$ transplantation, because the graft from the seronegative donor does not contain an-tigen-experienced HCMV-specific T-cells and the recipient has latent virus and simultane-ously receives immunosuppressive treatment to prevent GVHD. Of 11 D$^-$/R$^+$ transplan-tations, seven patients failed to develop HCMV-specific tetramer-positive T-cells, and the other four had very low levels of tetramer-positive cells that were not maintained, results which are in agreement with previous studies not based on tetramers (Aubert et al., 2001; Cwynarski et al., 2001; Gratama et al., 2001). Following D$^+$/R$^-$ transplantation, most of the recipients have very low or undetectable HCMV-specific T-cells unless the recipient develops acute GVHD (Cwynarski et al., 2001; Gandhi et al., 2003).

## Source of graft and manipulation

In cases of sibling allograft in which both the donor and recipient were HCMV seropositive (D$^+$/R$^+$) before allo-SCT, recovery of HCMV-specific CD8$^+$ T-cells was rapid. However, early reconstitution of tetramer-positive cells was not observed in recipients of matched unrelated donor (MUD) allo-SCT (Aubert et al., 2001; Cwynarski et al., 2001; Gratama et al., 2001). In order to reduce GVHD, MUD allo-SCT recipients receive T-cell-depleted grafts with the intention to avoid alloreactive T-cells: however this will also deplete antigen experienced T-cells including mature HCMV-specific cells and in part may account for the delay in reconstitution.

The origin of the HCMV-specific clones that reconstitute these transplant patients is of interest: does reconstitution occur from the expansion of donor-derived antigen ex-perienced T-cells or as the result of de novo naïve T-cell production and the mounting of a primary response? Staining with HCMV class I MHC tetramers does not distinguish between CD8$^+$ T-cells originating from the donor or from recipient, nor can tetramers dis-tinguish the individual T-cell clones that constitute the tetramer-positive population. These issues have been addressed by studying the reconstitution of HCMV-specific CD8$^+$ T-cells following allo-SCT at the level of individual clones, using clonotypic probing (Gandhi et al., 2003). Following D$^+$/R$^+$ allo-SCT, immunodominant donor HCMV-specific clones transferred in the allograft underwent early expansion in the recipient and were maintained

Elkington, R., Walker, S., Crough, T., Menzies, M., Tellam, J., Bharadwaj, M., and Khanna, R. (2003). Ex vivo profiling of CD8[+]-T-cell responses to human cytomegalovirus reveals broad and multispecific reactivities in healthy virus carriers. J. Virol. 77, 5226–5240.

Faint, J.M., Annels, N.E., Curnow, S.J., Shields, P., Pilling, D., Hislop, A.D., Wu, L., Akbar, A.N., Buckley, C.D., Moss, P.A., Adams, D. H., Rickinson, A. B., Salmon, M. (2001). Memory T-cells constitute a subset of the human CD8(+)CD45RA(+) pool with distinct phenotypic and migratory characteristics. J. Immunol. 167, 212–220.

Fletcher, J.M., Prentice, H.G., and Grundy, J.E. (1998). Natural killer cell lysis of cytomegalovirus (CMV)-infected cells correlates with virally induced changes in cell surface lymphocyte function-associated antigen-3 (LFA-3) expression and not with the CMV-induced down-regulation of cell surface class I HLA. J. Immunol. 161, 2365–2374.

Frankenberg, N., Pepperl-Klindworth, S., Meyer, R.G., and Plachter, B. (2002). Identification of a conserved HLA-A2-restricted decapeptide from the IE1 protein (pUL123) of human cytomegalovirus. Virology 295, 208–216.

Gamadia, L.E., Remmerswaal, E.B., Weel, J.F., Bemelman, F., van Lier, R.A., and Ten Berge, I.J. (2003). Primary immune responses to human CMV: a critical role for IFN-gamma-producing CD4[+] T-cells in protection against CMV disease. Blood 101, 2686–2692.

Gamadia, L.E., Rentenaar, R.J., Baars, P.A., Remmerswaal, E.B., Surachno, S., Weel, J.F., Toebes, M., Schumacher, T.N., ten Berge, I.J., and van Lier, R.A. (2001). Differentiation of cytomegalovirus-specific CD8(+) T-cells in healthy and immunosuppressed virus carriers. Blood 98, 754–761.

Gandhi, M.K., Wills, M.R., Okecha, G., Day, E.K., Hicks, R., Marcus, R.E., Sissons, J.G., and Carmichael, A.J. (2003). Late diversification in the clonal composition of human cytomegalovirus-specific CD8[+] T-cells following allogeneic hemopoietic stem cell transplantation. Blood 102, 3427–3438.

Gautier, N., Chavant, E., Prieur, E., Monsarrat, B., Mazarguil, H., Davrinche, C., Gairin, J.E., and Davignon, J.L. (1996). Characterization of an epitope of the human cytomegalovirus protein IE1 recognized by a CD4[+] T-cell clone. Eur. J. Immunol. 26, 1110–1117.

Gavin, M.A., Gilbert, M.J., Riddell, S.R., Greenberg, P.D., and Bevan, M.J. (1993). Alkali hydrolysis of recombinant proteins allows for the rapid identification of class I MHC-restricted CTL epitopes. J. Immunol. 151, 3971–3980.

Gillespie, G.M., Wills, M.R., Appay, V., O'Callaghan, C., Murphy, M., Smith, N., Sissons, P., Rowland-Jones, S., Bell, J.I., and Moss, P.A. (2000). Functional heterogeneity and high frequencies of cytomegalovirus-specific CD8(+) T-lymphocytes in healthy seropositive donors. J. Virol. 74, 8140–8150.

Goodrich, J.M., Boeckh, M., and Bowden, R. (1994). Strategies for the prevention of cytomegalovirus disease after marrow transplantation. Clin Infect Dis 19, 287–298.

Gratama, J.W., van Esser, J.W., Lamers, C.H., Tournay, C., Lowenberg, B., Bolhuis, R.L., and Cornelissen, J.J. (2001). Tetramer-based quantification of cytomegalovirus (CMV)-specific CD8[+] T-lymphocytes in T-cell-depleted stem cell grafts and after transplantation may identify patients at risk for progressive CMV infection. Blood 98, 1358–1364.

Hassan-Walker, A.F., Vargas Cuero, A.L., Mattes, F.M., Klenerman, P., Lechner, F., Burroughs, A.K., Griffiths, P.D., Phillips, R.E., and Emery, V.C. (2001). CD8[+] cytotoxic lymphocyte responses against cytomegalovirus after liver transplantation: correlation with time from transplant to receipt of tacrolimus. J. Infect. Dis.183, 835–843.

Hebart, H., Daginik, S., Stevanovic, S., Grigoleit, U., Dobler, A., Baur, M., Rauser, G., Sinzger, C., Jahn, G., Loeffler, J., Kanz, L., Rammensee, H. G., and Einsele, H. (2002). Sensitive detection of human cytomegalovirus peptide-specific cytotoxic T-lymphocyte responses by interferon-gamma-enzyme-linked immunospot assay and flow cytometry in healthy individuals and in patients after allogeneic stem cell transplantation. Blood 99, 3830–3837.

Jones, T.R., Hanson, L.K., Sun, L., Slater, J.S., Stenberg, R.M., and Campbell, A.E. (1995). Multiple independent loci within the human cytomegalovirus unique short region down-regulate expression of major histocompatibility complex class I heavy chains. J. Virol. 69, 4830–4841.

Jones, T.R., Wiertz, E., Sun, L., Fish, K.N., Nelson, J.A., and Ploegh, H.L. (1996). Human cytomegalovirus US3 impairs transport and maturation of major histocompatability complex class I heavy chains. Proc. Natl. Acad. Sci. USA 93, 11327–11333.

Karre, K., Ljunggren, H.G., Piontek, G., and Kiessling, R. (1986). Selective rejection of H-2-deficient lymphoma variants suggests alternative immune defence strategy. Nature 319, 675–678.

Kern, F., Bunde, T., Faulhaber, N., Kiecker, F., Khatamzas, E., Rudawski, I.M., Pruss, A., Gratama, J.W., Volkmer-Engert, R., Ewert, R., Reinke, P., Volk, H.D., Picker, L.J. (2002). Cytomegalovirus (CMV) phosphoprotein 65 makes a large contribution to shaping the T-cell repertoire in CMV-exposed individuals. J. Infect. Dis. *185*, 1709–1716.

Kern, F., Surel, I.P., Brock, C., Freistedt, B., Radtke, H., Scheffold, A., Blasczyk, R., Reinke, P., Schneider-Mergener, J., Radbruch, A., Walden, P., Volk, H.D. (1998). T-cell epitope mapping by flow cytometry. Nature Med. *4*, 975–978.

Kern, F., Surel, I.P., Faulhaber, N., Frommel, C., Schneider-Mergener, J., Schonemann, C., Reinke, P., and Volk, H.D. (1999). Target structures of the CD8(+)-T-cell response to human cytomegalovirus: the 72-kilodalton major immediate-early protein revisited. J. Virol. *73*, 8179–8184.

Khan, N., Cobbold, M., Keenan, R., and Moss, P.A. (2002a). Comparative analysis of CD8+ T-cell responses against human cytomegalovirus proteins pp65 and immediate-early 1 shows similarities in precursor frequency, oligoclonality, and phenotype. J. Infect. Dis. *185*, 1025–1034.

Khan, N., Shariff, N., Cobbold, M., Bruton, R., Ainsworth, J.A., Sinclair, A.J., Nayak, L., and Moss, P.A. (2002b). Cytomegalovirus seropositivity drives the CD8 T-cell repertoire toward greater clonality in healthy elderly individuals. J. Immunol. *169*, 1984–1992.

Khattab, B.A., Lindenmaier, W., Frank, R., and Link, H. (1997). Three T-cell epitopes within the C-terminal 265 amino acids of the matrix protein pp65 of human cytomegalovirus recognized by human lymphocytes. J. Med. Virol. *52*, 68–76.

Kondo, E., Akatsuka, Y., Kuzushima, K., Tsujimura, K., Asakura, S., Tajima, K., Kagami, Y., Kodera, Y., Tanimoto, M., Morishima, Y., and Takahashi, T. (2004). Identification of novel CTL epitopes of CMV-pp65 presented by a variety of HLA alleles. Blood *103*, 630–638.

Kuzushima, K., Hayashi, N., Kimura, H., and Tsurumi, T. (2001). Efficient identification of HLA-A*2402-restricted cytomegalovirus-specific CD8(+) T-cell epitopes by a computer algorithm and an enzyme-linked immunospot assay. Blood *98*, 1872–1881.

Lehner, P.J., Karttunen, J.T., Wilkinson, G.W., and Cresswell, P. (1997). The human cytomegalovirus US6 glycoprotein inhibits transporter associated with antigen processing-dependent peptide translocation. Proc. Natl. Acad. Sci. USA *94*, 6904–6909.

Li, C.R., Greenberg, P.D., Gilbert, M.J., Goodrich, J.M., and Riddell, S.R. (1994). Recovery of HLA-restricted cytomegalovirus (CMV)-specific T-cell responses after allogeneic bone marrow transplant: correlation with CMV disease and effect of ganciclovir prophylaxis. Blood *83*, 1971–1979.

Li Pira, G., Bottone, L., Ivaldi, F., Pelizzoli, R., Del Galdo, F., Lozzi, L., Bracci, L., Loregian, A., Palu, G., De Palma, R., Einsele, H., Manca, F. (2004). Identification of new Th peptides from the cytomegalovirus protein pp65 to design a peptide library for generation of CD4 T-cell lines for cellular immunoreconstitution. Int. Immunol. *16*, 635–642.

Longmate, J., York, J., La Rosa, C., Krishnan, R., Zhang, M., Senitzer, D., and Diamond, D.J. (2001). Population coverage by HLA class I restricted cytotoxic T-lymphocyte epitopes. Immunogenetics *52*, 165–173.

Manley, T.J., Luy, L., Jones, T., Boeckh, M., Mutimer, H., and Riddell, S.R. (2004). Immune evasion proteins of human cytomegalovirus do not prevent a diverse CD8+ cytotoxic T-cell response in natural infection. Blood *104*, 1075–1082.

Masuoka, M., Yoshimuta, T., Hamada, M., Okamoto, M., Fumimori, T., Honda, J., Oizumi, K., and Itoh, K. (2001). Identification of the HLA-A24 peptide epitope within cytomegalovirus protein pp65 recognized by CMV-specific cytotoxic T-lymphocytes. Viral Immunol. *14*, 369–377.

McLaughlin Taylor, E., Pande, H., Forman, S.J., Tanamachi, B., Li, C.R., Zaia, J.A., Greenberg, P.D., and Riddell, S.R. (1994). Identification of the major late human cytomegalovirus matrix protein pp65 as a target antigen for CD8+ virus-specific cytotoxic T-lymphocytes. J. Med. Virol. *43*, 103–110.

Miller, D.M., Cebulla, C.M., Rahill, B.M., and Sedmak, D.D. (2001). Cytomegalovirus and transcriptional down-regulation of major histocompatibility complex class II expression. Semin. Immunol. *13*, 11–18.

Parker, K.C., Bednarek, M.A., Hull, L.K., Utz, U., Cunningham, B., Zweerink, H.J., Biddison, W.E., and Coligan, J.E. (1992). Sequence motifs important for peptide binding to the human MHC class I molecule, HLA-A2. J. Immunol. *149*, 3580–3587.

Peggs, K., Verfuerth, S., and Mackinnon, S. (2001). Induction of cytomegalovirus (CMV)-specific T-cell responses using dendritic cells pulsed with CMV antigen: a novel culture system free of live CMV virions. Blood *97*, 994–1000.

Peggs, K., Verfuerth, S., Pizzey, A., Ainsworth, J., Moss, P., and Mackinnon, S. (2002). Characterization of human cytomegalovirus peptide-specific CD8(+) T-cell repertoire diversity following in vitro restimulation by antigen-pulsed dendritic cells. Blood *99*, 213–223.

Peggs, K.S., Verfuerth, S., Pizzey, A., Khan, N., Guiver, M., Moss, P.A., and Mackinnon, S. (2003). Adoptive cellular therapy for early cytomegalovirus infection after allogeneic stem-cell transplantation with virus-specific T-cell lines. Lancet 362, 1375–1377.

Petersen, J.L., Morris, C.R., and Solheim, J.C. (2003). Virus evasion of MHC class I molecule presentation. J. Immunol. 171, 4473–4478.

Podlech, J., Holtappels, R., Pahl-Seibert, M.-F., Steffens, H.-P., and Reddehase, M.J. (2000). Murine model of interstitial cytomegalovirus pneumonia in syngeneic bone marrow transplantation: persistence of protective pulmonary CD8-T-cell infiltrates after clearance of acute infection. J. Virol. 74, 7496–7507.

Polic, B., Hengel, H., Krmpotic, A., Trgovcich, J., Pavic, I., Lucin, P., Jonjic, S., and Koszinowski, U.H. (1998). Hierarchical and redundanT-lymphocyte subset control precludes cytomegalovirus replication during latent infection. J. Exp. Med. 188, 1047–1054.

Quinnan, G.V., Jr., Kirmani, N., Rook, A.H., Manischewitz, J.F., Jackson, L., Moreschi, G., Santos, G.W., Saral, R., and Burns, W.H. (1982). Cytotoxic T-cells in cytomegalovirus infection: HLA-restricted T-lymphocyte and non-T-lymphocyte cytotoxic responses correlate with recovery from cytomegalovirus infection in bone-marrow-transplant recipients. N. Engl. J. Med. 307, 7–13.

Rauser, G., Einsele, H., Sinzger, C., Wernet, D., Kuntz, G., Assenmacher, M., Campbell, J. D., and Topp, M.S. (2004). Rapid generation of combined CMV-specific CD4+ and CD8+ T-cell lines for adoptive transfer into recipients of allogeneic stem cell transplants. Blood 103, 3565–3572.

Reddehase, M.J. (2002). Antigens and immunoevasins: opponents in cytomegalovirus immune surveillance. Nature Rev. Immunol. 2, 831–844.

Reddehase, M.J., Mutter, W., Munch, K., Buhring, H.J., and Koszinowski, U.H. (1987). CD8-positive T-lymphocytes specific for murine cytomegalovirus immediate-early antigens mediate protective immunity. J. Virol. 61, 3102–3108.

Rentenaar, R.J., Gamadia, L.E., van DerHoek, N., van Diepen, F.N., Boom, R., Weel, J.F., Wertheim-van Dillen, P.M., van Lier, R.A., and ten Berge, I.J. (2000). Development of virus-specific CD4(+) T-cells during primary cytomegalovirus infection. J. Clin. Invest. 105, 541–548.

Retiere, C., Prod'homme, V., Imbert-Marcille, B.M., Bonneville, M., Vie, H., and Hallet, M. (2000). Generation of cytomegalovirus-specific human T-lymphocyte clones by using autologous B-lymphoblastoid cells with stable expression of pp65 or IE1 proteins: a tool to study the fine specificity of the antiviral response. J. Virol. 74, 3948–3952.

Reyburn, H.T., Mandelboim, O., Vales-Gomez, M., Davis, D.M., Pazmany, L., and Strominger, J.L. (1997). The class I MHC homologue of human cytomegalovirus inhibits attack by natural killer cells. Nature 386, 514–517.

Riddell, S.R., Watanabe, K.S., Goodrich, J.M., Li, C.R., Agha, M.E., and Greenberg, P.D. (1992). Restoration of viral immunity in immunodeficient humans by the adoptive transfer of T-cell clones. Science 257, 238–241.

Sester, M., Sester, U., Gartner, B., Kubuschok, B., Girndt, M., Meyerhans, A., and Kohler, H. (2002). Sustained high frequencies of specific CD4 T-cells restricted to a single persistent virus. J. Virol. 76, 3748–3755.

Solache, A., Morgan, C.L., Dodi, A.I., Morte, C., Scott, I., Baboonian, C., Zal, B., Goldman, J., Grundy, J.E., and Madrigal, J.A. (1999). Identification of three HLA-A*0201-restricted cytotoxic T-cell epitopes in the cytomegalovirus protein pp65 that are conserved between eight strains of the virus. J. Immunol. 163, 5512–5518.

Sutherland, C.L., Chalupny, N.J., and Cosman, D. (2001). The UL16-binding proteins, a novel family of MHC class I-related ligands for NKG2D, activate natural killer cell functions. Immunol. Rev. 181, 185–192.

Szmania, S., Galloway, A., Bruorton, M., Musk, P., Aubert, G., Arthur, A., Pyle, H., Hensel, N., Ta, N., Lamb, L., Jr., Dodi, T., Madrigal, A., Barrett, J., Henslee-Downey, J., van Rhee, F. (2001). Isolation and expansion of cytomegalovirus-specific cytotoxic T-lymphocytes to clinical scale from a single blood draw using dendritic cells and HLA-tetramers. Blood 98, 505–512.

Tomasec, P., Braud, V.M., Rickards, C., Powell, M.B., McSharry, B.P., Gadola, S., Cerundolo, V., Borysiewicz, L.K., McMichael, A.J., and Wilkinson, G.W. (2000). Surface expression of HLA-E, an inhibitor of natural killer cells, enhanced by human cytomegalovirus gpUL40. Science 287, 1031.

Tomasec, P., Wang, E.C., Davison, A.J., Vajtesek, B., Armstrong, M., Griffin, C., McSharry, B.P., Morris, R.J., Llewellyn-Lacey, S., Rickards, C., Nomoto, A., Sinzger, C., and Wilkinson, G.W. (2005).

Downregulation of natural killer cell-activating ligand CD155 by human cytomegalovirus UL141. Nat. Immunol. *6*, 181–188.

Utz, U., and Biddison, W.E. (1992). Presentation of three different viral peptides is determined by common structural features of the human lymphocyte antigen-A2.1 molecule. J. Immunother. *12*, 180–182.

Vales-Gomez, M., Browne, H., and Reyburn, H.T. (2003). Expression of the UL16 glycoprotein of Human Cytomegalovirus protects the virus-infected cell from attack by natural killer cells. BMC Immunol. *4*, 4.

van Leeuwen, E.M., Remmerswaal, E.B., Vossen, M.T., Rowshani, A.T., Wertheim-van Dillen, P.M., van Lier, R.A., and ten Berge, I.J. (2004). Emergence of a CD4+CD28- granzyme B+, cytomegalovirus-specific T-cell subset after recovery of primary cytomegalovirus infection. J. Immunol. *173*, 1834–1841.

van Lier, R.A., ten Berge, I.J., and Gamadia, L.E. (2003). Human CD8(+) T-cell differentiation in response to viruses. Nature Rev. Immunol. *3*, 931–939.

Waldrop, S.L., Davis, K.A., Maino, V.C., and Picker, L.J. (1998). Normal human CD4+ memory T-cells display broad heterogeneity in their activation threshold for cytokine synthesis. J. Immunol. *161*, 5284–5295.

Waldrop, S.L., Pitcher, C.J., Peterson, D.M., Maino, V.C., and Picker, L.J. (1997). Determination of antigen-specific memory/effector CD4+ T-cell frequencies by flow cytometry: evidence for a novel, antigen-specific homeostatic mechanism in HIV-associated immunodeficiency. J. Clin. Invest. *99*, 1739–1750.

Walter, E.A., Greenberg, P.D., Gilbert, M.J., Finch, R.J., Watanabe, K.S., Thomas, E.D., and Riddell, S.R. (1995). Reconstitution of cellular immunity against cytomegalovirus in recipients of allogeneic bone marrow by transfer of T-cell clones from the donor. N. Engl. J. Med. *333*, 1038–1044.

Wang, E.C., McSharry, B., Retiere, C., Tomasec, P., Williams, S., Borysiewicz, L.K., Braud, V.M., and Wilkinson, G.W. (2002). UL40-mediated NK evasion during productive infection with human cytomegalovirus. Proc. Natl. Acad. Sci. USA *99*, 7570–7575.

Weekes, M.P., Carmichael, A.J., Wills, M.R., Mynard, K., and Sissons, J.G. (1999a). Human CD28-CD8+ T-cells contain greatly expanded functional virus- specific memory CTL clones. J. Immunol. *162*, 7569–7577.

Weekes, M.P., Wills, M.R., Mynard, K., Carmichael, A.J., and Sissons, J.G. (1999b). The memory cytotoxic T-lymphocyte (CTL) response to human cytomegalovirus infection contains individual peptide-specific CTL clones that have undergone extensive expansion in vivo. J. Virol. *73*, 2099–2108.

Weekes, M.P., Wills, M.R., Sissons, J.G., and Carmichael, A.J. (2004). Long-Term Stable Expanded Human CD4+ T-cell Clones Specific for Human Cytomegalovirus Are Distributed in Both CD45RAhigh and CD45ROhigh Populations. J. Immunol. *173*, 5843–5851.

Wiertz, E.J., Jones, T.R., Sun, L., Bogyo, M., Geuze, H.J., and Ploegh, H.L. (1996a). The human cytomegalovirus US11 gene product dislocates MHC class I heavy chains from the endoplasmic reticulum to the cytosol. Cell *84*, 769–779.

Wiertz, E.J., Tortorella, D., Bogyo, M., Yu, J., Mothes, W., Jones, T.R., Rapoport, T.A., and Ploegh, H.L. (1996b). Sec61-mediated transfer of a membrane protein from the endoplasmic reticulum to the proteasome for destruction.. Nature *384*, 432–438.

Wills, M.R., Carmichael, A.J., Mynard, K., Jin, X., Weekes, M.P., Plachter, B., and Sissons, J.G. (1996). The human cytotoxic T-lymphocyte (CTL) response to cytomegalovirus is dominated by structural protein pp65: frequency, specificity, and T-cell receptor usage of pp65-specific CTL. J. Virol. *70*, 7569–7579.

Wills, M.R., Carmichael, A.J., Weekes, M.P., Mynard, K., Okecha, G., Hicks, R., and Sissons, J.G. (1999). Human virus-specific CD8+ CTL clones revert from CD45ROhigh to CD45RAhigh in vivo: CD45RAhighCD8+ T-cells comprise both naive and memory cells. J. Immunol. *162*, 7080–7087.

Wills, M.R., Okecha, G., Weekes, M.P., Gandhi, M.K., Sissons, P.J., and Carmichael, A.J. (2002). Identification of naive or antigen-experienced human CD8(+) T-cells by expression of costimulation and chemokine receptors: analysis of the human cytomegalovirus-specific CD8(+) T-cell response. J. Immunol. *168*, 5455–5464.

# Combat between Cytomegalovirus and Dendritic Cells in T-cell Priming

# 18

*Christian Davrinche*

## Abstract

Dendritic cells are the sentry cells of the immune system as they are located in most organs where they monitor entry of harmful pathogens such as viruses. Sensing of danger results in the innate response providing essential signals for priming of pathogen-specific T-cells to fight against infection. When viruses meet dendritic cells a battle is starting for each to ensure its own survival. Cytomegalovirus is considered as a spearhead in developing strategies to impair functions of infected dendritic cells. Successful containment of virus spreading and disease following primary infection of the immunocompetent host questions on forces engaged by the host to win the combat. In this chapter we argue in favor of a model involving acquisition of CMV antigens by non-infected dendritic cells from neighboring dead cells that were derived very early in the innate phase of the antiviral response. Phagocytosis of dying cells containing incoming and immediate-early CMV proteins could provide both appropriate costimulatory signals and antigenic material for cross-presentation to naïve $CD8^+$ T-cells in lymph nodes. Temporal coordination between (i) innate sensing through recognition by Toll-like receptors, (ii) apoptosis of infected cells in tissues and (iii) appropriate processing of antigens into the cross-presenting machinery should guarantee an efficient primary response against the virus able to bypass viral subversion.

## Cytomegalovirus sensing by dendritic cells and innate resistance to infection

The initial interaction of pathogens with dendritic cells (DCs) plays a central role in the establishment of an appropriate innate and adaptive response to avoid pathogen spreading and disease development. DC populations include plasmacytoid DCs found in lymphoid organs (thymus, bone marrow, spleen, tonsil and lymph nodes) and myeloid DCs located in the mucosa and in peripheral tissues which most viruses use to enter. In the skin, the epidermis is inhabited by Langerhans cells (LC) and the dermis by interstitial DCs. Throughout the text, DCs will refer to the myeloid DCs unless otherwise stated. According to their first-line position throughout the body, DCs are armed with conserved pattern-recognition receptors for the recognition and internalization of pathogens. Such receptors include the Toll-like-receptors (TLRs) and C-type lectin receptors (CLRs), the former being involved in cell activation and the latter in capture and antigen processing for presentation to T-cells (for reviews see, Larsson et al., 2004; van Kooyk et al., 2004). Besides their well-known function in sensing bacteria these receptors appeared more recently as crucial in recognition, internalization and cell to cell transmission of many viruses including HIV, Dengue virus, HCV, measles virus and Ebola viruses (Vaidya and Cheng, 2003). Human cytomegalovirus does not depart from the rule since it was recently shown to interact with TLR-2 ectopically expressed by human embryonic kidney (HEK) cells (Compton et al., 2003). Although no direct proof of HCMV interaction with TLR-2 on DCs was provided, data demonstrating the secretion of inflammatory cytokines following infection of PBMC with HCMV and the ability of virions to trigger inflammatory signals through TLR-2 in an NF-κB-dependent pathway support the notion that an HCMV–TLR2 interaction on DCs may result in activation of the innate response. Interestingly, recognition of HCMV by TLR-2 and subsequent signaling did not depend on viral gene transcription since dense bodies and UV-inactivated virus retained live virus properties in this respect. We can speculate that hyperimmediate recognition of HCMV by TLR-2 might be essential for innate resistance to infection through secretion by DCs of antiviral cytokines such as TNF-α and type I interferons (IFN-α/β), considered historically as the first line antiviral defense. Direct proof of TLR-2 signaling being crucial for an efficient innate defense against HCMV remains to be obtained experimentally. Evidence for TLR involvement in CMV sensing and in innate defense against the virus has been provided recently in an in vivo model of mouse cytomegalovirus (MCMV) infection (Tabeta et al., 2004). Lack of the TLR9-MyD88 pathway in infected mutant mice diminished the secretion of IFN-α/β, IFN-γ and IL-12, reduced the proportion of activated NK and NKT-cells, and had a dramatic effect on viral load and course of disease. Similar defects were observed in TLR3 knockout mice though to a lesser extent than in TLR9-deficient animals. These data suggest that sensing of MCMV by DCs through TLR9 and TLR3 is an essential step in innate defense against MCMV mediated by secretion of type I interferons and NK cell activation. Moreover, owing to the critical role of IL-12 in the induction of NK cells and the priming of effector T-cells, we can assume that recognition of HCMV by TLR-2 constitutes a crucial link between innate and adaptive immunity against the virus. Besides their known ability to activate NK cell cytotoxicity, DCs may acquire an own killing activity through either secretion of pro-apoptotic cytokines or direct cell contact. Indeed, it has been shown that immature DCs that were infected for 5 days with an endothelial cell-adapted HCMV strain acquired the ability to sensitize T-cells to apoptosis and to kill them through FasL- and Trail-dependent mechanisms (Raftery et al., 2001). Even though in this case there is no benefit for the host, this observation points out the ability of DCs to acquire a killing license. We can assume that in spite of this late immunosuppressive mechanism, immediate engagement of TLR and CLR by HCMV could make activated DCs cytotoxic, providing an efficient means of innate resistance to infection, but this remains to be demonstrated experimentally.

Capture of HCMV by DCs through the C-type lectin receptor DC-SIGN (DC-specific ICAM3 grabbing non-integrin) was demonstrated to involve ligation with the envelope glycoprotein gB (Halary et al., 2002) and to contribute to cis-enhancement of cell infection and to trans-cell to cell virus propagation. Beyond its role as a critical mediator of HCMV entry, gB seems to be sufficient to establish an antiviral response following interaction with its receptor (Boehme et al., 2004; Simmen et al., 2001) but it remains to be explored whether recognition of gB by DC-SIGN could initiate this process. By analogy to HIV (Moris et al., 2004), we also may ask whether internalization of HCMV by DCs through DC-SIGN could allow the delivery of viral antigens into MHC-I and MHC-II pathways for CD8$^+$ and CD4$^+$ T-cell activation, respectively. The physiological relevance of these exogenous pathways in response to viral immunosubversion will be discussed further in the section on adaptive immunity (see below).

Unfortunately, the innate response to virus recognition by DCs as described above (Figure 18.1) is not sufficient to win the battle against infection. It can rather be considered as a primary control and warning system to prepare DCs to fulfill their main task, which consists in acquisition and presentation of viral antigens to naïve T-cells. To this end, DCs have to acquire a mature phenotype and to migrate to lymph nodes where they encounter antiviral naïve T-cells. This adaptive response is closely regulated by virus-induced conditioning of DCs during the innate phase, and the outcome of the response is determined by the pattern of cytokines and chemokines they produce. It can thus be expected that a combat between DCs, trying to carry out their job in antiviral defense, and CMV, trying to ensure its survival and dissemination to tissues, starts from the very early steps of virus sensing and goes on throughout the infectious process in DCs.

## Subversion of dendritic cell functions by cytomegalovirus

Because of its central role in antiviral immunity, the DC is a favored cellular target for viruses to attack first. Direct arguments that DCs are real targets for HCMV in vivo are suggested for instance by the ability of the virus to infect hematopoietic progenitor cells (Hahn et al., 1998) and by the presence of viral DNA in purified DCs from viremic renal transplant patients (Beck et al., 2003). The option to use fibroblast- and endothelial cell-adapted HCMV strains for selective in vitro non-productive and productive infection of DCs, respectively, as well as the possibility to generate DCs from peripheral blood monocytes already provided a lot of information on the effects of HCMV infection on DC function. Infection of the mouse by MCMV provided valuable models to assess functional outcomes of DC infection in vivo. Both myeloid DCs and Langerhans cells (LCs) can be productively infected with endotheliotropic strains of HCMV (Hertel et al., 2003; Riegler

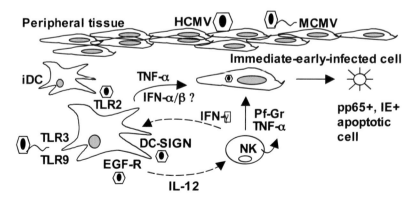

**Figure 18.1** Sensing of CMV by dendritic cells (DCs) involves recognition of HCMV by TLR2, DC-SIGN and EGFR (Wang et al, 2003). TLR3 and TLR9 take part in the innate response against MCMV. Innate resistance to infection may imply killing of immediate-early infected cells through both direct cytotoxicity by cell to cell contact with DCs or NK cells and secretion of pro-apoptotic cytokines. iDC, immature DC; Pf-Gr, perforin–granzyme.

Nevertheless, data from our laboratory (Mandron, Martin et al., unpublished data) have provided evidence that coculture of early-infected fibroblasts with immature DCs increased the number of apoptotic fibroblasts and that experimental blockade of apoptosis in donor cells by pan-caspase inhibitors impaired cross-presentation, suggesting that the presence of dying cells is important in the process. Findings of IE/E viral mechanisms interfering with TNF-R family-mediated apoptosis through either inhibition of mitochondrial permeability by UL37 (Goldmacher et al., 1999) or suppression of Caspase-8 (Flice) activity by UL36 (Skaletskaya et al., 2001) emphasizes the importance of the very immediate delivery of pp65 and expression of IE1 in infected donor cells as well as their sensitization to apoptosis before synthesis of these HCMV-encoded apoptosis inhibitors. Overall, this could explain why the CD8[+] T-cell response against HCMV is dominated by clones recognizing pp65- and IE1-derived epitopes. A requisite for DCs to promote immune activation rather than tolerance is the presence of danger signals such as inflammatory cytokines released by either neighboring cells or DCs. Besides data providing evidence that cross-presentation of cell-associated and particulate antigens may be up to $10^5$-fold more efficient than that of soluble antigens, recent findings pointed at the major role of the release of uric acid crystals by dead cells as a strong stimulatory signal (Shi et al., 2003). The authors showed that uric acid provided a danger signal to DCs with stronger stimulating properties than that exerted by LPS. We can assume that in CMV cross-presentation, apoptosis of infected peripheral cells may release uric acid in the vicinity of non-infected DCs providing at the same time viral antigens and a strong activation signal. Conclusively, even though it remains to be ascertained, one likely scenario is that non-infected DCs are licensed to cross-present CMV antigens thanks to signals provided in the innate phase of infection. Whatever processing pathway is taken by captured CMV antigens, cross-presentation to naïve CD8[+] T-cells is a rapid phenomenon which could overcome the subversive strategies of CMVs in the primary response.

Whether and how CMV might specifically impair cross-priming may be a senseless question since in our scenario non-infected DCs are not impaired in their ability to process viral antigens acquired from an exogenous cellular source. In that case, efficiency of cross-presentation may depend only on viral antigen availability and stability in the donor cell. This may explain why incoming pp65, which is stored in large amounts in dead donor cells, provides a major source of antigen for cross-presentation to prime a predominating anti-pp65 CD8[+] T-cell response. Irrespective of the possible viral effects, a study on factors affecting the efficiency of CD8[+] T-cell cross-presentation by DCs using exogenous HCMV antigens reported that the efficiency varied between individuals according to differences in the responding T-cell repertoire, the endocytic machinery and the HLA type of blood donor (Maecker et al., 2001). In light of this study, we can propose that cross-priming in vivo is similarly influenced and this may provide an explanation for the unresponsiveness of some donors to otherwise immunodominant peptides (Vaz-Santiago et al., 2001). Even though cross-presentation to CD4[+] T-cells is less constrained, their role in the primary response against CMV is critical as was demonstrated by studies showing that viral clearance is determined by the presence of IFN-γ-producing Th1 cells (Gamadia et al., 2003). We can assume that harmful variations in conditioning of DCs by CMV infection in the innate phase could promote polarization of CD4[+] T-cells into the Th2 subset and that Th2 skewing might explain impaired control of viral replication in some symptomatic transplantation recipients. Although we have no direct experimental evidence for CD4$^+$ T-cell help being critical in the cross-priming of CD8$^+$ T-cells, findings showing that they precede CD8$^+$ T-cells in the primary response (Gamadia et al., 2003) argue in favor of their contribution to the process. For instance, secretion of IFN-γ by CD4$^+$T-cells could induce up-regulation of cross-presentation through increased antigen uptake and delivery to the cytosol as well as enhanced proteasome and TAP activities (Gil-Torregrosa et al., 2004). Findings showing that the presence of CD4$^+$ T-cells is a critical point in the regulation of cross-presentation toward "cross-priming" versus "cross-tolerization" (Albert et al., 2001) supports the notion that CD4$^+$ T-cells take part in DC licencing to facilitate cross-priming of CD8$^+$ T-cells.

Finally, whatever the efficiency of cross-presentation is, the aim of priming T-cells against CMV is to equip the host with memory effector cells that are able to contain peripheral viral reactivation and re-infection. In vivo experiments in mice infected with MCMV showed that successful cross-priming of CD8$^+$ T-cells directed against a peptide classified as immunodominant did not guarantee a control of virus replication in spleen, lung and liver (Holtappels et al., 2004). Although we may ask whether viral peptides coming from cross-presentation can differ from those derived from endogenous antigens in infected tissues, these data do not challenge cross-presentation as being an important mechanism of T-cell priming. Rather, such examples invite us to take care in choosing epitopes in peptides-based vaccine design and cellular immunotherapy.

## Views and perspectives

In summary, a model can be proposed for T-cell cross-priming with special emphasis on the MHC-I-restricted CD8$^+$ subset based on the very early acquisition of CMV antigens contained in neighboring dead cells by non-infected DCs. This view is supported by the observation of an immediate sensitivity to death ligands of cells from peripheral tissues such as fibroblasts which contain mostly incoming proteins from the viral tegument such as for instance pp65. The notion of kinetic restriction for a successful priming is reinforced by (i) the early-late resistance of infected cells to apoptosis due to interfering UL proteins thus decreasing the availability of antigen-positive dead cells for phagocytosis by DCs, (ii) bystander inhibition of cross-presentation by early-late soluble factors secreted either by infected tissue cells, for instance TGF-β and cmvIL-10, or by DCs, for instance soluble CD83. Since chemokine-dependent recruitment of more DCs to the site of infection (Sallusto and Lanzavecchia, 2000) is likely to be essential to mount an efficient response, description of a competitive viral homolog of CC chemokine receptors encoded by US28 of HCMV (Bodaghi et al., 1998) provides an additional argument in favor of the idea that cross-presentation has to occur early after infection. Analysis of the dynamics of MCMV-DC interactions in vivo further supports this view since early after infection DCs efficiently prime naïve T-cells but fail later due to down-regulation of their functions (Mathys et al., 2003). More than does feeding of DCs with antigens, apoptosis of infected cells provides an appropriate microenvironment to trigger and improve cross-presentation, and the notion that dying cells have the ability to modulate innate and adaptive responses is an issue of recent debate (Albert, 2004). It may be that other activation signals than those provided by dead cells also modulate CMV-DC interaction in the innate phase, for instance type I interferons α/β resulting from activation through TLRs (Le Bon et al., 2003) and death ligands such as Trail, TNF-α and FasL. The in vivo relevance of these findings is supported

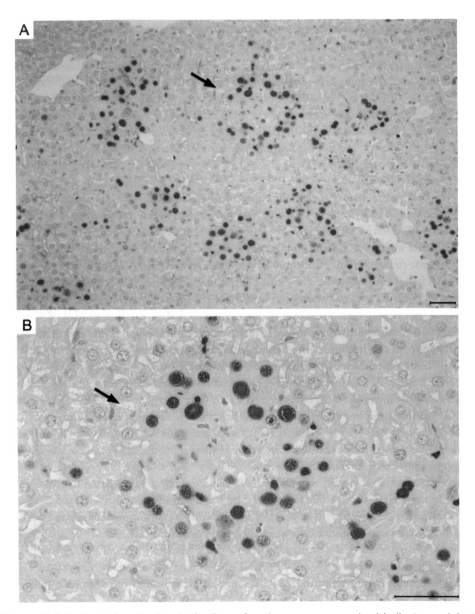

**Figure 19.2** Lytic viral replication in the liver of an immunocompromised indicator recipient. (A) Liver tissue literally resembling a sieve during viral hepatitis on day 14 after intraplantar infection, with numerous plaque-like lesions seen in the section. The arrow points to a virus plaque that is resolved to greater detail in B. (B) Section through a three-dimensional virus plaque in the liver. Note the lysed center of the plaque, which is devoid of hematoxylin-stainable nuclei, and the typical rim formed by more recently infected hepatocytes. Immunohistological staining of intranuclear viral IE1 protein pp76/89 (red) with hematoxylin counterstaining. Bars represent 50 μm.

in infected hepatocytes illustrates a massive CMV hepatitis in which the liver is literally riddled with holes, the three-dimensional organ equivalents of virus plaques well known from lytically infected cell monolayers in permissive cell cultures. When resolved to greater detail in Figure 19.2B, one can recognize the lysed center of a liver tissue plaque, devoid of nuclei, surrounded by a rim of more recently infected hepatocytes indicating the centrifugal expansion of the lesion. Besides the liver, many other organs are infected including the spleen, lungs, adrenal glands, kidney, salivary glands, and gastrointestinal tract (Grzimek et al., 1999) where petechial bleedings are prominent. It is worthy of notice that infection of stromal cells in the bone marrow leads to hemopoietin deficiency that inhibits endogenous as well as bone marrow transplantation-mediated hematopoietic reconstitution and thereby contributes to the maintenance of radiation-induced bone marrow aplasia (Mutter et al., 1988; Mayer et al., 1997; Steffens et al., 1998b). Infected under these conditions, immunocompromised mice can serve as "indicator recipients" to probe with high sensitivity for an antiviral protective function of ex vivo isolated immune cells or of viral epitope-specific CD8 T-cell lines.

In the original report (Reddehase et al., 1985), in vivo primed lymphocytes derived from draining lymph nodes of infected immunocompetent mice were depleted of either CD8-positive or CD4-positive T-cells. The resulting two cell populations were tested separately for their antiviral function by adoptive transfer into immunocompromised indicator recipients either on the day of infection (prophylactic transfer or "preemptive therapy" protocol) or on day 6 after infection, a time point at which infected tissue cells can already be detected in the organs of indicator recipients (therapeutic transfer protocol). In both protocols, albeit with higher cell numbers needed in the therapeutic transfer (Reddehase et al., 1987a), indicator recipients receiving CD8 T-cells showed reduced virus replication in all organs and recovered from moderate CMV disease, while recipients of CD4 T-cells plus all other CD8-negative lymph node cells (likely including NK-cells, NKT-cells, B-cells, possibly γ/δ T-cells, macrophages and immigrated dendritic cells) failed to control the infection and succumbed to CMV disease. In this first set of experiments based on a cell depletion approach, the CD8 T-cells were still accompanied by all the nondepleted CD4-negative cells. Later experiments using a short-term cultivation with IL-2 to selectively expand primed lymph node T-cells followed by CD4 T-cell depletion (Reddehase et al., 1987b) as well as more recent experiments performed with positively sort-purified ex vivo memory CD8 T-cells or with epitope-specific cytolytic CD8 T-lymphocyte lines (CTLL) have clearly shown that their protective antiviral function does not require a cooperation with any other co-transferred cell type. Moreover, a co-transfer experiment using purified memory CD8 T-cells mixed with graded numbers of purified memory CD4 T-cells in CD4/CD8 ratios varying from 0:1 to 4:1 did not reveal a helper function of CD4 T-cells for the antiviral in vivo function of transferred CD8 T-cells (Reddehase et al., 1988), which is an important finding for CD8 T-cell-based cytoimmunotherapy in the CD4 T-cell-deficient host. Taken together, there is firm evidence to conclude that CD8 T-cells expressing TCR α/β are essential and sufficient as direct antiviral effector cells for controlling acute infection and prevention of CMV disease in a cytoimmunotherapy approach. It must be noted however, that in these experiments the surviving recipients eventually reconstituted the complete immune system and that reconstituted CD4 T-cells may well be supportive in the long-term maintenance of protective CD8 T-cell memory.

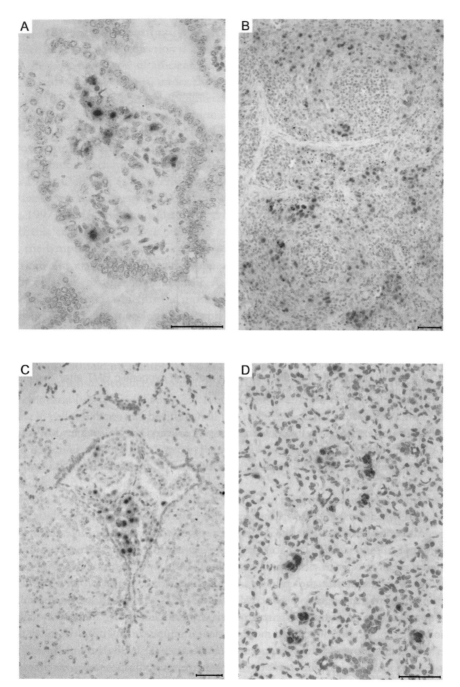

**Figure 19.5** Histopathology of multiple-organ CMV disease after depletion of CD8 T-cells in the model of syngeneic BMT. (A) Villus of small intestine; infected stromal cells in the lamina propria. (B) Spleen; infection of perifollicular stromal cells. (C) Brain; third ventricle in coronal section with infected liquor-producing choroid plexus epithelial cells. (D) Submandibular gland; infected acinar epithelial cells. Detection of the viral genome by virus-specific in situ DNA–DNA hybridization (red) with hematoxylin counterstaining. Tissue sections taken on day 21 (prefinal stage of lethal disease) after BMT and infection. Bars represent 50 μm.

2   CD4 T-cells are not essential for CD8 T-cell reconstitution, priming, recruitment to infected tissues, and antiviral effector function.

3   An efficient antiviral CD8 T-cell response is primed in the presence of an established productive multiple-organ infection, including infection of the spleen (Figure 19.5B). Thus, although conditions can be experimentally defined under which CMVs paralyze the antigen presentation function of DCs resulting in T-cell deletion and anergy (Andrews et al., 2001; Raftery et al., 2001; for greater detail see Chapter 18), such a mechanism may down-modulate but apparently does not prevent CD8 T-cell control of CMV replication in a more relevant model of in vivo infection.

4   Tissue-infiltrating, protective CD8 T-cells can recognize and lyse infected target cells in all three phases of the viral gene expression. As a consequence, virally encoded immune evasion proteins that interfere with the MHC class I pathway of antigen processing and presentation may down-modulate but apparently do not eventually prevent CD8 T-cell effector function in infected host tissues.

5   The protective consequences of CD8 T-cell function clearly prevail over possible immunopathological side-effects, at least in preemptive therapy at early stages of infection.

Altogether, the findings in the BMT model arouse hope that transfer of CD8 T-cells is an option for preemptive treatment of patients who are at risk of CMV disease.

## Bone marrow transplantation across an MHC class I disparity

As a further layer of complexity, a genetic difference between donor bone marrow cells and BMT recipients was introduced in the modular construction system of the model. Again, to keep the system manageable, we introduced a disparity at a single MHC class I locus, namely the $L^d$ locus (Alterio de Goss et al., 1998; Oettel, 2000; Podlech et al., 1998b). The inbred mutant mouse strain BALB/c-H-2$^{dm2}$ (briefly dm2) carries a genetic deletion that includes the gene coding for the $L^d$ molecule but is otherwise congenic with BALB/c. Thus, while the MHC class I genotype of BALB/c is $K^dD^dL^d$, the dm2 mutant can be described as $K^dD^d0$. This feature of the model allows to separate the two immunological complications of MHC disparity, namely graft-versus-host (GvH) and host-versus-graft (HvG) reactivity depending on the genetics of BMT donor and recipient:

| Donor | Recipient | Complication |
|---|---|---|
| BALB/c ($K^dD^dL^d$) | dm2 ($K^dD^d0$) | HvG |
| dm2 ($K^dD^d0$) | BALB/c ($K^dD^dL^d$) | GvH |

In both cases, the foreign target molecule is $L^d$. Besides a theoretically possible alloreaction against the $L^d$ molecule or $L^d$-presented endogenous peptides, the $L^d$ molecule is missing as a presenter of viral antigenic peptides on stromal and parenchymal tissue cells in the case of the HvG genetic constellation and on transplantable cells of hematopoietic origin – which include all major antigen-presenting cells (APCs) such as DCs, macrophages and B-cells – in the case of the GvH genetic constellation. In the absence of infection of the recipients, repopulation efficacy and survival rates were similar between both MHC-disparate BMT protocols and were also not significantly different to those in the two respective

previous literature must be reinvestigated in more depth with the knowledge of antigenic peptides presented by all three MHC class I molecules in the H-2$^d$ haplotype (Table 19.1) and with advanced methods to quantitate epitope-specific CD8 T-cells. The preclinical model of MHC class I disparate BMT (and future development by inclusion of further parameters in the modular construction system of the BMT model) bears a wealth of opportunities to address fundamental questions that are of clinical relevance, in particular in allogeneic HSCT.

## Antigens and immunodominance

As shown in the preceding section, missing information on the viral epitope specificity of the CD8 T-cells has hampered the interpretation of interesting findings made in otherwise highly elaborated disease models. Thus, although it is scientific journals' recent policy to no longer give credit for the identification of antigenic peptides, knowledge of the viral peptides that are presented by MHC class I molecules is fundamental to the understanding of disease control by CD8 T-cells. Without it, epitope-specific CD8 T-cells cannot be identified and quantitated by modern methods such as MHC-peptide multimer staining (MHC-Ig dimers, tetramers, pentamers, streptamers), ELISPOT assay or intracellular cytokine staining, and none of the pending questions in CMV immunology can be approached with any prospect.

The first antigenic peptide to be identified for a herpesvirus was the L$^d$-presented IE1 nonapeptide 168-YPHFMPTNL-176 (Reddehase and Koszinowski, 1984; Reddehase et al., 1989) that is derived by proteasomal processing of the regulatory intranuclear IE1 (ORFm123) protein pp76/89 of MCMV (for detailed reviews see Reddehase, 2000; 2002). After a pause of inspiration for one decade, only recently a number of further antigenic peptides have been identified in the H-2$^d$ haplotype (Table 19.1). While the H-2$^d$-restricted peptides of MCMV were identified one-by-one using sequence-based "reverse immunology" prediction algorithms (Nussbaum et al., 2003; Rammensee et al., 1997) and the testing of synthetic candidate peptides, systematic screening of a comprehensive library of ORF transfectants comprising the entire viral genome very recently led to a "quantum leap" of progress in the identification of H-2$^b$-restricted antigenic peptides of MCMV by author M.W.M. in the laboratory of Ann B. Hill (Table 19.2). The list includes three antigenic ORFs of MCMV (m04, M45, and m164) that are shared between the two haplotypes. Knowledge of the H-2$^b$-restricted peptides will greatly expedite CMV immunology, since a wealth of transgenic, knock-out, and knock-in mutants are on the C57BL/6 genetic background.

One result of quantitative analysis based on ELISPOT assay as well as MHC-peptide tetramer staining was the focus of the CD8 T-cell response in BALB/c mice on just two peptides, IE1 and m164 (Holtappels et al., 2002c), although the coding capacity of the virus, with ~ 170 ORFs to choose from (Rawlinson, 1996), predicts tens of thousands of potentially antigenic peptides matching MHC class I binding motifs. The phenomenon of a very focused immune response even to complex pathogens is not unique to CMVs but is generally known as "immunodominance" of antigenic peptides and is a main topic of ongoing research in basic immunology (Yewdell and Del Val., 2004). From a teleological point of view it makes evolutionary sense that a host facing numerous pathogens during its lifetime does not by luxury allocate all its T-cell receptor (TCR) repertoire and defensive resources

**Table 19.2** Antigenic peptides of MCMV in the H-2$^b$ haplotype

| ORF | Sequence | Restriction | CTL[a] |
|---|---|---|---|
| m04 | 1-MSLVCRLVL-9 | D | |
| M33 | 47-GGPMNFVVL-55 | D | |
| M36 | 213-GTVINLTSV-221 | D | |
| M38 | 38-STYTFVRT-45 | K | |
| M38 | 316-SSPPMFRV[PV]-325[b] | K | |
| M44 | 130-ACVHNQDII-138 | D | |
| M45 | 985-HGIRNASFI-993 | D | Clones 3; 55[c] |
| M57 | 816-SCLEFWQRV-824 | K | Clone 5[d] |
| M77 | 474-GCVKNFEFM-482 | D | |
| M78 | 8-VDYSYPEV-15 | K | |
| M86 | 1062-SQNINTVEM-1070 | D | |
| M97 | 210-IISPFPGL-217 | K | Clone 96[d] |
| M100 | 72-RIIDFDNM-79 | K | |
| M112 | 171-AAVQSATSM-179 | D | |
| m139 | 419-TVYGFCLL-426 | K | |
| m141 | 16-VIDAFSRL-23 | K | Clone 11[d] |
| m164 | 267-WAVNNQAIV-275 | D | |
| m164 | 283-GTTDFLWM-290 | K | |

[a]CTL clones used in published work prior to identification of the corresponding antigenic peptides.
[b]Determination of the C-terminal position is pending.
[c]Gold et al. (2002), Kavanagh et al. (2001).
[d]Kavanagh et al. (2001).

to the control of CMV but focuses on few peptides that suffice for the purpose of protection, with some security back-up to cope with the risk of antigenicity loss mutations in the peptide-coding sequences. With this reasoning, one may speculate that a highly focused response (as in BALB/c) reflects a good and robust protective quality of the epitope-specific CD8 T-cells and/or low peptide mutation rate, while a broader response (as in C57BL/6) reflects a lower success of each single epitope and/or higher mutation rate. Although the IE1 protein is a regulatory protein important for efficient virus replication in vivo (Ghazal et al., 2005), IE1 antigenicity loss mutations were found in virus isolates from feral mice (Lyons et al., 1996). Yet, as the natural variation of the recently identified D$^d$-restricted peptide that is derived from the non-essential ORFm164 protein (Holtappels et al., 2002c) and of the H-2$^b$-restricted antigenic peptides (Table 19.2) still await analysis, a comparison of the mutation rates to substantiate the hypothesis is not yet possible.

The molecular basis of immunodominance is currently not well understood, but is affected by many parameters of both the APC and the responding T-cells. The APC parameters include protein abundance and turnover, efficacy of proteasomal processing, resistance to cytosolic proteases, efficacy of TAP transport, N-terminal trimming, efficacy of class I molecule loading that includes peptide binding affinity, and MHC-peptide complex trafficking to the cell surface. Of these, peptide affinity for MHC is of central importance (for a review see Yewdell and Bennink, 1999). From the T-cell point of view, TCR affinity and

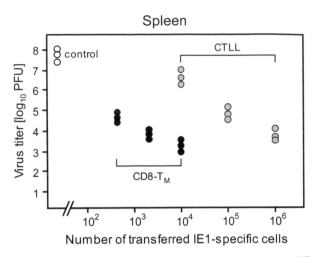

**Figure 19.9** CTLL and memory CD8 T-cells specific for the same epitope differ in their antiviral potential. Sort-purified IE1 epitope-specific memory CD8 T-cells (CD8-T$_M$; black circles) and IE1 epitope-specific effector CTL from a line selected by three rounds of restimulation with the IE1 peptide (CTLL; gray circles) were tested for their antiviral activity in the spleen by adoptive transfer of graded cell numbers. Control (open circles), no cell transfer. Virus titers were determined on day 12 after cell transfer and infection. Circles represent three individual adoptive transfer recipients.

for the high antiviral in vivo activity. From a TCR transgenic model there exists evidence to suggest that CD8-T$_{CM}$ have a higher proliferative potential and a greater capacity to persist in vivo (Wherry et al., 2003). Clearly, finding a method to select and expand epitope-specific CD8 T-cells in cell culture under conditions that either maintain or restore a CD8-T$_M$ – preferably a CD8-T$_{CM}$ – status is a rewarding goal and chance for a breakthrough in cytoimmunotherapy.

## Viral proteins interfering with MHC class I molecule trafficking

The data discussed so far have clearly revealed that destructive CMV replication is prevented by the immune system, by CD8 T-cells in particular, and that adoptive transfer of CD8 T-cells is a promising option for preemptive therapy of CMV disease in the immunocompromised host. In view of such a selective pressure exerted by CD8 T-cells it is not surprising that viral evolution has responded with mechanisms that interfere with CD8 T-cell priming and function, and the class I pathway of antigen processing and presentation is a key target for such a viral countermeasure. We owe the original observation of CMV "immune evasion" to MCMV and specifically to the IE phase-specific CTL clone IE1 (Reddehase et. al., 1986a), the prototype clone that years later aided the molecular identification of the IE1 peptide (Del Val et al., 1988; Reddehase et al., 1989; for more comprehensive referencing see Reddehase, 2000; 2002). Clone IE1 was found to recognize its target antigen in the IE phase of viral gene expression but failed to recognize target cells in the E phase despite continued presence of IE proteins (Reddehase et al., 1986b), a finding molecularly substantiated later with the same clone IE1 after its cognate L$^d$-restricted epitope was known (Del Val et al., 1989). In that study, prevention of IE1 peptide presentation in the E phase was found to be independent of the rate of IE1 protein synthesis, of protein amount, stability, and nuclear import, and, notably, was not associated with any down-modulation of MHC class I expression on the cell surface or reduced recognition of the cells by CTL clone B6αL$^d$ recognizing the L$^d$ molecule.

These early observations initiated an intense investigation into the molecular mechanisms, specifically the viral genes involved, and led to the identification of viral glycoproteins that are referred to as "immunoevasins" or as "VIPRs, viral proteins interfering with antigen presentation", because, as far as currently known, they serve the only purpose to interfere with the antigen presentation pathways. Though the involved genes and molecular details are unique for each virus, likely reflecting host adaptation, MCMV and HCMV show a phenotypic convergence in the fundamental concept of inhibiting the class I pathway. Numerous reports and review articles were dedicated to this hot topic, and it is beyond the scope of this chapter to re-review all the literature (for selected more recent reviews see Hengel et al., 1999; Kavanagh and Hill, 2001; Reddehase, 2002; Reddehase et al., 2004; Tortorella et al., 2000; Yewdell and Hill, 2002). With the focus on MCMV (see Chapter 17 for HCMV), the three glycoproteins m04/gp34 (which is simultaneously a source of antigenic peptides; see Tables 19.1 and 19.2), m06/gp48, and m152/gp40 have been shown to interfere with the trafficking of MHC class I molecules (Figure 19.10). Their tasks are apparently quite distinct: m152/gp40 appears to interact only transiently with peptide-loaded MHC class I complexes and catalyzes their retention in an ER-Golgi intermediate compartment (ERGIC), m06/gp48 binds to MHC class I molecules for degradation in the lysosome, and m04/gp34 rescues MHC class I molecules for presentation at the cell surface, thus rather opposing m06/gp48 and m152/gp40. The effects of the three molecules on MHC class I K$^d$ and K$^b$, D$^d$ and D$^b$, as well as L$^d$ cell-surface expression, singly and in all four possible combinations, have been deciphered recently by infecting cells with a complete set of viral immune evasion gene deletion mutants (Wagner et al., 2002; reviewed by Reddehase et al., 2004). In essence, with distinct differences between the MHC class I alleles, m06/gp48 and m152/gp40 each have a down-modulating effect with some cooperation when co-expressed, whereas m04/gp34 has no effect when expressed alone and has little effect when combined with m06/gp48. This can be explained by its later expression in the kinetics after the gp48-MHC class I complex is already degraded. Notably, m04/gp34 reverses the down-modulating effect of m152/gp40 and even leads to a class I cell-surface expression that is higher than normal. This can be explained again by its later expression in the kinetics after the transient interaction between m152/gp40 and MHC class I has led to an accumulation of peptide-loaded MHC class I complexes in the ERGIC. Thus, as far as MHC class I cell-surface expression is concerned, m04/gp34 should be expected to be an anti-immunoevasin/anti-VIPR. Its classification as an immunoevasin/VIPR is based on functional findings made with CTL clones in the lab of Ann B. Hill. Specifically, prevention of target cell lysis by K$^b$-restricted clones 5, 11, and 96 recognizing M57, m141, and M97 epitopes, respectively (Table 19.2), required a cooperation by m04/gp34, whereas such a help was not needed to prevent target cell lysis by the D$^b$-restricted clones 3 and 55 (Kavanagh et al., 2001), in the meantime both known to recognize the M45 epitope (Table 19.2). Of interest, the M97 epitope-specific clone 96 failed to lyse target cells infected with MCMV-WT, but recognized targets whenever any one of the three immunoevasins was deleted (LoPiccolo et al., 2003), indicating the need for cooperation between all three im-

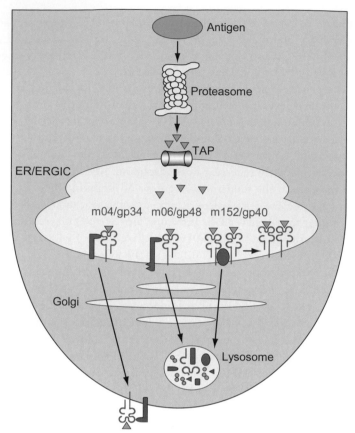

**Figure 19.10** Interference of the three MCMV-encoded immunoevasins/VIPRs with MHC class I molecule trafficking. (Adapted from Nature Rev. Immunol.; Reddehase, 2002.)

munoevasins. Collaborative effort required to prevent the presentation of K$^b$-restricted peptides is likely to relate to a less efficient retention of K$^b$, compared to D$^b$, by m152/gp40 in the ERGIC (Kavanagh et al., 2001).

Thus, being most evident for m04/gp34, the effect of an immunoevasin on antigenic peptide presentation is not mirrored by its effect on MHC class I cell-surface expression. In other words, class I cell-surface expression does not reliably predict peptide presentation. This brings us back to understand the original observation by Del Val and colleagues (1989), namely the finding that prevention of IE1 peptide presentation in the E phase is not associated with any detectable down-modulation of L$^d$ expression at the cell surface. As discussed in somewhat more detail in a recent review (Reddehase et al., 2004), there are good arguments to propose that effects on MHC class I molecules in peptide loading complexes in the ER or shortly after peptide loading are more relevant for antigen presentation than is the down-modulation of MHC class I cell-surface expression, which is likely to be only secondary to the function of the immunoevasins.

# Question marks concerning the in vivo relevance of immunoevasins

From the established protective function of CD8 T-cells in the adoptive transfer and genetic vaccination approaches discussed above, it was clear beforehand that the immunoevasins/VIPRs, even in concert, ultimately fail to prevent the presentation of antigenic peptides in infected host tissue cells (Reddehase, 2002). By using BAC-cloned MCMV-WT and the immune evasion gene triple-deletion mutant MCMV-Δm04+06+152 (Wagner et al., 2002), recent work by the group of Ann B. Hill has directly addressed the question of whether the presence of the three immune evasion genes has any impact on the acute or persisting CD8 T-cell response, the clearance of productive infection, and the establishment of latency in immunocompetent C57BL/6 mice (Gold et al., 2004). The somewhat surprising answer was: "no". So, based on these arguments, one may indeed question a biological relevance of immunoevasin/VIPR interference with the MHC class I pathway.

## Role of immunoevasins in the priming of naïve CD8 T-cells

It has previously been an idea that immunoevasins may contribute to the phenomenon of immunodominance by preventing the presentation of most epitopes, thereby restricting the CD8 T-cell response to a few epitopes. Accordingly, there was hope that deletion of the immune evasion genes in a candidate "vaccine virus" would improve immunogenicity by intensifying the presentation of peptides in general and in particular also by increasing the number of different epitopes presented, thus inducing a CD8 T-cell response both larger in size and of broader specificity.

The first evidence to abandon this hope came from work by Gold and colleagues (2002) who showed that deletion of m152/gp40 does not alter the immunodominance of the $D^b$-restricted M45 peptide (Table 19.2). This finding in the C57BL/6 model was later independently confirmed in the BALB/c model by infection of immunocompetent mice with mutant virus MCMV-Δm152 and its revertant MCMV- Δm152-rev, with the result that neither the response magnitude, i.e. the epitope-specific CD8 T-cell frequencies, nor the immunodominance hierarchy of epitopes IE1 ≈ m164 >> m04 > M83 > M84 was altered in any way (Reddehase et al., 2004). Since the first experiments did not take account for a possible contribution of m04/gp34 and m06/gp48, an analogous experiment was performed comparing the result of priming by infection with MCMV-WT bac and MCMV-Δm04+06+152 (Figure 19.11; and essentially corresponding results in the C57BL/6 model). Again, the deletions had no significant impact on response magnitude and immunodominance hierarchy. While we currently cannot exclude that a still undefined epitope may profit from the deletions, it is apparent from both the C57BL/6 and the BALB/c model that none of the many defined subordinate epitopes profited significantly.

An intriguing theory for explanation has been proposed by Gold and colleagues (2002) at the time of their original observation: priming of CD8 T-cells during CMV infection may occur primarily, if not exclusively, through uninfected DCs that take up viral proteins from infected cells that are destroyed as a consequence of lytic infection, a phenomenon known as cross-presentation or cross-priming (see Chapter 18). As the immunoevasins are not expressed in uninfected DCs, this theory elegantly explains the lack of an effect of immunoevasins on CD8 T-cell priming. This also somewhat deprives m04/gp34 of its mystique to be simultaneously an antigen and an immunoevasin (Holtappels et al., 2000b).

active replication in differentiated macrophages. Cellular factors that dictate macrophage differentiation must regulate CMV gene expression either directly, or indirectly by altering the cellular physiology of the host macrophage. In addition, differentiated macrophages are commonly exposed to stimulating and activating agents during inflammatory and cellular immune responses, and activated macrophages produce a plethora of intrinsic and extrinsic antiviral compounds. Therefore, the virus must adapt to this hostile environment, at least long enough to replicate and spread to neighboring cells. Challenges therefore exist to iden-tify which cellular factors associated with macrophage differentiation awaken the latent viral genome and promote virus replication, and which viral gene products serve to thwart antiviral activities of activated macrophages. However, identification of such factors is only the first step; deciphering the mechanisms is the ultimate goal.

The layers of complexity in macrophage cellular physiology provide perhaps the most voluminous challenge to characterizing the mechanisms of CMV regulation within this cell type. Macrophage differentiation and activation each represent a phenotypic and functional spectrum, and are not discrete entities. Phenotypically, there are at least two subpopulations of human peripheral blood monocytes (CD14hi and CD14low/CD16hi), and activated macrophages are now classified into at least three subpopulations depending on the induc-ing agent (Mosser, 2003) (Figure 20.1). Each subpopulation expresses different biological markers, secretory products, and intracellular antimicrobial compounds. In addition, mac-rophages residing within specific tissues are most assuredly physiologically distinct due to differences in their microenvironment, including the concentration of various chemokines, cytokines, and stimulating agents. It is likely that the complexity of CMV–macrophage interactions reflects such diversity. Because of common developmental pathways with en-dothelial cells and dendritic cells, many aspects of macrophage biology that influence CMV gene expression will likely be applicable to replication of the virus in these cell types as well. This chapter aims to highlight the complexities in the pivotal interactions between CMV and monocyte-derived cells, the progress that has been made to date, and the challenges that lie ahead in deciphering which interactions are key elements of CMV pathogenesis.

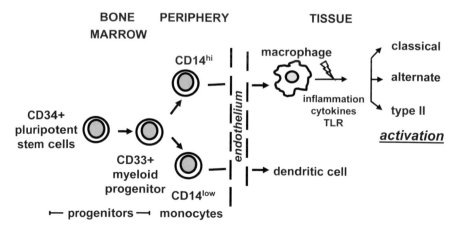

**Figure 20.1** Differentiation of monocytes into macrophages. Monocytes differentiate into tissue macrophages through the process of diapedesis across endothelial cell layers. A final stage in macrophage differentiation is activation in response to various stimuli via one of three pathways.

## Extent of CMV gene expression in monocytes, monocyte progenitors, and macrophages

Numerous studies have documented that CMV remains latent in bone marrow myeloid progenitors and peripheral blood monocytes (Sissons et al., 2002; Streblow and Nelson, 2003). Early studies documented that bone marrow CD34$^+$ precursors of monocytes and CD33$^+$ CD14$^+$ or CD33$^+$ CD15$^+$ monocytes harbor HCMV DNA (Hahn et al., 1998; Kondo et al., 1994; Larsson et al., 1998; Maciejewski et al., 1992; Mendelson et al., 1996; Taylor-Wiedeman et al., 1991; von Laer et al., 1995). It is estimated that during a natural infection, up to 0.01% of bone marrow-derived mononuclear cells harbor HCMV genome, with as many as 13 copies of genome per infected cell (Slobedman and Mocarski, 1999). A small percentage of latently infected cells also express viral transcripts that originate from the major immediate-early gene region but are either in the antisense orientation or are alternatively spliced compared to transcripts from this region that predominate during replication in permissive cells (Kondo et al., 1996; Slobedman and Mocarski, 1999).

It is important to keep in mind that the true definition of latency applies only to CMV DNA-containing cells that are capable of reactivating infectious virus upon some means of stimulation (Goodrum et al., 2002). This criterion is met for the CD14$^+$ monocytes and their CD34$^+$ progenitor cells (Goodrum et al., 2002; Hahn et al., 1998; Lee et al., 1999; Maciejewski et al., 1992; Movassagh et al., 1996; Taylor-Wiedeman et al., 1994; Zhuravskaya et al., 1997). Most recently, the CD34$^+$CD38- subpopulation of CD34$^+$ progenitor cells was specifically identified as sequestering latent CMV genome that could be reactivated (Goodrum, 2004).

Common to most all methods of reactivation from latency is cytokine-induced differentiation of progenitor cells into monocytes, or monocytes into macrophages (Ibanez et al., 1991; Soderberg-Naucler et al., 1997). Herein lies the key to defining cellular factors dictating tropism of CMV; however such factors have not as yet been clearly defined (see below). The observation that some populations of differentiated murine macrophages harbor latent MCMV DNA (Koffron et al., 1998; Pollock et al., 1997) suggests that differentiation per se may not be sufficient to restore the genome to a replicative form, but that a series of cytokine and TLR-mediated signals may be required to induce a permissive cell state. This hypothesis is supported by the finding that numerous cytokines, including interferon gamma (IFNγ), tumor necrosis factor alpha (TNFα), interleukin 4, and GM-CSF, have been identified as inducers of CMV reactivation from latency in cell culture systems (Hahn et al., 1998; Soderberg-Naucler et al., 1997; Soderberg-Naucler et al., 2001) and as inducers of the initial steps of reactivation in vivo in the murine model (Hummel et al., 2001; Simon et al., in press). The relative role of cytokine and TLR-mediated signaling in induction of host cell differentiation factors versus induction of viral gene expression (Hummel and Abecassis, 2002; Hummel et al., 2001; Kline et al., 1998; Lee et al., 2004; Ritter et al., 2000) is unknown; however, both mechanisms are likely involved in the switch from a latent to productive state of CMV infection.

Differentiated macrophages that are fully permissive for CMV replication nevertheless display differences in kinetics of viral replication, compared to fibroblast cells. Replication is delayed in macrophages, with viral transcripts detectable days after infection instead of

rophage activation. Therefore, they may function in dampening over-zealous inflammatory responses detrimental to the host. A third type of activation, called type II activation, ensues when macrophages engage their Fc gamma receptors via binding IgG and concomitantly receive a second signal through, for example, any of the TLR or CD40. Type II activated macrophages, like alternatively activated macrophages, secrete IL-10 and appear to have anti-inflammatory properties as well as a role in inducing type 2 cytokines from T-cells.

There are currently no data on CMV infection of alternatively or type II activated macrophages, or effects of virus infection on their function. Possible links however between CMV infection and the function of these cells come to mind when considering that human CMV encodes a functional IL-10 and/or induces cellular IL-10 (Chang et al., 2004; Jones et al., 2002; Kotenko et al., 2000; Nordoy et al., 2003; Spencer et al., 2002) and both human and murine CMV encode Fc gamma receptors (Atalay et al., 2002; Thäle et al., 1994). In addition, CMV infection of human macrophages functionally suppresses T-cell proliferative responses and MHC class II antigen presentation (Odeberg and Soderberg-Naucler, 2001; Redpath et al., 1999; Tomazin et al., 1999). Whether this results from suppression of classical macrophage activation by CMV infection, or promotion of the alternative or type II activation pathways is unknown.

### Effects of CMV infection on macrophage activation

Because differentiated macrophages are permissive for CMV replication and release infectious virus, it is possible that virion binding to these host cells or CMV gene expression within the cells may positively or negatively influence macrophage activation. For example, components of the CMV virion or replicative intermediates binding to TLR2 (Compton et al., 2003) or TRL9 (Tabeta et al., 2004), respectively, may promote or antagonize macrophage activation. Murine CMV infection of macrophages inhibits expression of TNFα receptors (Popkin and Virgin, 2003), potentially thwarting activation. Numerous studies have indeed documented that CMV infection blocks IFNγ-induced macrophage activation, resulting in suboptimal levels of MHC class II molecules and deficiencies in antigen presentation (Heise et al., 1998a; Heise et al., 1998b; Le Roy et al., 1999; Miller et al., 1998). Interestingly, the block in activation occurs at the level of IFNγ-induced chromosomal promoter assembly, thereby sparing basal levels of transcription and preserving expression of cellular genes required for virus replication and maintenance of the host cell integrity (Popkin et al., 2003). Additional viral gene products further contribute to the down-regulation of MHC class II expression to prevent antigen presentation (Odeberg and Soderberg-Naucler, 2001; Redpath et al., 1999; Tomazin et al., 1999).

In vivo, however, murine CMV infection promotes IFNγ and TNFα-mediated macrophage activation (Heise and Virgin, 1995), an important effector mechanism for controlling CMV infection and ensuring survival of the host required to harbor latent infection. Initially, CMV-induced macrophage activation in vivo may appear paradoxical to the finding that in vitro, virus infection of macrophages blocks IFNγ-induced activation. However, the majority of activated macrophages within the mouse during an acute infection are likely uninfected cells. Furthermore, it is plausible that only a subpopulation of infected macrophages, those infected prior to IFNγ exposure, are refractory to activation (Popkin et al., 2003). Macrophages exposed to IFNγ prior to infection are likely to be activated and to suppress CMV replication because this cytokine exerts potent antiviral activity in macrophages by a mechanism unique to this cell type and involving suppression of immediate-early gene expression (Presti et al., 2001). Under these circumstances, a balance between virus replication in a subpopulation of macrophages and preservation of the host through immune effector mechanisms is ensured.

## Intrinsic and extrinsic antiviral activities of activated macrophages

The need for CMV to suppress activation of a macrophage in which the virus has established residence may be related to the fact that activated macrophages produce a plethora of potent intrinsic and extrinsic antiviral agents. These agents are important cellular determinants of macrophage tropism, impacting on replication of CMV within the host macrophage (intrinsic) as well as neighboring cells (extrinsic). One such antiviral agent is nitric oxide (NO), which is produced by inducible nitric oxide synthase (iNOS2), an enzyme whose gene contains an ISRE within its promoter and is therefore induced in macrophages activated by IFNγ (Schroder et al., 2004). Several reports document inhibition of CMV replication by NO, both in vitro and in vivo (Bodaghi, 1999; Fernandez et al., 2000; Lee et al., 1999; Noda et al., 2001; Okada et al., 1999; Tanaka and Noda, 2001; Tay and Welsh, 1997) These studies made use of chemical inhibitors of iNOS2 or of mice genetically deficient in iNOS2 to demonstrate that in the absence of iNOS2, virus replication and copy numbers of latent CMV genomes are higher than in control cells or mice, although immunopathology, at least in the lungs, is reduced when the host is devoid of this antiviral cytokine.

Macrophage activation and MCMV infection of macrophages induces expression of the cytokine TNFα (Yerkovich et al., 1997), which exerts intrinsic antiviral activity in an autocrine manner through its binding to one of two structurally homologous and ubiquitous receptors (TNFR1 and 2) (Herbein and O'Brien, 2000). These same receptors also bind the ligand lymphotoxin alpha (LTα or TNFβ, produced only by T-cells). Several studies have documented that TNFα or LTα significantly influences HCMV or MCMV replication (Benedict and Ware, 2001). Initially, TNFα enhances HCMV and MCMV IE gene transcription (Hummel et al., 2001; Prösch et al., 1995) and differential splicing to yield the MCMV transactivator mRNA IE3 (Simon et al., in press). However, later in the virus replication cycle, TNFα confers antiviral functions by blocking MCMV DNA replication, subsequent late gene expression, and virion formation (Lucin et al., 1994). Neutralization of TNFα in vivo using antibodies to the cytokine or soluble TNFR enhances MCMV replication (Pavic et al., 1993; Yerkovich et al., 1997). Overcoming the antiviral activity of TNFα in macrophages is important for the pathogenesis of MCMV, and hence, as mentioned above, MCMV infection of macrophages inhibits expression of TNFα receptors (Popkin and Virgin, 2003).

Type 1 interferons alpha and beta, also produced by macrophages, have direct antiviral activity in reducing CMV replication (Delannoy et al., 1999; Gribaudo et al., 1995; Yeow et al., 1998). The major role of this antiviral cytokine in curtailing CMV infections in vivo, however, may actually be due to its immunoregulatory functions in inducing NK and CD8 T-cell antiviral functions (Dalod et al., 2003; Salazar-Mather et al., 2002). Interestingly, bone marrow macrophages from wild-type mice or knock-out mice devoid of interferon alpha/beta (IFNαβ) receptors, TNFR1R2 receptors, or iNOS all support low levels of

MCMV replication, with no significant enhancement of virus replication in the cells derived from knock-out mice (Presti et al., 2001). When these differentiated macrophages were activated with IFNγ, virus replication was equally suppressed in macrophages from wild-type and all three knock-out mice, indicating perhaps that the antiviral activities of IFNαβ, TNFα, and NO are functionally redundant. Another interpretation of these data, provided by the authors, is that IFNγ directly inhibits CMV replication in macrophages independent of these three antiviral agents (Presti et al., 2001).

Perhaps in attempts to thwart the antiviral activity of IFNα, human CMV blocks IFNα-induced signal transduction at numerous pathways, at least in endothelial cells and fibroblasts (Miller et al., 1999). For example, IFNα-stimulated antiviral protein MxA, IFN regulatory factor 1 (IRF-1), and 2′,5′-oligoadenylate synthetase (2′,5′-OAS) gene expression are suppressed by CMV infection of these cell types. It is apparent that a delicate balance between induction of IFNα-stimulated genes, whose products assist in CMV replication, and suppression of INFα-induced antiviral activities must be achieved to attain efficient replication of the virus in the host cell.

A fourth and newly discovered antiviral agent produced by macrophages in response to viral infections is the cellular protein viperin (*virus inhibitory protein, endoplasmic reticulum-associated, interferon-inducible*). Viperin is an ER-associated protein whose expression is induced by CMV infection, by binding of UV-inactivated virions, or by type I or II interferons in at least macrophages, fibroblasts, and dendritic cells (Boudinot et al., 2000; Chin and Cresswell, 2001; Zhu et al., 1997). Induction by CMV is not dependent upon type I interferon expression, but is similar to other IFN-inducible genes that are expressed upon virion gB binding (Chin and Cresswell, 2001; Zhu et al., 1997). Viperin suppresses CMV infection at an early or late stage of replication, sparing IE gene expression but squelching synthesis of early-late and late viral proteins (pp65, gB, pp28) (Chin and Cresswell, 2001). However, human CMV counteracts this antiviral effect by redistributing viperin from the ER to the Golgi and to cytoplasmic vacuoles associated with virion assembly (Chin and Cresswell, 2001). Further studies to determine the mechanism of how viperin subverts CMV replication and to identify the viral gene products that sabotage its antiviral function will surely contribute significantly to our understanding of CMV pathogenesis.

In order to assess the relative expression of viperin compared with iNOS, TNFα, and type I interferon in CMV-infected macrophages, we performed RT-PCR to detect expression of each gene product at various times after MCMV infection, exposure to noninfectious virions, or induction by type I interferon or LPS, another inducer of viperin (Boudinot et al., 2000). In the differentiated macrophage cell line IC-21, which is highly permissive for MCMV replication, TNFα was expressed constitutively and without regard to MCMV exposure (Figure 20.3A). In contrast, iNOS was expressed at early and late times during MCMV replication, with kinetics similar that of LPS-induced iNOS (Figure 20.3A and B). Viperin was minimally expressed in mock-infected cells, but was up-regulated within 4 h of exposure to UV-inactivated virus or of infection. Interestingly, progression into the late phase of virus replication resulted in suppressed levels of this antiviral agent, implicating perhaps a virus-encoded protein that is able to curtail expression of viperin. Induction of viperin protein coincided with RNA levels, with peak steady-state levels evident at 8 hr post infection (Figure 20.3C). In addition, the kinetics of expression of this gene was identical in primary peritoneal macrophage compared to the IC-21 cells (data not shown).

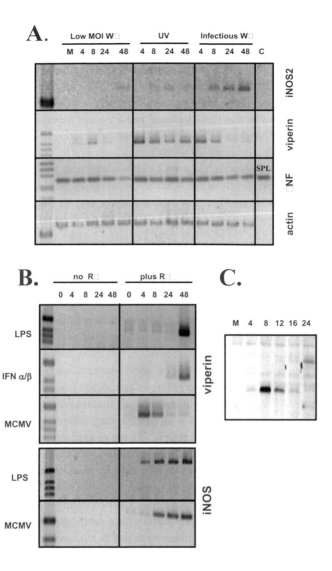

**Figure 20.3** Expression of intrinsic and extrinsic antiviral agents in IC-21 macrophages infected or exposed to WT MCMV. (A) IC-21 cells were infected (MOI = 5) with infectious MCMV or exposed to an equivalent amount of UV-inactivated virus. At the indicated times post infection (hours), RNA was isolated, reverse transcribed with hexamer primers (Superscript™) and PCR was performed using Failsafe™. PCR conditions were optimized for each product, and cycle numbers adjusted to insure a linear range of amplification. One-fifth of the PCR product was analyzed by agarose gel electrophoresis and ethidium bromide staining. M = mock; low MOI WT = infection with amount of infectious virus remaining in the UV prep; C = no DNA or no primer; SPL denotes control from infected spleen tissue. (B) Macrophages were infected with WT virus as above (MCMV) or stimulated with LPS or type I interferon for the indicated times (hours). RT-PCR was performed as above, with controls containing no reverse transcriptase (RT). PCR specific for each product was performed as in (A). (C) Western blot to detect expression of viperin (40 kDa) in MCMV-infected macrophages. Lysates of IC-21 macrophages infected as above with WT MCMV and harvested at the indicated times (hours) post infection were subjected to Western blotting using anti-viperin antibody (kindly provided by Dr. Peter Creswell, Yale University, New Haven, CT).

ficient virus replication in macrophages may necessitate anti-apoptotic, as well as antiviral, functions.

This discussion is based primarily on MCMV genes; identification of macrophage-tropic HCMV genes is restricted by the inability to maintain relevant and permissive human macrophage cell cultures in vitro. However, it is noteworthy that all of the MCMV genes listed in Table 20.1 have homology to HCMV genes (and hence are denoted with a capital M prefix). Current knowledge of the homologous HCMV genes will be discussed where appropriate. It is also noteworthy that four of the six genes listed in the table are members of the US22 family of genes. This gene family is found in all *Beta-herpesvirinae* and is characterized by the presence of one or more of four conserved motifs consisting of stretches of hydrophobic residues interspersed with charged amino acids (Chee et al., 1990; Efstathiou et al., 1992; Nicholas, 1996). Motifs I and II have consensus sequences, while motifs III and IV are less well defined but are rich in hydrophobic residues. The products of the US22 genes vary in kinetics of expression (either IE or E) and cellular localization (nuclear or cytoplasmic). Specific functions for many of the US22 family genes have not yet been defined, but they are likely to be diverse.

## M36

The MCMV M36 and HCMV UL36 immediate-early genes are members of the US22 family of genes and encode proteins of approximately 50–52 kDa that are dispensable for virus replication in fibroblasts (Dunn et al., 2003; McCormick et al., 2003; Menard et al., 2003; Patterson and Shenk, 1999). In human fibroblasts infected with clinical strains of HCMV, the UL36 gene product localizes to discrete but unidentified cytoplasmic structures that do not co-localize with markers specific for the Golgi, lysosomes, mitochondria, endosomes, the endoplasmic reticulum, or death effector filaments (Patterson and Shenk, 1999). Although pUL36 was originally reported to function as a transcriptional transactivator of viral and cellular promoters (Colberg-Poley, 1996), an anti-apoptotic function was recently identified for the UL36 gene product (Skaletskaya et al., 2001). This viral protein binds to procaspase-8, thereby inhibiting the enzyme's proteolytic activation required to trigger downstream events leading to apoptosis. The anti-apoptotic function of pUL36 was revealed in the context of fibroblasts that were induced to undergo apoptosis by stimulants and were expressing either WT or a mutant form of UL36. Skalestskaya and co-authors (2001) proposed that pUL36 has an important role in the pathogenesis of HCMV by either protecting infected target cells from cytolytic immune effector mechanisms, or preserving the integrity of a selective population of cell types in vivo that, once infected, would be particularly sensitive to apoptosis.

Studies by Menard et al. (2003) using MCMV have indeed proven that M36 is required for efficient replication of the virus selectively in macrophages. An M36 transposon insertion mutant of MCMV replicates unimpaired in fibroblasts and endothelial cells, but is significantly attenuated for replication in a relatively undifferentiated macrophage cell line (J774A1), a differentiated macrophage cell line (IC-21), and in stimulated primary peritoneal macrophages. Furthermore, dissemination of the M36 mutant virus to target organs is significantly impaired compared to WT virus, implying a compromise in monocyte tropism (U.H. Koszinowski, personal communication). These data indicate that M36 is likely a general determinant of macrophage tropism, independent of cellular differentiation.

Because of its homology to UL36, M36 was tested for anti-apoptotic functions in fibroblasts and macrophages treated with the pro-apoptotic stimulant, anti-Fas (Menard et al., 2003). Like pUL36, pM36 protected fibroblast cells from apoptosis. However, most striking was the protection rendered by WT M36, compared to mutant M36, in macrophages. Even in macrophages not exposed to a pro-apoptotic stimulant, M36 effectively inhibits caspase-8 activity by binding to procaspase-8 (Menard et al., 2003). Thus, M36 is required for a highly productive MCMV infection in the apoptosis-prone macrophage. Apparently, this viral gene product is required to thwart the antiviral activity of IFNγ and TLR signaling, which stimulate macrophages to secrete TNFα, an inducer of apoptosis (U.H. Koszinowski, personal communication). The M36 gene is the first general determinant of macrophage tropism for which a function has been defined.

## M45

The product of gene M45, which has homology to the ribonucleotide reductase R1 genes, is required for optimal replication of MCMV in the differentiated macrophage cell line IC-21 and some endothelial cell types, but not in fibroblasts, bone marrow stromal cells, or hepatocytes (Brune et al., 2001). The role of the 150–160 kDa M45 protein in virus replication in primary macrophages has not been reported to date; however, mutation of this gene renders MCMV highly attenuated for growth in SCID mice and in several target organs in vivo, including spleen, liver, kidneys, lungs, and salivary glands (Lembo et al., 2004). This phenotype may reflect a role for M45 in both macrophage and endothelial cell tropism, as well as in other cell types. In spite of M45's homology to the gene encoding the large subunit of ribonuclease reductase, the M45 product does not exert enzymatic function nor does it influence the activity of endogenous mouse ribonucleotide reductase (Lembo et al., 2004).

Two alternative functions have been proposed for the M45 gene product (Lembo et al., 2004). Because the protein is associated with the virion and expressed abundantly at late times post infection, it may function in virus maturation or at an immediate-early stage after virus entry. In addition, M45 has anti-apoptotic functions, especially evident in endothelial cells (Brune et al., 2001). Thus this gene product likely protects apoptosis-sensitive cells from premature death. Further studies are required to link the virion-associated nature of the M45 protein with its anti-apoptotic function and role in viral pathogenesis.

One complication in deciphering the function(s) of M45 in macrophage or endothelial cell tropism is the discrepancy between MCMV M45 and HCMV UL45 with regards to function. The latter gene has only a minimal role in protecting cells from apoptosis (Patrone et al., 2003) and it is dispensable for replication in at least human umbilical vein endothelial cells (Hahn et al., 2002). Expression of UL45 does nonetheless enhance replication of HCMV in fibroblasts at a low MOI (Patrone et al., 2003). These differences may reflect divergence between the human and mouse species of CMV and/or differences in the physiology of the cells employed.

## M78

The M78 gene, like UL78, encodes a predicted G-protein-coupled receptor (GPCR) with seven hydrophobic transmembrane domains and an amino-terminal extracellular domain characteristic of small ligand binding receptors (Oliveira and Shenk, 2001). The 53-kDa M78 protein accumulates during early and late phases of infection not at the plasma membrane, but in a perinuclear region that partially co-localizes with Golgi markers, consistent with the finding that this viral protein is assembled within the virion (Oliveira and Shenk, 2001). Although not yet proven to function as a GPCR, this protein apparently functions as a component of the virion by associating with the plasma membrane of the newly infected cell and generating a signal that enhances accumulation of IE mRNA, by perhaps stimulating transcription or promoting RNA stability (Oliveira and Shenk, 2001). Cells infected with MCMV deleted of M78 accumulate significantly less IE m123 RNA, early M54 RNA, and late M99 RNA (Oliveira and Shenk, 2001).

The potential macrophage-tropic nature of M78 lies in the fact that this viral protein has a greater impact upon virus replication in the macrophage cell line IC-21 compared to a fibroblast cell line, especially at a low MOI (Oliveira and Shenk, 2001). This lower rate of replication corresponds to significantly reduced steady-state levels of RNA accumulation in the macrophages. Deletion of the M78 gene also renders MCMV attenuated in lethality and growth in target organs, most notably in the salivary glands. This latter finding may reflect reduced dissemination of virus to this organ, a function attributed to virus-infected monocytes. However, once again, these data are not entirely reproducible in another species. While rat CMV deleted of R78 is attenuated with respect to lethality, its phenotype does not differ from WT virus with regards to replication in macrophages or in the salivary glands (Beisser et al., 1999). This poses the question as to whether M78, UL78, or R78 function as macrophage-tropic genes per se or whether they function in a broader sense to regulate virus replication in selective cell types or in cells in a specific physiological state.

## M139, M140, and M141

The US22 family member genes M139, M140, and M141 are appropriately discussed as a group due to molecular and biological similarities. These genes are transcribed at early times post infection, and their transcripts are 3′ co-terminal, sharing a polyadenylation signal within m138 (Hanson et al., 1999a). While single transcripts of 5.4 and 7.0 kb map to M140 and M141, respectively, M139 produces two transcripts of 4.0 and 3.0 kb, the latter likely derived from an alternative start site. The protein products of all three genes are abundantly expressed in both fibroblasts and macrophages at early times, with steady-state levels accumulating to a maximum about the time of DNA replication, with these levels sustained through the late phase of infection (Hanson et al., 2001). M139 encodes two proteins of 72–75 and 61 kDa (p75M139, p61M139), and M140 and M141 encode proteins of 56 and 52 kDa, respectively (pM140 and pM141) (Hanson et al., 2001). Most interestingly, the stability of each gene product is significantly compromised when one or two of the "partner" genes are deleted or mutated (Hanson et al., 2001; Menard et al., 2003), suggesting cooperative and perhaps physical interactions among the products of this gene region (see below).

The M139, M140, and M141 genes likely influence replication of MCMV in fully differentiated macrophages. They are required for optimal replication in the differentiated macrophage cell line IC-21 and in primary peritoneal macrophages differentiated in vivo by LPS (Hanson et al., 1999a; Hanson et al., 2001). These genes do not impact upon virus replication in less mature macrophages derived from bone marrow (Cavanaugh et al., 1996), from peritoneal cells stimulated with thioglycolate (Menard et al., 2003), or represented by the cell line J774A.1 (Menard et al., 2003). Therefore, at least in vitro, they are required for efficient MCMV replication in a cell differentiation-dependent manner, perhaps to overcome either an insufficiency or a molecular antagonist of viral replication acquired during the differentiation process.

The in vivo phenotype of recombinant MCMV mutated at one or more of these three US22 genes attests to the relevance of this gene region in the pathogenesis of the virus. Mutant MCMV deleted of M139, M140, and M141 is nonlethal for SCID mice, a host that is exquisitely sensitive to virus infection (Hanson et al., 1999a). The triple-deletion mutant virus fails to replicate in the spleen, a target organ rich in macrophages (red pulp, white pulp, marginal zone, and marginal metalophilic) (Cavanaugh et al., 1996; Hanson et al., 1999b). Deletion of M140 has the most detrimental effect on replication of the virus in vivo, followed by M141. Importantly, curtailed replication of the mutant virus in tissue macrophages, as opposed to other cell types such as stromal cells, was proven to be the factor limiting virus titers in the spleen. Selective in vivo depletion of splenic macrophages restored growth of the mutant to near WT levels (see Figure 20.4) by a mechanism independent of macrophage-mediated antiviral cytokine production (Hanson et al., 1999b). Notably, depletion of splenic macrophages also enhanced growth of WT MCMV; indicating that in vivo, macrophages have a net protective role. It appears that in vivo, macrophages support virus replication to an extent reflective of the replicative capacity of the virus and the action of intrinsic and extrinsic antiviral cytokines such that without splenic macrophages, virus has direct access to more highly permissive cell types in a less controlled manner.

Deciphering the function of each of the M139, M140, and M141 gene products is currently a challenge. Sequence analyses do not assist in defining potential functional domains, with the exception perhaps of a putative acidic domain in M139, nuclear localization signals in M139 and M141, and PEST sequences in M139 and M141. Unlike some other US22 genes, they do not appear to have an anti-apoptotic function (Menard et al., 2003). Electron microscopy indicates that significantly fewer virions are produced in macrophages infected with mutant virus deleted of M140 compared with WT virus; however, the mutant virions produced are as infectious as WT virus, assessed by comparing PFU/virion DNA ratios (J.S. Slater, unpublished data). These data indicate that the few mutant virions produced are properly assembled with the full repertoire of viral proteins. Although earlier studies implicated a defect in entry or penetration to explain the attenuated growth of the M139, M140, and M141 deletion mutant in macrophages (Hanson et al., 1999b), further analyses using gradient-purified virus selected for single-capsid virions to prevent entry due to phagocytosis of multi-capsid virions now sheds doubt on those findings. Neutralization of type I interferons does not restore growth of the triple-gene deletion mutant in macrophages, suggesting that these gene products do not play a major role in thwarting this antiviral activity. However, it is quite possible that this gene region functions to counteract

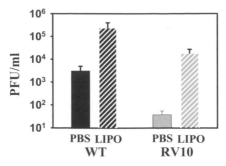

**Figure 20.4** Effects of splenic macrophage depletion on growth of WT and mutant MCMV in the spleen. BALB/c mice were injected i.v. with PBS or multilamellar liposomes encapsulating dichloromethylene-bisphosphonate (LIPO; kindly provided by Dr. Nico van Rooijen, Free University, Amsterdam). This treatment resulted in depletion of 80–90% of splenic macrophages, as determined by acid phosphatase staining of tissue sections. Two days later, the mice were infected i.v. with $3 \times 10^5$ PFU of WT MCMV or mutant virus deleted of genes M139, M140, and M141 (RV10). On days 1–3 post infection, the spleens of four or five individual mice in each group were harvested for virus titers. Titers on day 1 post infection were at or below the limits of detection (50 PFU/ml). Shown are virus titers on day 3 post infection.

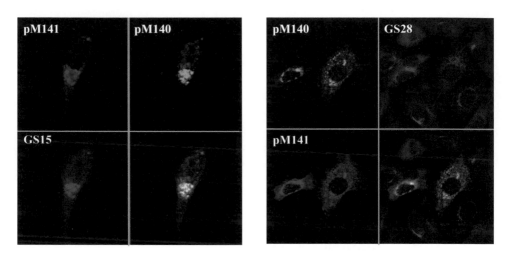

**Figure 20.5** Localization of MCMV proteins pM140 and pM141 to the Golgi region of the cell. NIH3T3 cells were co-transfected with GFP-M140 and myc-M141. Fixed cells were stained with rabbit anti-myc antibody and Cy5 conjugated goat anti-rabbit antibody, or mouse anti-GS15 or anti-GS28 antibody and Texas red conjugated goat anti-mouse serum. Convergence of all three labels appears white. Original magnification: 40×.

the intrinsic or extrinsic activity of other antiviral agents, a possibility currently under investigation.

Clues to the function of these important US22 gene products will likely be revealed through structural analyses of the gene products, as there is sufficient evidence to suggest that these proteins form complexes in virus infected cells. Functional evidence that these proteins physically interact is provided by that fact that, as stated above, they influence the stability of each other (Hanson et al., 2001; Menard et al., 2003). In addition, pM140 expressed alone localizes to the nucleus, but co-expression of pM141 redistributes pM140 to a perinuclear region of the cell where the products of all three genes converge (Karabekian et al., 2005). This site of convergence co-localizes with the Golgi-associated markers GS15 and GS28 (Figure 20.5), a cellular compartment where virion tegumentation and assembly occurs. Proof that the M139, M140, and M141 proteins physically interact is provided by data showing that the proteins co-immunoprecipitate from infected cells (Karabekian et al., 2005; Menard et al., 2003) and from cells transiently expressing these viral proteins (Karabekian et al., 2005). Furthermore, three physically distinct complexes containing these US22 proteins were identified by sucrose gradient sedimentation analyses of MCMV-infected cells (Karabekian et al., 2005). One complex clearly contains the products of all three genes, M139, M140, and M141. A second complex distinctively smaller in size and/or molecular weight contains an abundance of M139 gene products and pM140 and pM141 in significantly lower abundance. A third distinct complex, the smallest of the three, consists of pM140 and pM141. These complexes co-exist in steady-state conditions with monomeric (or perhaps multimeric) forms of each of the individual proteins. Defining the composition and stoichiometry of each of the structurally unique forms of these proteins will be a key step in ascribing function to these important US22 gene products.

## Acknowledgments

The authors gratefully acknowledge past and present members of the Campbell laboratory for their efforts in generating the data described in this chapter, especially Jacquelyn S. Slater, Victoria J. Cavanaugh, and Zarauhi Karabekian. We also sincerely thank Elizabeth R. Miller for her valuable contributions in creating figures and the table.

Work cited in the chapter was generated by funds from the Department of Health and Human Services, National Institutes of Health grant R01 CA41451 and the Thomas F. Jeffress and Kate Miller Jeffress Memorial Trust.

## References

Akira, S. (2003). Toll-like receptor signaling. J. Biol. Chem. *278*, 38105–38108.

Atalay, R., Zimmermann, A., Wagner, M., Borst, E., Benz, C., Messerle, M., and Hengel, H. (2002). Identification and expression of human cytomegalovirus transcription units coding for two distinct Fcgamma receptor homologs. J. Virol. *76*, 8596–8608.

Beisser, P.S., Grauls, G., Bruggeman, C.A., and Vink, C. (1999). Deletion of the R78 G protein-coupled receptor gene from rat cytomegalovirus results in an attenuated, syncytium-inducing mutant strain. J. Virol. *73*, 7218–7230.

Benedict, C.A., and Ware, C.F. (2001). Virus targeting of the tumor necrosis factor superfamily. Virology *289*, 1–5.

Bodaghi, B., Goureau, O., Zipeto, D., Laurent, L., Virelizier, J.L., and Michelson, S. (1999). Role of IFN-gamma-induced indoleamine 2,3 dioxygenase and inducible nitric oxide synthase in the replication of human cytomegalovirus in retinal pigment epithelial cells. J. Immunol. *162*, 957–964.

Boudinot, P., Riffault, S., Salhi, S., Carrat, C., Sedlik, C., Mahmoudi, N., Charley, B., and Benmansour, A. (2000). Vesicular stomatitis virus and pseudorabies virus induce a vig1/cig5 homologue in mouse dendritic cells via different pathways. J. Gen. Virol. *81*, 2675–2682.

Brune, W., Menard, C., Heesemann, J., and Koszinowski, U.H. (2001). A ribonucleotide reductase homolog of cytomegalovirus and endothelial cell tropism. Science *291*, 303–305.

Cavanaugh, V.J., Stenberg, R.M., Staley, T.L., Virgin, H.W., MacDonald, M.R., Paetzold, S., Farrell, H.E., Rawlinson, W.D., and Campbell, A.E. (1996). Murine cytomegalovirus with a deletion of genes spanning HindIII-J and – I displays altered cell and tissue tropism. J. Virol. *70*, 1365–1374.

Kopp, E., and Medzhitov, R. (2003). Recognition of microbial infection by Toll-like receptors. Curr. Opin. Immunol. 15, 396–401.

Kotenko, S.V., Saccani, S., Izotova, L.S., Mirochnitchenko, O.V., and Pestka, S. (2000). Human cytomegalovirus harbors its own unique IL-10 homolog (CMVIL-10). Proc. Natl. Acad. Sci. USA 97, 1695–1700.

Kurz, S.K., and Reddehase, M.J. (1999). Patchwork pattern of transcriptional reactivation in the lungs indicates sequential checkpoints in the transition from murine cytomegalovirus latency to recurrence. J. Virol. 73, 8612–8622.

Larsson, S., Soderberg-Naucler, C., Wang, F.Z., and Moller, E. (1998). Cytomegalovirus DNA can be detected in peripheral blood mononuclear cells from all seropositive and most seronegative healthy blood donors over time. Transfusion 38, 271–278.

Le Roy, E., Muhlethaler-Mottet, A., Davrinche, C., Mach, B., and Davignon, J.L. (1999). Escape of human cytomegalovirus from HLA-DR-restricted CD4(+) T-cell response is mediated by repression of gamma interferon-induced class II transactivator expression. J. Virol. 73, 6582–6589.

Lee, C.H., Lee, G.C., Chan, Y.J., Chiou, C.J., Ahn, J.H., and Hayward, G.S. (1999). Factors affecting human cytomegalovirus gene expression in human monocyte cell lines. Mol. Cells 9, 37–44.

Lee, Y., Sohn, W.J., Kim, D.S., and Kwon, H.J. (2004). NF-kappaB- and c-Jun-dependent regulation of human cytomegalovirus immediate-early gene enhancer/promoter in response to lipopolysaccharide and bacterial CpG-oligodeoxynucleotides in macrophage cell line RAW 264.7. Eur. J. Biochem. 271, 1094–1105.

Lembo, D., Donalisio, M., Hofer, A., Cornaglia, M., Brune, W., Koszinowski, U., Thelander, L., and Landolfo, S. (2004). The ribonucleotide reductase R1 homolog of murine cytomegalovirus is not a functional enzyme subunit but is required for pathogenesis. J. Virol. 78, 4278–4288.

Levine, A.J. (1984). Viruses and Differentiation: The Molecular Basis of Viral Tissue Tropisms. In Concepts in Viral Pathogenesis, A.L. Notkins and M.B.A. Oldstone, ed. (New York, Springer-Verlag), pp. 130–134.

Lucin, P., Jonjic, S., Messerle, M., Polic, B., Hengel, H., and Koszinowski, U.H. (1994). Late phase inhibition of murine cytomegalovirus replication by synergistic action of interferon-gamma and tumour necrosis factor. J. Gen. Virol. 75, 101–110.

Maciejewski, J.P., Bruening, E.E., Donahue, R.E., Mocarski, E.S., Young, N.S., and St Jeor, S.C. (1992). Infection of hematopoietic progenitor cells by human cytomegalovirus. Blood 80, 170–178.

McCormick, A.L., Smith, V.L., Chow, D., and Mocarski, E.S. (2003). Disruption of mitochondrial networks by the human cytomegalovirus UL37 gene product viral mitochondrion-localized inhibitor of apoptosis. J. Virol. 77, 631–641.

Menard, C., Wagner, M., Ruzsics, Z., Holak, K., Brune, W., Campbell, A.E., and Koszinowski, U.H. (2003). Role of murine cytomegalovirus US22 gene family members in replication in macrophages. J. Virol. 77, 5557–5570.

Mendelson, M., Monard, S., Sissons, P., and Sinclair, J. (1996). Detection of endogenous human cytomegalovirus in CD34+ bone marrow progenitors. J. Gen. Virol. 77 (Pt 12), 3099–3102.

Miller, D.M., Rahill, B.M., Boss, J.M., Lairmore, M.D., Durbin, J.E., Waldman, J.W., and Sedmak, D.D. (1998). Human cytomegalovirus inhibits major histocompatibility complex class II expression by disruption of the Jak/Stat pathway. J. Exp. Med. 187, 675–683.

Miller, D.M., Zhang, Y., Rahill, B.M., Waldman, W.J., and Sedmak, D.D. (1999). Human cytomegalovirus inhibits IFN-alpha-stimulated antiviral and immunoregulatory responses by blocking multiple levels of IFN-alpha signal transduction. J. Immunol. 162, 6107–6113.

Mosser, D.M. (2003). The many faces of macrophage activation. J Leukoc. Biol. 73, 209–212.

Movassagh, M., Gozlan, J., Senechal, B., Baillou, C., Petit, J.C., and Lemoine, F.M. (1996). Direct infection of CD34+ progenitor cells by human cytomegalovirus: evidence for inhibition of hematopoiesis and viral replication. Blood 88, 1277–1283.

Muller, W.A. (2002). Leukocyte-endothelial cell interactions in the inflammatory response. Lab. Invest. 82, 521–533.

Nicholas, J. (1996). Determination and analysis of the complete nucleotide sequence of human herpesvirus. J. Virol. 70, 5975–5989.

Noda, S., Tanaka, K., Sawamura, S., Sasaki, M., Matsumoto, T., Mikami, K., Aiba, Y., Hasegawa, H., Kawabe, N., and Koga, Y. (2001). Role of nitric oxide synthase type 2 in acute infection with murine cytomegalovirus. J. Immunol. 166, 3533–3541.

Noraz, N., Lathey, J.L., and Spector, S.A. (1997). Human cytomegalovirus-associated immunosuppression is mediated through interferon-alpha. Blood 89, 2443–2452.

Nordoy, I., Rollag, H., Lien, E., Sindre, H., Degre, M., Aukrust, P., Froland, S.S., and Muller, F. (2003). Cytomegalovirus infection induces production of human interleukin-10 in macrophages. Eur. J. Clin. Microbiol. Infect. Dis. 22, 737–741.

Odeberg, J., and Soderberg-Naucler, C. (2001). Reduced expression of HLA Class II molecules and interleukin-10- and transforming growth factor beta1-independent suppression of T-cell proliferation in human cytomegalovirus-infected macrophage cultures. J. Virol. 75, 5174–5181.

Okada, K., Tanaka, K., Noda, S., Okazaki, M., and Koga, Y. (1999). Nitric oxide increases the amount of murine cytomegalovirus-DNA in mice latently infected with the virus. Arch. Virol. 144, 2273–2290.

Oliveira, S.A., and Shenk, T.E. (2001). Murine cytomegalovirus M78 protein, a G protein-coupled receptor homologue, is a constituent of the virion and facilitates accumulation of immediate-early viral mRNA. Proc. Natl. Acad. Sci. USA 98, 3237–3242.

Patrone, M., Percivalle, E., Secchi, M., Fiorina, L., Pedrali-Noy, G., Zoppe, M., Baldanti, F., Hahn, G., Koszinowski, U.H., Milanesi, G., and Gallina, A. (2003). The human cytomegalovirus UL45 gene product is a late, virion-associated protein and influences virus growth at low multiplicities of infection. J. Gen. Virol. 84, 3359–3370.

Patterson, C.E., and Shenk, T. (1999). Human cytomegalovirus UL36 protein is dispensable for viral replication in cultured cells. J. Virol. 73, 7126–7131.

Pavic, I., Polic, B., Crnkovic, I., Lucin, P., Jonjic, S., and Koszinowski, U.H. (1993). Participation of endogenous tumour necrosis factor alpha in host resistance to cytomegalovirus infection. J. Gen. Virol. 74, 2215–2223.

Pollock, J.L., Presti, R.M., Paetzold, S., and Virgin, H.W. (1997). Latent murine cytomegalovirus infection in macrophages. Virology 227, 168–179.

Popkin, D.L., and Virgin, H.W. (2003). Murine cytomegalovirus infection inhibits tumor necrosis factor alpha responses in primary macrophages. J. Virol. 77, 10125–10130.

Popkin, D.L., Watson, M.A., Karaskov, E., Dunn, G.P., Bremner, R., and Virgin, H.W. (2003). Murine cytomegalovirus paralyzes macrophages by blocking IFN gamma-induced promoter assembly. Proc. Natl. Acad. Sci. USA 100, 14309–14314.

Presti, R.M., Popkin, D.L., Connick, M., Paetzold, S., and Virgin, H.W. (2001). Novel cell type-specific antiviral mechanism of interferon gamma action in macrophages. J. Exp. Med. 193, 483–496.

Prösch, S., Staak, K., Stein, J., Liebenthal, C., Stamminger, T., Volk, H.D., and Kruger, D.H. (1995). Stimulation of the human cytomegalovirus IE enhancer/promoter in HL-60 cells by TNFalpha is mediated via induction of NF-kappaB. Virology 208, 197–206.

Randolph, G.J. (2002). Is maturation required for Langerhans cell migration? J. Exp. Med. 196, 413–416.

Randolph, G.J., Beaulieu, S., Lebecque, S., Steinman, R.M., and Muller, W.A. (1998). Differentiation of monocytes into dendritic cells in a model of transendothelial trafficking. Science 282, 480–483.

Rao, K.M. (2001). MAP kinase activation in macrophages. J. Leukoc. Biol. 69, 3–10.

Reddehase, M.J., Podlech, J., and Grzimek, N.K. (2002). Mouse models of cytomegalovirus latency: overview. J. Clin. Virol. 25 Suppl 2, S23–36.

Redpath, S., Angulo, A., Gascoigne, N.R., and Ghazal, P. (1999). Murine cytomegalovirus infection down-regulates MHC class II expression on macrophages by induction of IL-10. J. Immunol. 162, 6701–6707.

Ritter, T., Brandt, C., Prösch, S., Vergopoulos, A., Vogt, K., Kolls, J., and Volk, H.D. (2000). Stimulatory and inhibitory action of cytokines on the regulation of hCMV-IE promoter activity in human endothelial cells. Cytokine 12, 1163–1170.

Ross, J.A. and Auger, M.J. (2002). The Biology of the Macrophage. In The Macrophage, B. Burke and C.E. Lewis, ed. (New York, Oxford University Press), pp. 1–72.

Salazar-Mather, T.P., Lewis, C.A., and Biron, C.A. (2002). Type I interferons regulate inflammatory cell trafficking and macrophage inflammatory protein 1alpha delivery to the liver. J. Clin. Invest. 110, 321–330.

Schroder, K., Hertzog, P.J., Ravasi, T., and Hume, D.A. (2004). Interferon-gamma: an overview of signals, mechanisms and functions. J. Leukoc. Biol. 75, 163–189.

Shi, S., Nathan, C., Schnappinger, D., Drenkow, J., Fuortes, M., Block, E., Ding, A., Gingeras, T.R., Schoolnik, G., Akira, S., Takeda, K., and Ehrt, S. (2003). MyD88 primes macrophages for full-scale activation by interferon-gamma yet mediates few responses to Mycobacterium tuberculosis. J. Exp. Med. *198*, 987–997.

Simon, C.O., Seckert, C.K., Dreis, D., Reddehase, M.J., and Grzimek, N.K.A. (2005). A role for tumor necrosis factor-α in murine cytomegalovirus transcriptional reactivation in latently infected lungs. J. Virol. *79*, in press

Sinclair, J.H., Baillie, J., Bryant, L.A., Taylor-Wiedeman, J.A., and Sissons, J.G. (1992). Repression of human cytomegalovirus major immediate-early gene expression in a monocytic cell line. J. Gen. Virol. *73 (Pt 2)*, 433–435.

Sissons, J.G., Bain, M., and Wills, M.R. (2002). Latency and reactivation of human cytomegalovirus. J. Infect. *44*, 73–77.

Skaletskaya, A., Bartle, L.M., Chittenden, T., McCormick, A.L., Mocarski, E.S., and Goldmacher, V.S. (2001). A cytomegalovirus-encoded inhibitor of apoptosis that suppresses caspase-8 activation. Proc. Natl. Acad. Sci. USA *98*, 7829–7834.

Slobedman, B. and Mocarski, E.S. (1999). Quantitative analysis of latent human cytomegalovirus. J. Virol. *73*, 4806–4812.

Slobedman, B., Stern, J. L., Cunningham, A.L., Abendroth, A., Abate, D.A., and Mocarski, E.S. (2004). Impact of human cytomegalovirus latent infection on myeloid progenitor cell gene expression. J. Virol. *78*, 4054–4062.

Smith, M.S., Bentz, G.L., Smith, P.M., Bivins, E.R., and Yurochko, A.D. (2004). HCMV activates PI(3)K in monocytes and promotes monocyte motility and transendothelial migration in a PI(3)K-dependent manner. J. Leukoc. Biol. *76*, 65–76.

Soderberg-Naucler, C., Fish, K.N., and Nelson, J.A. (1997). Interferon-gamma and tumor necrosis factor-alpha specifically induce formation of cytomegalovirus-permissive monocyte-derived macrophages that are refractory to the antiviral activity of these cytokines. J. Clin. Invest. *100*, 3154–3163.

Soderberg-Naucler, C., Streblow, D.N., Fish, K.N., Allan-Yorke, J., Smith, P.P., and Nelson, J.A. (2001). Reactivation of latent human cytomegalovirus in CD14(+) monocytes is differentiation dependent. J. Virol. *75*, 7543–7554.

Spencer, J.V., Lockridge, K.M., Barry, P.A., Lin, G., Tsang, M., Penfold, M.E., and Schall, T.J. (2002). Potent immunosuppressive activities of cytomegalovirus-encoded interleukin-10. J. Virol. *76*, 1285–1292.

Streblow, D.N., and Nelson, J.A. (2003). Models of HCMV latency and reactivation. Trends Microbiol. *11*, 293–295.

Tabeta, K., Georgel, P., Janssen, E., Du, X., Hoebe, K., Crozat, K., Mudd, S., Shamel, L., Sovath, S., Goode, J., Alexopoulou, L., Flavell, R.A., and Beutler, B. (2004). Toll-like receptors 9 and 3 as essential components of innate immune defense against mouse cytomegalovirus infection. Proc. Natl. Acad. Sci. USA *101*, 3516–3521.

Tanaka, K., and Noda, S. (2001). Role of nitric oxide in murine cytomegalovirus (MCMV) infection. Histol. Histopatho.l *16*, 937–944.

Tay, C.H., and Welsh, R.M. (1997). Distinct organ-dependent mechanisms for the control of murine cytomegalovirus infection by natural killer cells. J. Virol. *71*, 267–275.

Taylor-Wiedeman, J., Sissons, J.G., Borysiewicz, L.K., and Sinclair, J.H. (1991). Monocytes are a major site of persistence of human cytomegalovirus in peripheral blood mononuclear cells. J. Gen. Virol. *72 (Pt 9)*, 2059–2064.

Taylor-Wiedeman, J., Sissons, P., and Sinclair, J. (1994). Induction of endogenous human cytomegalovirus gene expression after differentiation of monocytes from healthy carriers. J. Virol. *68*, 1597–1604.

Thäle, R., Lucin, P., Schneider, K., Eggers, M., and Koszinowski, U.H. (1994). Identification and expression of a murine cytomegalovirus early gene coding for an Fc receptor. J. Virol. *68*, 7757–7765.

Tomazin, R., Boname, J., Hegde, N.R., Lewinsohn, D.M., Altschuler, Y., Jones, T.R., Cresswell, P., Nelson, J.A., Riddell, S.R., and Johnson, D.C. (1999). Cytomegalovirus US2 destroys two components of the MHC class II pathway, preventing recognition by CD4+ T-cells. Nat. Med. *5*, 1039–1043.

von Laer, D., Meyer-Koenig, U., Serr, A., Finke, J., Kanz, L., Fauser, A.A., Neumann-Haefelin, D., Brugger, W., and Hufert, F.T. (1995). Detection of cytomegalovirus DNA in CD34+ cells from blood and bone marrow. Blood *86*, 4086–4090.

Wang, X., Huong, S.M., Chiu, M.L., Raab-Traub, N., and Huang, E.S. (2003). Epidermal growth factor receptor is a cellular receptor for human cytomegalovirus. Nature *424*, 456–461.

Yeow, W.S., Lawson, C.M., and Beilharz, M.W. (1998). Antiviral activities of individual murine IFN-alpha subtypes in vivo: intramuscular injection of IFN expression constructs reduces cytomegalovirus replication. J. Immunol. *160*, 2932–2939.

Yerkovich, S.T., Olver, S.D., Lenzo, J.C., Peacock, C.D., and Price, P. (1997). The roles of tumour necrosis factor-alpha, interleukin-1 and interleukin-12 in murine cytomegalovirus infection. Immunology *91*, 45–52.

Yurochko, A.D., and Huang, E.S. (1999). Human cytomegalovirus binding to human monocytes induces immunoregulatory gene expression. J. Immunol. *162*, 4806–4816.

Zhu, H., Cong, J.P., and Shenk, T. (1997). Use of differential display analysis to assess the effect of human cytomegalovirus infection on the accumulation of cellular RNAs: induction of interferon-responsive RNAs. Proc. Natl. Acad. Sci. USA *94*, 13985–13990.

Zhuravskaya, T., Maciejewski, J.P., Netski, D.M., Bruening, E., Mackintosh, F.R., and St Jeor, S. (1997). Spread of human cytomegalovirus (HCMV) after infection of human hematopoietic progenitor cells: model of HCMV latency. Blood *90*, 2482–2491.

# Determinants of Endothelial Cell Tropism of Human Cytomegalovirus

# 21

*Margarete Digel and Christian Sinzger*

## Abstract

The vascular endothelium is an important target of human cytomegalovirus (HCMV) in vivo during acute infection. Infection of endothelial cells (EC) is assumed to contribute to replication and distribution of the virus, to modulate inflammatory responses in infected organs and to harbor virus during persistent infections. The endothelial cell tropism of HCMV is reflected in cell culture systems. Recent clinical isolates display a variable capacity of focal growth in endothelial cell culture. The inherent endothelial cell tropism of HCMV isolates is maintained during subsequent propagation in EC whereas it is greatly reduced during propagation in fibroblasts, resulting in 100- to 1000-fold differences in the endothelial cell tropism of the respective strains. Interstrain differences between highly endotheliotropic and low endotheliotropic strains dictate their reproductive and cytopathogenic potential and are determined by the efficiency of nuclear particle transport during the initial phase of replication. Multiple gene regions, including UL23–25 and UL128–132, contribute to successful infection of EC by HCMV. Which of these genes participate in the phenotypic diversity of naturally occurring variants and how they are linked to the interstrain differences in transport efficiency are questions to be resolved.

## Introduction

Hematogenous dissemination of human cytomegalovirus (HCMV) is assumed to be important for vertical transmission from mother to fetus and for organ manifestation during acute systemic infection of immunocompromised patients. Besides polymorphonuclear and mononuclear leukocytes, vascular EC have been identified as in vivo targets of HCMV that might participate in hematogenous dissemination. In addition, the strategic location at the interface between circulation and organ tissues has raised great interest in the potential contribution of infected vascular EC to the development of HCMV-associated disease. Infected endothelium was assumed to be involved in the modulation of inflammatory responses during acute infection, in the development of atherosclerosis and in persistence and reactivation of HCMV during phases of clinical latency. The endothelial cell tropism of HCMV has, therefore, been regarded as a potential virulence factor which might have implications for the development of an effective and safe live vaccine, once the molecular basis of HCMV–EC interactions is discovered. A thorough understanding of mechanisms and genes that govern the endothelial cell tropism of HCMV might also aid in the diagnostic evaluation of the virulence of a patient's isolate or in the design of an endotheliotropic HCMV-based vector for therapeutic purposes. This review describes the phenotypic manifestations of endothelial cell tropism of HCMV as well as recent insights into the mechanisms and genes involved in the expression of this phenotype.

## Endothelial cells as targets of HCMV infection

During acute infection, HCMV can be disseminated throughout the organism and can subsequently infect virtually any organ (Bissinger et al., 2002). The regular occurrence of viremia in HCMV-infected post-transplant patients strongly suggests a hematogenous route of virus dissemination (Wolf and Spector, 1993). Therefore, it was not surprising that EC were found to be infected in various organ tissues by immunohistochemical techniques. In particular, vascular EC of capillaries and venules were targeted by HCMV in lung and gastrointestinal tissues of immunocompromised patients (Sinzger et al., 1995). Infected EC expressed immediate-early, early, and late viral proteins (Figure 21.1A) and therefore seem to support the complete viral replication cycle. The assumption of a productive infection is further supported by the focal distribution of HCMV in infected vessels (Figure 21.1B) indicating that infected EC can pass the virus to adjacent cells by cell-to-cell spread (Myerson et al., 1984). Infected cells in the late stage of replication showed morphologic alteration including the characteristic cytomegalic appearance (Figure 21.1.A). Infected endothelium is often associated with infiltration of inflammatory cells and with HCMV infection in the surrounding stroma (Roberts et al., 1989; Sinzger et al., 1995). Although infected EC are a regular finding in tissues of post-transplant patients, AIDS patients, and congenitally infected newborns, the degree of endothelial involvement is variable as indicated by large interindividual differences in the absolute and relative frequency of HCMV-infected EC (Sinzger et al., 1995). Obviously, infected EC can detach from the vessel wall and enter the bloodstream, where they can be detected as circulating cytomegalic infected cells (CCIC) (Grefte et al., 1993b; Percivalle et al., 1993). The endothelial nature of these CCIC had been demonstrated by the detection of cell type-specific marker proteins. CCIC express late viral antigens and contain numerous viral capsids in their nucleus and cytoplasm, indicating that they are productively infected and may therefore be capable of spreading the virus (Grefte et al., 1993a; Grefte et al., 1994).

EC have also been suggested as sites of viral persistence during latency. Since the first reports of HCMV-DNA in vascular tissues raised this issue (Hendrix et al., 1991; Melnick et al., 1994), there has been controversy as to whether vascular EC are infected by HCMV during periods of latency. Attempts to detect viral genomes in situ in EC of atherosclerotic or nonatherosclerotic vessels have yielded contradictory results (Hendrix et al., 1997; Pampou et al., 2000). A recent report described the presence of viral genome as well as immediate-early antigen in EC of atherosclerotic lesions from trauma victims, whereas only DNA signals were found in nonatherosclerotic lesions (Pampou et al., 2000). However, the lack of a CMV-negative control group and particularly the cytoplasmic localization of genomes and immediate-early antigen raise doubts about the specificity of the results. In this context it is noteworthy that in vitro infections of explanted arterial tissues have demonstrated the susceptibility of the endothelial layer to HCMV and have confirmed the nuclear localization of viral immediate-early antigens in this cell type (Nerheim et al., 2004; Reinhardt et al., 2003).

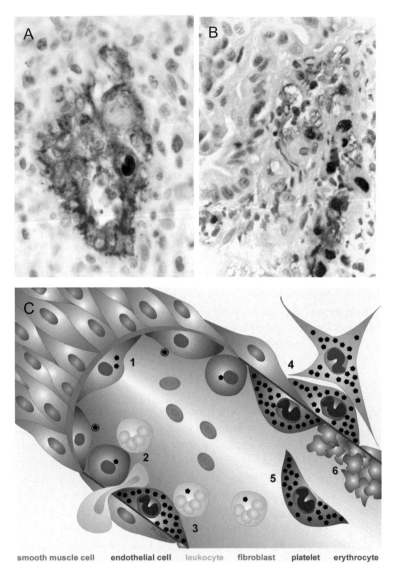

smooth muscle cell     **endothelial cell**     leukocyte     **fibroblast**     platelet     erythrocyte

**Figure 21.1** (A) Immunohistochemical detection of the late viral antigen pUL86 (brown) and endothelial cell marker F.VIII (red) in vascular endothelial cells of the colon of an AIDS patient. (B) Immunohistochemical detection of the immediate-early viral antigens pUL122/123 (brown) and endothelial cell marker F.VIII (red) in vascular endothelial cells of the stomach of a kidney transplant patient. (C) Hypotheses regarding the pathogenetic role of HCMV-infected endothelial cells. (1) The vascular endothelium has been suggested to be a site of viral persistence during periods of clinical latency. (2) Alteration of adhesion molecules on the surface of infected cells is assumed to facilitate leukocyte adhesion and subsequent transmigration into adjacent organ tissues. (3) Alternatively, leukocyte adhesion to productively infected endothelial cells might result in transmission of virus to leukocytes and subsequent hematogenous dissemination of the virus. (4) Productively infected endothelial cells might spread the virus to adjacent smooth muscle cells and fibroblasts. (5) Detachment of infected endothelial cells during the late phase of replication is also assumed to contribute to hematogenous distribution of HCMV. (6) Lesions in the endothelial layer due to detachment of infected cells or cytopathogenic effects of HCMV might initiate a procoagulant reaction and thus contribute to atherosclerotic events.

## Infected endothelial cells in the pathogenesis of HCMV infections

Since HCMV infection of EC in vivo has been described, the pathogenetic role of infected EC has been a matter of speculation (Figure 21.1C) particularly emphasizing (1) the immunomodulatory role of EC at the interface between circulation and organ tissue (Waldman, 1998), (2) the capacity of EC to produce and distribute viral progeny (Plachter et al., 1996), (3) the contribution of EC infection to the development of atherosclerosis (Degre, 2002), and (4) the role of infected EC for viral persistence and reactivation (Jarvis and Nelson, 2002).

During acute HCMV infection, viral replication in vascular EC is regularly associated with a mixed-cellular inflammatory infiltrate consisting of granulocytes, macrophages and lymphocytes (Sinzger et al., 1995). The perivascular distribution of the inflammatory cells has led to the assumption of an HCMV-induced vasculitis (Roberts et al., 1989) or endothelialitis (Koskinen et al., 1994). The in vivo findings showing increased expression of adhesion molecules ICAM-1 and VCAM-1 in infected tissues (Lautenschlager et al., 1996) and CD40 in individual infected EC (Maisch et al., 2002), together with cell culture data on the up-regulation of adhesion molecules in endothelial cell cultures (Sedmak et al., 1994; Shahgasempour et al., 1997), are consistent with the assumption of HCMV-induced leukocyte adhesion and subsequent transmigration into the adjacent tissues. Increased leukocyte adhesion to, and transmigration of leukocytes through, infected endothelial layers have been demonstrated in cell culture (Grundy et al., 1998; Shahgasempour et al., 1997), although the effects seem to be partly mediated by paracrine action on non-infected bystander cells (Dengler et al., 2000).

The formation of infected foci in the vascular endothelium (Figure 21.1B), the presence of viral structural proteins (Figure 21.1A) and the detection of viral progeny particles (Grefte et al., 1993a) in EC during acute infection suggests that they are productively infected and can distribute infection to adjacent ccells and into the bloodstream. Hematogenous spread of HCMV from infected EC might occur by release of viral progeny, by cell-to-cell transmission of progeny virus to transiently adhering leukocytes or by detachment of productively infected EC (CCIC, see above) from the vessel wall and dislocation by the bloodstream. Whereas cell-free infectivity in plasma or serum is a rare finding, polymorphonuclear leukocytes are well known to carry viral antigens and genomes, although the course of infection in these leukocytes is abortive (Grefte et al., 1994). Experimental evidence indicates that leukocytes acquire infectious virus by cell-to-cell transmission from productively infected EC. Although the underlying functional alterations are unclear, it is obvious that circulating cytomegalic EC, the CCIC, must have detached from the vascular wall, thus leaving a lesion in the endothelial layer of the respective vessel (Grefte et al., 1993a).

The damage caused by HCMV infection of EC might initiate a procoagulant inflammatory response, finally leading to atherosclerotic alterations of the vascular wall. While there is good evidence for a role of HCMV in the occurrence of coronary arteriosclerosis in heart transplant recipients (Valantine, 2004), the epidemiological and pathological data do not convincingly support a significant contribution of active HCMV infection to the development of classical atherosclerosis (Borgia et al., 2001). However, elevated anti-HCMV-IgG levels in cardiovascular disease patients might indicate a possible role of persistent HCMV infection.

Based on experimental data, infected vascular EC have been suggested as a source of persistent virus production during periods of clinical latency (Fish et al., 1998). However, analyses of tissue specimens from latently infected individuals argue against a major role of EC for persistence and reactivation of HCMV. In an in situ hybridization analysis of multiple organs of trauma victims, DNA signals were only found in cell types other than endothelial cells (Hendrix et al., 1997). More recently, using a highly sensitive PCR approach capable of detecting endogenous HCMV DNA in myeloid cells of seropositive individuals, HCMV genomes could not be found in vascular endothelial cells from seropositive bypass surgery patients (Reeves et al., 2004).

## HCMV infection of endothelial cells in cell culture

Successful infection of human umbilical vein endothelial cell (HUVEC) cultures with HCMV was first reported by Hirsch and co-workers, who noted that recent isolates might be more efficient in this cell type than laboratory strains (Ho et al., 1984). Under normal conditions, the infection efficiency of laboratory strains like AD169 or Towne in EC culture is limited to a few percent and it needs artificial conditions such as the addition of short fatty acids, the application of ultracentrifuged virus concentrates or centrifugal enhancement to significantly increase the fraction of infected cells (MacCormac and Grundy, 1999; Radsak et al., 1989; Wu et al., 1994). Despite these limitations, HUVEC have become the standard endothelial cell culture for HCMV infection and the use of this cell culture model was greatly facilitated by the introduction of endothelial propagated HCMV strains which enabled complete infection of HUVEC cultures (Sinzger et al., 1999; Waldman et al., 1991). HUVEC are preferentially used primarily because of their good availability, but they also seem to be a valid model due to the finding that EC of other origins show similar characteristics with regard to HCMV replication (Kahl et al., 2000; Knight et al., 1999). HCMV has been successfully used to infect venous, arterial and microvascular EC from umbilical vein, placenta, kidney, intestine and brain (Kahl et al., 2000; Knight et al., 1999; Lathey et al., 1990; Sindre et al., 2000; Ustinov et al., 1991) and recently even EC in explant cultures of arterial sections were found to be susceptible (Reinhardt et al., 2003).

Consistent with the immunohistochemical detection of late viral antigens in EC of infected tissues (Figure 21.1A), cultured EC support the full replication cycle of HCMV resulting in the production and release of viral progeny (Fish et al., 1998; Ho et al., 1984; Sinzger et al., 2000; Waldman et al., 1991). Recent clinical isolates are distinguished from cell culture adapted strains by the lack of release of cell-free infectious progeny from infected EC. Still they proved to be productive in endothelial monolayers as demonstrated by the detection of cell-to-cell spread resulting in focal expansion of the virus (Sinzger et al., 1999) (Figure 21.2A). Concomitant with productive infection, HCMV induces cytopathogenic effects in endothelial cell cultures of various origins, which include cell rounding, cell enlargement, formation of inclusion bodies and lysis of infected cells (Ho et al., 1984; Kahl et al., 2000; Waldman et al., 1991). Thus, HCMV infection of cultured EC widely resembles the well-known characteristics of fibroblast infection, except that they show a delay in kinetics of the replication cycle and an overall reduced infection efficiency (Kahl et al., 2000; Sinzger et al., 2000). In terms of pathogenesis, the finding that cultured EC permit lytic infection by HCMV supports the concept of HCMV-induced lesions within the vascular endothelium, which might provoke inflammatory responses and eventually contribute to the development of vasculitic or atherosclerotic processes in affected blood vessels.

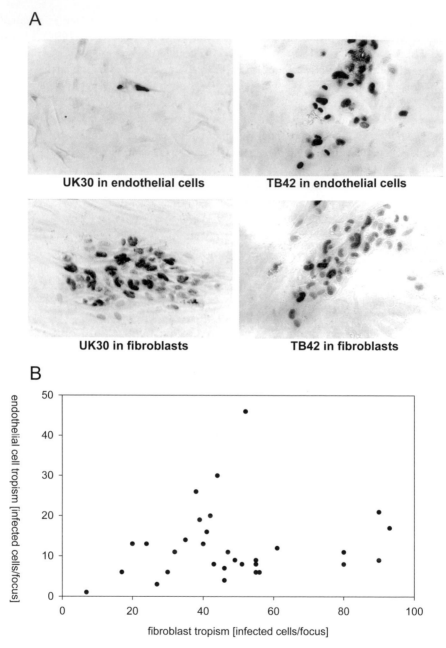

**Figure 21.2** Recent HCMV isolates differ from each other in the extent of focal spread in endothelial cell culture. (A) Focus formation of low endotheliotropic isolate UK30 and highly endotheliotropic isolate TB42 in endothelial cell monolayers and fibroblast monolayers as visualized by immunocytochemical detection of viral immediate-early antigen. (B) Graphical representation of the endothelial cell tropism and fibroblast tropism of 31 recent isolates indicates high interstrain variation and little correlation between growth properties in endothelial cells and fibroblasts.

Although HCMV infection of EC is productive and lytic in principle, persistent infection of aortic endothelial cultures has also been reported (Fish et al., 1998). As persistence of HCMV in individual EC has not been unequivocally demonstrated, the occurrence of persistence may be best explained by a balanced predator–prey relationship between cytopathogenic virus and proliferating uninfected bystanders (Kahl et al., 2000).

Beyond characterization of the viral replication cycle, endothelial cell cultures have been used as a model system for the analysis of functional aspects of HCMV infection. In particular, the effects of HCMV on EC–leukocyte interactions have been analyzed. Increased adhesion of polymorphonuclear cells, monocytes and lymphocytes to HCMV-infected endothelial cell cultures has been reported and has been attributed to increased surface expression of adhesion molecules. HCMV-induced elevation of ICAM-1 expression has been consistently found by many authors and appears to be mediated by binding of viral immediate early gene products to an SP1 binding site in the ICAM-1 promoter (Kronschnabl and Stamminger, 2003). Regarding induction of VCAM and E-selectin, contradictory findings have been reported (Knight et al., 1999; Shahgasempour et al., 1997) and some of these modulations may be due to paracrine effects on uninfected bystander cells (Dengler et al., 2000) since the infection efficiency in most of these studies was unknown or significantly below 100%. Regardless of whether HCMV induces these effects directly on infected cells or indirectly on noninfected bystanders, the findings indicate a modulatory role of the infected endothelium in the composition of the inflammatory response at sites of infection.

In addition, the alteration of adhesion molecule expression may have an impact on the hematogenous dissemination of HCMV. Whatever the exact mechanisms of interaction between HCMV-infected EC and leukocytes are, the degree of interaction appears to allow for transmission of infectious virus between both cell types. Productively infected endothelial monolayers have been demonstrated to transmit HCMV to mononuclear cells and to polymorphonuclear cells (Grundy et al., 1998; Revello et al., 1998; Waldman et al., 1995). Using co-cultures of infected EC and leukocytes, the clinical finding of pp65-antigenemia could thus be reproduced in an experimental model. Moreover, transmission did not only occur from infected EC to leukocytes but was bidirectional: monocytes or polymorphonuclear cells that have acquired HCMV from infected EC can also transmit infection to endothelial monolayers, thus resembling the hypothetical ways of hematogenous dissemination within the infected host (Revello et al., 1998; Waldman et al., 1995).

Interestingly, HCMV strains that efficiently infect endothelial cell cultures can also replicate in monocyte-derived macrophages or dendritic cells (Jahn et al., 1999; Riegler et al., 2000). Thus, increased efficiency in endothelial cell infection may also serve as a model system of a more generally extended cell tropism of certain HCMV strains as compared to laboratory strains AD169, Towne and Davis. However, further investigation into such interstrain differences requires the development of methods to quantify the degree of endothelial cell tropism.

## Endothelial cell tropism of HCMV strains – just a matter of definition?

HCMV strains that have been propagated in EC after primary isolation from clinical specimens maintain enhanced infectivity in EC when compared to fibroblast-propagated strains

(Waldman et al., 1991). This finding has initiated attempts to define the nature of such interstrain differences, and it has also initiated a debate about the qualitative or quantitative definition of endotheliotropism of HCMV.

The term "tropism" is generally used to describe movement of a living organism toward or away from a stimulus (positive or negative tropism, respectively). In virology, cell tropism means the capacity of a virus to enter and replicate in a certain cell type. From the cell's point of view, this implies susceptibility to infection by the virus and permissivity for a full viral replication cycle, finally resulting in production of infectious progeny. At first glance, this seems to be a clear definition. However, the interaction between the virus and its target cell is too complex to be described only in terms of susceptibility and permissivity. Definition of the term "tropism" becomes confusing when it is applied to describe aspects as diverse as (i) viral entry into the cell, (ii) activation events within the cell, (iii) viral gene expression, (iv) cytopathogenic effects caused by the virus, (v) replication of the virus by a cell type or (vi) establishment of viral latency and persistence.

In the context of HCMV, the term "endothelial cell tropism" has been used for the description of the cytopathogenicity in endothelial cell monolayers (Waldman et al., 1991), the production of viral progeny (Hahn et al., 2004; MacCormac and Grundy, 1999), the establishment of viral persistence in endothelial cell culture (Jarvis and Nelson, 2002), the long-term propagation of HCMV in endothelial cell cultures (Gerna et al., 2003; Sinzger et al., 1999), and for the discrimination of HCMV strains that are replicating more or less efficiently in a given type of endothelial cell culture (Kahl et al., 2000). It is evident that the diverse definitions of endothelial cell tropism will not be equivalent and are therefore prone to cause confusion.

The rationale behind a highly divergent usage of the term "tropism" becomes apparent when the pathogenetic impact of endothelial cell infection is considered. For instance: (i) in the context of HCMV-associated arteriosclerosis, virus-induced damage of the vascular endothelium is relevant regardless of whether it is due to abortive or permissive infection and whether cell damage is apoptotic or necrotic; (ii) in terms of viral replication it is relevant whether the endothelium is a site of virus amplification; (iii) for aspects of viral dissemination, in contrast, even low amounts of viral progeny may be sufficient and even a passive transport of endocytosed material might be effective; (iv) in terms of immunopathology the expression of cytokines, chemokines or adhesion molecules by infected endothelial cells is of interest – regardless of whether or not the infected cells are permissive for viral replication.

For reasons of clarity it would therefore be advantageous to specify whether "endothelial cell tropism of HCMV" means the capacity of HCMV strains to invade a given endothelial cell culture, causing functional alteration such as apoptosis or lysis of the target cell, whether it means generation of progeny virus in low amounts or in amounts that suffice to reproduce the parental generation, or whether it means any other aspect of interest. In a stricter sense "cell tropism" should be restricted to the cytopathogenic or reproductive potential of a virus in a certain cell type.

## Interstrain differences in endothelial cell tropism of HCMV

The capability to reproduce itself is an important characteristic of the relationship between a virus and its host cell, which will determine whether this cell type might serve as a site of viral replication within the host. At a reproduction ratio of >1, the virus will spread in the cell culture, whereas at a reproduction ratio of <1 the virus will eventually vanish. Hence, assays that quantify growth of HCMV in endothelial cell cultures allow for a dichotomy between endotheliotropic (reproduction ratio >1) and non-endotheliotropic (reproduction ratio <1) virus strains. Cell culture adapted strains, which release infectious progeny into the supernatant of infected endothelial cell cultures, can be analyzed by single-step or multi-step growth curves. If the amount of harvested infectivity exceeds the amount of input virus, the strain is regarded as being endotheliotropic. Commonly used laboratory strains such as AD169 and Towne fail to reproduce in EC and were therefore regarded non-endotheliotropic with respect to their reproductive capacity (Sinzger et al., 1999). In contrast, HCMV strains that have been propagated on EC after primary isolation, as it was the case with VHL/E or TB40/E, are endotheliotropic with regard to their reproductive capacity (Sinzger et al., 1999; Waldman et al., 1991). Recent isolates, which are strictly cell-associated, can be analyzed by focus expansion assays (Sinzger et al., 1997). If the fraction of infected cells within the endothelial cell culture increases over time, the isolate is regarded as endotheliotropic. Most clinical isolates displayed an endotheliotropic phenotype in such a focus expansion assay, although there was great variation in the degree of their *reproductive endothelial cell tropism* (Sinzger et al., 1999) (Figure 21.2). The advantage of reproduction assays is that they provide a clear definition of endothelial cell tropism.

However, in the relationship between virus and host cell, there are important aspects other than the reproductive capacity of a virus. From a pathogenetic point of view, a virus that causes cytopathogenic effects or functional alterations in EC will be regarded endotheliotropic despite failure of efficient reproduction. Even strains, such as AD169 and Towne, that fail to reproduce in EC, can infect a fraction of the endothelial cell culture, perform a complete replication cycle and finally induce lysis of individual infected cells (Sinzger et al., 2000). Hence, they show at least some degree of *cytopathogenic endothelial cell tropism*. However, again there are clear interstrain differences regarding the extent of endothelial cell damage induced by HCMV. Infection by EC-propagated strains like VHL/E and TB40/E finally results in complete lysis of the endothelial cell culture even when starting from low dose infection affecting only a minority of cells (Kahl et al., 2000). In contrast, there exist strains with a reproduction factor of <1 that cause an initial cytopathogenic effect in a fraction of cells but are eventually overgrown by proliferating noninfected bystander cells. If the reproduction factor of a certain HCMV strain in a certain endothelial cell culture is ~1, the virus will establish persistent infection with limited cytopathogenic effects (Kahl et al., 2000). Obviously, assays that quantify the cytopathogenic effect of HCMV strains in EC will not discriminate between endotheliotropic or non-endotheliotropic strains but will rather characterize HCMV strains as being low endotheliotropic or highly endotheliotropic.

Interstrain differences also reflect the course of viral replication in EC, which might be either productive or abortive at a certain stage of replication. However, such interstrain differences have not been reported. Once viral gene expression is initiated in individual EC, these appear to support the complete replicative cycle resulting in generation of viral progeny, irrespective of the HCMV strain used for infection (Sinzger et al., 2000). Interstrain differences rather appear to concern the ratio of cells in which gene expression is initiated and the extent of cytopathogenicity. This does not only apply to cell culture adapted strains but also to recent clinical isolates, which differ from each other by the efficiency of cell-associated focal expansion in endothelial cell cultures. Detection of viral immediate-early antigens in such assays revealed that low endotheliotropic isolates not only fail to cause cytolytic plaques around infected cells but are already inefficient at initiating expression of the earliest viral genes in adjacent cells (Sinzger et al., 1999). This indicates that the endotheliotropism of naturally occurring HCMV variants is determined by the efficiency of initial stages of infection rather than by the fate of infection at later time points.

Consequently, quantification of viral immediate-early gene expression in EC after infection at a defined multiplicity of infection has been used as a marker for the distinction between low endotheliotropic and highly endotheliotropic strains. Consistent with the above-mentioned aspects of propagation and cytolysis, HCMV strains with high reproductive and cytopathogenic endothelial cell tropism were also most efficient in viral gene expression. After normalization in fibroblast cultures, EC-propagated strains VHL/E and TB40/E turned out to be ~100- to 1000-fold more infectious in HUVECs when compared to fibroblast-propagated strains such as AD169, VHL/F and TB40/F (Sinzger et al., 2000) (Figure 21.3A). There has been concern that this "endothelial cell tropism" might reflect a very specific tropism for HUVECs rather than an endothelial cell tropism in a more general sense. However, commonly used EC-propagated HCMV strains, for instance VHL/E and TB40/E, have been shown to replicate well in endothelial cell cultures, irrespective of fetal or adult, of arterial, venous or microvascular origin (Kahl et al., 2000). Therefore, it seems justified to designate them "highly endotheliotropic", whereas their fibroblast-propagated counterparts VHL/F and TB40/F could not be amplified in any endothelial cell cultures and can therefore be regarded as "low endotheliotropic" in a more general sense.

In summary, aspects of cytopathogenicity and reproduction seem to be rather consistent when applied to interstrain comparisons of HCMV in endothelial cell culture: strains that were classified non-endotheliotropic or endotheliotropic in reproduction assays were classified low endotheliotropic or highly endotheliotropic, respectively, in assays measuring the extent of CPE or the extent of viral gene expression. Given that all HCMV strains display some endothelial cell tropism, but also taking into account the marked interstrain differences in its extent, we would suggest that HCMV strains should be classified as being low endotheliotropic or highly endotheliotropic, and that the term non-endotheliotropic should be avoided in the future.

## Molecular determinants of endothelial cell tropism of HCMV

In principle, the cell tropism of a virus can be influenced by both genetic and epigenetic factors. With respect to endothelial cell tropism of HCMV, there is strong evidence for a genetic contribution: long-term propagation of clinical isolates in fibroblast cultures results in selection of low endotheliotropic strains, while long-term endothelial-propagated strains of the same isolates retain both fibroblast tropism and endothelial cell tropism (Sinzger et al., 1999; Waldman et al., 1991). Such differentially propagated isolate pairs displayed genomic differences in restriction fragment length analyses (Sinzger et al., 1999); a finding which has been confirmed by others (Baldanti et al., 2003). Further evidence for a genetic determination of endothelial cell tropism was provided by transfection of purified DNA from various isolates into fibroblasts, showing that reconstituted virus retained the specific

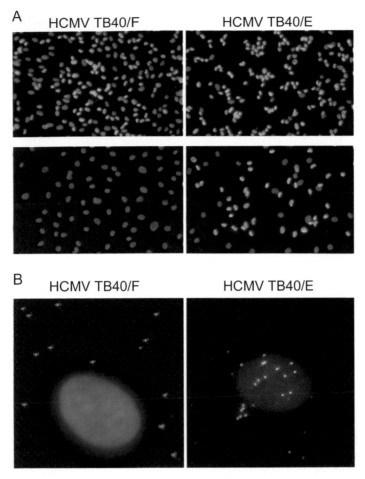

**Figure 21.3** Interstrain differences in the endothelial cell tropism of endothelial cell-propagated and fibroblast-propagated strains. (A) Infection efficiency in endothelial cells 24 h after infection with cell-free preparations of strains TB40/F and TB40/E at a multiplicity of infection of 0.5. Infected cells were detected by red immunofluorescence staining of viral immediate-early antigen. Nuclei are counterstained by DAPI (blue). (B) The low endotheliotropic strain TB40/F and the highly endotheliotropic strain TB40/E differ in the nuclear translocation efficiency of penetrated virus particles. Viral particles are visualized by green immunofluorescence staining of the capsid-associated tegument protein pUL32, and initiation of de novo viral gene expression is visualized by red immunofluorescence staining of viral immediate-early antigen. Nuclei are counterstained by DAPI (blue).

endothelial cell tropism of parental virus from which the DNA was isolated (Sinzger et al., 1999). Two lines of evidence strongly suggest that multiple gene regions contribute to interstrain differences in endothelial cell tropism of HCMV. Firstly, the extent of endothelial cell tropism in a collection of clinical HCMV isolates resembles a (skewed) normal distribution, which is more consistent with a polygenic than with a monogenic trait (Sinzger et al., 1999). Secondly, two low endotheliotropic strains can complement each other by bona fide genetic recombination to yield highly endotheliotropic recombinant progeny (Bolovan Fritts and Wiedeman, 2002; Sinzger et al., 1999).

Once a genetic contribution to the endothelial cell tropism of HCMV strains was established, there was increasing interest in the genetic determinants of this phenotype. In principle, there are various strategies to address this issue. Artificial knockouts of individual HCMV genes with subsequent phenotypic analyses in endothelial cell cultures allow identification of all genes that are necessary for efficient infection of endothelial cells. Similarly, genetic analyses of strain variants selected for by differential propagation in endothelial cell and fibroblast cultures allow the identification of those genes that determine the above mentioned interstrain differences in endothelial cell tropism. Also, correlation of naturally occurring polymorphisms of HCMV genes in clinical materials with endothelial cell tropism will aid the identification of genes that govern the endothelial cell tropism of HCMV quasispecies present in infected patients.

At present, most information about genetic determinants of HCMV cell tropism originates from the analysis of deletion mutants, which has revealed that a variety of genes contribute to successful replication of HCMV in endothelial cell culture.

## UL 23–25

In a comprehensive analysis of Towne strain deletion mutants it has recently been found that the UL24-deletion mutant was significantly defective in growth in EC, whereas it grew as well as wild-type virus in fibroblasts and epithelial cells (Dunn et al., 2003). pUL24 is a tegument protein of the US22 gene family, proteins which are assumed to participate in particle maturation (Adair et al., 2002). Intriguingly, such a finding has been predicted by analogy from the murine cytomegalovirus system where the US22 genes M140 and M141 function cooperatively and independently to regulate MCMV replication in macrophages (Hanson et al., 2001; see also Chapter 20). The authors assumed that the HCMV homologs of M140 and M141 (US23 and US24, respectively) might confer similar functions in regulating cell or tissue tropism of HCMV (Hanson et al., 2001). However, although these genes are analogous in MCMV and HCMV with respect to a function in regulation of cell tropism, they are distinct in the cell types that are affected. In MCMV, deletion of m140 and m141 only affects growth in macrophages but has no effect on infection efficiency in an endothelial cell line (Menard et al., 2003).

## UL45

Again, the first hint to an involvement of this gene in regulation of endothelial cell tropism came from the murine system. The MCMV homolog M45, which shares sequence homology to ribonucleotide reductase genes, appears to provide a cell-type specific antiapoptotic function. EC infected with M45-mutant viruses died by apoptosis in the early phase of viral replication resulting in aborted viral replication and spread within the endothelial cell culture; in contrast, growth in fibroblasts was unaffected (Brune et al., 2001). Surprisingly, deletion of UL45 in the context of HCMV Bacmid RVFIX did not affect the ability of the virus to replicate efficiently in EC, and an increase in levels of apoptotic death was not observed (Hahn et al., 2002). However, when apoptosis was induced by exposing fibroblasts to proapoptotic stimuli, cell survival was reduced by 50% in a UL45 deletion mutant on the AD169 background (Patrone et al., 2003). It will be of interest to see how the respective deletion mutants will behave in endothelial cell cultures under proapoptotic stimuli and on the background of other HCMV strains.

## UL128–131

Recently it was reported that the UL128–131 gene locus of HCMV is indispensable for productive infection of EC and that each of the genes within this locus is necessary for efficient growth in EC (Hahn et al., 2004). This gene region is highly polymorphic and, in particular, cell culture adapted strains were reported to contain disabling mutations in open reading frames UL128, UL130 and UL131A (Dolan et al., 2004).

### Other gene regions

A region located at the UL/b' (IRL) boundary, which is present in the Toledo strain but absent in AD169 (Cha et al., 1996), has been suggested to be involved in cell tropism of HCMV (Brown et al., 1995). However, in the AD169 background, a certain degree of endothelial cell tropism could be rescued by readaptation to EC cultures independent of the ULb' region (Gerna et al., 2003). Concordant with this finding, the Toledo isolate was reported to comprise both highly endotheliotropic and low endotheliotropic variants (MacCormac and Grundy, 1999). Using a low endotheliotropic Toledo strain variant in a "gain of function" approach, endothelial cell infectivity of this Toledo strain was rescued with AD169 cosmid sequences representing UL48–56 of AD169 (Bolovan Fritts and Wiedeman, 2002). Most genes within this region are essential for growth in fibroblasts and involved in DNA packaging and virus egress (Dunn et al., 2003). At present, functional data are missing that would indicate that the growth difference in EC between the low endotheliotropic Toledo strain variant and the rescued virus actually is manifested at the level of maturation and egress. Presumably, some of the genes that influence the cell tropism of HCMV are yet to be defined and particularly ORFs that are nonessential in fibroblasts might be first line candidates (Yu et al., 2003).

In conclusion, endothelial cell tropism of HCMV appears to be a complex phenotype which is governed by a number of different viral gene regions including UL23–25, UL48–56, UL128–131 and, under proapoptotic conditions, possibly UL45. In the near future, the continuation of genetic approaches is likely to identify additional genes which are nonessential in fibroblast cultures but contribute to growth in EC. Information about the genetic determinants of endothelial cell tropism will have an impact on both clinical and basic research. Firstly, it will facilitate a focused genetic analysis of clinical specimens in order to define naturally occurring polymorphisms that contribute to the observed interstrain variability of HCMV cell tropism and this may be exploited as virulence markers for prognostic purposes. Secondly, it will enable in-depth analyses of the virus-cell interactions by which the various genes influence the replication of HCMV in EC.

## Cell biological mechanisms underlying the endothelial cell tropism of HCMV

Speculation about the virus-cell interactions that determine the course of endothelial cell infection by HCMV strains are based on: (i) inference from analogous phenotypes of murine cytomegalovirus; (ii) putative functions of genetic loci involved in endothelial cell tropism; and (iii) experimental interstrain comparison of replication events in EC. Hypothetical mechanisms that determine interstrain differences in endothelial cell tropism include differential expression of antiapoptotic factors, the ability to induce microfusions between adjacent cells in order to promote cell-to-cell spread in endothelial cell monolayers, and interstrain differences regarding release of the viral genome from successfully penetrated virions in EC at initial time points of infection.

## Endothelial cell-specific apoptosis

In murine cytomegalovirus all available strains exhibit a natural endothelial cell tropism. In a whole genome screen using random transposon mutagenesis, deletion of the M45 gene has been reported to result in loss of focal growth in endothelial cell monolayers, despite initiation of viral gene expression at immediate-early times of replication (Brune et al., 2001). Infection by an M45 deletion mutant turned out to be aborted during the early phase of replication due to a failure in preventing apoptosis (Brune et al., 2001). Thus, viral inhibitors of apoptosis may contribute to endothelial cell tropism of cytomegaloviruses, although this mechanism is obviously different from the mechanisms underlying the interstrain differences observed with clinical HCMV isolates. The low endotheliotropic HCMV strains failed to express immediate-early antigens, and apoptosis was not observed to be the reason for inefficient growth in endothelial cell cultures (Sinzger et al., 1999). When the homologous UL45 gene of HCMV was deleted in the genetic background of the RVFIX virus, the level of endothelial cell infection was not reduced (Hahn et al., 2002). This could either mean that the UL45 gene is dispensable for growth of RVFIX in EC, possibly because of redundancy of antiapoptotic viral genes substituting for a loss of UL45, or that RVFIX is lacking a proapoptotic factor that might be present in murine cytomegalovirus and, possibly, in other HCMV strains. In addition, a hidden antiapoptotic function of UL45 or other HCMV genes might become apparent under proapoptotic conditions in the more complex environment of the infected host.

## Cell-to-cell spread in endothelial cells

Highly endotheliotropic HCMV strains and low endotheliotropic HCMV strains are not only discriminated by their behavior in endothelial cell cultures but also by their capacity to replicate in macrophages and dendritic cells and their efficiency of transmission to polymorphonuclear leukocytes (Gerna et al., 2002; Riegler et al., 2000). Although polymorphonuclear leukocytes are non-permissive for HCMV replication, the structural tegument protein pUL83 and the non-structural pUL122/123 proteins can be detected, indicating abortive infection of these cells (Grefte et al., 1994). This phenomenon also occurs in vivo and is exploited for diagnostic monitoring of post-transplant patients. Only highly endotheliotropic HCMV strains transmit virus from infected endothelial cell layers to attached polymorphonuclear cells, and specific fusions have been observed at intercellular attachment sites by ultrastructural analyses. These fusions were large enough to allow for transfer of HCMV capsids (Gerna et al., 2000). It is tempting to speculate that a similar mechanism might be involved in cell-to-cell spread of HCMV between endothelial cells. It has been suggested that polymorphisms in the UL128–131 gene region might contribute to interstrain differences in the induction of such fusion sites, either by a direct interference with the cellular membranes or indirectly by promoting cell–cell adhesions. This hypothesis is supported by the putative chemokine function of UL128 and UL130 gene products (Akter et al., 2003; Hahn et al., 2004). However, to date it has not been demonstrated that low endotheliotropic HCMV strains are actually less efficient in inducing such cell–cell fusions.

Initial events of replication

More direct insight into the mechanisms involved in endothelial cell tropism of HCMV originates from the comparative analysis of initial events of viral replication using low endotheliotropic and highly endotheliotropic HCMV strains. A prerequisite for the experimental dissection of replication events was the availability of highly endotheliotropic HCMV strains that release high titers of cell-free progeny. Most clinical HCMV isolates are endotheliotropic but are restricted to cell-to-cell spread, thus excluding highly efficient synchronized infections of endothelial cell cultures (Sinzger et al., 1999). Differential propagation of HCMV isolates in endothelial cell cultures and fibroblast cultures leads to selection of highly endotheliotropic and low endotheliotropic HCMV strains respectively, which are cell culture adapted in the sense that they produce high titers of free infectious virus in the culture supernatants (Sinzger et al., 1999; Waldman et al., 1991). This enables synchronized single-round infections and therefore allows for dissection of the individual replication steps that might determine the interstrain differences in endothelial cell tropism. All HCMV strains tested in that way could attach to and penetrate EC (Sinzger et al., 2000; Slobbe van Drunen et al., 1998). However, highly endotheliotropic HCMV strains VHL/E and TB40/E differed from their fibroblast-propagated counterparts by a 100-fold increase in efficiency of delivery of their genome into the nucleus of penetrated cells (Sinzger et al., 2000). Contradictory observations were published regarding initial events of the low endotheliotropic strain AD169 in EC: while in situ hybridization data had attributed the low endothelial cell tropism of AD169 to inefficient uncoating of the viral genome at the nucleus of EC (Slobbe van Drunen et al., 1998) (Figure 21.4B), ultrastructural data suggested an aberrant entry pathway by endocytosis and subsequent degradation as the underlying cause for low infectivity of AD169 in EC (Bodaghi et al., 1999) (Figure 21.4A). When single particles were detected by immunofluorescence of the capsid-associated tegument protein pUL32 (Figure 21.3B), it became evident that the nuclear translocation of penetrated particles of low endotheliotropic strains is almost completely blocked in EC, whereas it is efficient in fibroblasts (Sinzger et al., 2000). In contrast, capsids of highly endotheliotropic strains were transported efficiently in both cell types and subsequently delivered their genome and initiated viral gene expression, resulting in productive infection in either cell type (Figure 21.3B). The reason for interstrain differences in the translocation efficiency in EC is presently unknown. Similarly, there is no explanation for the differential behavior of EC and fibroblasts. The transport of penetrated virus particles from the cell periphery to the center of the cell is usually mediated by the dynein motor protein moving along microtubules toward the perinuclear microtubule organization center. This has also been demonstrated for the translocation of HCMV particles in fibroblasts (Ogawa Goto et al., 2003). It is therefore tempting to speculate that cell type-specific and virus strain-specific interaction of penetrated HCMV particles with the dynein-microtubule system determines the interstrain differences between highly endotheliotropic and low endotheliotropic HCMV strains (Figure 21.4C). This could be mediated by interstrain differences in access of the viral capsid to the motor protein, in activation of the cellular motor by viral structural proteins or in recruitment of cellular factors that promote binding and activation.

The cell tropism of many viruses is determined by the binding of virus particles to cell type-specific receptor molecules on the cell surface. The fact that both highly endotheliotropic and low endotheliotropic HCMV strains can bind to and penetrate EC with

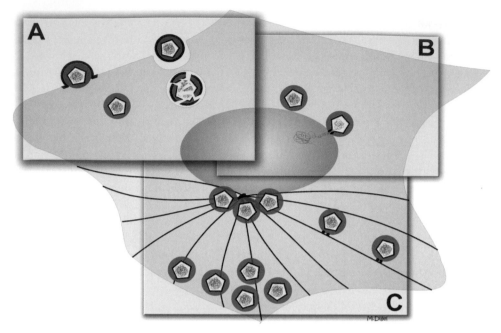

**Figure 21.4** The endothelial cell tropism of HCMV is determined by the efficiency of initial replication events. Hypothetical mechanisms mediating this phenotype are: (A) Strain-specific differences in the mode of viral entry. While fusion of the viral envelope is assumed to be the entry mode for successful infection, the low endotheliotropic strain AD169 was found in endocytic vesicles in endothelial cells, which might be an aberrant pathway leading to degradation and hence abortive infection. (B) Strain-specific differences regarding viral uncoating in endothelial cells. The low endotheliotropic strain AD169 has been shown to be inefficient in delivery of the viral genome into the nucleus. (C) Strain-specific differences in microtubule-based nuclear translocation of viral particles after penetration. Low endotheliotropic HCMV strains AD169, VHL/F and TB40/F have been found to differ from their highly endotheliotropic counterparts VHL/E and TB40/E by inefficient transport toward the nucleus despite successful penetration.

comparable efficiency argues against an interstrain difference on this level (Sinzger et al., 2000). This is consistent with the fact that cellular receptors for binding and entry such as HSPG (Compton et al., 1992) and EGFR (Wang et al., 2003) are expressed on EC. However, it cannot be ruled out that cell type-specific signaling events during the entry process, or differences in the route of entry determine the fate of penetrated particles in a strain-specific way.

In summary, there is growing evidence that early events prior to the initiation of viral gene expression determine the susceptibility of EC to naturally occurring HCMV variants. In particular, strain-specific interactions of penetrating virions with components of the cellular cytoskeleton appear to be critical and may be mediated by interstrain differences either in transport-activating signaling events or in direct virus–cytoskeleton interactions.

## Concluding remarks

It has become evident that interstrain differences in endothelial cell tropism are not an artifact of cell culture adaptation, but reflect a phenotype that is highly variable in recent clinical isolates. Interest in this phenotype has increased because it is possible that the degree of endothelial cell tropism might influence the virulence of a certain HCMV strain. Consistent with the finding that the endothelial cell tropism of HCMV is determined prior to viral gene expression, some of the genetic determinants identified so far encode viral structural proteins. Future work has to resolve how these genes are linked to the underlying mechanisms of endothelial cell tropism and which of the genes contribute to the variable endothelial cell tropism of clinical isolates. While any gene involved in endothelial cell tropism might be of relevance for the generation of an attenuated live vaccine, only genes that govern the variation in vivo are interesting for the genotypic diagnosis of the endotheliotropism of a patient's isolate. For the purpose of developing an endotheliotropic vector for gene therapy, it will be crucial to determine whether the endothelial cell tropism can be dissected from macrophage and dendritic cell tropism.

## Acknowledgments

We thank Dr. John Sinclair for critical reading of the manuscript. The contribution of Jutta Knapp to the analysis of recent HCMV isolates is greatly appreciated. The work from this laboratory was funded by the German Research Foundation.

## References

Adair, R., Douglas, E.R., Maclean, J.B., Graham, S.Y., Aitken, J.D., Jamieson, F.E., and Dargan, D.J. (2002). The products of human cytomegalovirus genes UL23, UL24, UL43 and US22 are tegument components. J. Gen. Virol. 83, 1315–1324.

Akter, P., Cunningham, C., McSharry, B.P., Dolan, A., Addison, C., Dargan, D.J., Hassan Walker, A.F., Emery, V.C., Griffiths, P.D., Wilkinson, G.W., and Davison, A.J. (2003). Two novel spliced genes in human cytomegalovirus. J. Gen. Virol. 84, 1117–1122.

Baldanti, F., Revello, M.G., Percivalle, E., Labo, N., and Gerna, G. (2003). Genomes of the endothelial cell-tropic variant and the parental Toledo strain of human cytomegalovirus are highly divergent. J. Med. Virol. 69, 76–81.

Bissinger, A.L., Sinzger, C., Kaiserling, E., and Jahn, G. (2002). Human cytomegalovirus as a direct pathogen: correlation of multiorgan involvement and cell distribution with clinical and pathological findings in a case of congenital inclusion disease. J. Med. Virol. 67, 200–206.

Bodaghi, B., Slobbe van Drunen, M.E., Topilko, A., Perret, E., Vossen, R.C., van Dam Mieras, M.C., Zipeto, D., Virelizier, J.L., LeHoang, P., Bruggeman, C.A., and Michelson, S. (1999). Entry of human cytomegalovirus into retinal pigment epithelial and endothelial cells by endocytosis. Invest. Ophthalmol. Vis. Sci. 40, 2598–2607.

Bolovan Fritts, C.A., and Wiedeman, J.A. (2002). Mapping the viral genetic determinants of endothelial cell tropism in human cytomegalovirus. J. Clin. Virol. 25 Suppl 2, S97–109.

Borgia, M.C., Mandolini, C., Barresi, C., Battisti, G., Carletti, F., and Capobianchi, M.R. (2001). Further evidence against the implication of active cytomegalovirus infection in vascular atherosclerotic diseases. Atherosclerosis 157, 457–462.

Brown, J.M., Kaneshima, H., and Mocarski, E.S. (1995). Dramatic interstrain differences in the replication of human cytomegalovirus in SCID-hu mice. J. Infect. Dis. 171, 1599–1603.

Brune, W., Menard, C., Heesemann, J., and Koszinowski, U.H. (2001). A ribonucleotide reductase homolog of cytomegalovirus and endothelial cell tropism. Science 291, 303–305.

Cha, T.A., Tom, E., Kemble, G.W., Duke, G.M., Mocarski, E.S., and Spaete, R.R. (1996). Human cytomegalovirus clinical isolates carry at least 19 genes not found in laboratory strains. J. Virol. 70, 78–83.

Compton, T., Nepomuceno, R.R., and Nowlin, D.M. (1992). Human cytomegalovirus penetrates host cells by pH-independent fusion at the cell surface. Virology 191, 387–395.

Degre, M. (2002). Has cytomegalovirus infection any role in the development of atherosclerosis? Clin. Microbiol. Infect. 8, 191–195.

Dengler, T.J., Raftery, M.J., Werle, M., Zimmermann, R., and Schoenrich, G. (2000). Cytomegalovirus infection of vascular cells induces expression of pro-inflammatory adhesion molecules by paracrine action of secreted interleukin-1beta. Transplantation 69, 1160–1168.

Dolan, A., Cunningham, C., Hector, R.D., Hassan Walker, A.F., Lee, L., Addison, C., Dargan, D.J., McGeoch, D.J., Gatherer, D., Emery, V.C., Emery V.C., Griffiths P.D., Sinzger C., McSharry B.P., Wilkinson G.W., and Davison A.J. (2004). Genetic content of wild-type human cytomegalovirus. J. Gen. Virol. 85, 1301–1312.

Dunn, W., Chou, C., Li, H., Hai, R., Patterson, D., Stolc, V., Zhu, H., and Liu, F. (2003). Functional profiling of a human cytomegalovirus genome. Proc. Natl. Acad. Sci. USA 100, 14223–14228.

Fish, K.N., Soderberg Naucler, C., Mills, L.K., Stenglein, S., and Nelson, J.A. (1998). Human cytomegalovirus persistently infects aortic endothelial cells. J. Virol. 72, 5661–5668.

Gerna, G., Percivalle, E., Baldanti, F., and Revello, M.G. (2002). Lack of transmission to polymorphonuclear leukocytes and human umbilical vein endothelial cells as a marker of attenuation of human cytomegalovirus. J. Med. Virol. 66, 335–339.

Gerna, G., Percivalle, E., Baldanti, F., Sozzani, S., Lanzarini, P., Genini, E., Lilleri, D., and Revello, M.G. (2000). Human cytomegalovirus replicates abortively in polymorphonuclear leukocytes after transfer from infected endothelial cells via transient microfusion events. J. Virol. 74, 5629–5638.

Gerna, G., Percivalle, E., Sarasini, A., Baldanti, F., Campanini, G., and Revello, M.G. (2003). Rescue of human cytomegalovirus strain AD169 tropism for both leukocytes and human endothelial cells. J. Gen. Virol. 84, 1431–1436.

Grefte, A., Blom, N., van der Giessen, M., van Son, W., and The, T.H. (1993a). Ultrastructural analysis of circulating cytomegalic cells in patients with active cytomegalovirus infection: evidence for virus production and endothelial origin. J. Infect. Dis. 168, 1110–1118.

Grefte, A., Harmsen, M.C., van der Giessen, M., Knollema, S., van Son, W.J., and The, T.H. (1994). Presence of human cytomegalovirus (HCMV) immediate-early mRNA but not ppUL83 (lower matrix protein pp65) mRNA in polymorphonuclear and mononuclear leukocytes during active HCMV infection. J. Gen. Virol. 75, 1989–1998.

Grefte, A., van der Giessen, M., van Son, W., and The, T.H. (1993b). Circulating cytomegalovirus (CMV)-infected endothelial cells in patients with an active CMV infection. J. Infect. Dis. 167, 270–277.

Grundy, J.E., Lawson, K.M., MacCormac, L.P., Fletcher, J.M., and Yong, K.L. (1998). Cytomegalovirus-infected endothelial cells recruit neutrophils by the secretion of C-X-C chemokines and transmit virus by direct neutrophil–endothelial cell contact and during neutrophil transendothelial migration. J. Infect. Dis. 177, 1465–1474.

Hahn, G., Khan, H., Baldanti, F., Koszinowski, U.H., Revello, M.G., and Gerna, G. (2002). The human cytomegalovirus ribonucleotide reductase homolog UL45 is dispensable for growth in endothelial cells, as determined by a BAC-cloned clinical isolate of human cytomegalovirus with preserved wild-type characteristics. J. Virol. 76, 9551–9555.

Hahn, G., Revello, M.G., Patrone, M., Percivalle, E., Campanini, G., Sarasini, A., Wagner, M., Gallina, A., Milanesi, G., Koszinowski, U., Baldanti F., and Gerna G. (2004). Human cytomegalovirus UL131–128 genes are indispensable for virus growth in endothelial cells and virus transfer to leukocytes. J. Virol. 78, 10023–10033.

Hanson, L.K., Slater, J.S., Karabekian, Z., Ciocco Schmitt, G., and Campbell, A.E. (2001). Products of US22 genes M140 and M141 confer efficient replication of murine cytomegalovirus in macrophages and spleen. J. Virol. 75, 6292–6302.

Hendrix, M.G., Daemen, M., and Bruggeman, C.A. (1991). Cytomegalovirus nucleic acid distribution within the human vascular tree. Am. J. Pathol. 138, 563–567.

Hendrix, R.M., Wagenaar, M., Slobbe, R. L., and Bruggeman, C.A. (1997). Widespread presence of cytomegalovirus DNA in tissues of healthy trauma victims. J. Clin. Pathol. 50, 59–63.

Ho, D.D., Rota, T.R., Andrews, C.A., and Hirsch, M.S. (1984). Replication of human cytomegalovirus in endothelial cells. J. Infect. Dis. 150, 956–957.

Jahn, G., Stenglein, S., Riegler, S., Einsele, H., and Sinzger, C. (1999). Human cytomegalovirus infection of immature dendritic cells and macrophages. Intervirology 42, 365–372.

Jarvis, M.A., and Nelson, J.A. (2002). Mechanisms of human cytomegalovirus persistence and latency. Front. Biosci. 7, 1575–1582.

Kahl, M., Siegel Axel, D., Stenglein, S., Jahn, G., and Sinzger, C. (2000). Efficient lytic infection of human arterial endothelial cells by human cytomegalovirus strains. J. Virol. 74, 7628–7635.

Knight, D.A., Waldman, W.J., and Sedmak, D.D. (1999). Cytomegalovirus-mediated modulation of adhesion molecule expression by human arterial and microvascular endothelial cells. Transplantation 68, 1814–1818.

Koskinen, P., Lemstrom, K., Bruggeman, C., Lautenschlager, I., and Hayry, P. (1994). Acute cytomegalovirus infection induces a subendothelial inflammation (endothelialitis) in the allograft vascular wall. A possible linkage with enhanced allograft arteriosclerosis. Am. J. Pathol. 144, 41–50.

Kronschnabl, M., and Stamminger, T. (2003). Synergistic induction of intercellular adhesion molecule-1 by the human cytomegalovirus transactivators IE2p86 and pp71 is mediated via an Sp1-binding site. J. Gen. Virol. 84, 61–73.

Lathey, J.L., Wiley, C.A., Verity, M.A., and Nelson, J.A. (1990). Cultured human brain capillary endothelial cells are permissive for infection by human cytomegalovirus. Virology 176, 266–273.

Lautenschlager, I., Hockerstedt, K., Taskinen, E., and von Willebrand, E. (1996). Expression of adhesion molecules and their ligands in liver allografts during cytomegalovirus (CMV) infection and acute rejection. Transpl. Int. 9 Suppl. 1, 213–215.

MacCormac, L.P., and Grundy, J.E. (1999). Two clinical isolates and the Toledo strain of cytomegalovirus contain endothelial cell tropic variants that are not present in the AD169, Towne, or Davis strains. J. Med. Virol. 57, 298–307.

Maisch, T., Kropff, B., Sinzger, C., and Mach, M. (2002). Up-regulation of CD40 expression on endothelial cells infected with human cytomegalovirus. J. Virol. 76, 12803–12812.

Melnick, J.L., Hu, C., Burek, J., Adam, E., and DeBakey, M.E. (1994). Cytomegalovirus DNA in arterial walls of patients with atherosclerosis. J. Med. Virol. 42, 170–174.

Menard, C., Wagner, M., Ruzsics, Z., Holak, K., Brune, W., Campbell, A.E., and Koszinowski, U.H. (2003). Role of murine cytomegalovirus US22 gene family members in replication in macrophages. J. Virol. 77, 5557–5570.

Myerson, D., Hackman, R.C., Nelson, J.A., Ward, D.C., and McDougall, J.K. (1984). Widespread presence of histologically occult cytomegalovirus. Hum. Pathol. 15, 430–439.

Nerheim, P.L., Meier, J.L., Vasef, M.A., Li, W.G., Hu, L., Rice, J.B., Gavrila, D., Richenbacher, W.E., and Weintraub, N.L. (2004). Enhanced cytomegalovirus infection in atherosclerotic human blood vessels. Am. J. Pathol. 164, 589–600.

Ogawa Goto, K., Tanaka, K., Gibson, W., Moriishi, E., Miura, Y., Kurata, T., Irie, S., and Sata, T. (2003). Microtubule network facilitates nuclear targeting of human cytomegalovirus capsid. J. Virol. 77, 8541–8547.

Pampou, S., Gnedoy, S.N., Bystrevskaya, V.B., Smirnov, V.N., Chazov, E.I., Melnick, J.L., and DeBakey, M.E. (2000). Cytomegalovirus genome and the immediate-early antigen in cells of different layers of human aorta. Virchows Arch. 436, 539–552.

Patrone, M., Percivalle, E., Secchi, M., Fiorina, L., Pedrali Noy, G., Zoppe, M., Baldanti, F., Hahn, G., Koszinowski, U.H., Milanesi, G., and Gallina, A. (2003). The human cytomegalovirus UL45 gene product is a late, virion-associated protein and influences virus growth at low multiplicities of infection. J. Gen. Virol. 84, 3359–3370.

Percivalle, E., Revello, M.G., Vago, L., Morini, F., and Gerna, G. (1993). Circulating endothelial gianT-cells permissive for human cytomegalovirus (HCMV) are detected in disseminated HCMV infections with organ involvement. J. Clin. Invest. 92, 663–670.

Plachter, B., Sinzger, C., and Jahn, G. (1996). Cell types involved in replication and distribution of human cytomegalovirus. Adv. Virus Res. 46, 195–261.

Radsak, K., Fuhrmann, R., Franke, R.P., Schneider, D., Kollert, A., Brucher, K.H., and Drenckhahn, D. (1989). Induction by sodium butyrate of cytomegalovirus replication in human endothelial cells. Arch. Virol. 107, 151–158.

Reeves, M.B., Coleman, H., Chadderton, J., Goddard, M., Sissons, J.G., and Sinclair, J.H. (2004). Vascular endothelial and smooth muscle cells are unlikely to be major sites of latency of human cytomegalovirus in vivo. J. Gen. Virol. 85, 3337–3341.

Reinhardt, B., Vaida, B., Voisard, R., Keller, L., Breul, J., Metzger, H., Herter, T., Baur, R., Luske, A., and Mertens, T. (2003). Human cytomegalovirus infection in human renal arteries in vitro. J. Virol. Methods 109, 1–9.

Revello, M.G., Percivalle, E., Arbustini, E., Pardi, R., Sozzani, S., and Gerna, G. (1998). In vitro generation of human cytomegalovirus pp65 antigenemia, viremia, and leukoDNAemia. J. Clin. Invest. 101, 2686–2692.

Riegler, S., Hebart, H., Einsele, H., Brossart, P., Jahn, G., and Sinzger, C. (2000). Monocyte-derived dendritic cells are permissive to the complete replicative cycle of human cytomegalovirus. J. Gen. Virol. *81*, 393–399.

Roberts, W.H., Sneddon, J.M., Waldman, J., and Stephens, R.E. (1989). Cytomegalovirus infection of gastrointestinal endothelium demonstrated by simultaneous nucleic acid hybridization and immunohistochemistry. Arch. Pathol. Lab. Med. *113*, 461–464.

Sedmak, D.D., Knight, D.A., Vook, N.C., and Waldman, J.W. (1994). Divergent patterns of ELAM-1, ICAM-1, and VCAM-1 expression on cytomegalovirus-infected endothelial cells. Transplantation 58, 1379–1385.

Shahgasempour, S., Woodroffe, S. B., and Garnett, H.M. (1997). Alterations in the expression of ELAM-1, ICAM-1 and VCAM-1 after in vitro infection of endothelial cells with a clinical isolate of human cytomegalovirus. Microbiol. Immunol. *41*, 121–129.

Sindre, H., Haraldsen, G., Beck, S., Hestdal, K., Kvale, D., Brandtzaeg, P., Degre, M., and Rollag, H. (2000). Human intestinal endothelium shows high susceptibility to cytomegalovirus and altered expression of adhesion molecules after infection. Scand J. Immunol. *51*, 354–360.

Sinzger, C., Grefte, A., Plachter, B., Gouw, A.S., The, T.H., and Jahn, G. (1995). Fibroblasts, epithelial cells, endothelial cells and smooth muscle cells are major targets of human cytomegalovirus infection in lung and gastrointestinal tissues. J. Gen. Virol. 76, 741–750.

Sinzger, C., Kahl, M., Laib, K., Klingel, K., Rieger, P., Plachter, B., and Jahn, G. (2000). Tropism of human cytomegalovirus for endothelial cells is determined by a post-entry step dependent on efficient translocation to the nucleus. J. Gen. Virol. *81*, 3021–3035.

Sinzger, C., Knapp, J., Plachter, B., Schmidt, K., and Jahn, G. (1997). Quantification of replication of clinical cytomegalovirus isolates in cultured endothelial cells and fibroblasts by a focus expansion assay. J. Virol. Methods 63, 103–112.

Sinzger, C., Schmidt, K., Knapp, J., Kahl, M., Beck, R., Waldman, J., Hebart, H., Einsele, H., and Jahn, G. (1999). Modification of human cytomegalovirus tropism through propagation in vitro is associated with changes in the viral genome. J. Gen. Virol. 80, 2867–2877.

Slobbe van Drunen, M.E., Hendrickx, A.T., Vossen, R.C., Speel, E.J., van Dam Mieras, M.C., and Bruggeman, C.A. (1998). Nuclear import as a barrier to infection of human umbilical vein endothelial cells by human cytomegalovirus strain AD169. Virus Res. 56, 149–156.

Ustinov, J.A., Loginov, R.J., Mattila, P.M., Nieminen, V.K., Suni, J.I., Hayry, P.J., and Lautenschlager, I.T. (1991). Cytomegalovirus infection of human kidney cells in vitro. Kidney Int. 40, 954–960.

Valantine, H.A. (2004). The role of viruses in cardiac allograft vasculopathy. Am. J. Transplant. 4, 169–177.

Waldman, W.J. (1998). Cytomegalovirus as a perturbing factor in graft/host equilibrium: havoc at the endothelial interface. In CMV-Related Immunopathology, M. Scholz, H.F. Rabenau, H.W. Doerr, and J.J. Cinatl, eds. (Basel, Karger), pp. 54–66.

Waldman, W.J., Knight, D.A., Huang, E.H., and Sedmak, D.D. (1995). Bidirectional transmission of infectious cytomegalovirus between monocytes and vascular endothelial cells: an in vitro model. J. Infect. Dis. *171*, 263–272.

Waldman, W.J., Roberts, W.H., Davis, D.H., Williams, M.V., Sedmak, D.D., and Stephens, R.E. (1991). Preservation of natural endothelial cytopathogenicity of cytomegalovirus by propagation in endothelial cells. Arch. Virol. *117*, 143–164.

Wang, X., Huong, S.M., Chiu, M. L., Raab Traub, N., and Huang, E.S. (2003). Epidermal growth factor receptor is a cellular receptor for human cytomegalovirus. Nature *424*, 456–461.

Wolf, D.G., and Spector, S.A. (1993). Early diagnosis of human cytomegalovirus disease in transplant recipients by DNA amplification in plasma. Transplantation 56, 330–334.

Wu, Q.H., Ascensao, J., Almeida, G., Forman, S.J., and Shanley, J.D. (1994). The effect of short-chain fatty acids on the susceptibility of human umbilical vein endothelial cells to human cytomegalovirus infection. J. Virol. Methods 47, 37–50.

Yu, D., Silva, M.C., and Shenk, T. (2003). Functional map of human cytomegalovirus AD169 defined by global mutational analysis. Proc. Natl. Acad. Sci. USA 100, 12396–12401.

# Myeloid Cell Recruitment and Function in Pathogenesis and Latency

# 22

*Edward S. Mocarski, Jr., Gabriele Hahn, Kirsten Lofgren White, Jiake Xu, Barry Slobedman, Laura Hertel, Shirley A. Aguirre, and Satoshi Noda*

## Abstract

Cytomegaloviruses have an association with myelomonocytic cells during acute infection, where they control viral dissemination, and latency, where progenitors are important sites for life-long viral genome residence. Recent efforts to understand the control of dissemination by virus-encoded chemokines have shown that immature myelomonocytic cells are recruited from the bone marrow to sites of viral infection. Investigations in mice that are deficient in CCR2 and CCL2 (MCP-1), a chemokine system responsible for monocyte recruitment from blood into tissues, exhibit no differences in susceptibility or dissemination patterns, suggesting that late myeloid progenitors rather than more mature monocytes or monocyte-derived macrophages, play key roles in MCMV dissemination. This, paired with understanding of HCMV pathogenesis from studies in naturally infected individuals, suggests that cytomegaloviruses depend on myeloid cell progenitors for both acute infection and latency. Latent infection by HCMV in granulocyte–macrophage progenitors is accompanied by the expression of specific latency-associated transcripts from the IE1/IE2 region whose gene products have not yet been found to affect experimental latent infection. A portion of the IE1/IE2 transcriptional enhancer-modulator that is dispensable for productive replication is important for genome maintenance in latency, predicting that HCMV has an origin of DNA replication in this region.

## Introduction

Cytomegaloviruses engage myeloid cells in the host as a component of dissemination and latency and share several characteristics that contribute to this interaction. Human CMV (HCMV) and the related beta-herpesvirus, murine CMV (MCMV), express potent immune modulators that facilitate escape from immune clearance and lead to persistence. The array of immunomodulatory functions encoded by these viruses likely increases the efficiency of primary infection, dissemination, and persistent infection of the host, measurable in certain cell types. These may help set up the host for efficient latency as well as sporadic reactivation throughout life and likely contribute to virulence of HCMV for the developing fetus and immunocompromised host. CMV-encoded proteins have been shown to modulate classical and nonclassical major histocompatibility complex protein function, host cell susceptibility to apoptosis, induction and activity of cytokines and interferons, and antibody defense mechanisms, together contributing to the success of this virus as a pathogen

(Mocarski 2002; 2004). One exploitive component of immunomodulation, the induction of migration and activation of leukocytes by virus-encoded chemokines, suggests that cytomegaloviruses benefit from encouraging a strong inflammatory response. Viral chemokines such as HCMV vCXCL-1 (UL146) and MCMV MCK-2 (m131-m129) attract myeloid cells. Studies on MCMV, in particular, revealed a role of the viral chemokine MCK-2 in enhancing systemic leukocyte mobilization from the bone marrow (BM) and recruitment to sites of infection via the peripheral blood (PB), as summarized below. Host leukocytes acquire virus at these sites and mediate specific dissemination patterns via PB in ways that may contribute to the reservoir of latent virus. Based on the distribution of viral DNA in host tissues and limited reports of reactivation from leukocytes, HCMV latency occurs in myeloid progenitors as primitive as stem cells and as fully differentiated as macrophages (MΦ) or dendritic cells (DC) (Jarvis and Nelson, 2002; Sissons et al., 2002). However, characterized latency-associated transcripts have only been detected in lineage-committed granulocyte–macrophage progenitors (GMPs) (Kondo et al., 1996; White et al., 2000) and not in more primitive or more differentiated cells. So far, the latency-associated genes and their products have been found to be dispensable for productive as well as latent infection in experimental infections (Kondo et al., 1996; White et al., 2000). The ability of proinflammatory viral functions expressed during productive infection to successfully manipulate the host response to infection is one of many diverse and surprising characteristics that ensure the success of this pathogen during evolution in the remarkably wide variety of different mammalian species where cytomegaloviruses are found.

## Control of host leukocyte migration by cytomegalovirus-encoded chemokines

Human CMV encodes at least four gene products (UL33, UL78, US27 and US28) that appear, based on sequence characteristics, to be 7-transmembrane spanning receptor homologs, one of which (US28) is a constitutively signaling receptor with broad specificity for CC and CX3C chemokines (Saederup and Mocarski, 2002) and another of which (UL33 and its MCMV homolog) supports ligand-independent signaling (Casarosa et al., 2003). HCMV encodes three chemokine homologs (UL128, UL146, UL147), one of which (UL146) encodes a ligand for host CXCR2, one of two IL-8 receptors (Saederup and Mocarski, 2002). UL146 varies up to 60% in amino acid composition from strain to strain, making it the most variable viral gene product known.

## Receptors

US28 and the adjacent gene, US27, are homologs found only in primate cytomegaloviruses and not in other beta-herpesviruses (Saederup and Mocarski, 2002). These gene products are encoded at late times during productive infection and not during latency. While HCMV has two family members, five US28 homologs have been identified in rhesus macaque CMV, where, like with HCMV, only one is functional (Penfold et al., 2003). Homologs of UL33 and UL78 are present in all beta-herpesviruses and although the UL33 and M33 gene products signal, none of these bind chemokines. US28 protein binds with very high affinity to many CC chemokines as well as to human fractalkine (CXC3L1). These interactions may activate and facilitate adhesion of virus-infected cells to endothelial cells. US28 protein is also an entry mediator for human immunodeficiency virus and binds Kaposi's sarcoma herpesvirus (KSHV, HHV-8) chemokine vMIP-2 with high affinity. US28 receptor does not bind to any of the known chemokines produced by human or animal cytomegaloviruses. Despite a strong impact on cells when expressed independent of viral infection, US28 signaling has little impact on infection, although growth defects unrelated to US28 deletion have been reported using a US28 mutant. Any consequence of US28 signaling remains imperceptible at late times of infection (Hertel and Mocarski, 2004). Initial predictions that US28 encodes a chemokine sink, have fallen away to models of cell behavior modification through expression of this potent receptor, though no experimental system has been used to fully evaluate this potential due to the substantial, US28-independent growth defects in the recombinant viruses that have been employed.

UL33 and UL78 and related homologs in other beta-herpesviruses appear to be important for infection (Saederup and Mocarski, 2002). M33 and R33 are both needed for dissemination to or replication in the salivary glands, a critical site for transmission to new hosts. M78 and R78 deletion mutants exhibit reduced replication in cultured cells as well as severe attenuation in the infected host, particularly in the spleen (Kaptein et al., 2003). Thus, these beta-herpesvirus-common homologs are important for infection, but the underlying mechanism remains unknown. There is no evidence that they function as receptors for any host- or virally encoded chemokines.

## Chemokines

Human CMV UL146 encodes vCXCL-1, a CXCR2-specific neutrophil-attracting viral chemokine (Penfold et al., 1999). This chemokine is conserved in primate cytomegaloviruses, which carry either two or three members of the UL146 family, and is absent from rodent CMV and other beta-herpesviruses. Furthermore, the IL-8 receptor in mice does not recognize vCXCL1, which limits experimental opportunities to study its impact on viral infection but stimulated the production of human CXCR2 transgenic mice (Sparer et al., 2004). In HCMV, the one other homolog, UL147 is encoded by the adjacent gene but has no activity. The specificity of vCXCL1 for CXCR2, one of two human IL-8 receptors, is surprising in light of the fact that the viral chemokine has virtually no sequence homology with IL-8. An amino terminal ELRCXC motif and a similar spacing of two conserved cysteines are the only features that vCXCL1 shares with IL-8. All studies to date have employed the UL146 gene from the Toledo strain of virus to study vCXCL1 function and these studies predict that the behavior of cells bearing CXCR2 would be affected. Neutrophils are considered the major population of CXCR2-bearing cells, but the other human PB cells, including myeloid progenitors as well as PB monocytes and granulocytes, bear this receptor as well. UL146 exhibits a remarkable level of sequence variability in different strains, up to 60% divergence, leaving open the possibility that other variants will have additional or different functions. Direct evaluation of vCXCL1 function in the context of the infected host remains very difficult given that this chemokine is only found in primate cytomegaloviruses.

Murine CMV encodes a chemokine homolog-1 (MCK-1) initially recognized by sequence characteristics (Saederup and Mocarski, 2002). As expressed naturally, the entire 63 amino acid MCK-1 chemokine domain is fused to a second exon (m129) that adds to the carboxyl-terminus to produce MCK-2. Expression of MCK-2 during infection drives greater inflammation and leads to enhanced dissemination (Saederup and Mocarski, 2002).

Following inoculation into BALB/c, C57BL/6, *scid/scid* mice or common γ-chain deficient mice, *mck* mutant viruses replicate at levels similar to wild type (WT) virus in most tissues and organs, but with reduced levels of peak viremia (day 5 post infection) and reduced titers of virus in salivary glands throughout the extended period of replication in this organ (day 7 to at least day 28 post infection) (Aguirre, Huang, Mocarski, in preparation). This temporal relationship suggests that the viral chemokine attracts cells that specifically facilitate dissemination to and seeding of this organ (Figure 22.1). Salivary glands are important for CMV pathogenesis because shedding in saliva is the source of virus for transmission to new hosts. This hypothesis has been reinforced by the demonstration that *mck* expression during viral infection (Saederup et al., 2001) or administration of recombinant MCK-2 (Noda et al., 2005) causes increased inflammation at sites of inoculation within the first few days of infection that is temporally related to dissemination. Increased levels of inflammation at sites of inoculation due to expression of MCK-2 have no impact on levels of replication or time course of clearance. These characteristics suggest that one role of MCK-2 is to attract leukocytes, initially thought to be antigen-presenting cells such as a Mφ or DC, to the inoculation site in order to assure dissemination via the bloodstream to salivary glands (Saederup and Mocarski, 2002). The virus-positive cell type responsible for disseminating MCMV via the PB was earlier shown to be a large, adherent and phagocytic mononuclear cell (MNC) consistent with some sort of Mφ or DC (Stoddart et al., 1994).

Most non-lymphoid MNCs in the PB are in the myeloid lineage, including BM-derived CD14+ leukocytes that, in uninfected mice are believed to give rise to tissue Mφ as well as to a subset of DC (Bruno et al., 2001; Geissmann et al., 2003). During viral infection an inflammatory insult leads to the mobilization of BM-derived progenitors and consequently an increase in PB MNCs that changes the composition of the PB. Mobilization and migration along blood vessels as well as transendothelial migration out of the bloodstream is controlled by sequential action of adhesion molecules and cytokines (Johnston and Butcher, 2002; Ley, 2003; Salazar-Mather and Hokeness, 2003; Schenkel et al., 2004). This response can involve leukocyte types at many stages of differentiation, including hematopoietic stem cells and a number of poorly characterized lineage-committed myeloid cell types (Christensen et al., 2004; Schenkel et al., 2004; Wagers and Weissman, 2004). In such settings, an immature (CD31+ Mac-1+ Gr-1+) myeloid cell predominates in the MNC fraction and is believed to represent a late stage progenitor of monocytes, DCs, Mφ or other MNCs (Biermann et al., 1999; Johnson et al., 1998; Serafini et al., 2004). Studies aimed at evaluating the migration of monocyte and Mφ have shown that the chemokine MCP-1 (CCL2) and its receptor CCR2 are critical for chemotaxis out of the bloodstream into tissues or the peritoneal cavity (Boring et al., 1997; Kurihara et al., 1997; Lu et al., 1998) and this response is critical for Mφ-mediated clearance of intracellular bacterial pathogens (Serbina et al., 2003a; Serbina et al., 2003b).

Mφ have long been held to be central to MCMV pathogenesis and dissemination in the host (Collins et al., 1994; Krmpotic et al., 2003; Stoddart et al., 1994). Monocytes and Mφ exhibit a differentiation-dependent susceptibility to MCMV replication (Brautigam et al., 1979; Collins et al., 1994; Hanson et al., 1999; Mims and Gould, 1978; Pollock et al., 1997; Stoddart et al., 1994) and appear to be one of the important sites of latency (Koffron et al., 1998; Mitchell et al., 1996; Pollock et al., 1997). Mφ have also been implicated in clearance via nitric oxide-enhanced mechanisms (Noda et al., 2001). Nonpermissive monocytes or

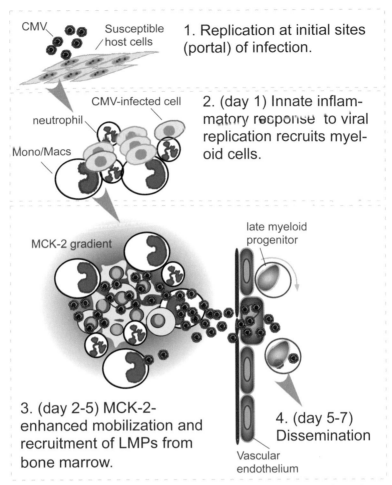

**Figure 22.1** Model of MCK-2 function in mobilization and recruitment of myelomonocytic progenitors for dissemination during MCMV infection. 1. Primary infection with MCMV leads to replication at the portal of entry over the first day of infection. 2. Replication induces an innate inflammatory response over the first day of infection that is independent of MCK-2, a very late viral gene product. 3. An MCK-2 gradient is produced and specifically mobilizes CD31$^+$Mac-1$^+$Gr-1$^{int}$ late myeloid progenitors (LMPs) starting over the second day and continuing at least until day 5 postinfection. Between day 2 and 5, late myeloid progenitors accumulate at sites of infection. 4. Recruited late myeloid progenitors acquire virus from infected vascular endothelium, leading to enhanced peak levels of viremia at day 5 and dissemination to salivary glands beginning at days 5–7 postinfection.

other MNCs are believed to disseminate infection and differentiate into permissive Mϕ upon entry into tissues. This pattern is similar to that proposed to explain HCMV latency in progenitors of Mϕ with reactivation under the appropriate cellular stimulation (Sissons et al., 2002). When three aspects of MCK-2 behavior were evaluated: (1) the cell type(s) that are mobilized by viral infection in response to MCK-2, (2) the impact of MCK-2 on mobilization, and (3) the importance of Mϕ in this migration, surprising results emerged (Noda et al., 2005). MCMV infection proceeds unhindered and efficiently in the absence of host CCL2 and/or CCR2 even though Mφ function is severely attenuated under these conditions. Consistent with these findings, MCK-2 induces an intense response of immature CD31⁺ Mac-1⁺ Gr-1^int BM-derived mononuclear leukocytes (Figure 22.2) that play key roles in dissemination. The influence of MCK-2 on accumulation of these late myeloid progenitors (LMPs), is dramatic, regardless of whether the chemokine is expressed from virus or injected as a recombinant protein (Noda et al., 2005). In the inoculated foot where cellular accumulation occurs, increased levels of LMPs are apparent by day 2 and continue to be present through day 5 post infection when peak viremia is attained with either *mck*-expressing WT or mutant virus. The situation in PB is similarly dramatic. Although the

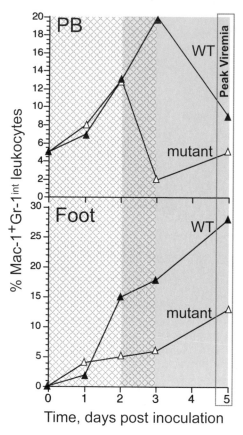

**Figure 22.2** Kinetics of mobilization of CD31⁺Mac-1⁺ Gr-1^int LMPs into PB and recruitment into the inoculated foot (Noda et al., 2005). (Top) Equal numbers of CD31⁺Mac-1⁺ Gr-1^int LMPs are mobilized into PB of mice inoculated with RQ461 (WT, *mck*-expressing) or RM461 (*mck* mutant) at one and two days postinfection (footpad inoculation). LMP levels continue to rise in WT virus infection at three days postinfection when *mck* mutant mobilization drops below uninfected mouse levels. Mobilization of LMPs into PB has resolved by five days postinfection when virus-positive PB leukocyte levels (viremia) peak (gray box) with either virus but at levels that are 10-fold higher with *mck*-expressing viruses (Saederup et al., 1999). (Bottom) Increased numbers of CD31⁺ Mac-1⁺ Gr-1^int LMPs accumulate in the foot of mice inoculated with WT virus compared to mutant virus starting at two days postinfection, temporally associated with higher levels of viremia with WT virus. The overlapping mobilization (cross-hatched area) and recruitment (gray area) stages of infection are indicated.

level of LMPs rises in a similar pattern over the first two days in either *mck*-expressing WT or mutant virus infection, *mck*-expressing WT virus is associated with a increased mobilization starting at day 3 (when 20% of PB leukocytes show characteristics of LMPs). In contrast, the levels of *mck* mutant virus drop below those found in uninfected animals between day two and three postinfection. These patterns provide evidence that MCK-2-responsive cells traverse the bloodstream and accumulate in the inoculated foot. This response is dependent upon functional BM and is reduced when antibody to Gr-1 is employed to deplete mobilized cells. Thus, monocytes and mature MΦ are neither vehicles for dissemination nor mediators of host innate immune control during experimental infection of mice. MCK-2 drives the increased mobilization and recruitment of immature LMPs to sites of infection and this appears to facilitate dissemination. Given the unlikelihood of reverse diapedesis to return virus-positive cells to the PB, we believe it is far more likely that MCMV replicates in vascular endothelial cells and infects immature cells passing through sites of initial infection. This allows for an MCK-2 impact on cells that never leave the bloodstream at all and accommodates all current information, including observations that MCMV gene products are specifically required for endothelial tropism. The intimate association of LMPs rolling along the endothelium may facilitate acquisition of virus and help explain the efficiency of dissemination in the presence or absence of MCK-2 (Figure 22.1).

## Persistence and latency of CMV in myeloid cells

CMV replicates in a wide variety of host cells during primary infection and is shed sporadically in saliva and other body fluids throughout life. The biology of CMV follows two characterized extremes (Figure 22.3) involving latent infection and productive replication. HCMV maintains a life-long latent infection with sporadic reactivation. MΦ and self-renewing myeloid progenitors that give rise to MΦ have received the most consistent supporting evidence as a site of HCMV latency (Jarvis and Nelson, 2002). MΦ and other BM-derived cells are also likely to be involved in MCMV latency (Koffron et al., 1998; Mitchell et al., 1996; Pollock et al., 1997). CMV latency studies have benefited from efforts to understand latency in other herpesviruses, and the establishment of important parameters that differentiate chronic infection from latency (Stevens, 1978). Many clinical investigators consider chronic, low level, productive (or lytic) infection to be a characteristic of CMV infection (Pass, 2001). The exceptionally high level of T-lymphocyte responsiveness to HCMV (Kern et al., 2002) lends support to the notion that frequent or persistent antigen stimulation may result from chronic infection. In addition to infection of hematopoietic cells, HCMV chronically infects ductal epithelial cells of the salivary glands or kidneys of healthy immunocompetent individuals and is regularly shed in saliva and/or urine. A wide variety of additional cell types are susceptible to viral infection in immunocompromised patients (Myerson et al., 1984).

Latency, a dormant relationship where the viral genome is deposited in cells but active viral replication ceases, is the favored persistence mechanism employed by herpesviruses. In studies of naturally infected host or in experimental models, the differentiation of latency from chronic infection or abortive infection remains a significant challenge for species-specific viruses like CMVs. Active replication must have ceased, the viral genome must remain and not be lost, and, importantly, infectious virus must be recoverable following reactivation of latently infected cells, and this may occur spontaneously or through the use of inducing

## CMV Latent Infection:

- Myelomonocytic cell.
- Low genome copy number.
- Restricted viral gene expression dependent on differentiation state.

## CMV Productive Infection:

- Differentiated cell types.
- Leads to cell death.
- Virus produced.

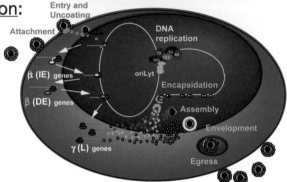

**Figure 22.3** Latent and productive infection by human CMV. (Top) Latent infection is shown, depicting stages of attachment, entry and uncoating in myelomonocytic granulocyte–macrophage progenitors. Entry results in the deposition of the viral genome as a plasmid in the nucleus and is accompanied by restricted, CMV latency-associated transcripts (CLTs). A nucleolar protein is encoded by ORF94 from sense CLTs, and a Golgi-localized protein is encoded by ORF152 (UL124) from antisense CLTs (Kondo et al., 1996; White et al., 2000). A putative DNA replication origin necessary for viral genome maintenance (oriP) is depicted as a replication bubble on the genome plasmid. (Bottom) Productive infection is shown, depicting stages of attachment, entry and uncoating, DNA replication, encapsidation, assembly, envelopment and egress (Mocarski and Courcelle, 2001). Entry into permissive fibroblasts, mesothelial, epithelial, endothelial or differentiated myeloid cells results in the deposition of the viral genome in the nucleus and is accompanied by a cascade of productive gene transcription and translation subdivided into the stages α or immediate-early (IE), β or delayed early (DE), and γ or late (L).

agents or immunosuppression. The HCMV genome is maintained as a circular plasmid in naturally infected peripheral blood myeloid cells (Bolovan-Fritts et al., 1999). Reactivation of HCMV has been tied to cell differentiation state as well as to proinflammatory cytokines and allogeneic stimulation (Hahn et al., 1998; Jarvis and Nelson, 2002; Soderberg-Naucler et al., 1997). In the naturally infected host, HCMV latency is likely to require cell-intrinsic and immune control components that restrict viral replication. Although transmission with blood remains infrequent (Roback, 2002), organ and hematopoietic cell transplantation has a high potential of transferring infection to naïve recipients (Ho, 1990). Natural latency in healthy adults is characterized by the presence of viral DNA (Mendelson et al., 1996; von Laer et al., 1995) and expression of latency-associated transcripts (Hahn et al., 1998; Kondo et al., 1996) in mononuclear cells. Reactivation of HCMV has been recently reported to follow CD34$^+$ progenitor cell differentiation into mature DC (Reeves et al., 2005). Quantitative analysis of natural latency of myelomonocytic cells from either BM or G-CSF-mobilized peripheral blood of seropositive donors (Slobedman and Mocarski, 1999) revealed a consistent percentage of approximately 0.01% CMV DNA-positive cells at a level of 2 to 10 genome equivalents per infected cell. Earlier studies had revealed latency-associated transcripts in myelomononuclear cells from seropositive donors (Kondo et al., 1996), but the frequency of transcript-positive cells is approximately 10-fold lower than DNA-positive cells (Hahn et al., 1998).

Recent attention to stem cells as a possible site of CMV latency (Goodrum et al., 2002; Jarvis and Nelson, 2002), has not been accompanied by any direct evaluation of viral DNA or transcripts in naturally infected stem cell-enriched populations. We analyzed mobilized peripheral blood mononuclear cells from three CMV-seropositive and two CMV-seronegative individuals undergoing hematopoietic stem cell transplantation using high speed fluorescence activated cell sorted cell populations (Vose et al., 2001). CD34$^+$ myelomononuclear cells lacking lineage markers (CD3$^-$ CD14$^-$ CD15$^-$ CD33$^-$) were batch-isolated from G-CSF-mobilized peripheral blood cells and further purified by fluorescence activated cell sorting into CD34$^+$ Thy-1/CD90$^+$ and CD34$^+$ Thy-1/CD90$^-$ populations. Both populations contained CMV DNA-positive cells in 6000-cell aliquots (Table 22.1), and in most but not all aliquots of 2000 or fewer cells (data not shown). Thus, the frequency of CMV DNA-positive cells is in the range of 0.05% in naturally infected stem cell-enriched populations. Lineage committed immature CD33$^+$ CD15$^+$ and more mature CD33$^-$ CD15$^+$ mononuclear cell populations were similarly purified and were also CMV DNA-positive at a similar frequency (Table 22.1). CMV DNA was detected in cells isolated from CMV-seropositive but not CMV-seronegative individuals. Thus, the natural distribution of the viral genome in myeloid cells includes primitive hematopoietic lineage cell types that are commonly referred to as stem cells as well as more highly differentiated cell types. Although CD34$^+$ myelomonocytic cell populations include multipotent and self-renewing stem cells, the level to which these and other cells contribute to long-term carriage of CMV remains unknown. Similarly, the level to which these progenitors remain quiescent or divide in natural, nontransplant settings remains to be established. A current

**Table 22.1** Detection of HCMV DNA and latency-associated transcripts in sorted populations of G-CSF-mobilized peripheral blood mononuclear cells

| Donor | CMV sero[a] | CD34$^+$Thy1$^-$ | CD34$^+$Thy1$^+$ | CD33$^+$CD15$^+$ | CD33$^-$CD15$^+$ |
|---|---|---|---|---|---|
| NHL1 | + | +[b] | + | +[c] | + |
| NHL2 | + | + | + | + | + |
| NHL3 | + | + | + | +[c] | + |
| NHL4 | − | − | − | − | − |
| NHL5 | − | − | − | − | − |

[a]CMV serological status.
[b]CMV-DNA PCR on 6 × 10$^3$ FACS-sorted cells. Nested CMV DNA-PCR was performed using primers IEP4BII/IEP2AII followed by IEP3B/IEP3A as described (Kondo et al., 1994).
[c]CMV sense latency-associated transcripts detected following analysis of multiple pools of 5 × 10$^3$ sorted cells as described (Hahn et al., 1998).

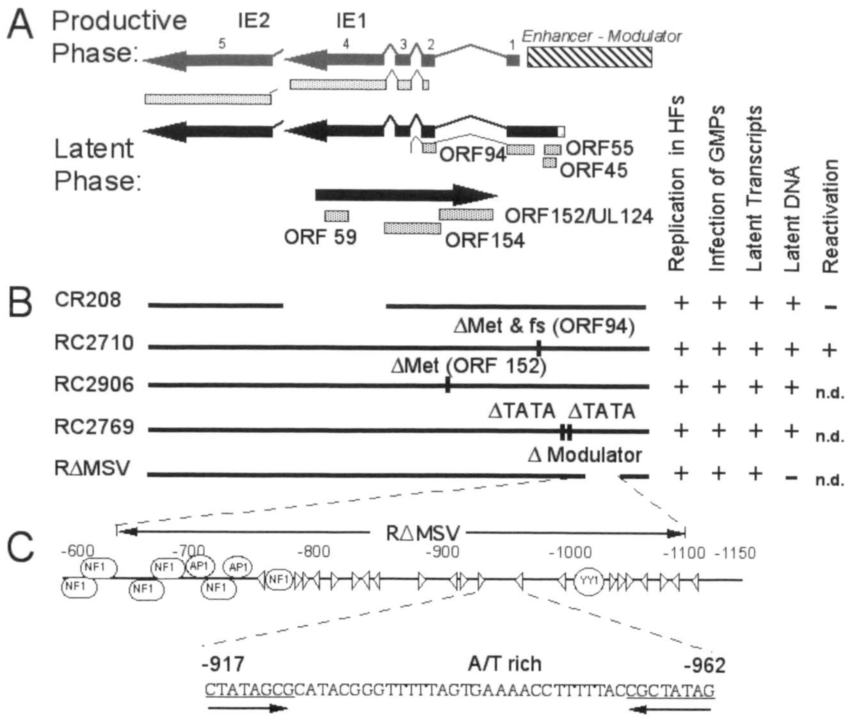

**Figure 22.5** Productive and latency-associated transcripts from the IE1/IE2 region of the human CMV genome. (A) Productive phase: enhancer-modulator region (hatched box) and productive phase transcripts (gray arrows), with splicing patterns indicated. Major productive phase IE proteins (gray stippled bars) shown below arrows. Latent Phase: sense and antisense latency-associated transcripts (black arrows), with splicing patterns indicated. Potential proteins encoded by sense (ORF94, ORF55, ORF45) and antisense (ORF59, ORF154, ORF152/UL124) transcripts are shown as stippled bars. (B) Depiction of IE1/IE2 region mutations in five HCMV (strain Towne) recombinants: exon 4 (IE1) mutant CR208 (Greaves and Mocarski, 1998), ORF94 mutant RC2710 (White et al., 2000), ORF152 mutant RC2906 (Xu and Mocarski, unpublished), sense transcript TATA box mutant RC2769 (White and Mocarski, unpublished) and modulator mutant RΔMSV (Meier and Stinski, 1997). Shown to the right of the mutants is a summary of replication efficiency (in human fibroblasts, HFs), presence of CMV DNA in GMPs, presence of sense latent transcripts in GMPs, stable maintenance of viral DNA during GMP culture (see Figure 22.6) and reactivation upon cocultivation of latent GMPs with permissive HFs (White et al., 2000). (C) Expanded modulator (−600 to −1150 relative to the IE1/IE2 transcription start site) showing the region deleted in RΔMSV (−640 to −1108), with positions of consensus NF1, AP1 and YY1 sites, as well as inverted repeat elements (triangles) indicated. Below, the nucleotide sequence of the A/T-rich region between −917 and −962 is shown bracketed by inverted repeats (arrows).

suppression consistent with transcriptional activity during latency. The IE1/IE2 enhancer-modulator region includes the cis-acting enhancer element important for expression of the major immediate-early genes (Mocarski and Courcelle, 2001), however the function of the modulator region upstream of the enhancer has remained an enigma. In order to evaluate the role of the modulator region, recombinant virus RΔMSV carrying a large deletion that removed the modulator region, was compared to WT Towne strain virus in a series of

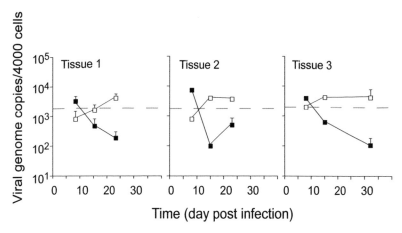

**Figure 22.6** Stable maintenance of human CMV DNA in GMPs requires the modulator region of the IE1/IE2 enhancer-modulator. Real time PCR analysis was used to assay aliquots of 4000 cells infected with RΔMSV (closed squares) or Towne strain (open squares) over a 24 to 33 day culture period using cultures derived from three independent fetal liver-derived GMPs (Tissue 1, Tissue 2, Tissue 3). GMP cultures infected with Towne strain maintain genome copy numbers of 2 to 13 genomes/cell (Slobedman and Mocarski, 1999). Dashed line indicates a level of 1 genome equivalent/cell. Standard deviation of three replicates is indicated by error bars and is within the symbol where not visible.

experiments in human fibroblasts (HFs) and GMPs. RΔMSV exhibits no obvious growth defects in permissive cells (Meier and Stinski, 1997) and establishes a latent infection in GMPs based on the detection of viral DNA and sense CLTs (Figure 22.5). In contrast to Towne strain, infection with RΔMSV results in the gradual loss of viral DNA over 3–5 weeks in culture (Figure 22.6), suggesting that genome maintenance may be affected by the deletion. Given that all of the cells start as viral DNA positive and cell growth is not affected in this experimental system (Kondo et al., 1996), the loss of viral DNA to levels below one genome copy per cell over time suggests that viral DNA is lost from cells that continue to grow during the assay period. When the region that is deleted in RΔMSV is examined for novel nucleotide sequence elements, an A:T-rich DNA element surrounded by inverted repeats in the modulator region is evident (between −917 and −962 relative to the IE1/IE2 transcription start site). While further investigation is necessary, these results open an alternative possibility for the large region upstream of the IE1/IE2 enhancer as a latent origin of DNA replication (oriP) for viral genome maintenance rather than in gene expression. Viral trans-acting functions that might be required for genome maintenance in GMPs have not emerged. Neither ORF94 (White et al., 2000), the only known nuclear protein, nor ORF152 (UL124), a Golgi body-localized protein, are required for any aspect of viral latency that has been tested (Figure 22.5). ORF152 mutant RC2906 carries a 4-bp deletion removing an NcoI site and the start codon, thus eliminating any chance that ORF152 could be translated. ORF94 is potentially encoded from either of two sense CLTs whose putative TATA boxes were predicted to lie an appropriate distance upstream of two transcription start sites (Kondo et al., 1996). Mutations introduced into latent start site (LSS)1 (GTATTTAC > GCGGCCGC) and LSS2 (CCCTATTGA > CGCGGCCGC) created NotI sites and viruses carrying one (RC2767 or RC2768, respectively) or two (RC2769)

NotI sites were isolated. Despite the disruption of these sequences, sense CLT expression was still detected in RC2769-infected GMPs (Figure 22.5). Thus, to date, CMV gene products that may play roles in latency have not emerged from candidate latency-associated gene products.

## Acknowledgments

We are indebted to Hideto Kaneshima, Systemix, Inc., and Karl Blume of the Stanford University School of Medicine Blood and Bone Marrow Transplantation Program for their encouragement. This work has been funded by USPHS grants AI30363 and CA49605.

## References

Bego, M.G., Maciejewski, J.P., Khaiboullina, S., Pari, G., and St Jeor, S. (2005). Characterization of a novel latency cytomegalovirus transcript. J. Virol. (in press)

Biermann, H., Pietz, B., Dreier, R., Schmid, K.W., Sorg, C., and Sunderkotter, C. (1999). Murine leukocytes with ring-shaped nuclei include granulocytes, monocytes, and their precursors. J. Leukoc. Biol. 65, 217–231.

Bolovan-Fritts, C.A., Mocarski, E.S., and Wiedeman, J.A. (1999). Peripheral blood CD14(+) cells from healthy subjects carry a circular conformation of latent cytomegalovirus genome. Blood 93, 394–398.

Boring, L., Gosling, J., Chensue, S.W., Kunkel, S.L., Farese, R.V., Jr., Broxmeyer, H.E., and Charo, I.F. (1997). Impaired monocyte migration and reduced type 1 (Th1) cytokine responses in C-C chemokine receptor 2 knockout mice. J. Clin. Invest. 100, 2552–2561.

Brautigam, A.R., Dutko, F.J., Olding, L.B., and Oldstone, M.B. (1979). Pathogenesis of murine cytomegalovirus infection: the macrophage as a permissive cell for cytomegalovirus infection, replication and latency. J. Gen. Virol. 44, 349–359.

Bruno, L., Seidl, T., and Lanzavecchia, A. (2001). Mouse pre-immunocytes as non-proliferating multipotent precursors of macrophages, interferon-producing cells, CD8alpha(+) and CD8alpha(-) dendritic cells. Eur. J. Immunol. 31, 3403–3412.

Casarosa, P., Gruijthuijsen, Y.K., Michel, D., Beisser, P.S., Holl, J., Fitzsimons, C.P., Verzijl, D., Bruggeman, C.A., Mertens, T., Leurs, R., Vink, C., and Smit, M.J. (2003). Constitutive signaling of the human cytomegalovirus-encoded receptor UL33 differs from that of its rat cytomegalovirus homolog R33 by promiscuous activation of G proteins of the Gq, Gi, and Gs classes. J. Biol. Chem. 278, 50010–50023.

Christensen, J.L., Wright, D.E., Wagers, A.J., and Weissman, I.L. (2004). Circulation and chemotaxis of fetal hematopoietic stem cells. PLoS Biol. 2, 368–377.

Collins, T.M., Quirk, M.R., and Jordan, M.C. (1994). Biphasic viremia and viral gene expression in leukocytes during acute cytomegalovirus infection of mice. J. Virol. 68, 6305–6311.

Diosi, P., Moldovan, E., and Tomescu, N. (1969). Latent cytomegalovirus infection in blood donors. Br. Med. J. 4, 660–662.

Geissmann, F., Jung, S., and Littman, D.R. (2003). Blood monocytes consist of two principal subsets with distinct migratory properties. Immunity 19, 71–82.

Goodrum, F.D., Jordan, C.T., High, K., and Shenk, T. (2002). Human cytomegalovirus gene expression during infection of primary hematopoietic progenitor cells: a model for latency. Proc. Natl. Acad. Sci. USA 99, 16255–16260.

Greaves, R.F., and Mocarski, E.S. (1998). Defective growth correlates with reduced accumulation of a viral DNA replication protein after low multiplicity infection by a human cytomegalovirus ie1 mutant. J. Virol. 72, 366–379.

Hahn, G., Jores, R., and Mocarski, E.S. (1998). Cytomegalovirus remains latent in a common precursor of dendritic and myeloid cells. Proc. Natl. Acad. Sci. USA 95, 3937–3942.

Hanson, L. K., Slater, J. S., Karabekian, Z., Virgin, H. W. t., Biron, C. A., Ruzek, M. C., van Rooijen, N., Ciavarra, R. P., Stenberg, R. M., and Campbell, A. E. (1999). Replication of murine cytomegalovirus in differentiated macrophages as a determinant of viral pathogenesis. J. Virol. 73, 5970–5980.

Hertel, L., Lacaille, V.G., Strobl, H., Mellins, E.D., and Mocarski, E.S. (2003). Susceptibility of immature and mature Langerhans cell-type dendritic cells to infection and immunomodulation by human cytomegalovirus. J. Virol. 77, 7563–7574.

Hertel, L., and Mocarski, E.S. (2004). Global analysis of host cell gene expression late during cytomegalovirus infection reveals extensive dysregulation of cell cycle gene expression and induction of Pseudomitosis independent of US28 function. J. Virol. *78*, 11988–12011.

Ho, M. (1990). Epidemiology of cytomegalovirus infections. Rev. Infect. Dis. *12 Suppl. 7*, 701–710.

Jarvis, M., and Nelson, J. (2002). Human cytomegalovirus persistence and latency in endothelial cells and macrophages. Curr. Opin. Microbiol. *5*, 403–407.

Johnson, B.D., Hanke, C.A., Becker, E.E., and Truitt, R.L. (1998). Sca1(+)/Mac1(+) nitric oxide-producing cells in the spleens of recipients early following bone marrow transplant suppress T-cell responses in vitro. Cell. Immunol. *189*, 149–159.

Johnston, B., and Butcher, E.C. (2002). Chemokines in rapid leukocyte adhesion triggering and migration. Semin. Immunol. *14*, 83–92.

Kaptein, S.J., Beisser, P.S., Gruijthuijsen, Y.K., Savelkouls, K.G., van Cleef, K.W., Beuken, E., Grauls, G.E., Bruggeman, C.A., and Vink, C. (2003). The rat cytomegalovirus R78 G protein-coupled receptor gene is required for production of infectious virus in the spleen. J. Gen. Virol. *84*, 2517–2530.

Kern, F., Bunde, T., Faulhaber, N., Kiecker, F., Khatamzas, E., Rudawski, I.M., Pruss, A., Gratama, J.W., Volkmer-Engert, R., Ewert, R., Reinke P., Volk H.D., Picker L.J. (2002). Cytomegalovirus (CMV) phosphoprotein 65 makes a large contribution to shaping the T-cell repertoire in CMV-exposed individuals. J. Infect. Dis. *185*, 1709–1716.

Koffron, A.J., Hummel, M., Patterson, B.K., Yan, S., Kaufman, D.B., Fryer, J.P., Stuart, F.P., and Abecassis, M.I. (1998). Cellular localization of latent murine cytomegalovirus. J. Virol. *72*, 95–103.

Kondo, K., Kaneshima, H., and Mocarski, E.S. (1994). Human cytomegalovirus latent infection of granulocyte–macrophage progenitors. Proc. Natl. Acad. Sci. USA *91*, 11879–11883.

Kondo, K., Xu, J., and Mocarski, E.S. (1996). Human cytomegalovirus latent gene expression in granulocyte–macrophage progenitors in culture and in seropositive individuals. Proc. Natl. Acad. Sci. USA *93*, 11137–11142.

Kondo, M., Wagers, A.J., Manz, M.G., Prohaska, S.S., Scherer, D.C., Beilhack, G.F., Shizuru, J.A., and Weissman, I.L. (2003). Biology of hematopoietic stem cells and progenitors: implications for clinical application. Annu. Rev. Immunol. *21*, 759–806.

Krmpotic, A., Bubic, I., Polic, B., Lucin, P., and Jonjic, S. (2003). Pathogenesis of murine cytomegalovirus infection. Microbes Infect. *5*, 1263–1277.

Kurihara, T., Warr, G., Loy, J., and Bravo, R. (1997). Defects in macrophage recruitment and host defense in mice lacking the CCR2 chemokine receptor. J. Exp. Med. *186*, 1757–1762.

Ley, K. (2003). The role of selectins in inflammation and disease. Trends Mol. Med. *9*, 263–268.

Lu, B., Rutledge, B.J., Gu, L., Fiorillo, J., Lukacs, N.W., Kunkel, S.L., North, R., Gerard, C., and Rollins, B.J. (1998). Abnormalities in monocyte recruitment and cytokine expression in monocyte chemoattractant protein 1-deficient mice. J. Exp. Med. *187*, 601–608.

Meier, J.L., and Stinski, M.F. (1997). Effect of a modulator deletion on transcription of the human cytomegalovirus major immediate-early genes in infected undifferentiated and differentiated cells. J. Virol. *71*, 1246–1255.

Mendelson, M., Monard, S., Sissons, P., and Sinclair, J. (1996). Detection of endogenous human cytomegalovirus in CD34+ bone marrow progenitors. J. Gen. Virol. *77*, 3099–3102.

Mims, C.A., and Gould, J. (1978). The role of macrophages in mice infected with murine cytomegalovirus. J. Gen. Virol *41*, 143–153.

Mitchell, B.M., Leung, A., and Stevens, J.G. (1996). Murine cytomegalovirus DNA in peripheral blood of latently infected mice is detectable only in monocytes and polymorphonuclear leukocytes. Virology *223*, 198–207.

Mocarski, E.S., Jr., (2002). Immunomodulation by cytomegaloviruses: manipulative strategies beyond evasion. Trends Microbiol. *10*, 332–339.

Mocarski, E.S., Jr., (2004). Immune escape and exploitation strategies of cytomegaloviruses: impact on and imitation of the major histocompatibility system. Cell. Microbiol. *6*, 707–717.

Mocarski, E.S., Jr., and Courcelle, C.T. (2001). Cytomegaloviruses and their replication. In Fields Virology, D.M. Knipe, P.M. Howley, D.E. Griffin, R.A. Lamb, M.A. Martin, B. Roizman, and S.E. Straus, eds. (Philadelphia, Lippincott Williams & Wilkins), pp. 2629–2673.

Myerson, D., Hackman, R.C., Nelson, J.A., Ward, D.C., and McDougall, J.K. (1984). Widespread presence of histologically occult cytomegalovirus. Hum. Pathol. *15*, 430–439.

Noda, S., Aguirre, S.A., BitMansour, A., Sparer, T.E., Brown, J., Huang, J., Bouley, D.M., and Mocarski, E.S. (2005). Cytomegalovirus MCK-2 controls mobilization and recruitment of myeloid progenitor cells to facilitate dissemination. Blood (in press).

Noda, S., Tanaka, K., Sawamura, S., Sasaki, M., Matsumoto, T., Mikami, K., Aiba, Y., Hasegawa, H., Kawabe, N., and Koga, Y. (2001). Role of nitric oxide synthase type 2 in acute infection with murine cytomegalovirus. J. Immunol. 166, 3533–3541.

Pass, R.F. (2001). Cytomegalovirus. In Fields Virology, D.M. Knipe, P.M. Howley, D.E. Griffin, R.A. Lamb, M.A. Martin, B. Roizman, and S.E. Straus, eds. (Philadelphia, Lippincott Williams & Wilkins), pp. 2675–2705.

Penfold, M.E., Dairaghi, D.J., Duke, G.M., Saederup, N., Mocarski, E.S., Kemble, G.W., and Schall, T.J. (1999). Cytomegalovirus encodes a potent alpha chemokine. Proc. Natl. Acad. Sci. USA 96, 9839–9844.

Penfold, M.E., Schmidt, T.L., Dairaghi, D.J., Barry, P.A., and Schall, T.J. (2003). Characterization of the rhesus cytomegalovirus US28 locus. J. Virol. 77, 10404–10413.

Pollock, J.L., Presti, R.M., Paetzold, S., and Virgin, H.W. (1997). Latent murine cytomegalovirus infection in macrophages. Virology 227, 168–179.

Reeves, B., MacAry, P.A., Lehner, P.J., Sissons, J.G., and Sinclair, J. (2005). Latency, chromatin remodelling and reactivation of human cytomegalovirus in the dendritic cells of healthy carriers. Proc. Natl. Acad. Sci. USA, in press.

Roback, J.D. (2002). CMV and blood transfusions. Rev. Med. Virol. 12, 211–219.

Saederup, N., Aguirre, S.A., Sparer, T.E., Bouley, D.M., and Mocarski, E.S. (2001). Murine cytomegalovirus CC chemokine homolog MCK-2 (m131–129) is a determinant of dissemination that increases inflammation at initial sites of infection. J. Virol. 75, 9966–9976.

Saederup, N., Lin, Y.C., Dairaghi, D.J., Schall, T.J., and Mocarski, E.S. (1999). Cytomegalovirus-encoded beta chemokine promotes monocyte-associated viremia in the host. Proc. Natl. Acad. Sci. USA 96, 10881–10886.

Saederup, N., and Mocarski, E.S. (2002). Fatal attraction: Cytomegalovirus-encoded chemokine homologs. Curr. Top. Microbiol. Immunol. 269, 235–256.

Salazar-Mather, T.P., and Hokeness, K.L. (2003). Calling in the troops: regulation of inflammatory cell trafficking through innate cytokine/chemokine networks. Viral. Immunol. 16, 291–306.

Schenkel, A.R., Mamdouh, Z., and Muller, W.A. (2004). Locomotion of monocytes on endothelium is a critical step during extravasation. Nature Immunol. 5, 393–400.

Serafini, P., De Santo, C., Marigo, I., Cingarlini, S., Dolcetti, L., Gallina, G., Zanovello, P., and Bronte, V. (2004). Derangement of immune responses by myeloid suppressor cells. Cancer Immunol. Immunother. 53, 64–72.

Serbina, N.V., Kuziel, W., Flavell, R., Akira, S., Rollins, B., and Pamer, E.G. (2003a). Sequential MyD88-independent and -dependent activation of innate immune responses to intracellular bacterial infection. Immunity 19, 891–901.

Serbina, N.V., Salazar-Mather, T.P., Biron, C.A., Kuziel, W.A., and Pamer, E.G. (2003b). TNF/iNOS-producing dendritic cells mediate innate immune defense against bacterial infection. Immunity 19, 59–70.

Sissons, J.G., Bain, M., and Wills, M.R. (2002). Latency and reactivation of human cytomegalovirus. J. Infect. 44, 73–77.

Slobedman, B., and Mocarski, E. S. (1999). Quantitative analysis of latent human cytomegalovirus. J. Virol. 73, 4806–4812.

Slobedman, B., Mocarski, E.S., Arvin, A.M., Mellins, E.D., and Abendroth, A. (2002). Latent cytomegalovirus down-regulates major histocompatibility complex class II expression on myeloid progenitors. Blood 100, 2867–2873.

Slobedman, B., Stern, J.L., Cunningham, A.L., Abendroth, A., Abate, D.A., and Mocarski, E.S. (2004). Impact of human cytomegalovirus latent infection on myeloid progenitor cell gene expression. J. Virol. 78, 4054–4062.

Soderberg-Naucler, C., Fish, K.N., and Nelson, J.A. (1997). Reactivation of latent cytomegalovirus by allogeneic stimulation of blood cells from healthy donors. Cell 91, 119–126.

Sparer, T.E., Gosling, J., Schall, T.J., and Mocarski, E.S. (2004). Expression of human CXCR2 in murine neutrophils as a model for assessing cytomegalovirus chemokine vCXCL-1 function in vivo. J. Interferon Cytokine Res. 24, 611–620.

Stevens, J.G. (1978). Latent characteristics of selected herpesviruses. Adv. Cancer Res. 26, 227–256.

Stoddart, C.A., Cardin, R.D., Boname, J.M., Manning, W.C., Abenes, G.B., and Mocarski, E.S. (1994). Peripheral blood mononuclear phagocytes mediate dissemination of murine cytomegalovirus. J. Virol. 68, 6243–6253.

Streblow, D.N., and Nelson, J. A. (2003). Models of HCMV latency and reactivation. Trends Microbiol. 11, 293–295.

von Laer, D., Meyer-Koenig, U., Serr, A., Finke, J., Kanz, L., Fauser, A.A., Neumann-Haefelin, D., Brugger, W., and Hufert, F.T. (1995). Detection of cytomegalovirus DNA in CD34$^+$ cells from blood and bone marrow. Blood 86, 4086–4090.

Vose, J.M., Bierman, P.J., Lynch, J.C., Atkinson, K., Juttner, C., Hanania, C.E., Bociek, G., and Armitage, J.O. (2001). Transplantation of highly purified CD34$^+$Thy-1+ hematopoietic stem cells in patients with recurrent indolent non-Hodgkin's lymphoma. Biol. Blood Marrow Transplant. 7, 680–687.

Wagers, A.J., and Weissman, I.L. (2004). Plasticity of adult stem cells. Cell 116, 639–648.

White, K.L., Slobedman, B., and Mocarski, E.S. (2000). Human cytomegalovirus latency-associated protein pORF94 is dispensable for productive and latent infection. J. Virol. 74, 9333–9337.

and gene activating signaling by inflammatory cytokines that are associated with a graft-versus-host (GvH) reaction. Regardless of the source of reactivating virus, donor or host or both, CMV recurrence can lead to graft failure and multiple organ disease, with the highest fatality rate being associated with interstitial pneumonia. Obviously, complications caused by CMV reactivation have a major health impact and can severely limit the success of therapy by transplantation.

Given this medical background, it is of paramount importance to understand the mechanisms that lead to virus latency, maintain latency, and cause reactivation from latency (Goodrum et al., 2002; for a research focus commentary, see Streblow and Nelson, 2003). Although latency and reactivation have been investigated since decades, there is probably no other issue in CMV research that has been on less firm ground and has been burdened with dogma for so long time. Only recently, in vitro models of HCMV latency and in vivo models of MCMV latency have allowed investigation into the molecular details, though we are still far from a generally accepted unifying concept. There is firm evidence to conclude that HCMV becomes latent in hematopoietic progenitor cells of the myeloid lineage, and it emerges that chromatin remodeling in the course of lineage differentiation as well as upon cell activation through cytokines is the key to virus reactivation (for HCMV latency, see Chapter 9 and Chapter 22). What appears to be less clear is whether "hematopoietic CMV latency" is the only type of latency that exists. Hematopoietic cells, as compared with organ biopsy specimens, are easier to obtain in humans, and this, not meant as a criticism, has directed the focus to latency in cells of the hematopoietic cell lineages. A hint to the existence of a non-hematopoietic second site (or even more sites) of latency is given by HCMV transmission rates in transplantations: while latent HCMV is infrequently transmitted from a seropositive (i.e. latently infected) donor to a seronegative recipient in hematopoietic transplantations, transmission is frequent in solid organ transplantations such as in liver, kidney, heart, lung or heart-lung transplantations (Emery, 1998; Goodrich et al., 1994). This is in accordance with murine models that have shown a lifelong maintenance of a high load of latent virus genomes in many organs, including liver, kidney, heart, lungs, spleen, adrenal glands, and salivary glands, after clearance of the viral DNA from hematopoietic and vascular compartments (Balthesen et al., 1993; Collins et al., 1993; Koffron et al., 1998; Kurz et al., 1997; 1999; Reddehase et al., 1994; and C.K.S., unpublished data). Thus, in agreement with early data on a stromal site of MCMV latency in the spleen (Mercer et al., 1988; Pomeroy et al., 1991), it is now established knowledge that stromal and/or parenchymal cellular sites of MCMV latency exist in most organs. We do not believe that this is a fundamental difference between human CMV latency and the mouse model; rather, although the message given by the mouse model is loud and clear, stromal/parenchymal latency is a largely unattended issue in human CMV latency research.

It is not the intention of this chapter to give a comprehensive and critical literature review of data obtained in the many different settings of mouse latency models. The "classical era" has been reviewed by Jordan (1983) and recent updates were given by Hummel and Abecassis (2002) and by our own group (Reddehase et al., 2002). We focus here on data obtained in a BALB/c mouse model of CMV latency and reactivation in the lungs, a key manifestation site of CMV disease. Specifically, we offer a hypothesis to explain latency and reactivation.

## The conservative and the provocative definition of latency

The classical definition of latency demands resolution of productive infection on the organismal level. Thus, as long as virus continues to replicate at a single site, latency is not established in the host. As the salivary glands, specifically the glandular epithelial cells, are a recognized site of a prolonged – so-called "persistent" – productive infection, cessation of virus replication in the salivary glands (Henson and Strano, 1972) can define the time point at which classical latency is established in the host. Though this definition makes medical/epidemiological sense in that it marks the end of the risk of virus transmission through saliva, it is not very helpful for the molecular understanding of latency, and insisting on this definition is dogmatic. The glandular epithelial cell is a highly specialized secretory cell type in which virion morphogenesis is different than in other cell types. Virus productivity is very high, and enormous numbers of virions become sequestered in huge vacuoles for secretion into the salivary duct using the cell's physiological secretory pathway (Jonjic et al., 1989). In addition, the glandular epithelial cell appears to be immunoprivileged in that virus production continues in the presence of a fulminant innate and adaptive immune response mediated by tissue infiltrating leukocytes (Cavanaugh et al., 2003). Unlike the situation in other organs, CD8 T-cells alone fail to resolve the infection of the salivary glands (Jonjic et al., 1989; Lucin et al., 1992). Long before virus replication discontinues in the glandular epithelial cells, productive infection is resolved everywhere else. So, there is no rationale to claim that latency is not established in the lungs or other organs if it is not yet established in the glandular epithelial cells. This becomes even more obvious in the reverse case, namely upon virus reactivation. As shown in a model of latency established after neonatal infection, virus recurrence induced by hematoablative γ-irradiation is a stochastic event that occurs independently in different organs (Reddehase et al., 1994). Specifically, out of 30 mice tested in this report, seven reactivated virus in the salivary glands but not in the lungs, six reactivated virus in the lungs but not in the salivary glands, and four reactivated virus in both organs simultaneously. This result was in accordance with the null hypothesis of independent distribution (Fisher's 2-by-2 contingency table test, $P = 0.55 > 0.05$; for the principle, see Figure 23.3). Thus, both organs are independent sites of latency where, accordingly, reactivation can occur also independently. In conclusion, latency in the lungs (or any other organ) and persistent productive infection in the salivary glands can coexist in an individual.

Notably, this fundamental principle of independent distribution proved to be applicable also when downscaled to latency within a particular organ, within the lungs in the specific case documented (Kurz and Reddehase, 1999). In this report, reactivation was induced by hematoablative γ irradiation, and virus recurrence was analyzed for individual pieces of the lungs that prior to the recurrence all carried a similar load of viral DNA. The result was a random pattern of often directly neighboring virus-positive and virus-negative lung pieces, which documented that sites of virus reactivation and sites of latency can coexist within the same tissue. We can now perform a thought experiment (Brooks, 1994) and downscale the principle to individual cells within a lung piece in which virus reactivation was observed. As Poisson distribution analysis gave an estimate of only one recurrence event per ~3 million lung cells and per ~ 6,000 copies of latent viral DNA (Kurz and Reddehase, 1999), a cell in which virus reactivated was likely just one amongst a majority of cells in which latency was maintained. If we go on with our thought experiment and downscale the principle of independent distribution to a single cell in which virus has reactivated we might even find a desilenced viral genome amongst still silenced "latent" viral genomes. Thus, admittedly a bit provocative, reactivation and latency on the level of viral genomes may coexist in the same cell nucleus. We were inspired with this idea by the "choreography of genome silencing and desilencing events" (see Chapter 7) that occur during acute infection of a cell and that decide upon lytic progression or latency of the genomes that enter the host cell's nucleus. Interestingly, as documented for a set of temperature-sensitive mutants, viral genomes that are unable to replicate under the non-permissive in vivo conditions due to mutations in essential genes are nevertheless maintained in the nonproductively infected cell in a silenced state and can even reactivate gene expression until the transcription stops by reaching the mutated gene (Bevan et al., 1996).

## Model of pulmonary CMV latency after bone marrow transplantation

The in vivo model of MCMV infection under the conditions of syngeneic BMT performed with BALB/c mice as bone marrow cell donors and recipients was used to establish a "high-load viral latency" in the lungs of the convalescent recipients (Figure 23.1, corresponding to Figure 19.3). The acute phase of pulmonary infection and immune control is explained in depth in Chapter 19. Briefly, after the hematoablative conditioning, repopulation of the recipients' bone marrow by transplanted hematopoietic stem- and progenitor cells initiates the reconstitution of CD8 T-cells that are primed during the ongoing infection and are recruited to the lungs, where they form focal infiltrates confining and finally resolving the productive infection. The antiviral CD8 T-cells in the focal infiltrates are activated $CD62L^{low}$ cytolytic effector cells (CTL) capable of lysing infected cells in all three temporal stages of viral gene expression referred to as immediate-early (IE), early (E) and late (L) phases (Holtappels et al., 1998; Podlech et al., 2000). Though infection and CD8 T-cell infiltration occur almost contemporaneously with a peak at ~ 3 weeks (Podlech et al., 2000), CTL activity peaks at about one week later, coincident with the decline in virus titers (Holtappels et al, 1998). After resolution of the productive lung infection, which is usually completed after ~ 6 weeks, CD8 T-cells persisted in the lungs for the lifespan of the recipients with only a slow decline in absolute numbers, but the focal character of the acute infiltrates was then replaced by a disseminated interstitial infiltrate (Podlech et al., 2000). Notably, CD8 T-cells specific for the immunodominant $L^d$-restricted IE1 epitope 168-YPHFMPTNL-176 were found to accumulate in latently infected lungs and to maintain an activated state as indicated by a $CD62L^{low}$ effector memory (CD8-$T_{EM}$) phenotype, suggesting frequent restimulation by presentation of the IE1 epitope and a role as "guardians of latency" (Holtappels et al., 2000). Thus, the immunological scenario that underlies the molecular characterization of latency in this particular model predicts latency-associated transcription from the major immediate-early (MIE) locus.

To facilitate a statistical approach, the lungs (composed of superior lobe, middle lobe, inferior lobe, postcaval lobe and left lung) were subdivided into 18 equally sized pieces numbered 1–18, and latent viral load as well as infectious virus were measured for each piece. Using an improved infectivity assay capable of detecting every infectious particle (Kurz et al., 1997), all 18 lung pieces proved to be free of infectious virus, whereas all pieces contained viral DNA with only little differences in quantity (Kurz et al., 1999). Between

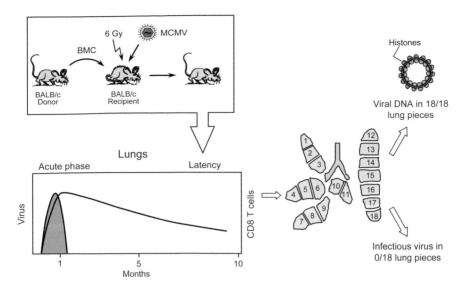

**Figure 23.1** Basic protocol of the establishment of latency in the BALB/c model of bone marrow transplantation and MCMV infection (see also Figure 19.3). For detection of latent viral genomes, viral transcripts, and infectious virions, lungs were subdivided into 18 tissue pieces to facilitate statistical analysis. Latent viral DNA is symbolized as a circular plasmid associated with histones in a postulated chromatin-like structure (see Chapter 9). BMC: bone marrow cells.

different transplantation experiments, the latent viral DNA load varied between ~ 2,000 and 9,000 copies per one million lung cells (Grzimek et al., 2001; Kurz et al., 1999; Simon et al., 2005). This variance likely reflects the efficacy in the reconstitution of antiviral CD8 T-cells resolving the productive infection. In accordance with this explanation, a combination of BMT with CD8 T-cell immunotherapy (see Chapter 19) revealed a direct relation between protective CD8 T-cells and decreased viral load in latency. Notably, the load could not be reduced to below ~ 1000 copies per one million lung cells, suggesting the existence of two types of latency characterized by differential susceptibility to control by CD8 T-cells (Steffens et al., 1998). Importantly, as documented in the same report, the load determines the risk of virus recurrence.

## Variegated expression of MIE locus genes during latency

The accumulation of IE1 epitope-specific CD8-$T_{EM}$ cells in latently infected lungs has prompted us to investigate the expression of MIE genes. The MIE locus of MCMV (Figure 23.2) is composed of a strong enhancer (Dorsch-Häsler et al., 1985; see Chapter 8 for CMV enhancers in general and Figure 8.1 for the mechanics of the MCMV enhancer), which governs the expression of flanking transcription units *ie1/3* and *ie2* that are transcribed in opposite direction. The MIE enhancer of MCMV is essential for virus growth (Ghazal et al., 2003), but, interestingly, can be replaced in enhancer-swap mutants of MCMV by the MIE enhancer of HCMV for virus growth in cell culture (Angulo et al., 1998) and in basically all tissues in vivo (Grzimek et al., 1999). Transcription unit *ie1/3* is driven by promoter $P^{1/3}$ and consists of five exons which, upon differential splicing, give rise

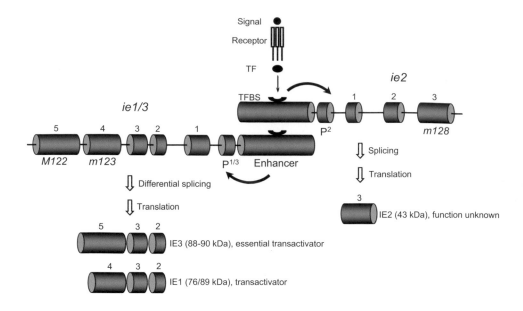

**Figure 23.2** MIE region of MCMV. Shown are the exon–intron structure of MIE genes and the resulting MIE proteins. TF, transcription factor(s). TFBS, transcription factor binding site(s) of the enhancer (see Chapter 8). P, core promoter(s).

to IE1 mRNA (exons 1/untranslated, 2, 3, and 4) and IE3 mRNA (exons 1/untranslated, 2, 3, and 5) (Keil et al., 1984; 1987a and b; Messerle et al., 1992). In acute infection, both mRNAs appear almost simultaneously within minutes, with the IE1 mRNA reaching an ~ 10-fold higher peak amount at ~ 60 minutes. Both mRNAs are fairly stable, with IE1 mRNA and IE3 mRNA declining by a factor of 10 and 100, respectively, within 24 hours (C.O.S., unpublished data). The resulting IE1 protein pp76/89 kDa (Keil et al., 1985) is a co-transactivator of downstream viral E genes and a transactivator of cellular genes of the nucleotide metabolism in quiescent cells (Gribaudo et al., 2000; Lembo et al., 2000). As shown in a recent study by Ghazal and colleagues (2005), a mutant MCMV in which *ie1* exon 4 is deleted replicates like MCMV-WT in different cultured cell types, indicating that the IE1 protein is dispensable for virus replication. However, growth of the IE1 deletion mutant was found to be severely attenuated in vivo, suggesting that the property of IE1 pp76/89 to activate nucleotide metabolism in quiescent cells in host tissues is a major pathogenicity determinant. In addition, as discussed above, the IE1 protein is the source of the IE1 epitope that is immunodominant in the H-2$^d$ haplotype (see Chapter 19). The 88–90 kDa IE3 protein, like its HCMV homolog IE2 (see Chapter 8), is the essential transactivator of viral E gene expression (Angulo et al., 2000; Messerle et al., 1992). Transcription unit *ie2*, for which no counterpart exists in HCMV, is driven by promoter P$^2$ and comprises three exons. Coding exon 3 specifies the 43 kDa IE2 protein (Messerle et al., 1991) for which no function could be defined to date (Cardin et al., 1995).

Using the strategy of piece-by-piece analysis (see above), IE1 transcripts were indeed detected in latently infected lungs. However, a variegated "on-or-off" pattern was found with lung tissue pieces positive or negative for IE1 transcripts, respectively, indicating focal events of transcriptional activity at the MIE locus during latency (Kurz et al., 1999). Detection of IE1 transcripts had occasionally been reported in earlier literature (discussed quite comprehensively by Bevan et al., 1996), but at that time the interpretation of this finding was controversial because IE1 transcription was generally regarded as a molecular marker for the productive viral cycle. For instance, Yuhasz et al. (1994) detected IE1 transcripts in the lungs but not in the spleen of latently infected mice. They took this finding as an argument against molecular latency in the lungs; although they failed to detect infectious virus, they proposed the persistence of a low-level productive infection specifically in the lungs. By contrast, Bevan et al. (1996) detected IE1 transcripts in the salivary glands, but not in the lungs. As these authors did not detect E phase transcripts E1 and M55/gB, they correctly argued that IE1 gene expression does not necessarily indicate persistent infection or transient reactivation. With optimized RT-PCRs specific for spliced IE1 and IE3 mRNAs and controlled for sensitivity with synthetic transcripts, Kurz and colleagues (1999) resolved the controversy by demonstrating the absence of IE3 mRNA in lung tissue pieces that contained IE1 mRNA. As the IE3 protein is essential as a transactivator of downstream E gene expression, the productive viral cycle was not started, and this explained the maintenance of latency despite MIE locus transcriptional activity. These findings thus showed that there exists a second molecular checkpoint of latency beyond $P^{1/3}$ core promoter activity and *ie1/3* transcription initiation, most likely involving the regulation of differential splicing. As shown recently, this checkpoint is surmounted after enhancement of MIE locus transcription through TNF-$\alpha$ (Simon et al., 2005) that signals to CMV enhancers by transcription factors NF-$\kappa$B and AP1 (Hummel et al., 2001; Prösch et al., 1995).

As the MIE enhancer governs also transcription unit *ie2* of the MIE locus (Figure 23.2), it was predicted that lung pieces in which the presence of IE1 indicated MIE locus activity simultaneously contain also IE2 transcripts. Surprisingly, though IE2 was indeed found to be expressed in latently infected lungs, the conjecture of a synchronous expression of the two enhancer flanking genes was not confirmed (Grzimek et al., 2001). The key result of this work is illustrated in Figure 23.3 for latently infected mouse (LIM) #1, a mouse that turned out to be representative of the group of mice analyzed by Grzimek and colleagues. RT-PCRs performed with poly(A)$^+$ RNA isolated from individual lung tissue pieces gave variegated "on-or-off" patterns of transcription for IE1 and IE2 transcripts. Yet, instead of the expected synchronized "on–on" and "off–off" events, the pattern comparison revealed also "on–off" and "off–on" events for IE1 and IE2 transcription, respectively. For statistical analysis, experimental data were arranged in a 2-by-2 contingency table, and the probability for independent distribution (null hypothesis) was calculated by using Fisher's exact probability test. Notably, the experimentally observed pattern matched precisely the pattern expected for independent transcription of IE1 and IE2. Detection of MIE transcripts in a lung tissue piece implies that the MIE locus was transcriptionally active in at least one latent viral genome present in a latently infected cell within a tissue piece consisting of ~ 3 million cells; however, the detected transcripts may be derived also from more than one latent genome present in one or in more than one latently infected cell. The Poisson distribution analysis is the statistical tool that allows, based on the proportion of transcriptionally silent tissue pieces, an estimation of the number of tissue pieces that contained one or two, or more, latent genomes from which IE1 or IE2 transcription occurred (see Appendix in Grzimek et al., 2001). The situation in the lungs of mouse LIM#1 is summarized in the

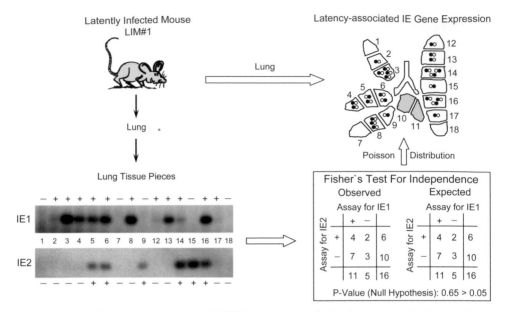

**Figure 23.3** Variegated expression of MIE locus genes during latency in the lungs. For a representative, latently infected mouse LIM#1, original data are reproduced in a rearranged form from the Journal of Virology (Grzimek et al., 2001). Shown are Southern blots of IE1 and IE2 RT PCR amplificates hybridized with the respective specific oligonucleotide probes. Statistical pattern analysis was performed by arranging the data in a 2-by-2 contingency table and applying Fisher's exact probability test. Independent distribution was confirmed by data from five mice LIM#1 through LIM#5 (P-value: 0.37 > 0.05). The topographical lung map shows the distribution of MIE locus transcription in the tissue pieces calculated by Poisson distribution analysis based on the fraction of negative pieces (see Appendix in Grzimek et al., 2001). Pieces 10 and 11 were used to determine the latent viral DNA load and did therefore not enter the statistical analysis. Closed and open circles represent presence and absence of transcripts, respectively. Left circle, IE1. Right circle, IE2.

topographical lung map shown in Figure 23.3. In essence, the message is that detection of IE1 and IE2 transcripts in the same tissue piece does not imply simultaneous transcription from a same latent viral genome but can be explained by the coincidence of transcriptionally active viral genomes of which at least one expressed IE1 and at least a second one expressed IE2. The data also revealed an imbalance of 2:1 between IE1 and IE2 transcriptional events. For reasons discussed above, only transcription from the *ie1/3* transcription unit bears the chance to proceed to virus recurrence.

During acute infection in cell culture, "statistical ensembles" of many viral genomes are studied, so that it is impossible to decide whether the enhancer-flanking MIE locus genes are transcribed simultaneously or whether the enhancer serves as a "flip-flop" switch between both as suggested in Figure 23.2. In latency, the frequency of MIE locus transcription is extremely low, so that one can observe transcription from literally single genomes. MIE locus transcription during latency supports a "flip-flop" model of enhancer action. However, it is a pending question whether the enhancer is at all involved in MIE locus transcription during latency. An alternative explanation not yet ruled out is that MIE locus transcription during latency reflects basal activities of the core promoters $P^{1/3}$ and $P^2$. A second pending question is why the transcription pulse is 2:1, as a comparison of $P^{1/3}$-enhancer and enhancer-$P^2$ strengths in a reporter gene assay did not reveal a higher activity of the $P^{1/3}$-enhancer construct. Finally, it is open to question which purpose IE2 is serving.

CMV "latency-associated" alternative sense and antisense MIE region transcripts (CLTs) as found in HCMV latency (see Chapter 22) were so far not described for MCMV, but this may relate in part to focus of investigation. So far, no function could be ascribed to MIE locus CLTs of HCMV, whereas IE1 in MCMV latency may indicate a first step in transcriptional reactivation. It is a matter of definition whether the IE1 and IE2 transcripts observed in latently infected lungs should be regarded as CLTs. We tend to say "no", as the vast majority of latent viral genomes is silenced at the MIE locus. To give an idea of orders of magnitude, data from Simon et al. (2005) based on a latent viral DNA load of ~ 2,000 copies per one million lung cells are summarized in Figure 23.4A. According to that, only ~ 20 viral genomes out of one million latent viral genomes expressed IE1 in a snapshot analysis of the transcriptional activity in latently infected lungs. As explained above, absence of IE3 as well as of downstream E gene transcripts verified the latent state.

## Role of signaling to the MIE enhancer in transcriptional reactivation

The MIE enhancer is a genetic switch to turn on the expression of MIE locus genes and kick-start the productive cycle. Accordingly, the enhancer is a prime checkpoint in viral latency, and signaling to the enhancer can be expected to enhance the frequency of tran-

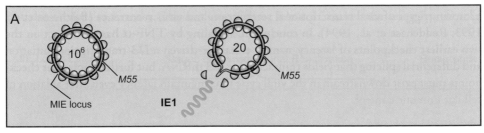

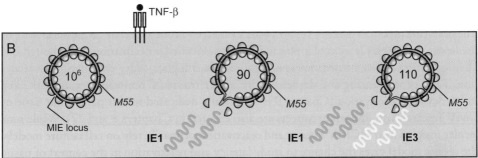

**Figure 23.4** Compilation of MIE locus transcription frequencies in latently infected lungs. (A) Basal expression. (B) Expression after induction with TNF-α. Transcription frequencies refer to 1 million viral genomes. The latent viral genome is symbolized as a circular plasmid in a chromatin-like higher order structure. Silenced and desilenced MIE locus is indicated by closed and open chromatin structure, respectively. Transcripts are symbolized by wavy lines. Numbers indicate transcription frequencies calculated on the basis of original data shown in the Journal of Virology (Simon et al., 2005).

lung pieces during reactivation revealed transcriptional events in three stages of reactivation: (stage I) IE1 transcripts, and likely also IE2 transcripts; (stage II) IE1 plus IE3 transcripts; and (stage III) IE1 plus IE3 plus M55/gB transcripts (Figure 23.5). So, obviously, stage III represented viral genomes with an open chromatin structure at the MIE locus and the M55/gB locus. Yet, viral genomes desilenced at these two loci but maintaining a closed chromatin structure at any other essential gene cannot complete the productive cycle. If one would test for not just three marker transcripts but for all transcripts in a global approach, many more stages of reactivation may be defined, but regardless of how many, full reactivation requires an open chromatin structure at all essential loci. As the expression of true late genes requires prior viral DNA replication by definition, one must propose that viral DNA replication and thus amplification precedes nucleocapsid assembly (Chapter 12) and virion morphogenesis (Chapter 13) in virus recurrence (Figure 23.5).

## Evidence for immune control of latency: the immune sensing hypothesis

Is there any need to propose a role for the immune system in the control of latency? The data that led to the silencing/desilencing hypothesis discussed above (Figure 23.5) were obtained after treatment by total-body γ irradiation that wipes out cellular immunity, and they were perfectly explained by viral chromatin remodeling. Clearly, the observed stages of transcriptional reactivation are defined by molecular events of gene regulation within latently infected cells. On the other hand, one cannot deny that loss of immune control is the common denominator in all the various experimental protocols for inducing virus recurrence (for a review see Reddehase et al., 2002). To give here the most instructive example, Polic and colleagues (1998) selectively depleted lymphocyte subsets in latently infected B-cell deficient $\mu^-/\mu^-$ mice and showed a contribution of CD8 and CD4 T-cells as well as of NK cells to the control of recurrent infection. As recurrence occurs also in the presence of neutralizing antibodies, humoral immunity is most likely not involved in the molecular control of intracellular virus reactivation and of virion recurrence. Yet, it has of course an impact on pathogenesis caused by recurrent virus in that antiviral antibodies limit virus dissemination from the recurrently infected cell to neighboring cells and thus suppress virus multiplication in subsequent rounds of productive infection (Jonjic et al., 1994; Reddehase et al., 1994). Likewise, if the experimental read-out parameter is the amount of recurrent virus, as it was the case in most previous studies, it is also difficult to distinguish between control of latently infected cells and control of recurrent virus multiplication by the effector cells of cellular immunity. Yet, T-cells and NK cells do have the potential to monitor stages of transcriptional reactivation in a latently infected cell if reactivated viral gene expression leads to a display at the cell surface. Specifically, CD8 T-cells may recognize reactivation-associated viral gene expression as soon as an antigenic peptide is presented by an MHC class I molecule (Chapter 19) and NK cells can screen for a down-modulation of MHC class I molecules (Chapter 15). On the other hand, reactivated expression of viral gene products that interfere with the MHC class I pathway of antigen presentation can promote virus recurrence by inhibition of CD8 T-cell effector function (Chapter 19). Similarly, reactivated viral gene products can inhibit NK cell recognition by down-modulating ligands of activating NK cell receptors and/or by up-regulating ligands of inhibitory NK cell receptors (Chapter 15).

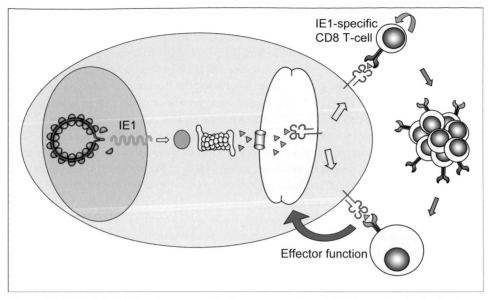

**Figure 23.6** Model of IE1 epitope-specific immune sensing of MIEP reactivation in the lungs of BALB/c mice. It is proposed that MIE locus activity during latency leads to the synthesis of the IE1 protein pp76/89 that is processed by the proteasome. Presentation of the IE1 peptide (green triangles) by MHC class I molecule $L^d$ at the cell surface (see Chapter 19) is recognized by lung-resident IE1 epitope-specific memory CD8 T-cells, which expand clonally, acquire an activated CD62L$^{low}$ CD8-$T_{EM}$ phenotype and develop into antiviral effector cells (based on data by Holtappels et. al., 2000).

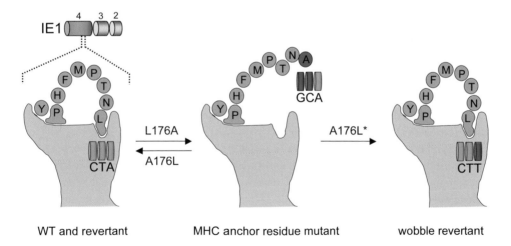

WT and revertant          MHC anchor residue mutant          wobble revertant

**Figure 23.7** Point mutagenesis of the C-terminal MHC anchor residue of the IE1 peptide. BAC mutagenesis of the viral genome (Chapter 4) was used to generate a recombinant mutant virus mCMV-IE1-L176A, the revertant virus mCMV-IE1-A176L, and a genetically marked phenotypic revertant virus mCMV-IE1-A176L* that carries a nucleotide point mutation in the wobble position of the Leu codon. "Rammensee-style" cartoons illustrate the IE1 peptide looping out from the binding groove of the presenting MHC class I molecule. The codons corresponding to the C-terminal amino acid residues are indicated with mutated nucleotides marked in red.

A combination of facts prompted an investigation into the role of IE1 epitope-specific CD8 T-cells in the control of latency in the lungs of BALB/c mice: (1) the IE1 gene is expressed during latency (Kurz et al., 1999); (2) processing of the IE1 protein yields the immunodominant IE1 peptide presented by the MHC class I molecule $L^d$ (Chapter 19); and (3) IE1 epitope-specific CD8 T-cells accumulate in the latently infected lungs and maintain an activated $CD62L^{low}$ CD8-$T_{EM}$ phenotype that suggests recent restimulation during latency (Holtappels et al., 2000). Though the frequency of transcription at the MIE locus in latently infected lungs is fairly low when viewed in a snapshot (Figures 23.3 and 23.4), expression at any single moment over a period of months can explain the maintenance of a pool of CD8-$T_{EM}$ cells in latently infected lungs (Figure 23.6). We are currently approaching this problem by studying MIE locus transcription in the lungs of BALB/c mice that are latently infected with a virus mutant mCMV-IE1-L176A in which the IE1 epitope is deleted by a point mutation Leu → Ala at the C-terminal MHC anchor residue (Figure 23.7), a mutation that prevents both cleavage after position 176 by the proteasome and efficient MHC binding. In accordance with the model proposed in Figure 23.6, the first results are promising to indicate that deletion of the IE1 epitope, and thus of IE1 epitope-specific CD8 T-cells, increases the frequency of latently infected cells that express IE1.

## Concluding hypothesis

Control of latency and reactivation by chromatin remodeling and sensing of reactivation stages by cells of the immune system, of CD8 T-cells in particular, are unified in the silencing/desilencing and immune sensing hypothesis (Figure 23.8). It is proposed that gene desilencing by viral chromatin opening is the prime event that leads to molecular reactivation of the viral transcriptional program. For virus recurrence to take place, local desilencing at the MIE locus, though essential to launch reactivation, is not sufficient. Instead, all essential genes must acquire an open viral chromatin structure, but not necessarily simultaneously. One can also envision a model of dynamic opening and closing that nevertheless fulfils the condition to provide the transcripts that encode the essential proteins for being present at the time when they are needed to move on in the productive cycle. The immune system senses reactivation-associated alterations at the cell surface to eliminate the reactivating cells or terminate reactivation by inhibitory lymphokines. The state at which CD8 T-cells interfere is dictated by the repertoire of viral epitopes presented in a particular MHC haplotype (see Chapter 19). Specifically, in BALB/c mice (H-$2^d$ haplotype), latently infected cells can present the immunodominant IE1 epitope by the MHC class I molecule $L^d$ and become visible for CD8 T-cells at the earliest stage of viral transcriptional reactivation. Accordingly, CD8 T-cells specific for E phase epitopes may check for reactivation stages that have evaded control by IE1 epitope-specific CD8 T-cells. As far as currently known, C57BL/6 (H-$2^b$ haplotype) mice do not present an epitope derived from the IE1 protein, as it is the case also for the BALB/c $L^d$ gene deletion mutant BALB/c-H-$2^{dm2}$. Likewise, the IE1 peptide is not presented in BALB/c mice infected with the IE1 peptide mutant mCMV-IE1-L176A. For these cases the model predicts immune sensing beyond the stage of IE1 expression. Generally, the hypothesis predicts that immunological checkpoints of latency and reactivation vary individually based on MHC immunogenetics. In conclusion, virus recurrence can take place only if both levels of control, gene silencing and immune sensing, are "out of order".

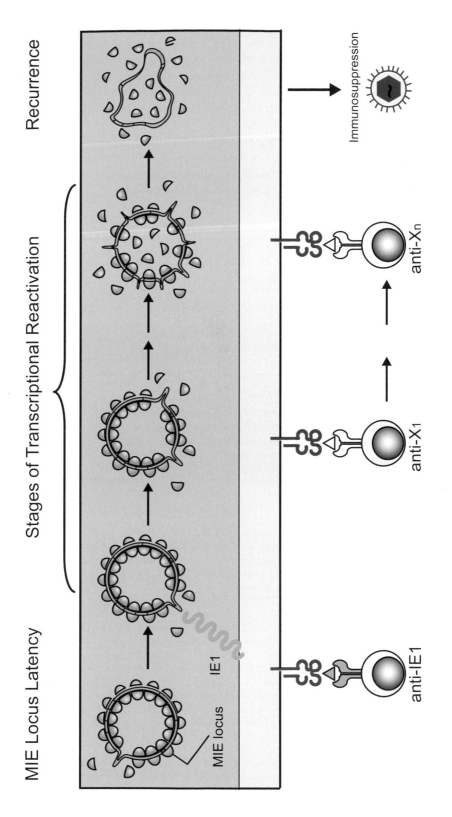

**Figure 23.8** Unifying hypothesis of chromatin remodeling and immune sensing in cytomegalovirus reactivation.

## Acknowledgments

The authors apologize to all colleagues whose work is not represented owing to the focus of this chapter or is not separately cited here because it is covered by cited recent review articles and other chapters of this book. We thank all current and previous members of the lab who have contributed to the experimental work. Critical reading of the manuscript by Rafaela Holtappels is greatly appreciated.

Current work of the authors is supported by the Deutsche Forschungsgemeinschaft, Sonderforschungsbereich (Collaborative Research Grant) 490 "Invasion and Persistence in Infections", individual project B1 "Immune Control of Cytomegalovirus Infection". This chapter includes the results of the funding period 2000–2004.

## References

Angulo, A., Ghazal, P., and Messerle, M. (2000). The major immediate-early gene ie3 of mouse cytomega-lovirus is essential for viral growth. J. Virol. 74, 11129–11136.

Angulo, A., Messerle, M., Koszinowski, U.H., and Ghazal, P. (1998). Enhancer requirement for murine cytomegalovirus growth and genetic complementation by the human cytomegalovirus enhancer. J. Virol. 72, 8502–8509.

Aggarwal, B.B. (2003). Signalling pathways of the TNF superfamily: a double-edged sword. Nature Rev. Immunol. 3, 745–756.

Balthesen, M., Messerle, M., and Reddehase, M.J. (1993). Lungs are a major organ site of cytomegalovirus latency and recurrence. J. Virol. 67, 5360–5366.

Benedict, C.A., Angulo, A., Patterson, G., Ha, S., Huang, H., Messerle, M., Ware, C.F., and Ghazal, P. (2004). Neutrality of the canonical NF-kappaB-dependent pathway for human and murine cytomega-lovirus transcription and replication in vitro. J. Virol. 78, 741–750.

Bevan, I.S., Sammons, C.C., and Sweet, C. (1996). Investigation of murine cytomegalovirus latency and re-activation in mice using viral mutants and the polymerase chain reaction. J. Med. Virol. 48, 308–320.

Brooks, D.H.M. (1994). The method of thought experiment. Metaphilosophy 25, 71–83.

Cardin, R.D., Abenes, G.B., Stoddart, C.A., and Mocarski, E.S. (1995). Murine cytomegalovirus IE2, an activator of gene expression, is dispensable for growth and latency in mice. Virology 209, 236–241.

Cavanaugh, V.J., Deng, Y., Birkenbach, M.P., Slater, J.S., and Campbell, A.E. (2003). Vigorous innate and virus-specific cytotoxic T-lymphocyte responses to murine cytomegalovirus in the submaxillary sali-vary gland. J. Virol. 77, 1703–1717.

Collins, T., Pomeroy, C., and Jordan, M.C. (1993). Detection of latent cytomegalovirus DNA in diverse organs of mice. J. Infect. Dis. 168, 725–729.

Dorsch-Häsler, K., Keil, G.M., Weber, F., Jasin, M., Schaffner, W., and Koszinowski, U. H. (1985). A long and complex enhancer activates transcription of the gene coding for the highly abundant immediate-early mRNA in murine cytomegalovirus. Proc. Natl. Acad. Sci. USA 82, 8325–8329.

Emery, V.C. (1998). Relative importance of cytomegalovirus load as a risk factor for cytomegalovirus dis-ease in the immunocompromised host. Monogr. Virol. 21, 288–301.

Fiering, S., Whitelaw, E., and Martin, D.I. (2000). To be or not to be active: the stochastic nature of en-hancer action. Bioassays 22, 381–387.

Ghazal, P., Messerle, M., Osborn, K., and Angulo, A. (2003). An essential role of the enhancer for murine cytomegalovirus in vivo growth and pathogenesis. J. Virol. 77, 3217–3228.

Ghazal, P., Visser, A.E., Gustems, M., Garcia, R., Borst, E., Sullivan, K., Messerle, M., and Angulo, A. (2005). Elimination of ie1 significantly attenuates murine cytomegalovirus virulence but does not alter replicative capacity in cell culture. J. Virol. 79, 7182–7194.

Goodrich, J.M., Boeckh, M., and Bowden, R. (1994). Strategies for the prevention of cytomegalovirus disease after marrow transplantation. Clin. Infect. Dis. 19, 287–298.

Goodrum, F.D., Jordan, C.T., High, K., and Shenk, T. (2002). Human cytomegalovirus gene expression during infection of primary hematopoietic progenitor cells: a model for latency. Proc. Natl. Acad. Sci. U.S.A. 99, 16255–16260.

Gribaudo, G., Riera, L., Lembo, D., De Andrea, M., Gariglio, M., Rudge, T.L., Johnson, L.F., and Landolfo, S. (2000). Murine cytomegalovirus stimulates cellular thymidylate synthase gene expression in quies-cent cells and requires the enzyme for replication. J. Virol. 74, 4979–4987.

Grzimek, N.K.A., Dreis, D., Schmalz, S., and Reddehase, M.J. (2001). Random, asynchronous, and asymmetric transcriptional activity of enhancer-flanking major immediate-early genes ie1/3 and ie2 during murine cytomegalovirus latency in the lungs. J. Virol. 75, 2692–2705.

Grzimek, N.K.A., Podlech, J., Steffens, H.P., Holtappels, R., Schmalz, S., and Reddehase, M.J. (1999). In vivo replication of recombinant murine cytomegalovirus driven by the paralogous major immediate-early promoter-enhancer of human cytomegalovirus. J. Virol. 73, 5043–5055.

Henson, D., and Strano, A.J. (1972). Mouse cytomegalovirus. Necrosis of infected and morphologically normal submaxillary gland acinar cells during termination of chronic infection. Am. J. Pathol. 68, 183–202.

Holtappels, R., Pahl-Seibert, M.F., Thomas, D., and Reddehase, M.J. (2000). Enrichment of immediate-early 1 (m123/pp89) peptide-specific CD8 T-cells in a pulmonary CD62L(lo) memory-effector cell pool during latent murine cytomegalovirus infection of the lungs. J. Virol. 74, 11495–11503.

Holtappels, R., Podlech, J., Geginat, G., Steffens, H.P., Thomas, D., and Reddehase, M.J. (1998). Control of murine cytomegalovirus in the lungs: relative but not absolute immunodominance of the immediate-early 1 nonapeptide during the antiviral cytolytic T-lymphocyte response in pulmonary infiltrates. J. Virol. 72, 7201–7012.

Hummel, M., and Abecassis, M.M. (2002). A model for reactivation of CMV from latency. J. Clin. Virol. 2, 123–136.

Hummel, M., Zhang, Z., Yan, S., DePlaen, I., Golia, P., Varghese, T., Thomas, G., and Abecassis, M.I. (2001). Allogeneic transplantation induces expression of cytomegalovirus immediate-early genes in vivo: a model for reactivation from latency. J. Virol. 75, 4814–4822.

Jonjic, S., Mutter, W., Weiland, F., Reddehase, M.J., and Koszinowski, U.H. (1989). Site-restricted persistent cytomegalovirus infection after selective long-term depletion of CD4+ T-lymphocytes. J. Exp. Med. 169, 1199–1212.

Jonjic, S., Pavic, I., Polic, B., Crnkovic, I., Lucin, P., and Koszinowski, U.H. (1994). Antibodies are not essential for the resolution of primary cytomegalovirus infection but limit dissemination of recurrent virus. J. Exp. Med. 179, 1713–1717.

Jordan, M.C. (1983). Latent infection and the elusive cytomegalovirus. Rev. Infect. Dis. 5, 205–215.

Keil, G.M., Ebeling-Keil, A., and Koszinowski, U.H. (1984). Temporal regulation of murine cytomegalovirus transcription and mapping of viral RNA synthesized at immediate-early times after infection. J. Virol. 50, 784–795.

Keil, G.M., Ebeling-Keil, A., and Koszinowski, U.H. (1987a). Immediate-early genes of murine cytomegalovirus: location, transcripts, and translation products. J. Virol. 61, 526–533.

Keil, G.M., Ebeling-Keil, A., and Koszinowski, U.H. (1987b). Sequence and structural organization of the murine cytomegalovirus immediate-early gene 1. J. Virol. 61, 1901–1908.

Keil, G.M., Fibi, M.R., and Koszinowski, U.H. (1985). Characterization of the major immediate-early polypeptides encoded by murine cytomegalovirus. J. Virol. 54, 422–428.

Koffron, A.J., Hummel, M., Patterson, B.K., Yan, S., Kaufman, D.B., Fryer, J.P., Stuart, F.P., and Abecassis, M.I. (1998). Cellular localization of latent murine cytomegalovirus. J. Virol. 72, 95–103.

Kurz, S., Rapp, M., Steffens, H.P., Grzimek, N.K., Schmalz, S., and Reddehase, M.J. (1999). Focal transcriptional activity of murine cytomegalovirus during latency in the lungs. J. Virol. 73, 482–494.

Kurz, S.K., and Reddehase, M.J. (1999). Patchwork pattern of transcriptional reactivation in the lungs indicates sequential checkpoints in the transition from murine cytomegalovirus latency to recurrence. J. Virol. 73, 8612–8622.

Kurz, S., Steffens, H.P., Mayer, A., Harris, J.R., and Reddehase, M.J. (1997). Latency versus persistence or intermittent recurrences: evidence for a latent state of murine cytomegalovirus in the lungs. J. Virol. 71, 2980–2987.

Lembo, D., Gribaudo, G., Hofer, A., Riera, L., Cornaglia, M., Mondo, A., Angeretti, A., Gariglio, M., Thelander, L., and Landolfo, S. (2000). Expression of an altered ribonucleotide reductase activity associated with the replication of murine cytomegalovirus in quiescent fibroblasts. J. Virol. 74, 11557–11565.

Lucin, P., Pavic, I., Polic, B., Jonjic, S., and Koszinowski, U.H. (1992). Gamma interferon-dependent clearance of cytomegalovirus infection in salivary glands. J. Virol. 66, 1977–1984.

Mercer, J.A., Wiley, C.A., and Spector, D.H. (1988). Pathogenesis of murine cytomegalovirus infection: identification of infected cells in the spleen during acute and latent infections. J. Virol. 62, 987–997.

Messerle, M., Bühler, B., Keil, G.M., and Koszinowski, U.H. (1992). Structural organization, expression, and functional characterization of the murine cytomegalovirus immediate-early gene 3. J. Virol. *66*, 27–36.

Messerle, M., Keil, G.M., and Koszinowski, U.H. (1991). Structure and expression of murine cytomegalovirus immediate-early gene 2. J. Virol. *65*, 1638–1643.

Podlech, J., Holtappels, R., Pahl-Seibert, M.F., Steffens, H.P., and Reddehase, M.J. (2000). Murine model of interstitial cytomegalovirus pneumonia in syngeneic bone marrow transplantation: persistence of protective pulmonary CD8-T-cell infiltrates after clearance of acute infection. J. Virol. *74*, 7496–7507.

Polic, B., Hengel, H., Krmpotic, A., Trgovcich, J., Pavic, I., Lucin, P., Jonjic, S., and Koszinowski, U.H. (1998). Hierarchical and redundanT-lymphocyte subset control precludes cytomegalovirus replication during latent infection. J. Exp. Med. *188*, 1047–1054.

Pomeroy, C., Hilleren, P.J., and Jordan, M.C. (1991). Latent murine cytomegalovirus DNA in splenic stromal cells of mice. J. Virol. *65*, 3330–3334.

Prösch, S., Staak, K., Stein, J., Liebenthal, C., Stamminger, T., Volk, H.D., and Krüger, D.H. (1995). Stimulation of the human cytomegalovirus IE enhancer/promoter in HL-60 cells by TNFalpha is mediated via induction of NF-kappaB. Virology *208*, 197–206.

Reddehase, M.J., Balthesen, M., Rapp, M., Jonjic, S., Pavic, I., and Koszinowski, U.H. (1994). The conditions of primary infection define the load of latent viral genome in organs and the risk of recurrent cytomegalovirus disease. J. Exp. Med. *179*, 185–193.

Reddehase, M.J., Podlech, J., and Grzimek, N.K.A. (2002). Mouse models of cytomegalovirus latency: overview. J. Clin. Virol. *2*, 23–36.

Simon, C.O., Seckert, C.K., Dreis, D., Reddehase, M.J., and Grzimek, N.K.A. (2005). Role for tumor necrosis factor alpha in murine cytomegalovirus transcriptional reactivation in latently infected lungs. J. Virol. *79*, 326–340.

Steffens, H.P., Kurz, S., Holtappels, R.,and Reddehase, M.J. (1998). Preemptive CD8 T-cell immunotherapy of acute cytomegalovirus infection prevents lethal disease, limits the burden of latent viral genomes, and reduces the risk of virus recurrence. J. Virol. *72*, 1797–1804.

Streblow, D.N., and Nelson, J.A. (2003). Models of HCMV latency and reactivation. Trends. Microbiol. *11*, 293–295.

Yuhasz, S.A., Dissette, V.B., Cook, M.L., and Stevens, J.G. (1994). Murine cytomegalovirus is present in both chronic active and latent states in persistently infected mice. Virology *202*, 272–280.

# The Rat Model for CMV Infection and Vascular Disease

# 24

*Suzanne J.F. Kaptein, Cornelis Vink, Cathrien A. Bruggeman, and Frank R.M. Stassen*

## Abstract

The rat cytomegalovirus (RCMV)/rat model has been exploited to address various aspects of the biology of cytomegaloviruses (CMVs). One of these aspects is the putative role of CMVs in the development and/or aggravation of vessel wall pathology. In this chapter, the findings regarding this issue will be reviewed and discussed.

## Introduction – rat cytomegalovirus (Maastricht strain)

The Maastricht strain of rat cytomegalovirus (RCMV) was first isolated in 1982 from the salivary glands of wild brown rats (*Rattus norvegicus*) and passed in rat embryo fibroblasts (REF) in vitro as well as in laboratory rats (Bruggeman et al., 1982). Sequencing of the entire 230-kbp genome of RCMV was completed nearly 20 years later (Vink et al., 2000). The RCMV genome is arranged as a single unique sequence flanked by terminal direct repeats and its sequence was found to be largely colinear with that of HCMV and MCMV (Chee et al., 1990a; Rawlinson et al., 1996; Vink et al., 1996; 2000). Moreover, the RCMV genome was predicted to encompass at least 170 open reading frames (ORFs), approximately half of which are conserved amongst the HCMV, MCMV and RCMV genomes and about two-thirds are conserved amongst both rodent CMV genomes (Vink et al., 2000). Interestingly, although the vast majority of the genes encode viral (structural) proteins, the genome was also found to encode proteins that show homology to host proteins which, thus, appear to be hijacked from the host genome by an ancestral virus. These include R33 and R78, both putatively encoding G protein-coupled receptors (Beisser et al., 1998; 1999), r144, encoding a major histocompatibility class I heavy chain homolog (Beisser et al., 2000), and r131 and r129, both encoding homologs of CC chemokines (Kaptein et al., 2004; Vink et al., 2000). Intriguingly, the RCMV genome also contains a unique ORF, r127, with similarity to the rep gene of parvoviruses as well as ORF U94 of human herpesvirus type 6A (HHV-6A) and 6B (HHV-6B) (van Cleef et al., 2004; Vink et al., 2000). Hitherto, counterparts of these ORFs have not been found in any of the other sequenced herpesvirus genomes.

growth factors and chemotactic factors, and the permeability of the vascular wall increases, leading to adhesion of platelets and monocytes/macrophages to the endothelium lining the vessel and extravasation of leukocytes (monocytes, macrophages and lymphocytes) into the subendothelium. Concomitantly, smooth muscle cells (SMCs) migrate from the media to the subendothelial layer, where they proliferate and secrete extracellular matrix proteins upon activation by growth factors, causing intimal thickening. In case of a hypercholesterolemic environment, macrophages and SMCs imbibe oxidized lipid (primarily cholesteryl ester droplets), generating the so-called 'foam cells', which successively progress into fatty streaks. Fatty streaks are flat intimal cell aggregations of predominantly lipid-laden macrophages, T-lymphocytes and SMCs. Subsequently, proliferating SMCs, macrophages and T-lymphocytes generate connective tissue, leading to the formation of fibro-fatty lesions that are characterized by the presence of a covering fibrous cap. Finally, fibro-fatty lesions gradually progress into atherosclerotic plaques as a result of ongoing cell replication, lipid accumulation, connective tissue formation and necrosis. Since the outer thinner parts of the fibrous cap of the plaque contain only relatively small amounts of connective tissue, cracks in these parts make the plaque unstable. Instability ultimately may cause rupture of the plaque and the emergence of acute vascular accidents.

## CMV and atherosclerosis

Over the years, herpesviruses have increasingly attracted attention with regard to their potential role in the process of atherosclerosis. In particular, the putative relation between CMV and atherosclerosis has been a favorite subject of investigation in both human and animal studies. In this respect, the RCMV/rat model system has proven to be a useful tool, shedding some light on the mechanisms by which CMV may contribute to atherosclerosis.

The existence of a possible association between CMV infection and atherosclerosis was originally based on a case-control study of cardiovascular patients undergoing surgery (Adam et al., 1987). This study showed a higher incidence of positive CMV antibodies in patients requiring vascular surgery for atherosclerosis than in control patients. Likewise, results from a cohort study performed by Nieto and coworkers (1996) suggested that CMV infection constitutes a risk factor for atherosclerosis. In this study, the level of carotid intimal-medial thickening (IMT) was compared with titers of antibodies directed against CMV, which were measured in sera obtained approximately 15 years earlier. Interestingly, subjects with higher levels of IMT also had higher mean CMV antibody titers than control subjects, suggesting that CMV might play a causal role in atherosclerosis. A potential association between CMV and atherosclerosis was further supported by sero-epidemiological studies, which reported a higher prevalence of CMV and higher levels of anti-CMV antibodies in patients suffering from atherosclerotic disease (Gattone et al., 2001: Musiani et al., 1990; Sorlie et al., 1994; Visseren et al., 1997). In contrast, in other studies, a correlation between CMV antibody levels and cardiovascular disease was not found (Adler et al., 1998; Ridker et al., 1998). It should be mentioned, however, that the study of Ridker et al. did not give consideration to the height of anti-CMV antibody titers, since only high but not low titers seem to be correlated with atherosclerotic disease (Gattone et al., 2001; Nieto et al., 1996).

In addition to data from sero-epidemiological studies, data from histopathological as well as in vitro studies render further support to the hypothesis that CMV appears to be implicated in the pathogenesis of atherosclerosis. Arterial SMCs derived from plaque samples of patients suffering from severe atherosclerotic disease were not only found to harbor CMV DNA, but, when cultured, were also demonstrated to express CMV antigens (Chen et al., 2003; Hendrix et al., 1989; 1990; Melnick et al., 1983; 1994; Petrie et al. 1987). More specifically, Hendrix et al. (1990) reported that CMV DNA was present in 90% of the arterial walls of patients with severe atherosclerotic changes (grade III) compared to 50% of those in control subjects with no or only small atherosclerotic changes (grade I at the most), as detected by PCR. Similar percentages of CMV DNA were observed by other investigators (Melnick et al., 1994; Qavi et al., 2000). However, Melnick et al. (1994) were not able to demonstrate a significant difference between the presence of CMV DNA in atherosclerotic plaque tissue and that in non-atherosclerotic regions within the same patient. On the other hand, atherosclerotic plaques with thrombosis from patients with carotid artery stenosis who required surgical endarterectomy were found to contain more CMV than plaques without thrombosis in carotid endarterectomy samples within the same patient (Chiu et al., 1997). As described earlier, experiments in infected rats revealed that CMV induced early vascular lesions, as exemplified by endothelial cell damage, increased adhesion of leukocytes to the endothelium followed by diapedesis through the endothelium into the subendothelial space, and presence of leukocytes (lymphocytes and macrophages) in the subendothelium (Span et al., 1992; 1993). Moreover, the CMV-induced early vascular lesions were characterized by the presence of lipophages as well as lipid accumulation in the endothelium and subendothelial space that consisted of extracellular lipid deposition and the location of subendothelial foam cells (Span et al., 1992; 1993). Additionally, CMV infection was shown to cause morphological alterations of the endothelium and subendothelial space of the large vessels in rats (Span et al., 1993). These changes included swollen endothelial cells with the formation of blebs, microvilli and craters on their surfaces. Moreover, CMV infection in rats led to the partial detachment of endothelial cells from the basement membrane, followed by an expansion of the space between the endothelium and the basement membrane and filling it with increasing amounts of basal lamina-like material mixed with collagen (Span et al., 1993).

The increased adherence of leukocytes (polymorphonuclear and monocytes) following a CMV infection was also supported by data from in vitro experiments (Span et al., 1991b). CMV infection of endothelial cells was accompanied by the expression of glycoproteins on their cell surface, promoting the adhesion of granulocytes to endothelial cells (Span et al., 1991b). In vitro as well as in vivo studies further revealed that apart from endothelial cells, a wide variety of cell types are permissive for CMV, including monocytes/macrophages, fibroblasts and SMCs (Bruggeman et al., 1985; Ibanez et al., 1991; Sinzger et al., 1995; Smiley et al., 1988; Tumilowicz et al., 1985; Vossen et al., 1996). Although the exact mechanism underlying the role of CMV in atherogenesis has not been elucidated yet, several studies did raise some ideas on how CMV might be involved. CMV was reported to induce SMC migration and proliferation, increase modified low-density lipoprotein (LDL) uptake in vascular SMCs, promote leukocyte influx, and induce an increased expression of adhesion molecules (e.g. ICAM-1), inflammatory cytokines (e.g. IL-6) and chemokines (e.g. MCP-1), ultimately contributing to the aggravation of the atherosclerotic process (Almeida et al.,

## Transplant vascular sclerosis

Transplant vascular sclerosis (TVS), also designated transplant-associated arteriosclerosis (TAA) or chronic rejection, has emerged as the major factor affecting long-term survival of transplanted organs (Barnart et al., 1987; Billingham, 1987). TVS has been identified as the most important cause of graft failure following the first post-transplant year (Paul, 1993). In cardiac allograft recipients the incidence of TVS is nearly 50% at 5 years after transplantation (Gao et al., 1988; Olivari et al., 1989). However, TVS is not restricted to cardiac allografts; it can develop after transplantation of a broad range of solid organs, including kidney, lung, liver and small bowel transplantation. In all transplanted solid organs the proliferative vascular changes have a similar appearance in TVS, being most prominent in the kidney and heart, whereas changes in the functional unit of the organ are dependent on organ specificity (Demetris et al., 1989). In TVS the manifestations are generalized, affecting the entire length of the vessel (arteries and arterioles), and allograft intimal thickening is concentric and develops rapidly, whereas the slowly progressing atherosclerotic process is characterized by focal, eccentric lesions containing cholesterol deposits (Billingham, 1987; Gao et al., 1988; Häyry et al., 1993; Olivari et al., 1989; Ross, 1986; 1992; Russell et al., 1993). Moreover, TVS is associated with a persistent perivascular and interstitial inflammation and a usually intact internal elastic lamina, whereas in atherosclerosis, perivascular inflammation is absent and the internal elastic lamina is disrupted (Billingham et al., 1987; Ross, 1986). Cardiac allografts undergoing chronic rejection showed vascular intimal thickening, due to proliferation of SMCs and infiltration of inflammatory cells (macrophages and activated T-lymphocytes), together with interstitial fibrosis (Billingham, 1989; Demetris et al., 1989; Uys and Rose, 1984). The pathophysiological mechanism of TVS is complicated and immunological as well as non-immunological factors, such as hyperlipidemia, pretransplant ischemia and CMV infection, are thought to be involved (Bieber et al., 1981; Eich et al., 1991; Gao et al., 1987; 1988; Grattan et al., 1989; McDonald et al., 1989; Olivari et al., 1989; Sharples et al., 1991; Stovin et al., 1991; Winters et al., 1990).

## CMV and transplant vascular sclerosis

Infection with CMV was hypothesized to be associated with an increased incidence of graft failure as well as increased recipient mortality rates following transplantation, indicating that CMV either induces or expedites TVS (Kendall et al., 1992; Light and Burke, 1979; Lopez et al., 1972). Indeed, various clinical studies suggested the existence of a relationship between TVS and CMV. Grattan and coworkers showed that allograft vasculopathy occurred more frequently as well as earlier in CMV-infected cardiac transplant recipients than in non-infected recipients (Grattan et al., 1989). Also, the rate of graft loss was significantly higher in CMV-infected patients than in non-infected patients (Grattan et al., 1989). Other investigators reported similar observations in cardiac transplant recipients infected with CMV (Loebe et al., 1990; McDonald et al., 1989). Additionally, prolonged CMV infection accompanied by viremia was found to correlate with the development of allograft vasculopathy in cardiac recipients (Everett et al., 1992). When allografts were examined for the presence of CMV nucleic acids using in situ hybridization with a CMV DNA probe, coronary arteries of allografted hearts with severe accelerated arteriosclerosis were found to harbor CMV nucleic acids (DNA and/or RNA), whereas these were absent in cardiac transplant patients showing no signs of allograft vasculopathy (Hruban et al., 1990; Wu et al., 1992). Correspondingly, Koskinen et al. (1993b; 1993c) found comparable results when using endomyocardial biopsies and angiography. These investigators also observed a link between CMV antigenemia and the induction of vascular cell adhesion molecules in the graft of cardiac transplant patients (Koskinen et al., 1993a). Finally, CMV infection putatively induces immune activation and subendothelial inflammation, subsequently contributing to graft arteriosclerosis and rejection (Koskinen et al., 1994a; 1994b). Besides cardiac allografts, CMV was reported to be implicated in TVS pathogenesis in a variety of allografts, including renal, kidney, liver and lung allografts (Boyce et al., 1988; Lautenschlager et al., 1997a; 1997b; Pouteil-Noble et al., 1993; Steinhoff et al., 1996).

The most compelling evidence that CMV might be involved in the development of TVS comes from various transplantation models in rats. Lemström and coworkers extensively studied the role of RCMV in transplant arteriosclerosis using the rat aortic transplant model. In this model, the thoracic aorta is transplanted between inbred rats in syngeneic and allogeneic combinations in the absence of immunosuppressive compounds (Mennader et al., 1991). Using this model, RCMV infection was shown to boost cell-mediated and humoral immune responses resulting in increased leukocyte influx, SMC and EC proliferation, and intimal thickening in the rat aortic as well as cardiac allografts (Lemström et al., 1993; 1995). Additionally, RCMV infection caused the induction of inflammation in the adventitia and subendothelium of the allografted vascular wall, as was evident by an increased infiltration of monocytes/macrophages and T-lymphocytes (Koskinen et al., 1994b; Lemström et al., 1993; Li et al., 1996a; 1996b). Similar findings were reported using the rat renal transplant model, in which RCMV infection was found to both accelerate and enhance the early immune response and the development of chronic rejection in renal allografts (Lautenschlager et al., 1999). Moreover, RCMV infection stimulated the induction of growth factors, the synthesis of collagen and the generation of fibrosis in rat renal allografts (Inkinen et al., 1999; Lautenschlager et al., 1999). In different kinds of allograft recipients, RCMV infection caused an increased expression of adhesion molecules VCAM-1 and ICAM-1 as well as increased numbers of inflammatory cells expressing activation markers LFA-1 and VLA-4 in the vascular endothelium, all contributing to the CMV-mediated acceleration of allograft arteriosclerosis (Kloover et al., 2000; Martelius et al., 1998; Steinhoff et al., 1995; Yilmaz et al., 1996). In addition, other antigens of which the expression was reported to be increased were MHC class I, MHC class II, IL-2 receptor, and several growth factors (e.g. PDGF-AA, PDGF-BB, TGF-β1, B-FDF), cytokines (e.g. TNF, IFN-γ) and chemokines (e.g. RANTES, MIP-1α, Fractalkine, IP-10, Lymphotactin) (Inkinen et al., 2003; Lemström et al., 1994a; Steinmüller et al., 1997; Streblow et al., 2003; Ustinov et al., 1994; van Dorp et al., 1989; von Willebrand et al., 1986; You et al., 1996). On the other hand, other studies failed to show a CMV-mediated up-regulation of MHC class II molecules (Sedmak et al., 1990; 1994; Ustinov et al., 1991; van Dorp et al., 1989). However, in all these cases only in vitro experiments were carried out, abrogating a possible indirect effect of CMV on MHC class II expression. Indeed, when endothelial cells were co-cultured with allogeneic CD4+ T-cells from subjects positive for CMV antibodies, CMV infection of cultured endothelial cells resulted in a substantial increase in MHC class II expression (Waldman et al., 1993). Moreover, this CMV-mediated increase in MHC class II expression could be abolished entirely by adding anti-IFN-γ antibodies to the culture medium, implicating that CMV indirectly, through IFN-γ, enhances MHC class II expression (Waldman et al., 1993).

CMV-mediated activation of the immune system, which mounts an additional inflammatory response to the alloresponse, is crucial in accelerated TVS pathogenesis, since treatment of animals with immunosuppressive compounds significantly diminished the CMV-induced exacerbation of transplant arteriosclerosis (Bruning et al., 1994; Lemström et al., 1994b; 1994c; 1994d; 1997). In addition, differences in MHC antigens between donor and recipient are also required, as enhanced graft-associated arteriosclerosis by CMV was rarely observed in syngeneic graft recipients (Cramer et al., 1989; Lemström et al., 1993; 1994c). Interestingly, although more vascular lesions were observed in the class II-mismatched allografts, substantial CMV-mediated neointima formation was only observed in allografts of rats with MHC class I disparity, probably due to CMV-linked increase in perivascular influx of monocytes/macrophages (Li et al., 1998). The importance of the alloreactive immune response in CMV acceleration of TVS was further supported by a study using bone marrow chimerism in a rat small bowel transplant model and a rat heart transplant model (Orloff et al., 2002). In this model, recipients lack an alloreactive response because of donor-specific tolerance induced by bone marrow chimerism, which was demonstrated to prevent transplant arteriosclerosis and chronic rejection (Orloff et al., 2000). Clearly, the alloreactive environment plays a crucial role in CMV-accelerated TVS, since allograft recipients given autologous bone marrow failed to develop TVS even when infected with RCMV (Orloff et al., 2002). In summary, CMV infection has the potential to modulate the allorecognition and alloreaction process, resulting in an intensified immune response.

## The role of endothelial cells in CMV-mediated vascular disease: endotheliotropism

The endothelium performs a pivotal role in the pathogenesis of vascular disease (Ross, 1993). Under normal conditions, the endothelium lining the vessel wall provides a protective barrier between the circulating blood and the vessel wall, maintaining an anticoagulant status. Moreover, the endothelial layer regulates vascular functions that offer resistance to a CMV invasion. Endothelial dysfunction constitutes a main event in the atherosclerotic disease process that may result from immunological destruction and clearance of CMV-infected ECs (Figure 24.1). Also, CMV infection is capable of modulating the status of ECs from an anticoagulant to a procoagulant one (van Dam-Mieras et al., 1992). Originally, permissiveness of endothelial cells for CMV was thought to be rather restricted, depending on the activation state of the cells, the virus strain used, and the origin of the endothelial cells (Slobbe-van Drunen et al., 1997; Smiley et al., 1988; Vossen et al., 1996; 1997; Waldman et al., 1989). Notably, in vitro, CMV has traditionally been propagated in fibroblasts. However, extensive virus propagation in fibroblasts was demonstrated to be associated with alterations in phenotypic characteristics presumably due to the occurrence of mutations in the viral genome during long-term propagation in fibroblasts, resulting in loss of tropism for various, otherwise natural host cells, including ECs (Cha et al., 1996; Sinzger et al., 1999; Waldman et al., 1989; 1991). Endothelial-based propagation, on the other hand, was demonstrated to preserve the natural endothelial cytopathogenicity of the original clinical isolate (Waldman et al., 1989; 1991). Intriguingly, results from unpublished studies indicate that the UL131-UL128 locus is required for endotheliotropism, although this locus most likely does not represent the sole determinant for EC tropism (Baldanti et al., International Cytomegalovirus Workshop 2003, Maastricht, The Netherlands; Hahn et al., International Cytomegalovirus Workshop 2003, Maastricht, The Netherlands).

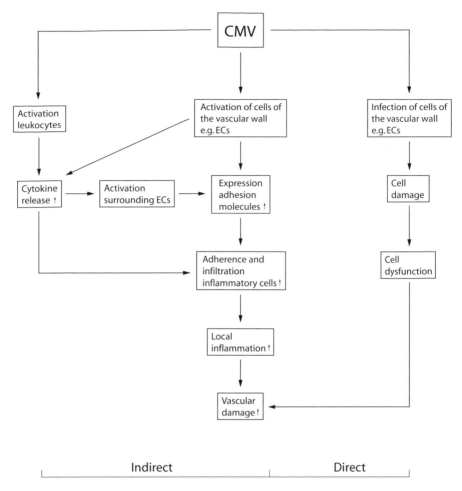

**Figure 24.1** Scheme of the major direct and indirect routes by which the endothelium might be implicated in aggravating CMV-mediated vascular diseases. CMV may enhance vascular damage through endothelial dysfunction as well as through (endothelial-mediated) modulation of the immune responses.

In addition to a direct detrimental effect to the endothelium, CMV may indirectly promote atherosclerosis via the release of inflammatory mediators (Figure 24.1). As stated earlier, CMV infection was found to up-regulate the expression of a variety of cytokines (e.g. IL-1, IL-2, IL-6, and IL-8) either directly by ECs or indirectly by other cells, e.g. leukocytes, which interact with the infected cells through the cytokine-mediated pathway (Almeida et al., 1994; Craigen and Grundy, 1996; Murayama et al., 1998; Waldman and Knight, 1996; Waldman et al., 1992; Woodroffe et al., 1993). Directly, CMV infection of ECs may increase the expression of adhesion molecules on the endothelial cell surface, such as VCAM-1, ICAM-1 and LFA-3, raising the endothelium to an activated, highly immunogenic state that can promote adherence and transendothelial migration of inflammatory cells, but also vascular damage (Craigen and Grundy, 1996; Sedmak et al., 1994; Steinhoff et al., 1995). Indirectly, CMV-infected, allogeneic ECs can activate T-lymphocytes (CD4[+]), resulting in proliferation and production of IL-2, which in turn stimulates the release of other T-cell-derived cytokines, such as IFN-γ (Waldman et al., 1992). Also, CMV-infected ECs may activate and stimulate allogeneic CD4$^+$ and CD8$^+$ T-lymphocytes to release cytokines (IFN-γ and TNF-α, respectively) that subsequently activate surrounding ECs and induce expression of adhesion molecules on their cell surface (Span et al., 1991; Waldman and Knight, 1996; Waldman et al., 1993). Finally, CMV infection of ECs may induce the up-regulation of chemokines (e.g. MCP-1), creating a chemokine gradient near the vessel wall and promoting leukocyte adherence and infiltration and, eventually, vascular disease (Taub et al., 1995). Interestingly, since CMV is capable of promoting vascular disease even in the absence of viral DNA in the vessel wall, the indirect pathway most likely represents the primary mechanism by which CMV augments atherogenesis.

Recently, a novel interesting mechanism in atherogenesis was reported that involves the interplay between leukocytes and endothelial cells and in which the proinflammatory cytokine IL-6, an important regulator of inflammation and immunity (e.g. activation of T-lymphocytes), plays a key role (Rott et al., 2003). It was postulated that T-lymphocytes that are clonally expanded in response to antigens presented by CMV infection, home and reside to sites of vascular damage via binding to adhesion molecules. Due to the presence of pathogen antigens in the plaque or cross-reaction with host protein homologs of pathogen antigens, T-lymphocytes release IL-6, which in turn triggers ECs in situ to release MCP-1. Thereupon, MCP-1, which is chemoattractant to monocytes and CD4$^+$ as well as CD8$^+$ T-lymphocytes, recruits even more monocytes and T-cells into the vascular wall, resulting in an aggravation of local inflammation and, hence, atherosclerosis (Rott et al., 2003; Taub et al., 1995). Although numerous pathways are involved in CMV-mediated exacerbation of vascular disease, this hypothesis sets a good example of the refined interplay between leukocytes and endothelial cells.

## The role of CMV-encoded GPCRs in vascular wall pathology and CMV disease

As is evident, chemokines and chemokine receptors play a prominent role in the development of vasculopathies. Chemokine receptors belong to the superfamily of the G protein-coupled receptors (GPCRs). Interestingly, the HCMV genome was found to encompass four genes encoding proteins with homology to host GPCRs, namely those encoded by UL33, UL78, US27 and US28 (Chee et al., 1990a; 1990b). Intriguingly, the HCMV US28-encoded protein (pUS28), comprises the first CMV-encoded protein reported to be putatively involved in SMC migration to sites of atherogenesis or stenosis and, hence, in the development of vascular disease. Streblow and coworkers (1999) found that HCMV infection of SMCs that were isolated from the aorta induced their migration. This migration was cell-type specific, not occurring in case of venous SMCs, ECs or fibroblasts. Since infection with a mutant HCMV strain lacking a functional US28 gene (HCMV-ΔUS28) failed to induce SMC migration, US28 was considered a prerequisite for HCMV-induced cellular migration. This notion was further supported by the observation that co-infection with adenoviruses expressing pUS28 was capable of rescuing the migration of SMCs that were infected with HCMV-ΔUS28. Moreover, expression of pUS28 lent SMCs the capability to migrate by chemokinesis as well as chemotaxis in the presence of CC chemokines, including RANTES or MCP-1 (Streblow et al., 1999). Altogether, these data may indicate that by virtue of pUS28 HCMV might induce SMC migration to sites of vascular inflammation, thereby contributing to the process of chronic vascular diseases (Streblow et al., 1999; 2001).

Although the US28 gene appears to be dispensable for virus replication in vitro (Vieira et al., 1998), this finding does not preclude the assumption that US28 plays an essential role in CMV disease in vivo. However, due to the species specificity as well as ethical considerations, it is practically impossible to study the role of pUS28 in HCMV pathogenesis in vivo. Likewise, the function of the HCMV-encoded US27 protein (pUS27) also remains speculative. Both the HCMV UL33 and UL78 gene, however, are well conserved in all currently known beta-herpesvirus genomes, including the RCMV genome (Beisser et al., 1998; 1999), enabling investigators to elucidate the in vivo function of homologs of the UL33- and UL78-encoded proteins (pUL33 and pUL78, respectively). Interestingly, the first pUL33-like GPCR that was demonstrated to be capable of signaling in a ligand-independent (i.e. constitutive) way was the receptor encoded by the RCMV R33 gene (Gruijthuijsen et al., 2002; 2004). In order to study the in vivo function of the R33 gene, recombinant viruses were generated in which the R33 gene is disrupted (Beisser et al., 1998). Subsequently, the replication characteristics of these null mutant strains were compared to those of wild-type (WT) RCMV both in vitro and in vivo. This study showed that the R33 gene was not required for the replication of RCMV in vitro. However, pronounced differences were found between animals infected with the R33-knockout viruses and those infected with WT RCMV. First, whereas WT RCMV was capable of establishing a high-titer infection in the salivary glands, the R33-knockout viruses were unable to replicate efficiently in this organ. This implied that R33 plays an important role in virus dissemination to or replication in the salivary glands. Second, a significantly lower mortality rate was observed amongst rats infected with the R33-knockout strains than amongst those infected with WT RCMV, demonstrating a more general role of R33 during in vivo infection. Similar findings were reported for the UL33-like gene present within the genome of murine CMV (MCMV), denoted M33 (Davis-Poynter et al., 1997), underlining the biological importance of the UL33-like genes in CMV pathogenesis.

An identical strategy was implemented to clarify the in vivo function of the GPCR encoded by the UL78-like gene within the RCMV genome (denoted R78), i.e. comparison of the replication characteristics of recombinant RCMV strains from which part of the R78 sequence was deleted to those of WT RCMV. In contrast to the R33 null mutants, the R78-knockout viruses displayed a diminished production of progeny virus in vitro (Beisser et al., 1999; Kaptein et al., 2003). Similar results were found for the UL78-like gene present in the MCMV genome, M78 (Oliveira and Shenk, 2001). Moreover, during infection of fibroblasts with the R78-knockout strains a conspicuous morphological phenomenon was observed in vitro. Unlike WT RCMV-infected fibroblasts, fibroblasts infected with the R78 null mutants developed a syncytium-like appearance (Beisser et al., 1999). The ability of these R78 knockout strains to induce syncytium formation in infected fibroblasts led to the suggestion that R78 might be implicated in stabilizing glycoprotein complexes in order to establish cell-to-cell contacts (Beisser et al., 1999). Similarly to the replication in vitro, replication of a recombinant R78 strain expressing an enhanced green fluorescent protein (EGFP)-tagged version of pR78 instead of native pR78 was also shown to be attenuated in vivo (Kaptein et al., 2003). While infectious virus was readily detected in the spleen of WT

RCMV-infected rats, virus could not be detected in the spleen of rats infected with the recombinant R78 strain. Correspondingly, recombinant M78 strains replicated less efficiently in the spleen as well as in the liver and salivary glands of infected mice than WT MCMV (Oliveira and Shenk, 2001). Summarizing, these data indicate that the UL78-like genes of RCMV and MCMV constitute non-essential genes that are, nevertheless, required for efficient CMV replication, at least in the spleen. Moreover, deletion of the R78 and M78 gene from the RCMV and MCMV genome, respectively, resulted in mutant strains with reduced virulence in vivo, implying that the UL78-like genes play an important role in the pathogenesis of CMV infection (Beisser et al., 1999; Oliveira and Shenk, 2001).

## Concluding remarks

The RCMV/rat model has proven to be suitable for studying the CMV-mediated development of vascular disease, particularly in atherosclerosis, arterial restenosis and transplant vascular sclerosis. The use of this model resulted in substantial data showing a stimulatory role of CMV in the pathogenesis of various types of vascular disease. RCMV was found to exert its detrimental effects both directly, through dysfunction of cells of the vascular wall, and indirectly, through modulation of the inflammatory and immune responses, which putatively represents the main mechanism of action. Also, CMV seems to affect all stages of the vascular disease process. However, this activity may not be unique for CMV. Several studies are indicative of the existence of a positive correlation between the number of infectious pathogens to which a subject has been exposed (e.g. CMV, *Chlamydia pneumoniae*, *Helicobacter pylori*, Epstein–Barr virus) and the degree of atherogenesis, giving rise to the so-called 'pathogen burden hypothesis' (Epstein et al., 2000; Espinola-Klein et al., 2002; Zhu et al., 2000). Nonetheless, more animal studies are required to elucidate the mechanisms of CMV-enhanced progression of vascular disease, which may ultimately provide us with new targets for prophylaxis and/or treatment.

Acknowledgment

C.V. is supported by a fellowship of the Royal Netherlands Academy of Arts and Sciences (KNAW).

References

Adam, E., Melnick, J.L., Probtsfield, J.L., Petrie, B.L., Burek, J., Bailey, K.R., McCollum, C.H., and DeBakey, M.E. (1987). High levels of cytomegalovirus antibody in patients requiring vascular surgery for atherosclerosis. Lancet 2, 291–293.

Adler, S.P., Hur, J.K., Wang, J.B., and Vetrovec, G.W. (1998). Prior infection with cytomegalovirus is not a major risk factor for angiographically demonstrated coronary artery atherosclerosis. J. Infect. Dis. 177, 209–212.

Almeida, G.D., Porada, C.D., St Jeor, S., and Ascensao, J.L. (1994). Human cytomegalovirus alters interleukin-6 production by endothelial cells. Blood 83, 370–376.

Barnart, G.R., Pascoe, E.A., Mills, A.C., Szentpetery, S., Eich, D.M., Mohanacumar, T., Hastillo, A., Thompson, J.A., Hess, M.L., and Lower, R.R. (1987). Accelerated coronary arteriosclerosis in cardiac transplant recipients. Transplant. Rev. 1, 31–46.

Beisser, P.S., Grauls, G.E.L.M., Bruggeman. C.A., and Vink, C. (1999). Deletion of the R78 G protein-coupled receptor gene from rat cytomegalovirus results in an attenuated, syncytium-inducing mutant strain. J. Virol. 73, 7218–7230.

Beisser, P.S., Kloover, J.S., Grauls, G.E.L.M., Blok, M.J., Bruggeman. C.A., and Vink, C. (2000). The r144 major histocompatibility complex class I-like gene of rat cytomegalovirus is dispensable for both acute and long-term infection in the immunocompromised host. J. Virol. 74, 1045–1050.

Beisser, P.S., Vink, C., van Dam, J.G., Grauls, G.E.L.M., Vanherle, S.J.V., and Bruggeman, C.A. (1998). The R33 G protein-coupled receptor gene of rat cytomegalovirus plays an essential role in the pathogenesis of viral infection. J. Virol. 72, 2352–2363.

Bieber, C.P., Hunt, S.A., Schwimm, D.A., Jamieson, S.A., Reaitz, B.A., Oyer, P.E., Shumway, N.E., and Stinson, E.B. (1981). Complications in long-term survivors of cardiac transplantation. Transplant. Proc. 13, 207–211.

Billingham, M.E. (1987). Cardiac transplant atherosclerosis. Transplant. Proc. 19 (Suppl.5), 19–25.

Billingham, M.E. (1989). Graft coronary disease: The lesions and the patients. Transplant. Proc. 21, 3665–3666.

Billingham, M.E. (1992). Histopathology of graft coronary disease. J. Heart Lung Transpl. 11, 38–44.

Blok, M.J., Savelkouls, K.G.M., Grauls, G.E.L.M., Bruggeman, C.A., and Vink, C. (2001). Immediate-early-1 mRNA expression and virus production are restricted during the acute phase of rat cytomegalovirus infection in immunocompetent rats. In Monitoring the course of CMV infection by detection of specific viral transcripts, M.J. Blok, ed. (Geleen, The Netherlands: ANDI Press), pp. 91–124.

Blum, A., Giladi, M., Weinberg, M., Kaplan, G., Pasternack, H., Laniado, S., and Miller, H. (1998). High anti-cytomegalovirus (CMV) IgG antibody titer is associated with coronary artery disease and may predict post-coronary balloon angioplasty restenosis. Am. J. Cardiol. 81, 866–868.

Boyce, N.W., Hayes, K., Gee, D., Holdsworth, S.R., Thomson, N.M., Scott, D., and Atkins, R.C. (1988). Cytomegalovirus infection complicating renal transplantation and its relationship to acute transplant glomerulopathy. Transplantation 45, 706–709.

Bruggeman, C.A. (1993). Cytomegalovirus and latency: an overview. Virchows Arch. B-cell Pathol. 64, 325–333.

Bruggeman, C.A., Bruning, J.H., Grauls, G., van den Bogaard, A.E., and Bosman, F. (1987). Presence of cytomegalovirus in brown fat. Study in a rat model. Intervirology 27, 32–37.

Bruggeman, C.A., Debie, W.H.M., Grauls, G.E.L.M., Majoor, G., and van Boven, C.P.A. (1983). Infection of laboratory rats with a new cytomegalo-like virus. Arch. Virol. 76, 189–199.

Bruggeman, C.A., Meijer, H., Bosman, F., and van Boven, C.P.A. (1985). Biology of rat cytomegalovirus infection. Intervirology 24, 1–9.

Bruggeman, C.A., Meijer, H., Dormans, P.H.J., Debie, W.H.M., Grauls, G.E.L.M., and van Boven, C.P.A. (1982). Isolation of a cytomegalovirus-like agent from wild rats. Arch. Virol. 73, 231–241.

Bruggeman, C.A., and van Dam-Mieras, M.C. (1991). The possible role of cytomegalovirus in atherogenesis. Prog. Med. Virol. 38, 1–26.

Bruning, J.H., Bruggeman, C.A., van Boven, C.P., and van Breda Vriesman, P.J. (1986). Passive transfer of latent rat cytomegalovirus by cardiac and renal organ transplants in a rat model. Transplantation 41, 695–698.

Bruning, J.H., Bruggeman, C.A., van Boven, C.P., and van Breda Vriesman, P.J. (1989). Reactivation of latent rat cytomegalovirus by a combination of immunosuppression and administration of allogeneic immunocompetent cells. Transplantation 47, 917–918.

Bruning, J.H., Persoons, M.C.J., Lemström, K.B., Stals, F.S., De Clercq, E., and Bruggeman, C.A. (1994). Enhancement of transplantation-associated atherosclerosis by CMV, which can be prevented by antiviral therapy in the form of HPMPC. Transplant. Int. 7(Suppl. 1), S365-S370.

Cha, T.A., Tom, E., Kemble, G.W., Duke, G.M., Mocarski, E.S., and Spaete, R.R. (1996). Human cytomegalovirus clinical isolates carry at least 19 genes not found in laboratory strains. J. Virol. 70, 78–83.

Chee, M.S., Bankier, A.T., Beck, S., Bohni, R., Brown, C.M., Cerny, R., Horsnell, T., Hutchinson III, C.A., Kouzarides, T., Martignetti, J.A., Preddie, E., Satchwell, S.C., Tomlinson, P., Weston, K.M., and Barrell, B.G. (1990a). Analysis of the protein-coding content of the sequence of human cytomegalovirus strain AD169. Curr. Top. Microbiol. Immunol. 154, 125–169.

Chee, M.S., Satchwell, S.C., Preddie, E., Weston, K.M., and Barrell, B.G. (1990b). Human cytomegalovirus encodes three G protein-coupled receptor homologues. Nature 344, 774–777.

Chen, R., Xiong, S., Yang, Y., Fu, W., Wang, Y., and Ge, J. (2003). The relationship between human cytomegalovirus infection and atherosclerosis development. Mol. Cell. Biochem. 249, 91–96.

Chiu, B., Viira, E., Tucker, W., and Fong, I.W. (1997). Chlamydia pneumoniae, cytomegalovirus and herpes simplex virus in atherosclerosis of the carotid artery. Circulation 96, 2144–2148.

Craigen, J.L., and Grundy, J.E. (1996). Cytomegalovirus induced up-regulation of LFA-3 (CD58) and ICAM-1 (CD54) is a direct viral effect that is not prevented by ganciclovir or foscarnet treatment. Transplantation 62, 1102–1108.

Cramer, D.V., Qian, S.Q., Harnaha, J., Chapman, F.A., Estes, L.W., Starzl, T.E., and Makowka, L.(1989). Cardiac transplantation in the rat. The effect of histoincompatibility differences on graft arteriosclerosis. Transplantation 47, 414–419.

Davis-Poynter, N.J., Lynch, D.M., Vally, H., Shellam, G.R., Rawlinson, W.D., Barrell, B.G., and Farrell, H.E. (1997). Identification and characterization of a G protein-coupled receptor homolog encoded by murine cytomegalovirus. J. Virol. 71, 1521–1529.

Danesh, J., Collins, R., and Peto, R. (1997). Chronic infections and coronary heart disease: is there a link? Lancet 350, 430–436.

Demetris, A.J., Zerbe, T., and Banner, B. (1989). Morphology of solid organ allograft arteriopathy: Identification of proliferating intimal cell populations. Transplant. Proc. 21, 3667–3669.

DeRodriguez, C.V., Fuhrer, J., and Lake-Bakaar, G. (1994). Cytomegalovirus colitis in patients with acquired immunodeficiency syndrome. J. R. Soc. Med. 87, 203–205.

Eich, D., Thompson, J.A., Ko, D., Hastillo, A., Lower, R., Katz, S., Katz, M., and Hess, M.L. (1991). Hypercholesterolemia in long-term survivors of heart transplantation; an early marker of accelerated coronary artery disease. J. Heart Lung Transplant. 10, 45–49.

Epstein, S.E., Speir, E., Zhou, Y.F., Guetta, E., Leon, M., and Finkel, T. (1996). The role of infection in restenosis and atherosclerosis: focus on cytomegalovirus. Lancet 348, S13-S17.

Epstein, S.E., Zhu, J., Burnett, M.S., Zhou, Y.F., Vercellotti, G., and Hajjar, D. (2000). Infection and atherosclerosis: potential roles of pathogen burden and molecular mimicry. Arterioscler. Thromb. Vasc. Biol. 20, 1417–1420.

Espinola-Klein, C., Rupprecht, H.J., Blankenberg, S., Bickel, C., Kopp, H., Rippin, G., Victor, A., Hafner, G., Schlumberger, W., and Meyer, J. (2002). Impact of infectious burden on extent and long-term prognosis of atherosclerosis. Circulation 105, 15–21.

Everett, J.P., Hershberger, R.E., Norman, D.J., Chou, S., Ratkovec, R.M., Cobanoglu, A., On, G.Y., and Hosenpud, J.D. (1992). Prolonged cytomegalovirus infection with viremia is associated with development of cardiac allograft vasculopathy. J. Heart Lung Transplant. 11, S133-S137.

Francis, N.D., Boylston, A.W., Roberts, A.H., Parkin, J.M., and Pinching, A.J. (1989). Cytomegalovirus infection in gastrointestinal tracts of patients with HIV-1 or AIDS. J. Clin. Pathol. 42, 1055–1064.

Gao, S.Z., Alderman, E.L., Schroeder, J.S., Silverman, J.F., and Hunt, S.A. (1988). Accelerated coronary vascular disease in the heart transplant patient: coronary arteriographic findings. J. Am. Coll. Cardiol. 12, 334–340.

Gao, S.Z., Schroeder, J.S., Alderman, E.L., Hunt, S.A., Silverman, J.F., Wiederholt, V., and Stinson, E.B. (1987). Clinical and laboratory correlates of accelerated coronary artery disease in the cardiac transplant patient. Circulation 76 (Suppl.4), 56–61.

Gattone, M., Iacoviello, L., Colombo, M., Castelnuovo, A.D., Soffiantino, F., Gramoni, A., Picco, D., Benedetta, M., and Giannuzzi, P. (2001). Chlamydia pneumoniae and cytomegalovirus seropositivity, inflammation markers, and the risk of myocardial infarction at a young age. Am. Heart J. 142, 633–640.

Glagov, S. (1994). Intimal hyperplasia, vascular modeling, and the restenosis problem. Circulation 89, 2888–2891.

Golden, M.P., Hammer, S.M., Wanke, C.A., and Albrecht, M.A. (1994). Cytomegalovirus vasculitis. Case reports and reviews of the literature. Medicine 73, 246–255.

Grattan, M.T., Moreno-Cabral, C.E., Starnes, V.A., Oyer, P.E., Stinson, E.B., and Shumway, N.E. (1989). Cytomegalovirus is associated with cardiac allograft rejection and atherosclerosis. JAMA 261, 3561–3566.

Gruijthuijsen, Y.K., Beuken, E.V.H., Smit, M.J., Leurs, R., Bruggeman, C.A., and Vink, C. (2004). Mutational analysis of the R33-encoded G protein-coupled receptor of rat cytomegalovirus: identification of amino acid residues critical for cellular localization and ligand-independent signaling. J. Gen. Virol. 85, 897–909.

Gruijthuijsen, Y.K., Casarosa, P., Kaptein, S.J.F., Broers, J.L.V., Leurs, R., Bruggeman, C.A., Smit, M.J., and Vink, C. (2002). The rat cytomegalovirus R33-encoded G protein-coupled receptor signals in a constitutive fashion. J. Virol. 76, 1328–1338.

Grundy, J.E., and Downes, K.L. (1993). Up-regulation of LFA-3 and ICAM-I on the surface of fibroblasts infected with cytomegalovirus. Immunology 78, 405–412.

Häyry, P., Mennander, A., Räisänen, A., Ustinov, J., Lemström, K., Aho, P., Yilmaz, S., Lautenschlager, I., and Paavonen, T. (1993). Pathophysiology of vascular wall changes in chronic allograft rejection. Transplant. Rev. 7, 1–20.

Hendrix, M.G., Dormans, P.H., Kitselaar, P., Bosman, F., and Bruggeman, C.A. (1989). The presence of CMV nucleic acids in arterial walls of atherosclerotic and non-atherosclerotic patients. Am. J. Pathol. 134, 1151–1157.

Hendrix, M.G., Salimans, M.M., van Boven, C.P., and Bruggeman, C.A. (1990). High prevalence of latently present cytomegalovirus in arterial walls of patients suffering from grade III atherosclerosis. Am. J. Pathol. 136, 23–28.

Ho, M. (1991). Cytomegalovirus. Biology and infection (New York: Plenum Medical Book Company).

Hruban, R.H., Wu, T.C., Beschorner, W.E., Cameron, D.E., Ambinder, R.F., Baumgartner, W.A., Reitz, B.A., and Hutchins, G.M. (1990). Cytomegalovirus nucleic acids in allografted hearts. Hum. Pathol. 21, 981–983.

Hsich, E., Zhou, Y.F., Paigen, B., Johnson, T.M., Burnett, M.S., and Epstein, S.E. (2001). Cytomegalovirus infection increases development of atherosclerosis in Apolipoprotein-E knockout mice. Atherosclerosis 156, 23–28.

Ibanez, C.E., Schrier, R., Ghazal, P., Wiley, C., and Nelson, J.A. (1991). Human cytomegalovirus productively infects primary differentiated macrophages. J. Virol. 65, 6581–6588.

Inkinen, K., Holma, K., Soots, A., Krogerus, L., Loginov, R., Bruggeman, C., Ahonen, J., and Lautenschlager, I. (2003). Expression of TGF-beta and PDGF-AA antigens and corresponding mRNAs in cytomegalovirus-infected rat kidney allografts. Transplant. Proc. 35, 804–805.

Inkinen, K., Soots, A., Krogerus, L., Bruggeman, C., Ahonen, J., and Lautenschlager, I. (1999). CMV increases collagen synthesis in chronic rejection in rat renal allograft. Transplant. Proc. 31, 1361.

Ishibashi, S., Brown, M.S., Goldstein, J.L., Gerard, R.D., Hammer, R.E., and Herz, J. (1993). Hypercholesterolemia in low density lipoprotein receptor knockout mice and its reversal by adenovirus-mediated gene delivery. J. Clin. Invest. 92, 883–893.

Kaptein, S.J.F., Beisser, P.S., Gruijthuijsen, Y.K., Savelkouls, K.G.M., van Cleef, K.W.R., Beuken, E., Grauls, G.E.L.M., Bruggeman, C.A., and Vink, C. (2003). The rat cytomegalovirus R78 G protein-coupled receptor gene is required for the production of infectious virus in the spleen. J. Gen. Virol. 84, 2517–2530.

Kaptein, S.J.F., van Cleef, K.W.R, Gruijthuijsen, Y.K., Beuken E.V.H., Van Buggenhout, L., Beisser, P.S., Stassen, F.R.M., Bruggeman, C.A., and Vink, C. (2004). The r131 gene of rat cytomegalovirus encodes a proinflammatory CC chemokine homolog which is essential for the production of infectious virus in the salivary glands. Virus Genes 29, 43–61.

Kendall, T.J., Wilson, J.E., Radio, S.J., Kandolf, R., Gulizia, J.M., Winters, G.L., Costanzo-Nordin, M.R., Malcom, G.T., Thieszen, S.L., Miller, L.W., and McManus, B.M. (1992). Cytomegalovirus and other herpes viruses: do they have a role in the development of accelerated coronary arterial disease in human heart allografts? J. Heart Lung Transplant. 11, 14–20.

Kikuchi, S., Umemura, K., Kondo, K., and Nakashima, M. (1996). Tranilast suppresses intimal hyperplasia after photochemically induced endothelial injury in the rat. Eur. J. Pharmac. 295, 221–227.

Kloover, J.S., Hillebrands, J.L., de Wit, G., Grauls, G., Rozing, J., Bruggeman, C.A., and Nieuwenhuis, P. (2000). Rat cytomegalovirus replication in the salivary glands is extensively confined to striated duct cells. Virchows Arch. 437, 413–421.

Kloover, J.S., Soots, A.P., Krogerus, L.A., Kauppinen, H.O., Loginov, R.J., Holma, K.L., Bruggeman, C.A., Ahonen, P.J., and Lautenschlager, I.T. (2000). Rat cytomegalovirus infection in kidney allograft recipients is associated with increased expression of intracellular adhesion molecule-1 vascular adhesion molecule-1, and their ligands leukocyte function antigen-1 and very late antigen-4 in the graft. Transplantation 69, 2641–2647.

Kloppenburg, G., de Graaf, R., Herngreen, S., Grauls, G., Bruggeman, C., and Stassen, F. (2005). Cytomegalovirus aggravates intimal hyperplasia in rats by stimulating smooth muscle cell proliferation. Microbes Infect. 7, 164–170.

Koskinen, P.K. (1993a). The association of the induction of vascular cell adhesion molecule-I with cytomegalovirus antigenemia in human allografts. Transplantation 53, 1103–1108.

Koskinen, P., Krogerus, L.A., Nieminen, M.S., Mattila, S.P., Häyry, P., and Lautenschlager, I.T. (1993b). Quantitation of cytomegalovirus infection-associated histologic findings in endomyocardial biopsies of heart allografts. J. Heart Lung Transplant. 12, 343–354.

Koskinen, P., Krogerus, L.A., Nieminen, M.S., Mattila, S.P., Häyry, P., and Lautenschlager, I.T. (1994a). Cytomegalovirus infection-associated generalized immune activation in heart allograft recipients: a study of cellular events in peripheral blood and endomyocardial biopsy specimens. Transplant. Int. 7, 163–171.

Koskinen, P., Lemström, K., Bruggeman, C., Lautenschlager, I., and Häyry, P. (1994b). Acute cytomegalovirus infection induces a subendothelial inflammation (endothelialitis) in the allograft vascular wall. A possible linkage with enhanced allograft arteriosclerosis. Am. J. Pathol. 144, 41–50.

Koskinen, P., Nieminen, M.S., Krogerus, L.A., Mattila, S.P., Häyry, P., and Lautenschlager, I.T. (1993c). Cytomegalovirus infection accelerates cardiac allograft vasculopathy: correlation between angiographic and endomyocardial biopsy findings in heart transplant patients. Transplant. Int. 6, 341–347.

Lai, I.R., Chen, K.M., Shun, C.T., and Chen, M.Y. (1996). Cytomegalovirus enteritis causing massive bleeding in a patient with AIDS. Hepatogastroenterology 43, 987–991.

Lautenschlager, I., Höckerstedt, K., Jalanko, H., Loginov, R., Salmela, K., Taskinen, E., and Ahonen, J. (1997a). Persistent CMV in liver allografts ending up with chronic rejection. Hepatology 25, 190–194.

Lautenschlager, I., Soots, A., Krogerus, L., Inkinen, K., Kloover, J., Loginov, R., Holma, K., Kauppinen, H., Bruggeman, C., and Ahonen, J. (1999). Time-related effects of cytomegalovirus infection on the development of chronic renal allograft rejection in a rat model. Intervirology 42, 279–284.

Lautenschlager, I., Soots, A., Krogerus, L., Kauppinen, H., Saarinen, O., Bruggeman, C., and Ahonen, J. (1997b). Effect of cytomegalovirus on an experimental model of chronic renal allograft rejection under triple-drug treatment in the rat. Transplantation 64, 391–398.

Lemström, K.B., Aho, P.T., Bruggeman, C.A., and Häyry, P.J. (1994a). Cytomegalovirus enhances mRNA expression of platelet-derived growth factor BB and transforming growth factor-beta 1 in rat aortic allografts. Possible mechanism for cytomegalovirus-enhanced graft arteriosclerosis. Arterioscler. Thromb. 14, 2043–2052.

Lemström, K.B., Bruning, J.H., Bruggeman, C.A., Koskinen, P.K., Aho, P.T., Yilmaz, S., Lautenschlager, I.T., and Häyry, P.J. (1994b). Cytomegalovirus infection-enhanced allograft arteriosclerosis is prevented by DHPG prophylaxis in the rat. Circulation 90, 1969–1978.

Lemström, K.B., Bruning, J.H., Bruggeman, C.A., Lautenschlager, I.T., and Häyry, P.J. (1993). Cytomegalovirus infection enhances smooth muscle cell proliferation and intimal thickening of rat aortic allografts. J. Clin. Invest. 92, 549–558.

Lemström, K.B., Bruning, J.H., Bruggeman, C.A., Lautenschlager, I.T., and Häyry, P.J. (1994c). Triple drug immunosuppression significantly reduces immune activation and allograft arteriosclerosis in cytomegalovirus-infected rat aortic allografts and induces early latency of viral infection. Am. J. Pathol. 144, 1334–1347.

Lemström, K.B., Bruning, J., Koskinen, P., Bruggeman, C., Lautenschlager, I., and Häyry, P.J. (1994d). Triple-drug immunosuppression significantly reduces chronic rejection in noninfected and RCMV-infected rats. Transplant. Proc. 26, 1727–1728.

Lemström, K., Koskinen, P., Krogerus, L., Daemen, M., Bruggeman, C., and Häyry, P. (1995). Cytomegalovirus antigen expression, endothelial cell proliferation, and intimal thickening in rat cardiac allografts after cytomegalovirus infection. Circulation 92, 2594–2604.

Lemström, K., Sihvola, R., Bruggeman, C., Häyry, P., and Koskinen, P. (1997). Cytomegalovirus infection-enhanced cardiac allograft vasculopathy is abolished by DHPG prophylaxis in the rat. Circulation 95, 2614–2616.

Li, F., Grauls, G., Yin, M., and Bruggeman, C. (1996a). Correlation between the intensity of cytomegalovirus infection and the amount of perivasculitis in aortic allografts. Transpl. Int. 9, S340–S344.

Li, F., Grauls, G., Yin, M., and Bruggeman, C. (1996b). Initial endothelial injury and cytomegalovirus infection accelerate the development of allograft arteriosclerosis. Transplant. Proc. 27, 3552–3554.

Li, F., Yin, M., van Dam, J.G., Grauls, G., Rozing, J., and Bruggeman, C.A. (1998). Cytomegalovirus infection enhances the neointima formation in rat aortic allografts: effect of major histocompatibility complex class I and class II antigen differences. Transplantation 65, 1298–1304.

Light, J.A., and Burke, D.S. (1979). Association of cytomegalovirus (CMV) infections with increased recipient mortality following transplantation. Transplant. Proc. 11, 79–82.

Loebe, M., Schüler, S., Zais, O., Warnecke, H., Fleck, E., and Hetzer, R. (1990). Role of cytomegalovirus infection in the development of coronary artery disease in the transplanted heart. J. Heart Lung Transplant. 9, 707–711.

Lopez, C., Simmons, R.L., Mauer, S.M., Park, B., Najarian, J.S., and Good, R.A. (1972). Virus infections may trigger rejection in immunosuppressed renal transplant recipients. Proc. Clin. Dial. Transplant. Forum 2, 107–111.

Martelius, T., Krogerus, L., Höckerstedt, K., Bruggeman, C., and Lautenschlager, I. (1998). Cytomegalovirus infection is associated with increased inflammation and severe bile duct damage in rat liver allografts. Hepatology 27, 996–1002.

Martelius, T., Scholz, M., Krogerus, L., Höckerstedt, K., Loginov, R., Bruggeman, C., Cinatl Jr, J., Doerr, H.W., and Lautenschlager, I. (1999). Antiviral and immunomodulatory effects of desferrioxamine in cytomegalovirus-infected rat liver allografts with rejection. Transplantation 68, 1753–1761.

McDonald, K., Rector, T.S., Braulin, E.A., Kubo, S.H., and Olivari, M.T. (1989). Association of coronary artery disease in cardiac transplant recipients with cytomegalovirus infection. Am. J. Cardiol. 64, 359-362.

Melnick, J.L., Adam, E., and DeBakey, M.E. (1990). Possible role of cytomegalovirus in atherogenesis. J. Am. Med. Assoc. 263, 2204–2207.

Melnick, J.L., Hu, C., Burek, J., Adam, E., and DeBakey, M.E. (1994). Cytomegalovirus DNA in arterial wall of patients with atherosclerosis. J. Med. Virol. 42, 170–174.

Melnick, J.L., Petrie, B.L., Dreesman, G.R., Burek, J., McCollum, C.H., and DeBakey, M.E. (1983). Cytomegalovirus antigen within human arterial smooth muscle cells. Lancet 2, 644–647.

Mennader, A., Tiisala, S., Harttunen, J., Yilmaz, S., Paavonen, T., and Häyry, P. (1991). Chronic rejection in rat aortic allografts: an experimental model for transplant arteriosclerosis. Arterioscler. Thromb. 11, 671–680.

Muldoon, J., O'Riordan, K., Rao, S., and Abecassis, M. (1996). Ischaemic colitis secondary to venous thrombosis. A rare presentation of cytomegalovirus vasculitis following renal transplantation. Transplantation 61, 1651–1653.

Murayama, T., Mukaida, N., Khabar, K.S.A., and Matsushima, K. (1998). Potential involvement of IL-8 in the pathogenesis of human cytomegalovirus infection. J. Leuk. Biol. 64, 62–67.

Musiani, M., Zurbini, M.L., Muscari, A., Puddu, G.M., Gentilomi, G., Gibellini, D., Gallinella, G., Puddu, P., and La Placa, M. (1990). Antibody patterns against cytomegalovirus and Epstein–Barr virus in human atherosclerosis. Microbiologica 13, 35–41.

Nieto, F.J., Adam, E., Sorlie, P., Farzadegan, H., Melnick, J.L., Comstock, G.W., and Szklo, M. (1996). Cohort study of cytomegalovirus infection as a risk factor for carotid intimal-medial thickening, a measure of subclinical atherosclerosis [see comments]. Circulation 94, 922–927.

Olivari, M.T., Homans, D.C., Wilson, R.F., Kubo, S.H., and Ring, W.S. (1989). Coronary artery disease in cardiac transplant patients receiving triple-drug immunosuppressive therapy. Circulation 80 (Suppl. III), 111–115.

Oliveira, S.A., and Shenk, T.E. (2001). Murine cytomegalovirus M78 protein, a G protein-coupled receptor homologue, is a constituent of the virion and facilitates accumulation of immediate-early viral mRNA. Proc. Natl. Acad. Sci. USA 98, 3237–3242.

Orloff, S.L., Streblow, D.N., Söderberg-Nauclér, C., Yin, Q., Kreklywich, C., Corless, C.L., Smith, P.A., Loomis, C.B., Mills, L.K., Cook, J.W., Bruggeman, C.A., Nelson, J.A., and Wagner, C.R. (2002). Elimination of donor-specific alloreactivity prevents cytomegalovirus-accelerated chronic rejection in rat small bowel and heart transplants. Transplantation 73, 679–688.

Orloff, S.L., Yin, Q., Corless, C.L., Orloff, M.S., Rabkin, J.M., and Wagner, C.R. (2000). Tolerance induced by bone marrow chimerism prevents transplant vascular sclerosis in a rat model of small bowel transplant chronic rejection. Transplantation 69, 1295–1303.

Paul, L.C. (1993). Chronic rejection of organ allografts: Magnitude of the problem. Transplant. Proc. 25, 2024–2025.

Persoons, M.C., Daemen, M.J., Bruning, J.H., and Bruggeman, C.A. (1994). Active cytomegalovirus infection of arterial smooth muscle cells in immunocompromised rats. A clue to herpesvirus-associated atherogenesis? Circ. Res. 75, 214–220.

Persoons, M.C., Daemen, M.J., van Kleef, E.M., Grauls, G.E., Wijers, E., and Bruggeman, C.A. (1997). Neointimal smooth muscle cell phenotype is important in its susceptibility to cytomegalovirus (CMV) infection: a study in rat. Cardiovasc. Res. 36, 282–288.

Persoons, M.C., Stals, F.S., van Dam-Mieras, M.C., and Bruggeman, C.A. (1998). Multiple organ involvement during experimental cytomegalovirus infection is associated with disseminated vascular pathology. J. Pathol. 184, 103–109.

Petrie, B.L., Melnick, J.L., Adam, E., Burek, J., McCollum, C.H., and DeBakey, M.E. (1987). Nucleic acid sequences of cytomegalovirus in cells cultured from human arterial tissue. J. Infect. Dis. *155*, 158–159.

Plump, A.S., Smith, J.D., Hayek, T., Aalto Setala, K., Walsh, A., Verstuyft, J.G., Rubin, E.M., and Breslow, J.L. (1992). Severe hypercholesterolemia and atherosclerosis in apolipoprotein E-deficient mice created by homologous recombination in ES cells. Cell *71*, 343–353.

Pouteil-Noble, C., Ecochard, R., Landrivon, G., Donia-Maged, A., Tardy, J.C., Bosshard, S., Colon, S., Betuel, H., Aymard, M., and Touraine, J.L. (1993). Cytomegalovirus infection – an etiological factor for rejection. A prospective study in 242 renal transplant patients. Transplantation *55*, 851–857.

Qavi, H.B., Melnick, J.L., Adam, E., and DeBakey, M.E. (2000). Frequency of coexistence of cytomegalovirus and Chlamydia pneumoniae in atherosclerotic plaques. Cent. Eur. J. Public Health *8*, 71–73.

Rawlinson, W.D., Farrell, H.E., and Barrell, B.G. (1996). Analysis of the complete DNA sequence of murine cytomegalovirus. J. Virol. *70*, 8833–8849.

Ridker, P.M., Hennekens, C.H., Stampfer, M.J., and Wang, F. (1998). Prospective study of herpes simplex virus, cytomegalovirus, and the risk of future myocardial infarction and stroke. Circulation *98*, 2796–2799.

Ross, R. (1986). The pathogenesis of atherosclerosis – An update. N. Eng. J. Med. *314*, 488–499.

Ross, R. (1993). The pathogenesis of atherosclerosis: a perspective for the 1990s. Nature *362*, 801–809.

Ross, R. (1999). Atherosclerosis – an inflammatory disease. N. Engl. J. Med. *340*, 115–126.

Rott, D., Zhu, J., Burnett, M.S., Zhou, Y.F., Wasserman, A., Walker, J., and Epstein, S.E. (2001). Serum of cytomegalovirus-infected mice induces monocyte chemoattractant protein-1 expression by endothelial cells. J. Infect. Dis. *184*, 1109–1113.

Rott, D., Zhu, J., Zhou, Y.F., Burnett, M.S., Zalles-Ganley, A., and Epstein, S.E. (2003). IL-6 is produced by splenocytes derived from CMV-infected mice in response to CMV antigens, and induces MCP-1 production by endothelial cells: a new mechanistic paradigm for infection-induced atherogenesis. Atherosclerosis *170*, 223–228.

Russell, M.E., Fujita, M., Masek, M.A., Rowan, R.A., and Billingham, M.E. (1993). Cardiac graft vascular disease. Non-selective involvement of large and small vessels. Transplantation *56*, 762–764.

Saniabadi, A.R., Umemura, K., Matsumoto, N., Sakuma, S., and Nakashima, M. (1995). Vessel wall injury and arterial thrombosis induced by a photochemical reaction. Thromb. Haemost. *73*, 868–872.

Sedmak, D.D., Knight, D.A., Vook, N.C., and Waldman, J.W. (1994). Divergent patterns of ELAM-1, ICAM-1, and VCAM-1 expression on cytomegalovirus-infected endothelial cells. Transplantation *58*, 1379–1385.

Sedmak, D.D., Roberts, W.H., Stephens, R.E., Buesching, W.J., Morgan, L.A., Davis, D.H., and Waldman, W.J. (1990). Inability of cytomegalovirus infection of cultured endothelial cells to induce HLA class II antigen expression. Transplantation *49*, 458–462.

Sharples, L.D., Caine, N., Mullins, P., Scott, J.P., Solis, E., English, T.A., Large, S.R., Schofield, P.M., and Wallwork, J. (1991). Risk factor analysis for the major hazards following heart transplantation – rejection, infection, and coronary occlusive disease. Transplantation *52*, 244–252.

Sinzger, C., Grefte, A., Plachter, B., Gouw, A.S.H., The, T.H., and Jahn, G. (1995). Fibroblasts, epithelial cells, endothelial cells and smooth muscle cells are major targets of human cytomegalovirus infection in lung and gastrointestinal tissues. J. Gen. Virol. *76*, 741–750.

Sinzger, C., Schmidt, J., Knapp, J., Kahl, M., Beck, R., Waldman, J., Hebart, H., Einsele, H., and Jahn, G. (1999). Modification of human cytomegalovirus tropism through propagation in vitro is associated with changes in the viral genome. J. Gen. Virol. *80*, 2867–2877.

Slobbe-van Drunen, M.E., Vossen, R.C., Couwenberg, F.M., Hulsbosch, M.M., Heemskerk, J.W., van Dam-Mieras, M.C., and Bruggeman, C.A. (1997). Activation of protein kinase C enhances the infection of endothelial cells by human cytomegalovirus. Virus Res. *48*, 207–213.

Smiley, M.L., Mar, E.C., and Huang, E.S. (1988). Cytomegalovirus infection and viral induced transformation of human endothelial cells. J. Med. Virol. *25*, 213–226.

Sorlie, P.D., Adam, E., Melnick, S.L., Folsom, A., Skelton, T., Chambless, L.E., Barnes, R., and Melnick, J.L. (1994). Cytomegalovirus/herpesvirus and carotid atherosclerosis: the ARIC study. J. Med. Virol. *42*, 33–37.

Span, A.H.M., Frederik, P.M., Grauls, G, van Boven, C.P.A., and Bruggeman, C.A. (1993). CMV induced vascular injury: an electron-microscopic study in the rat. In vivo *7*, 567–574.

Span, A.H.M., Grauls, G., Bosman, F., van Boven, C.P.A., and Bruggeman, C.A. (1992). Cytomegalovirus infection induces vascular injury in the rat. Atherosclerosis *93*, 41–52.

Span, A.H.M., Mullers, W., Miltenburg, A.H.M., and Bruggeman, C.A. (1991a). Cytomegalovirus-induced PMN adherence in relation to an ELAM-I antigen present on infected endothelial cell monolayers. Immunology *72*, 355–360.

Span, A.H.M., van Dam-Mieras, M.C.E., Mullers, W., Endert, J., Muller, A.D., and Bruggeman, C.A. (1991b). The effect of virus infection on the adherence of leukocytes or platelets to endothelial cells. Eur. J. Clin. Invest. *21*, 331–338.

Speir, E., Huang, E.S., Modali, R., Leon, M.B., Shawl, F., Finkel, T., and Epstein, E.S. (1995). Interaction of human cytomegalovirus with p53: possible role in coronary restenosis. Scand. J. Infect. Dis. Suppl. *99*, 78–81.

Speir, E., Modali, R., Huang, E.S., Leon, M.B., Shawl, F., Finkel, T., and Epstein, E.S. (1994). Potential role of human cytomegalovirus and p53 interaction in coronary restenosis. Science *265*, 391–394.

Stals, F.S., Bosman, F., van Boven, C.P.A., and Bruggeman, C.A. (1990). An animal model for therapeutic intervention studies of CMV infection in the immunocompromised host. Arch. Virol. *114*, 91–107.

Stals, F.S., de Clercq, E., and Bruggeman, C.A. (1991). Comparative activity of (S)-1-(3-hydroxy-2-phosphonylmethoxypropyl)cytosine and 9-(1,3-dihydroxy-2-propoxymethyl)-guanine against rat cytomegalovirus infection in vitro and in vivo. Antimicrob. Agents Chemother. 35, 2262–2266.

Stals, F.S., Wagenaar, S.S., and Bruggeman, C.A. (1994). Generalized cytomegalovirus (CMV) infection and CMV-induced pneumonitis in the rat: combined effect of 9-(1,3-dihydroxy-2-propoxymethyl)-guanine and specific antibody treatment. Antiviral Res. 25, 147–160.

Stals, F.S., Wagenaar, S.S., Kloover, J.S., Vanagt, W.Y., and Bruggeman, C.A. (1996). Combinations of ganciclovir and antibody for experimental CMV infections. Antiviral Res. 29, 61–64.

Stals, F.S., Zeytinoglu, A., Havenith, M., de Clercq, E., and Bruggeman, C.A. (1993). Rat cytomegalovirus-induced pneumonitis after allogeneic bone marrow transplantation: effective treatment with (S)-1-(3-hydroxy-2-phosphonylmethoxypropyl)cytosine. Antimicrob. Agents Chemother. 37, 218–223.

Steinhoff, G., You, X.M., Steinmüller, C., Bauer, D., Lohmann-Matthes, M.L., Bruggeman, C.A., and Haverich, A. (1996). Enhancement of cytomegalovirus infection and acute rejection after allogeneic lung transplantation in the rat. Transplantation *61*, 1250–1260.

Steinhoff, G., You, X.M., Steinmüller, C., Boeke, K., Stals, F.S., Bruggeman, C.A., and Haverich, A. (1995). Induction of endothelial adhesion molecules by rat cytomegalovirus in allogeneic lung transplantation in the rat. Scand. J. Infect. Dis. Suppl. *99*, 58–60.

Steinmüller, C., Steinhoff, G., Bauer, D., You, X.M., Denzin, H., Franke-Ullmann, G., Hausen, B., Bruggeman, C.A., Wagner, T.O., Lohmann-Matthes, M.L., and Emmendorffer, A. (1997). Analysis of leukocyte activation during acute rejection of pulmonary allografts in noninfected and cytomegalovirus-infected rats. J. Leukoc. Biol. *61*, 40–49.

Stovin, P.G.I., Sharples, L., Hutter, J.A., Wallwork, J., and English, T.A.H. (1991). Some prognostic factors for the development of transplant-related coronary artery disease in human cardiac allografts. J. Heart Lung Transplant. *10*, 38–44.

Streblow, D.N.., Kreklywich, C., Yin, Q., De La Melena, V.T., Corless, C.L., Smith, P.A., Brakebill, C., Cook, J.W., Vink, C., Bruggeman, C.A., Nelson, J.A., and Orloff, S.L. (2003). Cytomegalovirus-mediated up-regulation of chemokine expression correlates with the acceleration of chronic rejection in rat heart transplants. J. Virol. *77*, 2182–2194.

Streblow, D.N., Orloff, S.L., and Nelson, J.A. (2001). The HCMV chemokine receptor US28 is a potential target in vascular disease. Curr. Drug Targets Infect. Disord. *1*, 151–158.

Streblow, D.N., Söderberg-Nauclér, C., Vieira, J., Smith, P., Wakabayashi, E., Ruchti, F., Mattison, K., Altschuler, Y., and Nelson, J.A. (1999). The human cytomegalovirus chemokine receptor US28 mediates vascular smooth muscle cell migration. Cell *99*, 511–520.

Tanaka, H., Suzuki, A., Schwartz, D., Sukhova, G., and Libby, P. (1995). Activation of smooth muscle and endothelial cells following balloon injury. Ann. N. Y. Acad. Sci. *748*, 526–529.

Taub, D.D., Proost, P., Murphy, W.J., Anver, M., Longo, D.L., van Damme, J., and Oppenheim, J.J. (1995). Monocyte chemotactic protein-1 (MCP-1), protein-2 and protein-3 are chemotactic fot human T-lymphocytes. J. Clin. Invest. *95*, 1370–1376.

Ten-Napel, H.H., Houthoff, H.J., and The, T.H. (1984). Cytomegalovirus hepatitis in normal and immune compromised hosts. Liver *4*, 184–194.

Tumilowicz, J.J., Gawlik, M.E., Powell, B.B., and Trentin, J.J. (1985). Replication of cytomegalovirus in human arterial smooth muscle cells. J. Virol. *56*, 839–845.

Ustinov, J.A., Bruggeman, C.A., Häyry, P.J., and Lautenschlager, I.T. (1994). Cytomegalovirus-induced class II expression in rat kidney. Transplant. Proc. *26*, 1729.

Ustinov, J.A., Loginov, R.J., Mattila, P.M., Nieminen, V.K., Suni, J.I., Häyry, P.J., and Lautenschlager, I.T. (1991). Cytomegalovirus infection of human kidney cells in vitro. Kidney Int. 40, 954–960.

Uys, C., and Rose, A. (1984). Pathologic findings in long-term cardiac transplants. Arch. Pathol. Lab. Med. 108, 112–116.

van Cleef, K.W.R., Scaf, W.M.A., Maes, K., Kaptein, S.J.F., Beuken E., Beisser, P.S., Stassen, F.R.M., Grauls, G.E.L.M., Bruggeman, C.A., and Vink, C. (2004). The rat cytomegalovirus homologue of parvoviral *rep* genes, r127, encodes a nuclear protein with single- and double-stranded DNA-binding activity that is dispensable for virus replication. J. Gen. Virol. 85, 2001–2013.

van Dam-Mieras, M.C., Muller, A.D., van Hinsbergh, V.W., Mullers, W.J., Bomans, P.H., and Bruggeman, C.A. (1992). The procoagulant response of cytomegalovirus infected endothelial cells. Thromb. Haemost. 68, 364–370.

van Dorp, W.T., Jonges, E., Bruggeman, C.A., Daha, M.R., van Es, L.A., and van der Woude, F.J. (1989). Direct induction of MHC class I, but not class II, expression on endothelial cells by cytomegalovirus infection. Transplantation 48, 469–472.

Vanstapel, M.J., and Desmet, V.J. (1983). Cytomegalovirus hepatitis: a histological and immunohisto-chemical study. Appl. Pathol. 1, 41–49.

Vieira, J., Schall, T.J., Corey, L., and Geballe, A.P. (1998). Functional analysis of the human cytomeg-alovirus US28 gene by insertion mutagenesis with the green fluorescent protein gene. J. Virol. 72, 8158–8165.

Vink C., Beuken, E., and Bruggeman, C.A. (1996). Structure of the rat cytomegalovirus genome termini. J. Virol. 70, 5221–5229.

Vink C., Beuken, E., and Bruggeman, C.A. (2000). Complete DNA sequence of the rat cytomegalovirus genome. J. Virol. 74, 7656–7665.

Visseren, F.L., Bouter, K.P., Pon, M.J., Hoekstra, J.B., Erkelens, D.W., and Diepersloot, R.J. (1997). Patients with diabetes mellitus and atherosclerosis; a role for cytomegalovirus? Diabetes Res. Clin. Pract. 36, 49–55.

Vliegen, I., Duijvestijn, A., Grauls, G., Herngreen, S., Bruggeman, C., and Stassen, F. (2004). Cytomegalovirus infection aggravates atherogenesis in apoE knockout mice by both local and systemic immune activation. Microbes. Infect. 6, 17–24.

Vliegen, I., Stassen, F., Grauls, G., Blok, R., and Bruggeman, C. (2002). MCMV infection increases early T-lymphocyte influx in atherosclerotic lesions in apoE knockout mice. J. Clin. Invest. 25, S159-S171.

von Willebrand, E., Pettersson, E., Ahonen, J., and Häyry, P. (1986). CMV infection, class II antigen ex-pression, and human kidney allograft rejection. Transplantation 42, 364–367.

Vossen, R.C., Derhaag, J.G., Slobbe-van Drunen, M.E., Duijvestijn, A.M., van Dam- Mieras, M.C., and Bruggeman, C.A. (1996). A dual role for endothelial cells in cytomegalovirus infection? A study of cytomegalovirus infection in a series of rat endothelial cell lines. Virus Res. 46, 65–74.

Vossen, R.C., Persoons, M.C., Slobbe-van Drunen, M.E., Bruggeman, C.A., and van Dam-Mieras, M.C. (1997). Intracellular thiol redox status affects rat cytomegalovirus infection of vascular cells. Virus Res. 48, 173–183.

Waldman, W.J., Adams, P.W., Orosz, C.G., and Sedmak, D.D. (1992). T-lymphocyte activation by cy-tomegalovirus-infected, allogeneic cultured human endothelial cells. Transplantation 54, 887–896.

Waldman, W.J., and Knight, D.A. (1996). Cytokine-mediated induction of endothelial adhesion mol-ecule and histocompatibility leukocyte antigen expression by cytomegalovirus-activated T-cells. Am. J. Pathol. 148, 105–119.

Waldman, W.J., Knight, D.A., Adams, P.W., Orosz, C.G., and Sedmak, D.D. (1993). In vitro induc-tion of endothelial HLA class II antigen expression by cytomegalovirus-activated CD4+ T-cells. Transplantation 56, 1504–1512.

Waldman, W.J., Roberts, W.H., Davis, D.H., Williams, M.V., Sedmak, D.D., and Stephens, R.E. (1991). Preservation of natural endothelial cytopathogenicity of cytomegalovirus by propagation in endothe-lial cells. Arch. Virol. 117, 143–164.

Waldman, W.J., Sneddon, J.M., Stephens, R.E., and Roberts, W.H. (1989). Enhanced endothelial cy-topathogenicity induced by a cytomegalovirus strain propagated in endothelial cells. J. Med. Virol. 28, 223–230.

Winters, G.L., Kendall, T.J., Radio, S.J., Wilson, J.E., Constanzo-Nordin, M.R., Switzer, B.L., Remmenga, J.A., and McManus, B.M. (1990). Post-transplant obesity and hyperlipidemia predictors of severity of coronary arteriopathy in failed human heart allografts. J. Heart Lung Transplant. 9, 364–371.

Woodroffe, S.B., Garnett, H.M., and Danis, V.A. (1993). Interleukin-1 production and cell-activation response to cytomegalovirus infection of vascular endothelial cells. Arch. Virol. *133*, 295–308.

Wu, T.C., Hruban, R.H., Ambinder, R.F., Pizzorno, M., Cameron, D.E., Baumgartner, W.A., Reitz, B.A., Hayward, G.S., and Hutchins, G.M. (1992). Demonstration of cytomegalovirus nucleic acids in the coronary arteries of transplanted hearts. Am. J. Pathol. *140*, 739–747.

Yagyu, K., Steinhoff, G., Duijvestijn, A.M., Bruggeman, C.A., Matsumoto, H., and van Breda Vriesman, P.J. (1992). Reactivation of rat cytomegalovirus in lung allografts: an experimental and immunohisto-chemical study in rats. J. Heart Lung Transplant. *11*, 1031–1040.

Yagyu, K., van Breda Vriesman, P.J., Duijvestijn, A.M., Bruggeman, C.A., and Steinhoff, G. (1993). Reactivation of cytomegalovirus with acute rejection and cytomegalovirus infection with obliterative bronchiolitis in rat lung allografts. Transplant. Proc. 25, 1152–1154.

Yilmaz, S., Koskinen, P.K., Kallio, E., Bruggeman, C.A., Häyry, P.J., and Lemström, K.B. (1996). Cytomegalovirus infection-enhanced chronic kidney allograft rejection is linked with intercellular ad-hesion molecule-1 expression. Kidney Int. 50, 526–537.

You, X.M., Steinmüller, C., Wagner, T.O., Bruggeman, C.A., Haverich, A., Steinhoff, G. (1996). Enhancement of cytomegalovirus infection and acute rejection after allogeneic lung transplantation in the rat: virus-induces expression of major histocompatibility complex class II antigens. J. Heart Lung Transplant. *15*, 1108–1119.

Zhang, S.H., Reddick, R.L., Piedrahita, J.A., and Maeda, N. (1992). Spontaneous hypercholesterolemia and arterial lesions in mice lacking apolipoprotein E. Science 258, 468–471.

Zhou, Y.F., Guetta, E., Yu, Z.X., Finkel, T., and Epstein, S.E. (1996a). Human cytomegalovirus increases modified low density lipoprotein uptake and scavenger receptor mRNA expression in vascular smooth muscle cells. J. Clin. Invest. 98, 2129–2138.

Zhou, Y.F., Leon, M.B., Waclawiw, M.A., Popma, J.J., Yu, Z.X., Finkel, T., and Epstein, S.E. (1996b). Association between prior cytomegalovirus infection and the risk of restenosis after coronary atherec-tomy. N. Eng. J. Med. 335, 624–630.

Zhou, Y.F., Shou, M., Guetta, E., Guzman, R., Unger, E.F., Yu, Z.X., Zhang, J., Finkel, T., and Epstein, S.E. (1999a). Cytomegalovirus infection of rats increases the neointimal response to vascular injury without consistent evidence of direct infection of the vascular wall. Circulation *100*, 1569–1575.

Zhou, Y.F., Shou, M., Harrell, R.F., Yu, Z.X., Unger, E.F., and Epstein, S.E. (2000). Chronic non-vas-cular cytomegalovirus infection: effects on the neointimal response to experimental vascular injury. Cardiovasc. Res. 45, 1019–1025.

Zhou, Y.F., Yu, Z.X., Wanishsawad, C., Shou, M., and Epstein, S.E. (1999b). The immediate-early gene products of human cytomegalovirus increase vascular smooth muscle cell migration, proliferation, and expression of PDGF beta-receptor. Biochem. Biophys. Res. Commun. 256, 608–613.

Zhu, J.H., Quyyumi, A.A., Norman, J.E., Csako, G., Waclawiw, M.A., Shearer, G.M., and Epstein, S.E. (2000). Effects of total pathogen burden on coronary artery disease risk and C-reactive protein levels. Am. J. Cardiol. 85, 140–146.

Zhu, H., Shen, Y., and Shenk, T. (1995). Human cytomegalovirus IE1 and IE2 proteins block apoptosis. J. Virol. 69, 7960–7970.

(Lockridge et al., 1999; Tarantal et al., 1998), but the expense of these primates makes widespread use of this model impractical for vaccine studies. Mouse and rat cytomegaloviruses have been studied as models of CMV disease, but these CMVs do not cross the placentas of their respective host species, and are therefore not useful for vaccine studies which target prevention of congenital infection. Fortunately, the guinea-pig CMV provides a distinctly useful model which is very relevant to HCMV vaccine studies. In contrast to the cytomegaloviruses of other small mammals, GPCMV is unique in its ability to cross the placenta and cause fetal infection. Thus, there has been extensive interest over many years in the development of this model for the study of immunity and pathogenesis. This chapter provides an overview of research on GPCMV, a summary of the valuable features of this model, and an update on the recent advances in the study of this virus that have facilitated vaccine and pathogenesis studies.

## History of the GPCMV

GPCMV was first characterized approximately 85 years ago, and, remarkably, the history of the GPCMV parallels in many ways the description of the syndrome of congenital HCMV infection (reviewed by Bia et al., 1983). GPCMV was first recognized by histopathologic analysis, by virtue of the identification of classic viral inclusions in the guinea-pig salivary gland (Jackson, 1920). As was initially the case for HCMV, early investigators mistakenly identified the virus as a putative protozoan species. By the late 1920s, however, the viral etiology of these lesions was correctly appreciated, when it was noted that the agent produced histologic changes very similar to those observed in cells infected with herpes simplex virus (Cole and Kuttner, 1926). It was later shown that salivary gland homogenates containing this agent were infectious, and could produce interstitial pneumonitis following intratracheal instillation in guinea-pigs, establishing the importance of the salivary gland as a reservoir for infection (Kuttner and T'ung, 1935). Early on in the study of GPCMV, there was interest in the nature of the immune response to infection, and it was shown in a passive vaccination study that experimental immunity could be produced against the virus, which in turn could protect naïve animals from intracerebral inoculation with GPCMV (Kuttner 1927; Kuttner and Wang, 1934). It was quickly appreciated, however, that humoral immunity to the virus was not completely protective against establishment of infection. Although studies indicated that immune serum was capable of neutralizing infectivity in vitro, animals which were inoculated with virus and serum at separate sites on the thigh went on to develop salivary gland infection, indicating that serum did not protect against development of a latent infection (Andrewes, 1930). These early observations are still germane to the study of vaccination today, since they suggest that humoral responses alone are unlikely to be sufficient to ensure true "sterilizing immunity" against GPCMV in experimental models of infection and disease.

Shortly following these studies, the most important aspect of the biology of GPCMV, namely, its ability to infect the guinea-pig placenta, was described. In the first studies of this important characteristic of the virus, it was found that intraplacental inoculation of guinea-pig fetuses with GPCMV in utero resulted in the generation of viral inclusions and histopathology at the inoculation site. Moreover, intraplacental inoculation also produced a widespread infection, not only of the fetus, but also the pregnant dam (Markham and Hudson, 1936). The investigator who reported these findings astutely noted the similarity between the pathology observed in the guinea-pig and a disease which had been described in the human fetus. This human disease, which was associated with stillbirths, would later be designated as cytomegalic inclusion disease of the newborn. Demonstration of permissiveness of placenta for GPCMV infection following direct inoculation with virus was a landmark observation made by these investigators. However, it would be another forty years before it would be appreciated that maternal inoculation with GPCMV at a distant site could also, like direct placental inoculation, lead to widespread infection of mother and fetus, thus setting the stage for more detailed congenital infection and maternal vaccine studies.

## Pathogenesis of GPCMV infection

Virtually all studies of GPCMV pathogenesis have been conducted with the strain originally isolated by Hartley in 1957, from infected guinea-pig salivary glands, which was provided to the American Type Culture Collection (ATCC) as strain 22122 (Hartley et al., 1957). Viral workpools for animal inoculation may be purified either from tissue culture (GPCMV-TC), or from homogenates of salivary glands (GPCMV-SG): either preparation is capable of inducing disease in guinea-pigs, including congenital infection, but SG preparations are typically more virulent, and cause disease at a lower inoculum. The basis for the apparent increased virulence of SG passaged virus is not known. Guinea-pig pathogenesis experiments can be performed in outbred (Hartley) strain guinea-pigs, and also in inbred (strain 2, JY-9) guinea-pigs. An advantage of using inbred guinea-pig strains is that this allows, in principle, more meaningful comparison, within an experiment, of the cellular immune factors that may play a role in protection against intrauterine CMV infections (Griffith et al., 1986). The disease manifestations of GPCMV infection have been well-characterized in pregnant animals, but GPCMV is also useful for pathogenesis studies in non-pregnant guinea-pigs (described below). Viral inoculation can be performed by subcutaneous (s.c.), intraperitoneal (i.p.), intracardiac (i.c.), or intranasal (i.n.) routes. Although infection at a mucosal surface may be the most relevant with respect to natural modes of acquisition in humans, inoculation by the sc route is generally the most convenient approach from a technical perspective, and is generally less stressful to the pregnant animal. In addition, although placental and fetal infections are known to occur after infection of pregnant animals by all of these above-mentioned routes, the transfer of GPCMV to placentas and fetuses is more efficient in mothers inoculated by the sc route (Griffith et al., 1990).

Following experimental infection of guinea-pigs with GPCMV, viremia ensues, irrespective of the route of challenge, and this appears to be the mechanism by which end-organ disease (including infection of the placental-fetal unit) occurs. Viremia persists for approximately 10 days post-inoculation of nonpregnant animals, and generalized infection and disease involving the lungs, spleen, liver, kidney, thymus, pancreas, and brain can be demonstrated for approximately 3 weeks following acute infection (Hsuing et al., 1978). Salivary gland virus, first demonstrable at 7–10 days post infection, reaches maximal levels at 3 weeks, and persists at high level for at least 10 weeks. Salivary glands may be harvested during this time window for preparation of stocks for future in vivo experiments.

Although the greatest value of the GPCMV model is in the study of congenital infection, a number of models of GPCMV-induced disease have been described in non-pregnant animals. Of considerable interest with respect to GPCMV pathogenesis is the description of a mononucleosis-like syndrome associated with acute infection, a syndrome of interest in light of the ability of HCMV to cause so-called "heterophile negative" mononucleosis. Following s.c. inoculation of Hartley guinea-pigs with GPCMV-SG, atypical lympho-cytosis has been reported in peripheral blood, and a constellation of findings including splenomegaly, lymphadenopathy, anemia, and neutropenia has been observed (Diosi and Georgescu, 1974; Griffith et al., 1981). Interestingly, neutrophil dysfunction has been ob-served in this setting, suggesting an immunosuppressive effect of acute GPCMV infection (Yourtee et al., 1982). Inbred strains of guinea-pigs, in particular strain 2, are, in general, more prone to GPCMV disease than are outbred animals, and these animals are at risk for the development of severe pneumonitis (Bia et al., 1982). In guinea-pigs that are phar-macologically immunosuppressed, with agents such as cyclophosphamide or cyclosporin, extensive GPCMV-induced disease is noted, including pneumonitis, allowing study of an interesting model of disease which mimics the clinical manifestations of HCMV disease observed in immunosuppressed patients. This immunosuppression model provides a use-ful system in which to evaluate antiviral therapies (Aquino-de Jesus and Griffith, 1989). Unfortunately, GPCMV is resistant to ganciclovir at medically relevant doses, precluding meaningful study in guinea-pigs of the most important treatment available for HCMV infections (Matthews and Boehme, 1988). However, GPCMV is very susceptible to several of the more recently developed experimental antivirals, including cyclic HPMPC, and BAY 38–4766, and the efficacy of these compounds has been demonstrated in animal experi-ments (Bourne et al., 2000; Schleiss et al., 2005). In addition to the immunocompromised model, other useful models of GPCMV-induced disease which have been developed include a neuropathogenesis model, induced by direct intracerebral inoculation of virus, producing a glial nodule encephalitis (Booss et al., 1988), and a labyrinthitis model, induced by direct viral inoculation into the cochlea, leading to deafness (Keithley et al., 1988). Experiments have also been described based on infection of newborn guinea-pigs, which produces a gen-eralized, viremic infection, resulting in hepatitis, pneumonitis, and central nervous system involvement (Booss et al., 1989; Griffith et al., 1985a; Zheng et al., 1987). Although many of these models are quite valuable in the study of the pathogenesis and treatment of CMV disease, clearly the greatest strength of the guinea-pig model lies in the study of infection during pregnancy, based on the ability of GPCMV to infect the placenta, and the pup in utero (see below). Accordingly, those studies which focus on congenital transmission are arguably of the greatest significance in the study of GPCMV.

## Immune response to GPCMV infection

### Humoral immune response

As noted, the importance of immunity to GPCMV was first investigated over seventy years ago, when early studies of the virus-neutralizing effect of immune serum obtained from infected animals were performed. In spite of progress in characterization of the antibody response to GPCMV since these early reports, much still remains to be learned about the nature of humoral targets in the setting of acute infection. Early efforts in characterization of the humoral response focused on the development of GPCMV-specific polyclonal anti-bodies in rabbits, in order to differentiate GPCMV from a number of other related, endog-enous guinea-pig herpesviruses (Bia et al., 1980b). Although these studies enabled the development of ELISA and IFA assays, there was no information available at this time about the molecular identity of specific GPCMV proteins. In a subsequent study, immunoblot analyses of the humoral response to infection in both pregnant and non-pregnant guinea-pigs were performed, and defined the evolution of the antibody response to 12 GPCMV polypeptides during primary infection. Interestingly, immune responses were delayed, and were of lower magnitude, in pregnant animals compared to non-pregnant animals. This study provided the first information about candidate immunogenic GPCMV polypeptides, although the specific genes encoding these polypeptides remained unknown (Bu and Griffith, 1990). In another immunoblot study of sera from GPCMV infected, non-pregnant guinea-pigs, at least 18 GPCMV polypeptides were identified using immune sera. Some of these polypeptides appeared to have immunologic cross-reactivity with HCMV proteins, but, again, the molecular identity of these viral proteins was not explored (Kacica et al., 1990). Some progress was made in the characterization of specific GPCMV proteins through the generation of monoclonal antibodies. This approach allowed characterization of a virion structural protein of ~160–180 kDa, a 50 kDa nuclear nonstructural protein, and a 76 kDa matrix protein, although the precise identities of these proteins and their potential roles as immune targets was not examined (Jones et al., 1994; Nogami-Satake and Tsutsui, 1988; Tsutsui, et al., 1986). Ultimately, the use of cloned, recombinant technologies was required to assign a molecular identity to specific GPCMV proteins encoded by the viral genome, and these studies in turn have been applied to the testing of subunit vaccine candidates in the guinea-pig model (see below).

## Cellular immune response

As is the case for study of humoral responses, the contribution of cellular immune responses to infection with GPCMV has not been well explored, owing in part to the paucity of reagents for cell-mediated immune (CMI) assays in guinea-pigs. However, several lines of evidence support the concept that CMI plays an important role in clearance of GPCMV infection. In inbred strain 2 animals, peripheral blood mononuclear cell (PBMC)-mediated cytolytic activity was described against GPCMV-infected syngeneic and allogeneic targets, and was unaffected by T-cell depletion (Harrison and Myers, 1988). Other studies have examined the role of T-cells in antiviral responses to infection. An enhancement of cutaneous basophil responses was found after challenge with GPCMV in immune animals, and this occurred in association with stimulation of T-cell-dependent areas of lymph nodes, suggesting a role for T-cells in viral clearance (Griffith et al., 1982a). In non-pregnant animals, CD4$^+$ and CD8$^+$ immune responses have been demonstrated against specific GPCMV proteins, by FACS analysis of T-cells with specific monoclonal antibodies, recovered following intradermal challenge with viral antigens (Belkaid et al., 2002; Lacayo et al., unpublished data). Although CMI is presumed to play a role in viral clearance, some evidence suggests that cellular responses during acute GPCMV infection in pregnancy could play a role in potentiating injury of the fetus. In pregnant strain 2 guinea-pigs, a lack of suppression of cytolysis of GPCMV-infected syngeneic fetal cells was associated with poor pregnancy outcomes, including pup mortality, runting, and conceptus loss (Harrison and Myers, 1989). Similarly, enhanced NK cell activity during early gestation of GPCMV-infected dams was correlated with poor pregnancy outcomes (Harrison and Caruso, 2000). Much remains to be learned about cellular immunity in these models. Better definition of leukocyte subsets and lymphocyte populations in blood and lymphoid tissue of guinea-pigs should facilitate an improved understanding of cellular immunity in all of the guinea-pig models of infectious diseases, including the GPCMV model (Takizawa et al., 2004).

Acute GPCMV infection appears to induce diverse immunomodulatory and, potentially, immunoevasive effects on the guinea-pig immune response. During acute GPCMV mononucleosis, when guinea-pigs are viremic and neutropenic, mobilization of neutrophils to sites of inflammatory stimuli was impaired (Yourtee et al., 1982). Neutrophils from infected animals were noted to have diminished bacterial killing in response to a bacterial stimulus, suggesting that intercurrent GPCMV infection was immunosuppressive. These effects were further explored in a study of neutrophil migration, which demonstrated that acute GPCMV infection resulted in abnormalities of neutrophil-directed migration toward C5-derived chemotactic factors (Tannous and Myers, 1983). Other leukocyte subpopulations may be impaired in the setting of acute GPCMV infection. Alterations in macrophage function have also been described in the setting of experimental interstitial pneumonitis due to GPCMV (Miller et al., 1985). In another study, the nonspecific functional capacity of spleen cells, taken from female guinea-pigs with primary GPCMV infection, was assessed using lipopolysaccharide (LPS), a B-cell mitogen, and concanavalin A (Con A), a T-cell mitogen. Proliferative responses to the two mitogens were found to be significantly depressed in animals inoculated with GPCMV, as compared to controls (Griffith et al., 1984). These differences were also found to be present in guinea-pigs infected as newborn pups, and were associated with significant depletion of the T-cell population in the thymus (Zheng et al., 1987). Other examples of GPCMV-induced immune modulation have been described, such as down-regulation of cell-surface expression of MHC class I expression (Lacayo et al., 2003). This presumably facilitates evasion of host CTL responses, although the impact of class I down-regulation on other aspects of the guinea-pig immune response is unclear. A better understanding of the diverse viral function(s) presumably responsible for these immunomodulatory effects will facilitate improved vaccine and pathogenesis studies in the GPCMV model.

## Biology of the GPCMV virus

### GPCMV morphology and ultrastructure

The ultrastructure of the GPCMV, both in salivary gland and in tissue culture, was initially analyzed by electron microscopy (EM) in the late 1960s (Middelkamp et al., 1967). These studies were later extended, both in cell culture, and in salivary gland studies (Fong et al., 1979; Fong et al., 1980). An interesting feature of GPCMV ultrastructure is the presence of "dense bodies" (DBs), which do not contain virus capsids, but are otherwise generated and enveloped in a manner similar to virus particles. Immuno-EM of GPCMV morphogenesis has revealed that enveloped dense virions and dense bodies shared common envelope antigens (Fong et al., 1979). The finding that DBs are present in GPCMV-infected cells is an intriguing similarity between this virus and HCMV. It has been shown that GPCMV DBs can be purified away from enveloped virions using gradient centrifugation techniques similar to those developed for HCMV (Schleiss et al., 1999). These observations may enable study of relevant vaccine strategies in the GPCMV model, insofar as HCMV DBs have been proposed to be a potentially useful vaccine candidate, since they contain a full repertoire of viral proteins capable of eliciting a broad repertoire of antiviral immune responses (Pepperl-Klindworth et al., 2002). When GPCMV is reconstituted from infectious bacterial artificial chromosome (BAC) DNA following transfection in cell culture, DBs are also observed (Figure 25.1; McGregor and Schleiss, 2001a). This observation further authenticates the utility of using BAC-derived viruses for the study of pathogenesis of viral infection in vivo.

## GPCMV DNA and genome configuration

The first progress toward the characterization of the molecular biology of GPCMV came in the early 1980s, when density gradient centrifugation of GPCMV DNA, and restriction endonuclease analysis of the viral genome, were first described (Bia et al., 1980a). A comprehensive analysis of GPCMV DNA resulted in the generation of *Hind*III, *Eco*RI, and *Xba*I restriction maps of the viral genome, and a subset of these restriction fragments were cloned in the plasmid, pBR322 (Gao and Isom, 1984; Isom et al., 1984). Analysis of restriction endonuclease maps and cross-hybridization studies indicated that the viral genome consisted of a long unique sequence, with terminal repeat sequences. In contrast to HCMV, the genome did not contain internal repeat regions. Based on restriction endonuclease mapping of cloned fragments, the genome was estimated to be ~239 kbp, corresponding to a molecular mass of 158 kDa. Of interest, HCMV gene probes were found to

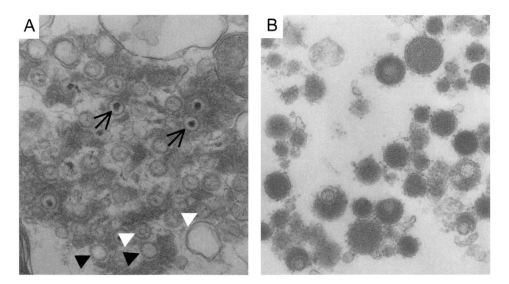

**Figure 25.1** Ultrastructural analysis of GPCMV reconstituted from infectious bacterial artificial chromosome (BAC). BAC DNA was transfected into guinea-pig lung fibroblasts and approximately 1 week following transfection was subjected to electron microscopy. Panel A (left), nucleocapsids include single-ring, type "A" capsids (black arrowhead) and double-ring "B" capsids (white arrowhead). Open arrow points to "C" capsids containing whorls of DNA. Panel B (right), dense bodies are reconstituted by BAC DNA transfection. The generation of these electron-dense particles, which do not contain capsids, represents an interesting similarity between the morphogenesis of GPCMV and HCMV. Phosphotungstic acid staining of BAC DNA-transfected cells: A, ×30,000; B, ×100,000.

cross-hybridize with GPCMV DNA fragments, resulting in the prediction that blocks of genes would be highly conserved between the two viruses.

One novel aspect of GPCMV genome structure which was concluded from these studies was the finding that GPCMV molecules exist in two forms in cell culture. In the predominant form, unit length genomes were noted to share sequence homology at the two termini, whereas in the less abundant genome isoform, one terminal fragment was noted to be smaller than (by ~0.7 kDa), and lacked homology with, the fragment at the other end of the genome (Gao and Isom, 1984). The mechanism(s) by which these genomic isoforms are generated in the setting of viral DNA replication have been elucidated in a recent series of elegant studies (McVoy et al., 1997). During generation of circularized GPCMV DNA during replication, two restriction fragments were identified which contained fused terminal sequences, and had sizes consistent with the presence of single or double terminal repeats. Cleavage to form the two genome types was found to occur at two sites, contained within a 64 bp element. Viral mutants were generated which authenticated the role of these *cis* elements in genomic cleavage. In a follow-up study, it was demonstrated that the double repeats that are formed by circularization of infecting genomes are rapidly converted to single repeats, such that the junctions between genomes within replicative concatemers formed late in infection contain single copies of the terminal repeat. However, although each cleavage event begins with a single repeat within a concatemer, two repeats are produced, one at each of the resulting termini, demonstrating that terminal repeat duplication occurs in conjunction with cleavage, and that both duplicative and nonduplicative cleavage events may occur concurrently (Nixon and McVoy, 2002). It remains to be elucidated whether these variations in GPCMV DNA replication, or the generation of multiple genome isoforms during replication, portends any implications for the pathogenesis of infection.

## GPCMV gene expression

As with other CMVs, GPCMV gene expression can be conveniently categorized into immediate-early (IE), early, or late gene expression, using metabolic blockade with cycloheximide and phosphonoacetic acid (Isom and Yin, 1990; Yin et al., 1990). IE gene expression has been extensively characterized. Mapping studies, utilizing cloned genomic fragments, showed that IE genes are located at three distinct sites on the viral genome. A total of 17 GPCMV IE transcripts were identified utilizing this approach, and 9 IE transcripts coded for by three specific regions of the genome (regions I, II, and III) were characterized in detail. By $^{35}$S pulse-label analysis, five major IE protein species were identified in GPCMV-infected cells, although the molecular identity of these proteins remains to be characterized (Isom and Yin, 1990).

Less is known about GPCMV early and late gene expression. Hybridization studies of cDNA probes with cloned genomic fragments indicated that early RNAs were encoded by 16 of 18 clones, but also that a plurality (~35%) of early transcripts hybridized with two cloned fragments, the *Hind*III 'N' and 'L' fragments, corresponding to the left and right termini of the genome, respectively. Late RNAs were found to be transcribed from cloned fragments representing ~99% of the viral genome (Yin et al., 1990). The kinetic class of several GPCMV genes has been determined by Northern analyses (see below).

## GPCMV gene products and status of genomic sequence

As noted above, although much information has been elucidated about the genome arrangement and structure of GPCMV by molecular cloning and cross-hybridization studies, until recently there has been little data available about the molecular identity of specific GPCMV gene products. However, over the past 10 years there have been significant advances in the molecular characterization of GPCMV. The first GPCMV gene to be identified was the glycoprotein B (gB) gene. The approach of low-stringency Southern blot hybridization of a HCMV gB probe with GPCMV restriction digests was used to pinpoint the region in the GPCMV genome which contained the gB homolog. This homolog was found to map to the *Hin*dIII "K" and "Q" fragments. This data allowed orientation of the GPCMV genome relative to other cytomegaloviruses, and facilitated the identification of other homologs. GPCMV gB is transcribed as an "early" gene (6.8 kb mRNA), and encodes a 901 aa protein with 42% homology to HCMV gB (Schleiss, 1994). The protein is cleaved into amino- and carboxy-terminal subunits of approximately 90 and 58 kDa. GPCMV gB is a major target of the neutralizing antibody response following natural GPCMV infection, and antibodies to gB are uniformly found in GPCMV seropositive animals (Schleiss, 2002; Schleiss and Jensen, 2003), making this an ideal candidate protein for subunit vaccine studies designed to prevent congenital GPCMV infection and disease (see below).

The characterization of the GPCMV gB homolog provided a framework for study of the organization of the viral genome, and for identification of other conserved genes. The glycoprotein H (gH; UL75 homolog) was found to map to the *Xba*I "T" region within the *Hin*dIII "A" fragment of the genome. This ORF is 723 aa in length, and the mature glycoprotein migrates at $M_r$ ~85 kDa in SDS-PAGE (Brady and Schleiss, 1996). The gH chaperone, glycoprotein L (gL; UL115 homolog) maps to a GPCMV genome region collinear with HCMV UL115, within the *Hin*dIII "B" region (Paglino et al., 1999). Both gH and gL of GPCMV are immunoreactive with anti-GPCMV antibodies, confirming that these conserved glycoprotein homologs are targets of immune response in natural GPCMV infection. ORFs with identity to HCMV UL74, the glycoprotein O member of the HCMV gCIII complex, and UL73 and UL100, the gN and gM proteins, have also been identified, although the role of these proteins in protective immunity in guinea-pigs, and their potential as vaccine candidates, remains to be assessed.

In addition to the conserved structural glycoproteins, other GPCMV genes, both structural and nonstructural, have been characterized. Sequence analysis of regions adjacent to the gB (GP55) homolog successfully identified the conserved DNA polymerase, GP54 (*pol*). GPCMV *pol* is transcribed as a 3.9 kb message with "early" kinetics, and encodes an ORF of 1094 amino acids (Schleiss, 1995). The GP97 phosphotransferase homolog has been mapped to *Hin*dIII "C" (Fox and Schleiss, 1997). A promoter with strong similarities, both in structure and *cis*-recognition sequences, to the HCMV major immediate-early promoter (MIEP) has been identified in the *Hin*dIII "E" region (Schleiss, 2002). Tegument phosphoprotein homologs have also been characterized, including a ~70 kDa phosphoprotein corresponding to the GPCMV UL83 homolog, identified in both virion and dense body fractions (Schleiss et al., 1999). Of particular interest are conserved and novel genes with putative immunomodulatory functions. A homolog of the HCMV-encoded G-protein coupled receptor homolog, UL33 (GP33), has been reported, and was found to be transcribed with "late" gene kinetics (Liu and Biegalke, 2001). Recently, a protein with

cinated animals showed acute viremia after similar challenge with virulent virus, but infection was less generalized than that in control animals, and GPCMV was not isolated from the fetuses of these vaccinated mothers. Virus-neutralizing titers were present in dams vaccinated by both strategies, but there was limited data concerning the qualitative aspects of the immune responses, because of the lack of any characterization of GPCMV structural proteins at the time these studies were performed. A subsequent vaccine study in non-pregnant strain 2 animals showed protection against pneumonitis, although once again the precise viral targets of protective immune responses were not known (Bia et al., 1982).

In a later vaccine study using an inbred model of congenital GPCMV infection, an immunoaffinity-purified glycoprotein was found to induce strong antibody and CMI responses when administered with Freund's adjuvant, and newborn pups were protected against congenital infection and disease (Britt and Harrison, 1994; Harrison et al., 1995). In another subunit vaccine study, immunity conferred by immunization with envelope glycoproteins resulted in substantial protection against pup mortality, with a reduction in pup mortality from 56% to 14% in the immunized group compared with controls ($P$ <0.001). Among live-born pups in this study, glycoprotein immunization also substantially reduced infection rates. GPCMV was isolated from 24 of 54 live-born pups born to immunized mothers, compared with 16 of 20 live-born pups born to controls, indicating that immunization significantly reduced in utero transmission in surviving animals ($P$ <0.01). Since a number of GPCMV glycoproteins were present in the glycoprotein eluate, it was difficult to be certain which protein(s) were key in protection (Bourne et al., 2001). Although both antibody and cellular responses have been noted in these adjuvanted glycoprotein studies, antibody appears to play the dominant role in protection of the fetus, as suggested by protection observed in pups following passive transfer of serum prior to GPCMV challenge (Bratcher et al., 1995; Chatterjee et al., 2001).

## Application of cloned, recombinant techniques to GPCMV vaccine studies

In order to optimize the study of vaccines in the GPCMV model, ongoing efforts have focused on both molecular characterization of the viral genome, and refinement of the congenital infection model. Key to the testing of subunit vaccines has been the cloning and characterization of GPCMV genes. As noted, the first GPCMV homolog identified was gB, which mapped to defined *Hind*III fragments of the genome (Schleiss, 1994). Subsequently, the GPCMV homolog of the HCMV UL83 tegument protein, GP83, was identified, and found to encode a 70 kDa phosphoprotein localized to virion and dense bodies (Schleiss et al., 1999). In light of the critical role of immune responses to these two proteins in control of CMV infection, these genes have been employed as the logical cornerstones of subunit vaccine studies using cloned, recombinant technologies.

## Protection against congenital CMV infection using GPCMV DNA vaccines

To examine the potential efficacy of DNA vaccines in the GPCMV model, the gB and GP83 ORFs were cloned in plasmid expression vectors, and immunogenicity evaluated, both in mice and guinea-pigs (Schleiss et al., 2000). These studies indicated optimal immunogenicity with a secreted form of gB, truncated upstream of the transmembrane domain. Dose range experiments addressed the effect of route of administration on immunogenicity, the effect of multiple doses on immune response, and the effect of dose of plasmid on immune responses. These data were used to direct design of efficacy studies for prevention of GPCMV disease in the congenital infection model.

Based on these immunogenicity comparisons, a four-dose series of plasmid DNA vaccine was administered epidermally, by "gene gun" approach, in a protection study. ELISA assay indicated that all animals ($n = 17$ with gB and $n = 14$ with GP83) seroconverted to GPCMV antigen, following the four-dose series of plasmid. The mean $log_{10}$ ELISA titer in gB vaccinated animals was 3.3, vs. 1.8 in GP83 immunized animals. In serum from guinea-pigs vaccinated with gB which became pregnant, complement-dependent neutralizing titers were determined. Neutralizing titers ranged from 1:80 to 1:1280, with a mean neutralizing titer of 2.55 $log_{10}$. To confirm that vaccination with recombinant gB plasmid induced antibodies capable of immunoprecipitating the native gB complex, RIP-PAGE analyses were performed. All immunized animals engendered antibody responses capable of precipitating gB species by RIP-PAGE at, ~150 kDa, ~90 kDa, and ~58 kDa, from [35]S-labeled tissue culture lysates.

Following vaccination, guinea-pigs were mated, pregnancy was established, and a challenge experiment was performed. A total of 17 guinea-pigs were vaccinated with gB plasmid, and 14 animals were immunized with GP83 plasmid. Among gB-vaccinated animals, 13 became pregnant and underwent GPCMV challenge. Among GP83-vaccinated animals, 11 became pregnant and underwent GPCMV challenge. Following third trimester challenge, pup mortality rates were compared. Of interest, significant differences in pup mortality were observed in high-ELISA responding dams in the gB vaccine group, compared to controls. In litters born to dams with a titer below 3.4 $log_{10}$ ($n = 8$), mortality was 14/28 (50%), a rate not statistically different from the negative control group. In contrast, for pups born to dams with a mean $log_{10}$ ELISA titer >3.4 $log_{10}$ ($n = 4$), no mortality was observed (0/13; $P < 0.005$ vs. dams with $\leq 3.4$ $log_{10}$ ELISA titer).

To analyze the impact of DNA vaccination on congenital GPCMV infection, liveborn pups were dissected within 72 hours of delivery, and organs (liver, spleen) were homogenized for tissue culture analysis and PCR-based detection of viral genome (Schleiss et al., 2003a). Among live-born pups, preconceptual vaccination with gB vaccine, but not UL83, resulted in a significant decrease in the incidence of congenital CMV infection (Table 25.1). Among 26 live-born pups in the control group, the total congenital infection rate was 77% (20/26). In contrast, in the gB group the total congenital infection rate was 41% (11/27; $P < 0.05$ vs. control group). Viral loads were compared among infected pups in gB and GP83 vaccine groups, and the control group, using quantitative competitive PCR (qcPCR). Mean viral loads in infected pups born to gB and GP83-immunized dams were in the 1.3–1.8 $log_{10}$/mg tissue range (Table 25.1). Thus, both vaccines resulted in a reduced viral load in infected animals, compared to controls. In infected pups born to the control (unimmunized) dams, the mean viral load in liver was 3.8 $log_{10}$ genomes/mg, and 4.0 $log_{10}$ genomes/mg in spleen (Table 25.1; $P < 0.005$). In summary, we have demonstrated that DNA vaccines protect against congenital CMV infection. Protection against pup mortality depended upon the antibody titer induced by immunization (Schleiss et al., 2003b). These data are the first reporting protection against a congenital CMV infection induced by a DNA vaccine.

Harrison, C.J., and Myers, M.G. (1989). Maternal cell-mediated cytolysis of CMV-infected fetal cells and the outcome of pregnancy in the guinea-pig. J. Med. Virol. 27, 66–71.

Hartley, J.W., Rowe, W.P., and Huebner, R.J. (1957). Serial propagation of the guinea-pig salivary gland virus in tissue culture. Proc. Soc. Exp. Biol. Med. 96, 281–285.

Ho, M. Cytomegalovirus, biology and infection, 2nd ed. Plenum Publishing Co., New York, 1992.

Hsuing, G.D., Choi, Y.C., and Bia, F. (1978). Cytomegalovirus infection in guinea-pigs. I. Viremia during acute primary and chronic persistent infection. J. Infect. Dis. 138, 191–196.

Isom, H.C., Gao, M., and Wigdahl, B. (1984). Characterization of guinea-pig cytomegalovirus DNA. J. Virol. 49, 426–436.

Isom, H.C., and Yin, C.Y. (1990). Guinea-pig cytomegalovirus gene expression. Curr. Top. Microbiol. Immuol. 154, 101–121.

Jackson, L. (1920). An intracellular protozoan parasite of the ducts of the salivary glands of the guinea-pig. J. Infect. Dis. 26, 347–350.

Johnson, K.P., and Connor, W.S. (1979). Guinea-pig cytomegalovirus, transplacental transmission. J. Exp. Med. 59, 263–267.

Jones, C.T., Keay, S.K., and Swoveland, P.T. (1994). Identification of GPCMV infected cells in vitro and in vivo with a monoclonal antibody. J. Virol. Methods. 48, 133–144.

Kacica, M.A., Harrison, C.J., Myers, M.G., and Bernstein, D.I. (1990). Immune response to guinea-pig cytomegalovirus polypeptides and cross reactivity with human cytomegalovirus. J. Med. Virol. 32, 155–159.

Keithley, E.M., Sharp, P., Woolf, N.K., and Harris, J.P. (1988). Temporal sequence of viral antigen expression in the cochlea induced by cytomegalovirus. Acta Otolaryngol. 106, 46–54.

Kumar, M.L., and Nankervis, G.A. (1978). Experimental congenital infection with cytomegalovirus, a guinea-pig model. J. Infect. Dis. 138, 650–654.

Kuttner, A.G. (1927). Further studies concerning the filterable virus present in the submaxillary glands of guinea-pigs. J. Exp. Med. 46, 935–956.

Kuttner, A.G., and T'ung, T. (1935). Further studies on the submaxillary gland viruses of rats and guinea-pigs. J. Exp. Med. 62, 805–822.

Kuttner, A.G., and Wang, S.H. (1934). The problem of the significance of the inclusion bodies found in the salivary glands of infants, and the occurrence of inclusion bodies in the submaxillary glands of hamsters, white mice, and wild rats. J. Exp. Med. 60, 773–791.

Lacayo, J., Sato, H., Kamiya, H., and McVoy, M.A. (2003). Down-regulation of surface major histocompatibility complex class I by guinea-pig cytomegalovirus. J. Gen. Virol. 84, 75–81.

Liu, Y., and Biegalke, B. (2001). Characterization of a cluster of late genes of guinea-pig cytomegalovirus. Virus Genes. 23, 247–256.

Lockridge, K.M., Sequar, G., Zhou, S.S., Yue, Y., Mandell, C.P., and Barry, P.A. (1999). Pathogenesis of experimental rhesus cytomegalovirus infection. J. Virol. 73, 9576–9583.

Markham, F.S., and Hudson, N.P. (1936). Susceptibility of the guinea-pig fetus to the submaxillary gland virus of guinea-pigs. Am. J. Path. 12, 175–181.

Matthews, T., and Boehme, R. (1988). Antiviral activity and mechanism of action of ganciclovir. Rev Infect Dis. 10, Suppl 3: 490–494.

McGregor, A., Liu, F., and Schleiss, M.R. (2004a). Identification of essential and non-essential genes of the guinea-pig cytomegalovirus (GPCMV) genome via transposome mutagenesis of an infectious BAC clone. Virus Res. 101, 101–108.

McGregor, A., Liu, F., and Schleiss, M.R. (2004b). Molecular, biological, and in vivo characterization of the guinea-pig cytomegalovirus (CMV) homologs of the human CMV matrix proteins pp71 (UL82) and pp65 (UL83). J. Virol. 78, 9872–9889.

McGregor, A., and Schleiss, M.R. (2001a). Recent Advances in Herpesvirus genetics using bacterial artificial chromosomes. Mol. Gen. Metab. 72, 8–14

McGregor, A., and Schleiss, M.R. (2001b). Molecular cloning of the guinea-pig cytomegalovirus (GPCMV) genome as an infectious bacterial artificial chromosome (BAC) in Escherichia coli. Mol. Gen. Metab. 72, 15–26.

McGregor, A., and Schleiss, M.R. (2004). Herpesvirus genome mutagenesis by transposon-mediated strategies. Methods Mol. Biol. 256, 281–302.

McVoy, M.A., Nixon, D.E., and Adler, S.P. (1997). Circularization and cleavage of guinea-pig cytomegalovirus genomes. J. Virol. 71, 4209–4217.

Medearis, D.N. (1964). Mouse cytomegalovirus infection. III. Attempts to produce intrauterine infections. Am. J. Hyg. *80*, 113–120.

Middelkamp, J.N., Patrizi, G., Reed, C.A. (1967). Light and electron microscopy studies of the guinea-pig cytomegalovirus. J. Ultrastruct. Res. *18*, 85–101.

Miller, S.A., Bia, F.J., Coleman, D.L., Lucia, H.L., Young, K.R., and Root, R.K. (1985). Pulmonary macrophage function during experimental cytomegalovirus interstitial pneumonia. Infect. Immun. *47*, 211–216.

Mocarski, E.S. (2004). Immune escape and exploitation strategies of cytomegaloviruses: impact on and imitation of the major histocompatibility system. Cell Microbiol. 6, 707–717.

Nixon, D.E., and McVoy, M.A. (2002). Terminally repeated sequences on a herpesvirus genome are deleted following circularization but are reconstituted by duplication during cleavage and packaging of concatemeric DNA. J. Virol. *76*, 2009–2013.

Nogami-Satake, T., and Tsutsui, Y. (1988). Identification and characterization of a 50K DNA-binding protein of guinea-pig cytomegalovirus. J. Gen. Virol. *69*, 2267–2276.

Paglino, J.C., Brady, R.C., and Schleiss, M.R. (1999). Molecular characterization of the guinea-pig cytomegalovirus glycoprotein L gene. Arch. Virol. *144*, 447–462.

Penfold, M., Miao, Z., Wang, Y., Haggerty, S., and Schleiss, M.R. (2003). A macrophage inflammatory protein homolog encoded by guinea-pig cytomegalovirus signals via CC chemokine receptor 1. Virology *316*, 202–212.

Pepperl-Klindworth, S., Frankenberg, N., and Plachter, B. (2002). Development of novel vaccine strategies against human cytomegalovirus infection based on subviral particles. J Clin Virol. *25*, Suppl 2: 75–85.

Plotkin, S.A. (1999). Vaccination against cytomegalovirus, the changeling demon. Pediatr Infect Dis J. *18*, 313–325.

Red-Horse, K., Drake, P.M., and Fisher, S.J. (2004). Human pregnancy, the role of chemokine networks at the fetal maternal interface. Expert Rev. Mol. Med. 1–14.

Reeves, M.M., Cui, X., Lacayo, J., McGregor, A., Schleiss, M., and McVoy, M. Guinea-pig cytomegalovirus (GPCMV) encodes three potential MHC class I homologs. In: International Cytomegalovirus and Betaherpesvirus Workshop, 11: 10.

Schleiss, M.R. (1994). Cloning and characterization of the guinea-pig cytomegalovirus glycoprotein B gene. Virology *202*, 173–185.

Schleiss, M.R. (1995). Sequence and transcriptional analysis of the guinea-pig cytomegalovirus DNA polymerase gene. J. Gen. Virol. *76*, 1827–1833.

Schleiss, M.R. (2002). Animal models of congenital cytomegalovirus infection: an overview of progress in the characterization of guinea-pig cytomegalovirus (GPCMV). J. Clin. Virol. *25, Suppl. 2*, 36–49.

Schleiss, M.R., Bourne, N., and Bernstein, D.I. (2003a). Preconception vaccination with a glycoprotein B (gB) DNA vaccine protects against cytomegalovirus (CMV) transmission in the guinea-pig model of congenital CMV disease. J. Infect. Dis. *188*, 1868–1874.

Schleiss, M.R., Bourne, N., Bravo, F.J., Jensen, N.J., and Bernstein, D.I. (2003b). Quantitative-competitive PCR monitoring of viral load following experimental guinea-pig cytomegalovirus infection. J. Virol. Methods. *108*, 103–110.

Schleiss, M.R., Bourne, N., Jensen, N.J., Bravo, F., and Bernstein, D.I. (2000). Immunogenicity evaluation of DNA vaccines that target guinea-pig cytomegalovirus proteins glycoprotein B and UL83. Virol. Immunol. *13*, 155–167.

Schleiss, M.R., Bourne, N., Stroup, G., Bravo, F.J., Jensen, N.J., and Bernstein, D.I. (2004). Protection against congenital cytomegalovirus infection and disease in guinea-pigs, conferred by a purified recombinant glycoprotein B vaccine. J. Infect. Dis. *189*, 1374–1381.

Schleiss, M.R., and Jensen, N.J. (2003). Cloning and expression of the guinea-pig cytomegalovirus glycoprotein B (gB) in a recombinant baculovirus: utility for vaccine studies for the prevention of experimental infection. J. Virol. Methods. *108*, 59–65.

Schleiss, M.R., McGregor, A., Jensen, N.J., Erdem, G., and Aktan, L. (1999). Molecular characterization of the guinea-pig cytomegalovirus UL83 (pp65) protein homolog. Virus Genes. *19*, 205–221.

Schleiss, M.R., Bernstein, D.I., McVoy, M.A., Stroup, G., Bravo, F., Creasy, B., McGregor, A., Henninger, K., and Hallenberger, S. (2005). The nonnucleoside antiviral, BAY 38–4766, protects against cytomegalovirus (CMV) disease and mortality in immunocompromised guinea-pigs. Antiviral Res. *65*, 35–43.

Staczek, J. (1990). Animal cytomegaloviruses. Microbiol. Rev. *54*, 247–265,

cal reactivity of human cytomegalovirus glycoprotein B (UL55) in seeds of transgenic tobacco. Vaccine *17*, 3020–3029.

Urban, M., Klein, M., Britt, W.J., Hassfurther, E., and Mach, M. (1996). Glycoprotein H of human cytomegalovirus is a major antigen for the neutralizing humoral immune response. J. Gen. Virol. *77*, 1537–1547.

Valantine, H.A., Luikart, H., Doyle, R., Theodore, J., Hunt, S., Oyer, P., Robbins, R., Berry, G., and Reitz, B. (2001). Impact of cytomegalovirus hyperimmune globulin on outcome after cardiothoracic transplantation: a comparative study of combined prophylaxis with CMV hyperimmune globulin plus ganciclovir versus ganciclovir alone. Transplantation *72*, 1647–1652.

Varnum, S.M., Streblow, D.N., Monroe, M.E., Smith, P., Auberry, K.J., Pasa-Tolic, L., Wang, D., Camp. D.G. 2nd, Rodland, K., Wiley, S., Britt, W., Shenk, T.E., Smith, R.D., and Nelson, J.A. (2004). Identification of proteins in human cytomegalovirus (HCMV) particles: the HCMV proteome. J. Virol. *78*, 10960–10966.

Walter, E.A., Greenberg, P.D., Gilbert, M.J., Finch, R.J., Watanabe, K.S., Thomas, E.D., and Riddell, S.R. (1995). Reconstitution of cellular immunity against cytomegalovirus in recipients of allogeneic bone marrow by transfer of T-cell clones from the donor. N. Engl. J. Med. *333*, 1038–1044.

Wang, Z., La, R.C., Maas, R., Ly, H., Brewer, J., Mekhoubad, S., Daftarian, P., Longmate, J., Britt, W.J., and Diamond, D.J. (2004a). Recombinant modified vaccinia virus Ankara expressing a soluble form of glycoprotein B causes durable immunity and neutralizing antibodies against multiple strains of human cytomegalovirus. J. Virol. *78*, 3965–3976.

Wang, Z., La, R.C., Mekhoubad, S., Lacey, S.F., Villacres, M.C., Markel, S., Longmate, J., Ellenhorn, J.D., Siliciano, R.F., Buck, C., Britt, W.J., and Diamond, D.J. (2004b). Attenuated poxviruses generate clinically relevant frequencies of CMV-specific T-cells. Blood *104*, 847–856.

Weller, T.H., Macauley, J.C., Craig, J.M., and Wirth, P. (1957). Isolation of intranuclear inclusion producing agents from infants with illnesses resembling cytomegalic inclusion disease. Proc. Soc. Exp. Biol. Med. *94*, 4–12.

Wills, M.R., Carmichael, A.J., Mynard, K., Jin, X., Weekes, M.P., Plachter, B., and Sissons, J.G. (1996). The human cytotoxic T-lymphocyte (CTL) response to cytomegalovirus is dominated by structural protein pp65: frequency, specificity, and T-cell receptor usage of pp65- specific CTL. J. Virol. *70*, 7569–7579.

Ye, M., Morello, C.S., and Spector, D.H. (2002). Strong CD8 T-cell responses following coimmunization with plasmids expressing the dominant pp89 and subdominant M84 antigens of murine cytomegalovirus correlate with long-term protection against subsequent viral challenge. J. Virol. *76*, 2100–2112.

Ye, M., Morello, C.S., and Spector, D.H. (2004). Multiple epitopes in the murine cytomegalovirus early gene product M84 are efficiently presented in infected primary macrophages and contribute to strong CD8$^+$ T-lymphocyte responses and protection following DNA immunization. J. Virol. *78*, 11233–11245.

Yeager, A.S., Grumet, F.C., Hafleigh, E.B., Arvin, A.M., Bradley, J.S., and Prober, C.G. (1981). Prevention of transfusion-acquired cytomegalovirus infections in newborn infants. J. Pediatr. *98*, 281–287.

# Current Perspectives in Vaccine Development

# 26

## Sandra Pepperl-Klindworth and Bodo Plachter

### Abstract

Prenatal transmission of HCMV is a frequent cause of mental retardation and hearing loss in children. Furthermore, infection with this virus is a severe threat to immunocompromised transplant recipients and AIDS patients. Consequently, development of a vaccine to prevent CMV disease has been identified as a first rank medical priority. Goals and target populations for such a vaccine have been named. With the availability of novel technologies, antigens to be targeted by vaccine-induced immune responses have been recently defined in greater detail. These studies indicated that only a subset of the more than 150 viral proteins may be sufficient to induce protective immunity. Using this information, strategies for the development of live virus vaccines as well as subunit vaccines can be refined.

## Introduction

Ever since the first isolation of the salivary gland virus from human material, the development of a vaccine to prevent the consequences of prenatal infection with this virus, later on termed cytomegalovirus, was a major focus of research in this field (Rowe et al., 1956; Smith, 1956; Weller et al., 1957). We have come a long way since those days. The number of publications about CMV stating, in the discussion section, that the results obtained in this work will serve as a basis for vaccine development are uncountable. Yet, still we stand lacking a licensed vaccine for the prevention of cytomegalovirus disease. One explanation for this could be that all of us being involved in one way or another in the project of CMV vaccine development have simply been unable to see obvious vaccine candidates which would be worthwhile testing. Alternatively, it might be that we have to deal here with an extremely smart and complicated virus that has learned to avoid attack by antiviral defenses to survive in an extremely hostile environment. The former explanation would be comfortable for us as the authors of this article, since we could stop writing at this point. Unfortunately, there are some clear indications that the latter is true and that we need to use all our knowledge about the molecular biology and immunology of CMV to design effective vaccine strategies.

This communication does not attempt to provide a complete coverage of the current status of CMV vaccine development, present and past. The reader is referred to recent excellent reviews on this subject (Britt, 1996; Plotkin, 2002, 2004). Rather, we will focus here on selected aspects of how the perspectives of vaccine development in this field could look like on the grounds of the body of information that has been gathered from previous studies on CMV candidate vaccines and on antigen targets of the virus-specific immune defense.

## Goals and target populations of a CMV vaccine

An important first step in the rationale for the design of any vaccine strategy should be to define the goals that a vaccine has to serve and to name target populations for a final formulation. This may have been somewhat easier in former times when diseases such as polio or smallpox, which affected large portions of the population, have set the immediate goal: to avoid infection and disease in virtually everybody. For CMV the situation is different in that only a fraction of the population will ever be affected and suffer from severe disease. Thus a vaccine must meet the demands imposed by the different clinical situations (Table 26.1).

**Table 26.1** Targets and goals of a CMV vaccine

| Setting | Target population | Goals |
|---|---|---|
| Prenatal care | CMV-seronegative girls/women | Induction of sterilizing or protective immunity Prevention of primary infection |
| | CMV-seropositive women | Boosting of preexisting immunity/induction of strain-independent immunity |
| | | Prevention of superinfection with a heterologous CMV strain |
| Transplantation of hematopoietic cells | D+ [a]/R+ [b] | Enhancement of CMV-specific immune reconstitution |
| | D− [c]/R+ | Prevention of CMV reactivation or superinfection/prevention of disease |
| | D+ /R− [d] | Induction of protective immunity |
| | | Prevention of primary infection/prevention of disease |
| Transplantation of solid organs | D+/R− | Induction of sterilizing or protective immunity Prevention of primary infection/prevention of disease |
| | D+/R+ | Boosting of preexisting immunity |
| | | Prevention of reactivation or superinfection/prevention of disease |
| AIDS | HIV-seropositive individuals | Boosting of preexisting immunity |
| Pediatric care | General population | Prevention or mitigation of CMV disease Induction of sterilizing immunity |
| | | Prevention of primary infection and latency |

[a]Donor CMV-antibody seropositive.
[b]Recipient CMV-antibody seropositive.
[c]Donor CMV-antibody seronegative.
[d]Recipient CMV-antibody seronegative.

The most important group from a public health point of view is certainly made up of newborns, who will be congenitally infected in 0.5% to 2.2% of cases. For the USA, an estimated annual number of 40,000 babies born with congenital CMV infection has been reported, resulting in roughly 8,000 cases of permanent central nervous system deficits (Fowler et al., 1992). This renders HCMV the most common cause of intrauterine infection (see also Chapter 1). Consequently, development of a vaccine to prevent congenital CMV disease has been ranked as top priority goal in vaccine development in the twenty-first century by the *Committee to Study Priorities for Vaccine Development* at the Institute of Medicine in Washington, D.C. (Stratton et al., 2001). The desirable goal of a vaccine would be to provide, in the first place, protection of the mother against primary infection with HCMV. However, from available data it remains uncertain whether this can ever be achieved by any immunization strategy. Consequently, a somewhat reduced goal may be to induce immune protection in a way that would prevent CMV transmission to and/or severe disease in the child. Of particular interest in this respect are recent findings that secondary CMV infection during pregnancy may lead to viral transmission and disease in the newborn in frequencies higher than previously anticipated (Boppana et al., 1999). This has been attributed to superinfections with heterologous HCMV strains rather than to reactivation of the endogenous latent CMV of the mother (Boppana et al., 2001). Thus, induction of strain-independent immunity is mandatory in this setting.

Recipients of hematopoietic transplants are a second group of patients severely endangered by CMV infection. Reactivation of endogenous latent CMV is a frequent cause of morbidity and mortality in patients undergoing allogeneic hematopoietic cell transplantations (reviewed in Boeckh et al., 2003). Although antiviral therapy has reduced the rates of complications in the immediate post-transplant period, mortality from late CMV disease now compromises the successful outcome of the transplantation strategy. In addition, improved and less toxic conditioning and treatment regimens after HSC-transplantation allow for a less stringent selection of transplant donors, thereby further enhancing the problems implemented by opportunistic infections. As a clear correlation between the ability to reconstitute antiviral immune responses, mediated primarily by CD4- and CD8-T-cells, and the outcome of viral reactivation after HSC transplantation has been documented, the requirement for an immunotherapeutic or immunoprophylactic strategy to prevent CMV reactivation or disease has been articulated frequently (Boeckh et al., 2003). In contrast to the situation in the prenatal care setting, strengthening of residual, transferred or newly developing antiviral cellular immunity would be a primary goal in these immunocompromised patients. As mature CMV-reactive T-lymphocytes can be transferred by the transplantation procedure, induction or boosting of antiviral immune responses by a vaccine may not only be effective in the recipient but also by immunization of the donor. The final goal of such a vaccination strategy would be to prevent reactivation of latent virus in the recipient or from the graft or to alleviate the consequences and reduce the levels of such reactivation.

In solid organ transplant recipients, cytomegalovirus remains the single most important pathogen, affecting 75% of patients in the first year after transplantation (reviewed in Pereyra and Rubin, 2004). Yet, whether infection or reactivation by CMV causes clinical symptoms is critically related to the mode of transplantation and the donor and recipient serostatus. In a setting where an organ from a CMV seropositive donor is transplanted to a seronegative recipient, prior immunization of the recipient should provide protection against reactivation from the graft. In contrast, pre-existing immunity should be boosted by a vaccine in seropositive recipients. In both instances, replication of CMV subsequent to reactivation needs to be controlled by vaccination in a way to prevent end-organ disease.

In contrast to the prenatal care and transplant settings, it remains unclear whether a CMV vaccine would be of benefit in patients suffering from AIDS. With the advent of highly active antiretroviral therapy, complications induced by CMV reactivation have drastically dropped in frequency (reviewed in Pass, 2001). Yet, deterioration of immune effector functions under antiretroviral therapy still results in severe CMV-mediated disease conditions. It remains to be determined whether patients would benefit from boosting of the CMV-specific immune response prior to the development of AIDS. Finally, it may be a matter of consideration whether a vaccine could be designed as a prophylactic strategy to prevent CMV infection in the general population and whether such a vaccine could be an integral part of a childhood vaccination program. Although this may sound disproportionate at a first glance, one has to remember that for prevention of congenital CMV infection, like in rubella, a large portion of the adolescent female population at the age of 12 to 14 years of age would have to be vaccinated. An effective general vaccine would cover this and concomitantly would prevent establishment of latent CMV infection in vaccinees. Such a strategy could thereby avoid the complications currently encountered with CMV at the later stages of life.

With that in mind, one faces a dilemma in the development of a CMV vaccine. Although prevention of prenatal infection is generally considered an obvious task, other target populations currently appear to be less well accepted from a public health interest perspective (Figure 26.1), despite the undeniable fact that CMV complications in transplant and AIDS patients can be just as devastating as the consequences of prenatal infection. Furthermore, whilst primary infections in healthy individuals do not frequently cause severe symptoms, they inevitably lead to the establishment of latency with the risk of virus recurrence and recrudescent disease. Thus, it should be of great public health interest to prevent latent CMV infection in the general population in the first place. It could be argued that once a vaccine for the prevention of prenatal infection is available, this could well be used in other groups. However, it is well acknowledged that clinical studies to prove tolerability and efficacy of a vaccine, as requested by health care authorities (Stratton et al., 2001), are difficult to perform in the setting of prenatal infection in terms of design, implementation, time frame and expenses. Clinical studies in transplant recipients, in contrast, would be easier to design and could provide a first step toward the goal of an effective vaccine for prevention of prenatal infection.

## Immune responses important for protection

As pointed out in the previous section, goals for CMV vaccine development may differ for the various target populations. In accordance with this, different branches of the immune system may have to be addressed by a vaccine in a certain setting. We will focus here on adaptive immune responses which are important for control of CMV infection and refer to Chapter 15 for a review of the interaction of CMVs with the innate immune system.

Humoral immune responses, particularly induction of virus-neutralizing antibodies (NT-Abs) appear to be crucial for protection against primary infection and, consequently,

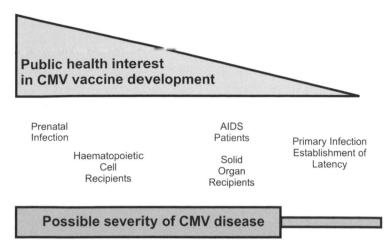

**Figure 26.1** Level of public health interest for the development of a CMV vaccine with respect to different patient populations and possible clinical consequences induced by infection. Highest rates of interest receive congenital infection and complications during hematopoietic transplantation. Less attention is directed towards CMV complications in AIDS-patients and recipients of solid organs, as improved therapeutic strategies have reduced frequencies of complications. However, severe and sometimes lethal complications are still seen in a fraction of the patients, thus warranting prophylactic intervention also in this group. In contrast, infection in healthy individuals is usually associated with mild symptoms or proceeds unnoticed. Yet these infections inevitably lead to viral latency, bearing a risk of reactivation later in life.

against congenital disease (see also Chapters 1 and 14). Transfer of maternal antibodies was reported to be protective against transfusion-associated infection in newborns in the immediate postnatal period (Yeager et al., 1981). Passive transfer of CMV-specific antibodies in multiply transfused preterm infants showed a tendency for reduction of CMV disease in these children (Snydman et al., 1995). In addition, several studies have shown that antibody seropositivity prior to pregnancy protects, though not completely, against prenatal infection and disease (Pass, 2001). The role of antibodies in the prevention of prenatal infection has also been emphasized in the guinea-pig CMV model (see Chapter 25). However, whether NT-Abs alone are protective in this setting remains debatable. Presence of NT-Abs have been correlated with slower progression toward CMV disease in AIDS patients (Boppana et al., 1995). Finally, kidney, cardiac, and liver transplant recipients appeared to benefit from passive transfer of CMV-specific antibodies in the post transplant period (Falagas et al., 1997; Pereyra and Rubin, 2004; Valantine et al., 2001). These findings, although admittedly circumstantial, indicate that antibodies are important for the prevention of infection. Thus a successful vaccine against CMV must be designed to address the humoral immune response and should particularly induce NT-Abs.

Strain differences have to be considered at this point, since it has been shown that sera from humans infected with one CMV strain may fail to neutralize a heterologous strain (Klein et al., 1999). These experimental data could serve as an explanation for CMV reinfections that have been reported to occur in otherwise healthy individuals. Such infections were seen in the presence of CMV-specific antibodies after exposure to secretions from individuals undergoing acute CMV infection (reviewed in Plotkin, 2002). More recently, severe congenital CMV infections have been reported in children born to antibody sero-positive mothers. This has been attributed to superinfection with an HCMV strain during pregnancy, which had not been previously present in the mother (Boppana et al., 2001).

Although antibodies appear to be involved in the prevention of CMV infection, it remains elusive whether they are sufficient. From the murine model of CMV infection, it became clear that cellular immune responses, in particular cytolytic CD8-positive T-lymphocytes (CTL) are protective against CMV disease and that these cells control viral reactivation from latency (Reddehase, 2002; Reddehase et al., 1985; see Chapters 19 and 23). In humans, Reusser and colleagues could show that reconstitution of antiviral CTL correlated with protection against CMV disease in bone marrow transplant recipients, a finding that has been confirmed on several occasions thereafter (Reusser et al., 1991). Making use of that knowledge, the same group provided evidence to suggest that in vitro expanded donor-derived CTL lines are protective against CMV disease after transfer into bone marrow transplant recipients (Walter et al., 1995). However, the use of this approach as a clinical routine has been impeded thus far by logistic limitations. Yet, what became clear from these studies was that CD4-positive T helper (Th) cells are important for long-term maintenance of antiviral CD8 T-cells. This work was thus pioneering for vaccine development in that it showed that also in humans, CTL are protective against CMV disease. Furthermore, it indicated that vaccination would have to address the Th-response as well in order to prime for a lasting CD8 T-cell memory. Subsequent studies confirmed that combined transfer of CD4$^+$ and CD8$^+$ T-cells for pre-emptive or prophylactic treatment of CMV reactivation in allogeneic transplant recipients was effective to generate a lasting CTL response specific for CMV (Einsele et al., 2002; Peggs et al., 2003).

## Antigen components for a CMV vaccine

Knowledge about the immune responses to be addressed should enable a rational design of the antigen composition for a CMV vaccine candidate. Viral proteins that are targeted by NT-Abs have to be combined with immunodominant antigens of the Th- and CTL responses. This should allow induction of a broad and lasting immunity. Although this may sound like a straightforward approach, it is hampered by the genetic complexity of CMV and by host HLA polymorphism. Initial DNA-sequence analyses of the viral genome demonstrated a coding capacity for the laboratory strain Ad169 of about 200 unique proteins, comprising the potential antigen repertoire for the antiviral response (Chee et al., 1990). Recent analyses revised this in that the putative number of genes contained in the Ad169 genome was reduced to 145 by means of comparison with the DNA-sequence of the chimpanzee cytomegalovirus (Davison et al., 2003). On the other hand genomic sequence analyses of recent HCMV isolates provided us with the surprising observation that large portions of the HCMV genome had been deleted during passage of laboratory strains in fibroblast culture over the years (Cha et al., 1996). Consequently, the genomes of clinical isolates encode additional proteins to be considered for vaccine design, making up a total of about 165 or even more unique proteins (Dolan et al., 2004; Murphy et al., 2003; see Chapter 3). Accordingly, the knowledge about important immunogenic proteins of HCMV is incomplete, to say the least. Most previous immunological studies focused on viral antigens that were abundantly detectable either in infected cells or in virus particles. Thus, our picture of "immunodominant" antigens of HCMV may be biased to some extent by our lack of knowledge about the immunogenicity of all the putative antigens of the virus. For the cellular immune response, the situation has changed only very recently by a genome-wide mapping of antigenic ORFs of HCMV for a cohort of 33 HLA-disparate HCMV-seropositive donors (work by Picker and colleagues, discussed in greater detail in Chapter 17). Yet, from quantitative analyses of the immune responses against particular viral proteins in infected individuals, information was already available about the antigens that should be included in a vaccine as a minimal requirement.

## Target antigens of the humoral immune response

The glycoprotein B (gB, gpUL55) has been identified as a dominant target antigen of the NT-Ab responses against HCMV (reviewed in Britt and Mach, 1996; see Chapter 14) (Table 26.2). The gB is an abundant CMV envelope glycoprotein. It is highly conserved both between HCMV isolates and in the family of herpesviruses. All infected individuals develop a humoral immune response against gB. Several studies have shown that a large portion of the virus-neutralizing capacity found in sera from infected individuals is directed against gB. The protein contains at least two predominant sites for the binding of NT-Ab. Because of these properties, it is generally accepted that gB would be an essential component for a vaccine against CMV. This is corroborated by the finding that antibodies against the gB homolog of GPCMV protect against prenatal infection in these animals (Schleiss et al., 2003). Since subtle changes in substructures of the gB molecule may lead to loss of NT-Ab binding, the molecule should be contained in a vaccine formulation in a close-to-native conformation.

The glycoprotein H (gH, gpUL75) is also abundant in the viral envelope as a component of the gH–gO–gL complex. As with gB, gH is conserved in herpesviruses. Antibodies are detectable in nearly 100% of cases following infection (reviewed in Britt and Mach, 1996; see Chapter 14). NT-Abs are synthesized against gH, and in some individuals this response may be dominant (Urban et al., 1996). NT-Abs directed against gH may be strain specific, since variations in an important amino-terminal epitope have been identified to be functional for antibody binding and virus neutralization. Clinical relevance of this phenomenon has been recently suggested. In women who encountered a CMV reinfection during pregnancy and transmitted the virus to their offspring, Boppana and colleagues detected newly synthesized strain-specific gH antibodies, indicating that these women were superin-

**Table 26.2** Viral proteins to be included in an HCMV vaccine

| Viral protein | Structure/Function | NT-Abs[a] | CTL target[b] | Th target[c] |
|---|---|---|---|---|
| gB (gpUL55) | Envelope glycoprotein/ attachment/fusion | Yes[d] | Yes | Yes |
| gH (gpUL75) | Envelope glycoprotein/ penetration | Yes[e] | Yes | Yes |
| pp65 (ppUL83) | Tegument | No | Yes | Yes |
| IE1 (ppUL123) | Non-structural/regulatory protein | No | Yes | Yes |

[a]Target antigen of the NT-Ab response against HCMV.
[b]Target antigen of the CTL response against HCMV.
[c]Target antigen of the Th response against HCMV.
[d]Binding of NT-Ab highly conformation dependent.
[e]NT-Ab response may be strain specific.

fected by an HCMV strain with another gH-type (Boppana et al., 2001). It is reasonable to assume that lack of antibodies directed against the gH of the newly infecting virus enabled intrauterine transmission of HCMV in women with pre-existing immunity. Consequently, for the design of a future CMV vaccine, strain specificity of target antigens for NT-Abs has to be considered and different "types" of gH may have to be included.

Information about further viral targets of the NT-Ab response is scarce. The complex of gN (gpUL73) and gM (gpUL100) has been reported to be targeted by NT-Abs. Antibody binding has been shown to be critically dependent on correct interaction of the two components (see Chapter 14). The gN–gM complex is highly abundant in the viral envelope and may exceed both gB and gH in copy number (Varnum et al., 2004; see Chapter 5) Although this renders the complex attractive for vaccine development, the well known hypervariability of gN may limit its value. Further information is needed about this complex to evaluate its applicability in vaccine design.

## CTL target antigens

As with antibody responses, information about the repertoire of target antigens for the CTL response is just emerging. Most studies on CTL against HCMV proteins so far focused on the tegument protein pp65 (pUL83) and the regulatory IE1-protein (pUL123) (reviewed in Reddehase, 2000; 2002; see also Chapter 17). Initial analyses have shown that CTL directed against pp65 are detectable with considerable frequency in seropositive individuals (Boppana and Britt, 1996; Wills et al., 1996). This observation has been confirmed in several subsequent studies both on healthy and immunocompromised individuals and has led to the conclusion that pp65 is the dominant CTL target antigen (see Chapter 17). Since the protein is highly conserved between viral strains, it indeed appears to be an ideal constituent for a vaccine. More recent analyses have shown that CTL from some individuals may also, or even preferentially respond to peptides from the IE1-protein (Gyulai et al., 2000; Kern et al., 1999). The IE1 was therefore also suggested as a component for a CMV vaccine (Plotkin, 2004). This may, however, be critically limited by the variability of the IE1 protein. Most CTL epitopes of IE1 mapped to date are located in regions of the protein that show at least some polymorphism between strains (Elkington et al., 2003; Prod'homme et al., 2003) (Figure 26.2). This applies in particular to some epitopes that have been classified as being immunodominant. It should be noted, however, that the term "immunodominance", as used in this context reflects frequencies of antigen-specific CTL found in the blood of infected individuals. It is worthwhile considering that frequency does not necessarily mean functionality in terms of protection. It still has to be shown for HCMV whether CTL frequency correlates with protection against reactivation or disease in humans. Subdominant epitopes from conserved parts of IE1 have been described at least for HLA-A1 and HLA-A2 (Elkington et al., 2003; Frankenberg et al., 2002). It remains to be seen whether subdominant antigens may become relevant inducers of an immune response after vaccination, as has been suggested by recent data about DNA vaccination in the murine model (Morello et al. 2005; Ye et al., 2002).

Besides pp65 and IE1, other viral antigens have been identified as CTL targets, including gB and gH. Genome wide peptide mapping has identified a number of additional viral proteins to be targeted by CTL (Elkington et al., 2003; see also Chapter 17). The relevance of these antigens for protection against HCMV disease and their usefulness to comple-

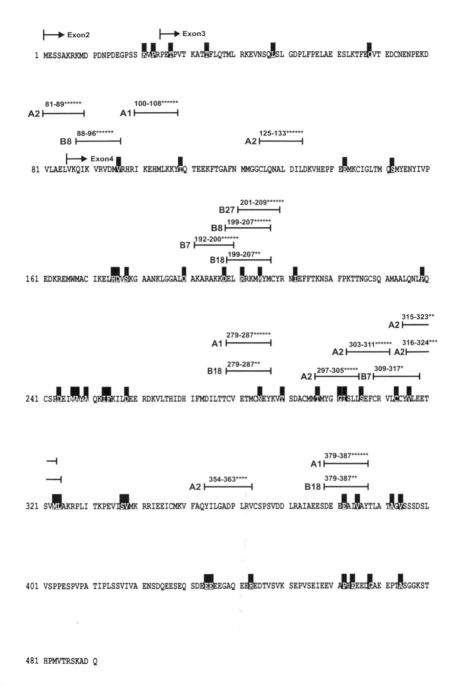

**Figure 26.2** Location of MHC class I presented peptides within the IE1-protein sequence of HCMV strain Ad169. Antigenic peptides are represented above the amino acid sequence by bars spanning from the N- to the C-terminal position with the presenting MHC-class I allomorph indicated at the left hand side of the bars. Exon boundaries are indicated by arrows. Black boxes mark variable positions. Peptide sequences published by: *Kern et al. (1999); **Retiere et al. (2000); ***Khan et al. (2002); ****Frankenberg et al. (2002); *****Gallez-Hawkins et al. (2003); ******Elkington et al. (2003).

ment pp65 and IE1 as components of a HCMV vaccine remain an open question at this point.

## Target antigens of the T-helper lymphocyte response

Infection with CMV is well known to induce a vigorous T helper (Th) response. Protection conferred by Th-lymphocytes has not been conclusively demonstrated in humans. However, as mentioned above, adoptive transfer experiments have convincingly shown that Th-responses are crucial, as expected, for mounting a lasting CTL response. Furthermore, it can be assumed that sustained antibody responses can only be achieved by a vaccine, if antigens for the induction of Th-lymphocytes are included in the formulation. Fortunately, the components that have to be used for vaccine purposes anyway, namely pp65, IE1, gB and gH, are also targeted by Th-cells (Beninga et al., 1995) (Table 26.2).

## Optimal composition

As alluded to above, a number of questions about the optimal composition of a candidate CMV vaccine still remain open. However, based on available data, a successful vaccine might require only a limited complexity in its antigenic components. Some of the proteins combine immunogenicity for the induction of more than one branch of the immune system. Furthermore, at least gB and pp65 appear to be immunodominant in a way that most infected individuals naturally respond to these antigens and, in the face of a competent immune system, are efficiently protected against disease. Thus it may be more than a good guess that the proteins listed in Table 26.2 are necessary and probably sufficient to induce a broad and lasting immunity after vaccination.

## Candidate vaccines

Several candidate vaccines have been developed and tested in preclinical and, for some of these, in clinical trials (Gonczol and Plotkin, 2001) (Table 26.3). It is interesting, though probably a bit unfair, to look into these approaches from the standpoint of our current knowledge about CMV immunology and CMV disease. The information gathered in all these studies is extremely helpful for current approaches to vaccine development and thus shall be briefly reviewed. We will not discuss work that has been conducted using antigen-loaded dendritic cells for vaccination here and refer to a recent review concerning this issue (Arrode and Davrinche, 2003; see also Chapter 18).

## Live virus vaccine

Live virus vaccines have been tremendously successful for battling virus infections. It is Stanley Plotkin and his colleagues, who merit the acknowledgment of having performed the pioneering work for CMV vaccine development generating a candidate, the Towne strain, and testing it in several clinical studies (Gonczol and Plotkin, 2001; Plotkin, 2002; 2004). Thus far, the Towne vaccine still is the best tested candidate available. It has been generated through serial passage on human fibroblasts and proved to be safe upon evaluation in phase I/II clinical studies. Although some of the genes of clinical isolates are missing from Towne, it proved to induce both humoral and cellular immune responses. As a matter of fact, all the antigens listed in Table 26.2 to be – according to the current state of knowledge – required for the induction of immunity, are abundantly synthesized by the Towne strain

**Table 26.3** Strategies and status of testing of available CMV candidate vaccines[a]

| | Preclinical testing | Clinical testing (phase I/II) | Clinical testing (phase III) | Outcome/perspectives |
|---|---|---|---|---|
| Live, attenuated (Towne) | Yes | Yes | Yes | Protective against disease<br>No prevention of infection |
| Live, recombinant | Yes | Yes | No | Well tolerated in healthy volunteers<br>Further clinical trials pending |
| Recombinant gB subunit | Yes | Yes | No | Induction of NT-Abs in volunteers<br>Rapid decline of humoral responses<br>Rapid restoration of NT-Ab titers subsequent to boosting<br>Further clinical trials pending |
| Viral vectors (canarypox) | Yes | Yes | No | CTL and Th responses in seronegative volunteers<br>Poor NT-Ab response<br>Candidate for prime boost approaches |
| Viral vectors (MVA, Adeno) | Yes | No | No | Induction of NT-Ab, CTL and Th in mice<br>Clinical trials pending |
| Peptides | Yes | No | No | Induction of CTL and Th responses in mice<br>Clinical trials in transplant patients pending |
| DNA – vaccines | Yes | No | No | Protection against challenge in the MCMV model using a prime boost approach<br>Induction of CTL- and NT-Ab- responses against HCMV antigens in mice<br>Clinical trials pending |
| Dense bodies | Yes | No | No | Induction of lasting NT-Ab responses in mice<br>Induction of CTL- and Th1-type responses in mice<br>Clinical trials pending |

[a]Listing according to published information available to the authors.

upon infection in cell culture. Thus it appears to be no surprise that protection against disease was afforded in renal transplant recipients. However, infection with CMV could not be prevented by prior immunization with the Towne vaccine in healthy individuals. This may be attributed to a limited immunogenicity of the formulation or the amount of virus used in these studies (Gonczol and Plotkin, 2001). Implicated in this argumentation is the assumption that simply the amount of virus expression in the vaccinees was insufficient to generate sterilizing immunity. Notwithstanding the discussion of whether such sterilizing immunity could ever be achieved by a CMV vaccine, limited replication of the Towne strain in vivo could have been a major drawback in this vaccine approach. Nonetheless, a lot has been learned from these studies, including the proof of principle that prevention of CMV disease by a vaccine is feasible. Importantly, these studies fostered other approaches in CMV vaccine development, particularly those aiming at the improvement of the replication and immunogenicity of a vaccine strain in vivo (Plotkin, 2004).

## Improvement of the live vaccine – Towne–Toledo chimeras

Based on the results of the clinical studies discussed above it may be assumed that the level of replication of the Towne vaccine strain in vivo limited its immunogenicity. Furthermore, Towne lacks large parts of genomic DNA sequences found in clinical isolates of HCMV, resulting in a limited repertoire of putative antigens (Cha et al., 1996). Attempts have thus been initiated to generate chimeric viruses. DNA fragments unique to recent clinical isolates, represented by the Toledo strain of HCMV but not found in the Towne strain, were transferred to Towne's genomic backbone to generate four different candidate vaccine strains (Kemble et al., 1996). The rationale behind this approach was to improve antigen expression of a candidate vaccine strain in vivo compared to the parental Towne strain and to broaden the repertoire of antigens expressed by the vaccine. Phase I/II clinical trials are currently performed with these viruses. It will be interesting to see whether the Towne–Toledo chimeras induce immune responses exceeding those obtained with the Towne vaccine and whether new antibody or T-lymphocyte specificities against the proteins encoded in the additional DNA sequences emerge.

## Recombinant gB subunit vaccine

As mentioned above, HCMV encodes proteins that have attracted scientific attention owing to their abundance and apparent immunodominance. The identification of the gB molecule as a major target of the NT-Ab response, in concert with the general agreement that the induction of NT-Ab responses is the foremost goal in vaccine development to prevent prenatal infection, supported the approach to use this protein alone as a recombinant subunit vaccine. For vaccine studies in humans, a modified version of the HCMV gB was expressed in Chinese hamster ovary cells. The protein was engineered in a way to allow secretion from cells into the culture supernatant without deleting the immunogenic parts of the molecule. The recombinant gB-vaccine proved to be well tolerated and effective in inducing antibody responses in healthy volunteers when administered with MF59, an oil-in-water emulsion of squalene as adjuvant (Pass et al., 1999). NT-Abs against gB were induced after three immunizations in titers comparable to those seen in natural infection. However, antibody levels declined substantially thereafter. A fourth dose resulted in a rapid secondary rise in anti-gB antibody and NT-Ab levels. This suggested that vaccinees would respond in a similar way when exposed to wild-type virus, thus being able to remount immunity against infection rapidly, which would reduce the risk of prenatal transmission. The gB vaccine also proved to be well tolerated and induced neutralizing antibodies in toddlers (Mitchell et al., 2002).

A vaccine based on gB has a good argument in terms of safety considerations. Furthermore, gB has been identified as a dominant NT-Ab antigen as well as a target, though likely a subdominant one, of both the CTL- and the Th-response. There is a fair chance that isolated induction of gB-specific NT-Abs could prevent diaplacental virus transmission in most cases. This is supported by using gB as a vaccine in an animal model of congenital infection (Schleiss et al., 2003; see also Chapter 25). However, recent data indicate that reinfections during pregnancy can occur in the face of a pre-existing NT-Ab response and that these reinfections can lead to prenatal transmission and disease (Boppana et al., 2001). Thus, there is reason to doubt that using one of the envelope glycoproteins alone as a vaccine would induce a broad-enough NT-Ab response to prevent intrauterine transmission in all instances. This view is supported by the finding obtained in the latter study that reinfection of pregnant women was mediated by a CMV strain that apparently showed specific alterations in an important neutralizing epitope in gH. Whether these strain specificities were the reason for reinfection and transmission or simply a means to detect such reinfections remains to be determined. Yet, there is evidence that both gB and gH may react with neutralizing antibodies in a strain-specific manner. Furthermore, recent proteome analyses revealed that other glycoproteins, such as gM, are far more abundant in the viral envelope (Varnum et al., 2004; see also Chapter 5) and thus may be required as antigens for induction of a broad NT-Ab response. Finally, a gB vaccine may be of limited use in transplant or AIDS patients, since a cellular immune response is considered to be crucial for protection against HCMV disease in these settings. Thus, although attractive because of its simplicity compared with other candidate vaccines, a stand-alone gB vaccine based on the recombinant protein may not be sufficient. However, combination of a gB-vaccine with other vaccine candidates remains a promising option to follow (see below).

## Canarypox recombinants

There is agreement that both humoral and cellular responses need to be induced for effective and lasting immunity. In most instances, however, vaccine candidates based on recombinant proteins proved to be poor inducers of cellular, particularly of CTL responses. This relates to the restrictions set by the MHC-class I processing/presentation pathway. Only those proteins which are either synthesized de novo or actively transduced in an antigen presenting cell gain sufficient access to the pathway to serve as antigens for priming of a CTL response. Recombinant proteins usually do not meet these criteria. Thus, viral expression systems have been designed for antigen synthesis to circumvent this inherent problem in the induction of cellular immune responses by recombinant vaccines. The canarypox virus is a versatile expression system in that it can accommodate large DNA fragments and infects mammalian cells without being able to fully replicate. Consequently, canarypox recombinants expressing gB and pp65 of HCMV have been constructed and tested both in preclinical and phase I/II clinical studies. In seronegative individuals, immunization with a pp65-expressing recombinant virus induced strong and lasting CTL responses, as well as Th- and humoral immune responses (Berencsi et al., 2001). In contrast to that, a gB-expressing recombinant virus failed to be superior to recombinant gB protein in priming an NT-Ab response (Adler et al., 1999; Bernstein et al., 2002). Consequently, canarypox recombinants may be useful as vaccine candidates in combination with other approaches like recombinant proteins.

## Adenovirus and vacciniavirus recombinants

Alternative viral vector systems expressing CMV proteins have been tested in preclinical models. Expression of HCMV gB and IE1 subsequent to infection with adenovirus recombinants has been shown to induce a cytotoxic T-cell response in mice (Berencsi et al., 1996). Furthermore, mucosal and systemic antibodies could be induced in mice after intranasal immunization with a replication-defective adenovirus vector, expressing MCMV gB (Shanley and Wu, 2003). Using the attenuated poxvirus strain, modified vaccine virus Ankara (MVA) expressing HCMV gB, Diamond and colleagues could induce sustained NT-Ab responses against multiple HCMV strains in mice (Wang et al., 2004a). Interestingly, pre-existing immunity against MVA did not interfere with antibody induction after gB-MVA immunization. Because of its lack of viral assembly and virulence in humans, the MVA strain appears to be particularly interesting for vaccine approaches in immunocompromised individuals. In an extension of their studies, these authors expressed the HCMV proteins IE1, pp65 and pp150 in MVA (Wang et al., 2004b). Robust and specific in vitro expansion of CTL could be achieved from PBMC of seropositive donors using autologous MVA-infected lymphoblastoid cell lines. Further to this, single immunization of HLA-A2 transgenic mice with MVA expressing pp65/pp150 or IE1 exon 4 induced strong CTL responses. These findings make MVA recombinants attractive candidates both for in vitro expansion of CMV-specific T-cells for clinical applications as well for use as a potential vaccine. With these properties in mind, MVA constructs expressing HCMV antigens await clinical testing.

## Peptide vaccines

Accumulation of information about antigenic peptides presented by different MHC class I and MHC class II allomorphs enables concepts to use synthetic peptides as immunogens for the stimulation or for boosting of CTL- and Th-responses in humans. This approach appears to be particularly attractive in the transplant setting, as both donors and recipients are HLA typed and specific combinations of peptides tailored for each vaccine could be designed. In one preclinical study, a dominant HLA-A*0201 presented nonamer peptide from pp65 fused to a Th-epitope peptide was able to induce significant CTL responses in the absence of adjuvant in HLA-A*0201 transgenic mice (La-Rosa et al., 2002). Use of a lipidated version of this peptide induced systemic responses in mice after intranasal application (BenMohamed et al., 2002). Again, clinical studies are warranted to investigate the usefulness of these approaches, e.g. for immunotherapy after transplantation. One limitation is that only conserved peptide epitopes can be used. Furthermore, peptides are not helpful for the induction of antiviral NT-Abs since epitopes in CMV glycoproteins targeted by the neutralizing immune response are usually conformation dependent and require the expression of full-length proteins in native conformation.

## DNA vaccines

DNA-based vaccines provide an alternative approach to viral vectors or protein/peptide-based formulations for the induction of immunity. In the model system of murine CMV infection, Spector and co-workers showed that co-immunization with plasmids expressing the IE1 protein pp89 of MCMV (equivalent to IE1 pp72 of HCMV) and the M84 gene product was sufficient to induce a protective and lasting CTL response (Ye et al., 2002). However, protection by this approach was incomplete as replication of MCMV in the salivary glands was not inhibited subsequent to challenge. Immunization with a plasmid-based vaccine expressing MCMV IE1, M84 and m04 together with a boost approach using inactivated virus particles induced both cellular and NT-Ab responses that led to prevention of challenge virus replication in all organs including the salivary glands (Morello et al., 2002). Using plasmids expressing HCMV gB or pp65, induction of NT-Abs and CTL against these HCMV antigens could also be demonstrated in mice (Endresz et al., 2001). Thus, a proof of principle has been provided that DNA-based vaccines can induce immune responses against CMVs and that, in the case of MCMV, these responses are protective. However, it remains to be determined whether a plasmid vaccine would be able to meet the criteria for safety and efficacy in clinical settings and would be superior to other CMV vaccine approaches, e.g. in the context of a prime-boost approach.

## Subviral particle vaccine – dense bodies

The knowledge that has been gathered both from vaccine studies as well as from analysis of the antiviral immune responses against CMV supports the notion that subunit vaccines, comprising a limited set of proteins out of the viral antigenic repertoire (Table 26.2), could be sufficient to provide protective immunity against infections in humans. The pp65 has been identified as a dominant CTL- as well as an important Th-antigen. Thus there is general agreement that a pp65-specific immune response should be induced. For neutralizing antibodies, gB and gH are considered mandatory and, more recently, also the abundant complex consisting of the viral proteins gM and gN has been suggested as a component of a vaccine (Plotkin, 2004). Especially for the NT-Ab response, however, native protein conformation appears to be a prerequisite for efficient induction (Britt and Mach, 1996). Thus, folding of these proteins in a way comparable to the situation in the viral envelope would be highly desirable.

Following infection with laboratory strains of HCMV, human fibroblast cultures release large amounts of subviral particles, so-called dense bodies (DBs), into the culture supernatant (see also Chapters 5 and 13). These particles do not contain genomic DNA but consist of an electron dense core enclosed by a lipid envelope. The protein content of these particles has been recently analyzed in a proteomic approach by mass spectrometry (Varnum et al., 2004; see also Chapter 5). Consistent with previous analyses, pp65 was found to be the dominant constituent of these particles. Compared to virions, only a subset of viral envelope glycoproteins was found in DBs. These glycoproteins are sufficient for adsorption and penetration of the particles into a variety of target cells (Pepperl et al., 2000; Pepperl-Klindworth et al., 2002, 2003). The properties of DBs, in particular the high content of the CTL and Th target pp65, the natural conformation of abundant viral surface glycoproteins in their envelope and the capacity for protein delivery into target cells renders these particles an attractive candidate for vaccine development. In preclinical evalu-

cal reactivity of human cytomegalovirus glycoprotein B (UL55) in seeds of transgenic tobacco. Vaccine *17*, 3020–3029.

Urban, M., Klein, M., Britt, W.J., Hassfurther, E., and Mach, M. (1996). Glycoprotein H of human cytomegalovirus is a major antigen for the neutralizing humoral immune response. J. Gen. Virol. *77*, 1537–1547.

Valantine, H.A., Luikart, H., Doyle, R., Theodore, J., Hunt, S., Oyer, P., Robbins, R., Berry, G., and Reitz, B. (2001). Impact of cytomegalovirus hyperimmune globulin on outcome after cardiothoracic transplantation: a comparative study of combined prophylaxis with CMV hyperimmune globulin plus ganciclovir versus ganciclovir alone. Transplantation *72*, 1647–1652.

Varnum, S.M., Streblow, D.N., Monroe, M.E., Smith, P., Auberry, K.J., Pasa-Tolic, L., Wang, D., Camp. D.G. 2nd, Rodland, K., Wiley, S., Britt, W., Shenk, T.E., Smith, R.D., and Nelson, J.A. (2004). Identification of proteins in human cytomegalovirus (HCMV) particles: the HCMV proteome. J. Virol. *78*, 10960–10966.

Walter, E.A., Greenberg, P.D., Gilbert, M.J., Finch, R.J., Watanabe, K.S., Thomas, E.D., and Riddell, S.R. (1995). Reconstitution of cellular immunity against cytomegalovirus in recipients of allogeneic bone marrow by transfer of T-cell clones from the donor. N. Engl. J. Med. *333*, 1038–1044.

Wang, Z., La, R.C., Maas, R., Ly, H., Brewer, J., Mekhoubad, S., Daftarian, P., Longmate, J., Britt, W.J., and Diamond, D.J. (2004a). Recombinant modified vaccinia virus Ankara expressing a soluble form of glycoprotein B causes durable immunity and neutralizing antibodies against multiple strains of human cytomegalovirus. J. Virol. *78*, 3965–3976.

Wang, Z., La, R.C., Mekhoubad, S., Lacey, S.F., Villacres, M.C., Markel, S., Longmate, J., Ellenhorn, J.D., Siliciano, R.F., Buck, C., Britt, W.J., and Diamond, D.J. (2004b). Attenuated poxviruses generate clinically relevant frequencies of CMV-specific T-cells. Blood *104*, 847–856.

Weller, T.H., Macauley, J.C., Craig, J.M., and Wirth, P. (1957). Isolation of intranuclear inclusion producing agents from infants with illnesses resembling cytomegalic inclusion disease. Proc. Soc. Exp. Biol. Med. *94*, 4–12.

Wills, M.R., Carmichael, A.J., Mynard, K., Jin, X., Weekes, M.P., Plachter, B., and Sissons, J.G. (1996). The human cytotoxic T-lymphocyte (CTL) response to cytomegalovirus is dominated by structural protein pp65: frequency, specificity, and T-cell receptor usage of pp65- specific CTL. J. Virol. *70*, 7569–7579.

Ye, M., Morello, C.S., and Spector, D.H. (2002). Strong CD8 T-cell responses following coimmunization with plasmids expressing the dominant pp89 and subdominant M84 antigens of murine cytomegalovirus correlate with long-term protection against subsequent viral challenge. J. Virol. *76*, 2100–2112.

Ye, M., Morello, C.S., and Spector, D.H. (2004). Multiple epitopes in the murine cytomegalovirus early gene product M84 are efficiently presented in infected primary macrophages and contribute to strong CD8$^+$ T-lymphocyte responses and protection following DNA immunization. J. Virol. *78*, 11233–11245.

Yeager, A.S., Grumet, F.C., Hafleigh, E.B., Arvin, A.M., Bradley, J.S., and Prober, C.G. (1981). Prevention of transfusion-acquired cytomegalovirus infections in newborn infants. J. Pediatr. *98*, 281–287.

# Antiviral Intervention, Resistance, and Perspectives

## 27

*Detlef Michel and Thomas Mertens*

## Abstract

Four drugs, ganciclovir/valganciclovir, cidofovir, foscarnet and fomivirsen, have been approved for the treatment of HCMV diseases. Partially they have also been used for prophylaxis or preemptive therapy of active infections. Some studies have shown a limited prophylactic effect of aciclovir (ACV) after organ transplantation, but conflicting results do exist. The first-line therapeutic nucleoside analog ganciclovir (GCV) but also ACV is activated by the viral UL97 protein (pUL97). Except for fomivirsen, all of these drugs share the same target molecule, the viral DNA polymerase. The compounds provoke significant drug-specific side-effects and the emergence of clinically relevant drug-resistant HCMV has been reported for all of them. Therefore, new compounds are urgently needed with less adverse effects, good oral bioavailability and possibly novel mechanisms of action to avoid cross-resistance. Consequently, new drugs and mechanisms of action as well as new molecular targets have to be identified. Benzimidazoles, phenylenediaminesulphonamides, and indolocarbazole protein kinase inhibitors are promising lead compounds for the development of even more specific inhibitors of HCMV. Inhibition of virus entry might be an imminent target for future antivirals. Since many symptoms of HCMV disease are quite unspecific, virologists have to provide reliable and fast methods for quantitative diagnosis of active systemic viral infection. Furthermore, for an optimum management of patients, monitoring of successful therapy, as well as early detection of drug-resistant HCMV emerging under therapy have to be performed.

## Introduction

Cytomegalovirus is well known as a major health problem for newborns, especially when infected during pregnancy after primary infection of the mother, as well as patients with innate, acquired or iatrogen-induced immunodeficiencies. More recently, it has been shown that HCMV reactivation does also preferentially occur in patients on intensive care units, e.g. individuals suffering from sepsis (Heininger et al., 2001). The clinical relevance of these reactivations with low antigenemias has still to be established. Clinical manifestations of HCMV disease show a very interesting but unexplained variation among different patient populations, depending on the type of underlying disease and the kind and intensity of immunosuppression, e.g. retinitis in AIDS

cidofovir are useful to treat HCMV diseases caused by GCV-resistant HCMV strains due to UL97 mutations. However, both substances are associated with nephrotoxicity.

## Antiviral prophylaxis

Antiviral prophylaxis should be safe and require a minimum of drug monitoring. Particularly, prophylaxis should be free of adverse interactions with other medications used in transplantation. However, current prophylaxis regimes vary widely among different transplant programs and also among different transplant centers. Accordingly, results of the mostly small single center studies are often difficult to reproduce, since differences in definitions, end points, viral surveillance, type of transplantation, immunosuppressive regime, and patient population are common. Consequently, the outcomes of these studies have revealed many discrepancies and ambiguous conclusions. For example, some retrospective studies appeared to document decreases in mortality from primary HCMV disease following oral aciclovir prophylaxis in high-risk patients. However, a randomized, placebo-controlled trial using high-dose oral ACV revealed only a moderate beneficial effect in preventing HCMV disease following kidney transplantation. Concerning liver transplantations, one study suggested a beneficial effect of ACV administered orally to HCMV-seropositive recipients while other studies reported decreases of HCMV infections and disease but no effect on survival or even no effect of oral ACV at all. Concerning GCV prophylaxis today it can be concluded that GCV (intravenous) or valganciclovir (oral) should be adapted to the risk of the transplant recipient to develop a severe HCMV manifestation, e.g. (i) for all HCMV-seropositive recipients and for HCMV-seronegative recipients receiving a liver or heart transplant from a seropositive donor or (ii) for seronegative renal transplant patients with an organ graft from an HCMV-seropositive donor.

Foscarnet and cidofovir are not generally used for prophylaxis in transplantation, particularly not in kidney transplant recipients, because of their nephrotoxicity.

A still open question is whether an antiviral prophylaxis may lead to a faster emergence of resistant virus during subsequent treatment of an HCMV-associated disease. The effect of prolonged prophylaxis on the emergence of resistant viral strains has not been extensively studied. Boivin et al. (2004) investigated the emergence of GCV-resistant HCMV in 301 high-risk solid-organ transplant recipients after oral prophylaxis with either valganciclovir or ganciclovir. The authors conclude that the prophylaxis did not promote the emergence of resistant virus. However, Alain et al. (2004) reported the detection of GCV-resistant HCMV after valaciclovir prophylaxis in renal transplant recipients with active HCMV infection. We could show that at least in vitro ACV selects precisely for those mutations in the UL97 gene that confer GCV resistance (Michel et al., 2001). In conclusion, a rational anti-HCMV prophylactic regime has still to be established.

## Preemptive therapy

Rapid diagnostic tests allow clinical virologists to detect viral replication prior to the onset of a symptomatic disease. This provides the opportunity of an early antiviral treatment following a laboratory-proven diagnosis of an active infection. This so-called "preemptive therapy" allows an early, effective, and often short-term treatment of individuals who are at risk of the development of serious HCMV disease. The advantage of this approach is to avoid overmedication by a general prophylaxis in patients who have no increased risk of HCMV disease and allows a more efficient early therapy, especially in patients at risk of interstitial pneumonia.

## Viral susceptibility testing and HCMV genotyping

The problem of clinically relevant drug-resistant HCMV has raised the need for methods of antiviral sensitivity testing for a fast identification of the emergence of resistant virus populations in treated patients. Resistance of isolated viruses can be determined phenotypically in an in vitro drug sensitivity assay mostly performed as focus reduction or plaque reduction assay (Landry et al., 2000; Michel et al., 2001). These assays are dependent on virus isolation and passage in cell culture, which requires up to 14 weeks. Comparative studies have been hampered by technical difficulties, lack of standardization of assays, and consequently by different definitions of GCV resistance. Resistance has been defined as an $ID_{50}$ (inhibitory dose) exceeding 6 µmol/l or 12 µmol/l GCV (Chou et al., 1999; Erice, 1999; Michel et al., 2001). For cidofovir and foscarnet it has been proposed to use an $ID_{50}$ of > 4 µmol/l and an $ID_{50}$ of >400 µmol/l, respectively, as cut-off values that define viral resistance to these compounds. The results of biological assays should be interpreted with some caution. Since mixed virus populations are involved, discordant results can be observed in genotypic and phenotypic assays. We as well as others have observed that a selection bias by passaging the virus in vitro before phenotypic susceptibility testing may underestimate the real impact of viral resistance in patients with failure of antiviral therapy.

## Biological function of pUL97

The UL97 protein does not show homology to any other known nucleoside kinases; rather it resembles protein kinases and bacterial phosphotransferases (Chee et al., 1989; Hanks et al., 1988). The protein is transported into the nucleus via a nuclear localization signal in the amino-terminal region (Michel et al., 1996). Prichard et al. (1999) generated an UL97-deficient virus which exhibited a severe replication deficiency in fibroblasts indicating that this gene plays an important role for virus replication in vitro. Is has been shown that pUL97 interacts with the DNA polymerase processivity factor pUL44, an essential component of the replication complex, and might be necessary for nuclear egress. However, data presented suggest that pUL97 plays roles in DNA replication, DNA encapsidation and/or nuclear egress but are in part conflicting (Krosky et al, 2003a; Krosky et al., 2003b; Marschall et al., 2003; Wolf et al., 2001). The region responsible for the interaction has been mapped between aa 366 and aa 459 of pUL97. The two proteins accumulate in the nucleus and are incorporated into replication centers. Treatment with cidofovir (CDV), an inhibitor of DNA synthesis, or with the indolocarbazoles NGIC-I and Gö6976, which are specific inhibitors of the pUL97 kinase, prevented the co-localization of pUL97 and pUL44 (Marschall et al., 2003). The cellular protein p32 appears to promote the accumulation of pUL97 at the nuclear lamina (Such et al., 2004). Thereby, p32 itself as well as lamin proteins are phosphorylated by pUL97.

The M97 protein (pM97) of mouse cytomegalovirus (MCMV) is the homolog of pUL97. We constructed an MCMV M97 deletion mutant and then replaced the

Chrisp, P., and Clissold, S.P. (1991). Foscarnet, A review of its antiviral activity, pharmacokinetic properties and therapeutic use in immunocompromised patients with cytomegalovirus retinitis. Drugs *41*, 104–129.

Cihlar, T., Fuller, M.D., and Cherrington J.M. (1998). Characterization of drug resistance-associated mutations in the human cytomegalovirus DNA polymerase gene by using recombinant mutant viruses generated from overlapping DNA fragments. J. Virol. *72*, 5927–5936.

Crumpacker, C.S. (1996). Ganciclovir. N. Engl. J. Med. *335*, 721–729.

Curran, M., and Noble, S. (2001). Valganciclovir. Drugs *61*, 1145–1150.

De Clercq, E., Sakuma, T., Baba, M., Pauwels, R., Balzarini, J., Rosenberg, I., and Holy, A. (1987). Antiviral activity of phosphonylmethoxyalkyl derivatives of purine and pyrimidines. Antiviral Res. *8*, 261–272.

Drew, W.L., Miner, R.C., Busch, D.F., Follansbee, S.E., Gullett, J., Mehalko, S.G., Gordon, S.M., Owen Jr., W.F., Matthews, T.R., Buhles, W.C., and DeArmond, B. (1991). Prevalence of resistance in patients receiving ganciclovir for serious cytomegalovirus infection. J. Infect. Dis. *163*, 716–719.

Emanuel, D., Cunningham, I., Jules-Elysee, K., Brochstein, J.A., Kernan, N.A., Laver, J., Stover, D., White, D.A., Fels, A., Polsky, B. (1988). Cytomegalovirus pneumonia after bone marrow transplantation successfully treated with the combination of ganciclovir and high-dose intravenous immune globulin. Ann. Intern. Med. *109*, 777–82.

Erice, C. (1999). Resistance of human cytomegalovirus to antiviral drugs. Clin. Microbiol. Rev. *12*, 286–297.

Hanks, S.K., Quinn, A.M., and Hunte, T. (1988). The protein kinase family: conserved features and deduced phylogeny of the catalytic domain. Science *241*, 42–52.

He, Z., He, Y., Kim, Y., Chu, L., Ohmstede, C., Biron, K.K., and Coen, D.M. (1997). The human cytomegalovirus UL97 protein is a protein kinase that autophosphorylates on serines and threonines. J. Virol. *71*, 405–411.

Heininger, A., Jahn, G., Engel, C., Notheisen, T., Unertl, K., and Hamprecht, K. (2001). Human cytomegalovirus infections in nonimmunosuppressed critically ill patients. Crit. Care Med. *29*, 541–547.

Jabs, D.A., Enger, C., Dunn, J.P., and Forman M. (1998a). Cytomegalovirus retinitis and viral resistance: ganciclovir resistance. J. Infect. Dis. *177*, 770–773.

Jabs D.A., Enger C., Forman M., and Dunn J.P. (1998b) Incidence of foscarnet resistance and cidofovir resistance in patients treated for cytomegalovirus retinitis. The Cytomegalovirus Retinitis and Viral Resistance Study Group. Antimicrob. Agents Chemother. *42*, 2240–2244.

John, G.T., Manivannan, J., Chandy, S., Peter, S., and Jacob, C.K. (2004). Leflunomide therapy for cytomegalovirus disease in renal allograft recepients. Transplantation *77*, 1460–1461.

Komazin, G., Ptak, R.G., Emmer, B., Townsend, L.B., and Drach, J.C. (2004) Resistance of human cytomegalovirus to the L-ribonucleoside maribavir maps to UL27. J. Virol. *77*, 11499–11506.

Krosky, P.M., Baek, M.C., and Coen, D.M. (2003). The human cytomegalovirus UL97 protein kinase, an antiviral drug target, is required at the stage of nuclear egress. J. Virol. *77*, 905–914.

Krosky, P.M., Baek, M.C., Jahng, W.J., Barrera, I., Harvey, R.J., Biron, K.K., Coen, D.M., and Sethna, P.B. (2003). The human cytomegalovirus UL44 protein is a substrate for the UL97 protein kinase. J. Virol. *77*, 7720–7727.

Krosky, P.M., Underwood, M.R., Turk, S.R., Feng, K.W., Jain, R.K., Ptak, R.G., Westerman, A.C., Biron, K.K., Townsend, L.B., and Drach, J.C. (1998). Resistance of human cytomegalovirus to benzimidazole ribonucleosides maps to two open reading frames: UL89 and UL56. J. Virol. *72*, 4721–4728.

Kruger, R.M., Shannon, W.D., Arens, M.Q., Lynch, J.P., Storch, G.A., and Trulock, E.P. (1999). The impact of ganciclovir-resistant cytomegalovirus infection after lung transplantation. Transplantation *68*, 1272–1279.

Lalezari, J.P., Aberg, J.A., Wang, L.H., Wire, M.B., Miner, R., Snowden, W., Talarico, C.L., Shaw, S., Jacobson, M.A., and Drew, W.L. (2002). Phase I dose escalation trial evaluating the pharmacokinetics, anti-human cytomegalovirus (HCMV) activity, and safety of 1263W94 in human immunodeficiency virus-infected men with asymptomatic HCMV shedding. Antimicrob. Agents Chemother. *46*, 2969–2976.

Landry, M.L., Stanat, S., Biron, K., Brambilla, D., Britt, W., Jokela, J., Chou, S., Drew, W.L., Erice, A., Gilliam, B., Lurain N., Manischewitz, J., Miner, R., Nokta, M., Reichelderfer, P., Spector, S., Weinberg, A., Yen-Lieberman, B., and Crumpacker, C. (2000). A standardized plaque reduction assay for determination of drug susceptibilities of cytomegalovirus clinical isolates. Antimicrob. Agents Chempther. *44*, 688–692.

Limaye, P., Raghu, G., Koelle, D.M., Ferrenberg, J., Huang, M.L., and Boeckh M. (2002). High incidence of ganciclovir-resistant cytomegalovirus infection among lung transplant recipients receiving pre-emptive therapy. J. Infect. Dis. *185*, 20–27.

Marschall, M. Freitag, M. Suchy, P. Romaker, D. Kupfer, R. Hanke, M., and Stamminger, T. (2003). The protein kinase pUL97 of human cytomegalovirus interacts with and phosphorylates the DNA polymerase processivity factor pUL44. Virology *311*, 60–71.

Mattes, F.M., Hainsworth, E.G., Geretti, A.M., Nebbia, G., Prentice, G., Potter, M., Burroughs, A.K., Sweny, P., Hassan-Walker, A.F., Okwuadi, S., Sabin, C., Amooty, G., Brown, V.S., Grace, S.C., Emery, V.C., Griffiths, P.D. (1004). A randomised, controlled trial comparing ganciclovir to ganciclovir plus foscarnet (each at half dose) for preemptive therapy of cytomegalovirus infection in transplant recipients. J. Infect. Dis. *189*, 1355–1361.

McGuigan, C., Pathirana, R.N., Snoeck, R., Andrei, G., De Clercq, E., and Balzarini, J. (2004). Discovery of a new family of inhibitors of human cytomegalovirus (HCMV) based upon lipophilic alkyl furano pyrimidine dideoxy nucleosides: action via a novel non-nucleosidic mechanism. J. Med. Chem. *47*, 1847–51.

McSharry, J.J., McDonough, A., Olson, B., Talarico, C., Davis, M., and Biron, K.K. (2001). Inhibition of ganciclovir-susceptible and -resistant human cytomegalovirus clinical isolates by the benzimidazole L-riboside 1263W94. Clin. Diagn. Lab. Immunol. *8*, 1279–1281.

Metzger, C, Michel, D. Schneider, K. Lüske, A. Schlicht, H.J., and Mertens, T. (1994). Human cytomegalovirus UL97 kinase confers ganciclovir susceptibility to recombinant vaccinia virus. J. Virol. *68*, 8423–8427.

Michel, D., Hohn, S., Haller, T., Jun, D., and Mertens, T. (2001). Aciclovir selects for ganciclovir-cross-resistance of human cytomegalovirus in vitro which is only in part explained by known mutations in the UL97 protein. J. Med. Virol. *65*, 70–76.

Michel, D., Kramer, S., Höhn, S., Schaarschmidt, P., Wunderlich, K., and Mertens, T. (1999). Amino acids of conserved kinase motifs of the cytomegalovirus protein UL97 are essential for autophosphorylation. J. Virol. *73*, 8898–8901.

Michel, D., Pavic, I., Zimmermann, A., Haupt, E., Wunderlich, K., Heuschmid, M., and Mertens T. (1996). The UL97 gene product of the human cytomegalovirus is an early-late protein with a nuclear localization but is not a nucleoside kinase. J. Virol. *70*, 6340–6347.

Michel, D., Schaarschmidt, P., Wunderlich, K., Heuschmid, M., Simoncini, L., Mühlberger, D., Zimmermann, A., Pavic, I., and Mertens, T. (1998). Functional regions of the human cytomegalovirus protein pUL97 involved in nuclear localization and phosphorylation of ganciclovir and pUL97 itself. J. Gen. Virol. *79*, 2105–2112.

Michel, D., Wunderlich, K., Michel, M., Wasner, T., Hauser, I., Just, M., Hampl, W., and Mertens, T. (2003). Fast genotypic identification and estimation of ganciclovir-resistant cytomegalovirus from clinical specimens, in: New Aspects of HCMV-Related Immunopathology. Scholz M. (ed), Monogr. Virol. Basel, Karger *24*, 160–170.

Mulamba, G., Hu, A., Azad, R., Anderson, K., and Coen, D. (1998). Human cytomegalovirus mutant with sequence-dependent resistance to the phosphorothioate oligonucleotide fomivirsen (ISIS2922). Antimcrob. Agents Chemother. *42*, 971–973.

Neyts, J., Balzarini, J., Andrei, G., Chaoyong, Z., Snoeck, R., Zimmermann, A., Mertens, T, Karlsson, A., and De Clercq, E. (1998). Intracellular metabolism of the N7-substituted acyclic nucleoside analog 2-amino-7-(1,3-dihydroxy-2-propoxymethyl)purine, a potent inhibitor of herpesvirus replication. Mol. Pharmacol. *53*, 157–65.

Perry, C.M. and Balfour J.A. (1999). Fomivirsen. Drugs *57*, 375–380.

Prichard, M.N., Gao, N., Jairath, S., Mulamba, G., Krosky, P., Coen, D.M., Parker, B.O., and Pari, GS. (1999). A recombinant human cytomegalovirus with a large deletion in UL97 has a severe replication deficiency. J. Virol. *73*, 5663–5670.

Reefschlaeger, J., Bender, W., Hallenberger, S., Weber, O., Eckenberg, P., Goldmann, S., Haerter, M., Buerger, I., Trappe, J., Herrington, J.A., Haebich, D., Ruebsamen-Waigmann, H. (2001). Novel nonnucleoside inhibitors of cytomegaloviruses (BAY 38–4766): in vitro and *in vivo* antiviral activity and mechanism of action. J. Antimicrob. Chemother. *48*, 757–767.

Scheffczik, H., Savva, C.G., Holzenburg, A., Kolesnikova, L., and Bogner, E. (2002). The terminase subunits pUL56 and pUL89 of human cytomegalovirus are DNA-metabolizing proteins with toroidal structure. Nucleic Acids Res. *30*, 1695–703.

For over a decade investigators have noted that recently isolated clinical strains of HCMV and some laboratory-maintained strains of HCMV, unlike the most commonly studied laboratory strain AD169, exhibit extended cellular tropism and replicate in human fibroblasts, endothelial cells, and monocyte-derived macrophages (MDM) (Gnann et al., 1988; Jahn et al., 1999; Kahl et al., 2000; Riegler et al., 2000; Sinclair and Sissons, 1996; Sinzger and Jahn, 1996; Sinzger et al., 1995; Soderberg-Naucler et al., 1997; Taylor-Wiedeman et al., 1991). Several reports have suggested that the extended tropism of clinical isolates can be explained by viral genes that are either deleted or their function is lost following mutations during serial passage of clinical viral isolates in human fibroblasts (Hahn et al., 2004). The rhesus CMV strain 68-1 replicates in fibroblasts derived from rhesus macaques (and human fibroblasts), endothelial cells and rhesus monocyte derived macrophages (Figure 28.1). These findings suggest that RhCMV expresses viral genes that are functionally homologous to genes of HCMV thought to be responsible for the expanded cellular tropism of recent clinical isolates of HCMV. Identification of these RhCMV genes, such as the Rh10 ORF, will help define their functions in in vitro cell culture systems and more importantly, help elucidate their role in the pathogenesis of RhCMV infection in rhesus macaques. Such experiments will be difficult to accomplish in currently used small animal models of HCMV pathogenesis.

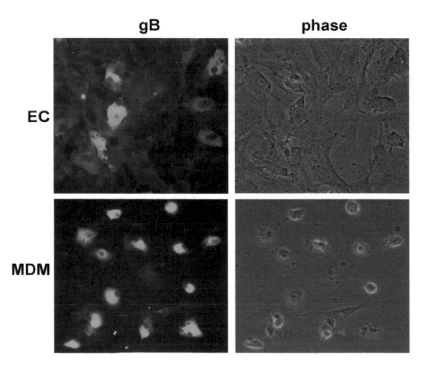

**Figure 28.1** Infection of endothelial cells and monocyte-derived macrophages by RhCMV. Primary rhesus macaque endothelial cells (EC) or monocyte-derived macrophages (MDM) were infected with RhCMV strain 68-1, and glycoprotein B (gB) expression was determined with a gB-specific mAb and FITC conjugated anti-mouse IgG. A phase-contrast photograph of the same field of cells is shown on the right. Photograph courtesy of Dr. A. Moses, Vaccine and Gene Therapy Institute, Oregon Health and Sciences University, Portland, OR, USA.

## The rhesus macaque model of HCMV infections: studies in viral pathogenesis

Although investigators have recognized the potential of the rhesus macaque model in the study of the pathogenesis of HCMV infections, a relatively small number of studies have been carried out utilizing this model (Baroncelli et al., 1997; Barry et al., 1997; Kaur et al., 1996; Kaur et al., 2002; Kaur et al., 2003; Lockridge et al., 1999; Sequar et al., 2002; Tarantal et al., 1998). In most rhesus macaque colonies, over 95% of animals are infected with RhCMV at the time of sexual maturity (Kessler et al., 1989; Laudenslager et al., 1999; Swack and Hsiung, 1982; Vogel et al., 1994). Interestingly, the near universal prevalence of RhCMV infection in macaque colonies has been reported for macaques housed in colonies in North and South America and in Europe, indicating that this virus spreads efficiently in animals housed in primate centers (Andrade et al., 2003; Swack and Hsiung, 1982; Vogel et al., 1994). One report has indicated that RhCMV infection is also common in free-ranging macaques, suggesting that the high incidence of RhCMV infection in primate colonies is not a consequence of containment but perhaps secondary to social behavior of these primates (Kessler et al., 1989). A near universal rate of infection with the other primate CMVs such as baboon CMV has been reported in other non-human primates housed in colonies (Blewett et al., 2001; Blewett et al., 2003). The routes of virus infection of rhesus macaques are not well documented but infection usually occurs during the first year of life (Andrade et al., 2003; Swack and Hsiung, 1982; Vogel et al., 1994). Although breast milk acquisition is a common route of HCMV transmission in humans, there is little evidence for this route of transmission in rhesus macaques and RhCMV spread within colonies of macaques is more likely to be related to social behavior such as fighting that occurs during sexual maturation. Following parenteral inoculation, RhCMV has been recovered from the saliva and bronchoalveolar lavage fluid as early as 7–14 days post virus inoculation and from urine by 2 months after infection (personal communication, P. Barry, University of California at Davis, Davis, CA, USA). Virus excretion following RhCMV infection can persist for well over 200 days and some anecdotal reports have suggested that viruria may persist for years, providing a durable reservoir for virus spread within a colony. Naturally acquired infections in individual animals are asymptomatic or at least do not manifest with findings consistent with a definable clinical syndrome. In contrast, experimental infections with RhCMV by an intravenous route can result in clinical and laboratory symptoms closely resembling those observed in humans infected with HCMV (Lockridge et al., 1999). A complete description of the clinical and laboratory findings associated with acute RhCMV infection in rhesus macaques has been reported by Lockridge et al. (1999), and the similarities between the response in rhesus macaques and humans following CMV infection are striking (Lockridge et al., 1999).

The incidence of congenital RhCMV infection in newborn macaques housed in primate colonies is unknown but symptomatic infection appears to be rare based on the lack of well documented cases. Thus, it remains unclear if macaques can be used to model congenital HCMV infection. It has been suggested that the high incidence of RhCMV immunity in sexually mature macaques significantly reduces the incidence of congenital infection in these primate populations. This explanation is not consistent with the epidemiology of congenital HCMV infection because the incidence of congenital HCMV infections increases as the prevalence of HCMV infection increases in human populations (Alford et al., 1990; Stagno et al., 1982). The limited data available from studies in macaques has not provided an explanation for this key difference between the natural history of CMV infections in human and macaques, but it is possible that the restricted genetic and antigenic variability of strains of RhCMV circulating in these colonies permits more efficient control of infection in pregnant macaques and, thus, prevents intrauterine transmission of RhCMV. In contrast, human populations could be expected to be exposed to a large number of genetically unrelated viral strains and as a result, virus replication and spread is less efficiently controlled by the host immune response. Studies utilizing direct inoculation of fetal monkeys have demonstrated histopathology similar to that observed in infants with congenital HCMV infection (Chang et al., 2002; London et al., 1986; Tarantal et al., 1998). A variety of CNS abnormalities were observed in infected fetal animals including loss of vascularity of the developing CNS, neuronal and glial migration deficits, direct cytopathic effects leading to loss of cellularity and calcifications (Chang et al., 2002; Tarantal et al., 1998). Interestingly, these infections were carried out in fetuses of dams with robust seroimmunity to RhCMV providing evidence that once the virus has crossed the placenta and infected the fetus early in gestation, the RhCMV immune status of the mother is of minimal importance in determining the outcome in infected offspring. Furthermore, infection transmitted in a retrograde fashion from fetus through placenta to the mother has been documented. This finding suggests only a limited role for maternal immunity in the control of established fetal infection (personal communication, P. Barry, University of California at Davis, Davis, CA, USA).

As was noted above, the development of AIDS-like syndrome in macaques inoculated with SIV lead to identification of a number of opportunistic infections in immunocompromised animals, including RhCMV infections. Necropsy of infected animals revealed evidence of productive RhCMV infection in several organ systems including gastrointestinal tract, lung, liver, genitourinary tract and in some animals, the CNS (Baroncelli et al., 1997; Kaup et al., 1998; Kaur et al., 2003; Kuhn et al., 1999; Sequar et al., 2002). Similar to findings in humans with HIV infection, the loss of immunological function in SIV-infected animals was associated with disseminated RhCMV infection and in some cases end-organ disease (Kaur et al., 2003; Sequar et al., 2002). These observations provided compelling evidence for the relevance of RhCMV infection in SIV-infected macaques as an informative model of HCMV infection in AIDS patients. Subsequently, other investigators have demonstrated that co-infection with SIV and RhCMV can lead to a more fulminant course of RhCMV infection with widespread dissemination, persistent viral DNAemia, and organ involvement (Kaur et al., 2003; Sequar et al., 2002). In these animals, a blunted anti-RHCMV antibody response was correlated with uncontrolled RhCMV replication (Kaur et al., 2003; Sequar et al., 2002). Gastrointestinal infections such as colitis are commonly observed disease manifestations of RhCMV infections in SIV-infected animals, but it is unclear whether RhCMV infection in the gastrointestinal tract in these animals is responsible for the chronic diarrhea and wasting syndrome or whether RhCMV is one of several opportunistic pathogens that contribute to gastrointestinal syndromes associated with SIV infection (Kaup et al., 1998). Finally, RhCMV retinitis has not been observed with any regularity in SIV-infected macaques and as a result, one of the most important manifestations of HCMV infection in AIDS patients cannot be easily modeled in this system.

The value of rhesus macaques as model for human disease caused by CMV is obvious from studies of SIV infections and from more limited studies of fetal infection with RhCMV. Yet the ultimate value of this model may be in the elucidation of mechanisms of vascular disease, including chronic graft rejection in macaques receiving allografts. The use of the macaque model for studies in allograft rejection and the development of allograft tolerance appear well accepted. However, there is only limited published data that has explored the role of RhCMV in accelerating vascular disease in these models. Because anti-rejection therapies can be more easily manipulated in these animals than rodents and rejection/graft function more readily monitored, grafts can be maintained for prolonged periods of time making this animal model an ideal system to experimentally define the role of RhCMV in allograft rejection. Moreover, the conservation of ORFs in HCMV and RhCMV will allow investigators to more accurately assign the role of specific viral genes in the development of disease, including those viral functions associated with vascular sclerosis and chronic rejection of transplanted organs.

## The rhesus macaque model of HCMV infection: immune responses to RhCMV

The development of assays and reagents to monitor immune responses in macaques infected with SIV has offered an opportunity to also define the role of host innate and adaptive immunological responses to RhCMV that are responsible for virus clearance and resistance to disease. Limited studies of immune responses to RhCMV have documented both cellular and antibody responses to virus-encoded proteins that appear to parallel those reported in humans infected with HCMV (Kaur et al., 1996; Kaur et al., 2002; Lockridge et al., 1999; Yue et al., 2003). Antibody responses to the RhCMV develop within 2 weeks following experimental infection and an IgM response to RhCMV can also be demonstrated (Lockridge et al., 1999). Antibody titers have been shown to increase with time and virus neutralizing antibodies are readily detectable. Specific antibody responses to the envelope glycoprotein gB have been demonstrated both in acute and chronic infections and a correlation has been made between the virus neutralizing antibodies and anti-gB antibodies, again suggesting that the antiviral antibody responses to RhCMV parallel those in humans infected with HCMV (Yue et al., 2003). Studies in SIV-infected macaques suggest that a failure to generate a robust antiviral antibody response is associated with RhCMV dissemination and persistent viremia (Kaur et al., 2003; Sequar et al., 2002). Similar findings have been reported in HIV-infected humans with disseminated HCMV infection (Boppana et al., 1995). Suppression of RhCMV-specific CD4$^+$ and CD8$^+$ T-lymphocyte responses in SIV-infected macaques have been correlated with levels of RhCMV replication and with time, virus dissemination and invasive infection in the majority of animals with dual infection (Kaur et al., 2002; Kaur et al., 2003). Little is known about specific virus-encoded targets of T-lymphocyte responses but T-lymphocyte responses specific for UL83 (pp65) homologs in RhCMV, Rh111 and Rh112, have been demonstrated in normal macaques (personal communication, L. Picker, Vaccine and Gene Therapy Institute, Oregon Health and Sciences Univ., Portland, OR, USA). Interestingly, the two RhCMV UL83 homologs share an approximate 32–35% identity with HCMV pp65 but share only 39% identity between themselves (Hansen et al., 2003). This finding argues that the two HCMV UL83 homologs in RhCMV may have separate functions, even if these two ORFs arose from a gene duplication event during RhCMV evolution. Together, these studies illustrate that detailed analysis of immunological responses in RhCMV-infected macaques can be expected to extend our understanding of the role of specific immune responses in the pathogenesis of RhCMV infections. Finally, and perhaps most importantly, the results of these studies can be directly translated into an additional understanding of the role of specific immune responses in the pathogenesis of human CMV infection as well.

## Models for HCMV infections: does the Rhesus macaque model lessen the relevance of rodent and other small animal models of HCMV infections?

The availability of a primate model of HCMV infection that utilizes a virus that is genetically closely related to HCMV has raised the question of whether studies performed in rodent and other small animal models can be translated into an increased understanding of the pathogenesis of HCMV infections. This question is timely because of the increasing emphasis on research with clearly defined translational objectives that can directly impact human health. Without question, studies in rodent models of HCMV infections have provided important insights into the pathogenesis of HCMV infections in transplant recipients, including the role of immune effector functions in the control of virus replication and dissemination. Yet it could be argued that definitive evidence for the role of specific immune responses leading to resistance to HCMV disease in these patients could have been gathered without studies in animal models such as murine CMV infection in immunocompromised mice. Similarly, the development of antiviral compounds could have proceeded without the need of rodent models of CMV infection, although toxicology studies would have still required the use of animal models. In the case of congenital HCMV infection, animal models have thus far only confirmed much of the information gathered from natural history studies in clinical populations and have added little insight into the pathogenesis of this maternal/fetal infection. Furthermore, it is likely that current small animal models underestimate the complexity of the maternal/fetal HCMV infection and may not address critical features of the pathogenesis of congenital HCMV infection. Thus, it is unclear if results from studies in these models will be readily translated to mechanisms of disease in humans. Lastly, the use of rodent models to study CMV infection in the post-transplant period is restricted by the acuity of graft rejection, thus limiting investigations into the role of CMV in chronic graft rejection. Thus, the value of small animal model systems for dissection of the pathogenesis of human disease associated with HCMV infection could become increasingly difficult to justify, especially with the availability of primate models.

Although the availability of reagents and the extensive genetic homology between HCMV and RhCMV argue convincingly for an increased utilization of the primate model for the study of the pathogenesis of HCMV infections, there are also several drawbacks that could potentially continue to limit its widespread use. The most obvious is cost and availability of experimental animals, particularly animals that are not infected with RhCMV. These factors have resulted in the restriction of most studies to investigators who are directly associated with primate centers, including the requirement that these investigators have access to facilities that can perform routine pathological studies as well as more sophisticated molecular analysis of tissue specimens. Furthermore, the demand for macaques in studies of SIV as a model of AIDS has outstripped the supply of animals

**Table 28.3** Advantages and disadvantages of primate models and small animal models

| Human disease | Primate models | | Small animal models | |
| --- | --- | --- | --- | --- |
| | Advantages | Disadvantages | Advantages | Disadvantages |
| Congenital infection | Primate organogenesis, immunity and CNS development similar to humans; RhCMV > 60% homologous to HCMV | Limited evidence that virus can efficiently cross placenta; high cost and limited availability of animals | Low cost; well-studied models of development; ease of manipulation of host genetic background, i.e., transgenics; guinea-pig CMV crosses placenta and establishes fetal infection; hearing damage secondary to guinea-pig CMV fetal infection | No evidence MCMV can cross placenta; limited data on CNS infection without direct inoculation; MCMV and HCMV distantly related, and guinea-pig CMV even more distantly related; limited availability of reagents for study of guinea-pig CMV |
| Infections in allograft recipients | Established model of allograft transplantation; long-term maintenance of graft; RhCMV homologous to HCMV; increasing availability of MHC matched and mismatched donor/recipients | High cost; limited availability of non-immune recipients; MHC matching limited at present | Low cost; genetics of donor and recipient easily controlled; reagents for definition of immune components of rejection response | Technical challenge for transplantation; rodent CMVs more distantly related to HCMV; limited value in studies of chronic rejection |
| Infection in HIV-infected hosts | Established models of SIV-induced disease and RhCMV-SIV co-infection; pathogenesis appears very similar to that seen in humans with HIV | Retinitis does not routinely develop; high cost and limited availability of non-immune animals; limited availability of animals for non-SIV studies | Low cost; availability of reagents and genetically defined animals and transgenic animals | Very limited data. In general, lack of adequate small animal models of HIV has limited usefulness for studies of co-infections |

available for studies of other virus infections such as RhCMV. The utilization of rhesus macaques as models of diseases associated with agents of bioterrorism remains undefined but if these animals are determined to be an informative model for pathogenesis studies of infectious agents of bioterrorism, then even fewer animals can be expected to be allocated for studies of RhCMV.

## Summary

Primate models appear to offer the most informative experimental model for the study of the pathogenesis of HCMV infections; however, it is also apparent that this model is not readily available to the majority of investigators studying the pathogenesis of HCMV. Furthermore, the availability and cost of rhesus macaques can be expected to continue to restrict utilization of this model, even in primate centers. For these reasons, small animal models of HCMV infections will almost certainly continue to be of great value in the definition of specific aspects of the pathogenesis of HCMV infections. In addition, the extensive collection of transgenic mice and recently developed murine models that permit organ specific expression of viral genes, have provided the investigator defined systems for characterization of the effects of specific viral functions in vivo as well as the importance of host responses to the outcome of an infection. Such elegant genetic systems are not available to investigators utilizing primate models. Even so, findings even in such sophisticated murine models will ultimately require validation in a more relevant primate model such as the rhesus macaque. Thus, primate and small animal models will continue to be essential in the definition of the pathogenesis of HCMV infections.

## Acknowledgments

This work was supported by grants from the HHS, NIH, NIAID and NICHD.

## References

Alford, C.A., Stagno, S., Pass, R.F., and Britt, W.J. (1990). Congenital and perinatal cytomegalovirus infections. Rev. Infect. Dis. 12, S745-S753.

Andrade, M.R., Yee, J., Barry, P., Spinner, A., Roberts, J.A., Cabello, P.H., Leite, J.P., and Lerche, N.W. (2003). Prevalence of antibodies to selected viruses in a long-term closed breeding colony of rhesus macaques (Macaca mulatta) in Brazil. Am. J. Primat. 59, 123–128.

Baroncelli, S., Barry, P.A., Capitanio, J.P., Lerche, N.W., Otsyula, M., and Mendoza, S.P. (1997). Cytomegalovirus and simian immunodeficiency virus coinfection: longitudinal study of antibody responses and disease progression. J. Acquir. Immune. Defic. Syndr. Hum. Retrovirol. 15, 5–15.

Barouch, D.H., Kunstman, J., Kuroda, M.J., Schmitz, J.E., Santra, S., Peyerl, F.W., Krivulka, G.R., Beaudry, K., Lifton, M.A., Gorgone, D.A., Montefiori, D.C., Lewis, M.G., Wolinsky, S.M.,and Letvin, N.L. (2002). Eventual AIDS vaccine failure in a rhesus monkey by viral escape from cytotoxic T-lymphocytes. Nature 415, 335–339.

Barry, P.A., Alcendor, D.J., Power, M.D., Kerr, H., and Luciw, P.A. (1996). Nucleotide sequence and molecular analysis of the rhesus cytomegalovirus immediate-early gene and the UL121–117 open reading frames. Virology 215, 61–72. (erratum appears in Virology (1996). 218, 296.).

Barry, P.A., Britt, W.J., Kravitz, R.H., and Tarantal, A.F. (1997). Developmental susceptibility of the fetal brain to rhesus cytomegalovirus infection. Paper presented at: The 22nd International Herpesvirus Workshop (La Jolla, California).

Blewett, E.L., Lewis, J., Gadsby, E.L., Neubauer, S.R., and Eberle, R. (2003). Isolation of cytomegalovirus and foamy virus from the drill monkey (Mandrillus leucophaeus) and prevalence of antibodies to these viruses amongst wild-born and captive-bred individuals. Arch. Virol. 148, 423–433.

Blewett, E.L., White, G., Saliki, J.T., and Eberle, R. (2001). Isolation and characterization of an endogenous cytomegalovirus (BaCMV) from baboons. Arch. Virol. 146, 1723–1738.

Boppana, S.B., Polis, M.A., Kramer, A.A., Britt, W.J., and Koenig, S. (1995). Virus specific antibody responses to human cytomegalovirus (HCMV) in human immunodeficiency virus type 1-infected individuals with HCMV retinitis. J. Infect. Dis. *171*, 182–185.

Chang, W.L., Tarantal, A.F., Zhou, S.S., Borowsky, A.D., and Barry, P.A. (2002). A recombinant rhesus cytomegalovirus expressing enhanced green fluorescent protein retains the wild-type phenotype and pathogenicity in fetal macaques. J. Virol. *76*, 9493–9504.

Davison, A.J., Dolan, A., Akter, P., Addison, C., Dargan, D.J., Alcendor, D.J., McGeoch, D.J., and Hayward, G.S. (2003). The human cytomegalovirus genome revisited: comparison with the chimpanzee cytomegalovirus genome. J. Gen. Virol. *84*, 17–28. (erratum appears in J. Gen. Virol. (2003). *84*, 1053.).

De La Melena, V.T., Kreklywich, C.N., Streblow, D.N., Yin, Q., Cook, J.W., Soderberg-Naucler, C., Bruggeman, C.A., Nelson, J.A., and Orloff, S.L. (2001). Kinetics and development of CMV-accelerated transplant vascular sclerosis in rat cardiac allografts is linked to early increase in chemokine expression and presence of virus. Transplant. Proc. *33*, 1822–1823.

Feinberg, M.B., and Moore, J.P. (2002). AIDS vaccine models: challenging challenge viruses. Nature Med. *8*, 207–210.

Gewurz, B.E., Wang, E.W., Tortorella, D., Schust, D.J., and Ploegh, H.L. (2001). Human cytomegalovirus US2 endoplasmic reticulum-lumenal domain dictates association with major histocompatibility complex class I in a locus-specific manner. J. Virol. *75*, 5197–5204.

Gnann, J.W., Jr., Ahlmen, J., Svalander, C., Olding, L., Oldstone, M.B., and Nelson, J.A. (1988). Inflammatory cells in transplanted kidneys are infected by human cytomegalovirus. Am. J. Pathol. *132*, 239–248.

Hahn, G., Revello, M.G., Patrone, M., Percivalle, E., Campanini, G., Sarasini, A., Wagner, M., Gallina, A., Milanesi, G., Koszinowski, U., Baldanti, F., Gerna, G. (2004). Human cytomegalovirus UL131–128 genes are indispensable for virus growth in endothelial cells and virus transfer to leukocytes. J. Virol. *78*, 10023–10033.

Hansen, S.G., Strelow, L.I., Franchi, D.C., Anders, D.G., and Wong, S.W. (2003). Complete sequence and genomic analysis of rhesus cytomegalovirus. J. Virol. *77*, 6620–6636.

Horton, H., Vogel, T.U., Carter, D.K., Vielhuber, K., Fuller, D.H., Shipley, T., Fuller, J.T., Kunstman, K.J., Sutter, G., Montefiori, D.C., Erfle, V., Desrosiers, R.C., Wilson, N., Picker, L.J., Wolinsky, S.M., Wang, C., Allison, D.B., and Watkins, D.I. (2002). Immunization of rhesus macaques with a DNA prime/modified vaccinia virus Ankara boost regimen induces broad simian immunodeficiency virus (SIV)-specific T-cell responses and reduces initial viral replication but does not prevent disease progression following challenge with pathogenic SIVmac239. J. Virol. *76*, 7187–7202.

Huang, E.S., Kilpatrick, B., Lakeman, A., and Alford, C.A. (1978). Genetic analysis of a cytomegalovirus-like agent isolated from human brain. J. Virol. *26*, 718–723.

Jahn, G., Stenglein, S., Riegler, S., Einsele, H., and Sinzger, C. (1999). Human cytomegalovirus infection of immature dendritic cells and macrophages. Intervirology *42*, 365–372.

Jones, T.R., and Sun, L. (1997). Human cytomegalovirus US2 destabilizes major histocompatibility complex class I heavy chains. J. Virol. *71*, 2970–2979.

Jones, T.R., Wiertz, E.J., Sun, L., Fish, K.N., Nelson, J.A., and Ploegh, H.L. (1996). Human cytomegalovirus US3 impairs transport and maturation of major histocompatibility complex class I heavy chains. Proc. Natl. Acad. Sci. USA *93*, 11327–11333.

Kahl, M., Siegel-Axel, D., Stenglein, S., Jahn, G., and Sinzger, C. (2000). Efficient lytic infection of human arterial endothelial cells by human cytomegalovirus strains. J. Virol. *74*, 7628–7635.

Kaup, F., Matz-Rensing, K., Kuhn, E., Hunerbein, P., Stahl-Hennig, C., and Hunsmann, G. (1998). Gastrointestinal pathology in rhesus monkeys with experimental SIV infection. Pathobiology *66*, 159–164.

Kaur, A., Daniel, M.D., Hempel, D., Lee-Parritz, D., Hirsch, M.S., and Johnson, R.P. (1996). Cytotoxic T-lymphocyte responses to cytomegalovirus in normal and simian immunodeficiency virus-infected rhesus macaques. J. Virol. *70*, 7725–7733.

Kaur, A., Hale, C.L., Noren, B., Kassis, N., Simon, M.A., and Johnson, R.P. (2002). Decreased frequency of cytomegalovirus (CMV)-specific CD4[+] T-lymphocytes in simian immunodeficiency virus-infected rhesus macaques: inverse relationship with CMV viremia. J. Virol. *76*, 3646–3658.

Kaur, A., Kassis, N., Hale, C.L., Simon, M., Elliott, M., Gomez-Yafal, A., Lifson, J.D., Desrosiers, R.C., Wang, F., Barry, P., Mach, M., Johnson, R.P. (2003). Direct relationship between suppression of virus-specific immunity and emergence of cytomegalovirus disease in simian AIDS. J. Virol. *77*, 5749–5758.

Kessler, M.J., London, W.T., Madden, D.L., Dambrosia, J.M., Hilliard, J.K., Soike, K.F., and Rawlins, R.G. (1989). Serological survey for viral diseases in the Cayo Santiago rhesus macaque population. Puerto Rico Health Sci. J. 8, 95–97.

Kropff, B., and Mach, M. (1997). Identification of the gene coding for rhesus cytomegalovirus glycoprotein B and immunological analysis of the protein. J. Gen. Virol. 78, 1999–2007.

Kuhn, E.M., Stolte, N., Matz-Rensing, K., Mach, M., Stahl-Henning, C., Hunsmann, G., and Kaup, F.J. (1999). Immunohistochemical studies of productive rhesus cytomegalovirus infection in rhesus monkeys (Macaca mulatta) infected with simian immunodeficiency virus. Vet. Pathol. 36, 51–56.

Laudenslager, M.L., Rasmussen, K.L., Berman, C.M., Lilly, A.A., Shelton, S.E., Kalin, N.H., and Suomi, S.J. (1999). A preliminary description of responses of free-ranging rhesus monkeys to brief capture experiences: behavior, endocrine, immune, and health relationships. Brain Behav. Immun. 13, 124–137.

Lautenschlager, I., Soots, A., Krogerus, L., Kauppinen, H., Saarinen, O., Bruggeman, C., and Ahonen, J. (1997). CMV increases inflammation and accelerates chronic rejection in rat kidney allografts. Transplant. Proc. 29, 802–803.

Li, F., Yin, M., Van Dam, J.G., Grauls, G., Rozing, J., and Bruggeman, C.A. (1998). Cytomegalovirus infection enhances the neointima formation in rat aortic allografts: effect of major histocompatibility complex class I and class II antigen differences. Transplantation 65, 1298–1304.

Lockridge, K.M., Sequar, G., Zhou, S.S., Yue, Y., Mandell, C.P., and Barry, P.A. (1999). Pathogenesis of experimental rhesus cytomegalovirus infection. J. Virol. 73, 9576–9583.

Lockridge, K.M., Zhou, S.S., Kravitz, R.H., Johnson, J.L., Sawai, E.T., Blewett, E.L., and Barry, P. A. (2000). Primate cytomegaloviruses encode and express an IL-10-like protein. Virology 268, 272–280.

London, W.T., Martinez, A.J., Houff, S.A., Wallen, W.C., Curfman, B.L., Traub, R.G., and Sever, J. L. (1986). Experimental congenital disease with simian cytomegalovirus in rhesus monkeys. Teratology 33, 323–331.

Martelius, T.J., Blok, M.J., Inkinen, K.A., Loginov, R.J., Hockerstedt, K.A., Bruggeman, C.A., and Lautenschlager, I.T. (2001). Cytomegalovirus infection, viral DNA, and immediate-early-1 gene expression in rejecting rat liver allografts. Transplantation 71, 1257–1261.

Matano, T., Kobayashi, M., Igarashi, H., Takeda, A., Nakamura, H., Kano, M., Sugimoto, C., Mori, K., Iida, A., Hirata, T., Hasegawa, M., Yuasa, T., Miyazawa, M., Takahashi, Y., Yasunami, M., Kimura, A., O'Connor, D.H., Watkins, D.I., and Nagai, Y.. (2004). Cytotoxic T-lymphocyte-based control of simian immunodeficiency virus replication in a preclinical AIDS vaccine trial. J. Exp. Med. 199, 1709–1718.

Nigida, S.M., Jr., Falk, L.A., Wolfe, L.G., Deinhardt, F., Lakeman, A., and Alford, C.A. (1975). Experimental infection of marmosets with a cytomegalovirus of human origin. J. Infect. Dis. 132, 582–586.

Orloff, S.L., Streblow, D.N., Soderberg-Naucler, C., Yin, Q., Kreklywich, C., Corless, C.L., Smith, P.A., Loomis, C.B., Mills, L.K., Cook, J.W., Bruggeman, C.A., Nelson, J.A., and Wagner, C.R. (2002). Elimination of donor-specific alloreactivity prevents cytomegalovirus-accelerated chronic rejection in rat small bowel and heart transplants. Transplantation 73, 679–688.

Penfold, M.E., Schmidt, T.L., Dairaghi, D.J., Barry, P.A., and Schall, T.J. (2003). Characterization of the rhesus cytomegalovirus US28 locus. J. Virol. 77, 10404–10413.

Riegler, S., Hebart, H., Einsele, H., Brossart, P., Jahn, G., and Sinzger, C. (2000). Monocyte-derived dendritic cells are permissive to the complete replicative cycle of human cytomegalovirus. J. Gen. Virol. 81, 393–399.

Sequar, G., Britt, W.J., Lakeman, F.D., Lockridge, K.M., Tarara, R.P., Canfield, D.R., Zhou, S.S., Gardner, M.B., and Barry, P.A. (2002). Experimental coinfection of rhesus macaques with rhesus cytomegalovirus and simian immunodeficiency virus: pathogenesis. J. Virol. 76, 7661–7671.

Shiver, J.W., Fu, T.M., Chen, L., Casimiro, D.R., Davies, M.E., Evans, R.K., Zhang, Z.Q., Simon, A.J., Trigona, W.L., Dubey, S.A., Huang, L., Harris, V.A., Long, R.S., Liang, X., Handt, L., Schleif, W.A., Zhu, L., Freed, D.C., Persaud, N.V., Guan, L., Punt, K.S., Tang, A., Chen, M., Wilson, K.A., Collins, K.B., Heidecker, G.J., Fernandez, V.R., Perry, H.C., Joyce, J.G., Grimm, K.M., Cook, J.C., Keller, P.M., Kresock, D.S., Mach, H., Troutman, R.D., Isopi, L.A., Williams, D.M., Xu, Z., Bohannon, K.E., Volkin, D.B., Montefiori, D.C., Miura, A., Krivulka, G.R., Lifton, M.A., Kuroda, M.J., Schmitz, J.E., Letvin, N.L., Caulfield, M.J., Bett, A.J., Youil, R., Kaslow, D.C., and Emini, E.A.. (2002). Replication-incompetent adenoviral vaccine vector elicits effective anti-immunodeficiency-virus immunity. Nature 415, 331–335.

Sinclair, J., and Sissons, P. (1996). Latent and persistent infections of monocytes and macrophages. Intervirology 39, 293–301.

Sinzger, C., Grefte, A., Plachter, B., Gouw, A.S., The, T.H., and Jahn, G. (1995). Fibroblasts, epithelial cells, endothelial cells and smooth muscle cells are major targets of human cytomegalovirus infection in lung and gastrointestinal tissues. J. Gen. Virol. 76, 741–750.

Sinzger, C., and Jahn, G. (1996). Human cytomegalovirus cell tropism and pathogenesis. Intervirology 39, 302–319.

Soderberg-Naucler, C., Fish, K.N., and Nelson, J.A. (1997). Interferon-gamma and tumor necrosis factor-alpha specifically induce formation of cytomegalovirus-permissive monocyte-derived macrophages that are refractory to the antiviral activity of these cytokines. J. Clin. Invest. 100, 3154–3163.

Stagno, S., Pass, R.F., Dworsky, M.E., Henderson, R.E., Moore, E.G., Walton, P.D., and Alford, C.A. (1982). Congenital cytomegalovirus infection: The relative importance of primary and recurrent maternal infection. N. Engl. J. Med. 306, 945–949.

Streblow, D.N., Kreklywich, C., Yin, Q., De La Melena, V.T., Corless, C.L., Smith, P.A., Brakebill, C., Cook, J.W., Vink, C., Bruggeman, C.A., Nelson, J.A., and Orloff, S.L., (2003). Cytomegalovirus-mediated up-regulation of chemokine expression correlates with the acceleration of chronic rejection in rat heart transplants. J. Virol. 77, 2182–2194.

Swack, N.S., and Hsiung, G.D. (1982). Natural and experimental simian cytomegalovirus infections at a primate center. J. Med. Primatol. 11, 169–177.

Tarantal, A.F., Salamat, M.S., Britt, W.J., Luciw, P.A., Hendrickx, A.G., and Barry, P.A. (1998). Neuropathogenesis induced by rhesus cytomegalovirus in fetal rhesus monkeys (Macaca mulatta). J. Infec. Dis. 177, 446–450.

Taylor-Wiedeman, J., Sissons, J.G., Borysiewicz, L.K., and Sinclair, J.H. (1991). Monocytes are a major site of persistence of human cytomegalovirus in peripheral blood mononuclear cells. J. Gen. Virol. 72, 2059–2064.

Tikkanen, J., Kallio, E., Pulkkinen, V., Bruggeman, C., Koskinen, P., and Lemstrom, K. (2001). Cytomegalovirus infection-enhanced chronic rejection in the rat is prevented by antiviral prophylaxis. Transplant. Proc. 33, 1801.

Tomazin, R., Boname, J., Hegde, N.R., Lewinsohn, D.M., Altschuler, Y., Jones, T.R., Cresswell, P., Nelson, J.A., Riddell, S.R., and Johnson, D.C. (1999). Cytomegalovirus US2 destroys two components of the MHC class II pathway, preventing recognition by CD4+ T-cells. Nature Med. 5, 1039–1043.

Vogel, P., Weigler, B.J., Kerr, H., Hendrickx, A.G., and Barry, P.A. (1994). Seroepidemiologic studies of cytomegalovirus infection in a breeding population of rhesus macaques. Lab. Anim. Sci. 44, 25–30.

Warren, J. (2002). Preclinical AIDS vaccine research: survey of SIV, SHIV, and HIV challenge studies in vaccinated nonhuman primates. J. Med. Primat. 31, 237–256.

Whitney, J.B., and Ruprecht, R.M. (2004). Live attenuated HIV vaccines: pitfalls and prospects. Curr. Opin. Infect. Dis. 17, 17–26.

Yue, Y., Zhou, S.S., and Barry, P.A. (2003). Antibody responses to rhesus cytomegalovirus glycoprotein B in naturally infected rhesus macaques. J. Gen. Virol. 84, 3371–3379.

Zhu, H., Cong, J.P., Bresnahan, W.A., Shenk, T.E. (2002). Inhibition of cyclooxygenase 2 blocks human cytomegalovirus replication. Proc. Natl. Acad. Sci. USA 99, 3932–3937.

# Résumé and Visions: From CMV Today to CMV Tomorrow

*Ulrich H. Koszinowski*

This book expertly describes the state of the art in clinical and experimental cytomegalovirus (CMV) research. It is written by distinguished scholars in the field, several of whom have contributed to CMV research for decades. For the expert it is both fun and stimulating to read, and for the novice in the field it is an appetizer and entry to the manifold CMV problems and research opportunities.

Having a perspective or having an eye for the detail: these are advantages and disadvantages of being close or distant to a subject. Sometimes detail matters, sometimes perspective. When we start on a tour in the mountains we better go for the safe detail and carefully mind each of our footsteps. After some time, our feet find their way with little optical support. We lift our head and enjoy the perspective. Exploring CMV is more complex than exploring a mountain range. Weller once described cytomegaloviruses as ubiquitous agents with protean clinical manifestations (Weller, 1971a; b). The genetic shape of CMV is molded by co-evolution over millions of years with their mammalian hosts. We can only speculate about the driving forces that caused the present status. CMV from different species share many aspects, but do they differ in detail as much as their hosts do? For understanding cytomegalovirus disease and pathology we need to know under which conditions which detail matters.

This book, with its 28 chapters on different aspects of CMV research, provides different points of view, and this is a particular attractive and telling aspect of the editor's assemblage. The richness and also the diversity in thinking and in perspectives are perceived best by reading the complete book, not only one or the other chapter. Only in some contributions is there an attempt at a historical perspective and – with purity of the heart – authors have a superior memory for own contributions. Some authors explain details of their work that would remain unnoticed in a normal paper. Others are not so close to the field and thus do not have to defend an authority. This can also make very good reading because of a balanced view. A reader like me, who knows the majority of the authors, profits twice: first, by sensing the authenticity of the esteemed colleague and, second, by getting a well-adjusted view on the various areas of research in the field at this time. Is the book complete? I would have estimated the contribution of one or the other additional author in this concert. The picture of CMV would have acquired even more depth, but without these contributions the picture is not distorted. The result is clear: there is no consensus and no shared general view on the CMV problem.

We live in an anthropocentric world which is expecting the improvement of the conditio humana. Do we really need all these research efforts in the numerous experimental animal CMV models in the future? Or do they merely profit from the high costs and the poor availability of primate models (Chapter 28)? At the beginning of the 1980s, the CMV of the different species were just viruses of the same family giving rise to similar infections and similar pathology in their respective hosts. Now we have the genomic sequences and realize that there is a high degree of homology only in a core set of CMV genes. As we look more to the ropes' ends there is a continuous decrease of sequence homology and functional identity until we end up with genes found only in the cytomegalovirus infecting a given species. There has been a wealth of findings relating to virus–host interactions in the animal models. Unfortunately, the genes that apparently govern the balance of the virus–host interaction are also the most variable between CMV. This is an irritating fact and a problem that is hard to resolve in a short period of time.

Is there a prototypic model for human CMV disease? The various and variable viral gene products act on (in the majority unknown) host proteins. Therefore, host genetics also matters. Lacking knowledge on the effectors and targets we cannot decide to date which genes govern pathology under which disease conditions in which species. We somehow sense – through common sense – that, most probably, a primate CMV should be closer to the human situation. Doubts creep in after taking a look at the consensus factor binding motifs in the enhancers of the different CMV (Chapter 8), the region pivotal for CMV gene regulation. No clear-cut picture emerges. The beads of this abacus reveal basic elements of a sort of a counting instrument, but the mathematics involved does not seem to display a definite species-specific signature.

Defects, immaturity, ablation, exhaustion or the irritation of the immune system is the hallmark of CMV disease, reactivation, recrudescence, and pathology. Therefore, the desire to understand the immune control of CMV has many sources. This extends by far the motivation to develop vaccines or to provide passive immunization to the immunodeficient patient. Actually, the latter aspect is quite advanced and apparently close to translation efforts into manageable and economical medical application. This is the area where the rodent models have provided and still provide many valuable concepts for testing. In the chapters on innate immunity, chemokines, interferons, NK cells and T cells the reader gets a feeling that the field of virology gains from knowledge in cell biology and immunology and at the same time contributes to basic science (Chapters 15–25). Whenever a new immunological principle is detected it does not take long to discover that CMV has already taken care. This is the mammalian virus with the largest genome and it shows. Work of the last 20 years has brought the information that all viruses –to plagiarize Shakespeare's *Hamlet* – "take arms against a sea of troubles" caused by the host immune response, and by opposing though not always end but usually modulate or channel them, or even profit from them. The sheer size of the CMV genome provides many genes for the art of self-defense, resistance, and counteraction welded by evolution. We know this armory only by the inspection of its parts so far. We should like to see the components at their specific and important moment of action.

In the analysis of host defenses to CMV the antibody response has had the head start (Chapter 14). Many aspects of virus–host interactions have been considered and are still considered as a reflection of providing enough and effective enough antibodies. Then T cells appeared at the scene, antigen-presenting cells, chemokines and cytokines, NK cells, and innate immunity signal transduction cascades; and this list of relevant pathways and functions has certainly not come to an end. It clearly emerges from this book that we know many elements of the antiviral host defense. None of the functions has been falsified. As we usually have a very specific expertise, we know our favored defense mechanism, but we disagree on the major players and their hierarchical interaction.

The infection of the fetus and the newborn remains the most burning CMV problem. What do we know? Antibodies are powerful tools. They discriminate antigenic properties in proteins, and antibodies against specific antigens can even neutralize virus suspensions. Usage of the discriminative power of antibodies brought the sobering information that pregnant women – in the face of a robust immunity following primary infection – can be superinfected and that this secondary infection can be transmitted to the fetus. Which of our vaccines can claim to protect superior to natural infection? Is this the end of all vaccine attempts? Well, the frequency of this type of transmission is probably low. Is it linked to the lack of cross-neutralizing antibodies? Does a vaccine have to represent as many as possible, if not all, of the variants of glycoprotein B and other proteins against which neutralizing antibodies can be produced (Chapter 26)? Perhaps yes, provided that the protective effect of antibodies in vivo is really based on neutralization of free virus in vivo. But what is the contribution of Fc receptor-bearing cells which specifically recognize bound antibodies? A robust antibody titer often correlates with protection against disease. This conclusion may be premature. Are neutralizing antibodies really an unambiguous marker for protection against superinfection? Or do neutralizing antibodies represent merely a correlate but not a surrogate – a reliable and consistent indicator – for protection against superinfection? Absence of cross-neutralization proved the fact of superinfection but not its principle. Is the virus vulnerable to neutralizing antibodies during transmission? In the guinea-pig, the virus can replicate in the placenta despite a robust antibody response (Chapter 25) Is the virus transported or even transmitted by infected cells (Chapter 22)? Is there a possibility for virus transmission from cell to cell in the absence of glycoproteins (Silva et al., 2005)? Clearly, a robust antibody response reduces the risk of fetal infection. However, if other less frequent modes of transmission obey different principles, they cannot be shut this way. It may be that we have to broaden our view beyond antibodies. Other principles of innate and adaptive immunity that are modulated by the virus may play a role as well. Is it really justified to consider the guinea-pig model as outdated as long as we do not know the consistent and reliable marker of protection? Which, then, is the reliable surrogate for immunity against fetal transmission that we only need to confirm in the primate model?

What is the focus of CMV research of tomorrow? As Karl Valentin (1882–1948), a famous comedian from Munich, Bavaria, used to say: "Predictions are generally difficult, in particular if they concern the future". Without a vaccine in sight, efficient immune and chemotherapy with few side-effects is one option. Chemotherapeutics and their problems are described in Chapter 27. Which type of approach provides new targets for specific therapy? Morphogenesis, assembly, and egress (Chapters 12 and 13) have in the past been areas of research slowed down by considerable technical difficulty. New technical developments reported in Chapters 3, 4, and 5, provide new approaches and new chances of increasing the resolution of analysis to identify new target candidates for therapy.

# Open reading frames index

## MCMV